Advances in Biochemistry in Health and Disease

Volume 34

Advances in Biochemistry in Health and Disease focus on the latest developments in biochemical research with implications for health and disease. This book series consists of original edited volumes and monographs, presented by leading experts in the field and provides an up to date and unique source of information for all those interested in the fundamental, biochemical processes of the latest and emerging topics and techniques.

Covering a wide variety of topics, this book series is a valuable source of information from those at the lab bench through to the Health Care workers.

Surajit Pathak • Antara Banerjee
Editors

MicroRNAs: Role in Cancer Progression, Diagnostics and Therapy

Springer

Editors
Surajit Pathak
Faculty of Allied Health Sciences,
Medical Biotechnology Lab
Chettinad Academy of Research and
Education (CARE), Chettinad Hospital
and Research Institute
Chennai, Tamil Nadu, India

Antara Banerjee
Faculty of Allied Health Sciences,
Medical Biotechnology Lab
Chettinad Academy of Research and
Education (CARE), Chettinad Hospital
and Research Institute
Chennai, Tamil Nadu, India

ISSN 2512-2142 ISSN 2512-2150 (electronic)
Advances in Biochemistry in Health and Disease
ISBN 978-3-032-08823-9 ISBN 978-3-032-08824-6 (eBook)
https://doi.org/10.1007/978-3-032-08824-6

This Springer imprint is published by the registered company Springer Nature Switzerland AG
The registered company address is: Gewerbestrasse 11, 6330 Cham, Switzerland

Preface

In recent decades, the landscape of cancer biology has undergone a dramatic transformation, driven not just by advancements in genomic and proteomic technologies but by the discovery of a class of small, noncoding RNAs known as microRNAs (miRNAs). Once regarded as molecular noise, miRNAs have emerged as critical regulators of gene expression, orchestrating cellular pathways that govern proliferation, apoptosis, metastasis, immune modulation, and resistance to therapy. Their profound influence on cancer biology has positioned them at the center of diagnostic, prognostic, and therapeutic strategies across a wide spectrum of malignancies.

The book *MicroRNAs: Role in Cancer Progression, Diagnostics and Therapy* is a comprehensive exploration of the multifaceted roles of miRNAs in cancer, from their basic biogenesis and mechanisms of action to their translational relevance in clinical oncology. The chapters are contributed by a diverse and international group of researchers who bring forward their expertise in molecular biology, oncology, bioinformatics, and translational medicine. Each chapter delves into a distinct facet of miRNA function, highlighting their involvement in specific cancer types, their interaction with the tumor microenvironment, and their emerging role in epigenetic regulation, liquid biopsy-based diagnostics.

This book also places special emphasis on cutting-edge advancements such as the use of artificial intelligence to decode miRNA datasets, and the therapeutic targeting of exosomal miRNAs and miRNA modulators. Topics like miRNA roles in gastrointestinal, breast, lung, gynecological, and prostate cancers are explored in depth, along with specialized focus areas such as bone metastasis, radiotherapy resistance.

By bridging foundational concepts with advanced applications, this book is intended for a diverse audience, including oncologists, cancer researchers, pharmaceutical scientists, and students seeking a comprehensive understanding of the evolving role of microRNAs in cancer biology and therapy. It serves as a vital resource for those engaged in drug discovery, molecular diagnostics, and the development of RNA-based therapeutic strategies. With in-depth discussions on miRNA biogenesis, mechanisms of action, delivery systems, and the complexities of clinical translation, this volume offers valuable insights for both academic and industry professionals. It encourages readers to look beyond conventional paradigms and explore microRNAs as powerful clinical tools, capable of transforming cancer care from reactive treatments to predictive, precise, and personalized interventions.

We hope this volume contributes meaningfully to the ongoing conversation in cancer research and inspires future studies that push the boundaries of our understanding and application of microRNA biology in oncology. We sincerely thank all the contributors, researchers, and experts whose work and insights have made this book possible.

Chennai, Tamil Nadu, India

Surajit Pathak
Antara Banerjee

Contents

About the Editors

Surajit Pathak, is a leading researcher and academcian, who is currently a Professor at Chettinad Academy of Research and Education in Chennai. He has previously held esteemed positions as Principal Research Scientist at Linkoping University, Sweden. With a Ph.D from India (2007) and an extensive post-doctoral journey at renowned institutions like the University of Alabama (USA), University of Padova (Italy), and University of Linkoping (Sweden), his research focuses on microRNAs for early colorectal cancer detection. He has published over 150 research and review articles, earning more than 5500 citations (google scholar) and an h-index of 40. He has received several grants from various funding agencies and is a member of national research societies.

Antara Banerjee, is currently working as a Professor at Chettinad Academy of Research and Education (CARE), Chennai. She pursued her post-doctoral research on Cancer Biology and Regenerative Medicine at the University of Padova, Italy, and Linkoping University, Sweden. Her research interest covers areas on tissue regeneration and analyzing tumor specific niche in cancer. She has published over 120 articles with an h-index of 35 and 4296 citations (Google scholar) with 2 patents awarded. She is affiliated with professional research bodies and is a member of several national research societies.

Chapter 1
An Overview of MicroRNA Biology: From Biogenesis to Therapeutic Implications

Sibin Nambidi, Subhamay Adhikary, Sneha Pallatt, Amit Dey, Antara Banerjee, and Surajit Pathak

Abstract MicroRNAs (miRNAs) are short, noncoding RNAs that are key players in regulating genes after they are transcribed, by targeting mRNAs for degradation or inhibiting translation. The biogenesis of miRNAs is a process which occurs in multiple steps, first involving transcription into the genome, followed by processing by the Drosha and Dicer enzymes along the way, and finally by incorporation into the RNA-induced silencing complex (RISC) in order to carry out their regulatory functions. miRNAs are important for a variety of biological activities, including cell differentiation, proliferation, apoptosis, and immune response. Dysregulation of miRNA expression and function is linked to multiple diseases, particularly cancer, neurodegenerative diseases, and metabolic syndromes. Advances in sequencing technologies and bioinformatics have led to a better understanding of miRNA biogenesis and function, as well as their potential as both biomarkers and therapeutic targets. However, some challenges remain, for example, the development of efficient delivery systems, off-target effects, and the complexities of miRNA-mRNA interactions. This chapter will highlight important concepts regarding miRNA biogenesis and its implications for health and disease, as well as future opportunities for miRNA-based diagnostics and therapeutics.

Keywords MicroRNA · Biogenesis · Posttranscriptional regulation · Therapeutics · Biomarkers · RNA interference

S. Nambidi · S. Adhikary · S. Pallatt · A. Dey · A. Banerjee · S. Pathak (✉)
Faculty of Allied Health Sciences, Medical Biotechnology Lab, Chettinad Academy of Research and Education (CARE), Chettinad Hospital and Research Institute, Chennai, Tamil Nadu, India
e-mail: drsurajitpathak@care.edu.in

S. Pathak, A. Banerjee (eds.), *MicroRNAs: Role in Cancer Progression, Diagnostics and Therapy*, Advances in Biochemistry in Health and Disease 34,
https://doi.org/10.1007/978-3-032-08824-6_1

Introduction to MicroRNAs

miRNAs are short, noncoding RNAs involved in posttranscriptional gene regulation that can affect a variety of cellular functions including growth, differentiation, development, and apoptosis [1]. In general, regulatory noncoding RNA molecules originate from DNA sequences as primary miRNAs (pri-miRNAs), and then become processed to precursor miRNA (pre-miRNA) while also being solved to working copies of mature miRNAs [2]. miRNAs usually exert effects by binding to target mRNAs within the 3′ untranslated region (3′ UTR), which in turn induces either mRNA degradation or translational repression. In recent years, however, miRNAs have demonstrated additional interactions with other regions aside from the 3′ UTR, including the 5′ UTR, coding sequence, and gene promoter which further extends nonlinearity potential for regulation [2].

In some cases, miRNAs have also demonstrated translation activation or transcription modulation capabilities, pointing to additional functional diversity in how miRNAs act in normal physiology, and potentially, how they contribute to pathophysiology. Therefore, over the last several years there has been increased attention towards understanding how miRNAs contribute to normal physiologic function, and underlying physiological function pathophysiology. miRNAs serve an important regulatory function in biology under relatively homeostatic situations, in topics such as cell proliferation, cell's structural differentiation to functional state, and apoptosis [3]. miRNAs often act in large feedback loops to support conditions that favor cellular homeostasis. However, the deregulation of even a single miRNA or a small subset of miRNAs can trigger widespread alterations in gene expression, potentially driving cellular transformation and disease progression. The first disease-associated miRNAs, miR-15 and miR-16, located at the 13q14 chromosomal region, were initially identified in chronic lymphocytic leukemia (CLL), marking a pivotal discovery in miRNA research. Since then, tumor-associated miRNAs have been detected in the serum of patients with various malignancies, including those with diffuse large B-cell lymphoma (DLBCL), further underscoring the clinical significance of miRNAs in cancer diagnostics and therapeutics [2, 3].

MiRNAs are processed in multiple steps, starting with transcription and primary processing in the nucleus, followed by the maturation of the precursor and activation of the functional miRNA in the cytoplasm. In the cytoplasm, mature miRNAs not only perform their regulatory functions, but their levels can also be adjusted by degradation processes to establish precise control of gene expression [4]. In the early 1990s, researchers demonstrated that two small RNAs, instead of a protein, represented the functional products of the lin-4, which regulated the timing of larval development in *Caenorhabditis elegans* (*C. elegans*). The longer of the two RNAs (~70 nucleotides) was shown to be the precursor of the shorter RNA (~22 nucleotides), which we now know was one of the first identified members of the miRNA family. With the increasing evidence of the human genome and its functions, the roles of miRNAs in various cellular processes are becoming much clearer [5]. Along with miRNAs, several other noncoding RNAs (ncRNAs), including Air, H19, Ipw,

NTT, Tsix, and XIST, have been shown to be important regulators of gene expression. Noncoding RNAs are involved in a wide range of biological functions, from genomic imprinting to X-chromosome inactivation in mammals, revealing their importance in epigenetic regulation [5].

Since this finding, hundreds of miRNAs have been identified in nearly all metazoan genomes from worms, flies, and plants to mammals. These small regulatory RNAs are diverse in expression and arc known to be critical in regulating various developmental and physiological processes. The discovery of microRNAs has added further complexity to our understanding of gene regulatory networks, revealing an additional level of posttranscriptional regulation that modulates gene expression in a variety of biological contexts [6]. Over 1900 miRNAs have been identified in the human genome to date and are associated with a variety of common human diseases. The recent advances in high-throughput genomics, bioinformatics, and traditional molecular biology techniques and animal studies have progressed the scientific field exploring role of miRNAs greatly [7]. This research is now poised for translational work in clinical settings where miRNA may represent an important new avenue of research to improve public health as a foundation for diagnostics and therapeutics. In fact, miRNA-based biomarkers have been considered and investigated as tools for early detection of diseases related to cancer, cardiovascular diseases, and neurodegenerative disorders, for example [8]. The development of miRNA mimics and inhibitors also has great potential as therapeutic agents for targeted gene regulation and exploring miRNA may lead to opportunities within the framework of personalized medicine soon [7, 8].

In this chapter, we will discuss the complex processes involved in miRNA biogenesis and the various mechanisms of miRNA-mediated gene regulation. By discussing the role of miRNA in normal cellular contexts and also pathological states, this chapter aims to provide the reader with a thorough overview of the importance of miRNA in developmental processes, homeostasis, and potential therapeutic applications.

Biogenesis of MicroRNAs

The biogenesis of miRNAs starts with transcription from miRNA genes or miRNA gene clusters that can be transcribed as a single transcript or polycistronic transcripts. The synthesis of miRNA transcripts is performed primarily by RNA polymerase II or III and subject to co-transcriptional or posttranscriptional processing [9]. Approximately half of currently encoded miRNAs are classified as intragenic, primarily derived from protein-coding introns and less commonly from protein-coding exons. The remaining intergenic miRNAs are encoded independent of a host gene and rely on their own promoters for transcription, serving as transcriptional units [2, 9]. The understanding of miRNA biogenesis has developed rapidly in recent years. However, knowledge of the transcriptional regulation of miRNA genes is still limited, even though transcription regulation represents an important step in miRNA biogenesis. Identifying core transcriptional machinery has important implications for understanding miRNA gene regulation and can ultimately provide additional insights into miRNA function during development, homeostasis, and disease [10].

miRNAs can arise intragenically, derived from an intron and exon of nuclear DNA, or intergenically, generated via transcription independent of the original gene. Their biogenesis entails the transcription of primary miRNAs (pri-miRNAs), which can occur via RNA polymerase II or III, resulting in the generation of precursor miRNAs (pre-miRNAs) via either the canonical or noncanonical pathway. In the canonical pathway, pri-miRNA is then processed by Drosha in association with Dgcr8, resulting in a hairpin-shaped ~60–100 nucleotide pre-miRNA [11]. This pre-miRNA is then bound by Exportin-5 and transported to the cytoplasm via a Ran-GTP-dependent mechanism, where Dicer removes the terminal loop from the pre-miRNA, resulting in a ~ 22 nucleotide double-stranded miRNA duplex containing the guide strand (the mature miRNA) and the passenger strand, which is subsequently degraded [12]. The guide strand is then loaded into the RNA-induced silencing complex (RISC), which contains Argonaute (Ago) proteins that play a crucial role in miRNA-based gene silencing. miRNAs work to guide RISC to complementary sequences on target mRNAs and are responsible for translational repression and decay. The cooperation of the Drosha, Dgcr8, and Dicer collaboration are required in order to produce canonical miRNAs from pri-miRNAs, rendering miRNA biogenesis an intricate and highly regulated process [11, 12].

While significant advances have been made in understanding the basic steps of miRNA maturation, how miRNAs are regulated to control their specific expression is still unknown. Some studies suggest that not all miRNAs go through the same processing steps, and it is understood that there are several regulatory mechanisms that modulate the biogenesis of miRNAs at different stages. It has been established that miRNAs regulate gene expression posttranscriptionally, but we are still learning about the precise molecular mechanisms [13]. This is a rapidly-evolving area of research that is also highly relevant to clinical medicine and disease research. Changes in the expression of miRNAs can be associated with diseases, including cancer, and many can directly affect tumor initiation and progression. Even with many limitations, the profiling of miRNA expression is showing promise for development in diagnostics, prognostic, and therapeutic applications in multiple areas of medicine [12, 13].

In general, miRNAs are transcribed by RNA Polymerase II, and then subsequently processed by Drosha and Dicer, and eventually loaded into RISC for gene expression regulation. This regulation of gene expression most commonly occurs via specific binding to the 3′ UTR of an mRNA molecule which can lead to inhibition of translation and or degradation of the transcript. Recent publications have identified target-mediated decay of miRNA and a broader range of regulatory molecular function outside of standard binding to an mRNA target suggesting further regulatory implications [14]. Although our general understanding of producing a functional miRNA is broad and diverse beyond the classical, canonical pathway, the classical canonical pathway that has been clearly presented in academic literature is widely communicated. This pathway has established interest, which has led to the identification and characterization of noncanonical miRNA biogenesis pathways [15]. The consequence of recent studies using modern molecular investigations are studying (and characterizing) noncanonical biogenesis pathways for miRNAs that normally are considered canonical. Noncanonical pathways of miRNA biogenesis deviate from the classical Drosha–Dicer processing route. While most pre-miRNAs are generated through Drosha cleavage, certain alternative pathways bypass Drosha, and only a few miRNAs have been shown to become functional without Dicer processing. In addition, many miRNAs are organised in clusters, and approximately half of the miRNA genes are conserved across the human genome, often showing coordinated expression within these clusters. A number of miRNA are characterized to cluster, and approximately half of the miRNA genes are conserved across the human genome with different series of coupling to each miRNA [15, 16]. A study done by Havens et al. [17] demonstrated that miRNAs are typically processed by Drosha, Dgcr8, and Dicer, but some bypass these steps. This study identifies simtrons, a new class of miRNAs that do not require Dgcr8, Dicer, Exportin-5, or Argonaute 2 for biogenesis. Instead, they are directly processed by Drosha and still function in gene silencing via the RISC complex. These findings highlight a noncanonical miRNA biogenesis pathway, expanding our understanding of miRNA regulation [17]. Figure 1.1 depicts the biogenesis and function of microRNA, beginning with the transcription of pri-miRNA from DNA, its subsequent processing into pre-miRNA by the Drosha complex, transport by Exportin-5, cleavage by Dicer in the cytoplasm of the miRNA duplex, loading the mature miRNA into the RISC complex, and its ultimate function of regulating target mRNA by cleavage, translational repression, or degradation.

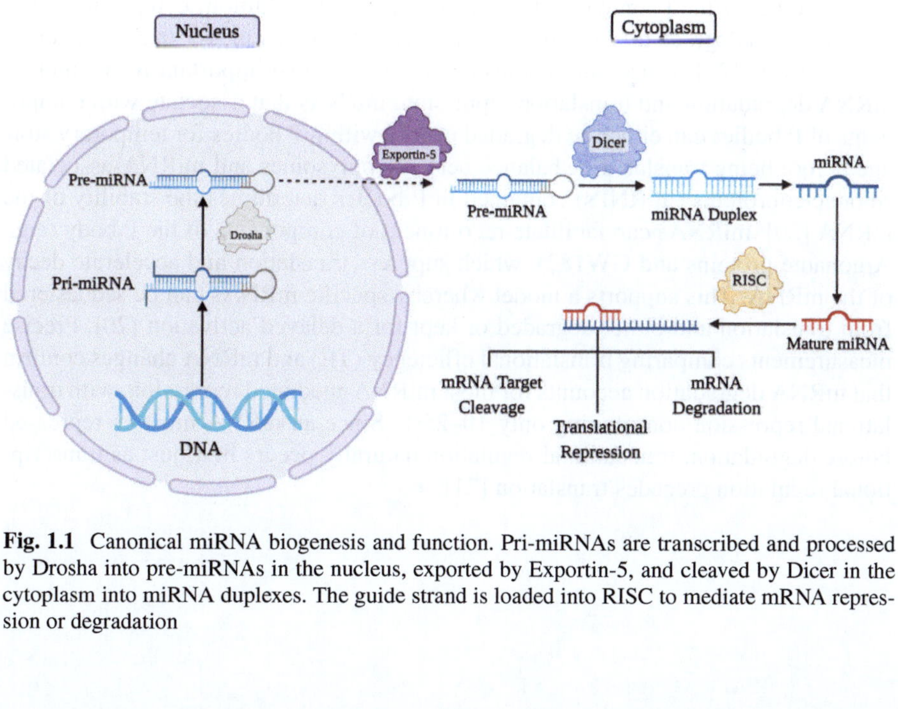

Fig. 1.1 Canonical miRNA biogenesis and function. Pri-miRNAs are transcribed and processed by Drosha into pre-miRNAs in the nucleus, exported by Exportin-5, and cleaved by Dicer in the cytoplasm into miRNA duplexes. The guide strand is loaded into RISC to mediate mRNA repression or degradation

Mechanisms of miRNA Action

miRNAs can regulate gene expression by degrading mRNA, repressing translation, or, in some cases, activating gene expression. The way miRNAs regulate gene expression is by binding to complementary sequences in target mRNAs to regulate the expression of the mRNA. The principal way that miRNAs regulate gene expression is posttranscriptionally [18]. miRNAs bind to sequences typically in the 3′ UTR of target mRNA to prevent protein translation. In addition, miRNAs can also regulate gene expression by binding to gene promoters, coding regions, or 5′ UTRs to regulate translation. Coding region binding may represent an alternative pathway of regulation especially when the 3′ UTR is too short or when it is necessary to control the protein levels of an alternative splice variant [18].

mRNA Degradation

miRNAs are critical gene regulators that repress expression at multiple levels, including mRNA stability and translation. miRNAs repress expression by binding to the 3′ untranslated region (3′UTR) of target mRNAs, which either promotes suppression of protein synthesis at the various phases of translation or promotes polypeptide degradation directly [18, 19]. The mRNA degradation begins with removal of the poly(A) tail at the 3′ end of the mRNA and/or removal of the 5′ cap made by enzymes such as DCP1 and DCP2. Without both the poly(A) tail and 5′ cap, the remaining RNA is subject to degradation by exonucleases, with the most likely candidate being the exonucleolytic enzyme Xrn1p. In addition, cytoplasmic exonucleases can degrade mRNAs missing a poly(A) tail by using a 3′ to 5′ exonucleolytic pathway [20]. Processing bodies (P bodies) are also important for regulating mRNA degradation and translation repression. mRNAs that associate with components of P bodies can either be degraded or stay within P bodies for temporary storage before being translated. A balance between polysomes and miRNA-associated ribonucleoproteins (miRNPs) contained in P bodies determines the stability of the mRNA [20]. miRNAs can facilitate recruitment of components of the P body (e.g., Argonaute proteins and GW182), which suppress translation and accelerate decay of the mRNA. This supports a model whereby specific mRNAs can be sequestered from translation and either degraded or kept for a delayed activation [20]. Precise measurements comparing translational efficiency (TE) and mRNA changes confirm that mRNA degradation accounts for most miRNA-mediated repression, with translational repression contributing only 10–25%. Since an mRNA must be repressed before degradation, translational regulation naturally occurs first, just as transcriptional regulation precedes translation [21].

Translational Repression

Shortly after recognition, miRNAs govern the expression of a multitude of protein-coding mRNAs through translational repression. Translational repression is a reversible process that allows accelerated changes to internal or external environmental cues. mRNA degradation cannot be reversed; however, miRNAs generally inhibit translation before they initiate mRNA decay by means of de-adenylation and decapping, whereupon mRNAs are rendered vulnerable to exonuclease activity. The current view of miRNA-mediated repression has transitioned from the idea that miRNAs inhibit the formation of 80S complexes or elongation, to the understanding that miRNAs primarily inhibit cap-dependent initiation [22]. This research describes an example when miRNA-mediated CAT-1 mRNA repression was observed to be reversed in stress conditions in a human hepatocarcinoma cell line. Specifically, once CAT-1 mRNA was released from P bodies, the repression of CAT-1 was lifted by *miR-122*, allowing CAT-1 mRNA to be recruited to polysomes where translation was resumed. This process was facilitated by the actions of HuR, an RNA-binding protein that interacted with the CAT-1 3′ UTR. Overall, these findings highlight the ability of RNA-binding proteins to control miRNA activity and support the ability to resume mRNA translation under stress conditions [23]. New evidence has shown that miRNA-induced gene repression occurs in stages, including translational repression, followed by mRNA de-adenylation, then degradation. The importance of the 5′ cap structure for translational repression is well established, while there remains divergence on the role of the poly(A) tail, with studies citing necessity on the one hand and minor impacts on the other. There is now solid evidence that translational repression occurs prior to mRNA deadenylation and decay, and through the Renilla luciferase reporter (RRL model), endogenous miRNAs are shown to successfully repress translation and initiate siRNA responses [24]. Figure 1.2 illustrates the mechanism of gene regulation via miRNAs. miRNAs are transcribed from DNA, and these miRNAs with respect to the RISC complex scan target mRNAs prior to moving forward with either translational repression (partial complementarity) or mRNA degradation (complete complementarity).

miRNA-Mediated Transcriptional Activation or Repression

miRNAs are mainly known for downregulating target mRNAs (thereby inactivating them) through binding to 3′ untranslated regions (UTRs). This process can result in either mRNA degradation or repression of translation. In addition to their well-characterized role in regulating gene expression posttranscriptionally, there is now a body of work to suggest that miRNAs interact with gene promoters to affect transcriptional regulation. This dual function allows miRNAs to activate or silence gene expression, referred to as transcriptional activation of gene (TAG) and transcriptional gene silencing (TGS), respectively. Furthermore, there is growing evidence

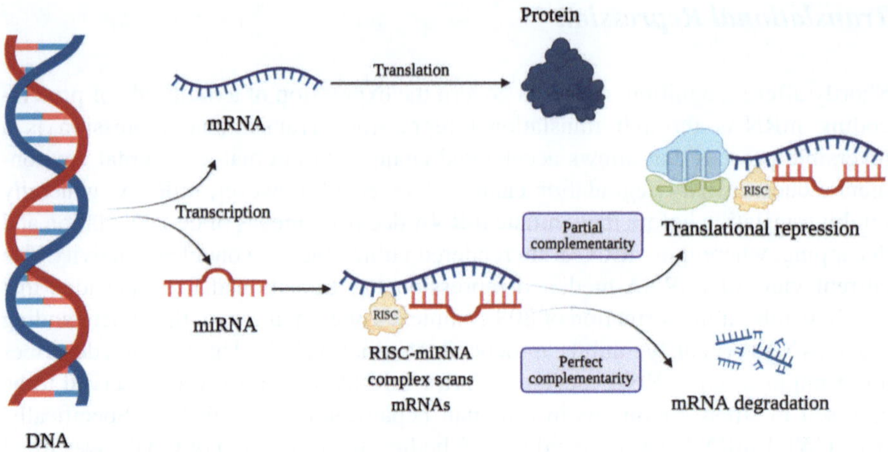

Fig. 1.2 Mechanisms of miRNA-mediated gene silencing. miRNAs guide the RISC complex to target mRNAs. Perfect complementarity leads to mRNA degradation, while partial complementarity results in translational repression, reducing protein synthesis

that miRNAs that locate in the nucleus can induce epigenetic changes to chromatin structure and affect gene expression, demonstrating their importance in tumorigenesis and disease [25, 26].

The miRNAs can either induce translational repression (blocking protein synthesis) or induce degradation of the mRNA to decrease the expression of the targeted gene. This process is mediated by the miRNA-induced silencing complex (miRISC) that consists of miRNA and Argonaute proteins to either block translation or target the mRNA for degradation. In addition to their known posttranscriptional functions, miRNAs can perform noncanonical functions [27]. Evidence is emerging which suggests that miRNAs act as transcriptional repressors or transcriptional activators by interacting with gene promoters and other regulatory regions. This is called miRNA-mediated RNA activation (RNAa) and involves direct recruitment of transcription factors and RNA polymerase II that promotes gene transcription. Nuclear miRNAs can also regulate transcription through effects on chromatin structure to either promote an open chromatin state that activates the expression of the gene, or a closed chromatin state that represses gene expression [25, 27].

For instance, *miR*-373 is a notable example of miRNA-induced transcriptional activation, as a recent study demonstrated that *miR-373* directly increased expression of both E-cadherin (CDH1) and cold shock domain-containing protein C2 (CSDC2) by targeting their promoters. In addition to their regulatory role in transcription, miRNAs can affect translation by binding to the 5′ UTR (untranslated region) of target mRNAs, to protect the target transcript from degradation while promoting ribosome recruitment for enhanced protein translation. These emerging mechanisms demonstrate the diverse roles miRNAs may play in modulating gene expression and provide insight into the complicated function of miRNAs in cellular biology, tumorigenesis, and disease [26].

miRNA Sponging by Long Noncoding RNAs and Circular RNAs

Long noncoding RNAs (lncRNAs) modulate gene expression by serving as sponges for miRNAs, a method called miRNA sponging. When lncRNAs bind to particular miRNAs, they prevent the miRNAs from binding to and regulating the message of target mRNAs, thereby modifying gene expression. LncRNAs are noncoding RNA molecules longer than 200 nucleotides that, while they do not encode protein-protein coding potential, play crucial roles in gene regulation [28]. In contrast, miRNAs are small RNA molecules that limit gene expression by binding to and pairing with complementary sequences on the mRNA targets, ultimately leading to either translational repression or degradation. Some lncRNAs contain sequence motifs that facilitate binding to a specific miRNA, allowing the lncRNAs to sequester the miRNAs from their target mRNAs. In this way, miRNA sponging regulates gene expression to affect cellular functions including development, differentiation, and disease pathology. Based on context, lncRNAs can either promote or inhibit gene expression, highlighting their importance in regulation of diverse biological processes [29, 30]. The dysregulation of lncRNA-mediated miRNA sponging contributes to aberrant gene expression and tumor progression in pathological conditions such as cancer [31]. There are several well-characterized lncRNAs that are known to function as miRNA sponges, impacting several cellular pathways, MALAT1, MACC1-AS1, and ANRIL are examples of lncRNAs that function as miRNA sponges. In head and neck squamous cell carcinoma, for example, ANRIL acts as a sponge for *miR-125a-3p*, which impacts tumor progression [32]. As research continues to unveil the complex interactions between lncRNAs and miRNAs, it has become increasingly apparent their contribution to pathology in disease contexts such as cancer, neurodegenerative diseases, and metabolic diseases. The opportunity to increase the understanding these interactions will likely have a great impact on the strategy for novel biomarkers and therapeutics. New approaches that impact precision medicine utilizing their interactions with miRNA will become available [32–34].

Circular RNAs (circRNAs) also regulate gene expression via screening activities to miRNAs, preventing miRNAs from binding to and repressing their target mRNAs. Therefore, circRNA modulation of miRNA is a major contributor to the regulation of several cellular processes and biological pathways. Unlike linear RNAs, circRNAs have a unique circular structure, allowing for greater stability and degradation resistance. The addition of bonding a specific miRNA means circRNAs can sequester miRNAs and limit their activities regulating gene expression. The sponging of miRNAs reminisces of antiquity to modulate gene expression, which influences several cellular processes that can affect the progression of disease [35–37]. ciRS-7 (CDR1as) is a famous and studied circRNA that sponges *miR-7*, while circNRIP1 and circUBE2J2 sponge *miR-149-5p* and *miR-370-5p*, respectively. These observations become important in the disease context, such as with cancer, where circRNA-miRNA networks contribute to growth and malignancy [38–40].

circRNAs represent an important implication not only for altering disease mechanisms, but circRNAs may also serve as biomarkers for disease diagnosis and prognosis, and may be an important target for future therapies. Future work in this area and increasing biological understanding to target circRNA-miRNA interactions hold much promise for therapeutic precision medicine and unique treatment strategies [41, 42].

Regulatory Roles of miRNAs in Cellular Processes

The miRNAs are crucial for regulating important cellular processes such as proliferation, differentiation, apoptosis, and development through precise regulation of gene expression. However, miRNAs can also target regions other than the 3′ UTR, such as the 5′ UTR, coding sequence, and even gene promoters, leading to a range of potential regulatory outcomes [43, 44].

miRNAs oversee cell cycle progression in cell proliferation and differentiation by either supporting or antagonizing growth, influencing tissue formation in the process. They are also involved in the regulation of apoptosis, thereby either promoting or initiating programmed cell death and determining cell survival, which are also vital to tissue homeostasis [45]. miRNAs also influence metabolism via glucose balance, lipid metabolism, energy production, and other pathways. They regulate various aspects of the immune system from the development of immune cells to regulatory T cell roles in the immune response. The role of dysregulated miRNAs in cancer, heart disease, neurodegenerative diseases, and metabolic diseases establishes their importance in disease pathogenesis [44, 46]. miRNA actions take place through a multistep process that always begins with miRNA biogenesis in which precursor miRNAs are transcribed, processed, and ultimately loaded into the RNA-induced silencing complex (RISC) [47]. Mature miRNAs bind to specific mRNAs via their seed-sequences enabling posttranscriptional regulation to occur. In addition to acting as intracellular regulators, miRNAs can also be secreted into extracellular spaces where they act as intercellular mediators to influence the cellular activities of neighboring cells in a paracrine or systemic manner [2].

Numerous miRNAs have been recognized for having important functions in particular pathways. miR-128 functions in neurodevelopment, neuronal repair, and brain activity, and the *miR-17* family promotes proliferation in myeloid cells by targeting sequestome-1, an important regulator of protein degradation through autophagy [43, 48]. In stem cell biology, the clusters *miR-200b, miR-200c, miR-106a-363*, and *miR-302-367*, as well as the *miR-93/106b* family, participate in mesenchymal-to-epithelial transition during iPSC reprogramming, while *miR-19*, *miR-17*, and both *miR-290* and the *miR-8* family are necessary for the activation and maintenance of pluripotency [49]. *miR-135b* modulate Hippo signaling pathway members LATS2, BTRC, NDR2, and LZTS, while *miR-101*, *miR-129-5p*, and *miR-221* all target the 3′ UTR of FMR1 and modulate its levels. Overall, miRNAs act to regulate gene expression, impacting biological processes and diseases.

Understanding these functions and mechanisms have important implications for therapeutic opportunities, including miRNA based therapeutic initiatives in production of therapies with intended outcomes for disease treatment [50].

Experimental Techniques for miRNA Study

To study miRNAs, many experimental methodologies are utilized, each having its own benefits when profiling and examining miRNA expression and function. These primary experimental methodologies for studying miRNAs include quantitative reverse transcription PCR (qRT-PCR), hybridization-based microarrays, and high-throughput sequencing (RNA-seq), among others [51]. qRT-PCR is a method for examining miRNA expression that is also high in sensitivity and accuracy. In this procedure, the total RNA is isolated, including the small RNA fraction, next reverse transcription produces the miRNA complementary DNA (cDNA), and finally, PCR amplification is carried out using an assay specific for the miRNA with appropriate probes [52]. qRT-PCR quantitates the level of specific miRNAs and is the preferred assay for targeted miRNA studies. However, the necessary normalization, such as using the spike-in controls, must be performed for valid comparisons across samples. A common approach is to employ the stem-loop RT-based assay, which provides improved specificity for detection of a miRNA and the ability to differentiate a closely related miRNA species differing by a single nucleotide [52].

Hybridization-based techniques like microarrays have emerged as a high-throughput methodology to profile miRNA expression in different samples. In brief, labeled RNA obtained from tissue or cell sample is hybridized to arrays that contain probes for all known mature miRNAs of a given species. Microarrays have the capacity to simultaneously analyze many miRNAs, making them very suitable for training samples analysis at a large scale. The technology is built upon well-established, optimized protocols that make miRNA analysis feasible, effective and affordable [53]. High-throughput sequencing (RNA-seq) allows for an integrated approach to studying miRNAs. It enables one to quantify expression of known miRNAs as well as discover novel miRNAs. The RNA-seq process entails sequencing small RNA molecules and subsequently analyzing the sequence data through various bioinformatics analysis methods to identify miRNAs and determine expression. RNA-seq is beneficial for detecting low abundance miRNAs and is effective for assessing global miRNA profiles. This technique, however, requires specialized sequencing equipment and more advanced bioinformatics capabilities and software for data processing and analysis [54].

Along with these primary methods, there are other individual methods for specific miRNA studies. Northern blotting, which is a classical method, can identify both mature miRNAs and precursor miRNAs, but it requires a relatively large amount of RNA and is time-consuming. In situ hybridization enables spatial localization of miRNA expression in tissues and provides insights into the functional roles of miRNAs in specific cellular contexts. Nanomaterial-based detection approaches utilize advanced materials, such as gold nanoparticles, magnetic nanoparticles, silver nanoclusters, and quantum dots, for more sensitive and specific detection of miRNAs [55–57]. The functional validation of putative miRNA targets is usually done by luciferase reporter assays, which measure changes in luciferase activity that result from miRNA binding to a target sequence. Western blotting and proteomic methods also determine the downstream effects of miRNA regulation, such as changes in protein expression patterns. Overall, these approaches lead to an integrated understanding of miRNA biology, from expression profiling to functional validation, leading to potential applications in disease diagnostics or therapeutics [58, 59].

Role of miRNAs in Cancer

miRNAs are important regulators of gene expression and are often dysregulated in cancers, also function as either oncogenes (driving tumor formation) or tumor suppressors (preventing tumor growth). miRNAs modulate different cellular pathways to ultimately contribute to hallmarks of cancer, such as sustaining proliferative signaling by regulating cell growth and division, and avoiding growth suppressors by modulating factors that normally inhibit uncontrolled cell proliferation [60]. miRNAs also participate in resisting cell death such as preventing apoptosis, stimulating invasion and metastasis that enhance cancer cell migration, and facilitate angiogenesis by inducing new blood vessel growth to support tumor growth [61].

Due to their critical role in the induction and progression of cancer, miRNAs are being explored as potential therapeutic targets for cancer therapy. Moreover, altered miRNA expression can act as a biomarker for the detection, diagnosis, and prognosis of cancer, providing an exciting avenue for early identification of disease and patient stratification [62]. For example, *miR-21* is overexpressed in various cancers and has been linked to poor prognosis, cell survival, and metastasis. *miR-10b* has also been implicated in ontogeny and metastasis of breast cancer to the lung. In further example, *miR-34* is a well-established tumor suppressor, which has been shown to be downregulated in various cancers to promote tumor immunity by targeting PD-L1. The Let-7 family, another example of tumor suppressor miRNA, is silenced in numerous cancers to promote malignant transformation [63, 64]. Distinct miRNA profiles have been identified in different cancers, and specific signatures have also been proposed for specific cancers to further classify and diagnostically address cancer. Importantly, miRNAs are stable in body fluids (e.g., blood, saliva, and urine) and could be considered preferably for development of noninvasive diagnostics [65]. Since in cancers, detection of disease through miRNA profiling could lead to earlier diagnosis and potentially better therapeutic outcomes through timely intervention, this underscores the impact of identifying early disease. Further, miRNA expression levels can be predictive of disease progression and therapeutic response to assist physicians' therapeutic decision-making for individual patients [66].

Multiple miRNAs have developed into valuable biomarkers for the clinic. To illustrate, *miR-21, miR-155*, and *miR-221/222* are involved in tumor proliferation and angiogenesis in breast cancer; circulating *miR-332* and *miR-195* can differentiate between local breast cancer and metastatic breast cancer; and many specific miRNAs can differentiate between castration-sensitive and castration-resistance prostate cancer and offer prognostic information. The expanding knowledge of miRNA biology is highly promising for the development of new diagnostic and therapeutic approaches for cancer care and paves their way to precision oncology [67]. As reported by Malayaperumal et al. [1], miR-122 expression is significantly reduced in colon cancer cells and is involved in modulating both apoptotic pathways and inflammatory cytokine levels. When *miR-122* is overexpressed in HCT116, SW480, and SW620 cell lines, there is a marked increase in the levels of pro-apoptotic proteins Caspase-3, Caspase-9, and BAX, alongside a decrease in the anti-apoptotic protein BCL2. This shift promotes apoptosis by altering the balance between cell survival and programmed cell death [1]. Figure 1.3 illustrates how mutations in oncogenic or tumor suppressor miRNAs can promote tumorigenesis through increased oncomiR expression or decreased tumor suppressor miRNA levels.

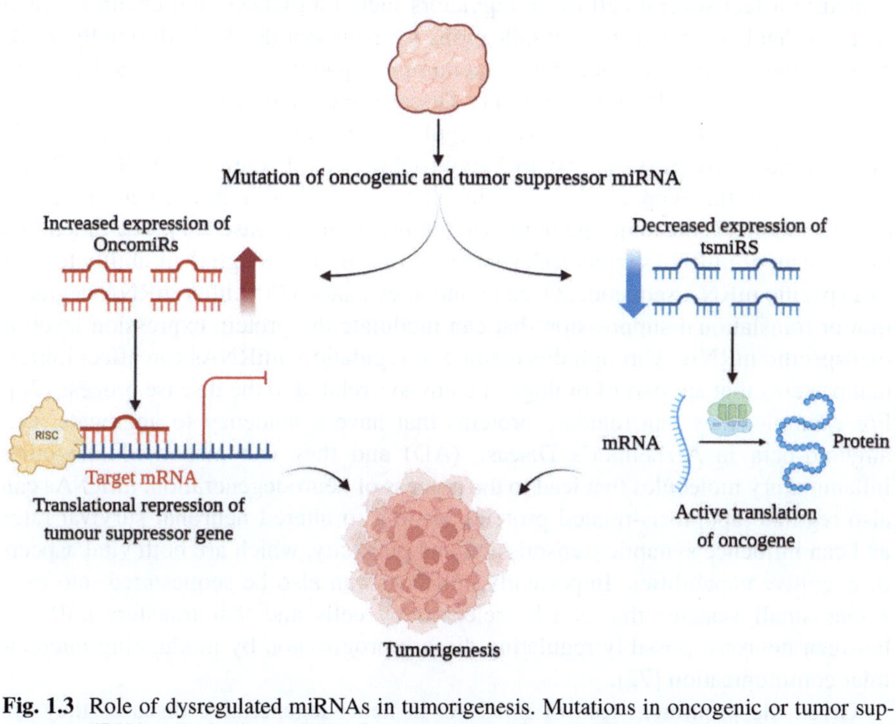

Fig. 1.3 Role of dysregulated miRNAs in tumorigenesis. Mutations in oncogenic or tumor suppressor miRNAs can lead to cancer. Overexpression of oncomiRs represses tumor suppressor genes, while reduced expression of tumor-suppressive miRNAs (tsmiRs) allows active translation of oncogenes, both contributing to tumor development

Role of miRNAs in Neurodegenerative Diseases

miRNAs have a significant role in posttranscriptional regulation of gene expression, and misregulated miRNA has been shown to be associated with developing and advancing neurodegenerative disorders, specifically Alzheimer's disease (AD), Parkinson's disease (PD), Huntington's disease (HD), and Amyotrophic lateral sclerosis (ALS) [68]. Studies indicate the pattern of miRNA expression is dramatically altered in the brains of patients afflicted with these disorders, which in turn can affect the downstream impact on key cellular mechanisms such as neuronal function, metabolism, and survival. Altered expression of specific miRNAs may function in disease processes through modulation of genes associated with protein aggregation, inflammation, neuronal death, and cholesterol homeostasis—all of which are hallmark mechanisms in neurodegeneration [69]. There are specific miRNAs identified as contributing substantially to neuronal health. For example, *miR-132* is known to contribute to neuronal axon growth, neural migration, and neuroplasticity. In AD, *miR-107* and *miR-34* contribute to disease processes, with *miR-107* regulating the cell cycle and neuronal survival and *miR-34* regulating neuronal apoptosis and inflammation [70].

In Parkinson's Disease (PD), the downregulation of *miR-34* has been demonstrated to affect several cell cycle regulators including CDK4 and cyclin D2, and such mechanisms would have implications for neuronal death. Additionally, miRNAs regulate mitochondrial function and immune pathways, both of which are significantly altered in PD. In Huntington's Disease (HD), molecules such as *miR-100, miR-151-3p,* and *miR-219-2-3p* are upregulated, suggesting they could have a role in the disease's progression [70]. In Amyotrophic Lateral Sclerosis (ALS), miRNAs also modulate the expression of proteins that are involved in motor neuron degeneration and this is the major pathological feature of this disease [71]. The functional mechanism of miRNAs in neurodegenerative disease is through their ability to bind to a specific mRNA sequence. Once bound, they can lead to either mRNA degradation or translational suppression that can modulate the protein expression level of that specific mRNA. Through this method of regulation, miRNAs can affect important proteins that are part of biological pathways related to the disease process [72]. For example, they can regulate proteins that have a tendency to aggregate, i.e., amyloid-beta in Alzheimer's Disease (AD) and they can also modulate other inflammatory molecules that lead to the process of neurodegeneration. miRNAs can also regulate apoptosis-related proteins leading to altered neuronal survival rates and can influence synaptic transmission and plasticity, which are both vital aspects of cognitive capabilities. Importantly, miRNAs can also be sequestered into exosomes small vesicles that can be released by cells and that transfers miRNAs between neurons, possibly regulating disease progression by modulating intercellular communication [72].

Given the prominent role of miRNAs in neurodegenerative diseases they are being pursued as both therapeutic targets and biomarkers. Researchers are exploring treatment options that involve restoring normal function of beneficial miRNAs via miRNA mimics or inhibiting the expression of harmful miRNAs [73]. Several delivery methods such as viral vectors, nanoparticles, and other exosome-based approaches for targeting miRNAs while increasing effectiveness and limiting off-target side effects. Delivery methods will expand research into miRNA-based therapies into more viable treatment options for reversing, or slowing neurodegenerative disease progression by targeting the underlying biological mechanisms [73].

Role of miRNAs in Cardiovascular Diseases

MicroRNAs are critical components of cardiovascular diseases (CVD). miRNAs at times serve as molecular switches that bind to the messenger RNA (mRNA) responsible for making proteins and regulate the stability of the mRNA, along with the translation of mRNA into proteins, which impacts a number of cellular activities including cell growth, differentiation, apoptosis, and inflammation, all relevant to cardiovascular pathology [74]. In the cardiovascular system, miRNAs regulate the behavior and function of key cell types including cardiomyocytes, endothelial cells, and vascular smooth muscle cells. Additionally, modifications of key miRNAs have been shown to be related to several cardiovascular diseases [75]. For example, during atherosclerosis, key miRNAs have been found to contribute to endothelial dysfunction, inflammatory response, and plaque formation [76]. In heart failure, changes in miRNA expression are directly responsible for driving cardiac remodeling, hypertrophy, and fibrosis. Certain miRNAs can also influence the regulation of the hearts rhythm, change in miRNA concentration can result in arrhythmias. miRNAs can also be released from damaged cardiac tissue during cardiac injury, especially in conditions such as myocardial infarction, which can circulate in the blood allowing miRNAs to be studied as biomarkers for diagnosis and to assess heart damage and severity of disease. Therefore, miRNAs are being studied as biomarkers for diagnosis, prognosis, along with therapeutic modulation of the miRNAs [77]. For instance, *miR-21* has been investigated in relation to a wide variety of contexts pertaining to vascular smooth muscle cells and cardiomyocyte growth, apoptosis, and cardiac fibroblasts; thus, it has been described as a potential therapeutic target in different surroundings related to CVD [78].

Role of miRNAs in Immunity

MicroRNAs are essential regulators of both the innate and adaptive immune systems, as they modulate the development, function, and responses of immune cells to pathogenesis and are implicated in immune-mediated diseases, and are thus key regulators of immune homeostasis [79]. In innate immunity, miRNAs regulate the differentiation and function of a variety of crucial immune cells, including macrophages, dendritic cells (DCs), natural killer (NK) cells, and neutrophils. In adaptive immunity, miRNAs modulate the development and function of T and B cells, which are integral to antigen-specific immunity [79, 80]. For example, *miRNA-155* can modulate the differentiation of T helper cells as well as the germinal center reaction that is required for antibody secretion. *miRNA-124* also plays an important role in the development of the immune system, regulating immune responses and the pathogenesis of inflammatory diseases. *miRNA-181c-5p* enhances the phagocytic capacity of bone marrow-derived macrophages, and *miRNA-155* is upregulated upon TCR-mediated T cell activation, regulating CD8 T cell proliferation by targeting SOCS1. miRNAs function in immune surveillance by inducing effective pathogen clearance, while maintaining immune system homeostasis [81–83].

In addition to immune cell development, miRNAs can also regulate immune responses by modulating inflammation, cytokine production, and immune checkpoint expression. They play essential roles in the regulation of inflammatory pathways, in the context of infection, and chronic inflammatory conditions. By controlling cytokine levels, miRNAs directly influence the intensity and duration of immune responses and limit excessive inflammation that can contribute to tissue damage. Furthermore, miRNAs regulate the expression of immune checkpoint molecules that are critical for maintaining self-tolerance and for preventing autoimmune diseases [84, 85].

Disruption of miRNAs has been implicated in numerous pathological processes, specifically, autoimmune disorders, cancer, and several other immunological pathologies. In the context of autoimmune disease, deregulation of miRNA expression can lead to hyperactivation of the immune response/barrier, resulting in loss of tolerance that could induce disease progression. In cancer, miRNAs could operate as either oncogenes or tumor suppressors, which could affect tumor immune evasion, proliferation, and metastasis. Additionally, deregulation of miRNAs has been shown to be associated with several inflammatory and allergic disorders, demonstrating their importance in maintaining immune homeostasis [86, 87]. miRNAs have become attractive targets for the treatment of immune-related disorders because of their profound effects on immune regulation. miRNA mimic and antisense oligonucleotides directed to specific miRNAs have been developed to prevent the loss of beneficial miRNAs and inhibit the action of pathogenic miRNAs. Overall, advances in miRNA modification of immune function may lead to new treatment strategies in autoimmune disease, cancer immunotherapy, and in the realm of inflammatory disorders [88].

Therapeutic Potential of miRNAs in Other Diseases

miRNAs are novel intriguing therapeutic targets since they can regulate gene expression and alter many important cellular processes relevant to human diseases (for example, cancer or neurological disorders) [89]. As with many diseases, miRNAs have an important role in the pathogenesis for a variety of the so-called non-communicable diseases like cancer. In this spirit, miRNAs can behave as either an oncogene or a tumor suppressor. Oncogenic miRNAs promote tumor growth, enhancing the proliferation, invasion, and metastasis of malignant cells, while cancer cell growth is inhibited by tumor suppressor miRNAs; hence, downregulation might promote the growth of tumors [90]. In addition to cancer, the roles of miRNAs are versatile in neurological disorders, where miRNAs are often dysregulated, which impacts neuronal function and survival in diseases such as neurodegenerative disease, developmental disorders, and stroke. miRNAs also have relevance in other issues, including metabolic disorders (for example, type 2 diabetes) and tissue repair (for instance, tendon injuries) [91, 92].

Due to their roles in numerous diseases, miRNAs have been described as promising therapeutic applications. When it comes to cancer treatment, miRNA mimics can restore the function of cancer suppressor miRNAs that are downregulated in cancers, thereby inhibiting the growth of cancer cells, whereas antisense oligonucleotides, called antimiRs, can inhibit oncogenic miRNAs to enhance their tumor-promoting effects [93]. In neurodegenerative diseases, miRNA therapies focus on restoring or inhibiting altered miRNAs, which could improve neuronal function and treatment outcomes. Within the context of altered cellular processes, miRNA-targeted treatment strategies are being developed to regulate cell behavior, such as inflammation/inflammation resolution, apoptosis, or tissue repair, where any of these processes may be exploited as therapeutic strategies [93, 94].

Nevertheless, substantial obstacles exist that must be overcome before miRNA-based treatments can achieve mainstream acceptance. Off-target effects are a possibility, since miRNAs often regulate multiple genes at the same time. Targeted and efficient delivery systems are also an obstacle, as miRNA-based medicines must reach the target tissue while remaining stable and active. Extensive research and validation are needed to assess the safety and adequacy of miRNA therapeutics for clinical use [95, 96]. Future research in miRNA therapeutics will attempt to create more specific therapeutic schemes; enhance delivery systems through nanoparticles, exosomes, and other innovative systems; and additionally implement miRNA profiling as a personalized medicine tool. Patients' miRNA signatures can be identified to create personalized targeted therapies in the future that maximize target effects while minimizing potential adverse effects, and create a paradigm shift in the use of miRNA-based therapies in modern medicine [95, 96]. Figure 1.4 shows therapeutic strategies targeting miRNAs, including miRNA mimics to restore tumor suppressor miRNAs and anti-miRNA approaches such as AMOs, masking, and sponges to inhibit oncogenic miRNAs.

Conclusion

MiRNAs are critical regulators of gene expression, impacting many cellular activities, such as development, differentiation, and disease progression. Their production is a highly coordinated process involving several enzymatic steps, from transcription to maturation and recognition of targets. Altered processing and function of

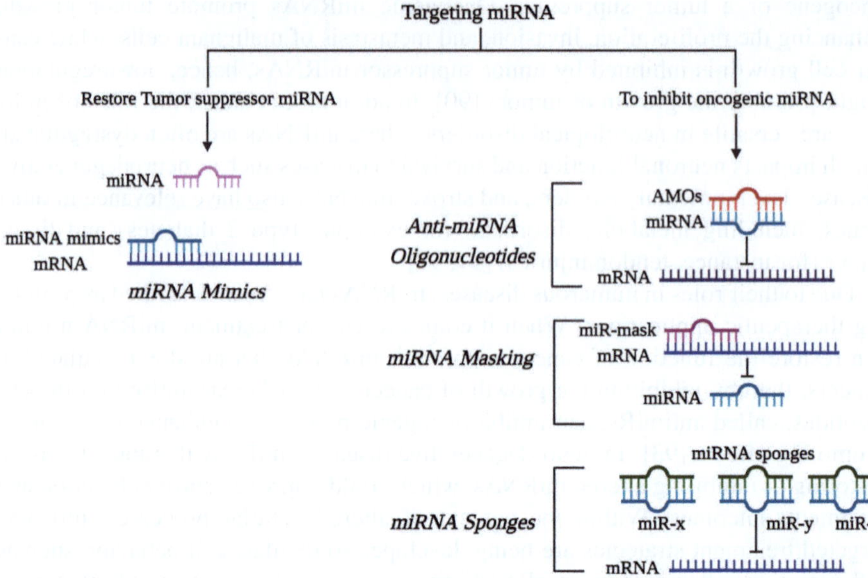

Fig. 1.4 Therapeutic strategies for targeting miRNAs. miRNA mimics restore tumor suppressor miRNA activity, while anti-miRNA oligonucleotides (AMOs), miRNA masking, and miRNA sponges inhibit oncogenic miRNAs. These approaches modulate gene expression for therapeutic benefit

miRNAs have been implicated in many ailments, with special emphases on cancer, neurodegenerative diseases, and metabolic disorders. These studies attest to the importance of miRNAs in normal physiology and pathology. Despite advances in miRNA biogenesis studies, several challenges remain. For instance, the complexity of miRNA-mRNA interactions, with potential for posttranscriptional modifications of both, as well as other regulatory components that could engage and influence miRNA function remain areas of active investigation. Moreover, as promising new sequencing technologies, bioinformatics tools, and single-cell approaches become available, they will enable us to quantify miRNA dynamics in greater detail for future translational applications.

Although the future of miRNA-mediated therapies is promising, it requires further optimization. The development of efficient delivery mechanisms such as nanoparticles and exosomes is vital for improving the stability and specificity of miRNA therapeutics. Personalized approaches to miRNA-based medicine, leveraging the study of miRNA profiles in the clinic may offer a pathway to targeted therapies. Future research will also entail elucidation of novel regulatory mechanisms controlling miRNA biogenesis and function as the groundwork for new therapeutic targets. Overall, miRNA biology is an exciting and rapidly expanding discipline with novel ideas related to gene regulation and therapeutic innovations.

This chapter has provided us a comprehensive overview of the steps that involves in miRNA biogenesis, transcription and nuclear processing, cytoplasmic maturation, and incorporation into the RISC. This chapter also highlights the various mechanisms by which miRNAs function, through mRNA degradation, translational repression or in some case, transcriptional regulation, as well as the miRNA sponging actions of lncRNAs and circRNAs. Building upon this foundation, the present book offers a comprehensive and multidisciplinary overview of miRNAs in cancer biology and translational medicine. The following chapters examine specific topics such as the role of miRNAs in cancer stem cell maintenance, cell cycle regulation, and epigenetic modulation. As well, organ- and cancer-specific insights are provided with focused discussions on the role of miRNAs in lung, breast, liver, gastrointestinal, prostate, oral, and cervical cancers. The role of miRNAs in processes such as radiotherapy resistance, epithelial–mesenchymal transition (EMT), and tumor microenvironment modulation is also explored. It is important to acknowledge the clinical and technological frontiers discussed in the book, which includes the potential of circulating and exosomal miRNAs for diagnostic and prognostic purposes, and therapeutic approaches utilizing miRNA mimics and inhibitors. With regard to AI and bioinformatics used for miRNA profiling and target prediction, the evolving role of computational tools in precision oncology. Lastly, the later part of the book discusses the translational aspects of how miRNAs move from their potential utility in early detection, prognosis, and personalized cancer therapy. Overall, the book provides a overview of how miRNAs have many different functions, as well as their regulatory features and clinical utility, making it a useful resource for the research and clinical application in molecular biology, oncology, and translational medicine.

Acknowledgments We want to thank the Chettinad Academy of Research and Education, Chettinad Hospital, and Research Institute for providing the facilities. The authors are also thankful to Department of Biotechnology, Ministry of Science & Technology, and Government of India for providing support to Mr. Subhamay Adhikary (Fellowship IDDBT/2021-22/CARE/1592).

Conflict of Interest All authors declare that they have no conflicts of interest.

Authors' Contributions Antara Banerjee and Surajit Pathak conceptualized and designed the study. Sibin Nambidi, Subhamay Adhikary, Sneha Pallatt, Amit Dey, Antara Banerjee, and Surajit Pathak were involved in the manuscript preparation. Sibin Nambidi and Surajit Pathak designed the images. Surajit Pathak and Antara Banerjee critically reviewed the chapter. All authors approved the final version of the chapter.

References

1. Malayaperumal S, Sriramulu S, Jothimani G, Banerjee A, Zhang H, Mohammed Rafi ST, Ramachandran I, Nr RK, Sun X-F, Pathak S (2025) MicroRNA-122 overexpression suppresses the colon cancer cell proliferation by downregulating the astrocyte elevated gene-1/metadherin oncoprotein. Ann Med 57(1):2478311. https://doi.org/10.1080/07853890.2025.2478311
2. O'Brien J, Hayder H, Zayed Y, Peng C (2018) Overview of MicroRNA biogenesis, mechanisms of actions, and circulation. Front Endocrinol 9(402). https://doi.org/10.3389/fendo.2018.00402
3. Reddy KB (2015) MicroRNA (miRNA) in cancer. Cancer Cell Int 15(1). https://doi.org/10.1186/s12935-015-0185-1

4. Bofill-De Ros X, Andersson U (2023) Recent progress in miRNA biogenesis and decay. RNA Biol 21(1):1–8. https://doi.org/10.1080/15476286.2023.2288741
5. Ardekani AM, Naeini MM (2024) The role of MicroRNAs in human diseases. Avicenna J Med Biotechnol 2(4):161
6. He L, Hannon GJ (2004) MicroRNAs: small RNAs with a big role in gene regulation. Nat Rev Genet 5(7):522–531. https://doi.org/10.1038/nrg1379
7. Li Y, Kowdley KV (2012) MicroRNAs in common human diseases. Genomics Proteomics Bioinformatics 10(5):246–253. https://doi.org/10.1016/j.gpb.2012.07.005
8. Song R, Zhang L (2024) MicroRNAs and therapeutic potentials in acute and chronic cardiac disease. Drug Discov Today 29(11):104179. https://doi.org/10.1016/j.drudis.2024.104179
9. Annese T, Tamma R, De Giorgis M, Ribatti D (2020) microRNAs biogenesis, functions and role in tumor angiogenesis. Front Oncol 10. https://doi.org/10.3389/fonc.2020.581007
10. Lee Y, Kim M, Han J, Yeom K-H, Lee S, Baek SH, Kim VN (2004) MicroRNA genes are transcribed by RNA polymerase II. EMBO J 23(20):4051–4060. https://doi.org/10.1038/sj.emboj.7600385
11. de Mello AS, Ferguson BS, Shebs-Maurine EL, Giotto FM (2024) MicroRNA biogenesis, gene regulation mechanisms, and availability in foods. Non-Coding RNA 10(5):52. https://doi.org/10.3390/ncrna10050052
12. Davis BN, Hata A (2009) Regulation of MicroRNA biogenesis: a miRiad of mechanisms. Cell Commun Signal 7(1). https://doi.org/10.1186/1478-811x-7-18
13. MacFarlane L-A, Murphy R, P. (2010) MicroRNA: biogenesis, function and role in cancer. Curr Genomics 11(7):537–561. https://doi.org/10.2174/138920210793175895
14. Stavast C, Erkeland S (2019) The non-canonical aspects of MicroRNAs: many roads to gene regulation. Cells 8(11):1465. https://doi.org/10.3390/cells8111465
15. Abdelfattah AM, Park C, Choi MY (2014) Update on non-canonical microRNAs. Biomol Concepts 5(4). https://doi.org/10.1515/bmc-2014-0012
16. Park S, Kang I, Shin C (2021) MicroRNA clustering on the biogenesis of suboptimal microRNAs. Appl Biol Chem 64(1). https://doi.org/10.1186/s13765-021-00624-3
17. Havens MA, Reich AA, Duelli DM, Hastings ML (2012) Biogenesis of mammalian microRNAs by a non-canonical processing pathway. Nucleic Acids Res 40(10):4626–4640. https://doi.org/10.1093/nar/gks026
18. Sadakierska-Chudy A (2020) MicroRNAs: diverse mechanisms of action and their potential applications as cancer epi-therapeutics. Biomolecules 10(9):1285. https://doi.org/10.3390/biom10091285
19. Valinezhad Orang A, Safaralizadeh R, Kazemzadeh-Bavili M (2014) Mechanisms of miRNA-mediated gene regulation from common downregulation to mRNA-specific upregulation. Int J Genomics 2014:1–15. https://doi.org/10.1155/2014/970607
20. Eichhorn SW, Guo H, McGeary SE, Rodriguez-Mias RA, Shin C, Baek D, Hsu S, Ghoshal K, Villén J, Bartel DP (2014) mRNA destabilization is the dominant effect of mammalian MicroRNAs by the time substantial repression ensues. Mol Cell 56(1):104–115. https://doi.org/10.1016/j.molcel.2014.08.028
21. Naeli P, Winter T, Hackett AP, Alboushi L, Jafarnejad SM (2022) The intricate balance between microRNA-induced mRNA decay and translational repression. FEBS J 290(10). https://doi.org/10.1111/febs.16422
22. Bhattacharyya SN, Habermacher R, Martine U, Closs EI, Filipowicz W (2006) Relief of microRNA-mediated translational repression in human cells subjected to stress. Cell 125(6):1111–1124. https://doi.org/10.1016/j.cell.2006.04.031
23. Ricci EP, Limousin T, Soto-Rifo R, Rubilar PS, Decimo D, Ohlmann T (2012) miRNA repression of translation in vitro takes place during 43S ribosomal scanning. Nucleic Acids Res 41(1):586–598. https://doi.org/10.1093/nar/gks1076
24. Ramchandran R, Chaluvally-Raghavan P (2017) miRNA-mediated RNA activation in mammalian cells. RNA Activation:81–89. https://doi.org/10.1007/978-981-10-4310-9_6

25. Liu J, Yang T, Huang Z, Chen H, Bai Y (2022) Transcriptional regulation of nuclear miRNAs in tumorigenesis (review). Int J Mol Med 50(1). https://doi.org/10.3892/ijmm.2022.5148
26. Wilczynska A, Bushell M (2014) The complexity of miRNA-mediated repression. Cell Death Differ 22(1):22–33. https://doi.org/10.1038/cdd.2014.112
27. Ma B, Wang S, Wu W, Shan P, Chen Y, Meng J, Xing L, Yun J, Hao L, Wang X, Li S, Guo Y (2023) Mechanisms of circRNA/lncRNA-miRNA interactions and applications in disease and drug research. Biomed Pharmacother 162:114672. https://doi.org/10.1016/j.biopha.2023.114672
28. Schwarzenbach H, Gahan PB (2023) Interplay between LncRNAs and microRNAs in breast cancer. Int J Mol Sci 24(9):8095. https://doi.org/10.3390/ijms24098095
29. Entezari M, Taheriazam A, Orouei S, Fallah S, Sanaei A, Hejazi ES, Kakavand A, Rezaei S, Heidari H, Behroozaghdam M, Daneshi S, Salimimoghadam S, Mirzaei S, Hashemi M, Samarghandian S (2022) LncRNA-miRNA axis in tumor progression and therapy response: an emphasis on molecular interactions and therapeutic interventions. Biomed Pharmacother 154:113609. https://doi.org/10.1016/j.biopha.2022.113609
30. Paraskevopoulou MD, Hatzigeorgiou AG (2016) Analyzing MiRNA–LncRNA interactions. Long Non-Coding RNAs:271–286. https://doi.org/10.1007/978-1-4939-3378-5_21
31. Wu X, Wang F, Li HF, Hu Y, Jiang L, Zhang F, Li ML, Wang XA, Jin Y, Zhang YJ, Lu W, Wu W, Shu Y, Weng H, Yang C, Bao RF, Liang HB, Wang Z, Gong W, Li Z (2017) LncRNA-PAGBC acts as a microRNA sponge and promotes gallbladder tumorigenesis. EMBO Rep 18(10):1837–1853. https://doi.org/10.15252/embr.201744147
32. Zhang X, Zhou Y, Chen S, Li W, Chen W, Gu W (2019b) LncRNA MACC1-AS1 sponges multiple miRNAs and RNA-binding protein PTBP1. Oncogene 8(12). https://doi.org/10.1038/s41389-019-0182-7
33. Nambidi S, Shruthi KB, Kanimozhe SP, Banerjee A, Kondaveeti SB, Duttaroy AK, Pathak S (2025) A review of circular RNAs in colorectal cancer: insights into biomarker discovery and therapeutic targeting. Front Oncol 15. https://doi.org/10.3389/fonc.2025.1549046
34. Yin J, Zeng X, Ai Z, Yu M, Wu Y, Li S (2020) Construction and analysis of a lncRNA-miRNA-mRNA network based on competitive endogenous RNA reveal functional lncRNAs in oral cancer. BMC Med Genet 13(1). https://doi.org/10.1186/s12920-020-00741-w
35. Jarlstad Olesen MT, Kristensen LS (2021) Circular RNAs as microRNA sponges: evidence and controversies. Essays Biochem. https://doi.org/10.1042/ebc20200060
36. Xiao J, Joseph S, Xia M, Teng F, Chen X, Huang R, Zhai L, Deng W (2022) Circular RNAs acting as miRNAs' sponges and their roles in stem cells. J Clin Med 11(10):2909. https://doi.org/10.3390/jcm11102909
37. Yu C-Y, Kuo H-C (2019) The emerging roles and functions of circular RNAs and their generation. J Biomed Sci 26(1). https://doi.org/10.1186/s12929-019-0523-z
38. Hashemi M, Khosroshahi EM, Daneii P, Hassanpoor A, Eslami M, Koohpar ZK, Asadi S, Zabihi A, Jamali B, Ghorbani A, Nabavi N, Memarkashani MR, Salimimoghadam S, Taheriazam A, Tan SC, Entezari M, Farahani N, Hushmandi K (2024) Emerging roles of CircRNA-miRNA networks in cancer development and therapeutic response. Non-Coding RNA Res 10:98–115. https://doi.org/10.1016/j.ncrna.2024.09.006
39. Zhang L, Liu Y, Tao H, Zhu H, Pan Y, Li P, Liang H, Zhang B, Song J (2021) Circular RNA circUBE2J2 acts as the sponge of microRNA-370-5P to suppress hepatocellular carcinoma progression. Cell Death Dis 12(11). https://doi.org/10.1038/s41419-021-04269-4
40. Zhang X, Wang S, Wang H, Cao J, Huang X, Chen Z, Xu P, Sun G, Xu J, Lv J, Xu Z (2019a) Circular RNA circNRIP1 acts as a microRNA-149-5p sponge to promote gastric cancer progression via the AKT1/mTOR pathway. Mol Cancer 18(1). https://doi.org/10.1186/s12943-018-0935-5
41. Gurtan AM, Sharp PA (2013) The role of miRNAs in regulating gene expression networks. J Mol Biol 425(19):3582–3600. https://doi.org/10.1016/j.jmb.2013.03.007

42. Huang J, Yu S, Ding L, Ma L, Chen H, Zhou H, Zou Y, Yu M, Lin J, Cui Q (2021) The dual role of circular RNAs as miRNA sponges in breast cancer and colon cancer. Biomedicine 9(11):1590. https://doi.org/10.3390/biomedicines9111590
43. Kunze-Schumacher H, Krueger A (2020) The role of MicroRNAs in development and function of regulatory T cells – lessons for a better understanding of MicroRNA biology. Front Immunol 11. https://doi.org/10.3389/fimmu.2020.02185
44. Peng Y, Croce CM (2016) The role of MicroRNAs in human cancer. Signal Transduct Target Ther 1(1). https://doi.org/10.1038/sigtrans.2015.4
45. Mathieu J, Hannele Ruohola-Baker (2013) Regulation of stem cell populations by microRNAs. Adv Exp Med Biol:329–351. https://doi.org/10.1007/978-94-007-6621-1_18
46. Shrivastav D, Singh DD (2024) Emerging roles of microRNAs as diagnostics and potential therapeutic interest in type 2 diabetes mellitus. World J Clin Cases 12(3):525–537. https://doi.org/10.12998/wjcc.v12.i3.525
47. Adlakha YK, Saini N (2014) Brain microRNAs and insights into biological functions and therapeutic potential of brain enriched miRNA-128. Mol Cancer 13(1):33. https://doi.org/10.1186/1476-4598-13-33
48. Afshari A, Yaghobi R, Rezaei G (2021) Inter-regulatory role of microRNAs in interaction between viruses and stem cells. World J Stem Cells 13(8):985–1004. https://doi.org/10.4252/wjsc.v13.i8.985
49. Pu M, Chen J, Tao Z, Miao L, Qi X, Wang Y, Ren J (2018) Regulatory network of miRNA on its target: coordination between transcriptional and post-transcriptional regulation of gene expression. Cell Mol Life Sci 76(3):441–451. https://doi.org/10.1007/s00018-018-2940-7
50. Svoronos AA, Engelman DM, Slack FJ (2016) OncomiR or tumor suppressor? The duplicity of MicroRNAs in cancer. Cancer Res 76(13):3666–3670. https://doi.org/10.1158/0008-5472.CAN-16-0359
51. Reyes-Calderón A, Gutiérrez-García C, Urióstegui-Pena AG, Srivastava A, Aguilar-Marcelino L, Hafiz A, Paul S, Sharma A (2023) Identification of cumin (Cuminum cyminum) MicroRNAs through deep sequencing and their impact on plant secondary metabolism. Plants 12(9):1756–1756. https://doi.org/10.3390/plants12091756
52. Faraldi M, Gomarasca M, Sansoni V, Perego S, Banfi G, Lombardi G (2019) Normalization strategies differently affect circulating miRNA profile associated with the training status. Sci Rep 9(1). https://doi.org/10.1038/s41598-019-38505-x
53. Siddika T, Heinemann IU (2021) Bringing MicroRNAs to light: methods for MicroRNA quantification and visualization in live cells. Front Bioeng Biotechnol 8. https://doi.org/10.3389/fbioe.2020.619583
54. Benesova S, Kubista M, Valihrach L (2021) Small RNA-sequencing: approaches and considerations for miRNA analysis. Diagnostics 11(6):964. https://doi.org/10.3390/diagnostics11060964
55. Bernardo BC, Charchar FJ, Lin RC, McMullen JR (2012) A microRNA guide for clinicians and basic scientists: background and experimental techniques. Heart, Lung Circulation 21(3):131–142
56. Kuhn DE, Martin MM, Feldman DS, Terry AV, Nuovo GJ, Elton TS (2008) Experimental validation of miRNA targets. Methods 44(1):47–54. https://doi.org/10.1016/j.ymeth.2007.09.005
57. Pritchard CC, Cheng HH, Tewari M (2012) MicroRNA profiling: approaches and considerations. Nat Rev Genet 13(5):358–369. https://doi.org/10.1038/nrg3198
58. Thomson DW, Bracken CP, Goodall GJ (2011) Experimental strategies for microRNA target identification. Nucleic Acids Res 39(16):6845–6853. https://doi.org/10.1093/nar/gkr330
59. Ye J, Xu M, Tian X, Cai S, Zeng S (2019) Research advances in the detection of miRNA. J Pharm Anal 9(4):217–226. https://doi.org/10.1016/j.jpha.2019.05.004
60. Di Leva G, Garofalo M, Croce CM (2014) MicroRNAs in cancer. Ann Rev Pathol Mech Dis 9(1):287–314. https://doi.org/10.1146/annurev-pathol-012513-104715
61. Kim T, Croce CM (2023) MicroRNA: trends in clinical trials of cancer diagnosis and therapy strategies. Exp Mol Med 55(7):1314–1321. https://doi.org/10.1038/s12276-023-01050-9

62. Radhika T, Gopalakrishnan S, Muthukumar RS, Arulpari M, Kumar BS, Hari R, Jeyaraman M, Jeddy N (2023) Role of serum and salivary microRNAs as diagnostic biomarkers in gastric cancer. J Orofacial Sci 15(2):182–188. https://doi.org/10.4103/jofs.jofs_228_23
63. Jothimani G, Das D, Pathak S, Malayaperumal S, Zhang H, Sun X-F, Banerjee A (2025) Unraveling the mechanism of microRNA-134 in colon cancer progression: targeting KRAS and PIK3CA for cell cycle control and histone deacetylase regulation. Exp Cell Res 444(2):114385. https://doi.org/10.1016/j.yexcr.2024.114385
64. Radhika T, Gopalakrishnan S, Muthukumar RS, Hari R, Jeyaraman M, Vijetha G (2025) Validation of salivary microRNA 301a as a potential non-invasive diagnostic biomarker in gastric carcinoma. J Oral Maxillof Pathol 29(1):104–108. https://doi.org/10.4103/jomfp.jomfp_39_25
65. Metcalf GAD (2024) MicroRNAs: circulating biomarkers for the early detection of imperceptible cancers via biosensor and machine-learning advances. Oncogene:1–8. https://doi.org/10.1038/s41388-024-03076-3
66. Galvão-Lima LJ, Morais AHF, Valentim RAM, Barreto EJSS (2021) miRNAs as biomarkers for early cancer detection and their application in the development of new diagnostic tools. Biomed Eng Online 20(1). https://doi.org/10.1186/s12938-021-00857-9
67. Nguyen TPN, Kumar M, Fedele E, Bonanno G, Bonifacino T (2022) MicroRNA alteration, application as biomarkers, and therapeutic approaches in neurodegenerative diseases. Int J Mol Sci 23(9):4718. https://doi.org/10.3390/ijms23094718
68. Junn E, Mouradian MM (2012) MicroRNAs in neurodegenerative diseases and their therapeutic potential. Pharmacol Ther 133(2):142–150. https://doi.org/10.1016/j.pharmthera.2011.10.002
69. Mu C, Gao M, Xu W, Sun X, Chen T, Xu H, Qiu H (2024) Mechanisms of microRNA-132 in central neurodegenerative diseases: a comprehensive review. Biomed Pharmacother 170:116029–116029. https://doi.org/10.1016/j.biopha.2023.116029
70. Roy B, Lee E, Li T, Rampersaud M (2022) Role of miRNAs in neurodegeneration: from disease cause to tools of biomarker discovery and therapeutics. Genes 13(3):425. https://doi.org/10.3390/genes13030425
71. Li S, Lei Z, Sun T (2022) The role of microRNAs in neurodegenerative diseases: a review. Cell Biol Toxicol 39(1):53–83. https://doi.org/10.1007/s10565-022-09761-x
72. Shaheen N, Shaheen A, Osama M, Nashwan AJ, Bharmauria V, Flouty O (2024) MicroRNAs regulation in Parkinson's disease, and their potential role as diagnostic and therapeutic targets. NPJ Parkinson's Dis 10(1). https://doi.org/10.1038/s41531-024-00791-2
73. Chandan K, Gupta M, Sarwat M (2020) Role of host and pathogen-derived MicroRNAs in immune regulation during infectious and inflammatory diseases. Front Immunol 10. https://doi.org/10.3389/fimmu.2019.03081
74. Cordes KR, Srivastava D (2009) MicroRNA regulation of cardiovascular development. Circ Res 104(6):724–732. https://doi.org/10.1161/circresaha.108.192872
75. Hata A (2013) Functions of MicroRNAs in cardiovascular biology and disease. Annu Rev Physiol 75(1):69–93. https://doi.org/10.1146/annurev-physiol-030212-183737
76. Li Z, Zhao Y, Suguro S, Suguro R (2023) MicroRNAs regulate function in atherosclerosis and clinical implications. Oxidative Med Cell Longev 2023:1–12. https://doi.org/10.1155/2023/2561509
77. Divakaran V, Mann DL (2008) The emerging role of MicroRNAs in cardiac remodeling and heart failure. Circ Res 103(10):1072–1083. https://doi.org/10.1161/circresaha.108.183087
78. Cheng Y, Zhang C (2010) MicroRNA-21 in cardiovascular disease. J Cardiovasc Transl Res 3(3):251–255. https://doi.org/10.1007/s12265-010-9169-7
79. Cortez MA, Anfossi S, Ramapriyan R, Menon H, Atalar SC, Aliru M, Welsh J, Calin GA (2019) Role of miRNAs in immune responses and immunotherapy in cancer. Genes Chromosom Cancer 58(4):244–253. https://doi.org/10.1002/gcc.22725
80. Ha T-Y (2011) The role of MicroRNAs in regulatory T cells and in the immune response. Immune Netw 11(1):11–41. https://doi.org/10.4110/in.2011.11.1.11

81. Nejad C, Stunden HJ, Gantier MP (2018) A guide to miRNAs in inflammation and innate immune responses. FEBS J 285(20):3695–3716. https://doi.org/10.1111/febs.14482
82. Qin Z, Wang P-Y, Su D-F, Liu X (2016) miRNA-124 in immune system and immune disorders. Front Immunol 7. https://doi.org/10.3389/fimmu.2016.00406
83. Gaál Z (2024) Role of microRNAs in immune regulation with translational and clinical applications. Int J Mol Sci 25(3):1942–1942. https://doi.org/10.3390/ijms25031942
84. Mehta A, Baltimore D (2016) MicroRNAs as regulatory elements in immune system logic. Nat Rev Immunol 16(5):279–294. https://doi.org/10.1038/nri.2016.40
85. Yu H-R, Huang L-H, Li S-C (2018) Roles of microRNA in the immature immune system of neonates. Cancer Lett 433:99–106. https://doi.org/10.1016/j.canlet.2018.06.014
86. Wu C-J, Lu L-F (2017) MicroRNA in immune regulation. Curr Top Microbiol Immunol:249–267. https://doi.org/10.1007/82_2017_65
87. Zhang Z, Huang Q, Yu L, Zhu D, Li Y, Xue Z, Hua Z, Luo X, Song Z, Lu C, Zhao T, Liu Y (2022) The role of miRNA in tumor immune escape and miRNA-based therapeutic strategies. Front Immunol 12. https://doi.org/10.3389/fimmu.2021.807895
88. Esau CC, Monia BP (2007) Therapeutic potential for microRNAs. Adv Drug Deliv Rev 59(2–3):101–114. https://doi.org/10.1016/j.addr.2007.03.007
89. Thorsen SB, Obad S, Jensen NF, Stenvang J, Kauppinen S (2012) The therapeutic potential of MicroRNAs in cancer. Cancer J 18(3):275–284. https://doi.org/10.1097/ppo.0b013e318258b5d6
90. Hutchison ER, Okun E, Mattson MP (2009) The therapeutic potential of microRNAs in nervous system damage, degeneration, and repair. NeuroMolecular Med 11(3):153–161. https://doi.org/10.1007/s12017-009-8086-x
91. Palihaderu P, Mendis B, Premarathne J, Dias W, Yeap SK, Ho WY, Dissanayake A, Rajapakse I, Karunanayake P, Senarath U, Satharasinghe D (2022) Therapeutic potential of miRNAs for type 2 diabetes mellitus: an overview. Epigenet Insights 15:251686572211300. https://doi.org/10.1177/25168657221130041
92. Shah MY, Ferrajoli A, Sood AK, Lopez-Berestein G, Calin GA (2016) microRNA therapeutics in cancer – an emerging concept. EBioMedicine 12:34–42. https://doi.org/10.1016/j.ebiom.2016.09.017
93. Reda El Sayed S, Cristante J, Guyon L, Denis J, Chabre O, Cherradi N (2021) MicroRNA therapeutics in cancer: current advances and challenges. Cancer 13(11):2680. https://doi.org/10.3390/cancers13112680
94. Diener C, Keller A, Meese E (2022) Emerging concepts of miRNA therapeutics: from cells to clinic. Trends Genet 38(6):613–626. https://doi.org/10.1016/j.tig.2022.02.006
95. Hasan A, Ardizzone A, Giosa D, Scuderi SA, Calcaterra E, Esposito E, Capra AP (2025) The therapeutic potential of MicroRNA-21 in the treatment of spinal cord injury. Curr Issues Mol Biol 47(2):70. https://doi.org/10.3390/cimb47020070
96. Seyhan AA (2024) Trials and tribulations of MicroRNA therapeutics. Int J Mol Sci 25(3):1469. https://doi.org/10.3390/ijms25031469

Chapter 2
The Role of miRNAs in Cancer: From Molecular Insights to Therapeutic Potential

Sibani Sarkar, Sunny Kumar, Malini Basu, and Mrinal K. Ghosh

Abstract MicroRNAs (miRNAs) control gene expression and are important for the initiation and spread of cancer. By functioning as either tumor suppressive or oncogenic, miRNAs dysregulation can influence one or more cancer hallmarks, viz., angiogenesis, metastasis, apoptosis resistance, and proliferation. Circulating miRNAs also show promise as noninvasive indicators for the diagnosis and prognosis of cancer. miRNAs are being investigated as therapeutic targets due to their critical roles in gene expression. Modern techniques include the suppression of oncogenic miRNAs, miRNA replacement treatment, and sophisticated delivery systems using viral vectors and nanoparticles. miRNA-based cancer therapies are progressing due to advancements in RNA-based therapeutics, despite obstacles like the efficiency of delivery systems and their off-target effects. This chapter opens the door for precision medicine approaches in oncology by highlighting the mechanistic insights of miRNAs in cancer and their developing therapeutic potentials.

Keywords MicroRNAs (miRNAs) · Cancer therapeutics · Gene regulation · Precision medicine

Introduction

Noncoding RNAs (ncRNAs) and their pathological significance in multiple diseases, including cancer, have been the subject of much discussion over the last 20 years. It has been observed that in the human genome, only 1.5% is protein-coding sequence, whereas the remaining 98.5% is noncoding [1]. They are a

S. Sarkar · S. Kumar · M. K. Ghosh (✉)
Cancer Biology and Inflammatory Disorder Division, Council of Scientific and Industrial Research-Indian Institute of Chemical Biology (CSIR-IICB), Kolkata, West Bengal, India

M. Basu
Department of Microbiology, Dhruba Chand Halder College, Paraganas, West Bengal, India

S. Pathak, A. Banerjee (eds.), *MicroRNAs: Role in Cancer Progression, Diagnostics and Therapy*, Advances in Biochemistry in Health and Disease 34,
https://doi.org/10.1007/978-3-032-08824-6_2

heterogeneous group of RNA molecules that perform distinct biological functions, biogenesis, structure, and length. According to the nucleotide length, these molecules are divided into small (sncRNAs) (<200 nt) and long (lncRNAs) (>200 nt). The spatial and temporal profiles of lncRNAs which are broadly expressed in various tissues and cell lines as well as their interaction with DNA, RNA, and proteins showed special functions. They regulate cell cycle, proliferation, differentiation, and other physiological processes, viz., development, aging, immunity, and are connected with multiple pathological processes such as cancer, cardiovascular diseases and diabetes [2, 3]. Regulation of transcription through targeting either in cis genes that regulate the expression of neighboring genes or in trans form that is far from their loci where they interact with chromatin at different locations of multiple chromosomes. Another type of ncRNA, i.e., sncRNA, is called microRNA (miRNA), which has become significantly popular over the last two decades and extensively functions to regulate health and diseases. miRNAs, having a length of approximately 19–25 nucleotides, induce mRNA degradation and translational suppression by binding to the 3′ untranslated region (3′ UTR) of mRNA, though it has been reported later that miRNAs can also induce mRNA translation. Regulatory network of simultaneous targeting of both miRNA and mRNA is aggravated by the involvement of long noncoding RNAs (lncRNA), which binds to endogenous RNAs, i.e., ceRNAs or miRNA sponges, resulting in depletion of the regulatory effect of miRNAs on their target mRNAs. Regulation of different biological processes occurs through the crosstalk of lncRNA and miRNA, and also various pathological conditions are linked with aberrant expression of miRNAs [4] (Fig. 2.1).

Cancer is one of the leading causes of morbidity and mortality worldwide, drawing significant attention from clinicians and researchers alike. Thus, the expression of ncRNAs and their downstream targets can be modified by the intake of natural compounds such as curcumin, resveratrol, epigallocatechin-3-gallate with simultaneous reduction of cancer growth, apoptosis induction, and enhanced efficacy of conventional cancer therapies [4]. This discovery has potential therapeutic implications based on the molecular mechanisms of ncRNA on the biochemical pathways to identify novel druggable targets.

Molecular Insights into miRNAs in Cancer

Genetic, epigenetic, and environmental factors may play a major role in the formation of cancer in the human body which involves abnormal cell growth and invasion or spread to other organs. The exact mechanism of changing the tissues and cell surface behavior of the tumor is unknown. Globally, cardiovascular diseases (including heart disease and stroke) are the primary cause of death, followed by cancer, which accounts for nearly 10 million deaths per year—approximately one in every six deaths worldwide. According to the American Cancer Society's *Colorectal Cancer Facts & Figures 2023–2025*, the proportion of colorectal cancer cases diagnosed at regional or distant stage (indicating spread beyond the primary site)

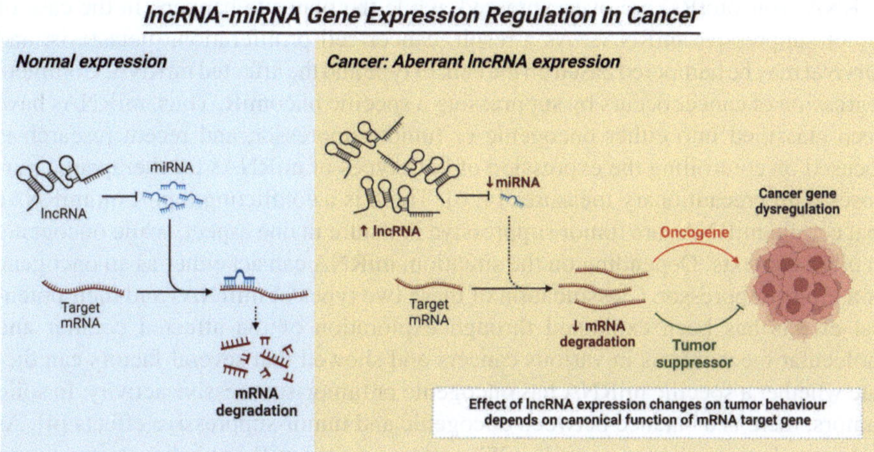

Fig. 2.1 Schematic representation of lncRNA-miRNA crosstalk in cancer. Schematic representation of the interplay between long noncoding RNAs (lncRNAs) and microRNAs (miRNAs) under normal physiological conditions and in cancer. In normal cells, lncRNAs and miRNAs maintain cellular homeostasis through tightly regulated interactions, including miRNA sponging, transcriptional modulation, and feedback loops. In cancer, dysregulated expression of lncRNAs and miRNAs disrupts these interactions, contributing to oncogenic signaling, tumor progression, and metastasis

increased from 52% in the mid-2000s to approximately 60% by 2019 [5]. The relation between the time of diagnosis and life expectancy is very high, but the mortality rate is reduced if diagnosed early. Small ncRNAs generally target mRNAs and manage gene expression at the posttranscriptional level. The new one is "oncomiRs," or tumor-suppressive miRNAs, which play an important role in cancer development and progression. One miRNA can act on multiple genes; otherwise, multiple miRNAs can also control one gene. Thus, various cellular processes, viz., apoptosis, growth, and proliferation, can be controlled by a specific miRNA. miRNAs are detected as a brand new biomarker for various diseases including cancer diagnosis with high sensitivity and specificity by observing their irregular expression profiles, which show differences between normal and tumor tissues [4]. Thus, different types of miRNAs are discussed below with respect to their activity in cancer.

Oncogenic and Tumor Suppressor miRNAs

miRNAs make mRNA unstable by inhibiting its translation and controlling gene expression. Thus, the modification of miRNA expression level has been used to treat cancer. But many miRNAs either activate or inhibit the cancer cell characteristics by modifying the expression of tumor suppressor or oncogenes. In cancer, oncogenic miRNAs (oncomiRs) are overexpressed, while the opposite happens in the case of tumor-suppressive miRNAs. As a result, cancer cell proliferation, metastasis, and survival may be hampered based on the cancer type and the affected miRNA. Complete regression of cancer occurs by suppressing a specific oncomiR. Thus, miRNAs have been classified into either oncogenic or tumor suppressor, and recent research is focused on controlling the expression of both types of miRNAs for therapeutic purposes with precautionary measures [4, 6]. There is a conflicting report on miRNAs that certain miRNAs are tumor suppressive in nature in one aspect, while oncogenic in other contexts. Depending on the situation, miRNA can act either as an oncogene or a tumor suppressor. Classification of these two types of miRNAs and their potential effects has been explained through exploration of the affected cellular and molecular mechanisms in various cancers and showed that several factors can dictate whether a specific miRNA has oncogenic or tumor-suppressive activity. In solid tumors, there is a balance between oncogenic and tumor-suppressive effects [4]. As in hematologic malignancy, miR-125b acts as an oncomiR but behaves as a tumor suppressor in multiple solid tumors. This inconsistency can be attributed to the fact that a single miRNA molecule targets 10–100's of different mRNA, amongst them some are functionally opposing tumor suppression and oncogenesis. In miR-125b, all the factors (antiapoptotic, proapoptotic, and pro-proliferative) are involved in targeting mRNAs and creating a balance in the expression of oncogenic or tumor-suppressive genes to determine whether miR-125b has an overall effect on any cancer individual [4, 6].

miRNA Biogenesis and Dysregulation

Production of miRNA occurs canonically through three main events, namely cropping, export, and dicing, which start in the nucleus and end in the cytoplasm. Transcription of miRNA genes occurs in primary capped and polyadenylated precursors of miRNA (pri-miRNA) by RNA polymerase II. Stem loop structure with long transcripts of mature miRNA stays in the 53 or 33 half of the stem and sheared in the nucleus by the nuclear Drosha/DGCR8 heterodimer and split at the stem of pri-miRNA to release a hairpin-structured pre-miRNA of 60-100 nt. DGCR8, a dsRNA binding protein, recognizes the -10 bp of the stem of the pri-miRNA hairpin, which is the catalytic site of the RNAse 3 enzyme, Drosha. In the cytoplasm, the RNAse 3 enzyme, Dicer, broke pre-miRNA to release th-22 nt miRNA duplexes [4, 6] (Fig. 2.2).

After the identification of miRNA, it has been observed that an individual miRNA can target multiple mRNA and each one is suppressed significantly by different miRNAs, suggesting that the miRNA biogenesis pathway plays an important role in gene regulation and anti-tumor action development [4]. Generally, miRNAs show dual action either oncogenic or tumor-suppressive functions, and their expression is dysregulated in tumors but not in normal tissues. So, miRNA biogenesis may be modulated in cancer cells. miRNA processing is low as Ribonuclease III, DROSHA,

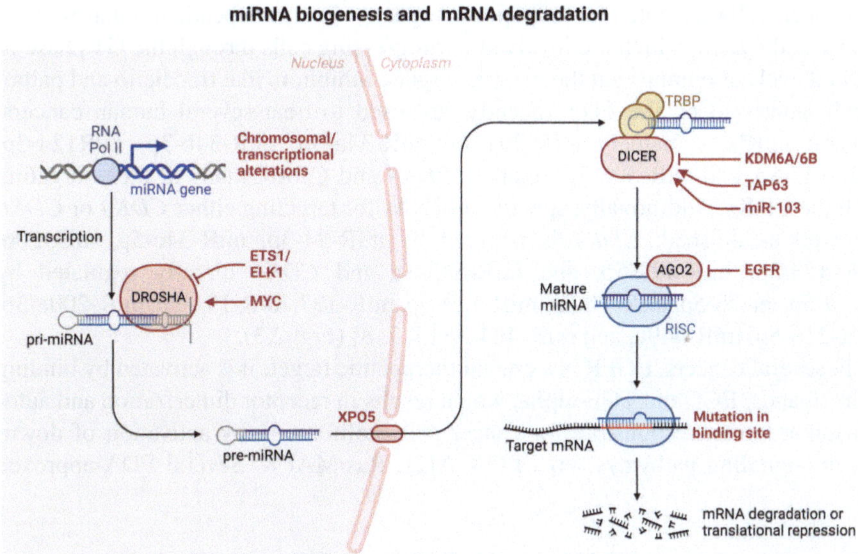

Fig. 2.2 MicroRNA biogenesis and degradation**.** RNA polymerase II (Pol II) transcribes miRNA genes into primary miRNAs (pri-miRNAs) to precursor miRNAs (pre-miRNAs) via the microprocessor complex (DROSHA/DGCR8). Pre-miRNAs are translocated to the cytoplasm by exportin 5 (XPO5), where DICER1 cuts them into mature miRNAs. To direct mRNA silencing by degradation or translational repression in processing bodies (P-bodies), the guide strand is integrated into the miRNA-induced silencing complex (miRISC) with Argonaute (AGO) proteins

and DICER 1 expression levels are decreased in lung and ovarian cancers [6]. Moreover, low DROSHA and DICER1 expression levels are observed in advanced tumor stages with poor clinical outcomes in cancer patients and can accelerate tumor growth due to the genetic deficiency of miRNA biogenesis components. Thus, this pathway plays an important role in cancer progression [4, 6]. The mutations and dysregulation of biogenesis pathway constituents exert their pathophysiological relevance in the tumor microenvironment. Moreover, it is also observed that disruption of miRNA biogenesis pathway in various pediatric and adult cancers occurs by destruction of recently discovered molecules and cellular mechanisms which control biogenesis machinery [6].

miRNAs in Hallmarks of Cancer

Several miRNAs are involved in cancer hallmarks, and proper initiatives have been taken for targeting them therapeutically, though no specific one has been used in a clinical trial. Only MRX34 was used clinically previously but was terminated due to some adverse effects. Thus, we will be focusing on many genes and proteins, especially kinases and kinase receptors, for regulating cell proliferation and growth of cancer cells as potential therapeutic agents. Cyclin-dependent kinases, viz., *CDK4* and *CDK6*, which are involved in progressing cells through the G1 phase of the cell cycle, are important therapeutic targets. Inhibitors like ribociclib and palbociclib, approved by the FDA recently, are used to treat several human cancers. Several miRNAs, namely miR-29a-3p, miR-34a-5p, miR-34b-3p, miR124-3p, miR-145-5p, and miR-506-3p, regulate *CDK4* and *CDK6*, many of them targeting both the CDKs. Additionally, specific miRNAs for targeting either *CDK4* or *CDK6* are well established. *CDK4* is targeted by miR-24-3p, miR-34c-5p, miR-206, miR-483-3p, miR-497-5p, and miR-545-3p and *CDK6* directly regulated by let-7a-5p, let-7b-5p, miR-16-5p, miR-129-5p, miR-137, miR-191-5p, miR-200a-3p, miR-218-5p, miR-449a, and miR-494-3p [4, 7, 8] (Fig. 2.3).

In several cancers, EGFR is a crucial therapeutic target. It is activated by binding to its ligands, EGF and TGF-alpha, which results in receptor dimerization and autophosphorylation, responsible for cancer cell proliferation by activation of downstream signaling pathways, viz., PI3K-AKT, Ras/MAPK. Several FDA-approved

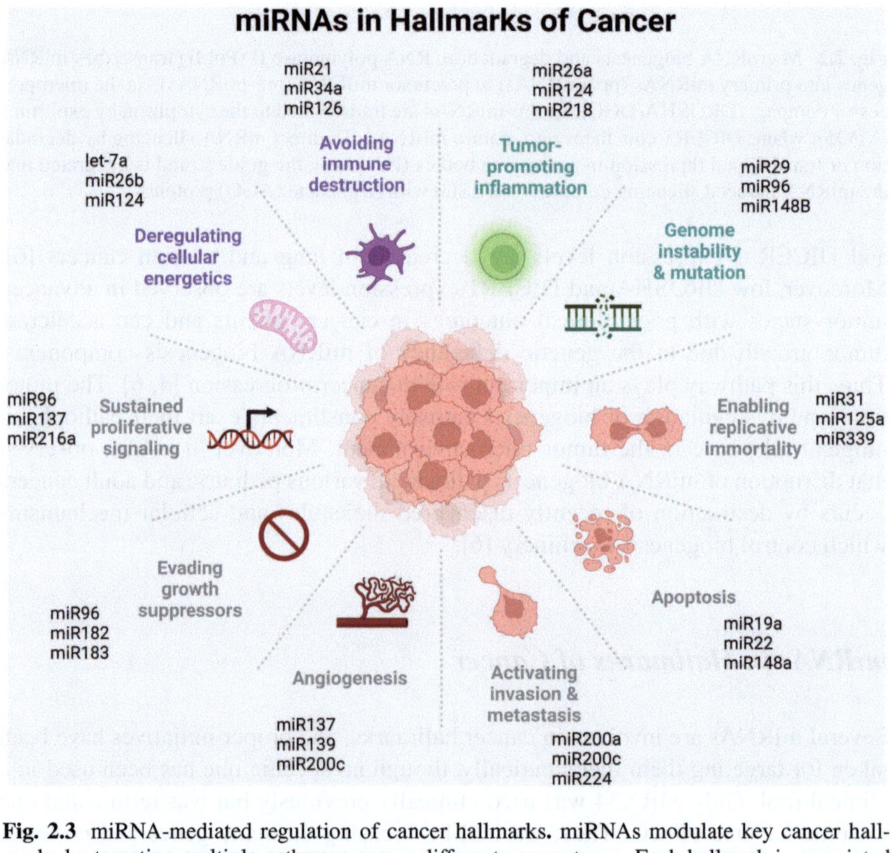

Fig. 2.3 miRNA-mediated regulation of cancer hallmarks. miRNAs modulate key cancer hallmarks by targeting multiple pathways across different cancer types. Each hallmark is associated with three representative microRNAs that influence the respective cellular function

inhibitors, viz., gefitinib, erlotinib, cetuximab, necitumumab, osimertinib, vandetanib, and so on, are used to target EGFR for the treatment of different cancers. Many miRNAs have also been used to target EGFR and control the proliferation and growth of cancer cells, namely miR-7-5p, miR-27a-3p, miR-125a-5p, miR-133a-3p, miR-1455p, miR-219-5p, miR-542-5p, etc. [9].

Another cancer hallmark for the evasion of programmed cell death has been studied. Anti-apoptotic protein *BCL-2* is also an important target for more than 50 miRNAs. 13q14 clustered miR-15a and miR-16 have been used to target both anti-apoptotic oncogenes Bcl-2 and MCL-1, the first noncoding RNAs, used in human diseases, specifically in CLL. PARP enzyme is involved in DNA repair and apoptosis. Its activity is necessary for base excision repair of single-stranded breaks. Therefore, PARP inhibition attenuates DNA damage repair and hence can cause apoptosis of cancer cells. Several FDA-approved inhibitors like Niraparib, Olaparib, and Rucaparib have been used, but till now no cancer-associated miRNA-PARP 2 interaction interrupter molecule has been established yet, although *PARP1* was targeted by miR-124, miR-216b, miR-223-3p, miR-335, and miR-708-5p. Thus, all protein-coding genes, regulated by miRNAs, could be targeted by therapeutic agents. Thus, miRNAs themselves form interesting druggable targets [7, 10, 11].

Circulating miRNAs as Biomarkers

miRNAs present in the circulation and other body fluids is known as circulating miRNA. Many of these miRNAs associated with various cancers are described below:

Lung Cancer (LC)

Discrimination between early-stage LC patients and nonsmoking healthy individuals can be possible by observing the expression pattern of circulating miR-21 along with miR-210 which were elevated in LC patients. Moreover, in the plasma of malignant patients' solitary pulmonary nodules (SPNs), circulating miR-486-5p expression was decreased significantly. LC patients are differentiated from healthy individuals by using these three circulating miRNAs, and they are used as a noninvasive biomarker for the detection and prognosis of LC patients. Additionally, Yan and colleagues reported that the expression pattern of circulating miR-31 was increased significantly in LC patients than that of healthy individuals, and the average survival of these patients was less (25–23 months) than that of the others (38–44 months). Moreover, miR-34 family can also be used in LC diagnosis where overexpressed miR-34a and miR-34c were linked with longer disease-free survival. The expression pattern of miR-944 and miR-3662 was enhanced in the advanced stage. For early detection of LC, miR-448 and miR-4478 have been used as potential noninvasive biomarkers. Their expressions in patients' plasma were upregulated in LC compared to those of normal individuals [12–16].

Gastric Cancer (GC)

GC metastasis is closely correlated with the expression pattern of miR-203 and miR-218. Tsai and colleagues established circulating miR-196a as a biomarker to discriminate GC by observing its expression pattern, which is reduced significantly after the surgical resection and is more specific than the carbohydrate-antigen method. Zhao and colleagues showed that the circulating miR-21, miR-93, miR-106a, and miR-106b were also higher in GC patients. The removal of lymph node metastasis, which is a crucial prognostic factor of GC and patients' survival, has been increased significantly. Serum circulating miR-143-3p, miR-146a, miR-451a, and miR-501-3p has been used to detect and provide prognosis information to LNM developing GC patients. Circulating miR-21, the most studied oncomiRs, has been used as a potential GC biomarker, also linked with other cancers. Thus, circulating miR-21 can be regarded as a holistic biomarker for many cancers [17–21].

Breast Cancer (BC)

Generally, it has been observed that ~30% of early-stage BC patients are reported with distant metastasis and recurrence. Circulating miR-376c-3p, miR-382-5p, and miR-411-5p expression was limited in recurrent patients than in nonrecurrent ones. In TNBC patients, circulating miR-34c expression was used clinically very little to correlate with tumor grades and distant metastasis. Lower expression of circulating miR-34c has been linked with tumor progression. Thus, these miRNAs can be designated as a noninvasive biomarker to monitor BC recurrence. Neo-adjuvant chemotherapy (NACT) was used to treat BC by reducing the complete auxiliary lymph node dissection and improving the breast cancer surgery rates. The expression of circulating miR-451 was also significantly lower in BC patients with higher NACT resistivity [22–26].

Colorectal Cancer (CRC)

Circulating miRNAs have been used in the early detection of many CRC studies. Chan and colleagues have observed that circulating miR-141 has been linked with stage IV CRC. Moreover, it has also been observed that the combination of circulating miR-141 with carcinoembryonic antigen (CEA), a marker for CRC, has been improved for CRC diagnosis, though it is also associated with other cancers. Maria and colleagues observed that a distinct increase in miR-21 for a period of 3 years was required for proper diagnosis. In another study, it has been observed that the expression of miR-21 was significantly higher in metastatic tissues, primary tumor tissues, and plasma-derived exosomes. Additionally, it has also been shown that the change in temporal expression pattern of circulating miR-21 in different stages of CRC was linked with poorer disease-free survival (DFS) and overall survival (OS).

It has also been observed that exosomal miR-21, which is an independent prognostic factors for DFS and OS, is compared to the patients with lower circulating miR-21 in TNM (T for tumor size and spread, N for lymph node involvement, and M stands for metastasis) stage II or III, and for OS in stage IV patients. Thus, miR-21 was designated as a potential biomarker for early detection, prediction of recurrence, and poor prognosis of CRC [27–32].

Therapeutic Potential of miRNAs in Cancer

Nowadays, many miRNA-based cancer therapies have been implemented to increase the survival rate of cancer patients. miRNAs can control multiple targets in the entire network of cancer cascade to act therapeutically, and it is an attractive way to develop a more practical strategy for use of miRNAs in cancer treatment. The miRNA-based cancer therapies are popularly known as replacement, restoration, or overexpression, reduction, inhibition, or downregulation therapy using miRNAs, which are known as antagomiRs (Fig. 2.4).

Replacement Therapy

There are two types of miRNAs, viz., tumor suppressor and oncogenic miRNAs, which are cancer related. In cancer, upregulation of oncogenic microRNAs and downregulation of tumor suppressor microRNAs have been observed. miRNA-based therapy, i.e., miRNA inhibition and replacement, is a promising approach in cancer therapy. miRNA replacement therapy is a modern approach in treating cancer by using synthetic miRNA mimics designed to resemble natural miRNA when introduced into cells restoring the function of downregulated or deleted suppressor miRNA in cancer, leading to the overexpression of the target oncogenes, thereby inhibiting tumor growth and affecting cancer stem cells [33]. These restored miRNAs inhibit the expression of key oncoproteins of miRNAs. Replacement of the lost miRNA level happens by ectopic expression of tumor-suppressive miRNAs for inhibition of cellular pathways to support oncogenesis. Tumor-suppressing miRNAs restore their functions by using mimics miRNA, which are synthetic versions of natural miRNA replacing the loss of tumor-suppressive miRNA and also restoring miRNA action, viz., let 7, miR-34a, and miR-29b [34], which are suppressed, induced apoptosis in malignant cancer. There are many commercially available miRNA mimics such as pre-miR miRNA precursors (Ambion) and miRIDIAN mimic hsa-miR-17 (Thermo Scientific, Dharmacon) [35–37].

In pancreatic cancer, restoring miR-34 downregulates Bcl-2 and notch-1/2 expression which inhibits angiogenesis. Decreased expression of miR-34 also plays a role in the progression of the most advanced epithelial ovarian cancer along with the pathogenesis. In hepatocellular carcinoma, the expression of both miR-101 and

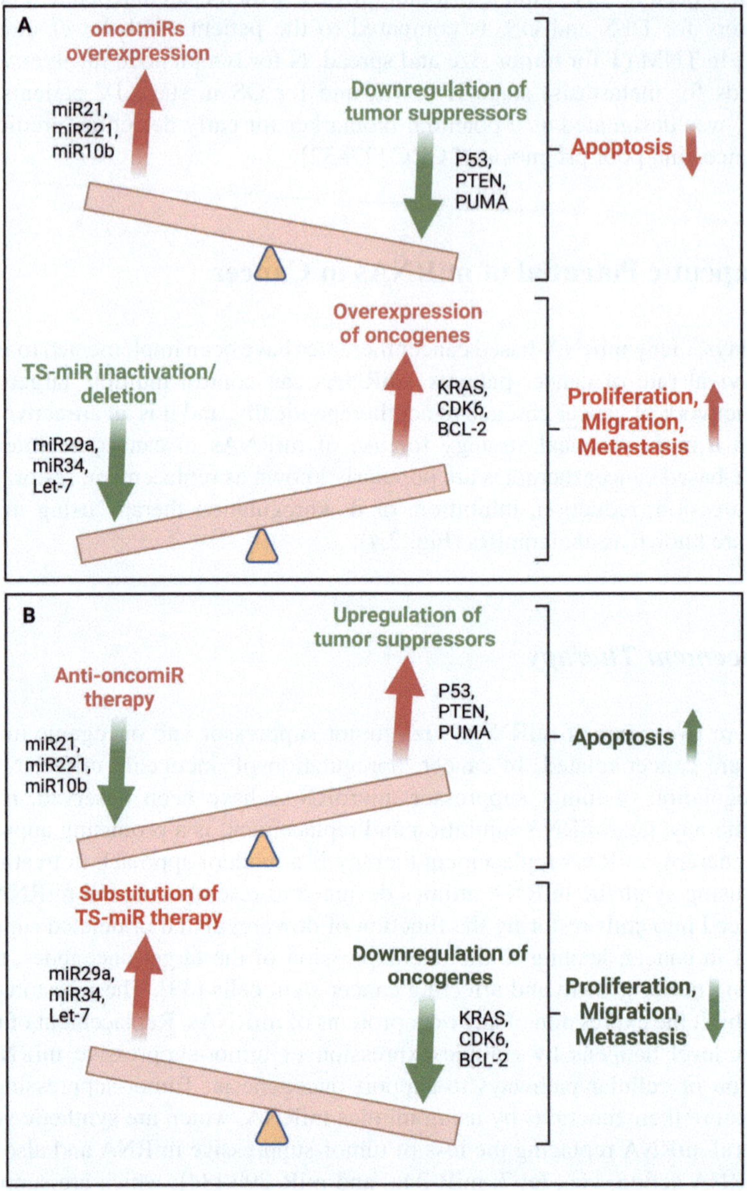

Fig. 2.4 miRNAs in cancer therapy. (**a**) Tumors exploit miRNA dysregulation by upregulating oncomiRs, which suppress tumor suppressor genes, and silencing TS-miRs, leading to oncogene overexpression. This imbalance fuels tumor growth, invasion, metastasis, and resistance to cell death. (**b**) Anti-miRNA therapies block oncomiRs, restoring tumor suppressor gene expression, while TS-miR replacement suppresses oncogenes. Together, these strategies promote cell death and inhibit tumor progression

miR-122 is present in a downregulated state. miR-101 induced HepG2 cells apoptosis, whereas miR-122 inhibits cancer cell proliferation and migration. Sometimes, miRNA can function as a tumor suppressor in one cancer type and an oncogene in others. miR-22 is an example of mostly studied tumor suppressive miRNA. However, its elevated level has been studied in BC and CRC [37, 38].

Inhibition Therapy

In human cancers, it is necessary to inhibit overexpressed oncogenic miRNAs to restore their expression and function. These are complementary single-stranded oligonucleotide segregated endogenous miRNAs in other unrecognized conformation resulting in RISC that cannot be processed to mature miRNA and eliminated from RISC. These are anti-miR oligonucleotides (AMOs), locked nucleic acid (LNA), anti-miRNAs, antagomiRs, miRNA sponges, miRNA masks, and small molecule inhibitors of miRNAs [39–42].

AMOs are designed to be complementary to the selected miRNA and are single-stranded 17–20 nt in length, chemically modified antisense oligonucleotides bind to the target mature miRNAs and act as competitive inhibitors of miRNAs. They inhibit the interaction of miRNA with their targets. Thus, targeted inhibition of the specific miRNAs results in subsequent upregulation of the targets and produces a mechanistically ASO-miRNA duplex through Watson-Crick binding that leads to RNAse H-mediated cleavage of the targeted miRNA gene occurs [4, 40].

The antagomiRs are cholesterol-conjugated, chemically modified single-stranded 23 nt RNA molecules complementary to the targeted miRNAs. It protects against degradation through modification and increases the stability of RNA. AntagomiR-122, when injected intravenously in mice, induced a marked, specific, and persistent reduction of endogenous miR122 in the liver, kidney, lung, intestine, heart, skin, bone marrow, ovary, etc. A similar observation occurs in the case of antagomiR-16. Other physiological effects were observed by silencing miRNAs by using these new agents, for example, antagomiR-122 can decrease the plasma cholesterol level [4, 43].

Multiple miRNA binding sites found in miRNA sponges compete with the endogenous target miRNA. Using the miRNA that prevents metastasis, *CDH1* is targeted by blocking miR-9, which is increased in breast cancer cells. miRNA masks are new gene-specific miRNAs that mask a particular mRNA from the endogenous miRNA by selectively preventing the target miRNA from interacting with that mRNA. They can therefore stop it from being suppressed. Therefore, one effective tactic in cancer treatment is to block overexpressed oncogenic miRNAs or replace them with tumor-suppressive miRNAs [4, 40].

Delivery Systems for miRNA Therapeutics

Clinical implementation of microRNA is difficult due to the lack of proper delivery systems. They are not biologically stable and cannot diffuse through the lipid membranes, though they are small fragments of nucleotides. Thus, perfect delivery systems are necessary to reduce toxicity and stability of miRNAs in circulation. There are several strategies to deliver miRNAs into the cancer cells therapeutically. Currently, they are categorized as nonviral and viral vectors, for efficient delivery of miRNA. The nonviral includes organic, inorganic, or polymer-based carriers, whereas viral-based delivery uses lentivirus, retrovirus, adenovirus, and adeno-associated viruses [44–46].

By transporting endogenous miRNA or its expressing vectors, nonviral delivery techniques can stop nuclease-mediated degradation. It is generally less toxic and immunogenic but has low efficiency. This delivery system uses modern techniques like electroporation, gene gun, ultrasound, or chemically using organic/inorganic or polymer-based carriers. Organic carriers, such as lipid-nucleic acid complexes called liposomes, exhibit therapeutic effects when applied clinically and preclinically in cancer research of miRNA. Wu et al. [47] developed a cationic liposome containing miR-29 for restoring tumor-suppressive microRNA in non-small cell lung cancer cells to inhibit tumor growth. High-density lipoprotein nanoparticles (HDL-NPs) were successfully delivered miR-34a to head and neck cancer cells, resulting in a reduction of metabolic activity, increased apoptosis, and disruption of cell cycle [48–50].

Inorganic materials include gold nanoparticles, Fe_3O_4 nanoparticles, and silica-based nanoparticles to deliver miR-29b in myeloid leukemia and miR-30b in multiple myeloid cells. But the successful delivery of miR-29b occurs when using a cell-penetrating peptide (CPP) in the case of osteogenic stem cells. Moreover, polyethylene amine (PEI)-based miR nanoparticles showed less toxicity and delivered miR-145 and miR-33a to colon cancer by damaging the cell membrane. PLGA is a biodegradable polymer, and PLGA-based nanoparticles with poly-plexed PEI coating delivered miR-26a in HepG2 cells. Polyamidoamines (PAMAMs) are positively charged with high transfection efficiency, which can deliver anti-miR-21 in glioblastoma cells for inhibition of miR-21 activity [51, 52].

Recombinant Viruses and Viral Vectors in miRNA Delivery System

It is safe to utilize genetically engineered viruses since they can deliver desired oligonucleotides to increase expression or anti-miR to decrease target miRNA and prevent replication. These delivery systems often use lentivirus, retrovirus, adenovirus, and adeno-associated virus. The expression of miR-138 in mouse embryonic fibroblasts was induced using retroviral vectors. miR-15a/16 decreased malignancy in a rat with chronic lymphocytic leukemia when administered lentivirally. Because of their great loading capacity, excellent stability, longer half-life in circulation, gradual degradation of the miRNA cargo, and low toxicity, these miRNA delivery vehicles are the most favored. Viral vectors have high immunogenicity, cytotoxicity, and transduction efficiencies. Otherwise, nonviral vectors have low transfection efficiency, but it can be overcome by chemical modifications such as positively charged PAMAM dendrimers, which show higher transfection efficiency and lower cytotoxicity compared to other polymers. Their half-life can also be increased by including PEG [49].

Future Directions

In recent years, the study of miRNA in cancer has been developing quickly, and its future is being shaped by a number of exciting avenues. By revealing the diversity of miRNA expression inside tumors, single-cell and spatial transcriptomics are illuminating their functions in treatment resistance, metastasis, and the progression of cancer. miRNA-mediated variations in glioma subpopulations have already been identified by technologies such as single-cell RNA sequencing (scRNA-seq), providing information on treatment resistance. Since integrating miRNA profiling with genomic, transcriptomic, and proteomic data improves biomarker identification and patient stratification for precision medicine, multi-omics integration is another important area of attention. To find new treatment targets, large-scale initiatives [viz., The Cancer Genome Atlas (TCGA)] are combining with miRNA datasets. Furthermore, by enhancing miRNA-target prediction, optimizing treatment design, and boosting biomarker selection, artificial intelligence (AI) and machine learning (ML) can also be used for transforming miRNA research. Deep learning techniques, in particular, drastically lower false positives in miRNA-mRNA interaction networks. However, effective delivery mechanisms are one of the key obstacles in miRNA therapies. To increase tissue selectivity and stability, novel strategies including RNA nanotechnology and exosome-based delivery are being developed. Exosomes have the potential to penetrate biological barriers like the blood-brain barrier (BBB). As demonstrated by MRX34 (a miR-34 mimic), the first miRNA-based treatment to enter clinical trials, which experienced toxicity-related difficulties, clinical translation and regulatory approval continue to be critical obstacles despite promising preclinical outcomes. To enhance the therapeutic outcome, future clinical trials must improve through patient selection and dosage techniques. Combination therapy is another intriguing approach, in which miRNA regulation is being investigated in conjunction with currently used therapies like immunotherapy, chemotherapy, and targeted therapy. According to recent research and development, miRNAs can improve the effectiveness of immune checkpoint inhibitors in malignancies such as lung cancer and melanoma, indicating a possibility for synergy in overcoming treatment resistance. miRNA-based treatments have enormous potential to transform cancer detection and treatment as these developments are rapidly progressing for opening the door for customized medicine.

Conclusion

miRNAs are important regulators of gene expression and play a significant role in cancer development. They are useful biomarkers and therapeutic targets because their dysregulation promotes tumor development, metastasis, and may cause treatment resistance. Although there are still issues with delivery, specificity, and safety in developments in miRNA-based treatments, such as replacement and inhibitory techniques, they hold great promise. The entire potential of miRNA therapies in precision oncology will be ascertained through ongoing research, which is fueled by clinical trials and technical advancements. In the age of personalized medicine, miRNAs have the potential to revolutionize cancer diagnosis, prognosis, and treatment as we deepen our understanding of miRNA biology and create more potent therapeutic approaches.

Conflicts of Interests There is no competing interest.

Funding This work is jointly supported by the Department of Science and Technology (NanoMission: DST/NM/NT/**2018/105**(G); SERB: EMR/**2017/000992**) and Focused Basic Research (FBR), HCT, OLP-**121**, CSIR, Govt. of India.

References

1. Lander ES, Linton LM, Birren B, Nusbaum C, Zody MC, Baldwin J, Devon K, Dewar K, Doyle M, FitzHugh W, Funke R, Gage D, Harris K, Heaford A, Howland J, Kann L, Lehoczky J, LeVine R, McEwan P, McKernan K et al (2001) Initial sequencing and analysis of the human genome. Nature 409(6822):860–921. https://doi.org/10.1038/35057062
2. Leung A, Natarajan R (2018) Long noncoding RNAs in diabetes and diabetic complications. Antioxid Redox Signal 29(11):1064–1073. https://doi.org/10.1089/ars.2017.7315
3. Fu P, Guo X, Cheung FMH, Yung KKL (2019) The association between PM2. 5 exposure and neurological disorders: a systematic review and meta-analysis. Sci Total Environ 655:1240–1248
4. Szczepanek J, Skorupa M, Tretyn A (2022) MicroRNA as a potential therapeutic molecule in cancer. Cells 11(6):1008
5. Primo WQSP (2023) National Cancer Institute and the 2023-2025 estimate–cancer incidence in Brazil. Rev Bras Ginecol E Obstetrícia 45(1):01–02
6. Weng YT, Chang YM, Chern Y (2023) The impact of dysregulated microRNA biogenesis machinery and microRNA sorting on neurodegenerative diseases. Int J Mol Sci 24(4):3443
7. Ravi S, Alencar AM Jr, Arakelyan J et al (2022) An update to hallmarks of cancer. Cureus 14(5)
8. Kumar S, Basu M, Ghosh MK (2021) Chaperone-assisted E3 ligase CHIP: a double agent in cancer. Genes Dis 9(6). https://doi.org/10.1016/j.gendis.2021.08.003

9. Normanno N, De Luca A, Bianco C et al (2006) Epidermal growth factor receptor (EGFR) signaling in cancer. Gene 366(1):2–16
10. Van Roosbroeck K, Calin GA (2017) Cancer hallmarks and MicroRNAs: the therapeutic connection. Adv Cancer Res 135:119–149
11. Kumar S, Chatterjee M, Ghosh P, Ganguly KK, Basu M, Ghosh MK (2023) Targeting PD-1/PD-L1 in cancer immunotherapy: an effective strategy for treatment of triple-negative breast cancer (TNBC) patients. Genes Dis. 10(4):1318–1350
12. Moretti F, D'Antona P, Finardi E et al (2017) Systematic review and critique of circulating miRNAs as biomarkers of stage I-II non-small cell lung cancer. Oncotarget 8(55):94980
13. Zhang YH, Jin M, Li J, Kong X (2020) Identifying circulating miRNA biomarkers for early diagnosis and monitoring of lung cancer. Biochim et Biophys Acta (BBA) 1866(10):165847
14. Wozniak MB, Scelo G, Muller DC, Mukeria A, Zaridze D, Brennan P (2015) Circulating microRNAs as non-invasive biomarkers for early detection of non-small-cell lung cancer. PLoS One 10(5):e0125026
15. Müller S, Janke F, Dietz S, Sültmann H (2020) Circulating MicroRNAs as potential biomarkers for lung cancer. In: Schaffner F, Merlin JL, Von Bubnoff N (eds) Tumor liquid biopsies. Vol 215. Recent results in cancer research. Springer International Publishing, pp 299–318. https://doi.org/10.1007/978-3-030-26439-0_16
16. Herath S, Sadeghi Rad H, Radfar P et al (2022) The role of circulating biomarkers in lung cancer. Front Oncol 11:801269
17. Shekari N, Baradaran B, Shanehbandi D, Kazemi T (2018) Circulating MicroRNAs: valuable biomarkers for the diagnosis and prognosis of gastric cancer. Curr Med Chem 25(6):698–714. https://doi.org/10.2174/0929867324666171003123425
18. Lindner K, Haier J, Wang Z, Watson DI, Hussey DJ, Hummel R (2015) Circulating microRNAs: emerging biomarkers for diagnosis and prognosis in patients with gastrointestinal cancers. Clin Sci 128(1):1–15
19. Kim SY, Jeon TY, Choi CI et al (2013) Validation of circulating miRNA biomarkers for predicting lymph node metastasis in gastric cancer. J Mol Diagn 15(5):661–669
20. Mirzaei H, Khataminfar S, Mohammadparast S et al (2016) Circulating microRNAs as potential diagnostic biomarkers and therapeutic targets in gastric cancer: current status and future perspectives. Curr Med Chem 23(36):4135–4150. https://doi.org/10.2174/0929867323666160818093854
21. Tsujiura M, Ichikawa D, Komatsu S et al (2010) Circulating microRNAs in plasma of patients with gastric cancers. Br J Cancer 102(7):1174–1179
22. Cortez MA, Welsh JW, Calin GA (2012) Circulating MicroRNAs as noninvasive biomarkers in breast cancer. In: Ignatiadis M, Sotiriou C, Pantel K (eds) Minimal residual disease and circulating tumor cells in breast cancer. Vol 195. Recent results in cancer research. Springer, Berlin/Heidelberg, pp 151–161. https://doi.org/10.1007/978-3-642-28160-0_13
23. Hamam R, Hamam D, Alsaleh KA et al (2017) Circulating microRNAs in breast cancer: novel diagnostic and prognostic biomarkers. Cell Death Dis 8(9):e3045–e3045
24. Heneghan HM, Miller N, Lowery AJ, Sweeney KJ, Newell J, Kerin MJ (2010) Circulating microRNAs as novel minimally invasive biomarkers for breast cancer. Ann Surg 251(3):499–505
25. Zhang L, Xu Y, Jin X et al (2015) A circulating miRNA signature as a diagnostic biomarker for non-invasive early detection of breast cancer. Breast Cancer Res Treat 154:423–434
26. Ng EK, Li R, Shin VY et al (2013) Circulating microRNAs as specific biomarkers for breast cancer detection. PLoS One 8(1):e53141
27. Clancy C, Joyce MR, Kerin MJ (2015) The use of circulating microRNAs as diagnostic biomarkers in colorectal cancer. Cancer Biomark 15(2):103–113
28. Giráldez MD, Lozano JJ, Ramírez G et al (2013) Circulating microRNAs as biomarkers of colorectal cancer: results from a genome-wide profiling and validation study. Clin Gastroenterol Hepatol 11(6):681–688
29. Rapado-González Ó, Álvarez-Castro A, López-López R, Iglesias-Canle J, Suárez-Cunqueiro MM, Muinelo-Romay L (2019) Circulating microRNAs as promising biomarkers in colorectal cancer. Cancer 11(7):898

30. Nassar FJ, Msheik ZS, Itani MM et al (2021) Circulating miRNA as biomarkers for colorectal cancer diagnosis and liver metastasis. Diagnostics 11(2):341
31. Aravalli RN, Steer CJ (2015) Circulating microRNAs: novel biomarkers for early detection of colorectal cancer. Transl Res 166(3):219–224
32. Sarkar S, Kumar S, Saha G, Basu M, Ghosh MK (2024) Glioma nanotherapy: unleashing the synergy of dual-loaded DIM and TMZ. Int J Pharm 665:124697
33. Martino MTD, Tagliaferri P, Tassone P (2025) MicroRNA in cancer therapy: breakthroughs and challenges in early clinical applications. J Exp Clin Cancer Res 44(1):126. https://doi.org/10.1186/s13046-025-03391-x
34. Botta C, Cucè M, Pitari MR et al (2018) MiR-29b antagonizes the pro-inflammatory tumor-promoting activity of multiple myeloma-educated dendritic cells. Leukemia 32(4):1003–1015
35. Mollaei H, Safaralizadeh R, Rostami Z (2019) MicroRNA replacement therapy in cancer. J Cell Physiol 234(8):12369–12384. https://doi.org/10.1002/jcp.28058
36. Henry JC, Azevedo-Pouly ACP, Schmittgen TD (2011) MicroRNA replacement therapy for cancer. Pharm Res 28:3030–3042
37. Hosseinahli N, Aghapour M, Duijf PHG, Baradaran B (2018) Treating cancer with microRNA replacement therapy: a literature review. J Cell Physiol 233(8):5574–5588. https://doi.org/10.1002/jcp.26514
38. Bader AG, Brown D, Winkler M (2010) The promise of microRNA replacement therapy. Cancer Res 70(18):7027–7030
39. Rothschild SI (2014) microRNA therapies in cancer. Mol Cell Ther 2(1):7. https://doi.org/10.1186/2052-8426-2-7
40. Spizzo R, Rushworth D, Guerrero M, Calin GA (2009) RNA inhibition, microRNAs, and new therapeutic agents for cancer treatment. Clin Lymphoma Myeloma 9:S313–S318
41. Ghosh MK, Kumar S, Begam S, Ghosh S, Basu M (2024) GBM immunotherapy: exploring molecular and clinical frontiers. Life Sci 356:123018
42. Kumar S, Basu M, Ghosh MK (2024) E3 ubiquitin ligases and deubiquitinases in colorectal cancer: emerging molecular insights and therapeutic opportunities. Biochim et Biophys Acta (BBA) 1871:119827
43. Wen D, Danquah M, Chaudhary AK, Mahato RI (2015) Small molecules targeting microRNA for cancer therapy: promises and obstacles. J Control Release 219:237–247
44. Ahmadzada T, Reid G, McKenzie DR (2018) Fundamentals of siRNA and miRNA therapeutics and a review of targeted nanoparticle delivery systems in breast cancer. Biophys Rev 10:69–86
45. Forterre A, Komuro H, Aminova S, Harada M (2020) A comprehensive review of cancer MicroRNA therapeutic delivery strategies. Cancer 12(7):1852
46. Dasgupta I, Chatterjee A (2021) Recent advances in miRNA delivery systems. Methods Protoc 4(1):10
47. Wu Y, Crawford M, Mao Y et al (2013) Therapeutic delivery of microRNA-29b by cationic lipoplexes for lung cancer. Mol Ther Nucleic Acids 2. Accessed June 19, 2025. https://www.cell.com/molecular-therapy-family/nucleic-acids/fulltext/S2162-2531(16)30145-7:e84
48. Wang H, Jiang Y, Peng H, Chen Y, Zhu P, Huang Y (2015) Recent progress in microRNA delivery for cancer therapy by non-viral synthetic vectors. Adv Drug Deliv Rev 81:142–160
49. Yang N (2015) An overview of viral and nonviral delivery systems for microRNA. Int J Pharm Investig 5(4):179
50. Labatut AE, Mattheolabakis G (2018) Non-viral based miR delivery and recent developments. Eur J Pharm Biopharm 128:82–90
51. Sousa DP, Conde J (2022) Gold Nanoconjugates for miRNA modulation in cancer therapy: from miRNA silencing to miRNA mimics. ACS Mater Au 2(6):626–640. https://doi.org/10.1021/acsmaterialsau.2c00042
52. Choi KY, Liu G, Lee S, Chen X (2012) Theranostic nanoplatforms for simultaneous cancer imaging and therapy: current approaches and future perspectives. Nanoscale 4(2):330–342

Chapter 3
Impact of MicroRNAs on Tumor Microenvironment and Their Therapeutic Implications

Kruthika Prakash, Srisri Satishkartik, Raksa Arun, T. Sasitharan, Janani Balaji, Sanjana Dhayalan, Surya Babu, Shanmuga Priya, M. Ezhilarasi, Satish Ramalingam, and K. N. Aruljothi

Abstract MicroRNAs (miRNAs) are small, noncoding RNAs that perform essential functions in posttranscriptional gene regulation. They are involved in essential cellular processes such as proliferation, apoptosis, and differentiation, and have been increasingly implicated in cancer development and progression. While the roles of miRNAs in cancer cells are well established, their regulatory influence on the tumor microenvironment (TME), including immune modulation, stromal remodeling, and intercellular communication, remains poorly understood. Moreover, the therapeutic application of miRNA modulation to target cancer cells and the surrounding TME remains in its early stages. This study highlights the multifaceted roles of miRNAs within the TME and evaluates their potential as diagnostic biomarkers and therapeutic targets in cancer. miRNAs regulate tumor stroma interactions by modulating immune responses, angiogenesis, and extracellular matrix dynamics. Tumor-derived miRNAs can be secreted via extracellular vesicles to influence the behavior of immune and stromal cells, thereby contributing to immune evasion and metastasis. Specific miRNA expression profiles are also emerging as promising noninvasive detection and treatment monitoring biomarkers. A deeper understanding of miRNA-mediated crosstalk within the TME may pave the way for innovative, targeted therapies. Therapeutic strategies involving miRNA mimics or inhibitors hold significant promise for overcoming the limitations of conventional cancer treatments and advancing personalized oncology.

Keywords MicroRNAs · Tumor microenvironment · Immune modulation · Biomarkers · Therapeutic targets

K. Prakash · S. Satishkartik · R. Arun · T. Sasitharan · J. Balaji · S. Dhayalan · S. Babu · S. Priya · M. Ezhilarasi · S. Ramalingam (✉) · K. N. Aruljothi (✉)
Department of Genetic Engineering, College of Engineering and Technology, SRM Institute of Science and Technology, Chennai, Tamil Nadu, India
e-mail: satishr@srmist.edu.in; aruljotn@srmist.edu.in

S. Pathak, A. Banerjee (eds.), *MicroRNAs: Role in Cancer Progression, Diagnostics and Therapy*, Advances in Biochemistry in Health and Disease 34,
https://doi.org/10.1007/978-3-032-08824-6_3

Introduction

MicroRNAs, also known as miRNAs, are a family of noncoding RNA molecules that are small and single-stranded. They exhibit a crucial function in the procedure of posttranscriptional gene regulation. Generally varying in length from 15 to 30 nucleotides, they exercise their regulatory effect via the binding to complementary sequences within target messenger RNAs (mRNAs). This involvement results in either translational repression or the degradation of messenger RNA, which in turn modulates the mechanism of protein synthesis and the function of the cell. The significance of miRNAs extends to virtually all aspects of cellular physiology, including proliferation, apoptosis, differentiation, and developmental processes. Their capacity to precisely control gene expression allows for precise control over cellular pathways, disease prevention, immune response modulation, and adaptation to environmental change. However, dysregulation of miRNA expression can disrupt these balances, contributing to the pathogenesis of diseases, particularly cancer [1].

The process of miRNA biogenesis begins with RNA polymerase II or III transcription, which results in the synthesis of primary miRNAs (pri-miRNAs) that form a hairpin structure. The pri-miRNAs are then processed by the Drosha-DGCR8 complex, forming precursor miRNAs (pre-miRNAs) consisting of roughly 70 nucleotides. Pre-miRNAs are exported to the cytoplasm and then sliced by the RNase III enzyme Dicer, resulting in a mature miRNA duplex of 21 to 23 nucleotides [2]. One of the duplex strands, the guide miRNA, is specifically incorporated into the RNA-induced silencing complex (RISC), whilst the other strand, the passenger miRNA, is subjected to degradation. The RISC, containing the mature miRNA, then targets specific mRNAs based on sequence complementarity, leading to gene silencing. The degree of complementarity between the miRNA and its target mRNA, leads to gene silencing. Ideal or almost ideal complementarity destroys mRNA, while partial complementarity induces translational inhibition [3].

In oncology, miRNAs have acquired significant focus owing to their multifaceted roles in cancer development and progression. They function as oncogenes (oncomiRs), promoting malignant transformation and tumor growth or as tumor suppressors, inhibiting oncogenic processes and maintaining cellular integrity by regulating survival, apoptosis, and angiogenesis. This dynamic regulatory function highlights the involvement of miRNAs in the initiation, development, and metastasis of diverse types of cancer [4]. Dysregulated miRNA expression is a characteristic feature of numerous malignancies, contributing to disrupting critical signaling pathways and acquiring cancer hallmarks. The oncogenic miR-17-92 cluster promotes the cell proliferation and inhibits apoptosis across multiple cancer types. Conversely, tumor suppressor miRNAs such as miR-15a and miR-16-1 contribute to disease progression in chronic lymphocytic leukemia (CLL), where they are frequently lost due to genetic deletions at chromosome 13q14. In addition to such deletions, their downregulation may also result from epigenetic silencing mechanisms or disruptions in miRNA processing pathways [5].

Beyond their direct effects on cancer cells, miRNAs also shape the tumor microenvironment. Their ability to influence the behavior of immune cells, stromal cells, and endothelial cells modulates the complex interplay between the tumor and its environment. miRNAs derived from tumors can be secreted into the TME via exosomes, influencing the differentiation and fibroblast activation, endothelial cells, and immune cells, facilitating angiogenesis, immunosuppression, and metastasis [6].

miRNAs are being recognized as significant diagnostic and predictive indicators in oncology. Patterns of tissue-specific expression stability in biological fluids and involvement in key oncogenic pathways make them valuable tools. They are engaged in essential cellular processes such as proliferation, apoptosis, and differentiation and have been increasingly implicated in cancer development and progression, and surveillance of illness advancement. Specific miRNA signatures can distinguish between cancer subtypes, predict patient survival, and monitor treatment efficacy. Therapeutic strategies targeting miRNAs are being explored, including antagomiRs to block oncomiRs and miRNA mimics to replace tumor suppressor miRNAs. These miRNA-based therapies offer a promising path for personalized cancer care, potentially facing the limitations of conventional therapies [7].

Tumor Microenvironment (TME): Structure and Components

The tumor microenvironment (TME) comprises an intertwined network of tumor cells, stromal cells, extracellular matrix (ECM), signaling molecules, and blood vessels. This dynamic ecosystem is a pivotal part in cancer metastasis, progression, and sustained tumor survival. The tumor microenvironment (TME) facilitates tumor proliferation and enables evasion from the immune response to chemotherapy and radiotherapy, constituting it a critical factor in cancer biology [8]. The TME promotes angiogenesis, suppresses antitumor immunity, and facilitates invasion through ECM remodeling, all supporting tumor growth. Moreover, it contributes to therapy resistance by activating signaling pathways that protect cancer cells from treatment-induced apoptosis. The reciprocal interactions between tumor cells and their microenvironment create a feedback loop that reinforces malignancy, complicating the development of effective therapeutic strategies [9].

A deeper understanding of the TME requires an examination of its key components. The following sub-sections will examine the function of immune cells in tumor immunity, the contribution of stromal elements such as cancer-associated fibroblasts and endothelial cells in tumor support, and the impact of noncellular factors like growth factors, extracellular matrix, and cytokine remodeling in shaping the tumor microenvironment.

Immune Cells

The immune cells within the TME play a bifunctional role, suppressing or promoting tumor growth. The key immune cell types in the TME include tumor-associated macrophages, T cells, tumor-associated macrophages, and regulatory T cells. These immune cells engage dynamically with tumor cells, affecting cancer growth and treatment resistance.

Tumor-associated macrophages (TAMs) are one of the prevalent immune cell types in the tumor microenvironment. Tumor-derived signals polarize TAMs, and their high infiltration is often associated with poor prognosis and reduced survival rates [10, 11]. TAMs are broadly classified into M1 (classically activated) or M2 (alternatively activated) macrophages, each with functions. M1 macrophages, activated by pro-inflammatory signals, secrete cytokines like interleukin- 12 (IL-12) and interleukin- 23 (IL-23) and present major histocompatibility complex class II (MHC-II) molecules, stimulating antitumor immune responses. Conversely, M2 macrophages, often induced by tumor-derived factors, promote tumor progression by secreting matrix metalloproteinases (MMPs), which cause the degradation of the ECM, facilitating tumor invasion and metastasis. Additionally, M2 macrophages contribute to immunosuppression, inhibit T cell activity, and secrete angiogenic factors like vascular endothelial growth factor (VEGF), IL-8, and Interleukin-6 (IL-6), further supporting tumor growth. The phenotypic plasticity of TAMs makes them challenging therapeutic targets [12, 13].

T cell suppression is a fundamental method through which cancers circumvent immune surveillance. Myeloid-derived suppressor cells (MDSCs), regulatory T cells (Tregs), and TAMs contribute to this suppression by suppressing T cell activation. The immunosuppressive factors in the TME impact the efficacy of T-cell-based immunotherapies like CAR T cell therapy, as the suppressive environment can lead to T cell exhaustion [14]. In tumors such as classic Hodgkin Lymphoma, Pancreatic Ductal Adenocarcinoma, and Glioblastoma Multiforme, T cells are often excluded or rendered dysfunctional due to the presence of Tregs and MDSCs, which hinder their activation and migration [15]. Furthermore, immune checkpoint molecules such as cytotoxic T-lymphocyte antigen-4 (CTLA-4) and programmed cell death protein 1(PD-1), along with cytokine-induced metabolic changes, further dampen T cell responses, impairing their ability to recognize and eliminate malignant cells [16].

Dendritic cells (DCs), which are essential for antigen presentation and initiating T cell responses, are also significantly affected by the TME. The suppressive environment hinders DC maturation and function. Immunosuppressive factors such as IL-6, interleukin- 10 (IL-10), and macrophage colony-stimulating factor (M-CSF) suppress DC development, while cytokines such as interleukin- 2 (IL-2), interleukin-15 (IL-15), and granulocyte-macrophage colony-stimulating factor (GM-CSF) promote DC maturation. Immature DCs in the TME exhibit reduced antigen presentation capability, which leads to impaired activation of T cells and weakened immune surveillance. Tumor-secreted factors like VEGF, IL-10, and transforming

Growth Factor-beta (TGF-β) further disrupt DC differentiation and function, ultimately undermining the immune response [17, 18].

This immunosuppressive landscape enables tumors to evade immune detection, highlighting the complexity of immune cell interactions within the TME and their implications for cancer progression and immunotherapy resistance.

Extracellular Matrix (ECM)

Following the disruption and manipulation of immune cells, tumor cells also remodel their physical surroundings to support growth and survival. The ECM is a physical barrier and a guiding structure for cell migration. Its physical and biochemical characteristics provide crucial regulatory cues that influence cellular states. ECM remodeling, a hallmark of cancer, is far more pronounced in normal ones [19]. Beyond its role as a structural scaffold, the ECM acts as a reservoir for cytokines and growth factors, regulating key pathological processes such as angiogenesis [20]. This dynamic structure undergoes MMP-dependent modifications, facilitating carcinogenic progression. A reciprocal interaction between cells and the ECM ensures that both influence each other's behavior. While cells remodel ECM composition, the ECM transmits signals that regulate cellular functions [21]. Moreover, the ECM is essential role in maintaining stem cell niches, making it a fundamental component of tumor development [22].

Cancer-associated fibroblasts (CAFs), key cell types interacting with the ECM, contribute significantly to its dysregulation. These fibroblasts produce excess ECM components, increasing the tensile strength of tissues. Overproduction of ECM proteins like collagen, and cross-linking of structural fibers by enzymes namely lysyl oxidase (LOX) results in denser and stiffer fibrils, facilitating tumor cell migration [23]. One of the primary hallmarks of cancer progression is metastasis, which is driven by ECM degradation through MMP family enzymes. Notably, the fragments generated by ECM proteolysis can inhibit the focal adhesion kinase (FAK)/ phosphoinositide 3-kinase (PI3K)/ protein kinase B (Akt)/ mechanistic target of rapamycin complex 1 (mTORC1) pathway, reducing cancer cell proliferation and invasion. These ECM-derived fragments, secreted into the circulation, may also be biomarkers for cancer detection and progression monitoring [24, 25].

The rigidity and density of the ECM pose a significant challenge for cancer treatment. Tumor cells, encapsulated within the stiff ECM composition, are shielded from therapeutic drugs, leading to reduced treatment efficacy [26]. Furthermore, ECM remodeling induces hypoxia, activating anti-apoptotic pathways and promoting drug resistance. In breast cancer, for instance, increased ECM stiffness drives epithelial-mesenchymal transition (EMT), significantly enhancing resistance to paclitaxel [27]. These ECM-induced changes, combined with genetic instability, are strongly linked to tumorigenesis and cancer heterogeneity, making the ECM a critical factor in both disease progression and therapeutic resistance [28].

Cancer-Associated Fibroblasts (CAFs)

Under normal conditions, fibroblasts serve a vital function in maintaining tissue integrity by producing collagen and supporting the connective tissue structure. When wound healing is required, quiescent fibroblasts are activated, a property that extends to healing cancerous tumors [29]. The fibroblasts located near tumors, termed Cancer-Associated Fibroblasts (CAFs), exhibit persistent activation unlike their regular counterparts. CAFs secrete many cytokines and growth factors while synthesizing excess ECM, ultimately promoting tumor growth and metastasis [30]. This perpetual activation is maintained by continuous signaling from cancer cells, keeping CAFs in a chronic wound-like state that prevents routine healing. CAFs can derive from native fibroblasts, bone marrow-derived mesenchymal stem cells (BM-MSCs), macrophages via macrophage-myofibroblast transition, pericytes, or adipocytes [31].

Within the TME, CAFs are key regulators that interact with tumor cells, influencing angiogenesis, immune evasion, and tumor progression processes. One of their primary functions is promoting tumor angiogenesis, facilitated through the overexpression of pro-angiogenic factors like VEGFA, platelet-derived growth factor C (PDGFC), and colony-stimulating factor 3 (CSF3) [32]. CAFs also recruit myeloid cells and form vascular mimicry networks, which can develop into functional blood vessels, further supporting tumor survival. Beyond angiogenesis, CAFs contribute to immune evasion by the secretion of immunosuppressive cytokines and chemokines, such as C-X-C motif chemokine ligand 12 (CXCL12) and TGF-β, which inhibit T cell activation and recruitment at tumor sites. Additionally, CAFs play a role in fostering an immunosuppressive microenvironment that facilitates the differentiation of CD4+ and CD25+ cells into Tregs, further enhancing tumor growth by suppressing the antitumor immune response [33]. Given their diverse roles in tumor progression, CAFs have become a crucial target for therapeutic interventions focusing on disrupting the TME's supportive functions. However, designing effective CAF-targeted therapies remains challenging due to their phenotypic plasticity and heterogeneity.

Growth Factors and Cytokines

Growth factors and cytokines, disseminated by tumor cells and adjacent stromal and immune cells, are pivotal in influencing the TME. These signaling molecules directly influence cancer progression, metastasis, and therapy resistance by regulating angiogenesis, immune suppression, and cellular proliferation. Their persistent production creates an inflammatory and immunosuppressive milieu, ultimately contributing to poor therapeutic outcomes [11].

Among the most significant growth factors, VEGF is secreted by both tumor cells and tumor-associated macrophages, initiating angiogenesis to guarantee a sufficient provision of oxygen and nutrients to the tumor. Many cancer therapies, including VEGF inhibitors, target this pathway to restrict tumor vascularization. Fibroblast growth factor also promotes tumor expansion by stimulating cellular proliferation and migration, aiding metastasis. TGF-β exhibits a dual role depending on the stage of cancer. In the early cancer stage, it functions like a tumor suppressor by preventing the proliferation of cells. Nevertheless, TGF-β stimulates EMT, invasion, and immune evasion in advanced cancers. It stimulates the expression of EMT-related transcription factors such as *Slug, Snail*, and *Twist* while suppressing the activation of T cells, enabling the tumor to elude immune destruction [34].

Cytokines in the TME can be broadly classified as pro-tumorigenic or antitumorigenic based on their impact on tumor behavior and immune response. Pro-tumorigenic cytokines like IL-6 promote tumor development by enhancing cell proliferation, apoptosis resistance, and angiogenesis. It exerts its effects through the janus kinase (JAK)/signal transducers and activator of transcription 3 (STAT3) pathways, leading to the transcription of required genes that enhance tumor survival. Additionally, IL-6 recruits MDSCs, which create an immunosuppressive environment, inhibiting antitumor T cell activity. Similarly, interleukin-8 (IL-8) contributes to tumor progression by recruiting neutrophils to the TME, further amplifying inflammation and angiogenesis [35].

Conversely, antitumorigenic cytokines, namely interferons (IFNs), particularly IFN-γ, activate macrophages and cytotoxic T lymphocytes (CTLs), improving the immune system's ability to eradicate tumor cells. However, tumors counteract this by creating an immunosuppressive microenvironment dominated by Tregs and MDSCs. These tumor cells release IL-10 and TGF-β, which suppress effector T cell responses, allowing the tumor to evade immune attack and continue proliferating [36].

Dynamics of Tumor Microenvironment and the Tumor Cells in Cancer Progression

All components of the TME, from the ECM to soluble factors like cytokines and growth factors, intricately interact to assist tumor development, proliferation, as well as metastasis. These components form a dynamic and highly adaptive network that evolves in response to physiological and therapeutic pressures through complex feedback mechanisms. One of the most significant features of the TME is its capacity to circumvent the host's immune surveillance. This is primarily achieved through the secretion of immunosuppressive cytokines and by recruiting specialized immune cells, like TAMs, Tregs, and MDSCs, all inhibiting antitumor immune responses. These immunosuppressive mechanisms establish a permissive environment for tumor development and metastasis [37].

The tumor cells and their surrounding microenvironment interaction is bidirectional and highly coordinated, as a foundation for immune evasion, metastasis, and resistance to therapies. Tumor cells remodel their microenvironment by secreting chemokines and cytokines that recruit and polarize stromal and immune cells. For example, tumor-derived signals differentiate TAMs into the M2 phenotype, which supports tumor growth, motility, and immunosuppression [38]. M2 macrophages secrete anti-inflammatory and pro-tumorigenic cytokines that inhibit T cell function and influence the ECM to favor invasion. Similarly, once activated by tumor signals, cancer-associated fibroblast remains perpetually active and secretes factors such as TGF-β, further promoting angiogenesis, immune evasion, and ECM remodeling [39].

Conversely, the TME influences tumor cell behavior through mechanical forces and biochemical signals. These include interactions with the ECM, hypoxia-induced signaling cascades, and crosstalk with immune and stromal cells. A dysregulated ECM or altered signaling pathways within the TME can enhance tumor cell survival, particularly under stress conditions like hypoxia. The hypoxic environment triggers changes in tumor cell metabolism that facilitate adaptation and resistance to therapy. Furthermore, the upregulation of inhibitory molecules of the immune checkpoint, like CTLA-4 and PD-1, within the TME effectively inhibits antitumor immune responses, enabling the persistence of cancer cells [40].

Therefore, targeting only tumor cells is insufficient for effective cancer therapy. Comprehensive treatment strategies must also address the dynamic components of the TME. By simultaneously modulating tumor cell activity and reprograming the surrounding microenvironment, it is possible to enhance therapeutic outcomes and improve patient prognosis.

miRNA-Mediated Modulation of the Tumor Microenvironment

The last phase of cancer progression, metastasis, is marked through migration and colonization of neoplastic cells in remote tissues. This process is a primary contributor to cancer-related mortality and is frequently linked to unfavorable treatment outcomes. The metastatic cascade involves two critical stages: the spread of cancerous cells from the original tumor and their successful colonization in secondary tissues. The capability of circulating tumor cells (CTCs) to establish colonies in distant organs is heavily influenced by the local microenvironment of those organs. Interestingly, primary tumors can actively prepare these distant sites by creating pre-metastatic niches by the release of soluble factors and via dynamic interactions with stromal and immune cells [41].

Among the key mediators of this process are exosomes, which are small extracellular vesicles which transport a wide array of molecular cargo, such as miRNAs, proteins, and RNAs. These vesicles facilitate long-range intercellular communication and are instrumental in remodeling the tumor microenvironment. Cancer-derived exosomes often exhibit aberrant miRNA signatures, critical in promoting tumor progression. These miRNAs modulate gene expression in recipient cells, affecting key regulatory networks that control oncogene activation, tumor suppressor silencing, immune modulation, and angiogenesis [42, 43].

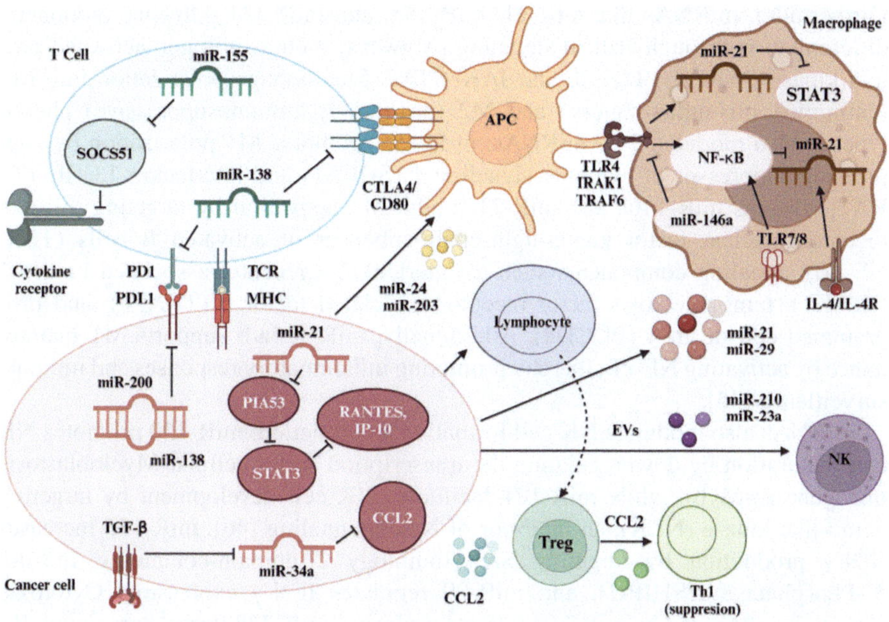

Fig. 3.1 miRNAs play a crucial modulatory role in shaping the tumor microenvironment by regulating interactions among immune cells, cancer cells, and signaling molecules

This section has illuminated the complex roles of miRNAs in shaping the TME through precise modulation of immune cell activity, fibroblast activation, angiogenesis, and ECM remodeling, thereby supporting tumor initiation, progression, and metastasis, as shown in Fig. 3.1. These small but powerful molecules have identified as essential regulators of cancer dynamics and signify prospective treatment targets for future methods aimed at disrupting tumor progression through modulation of the TME.

Regulation of Tumor-Associated Immune Cells

The TME is heavily influenced by the composition and activity of infiltrating immune cells. MicroRNAs are pivotal regulators in orchestrating this immune landscape, modulating differentiation, polarization, and effector functions of diverse immune cell populations. Deciphering miRNA-mediated mechanisms offers promising insights into restoring antitumor immunity.

miRNAs regulate the maturation and differentiation of monocytes by modulating transcription factors. The downregulation of the miR-17-5p, miR-20a, and miR-106a cluster leads to elevated expression of the M-CSF receptor, facilitating monocytic differentiation and supporting the survival of tumor-associated macrophages [44].

Furthermore, miRNAs like miR-21, miR-155, and miR-424 influence monocytic differentiation through critical signaling pathways, such as mitogen-activated protein kinase (MAPK), TGF-β, and JAK–STAT. Macrophage polarization into M1 (antitumor, pro-inflammatory) and M2 (pro-tumor, immunosuppressive) phenotypes is also modulated by miRNAs. miR-155 promotes M1 polarization by suppressing suppressor of cytokine signaling 1 (SOCS1) and interleukin-13 (IL-13), RA1, whereas miR-146a and miR-21 inhibit M1 activation by targeting toll-like receptor/ nuclear factor kappa-light-chain-enhancer of activated B cells (TLR/ NF-κB) signaling components such as interleukin-1 receptor-associated kinase 1 (IRAK1), tumor necrosis factor receptor-associated factor 6 (TRAF6), and programmed cell death 4 (PDCD4). Additionally, miR-125a/b supports M1 maintenance by activating NF-κB, thereby promoting inflammatory responses and immune surveillance [45].

miRNAs also modulate NK cell formation and function. miR-150 promotes NK cell maturation by downregulating the transcription factor cellular Myeloblastosis oncogene (c-Myb), while miR-181 facilitates NK cell development by targeting nemo-like kinase (NLK), an inhibitor of Notch signaling [46]. miR-155 increases IFN-γ production by targeting Src Homology 2 domain-containing Inositol 5'-Phosphatase 1(SHIP-1), and miR-29 regulates IFN-γ expression. Cytotoxic activity is influenced by miR-30e (targets perforin), miR-378 (targets granzyme B), and miR-27a-5p (targets both perforin and granzyme B) [47]. The miR-17-92 cluster facilitates T follicular helper (TFH) cell differentiation by targeting phosphatase and tensin homolog (PTEN) and Retinoic acid receptor-related Orphan Receptor alpha (RORα), thereby preventing commitment to T helper 17 cells (Th17) or T helper 17 cells (Th22) lineages [42]. This cluster also enhances CD8+ T cell function by downregulating TGFβRII, promoting IFN-γ production, cytotoxicity, and memory formation. miR-155, induced by Foxp3, suppresses SOCS1 to augment IL-2 signaling and signal transducer and activator of transcription 5 (STAT5) activation, promoting Treg proliferation. Conversely, miR-15a and miR-16 suppress forkhead box P3 (Foxp3) and cytotoxic T-lymphocyte antigen 4 (CTLA4), reducing Treg-mediated immunosuppression [48]. miR-146a targets STAT1 to inhibit Th1 differentiation and enhance Treg functionality. In CD8+ T cells, miR-155 regulates proliferation via SOCS1 suppression, while miR-21 and miR-30b target B cell lymphoma 6 (BCL-6) and dual specificity phosphatase 10 (DUSP10). miR-29 limits effector functions by inhibiting IFN-γ expression [49, 50].

Impact on Cancer-Associated Fibroblasts

Cancer-associated fibroblasts (CAFs), a dominant member of the TME, support tumor progression by secreting growth factors, cytokines, and extracellular matrix components. miRNAs regulate CAF activation and maintenance, modulating their tumor-promoting functions.

Mutant p53-expressing colorectal cancer cells release exosomes containing miR-21-3p and miR-769-3p, activating fibroblasts via TGF-β/Smad signaling, enhancing metastasis and EMT. Unlike normal fibroblasts, CAFs support tumor growth independently of continuous tumor-derived signals, mainly due to sustained epigenetic reprogramming, including miRNA dysregulation [51]. In ovarian cancer, overexpression of miR-155 and downregulation of miR-31 and miR-214 in CAFs are critical for their pro-tumorigenic phenotype. Restoring miR-214/miR-31 or inhibiting miR-155 reverses the CAF phenotype and reduces the secretion of chemokines like C–C motif chemokine ligand 5 (CCL5), C–C motif chemokine ligand 20 (CCL20), and C–C motif chemokine ligand 5 (CXCL8). Downregulation of miR-31 in endometrial cancer elevates AT-rich sequence-binding protein 2 (SATB2) expression, promoting cell motility. Similarly, decreased miR-148a in CAFs enhances tumor growth [52].

Another regulatory axis, PTEN-miR-320, plays a key tumor-suppressive role in CAF biology. Stromal cells lacking PTEN show altered miRNA profiles, with miR-320 acting as a downstream tumor-suppressive mediator.

Modulation of Angiogenesis and Regulation of Hypoxia in the TME

Angiogenesis and hypoxia are hallmarks of the TME, facilitating tumor progression. miRNAs regulate vascular remodeling and cellular responses to low oxygen levels. miR-320 downregulation enhances the release of pro-angiogenic molecules such as MMP9, BMP1, and LOXL2, that support tumor invasion and neovascularization. Expression patterns of miR-320 correlate with patient prognosis and can differentiate malignant from normal stroma [53].

miRNAs including miR-205 and miR-200 regulate angiogenesis as well as EMT via VEGF-A signaling. Additional angiogenesis-regulating miRNAs include miR-210, miR-126, and miR-296, which modulate VEGF pathways. Hypoxia-inducible miRNAs, including miR-210 and miR-1246, reprogram stromal and tumor cells to adapt to hypoxia, often enhancing invasion and survival [52].

Control of Metastasis and ECM Remodeling

Metastasis relies on ECM remodeling, cell migration, and invasion. The ECM offers a structural framework for the TME, and its remodeling is crucial for tumor progression. miRNAs influence these processes by modulating ECM components and cellular responses to external cues.

Lysophosphatidic acid (LPA) in the ECM promotes tumor metastasis by activating G-protein coupled receptors and upregulating miR-21, suppressing anti-metastatic targets like PTEN [54]. Cytokine-driven miRNA expression further modulates signaling to enhance tumor proliferation and immune evasion. CircRNAs such as circ_0000854 act as sponges for miRNAs like miR-1294. In hepatocellular carcinoma (HCC), downregulation of circ_0000854 leads to miR-1294 silencing, promoting tumor growth. Rebalancing miR-1294 levels can thus offer a novel therapeutic strategy for HCC [55]. Also, miR-181a modulates γδ T cell development, which is essential for antitumor immunity. Overexpression suppresses cytotoxicity and downregulates natural killer group 2, member D (NKG2D), a crucial activating receptor. Thus, targeting miR-181a can potentiate γδ T-cell-based immunotherapy. Additionally, miR-181a modulates γδ T cell development, essential for antitumor immunity. Overexpression suppresses cytotoxicity and downregulates NKG2D, a key activating receptor. Thus, targeting miR-181a can potentiate γδ T-cell-based immunotherapy. miR-181a has the potential to be used in next-generation immunotherapies since it can rewire γδ T cells toward an active effector phenotype. These results underscore the dual function of miRNAs in the TME as regulators and therapeutic targets, providing encouraging treatment options for cancer [56].

miRNAs orchestrate diverse TME functions such as immune modulation, angiogenesis, ECM remodeling, and tumor metastasis. In colorectal cancer, miR-27a reduces T cell infiltration and impairs antigen presentation, facilitating immune evasion and poor prognosis. Tumor-derived exosomal miRNAs suppress dendritic cell activation, promote M2 macrophage polarization, and inhibit T helper cell development [57]. Dysregulated miRNAs affect immune checkpoint pathways, including PD-L1/PD-1 and CD80/CTLA-4, blunting immune responses. miRNAs like miR-155 also regulate immune cell metabolism and function via non-cell-autonomous mechanisms. These insights underscore the potential of miRNA-based cancer therapies, including ex vivo immune cell engineering and in vivo targeted delivery. By targeting the TME and metastasis-related pathways, these methodologies present intriguing pathways for next-generation cancer therapy [58].

Mechanisms of miRNA Dysregulation in Tumor Microenvironment

MicroRNAs are essential posttranscriptional regulators of gene expression that perform critical functions in various cellular processes. Dysregulation of miRNAs in the TME may be attributed to various variables, including transcriptional control, epigenetic modifications, inflammatory responses, and extracellular vesicle-mediated transfer (Table 3.1). Upon understanding these pathways, gaining significant insights into miRNA-driven oncogenesis and prospective therapeutic interventions is possible.

Table 3.1 Key miRNAs involved in cancer and the tumor microenvironment, highlighting their regulatory mechanisms, target genes, and functional roles

miRNA	Regulatory mechanism	Target genes	Pathways involved	Role in cancer and TME	References
miR-21	Upregulated due to activation of NF-κB	*PTEN, PDCD4, TGF-β*	TGF-β, PI3K/AKT	OncomiR Downregulates tumor suppressor genes	[60, 64]
miR-29	Downregulated in lung cancer, breast cancer, pancreatic cancer, colorectal cancer, and leukemia, epigenetically silenced by DNMTs	*DNMT3, DNMT3B, TCL1*	Epigenetic regulation	Tumor-suppressive miRNA Induces apoptosis and inhibits proliferation Reduces DNA methylation	[60]
miR-34	Activated by p53	*SIRT1, NOTCH1, BCL-2*	p53, NOTCH	Tumor-suppressive miRNA Induces apoptosis and senescence Suppresses stemness and EMT	[61, 62]
miR-155	Upregulated by NF-κB and STAT3 in immune cells	*SOCS1, TP53,*	PI3K/AKT, JAK/STAT	Enhances tumor cell survival and angiogenesis Increases inflammation and immune evasion	[67, 71]
miR-210	Induced by HIF-1α during hypoxia	*ISCU, SDHD, RAD52*	hypoxia response, DNA repair	Promotes angiogenesis and metabolic adaptation Alters mitochondrial function Reduces DNA repair capacity	[64]
miR-146a-5p	Activated by NF-κB in immune cells	*TRAF6, IRAK1*	NF-κB signaling	Reduces excessive inflammation Inhibits immune activation in TME Modulates cytokine production	[61]

(continued)

Table 3.1 (continued)

miRNA	Regulatory mechanism	Target genes	Pathways involved	Role in cancer and TME	References
miR-124	Silenced through promoter methylation by DNMTs	*CDK6, STAT3, EZH2*	Cell cycle, epigenetic regulation	Tumor-suppressive miRNA Restoring miR-124 inhibits tumor growth Loss of miR-124 leads to uncontrolled proliferation	[65]
miR-223	Selectively packed into exosomes via hnRNPA2B1	*PTEN, FBXW7*	NF-κB, PI3K/AKT	Promotes tumor growth and immune suppression Transfers from TAMs to cancer cells via exosomes Suppresses PTEN to activate PI3K/AKT	[64]
miR-101	Downregulated in various cancers like breast cancer, prostate cancer, lung cancer, etc. Targets EZH2 for chromatin remodeling	*EZH2*	Epigenetic silencing	Tumor-suppressive miRNA Reduces metastasis and epigenetic silencing Inhibits cell cycle progression	[62]
miR-17-92	Upregulated by c-Myc and E2F transcription factors	*PTEN, BIM, E2F1*	PI3K/AKT, cell cycle regulation	Oncogenic miRNA cluster Promotes proliferation and inhibits apoptosis Enhances tumor angiogenesis	[91]

Transcriptional and Epigenetic Regulation of miRNA Expression in the TME

In the TME, miRNA expression is tightly controlled by transcription factors and epigenetic modifiers, shaping oncogenic and tumor-suppressive pathways. One of the most well-documented cases is the overexpression of the oncogenic miRNA, miR-21, in ovarian and colorectal cancers. This dysregulation is primarily driven by the transcription factor NF-κB, which becomes activated under inflammatory conditions such as exposure to TNF-α or IL-1β. Upon activation, the transcription factor NF-κB undergoes a process of translocation to the nucleus, where it interacts with the κB response elements. Chromatin immunoprecipitation (ChIP) assays have confirmed that the p65 subunit of NF-κB recruits histone acetyltransferase (HAT) p300/CBP that subsequently acetylates histone H3 at lysine 27 (H3K27Ac), leading to increased miR-21 transcription [59].

Apart from NF-κB, other transcription factors regulate miRNA expression, promoting tumorigenesis or acting as tumor suppressors. The tumor suppressor p53, in reaction to DNA damage, activates miR-34a transcription, which induces cell-cycle arrest and apoptosis. Conversely, the oncogenic transcription factor c-Myc binds to E-box motifs (CACGTG) in the promoters of miRNA clusters such as miR-17-92, which drive tumor proliferation. These transcription factors often function in coordination with coactivators that integrate various signaling pathways, further influencing the miRNA regulatory network [60].

Epigenetic alterations significantly contribute to miRNA dysregulation within the TME. DNA methyltransferases (*DNMT1, DNMT3A*, and *DNMT3B*) silence tumor-suppressor miRNAs such as miR-124 and miR-101 by methylating CpG islands within their promoters. Methyl-binding proteins like MeCP2 recognize these methylated regions and recruit histone deacetylases (*HDAC1* and *HDAC2*), leading to chromatin condensation and transcriptional repression [60]. Conversely, hypomethylation of oncogenic miRNAs, such as miR-21, increases their overexpression and contributes to tumor progression. In many cancers, the hypermethylation of tumor-suppressive miRNA promoters (e.g., miR-124 and miR-101) leads to their silencing [61].

A complex feedback loop exists between miRNAs and epigenetic regulators. For instance, miR-29 targets *DNMT3A* and *DNMT3B*, leading to their downregulation and a subsequent decrease in global DNA methylation levels. Depending on the miRNAs involved, such feedback mechanisms can reinforce either a tumor-promoting or tumor-suppressive state. Functional studies further confirm the significance of miRNA dysregulation in oncogenesis [60, 62]. Luciferase reporter assays showed miR-21 binds directly to the 3′ UTR of *PTEN* mRNA, resulting in downregulation. The removal of *PTEN* function inhibits the control over the PI3K/AKT signaling pathway, thereby enhancing the survival of cells and proliferation. The regulatory interplay of transcription factors and epigenetic mechanisms determines the miRNA landscape in the TME, significantly impacting oncogenic pathways [60, 62]. Targeting these regulatory networks presents promising therapeutic strategies for reversing miRNA-mediated tumor progression. Comprehending such molecular interactions is essential for fostering miRNA-based treatment interventions designed to modify tumor progression and response to therapy.

Impact of Hypoxia and Inflammatory Signaling on miRNA Biogenesis and Function

The harsh conditions of the tumor microenvironment, particularly hypoxia and chronic inflammation, profoundly influence miRNA biogenesis, from transcription to maturation. Hypoxia, a hallmark of solid tumors, maintains hypoxia-inducible factor 1-alpha (HIF-1α), a crucial transcriptional regulator adaptive response. Under normoxic conditions, HIF-1α undergoes hydroxylation at Proline at position 402 (Pro402) and Proline at position 564 (Pro564) by prolyl hydroxylase domain enzymes (PHD2), marking it to ubiquitination by the Von Hippel–Lindau (VHL) protein and ensuing proteasomal degradation [63]. However, under hypoxic conditions, PHD2 activity is suppressed due to oxygen deprivation, allowing HIF-1α to accumulate in the cell. The HIF-1α that has been stabilized moves to the nucleus, where it dimerizes with the HIF-1β and binds to hypoxia-responsive elements (HREs) in the promoters of target genes, including miR-210. Chromatin immunoprecipitation assays have confirmed direct HIF-1α binding to the miR-210 promoter, and the recruitment of coactivators such as p300/CBP further enhances transcription. Functionally, miR-210 modulates mitochondrial respiration, promotes angiogenesis by targeting Ephrin-A3, and regulates the cell cycle through interactions with multiple downstream mRNAs [64].

Beyond transcription, hypoxia and inflammation modulate miRNA maturation at multiple levels. The initial step in miRNA processing involves the cleavage of pri-miRNAs into pre-miRNAs, which is executed by the microprocessor complex that includes Drosha and its cofactor DGCR8. Under inflammatory conditions, post-translational modifications of Drosha, such as phosphorylation by p38 MAPK, have been reported to alter its activity, selectively influencing the maturation efficiency of specific miRNAs. Similarly, hypoxia can regulate the subcellular localization of Exportin-5 (XPO5), affecting the nuclear export rates of specific pre-miRNAs [61, 64].

Further processing of pre-miRNAs into mature 22-nucleotide miRNA duplexes is carried in the cytoplasm by Dicer, an essential RNase III enzyme. This process is assisted by cofactors such as TAR RNA-binding protein (TRBP) and the interferon-induced protein kinase (PACT) protein activator. Under stress conditions, modifications to Dicer or fluctuations in TRBP levels can shift the miRNA repertoire, influencing tumor cell survival. For instance, altered Dicer activity under hypoxia has been associated with the differential processing of miRNAs that regulate key oncogenic pathways [64].

These molecular alterations collectively reshape the miRNA landscape within the TME, promoting critical adaptive responses namely angiogenesis, metabolic reprogramming, immune evasion, as well as enhanced tumor cell survival. Advanced biochemical techniques, including ChIP, luciferase reporter assays, and in vitro processing assays, have delineated these molecular interactions, underscoring their significance in cancer progression [65].

Exosomal Communication and Intercellular Transfer of miRNAs in the TME

Exosomes, a class of specialized extracellular vesicles (30–150 nm), mediate intercellular interactions within the TME. By selectively packaging and delivering miRNAs, exosomes influence key oncogenic mechanisms including immune evasion, angiogenesis, metabolic adaptation, and therapy resistance. The categorization of specific miRNAs into exosomes is a meticulously regulated procedure controlled by RNA-binding proteins like heterogeneous nuclear ribonucleoprotein A2B1 (hnRNPA2B1) and Y-box binding protein 1 (YBX1). Sumoylated hnRNPA2B1 recognizes EXO-motifs within miRNA sequences, such as those found in miR-223 and miR-21, directing their incorporation into exosomal cargo. Similarly, YBX1 binds to structured regions of miRNAs, facilitating selective sorting, as observed in macrophage-derived exosomes enriched in miR-223 [64, 66].

Beyond selective packaging, exosomes enable horizontal transfer of miRNAs between various cell types within the TME, altering gene expression in recipient cells. A well-characterized example is the delivery of miR-223 from TAMs to epithelial ovarian cancer cells. Under hypoxic conditions, macrophages polarize into a TAM-like phenotype and release exosomes highly enriched in miR-223. Upon uptake by ovarian cancer cells through receptor-mediated endocytosis, miR-223 specifically targets the 3′ untranslated region (UTR) of PTEN mRNA, leading to its downregulation. PTEN, a lipid phosphatase that converts phosphatidylinositol-3,4,5-trisphosphate (PIP3) to phosphatidylinositol-4,5-bisphosphate (PIP2), is a critical regulator of the PI3K/AKT signaling pathway. Consequently, PTEN loss results in PI3K/AKT hyperactivation, enhancing cell survival, proliferation, and chemoresistance. This regulatory mechanism has been validated through dual-luciferase reporter assays and in vivo xenograft models [64, 67].

Similarly, tumor cells undergo hypoxia-induced alterations in their exosomal miRNA cargo, impacting recipient cell behavior. Hypoxic conditions in esophageal squamous cell carcinoma (ESCC) lead to a distinct shift in exosomal miRNA profiles. High-throughput sequencing has revealed the overexpression of microRNAs such as miR-128-3p, miR-140-3p, and miR-340-5p in exosomes derived from hypoxic TE-13 cells. Gene ontology and KEGG pathway analyses demonstrate that these miRNAs regulate key pathways involved in focal adhesion, autophagy, and AGE-RAGE signaling, all of which facilitate tumor invasion and metastasis [64, 68]. Once secreted, exosomal miRNAs reach target cells via multiple uptake mechanisms, including receptor-mediated endocytosis, direct membrane fusion, and phagocytosis. Following internalization, exosomal membranes disassemble, releasing miRNAs into the cytoplasm. These miRNAs then integrate into the RISC, where they guide posttranscriptional gene silencing by targeting complementary mRNAs [69].

Beyond their direct influence on tumor cells, exosomal miRNAs additionally serve a pivotal role in reprogramming stromal components of the TME. Exosomes generated from cancer cells convert normal fibroblasts into cancer-associated fibroblasts by transferring of miRNAs that downregulate tumor suppressor genes. Additionally, endothelial cells receiving exosomes enriched in miR-21 and miR-210 exhibit increased VEGF expression and enhanced tube formation, driving angiogenesis. These results underscore the complex role of exosomal miRNAs in shaping the tumor microenvironment [67].

Given their ability to modulate key oncogenic pathways, exosomal miRNAs are promising targets for cancer therapeutic intervention in cancer. Comprehending the molecular mechanisms that regulate miRNA sorting, transfer, and functional impact may provide innovative approaches to interrupt tumor-enhancing exosomal communication.

Molecular Signatures of Key OncomiRs: In-depth Analysis of miR-21, miR-146a-5p, and miR-155

The decreased activity of specific miRNAs is pivotal in the malignant transformation of the tumor microenvironment. Among them, miR-21, miR-146a-5p, and miR-155 have been extensively studied for their well-defined molecular targets and contributions to cancer progression. These oncomiRs drive tumorigenesis via altering essential signaling pathways implicated in cellular survival, inflammation, and immune evasion, as shown in Figure 3.2.

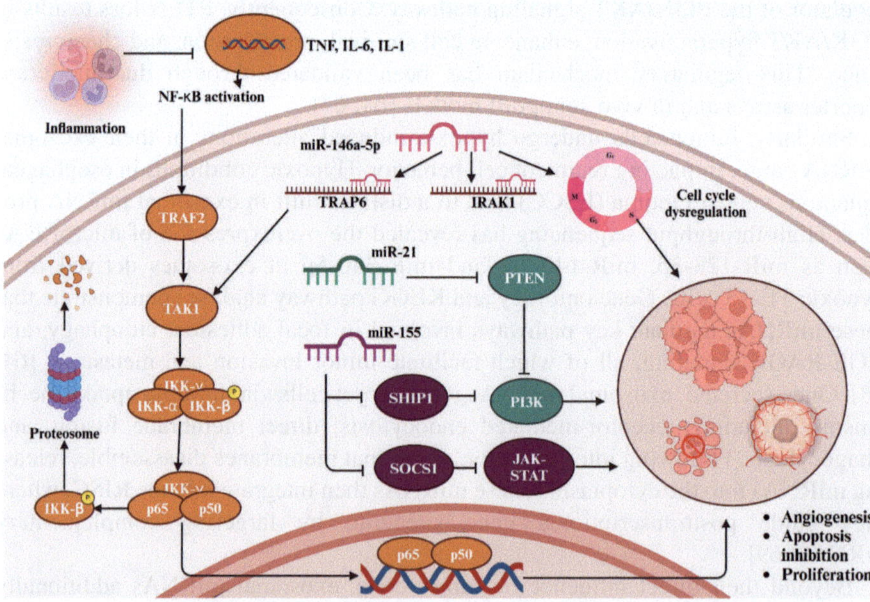

Fig. 3.2 Role of different miRNAs in regulating pathways associated with tumorigenesis by influencing signaling networks that drive cell cycle dysregulation, proliferation, and survival

miR-21 is among the most consistently elevated microRNAs across multiple cancers, including ovarian and colorectal malignancies. Its oncogenic role primarily stems from directly suppressing *PTEN*, a tumor suppressor that inhibits the PI3K/AKT pathway. miR-21 binds to the 3′ untranslated region (UTR) of *PTEN* mRNA, promoting its degradation or translational inhibition. *PTEN* dephosphorylates phosphatidylinositol-3,4,5-trisphosphate (PIP3) to phosphatidylinositol-4,5-bisphosphate (PIP2), thereby counteracting PI3K/AKT signaling. However, miR-21 overexpression leads to PTEN suppression, resulting in sustained AKT activation that promotes cellular proliferation, survival, and resistance to apoptosis. This dysregulation fosters an aggressive tumor phenotype, contributing to therapeutic resistance [61, 70].

miR-146a-5p is pivotal in modulating inflammation-induced carcinogenesis through its targeting components of the Toll-like receptor (TLR) signaling pathway, including *TRAF6* and interleukin-1 receptor-associated kinase 1 (*IRAK1*). By binding to their 3′ UTRs, miR-146a-5p suppresses nuclear factor-kappa B (NF-κB) activation, modulating inflammatory responses closely linked to cancer progression. This negative feedback loop is essential for preventing chronic inflammation, a recognized catalyst of tumorigenesis. Beyond inflammation, miR-146a-5p regulates cell cycle progression and apoptosis, with its dysregulation being associated with altered cytokine production and tumor growth. Notably, its expression is tightly controlled by NF-κB itself, creating a complex regulatory network [71].

miR-155, primarily known for its role in hematological malignancies, also contributes to tumorigenesis in solid tumors. It is derived from the B cell integration cluster (BIC) gene and is upregulated in response to inflammatory stimuli and oncogenic stress. miR-155 exerts its oncogenic effects by targeting tumor suppressors such as Src homology 2 domain-containing inositol-5-phosphatase 1 (SHIP1) and SOCS1. *SHIP1*, like *PTEN*, dephosphorylates PIP3, thus modulating PI3K/AKT signaling, while SOCS1 inhibits the JAK/STAT pathway. The loss of these tumor suppressors due to miR-155 overexpression leads to enhanced proliferative and anti-apoptotic signaling, promoting tumor progression. Elevated miR-155 expression is clinically linked to unfavorable prognosis and chemoresistance, especially in lymphomas and solid malignancies [72].

Together, these miRNAs orchestrate a molecular landscape that supports tumor progression by modulating distinct yet interconnected signaling pathways. miR-21 suppresses *PTEN*, leading to PI3K/AKT activation and enhanced tumor cell survival. miR-146a-5p inhibits *TRAF6* and *IRAK1*, attenuating NF-κB-driven inflammation. miR-155 downregulates *SHIP1* and SOCS1, amplifying PI3K/AKT and JAK/STAT signaling to drive tumor growth. This intricate regulatory network underscores the central role of oncomiRs in shaping the TME. Focusing on these miRNAs is a viable therapeutic approach, offering new avenues for precision oncology interventions to disrupt tumor-promoting signaling cascades [73]. This intricate regulatory network underscores the central role of oncomiRs in shaping the TME. Targeting these miRNAs offers a possible therapeutic approach, offering new avenues for precision oncology interventions aimed at disrupting tumor-promoting signaling cascades.

Therapeutic Implications of miRNAs Targeting the Tumor Microenvironment

The TME, a dynamic ecosystem of endothelial cells, fibroblasts, immune cells, and other stromal components, significantly influences cancer development and progression. miRNAs are pivotal in modulating the biology and function of these stromal cells. The abnormal regulation of miRNAs is connected virtually with every aspect of cancer, from initiation as well as growth to metastasis and drug resistance. The following subsection will offer a comprehensive breakthrough of the most recent developments in comprehending the functions of miRNAs in TME remodelling, their potential as diagnostic, prognostic, and therapeutic biomarkers, and the challenges associated with their clinical application [74].

miRNAs as Molecular Biomarkers for Diagnosis and Detection

miRNAs function as valuable biomarkers for cancer detection. Techniques like real-time transcription-mediated amplification have demonstrated the focused acquisition and identification of mature miR-221 from intricate samples, including human prostate cancer cells and xenografts. This high-throughput approach offers excellent analytical sensitivity and specificity for miRNA profiling in clinical specimens. Additionally, integrative molecular bioinformatics tools such as PicTar, TargetScan, and miRBase can predict miRNA-mRNA interactions, enhancing biomarker discovery [75].

Comparative analyses of delta cycle threshold (Ct) values for specific miRNAs also facilitate cancer classification. For instance, differential expression of miR-503 and miR-511 with high accuracy can distinguish adrenocortical cancers from benign adenomas. Moreover, miRNA analysis in formalin-fixed tissues, the standard histological procedure, broadens their clinical utility. Studies have consistently identified dysregulated miRNA panels in breast cancer, such as the upregulation of miR-21, miR-155, miR-191, and miR-196a, alongside the downregulation of miR-125b and miR-221. These results underscore the possibilities of miRNA profiling for cancer detection and risk stratification [76].

miRNAs as Prognostic Biomarkers: Predicting Treatment Response, Metastasis, Progression, and Survival

MicroRNAs have emerged as robust prognostic biomarkers in various cancers, offering insights into treatment response, metastatic potential, disease progression, and patient survival. Their ability to regulate gene expression makes them sensitive indicators of tumor behavior. For example, the cluster of microRNAs such as miR-17-92 is overexpressed in patients with small cell lung cancer (SCLC) with inactivated retinoblastoma (RB) may modulate DNA damage response, suggesting that its expression levels predict the effectiveness of DNA-damaging chemotherapeutics. Similarly, elevated miR-196b in mixed lineage leukemia (MLL) could be a marker of the progression of aggressive disease and potentially poor response to standard therapies, given its function in promoting the proliferation and persistence of leukemic cells [77].

In chronic lymphocytic leukemia (CLL), miR-92-1 targets the VHL transcript, stabilizing HIF-1 and activating VEGF expression. This suggests that high miR-92-1 levels may predict increased angiogenesis and a heightened risk of disease progression. Furthermore, miR-21, a widely studied oncomiR, exhibits variable expression across different cancers and stages. In tongue squamous cell carcinoma (TSCC), its inverse correlation with TPM1 suggests that higher miR-21 levels may indicate reduced apoptosis and poorer survival outcomes. In laryngeal carcinoma, its suppression of BTG2 is associated with aggressive tumor behavior and metastasis [78]. Conversely, tumor-suppressor miRNAs also serve as valuable prognostic markers. The downregulation of miR-449a in prostate cancer leads to the initiation of apoptosis as well as the arrest of the cell cycle when re-expressed, suggesting that low miR-449a levels may indicate aggressive disease and therapy resistance. Similarly, in acute myeloid leukemia (AML), miR-34b downregulation correlates with increased CREB expression, which is linked to uncontrolled proliferation of cells and poor prognosis. In non-small cell lung cancer (NSCLC), the downregulation of miR-145 in EGFR-mutant tumors suggests an increased risk of tumor progression and resistance to EGFR-targeted therapies [79].

Some miRNA dual role oncogenes and tumor suppressors highlight their complexity and prognostic potential. miR-29a, for instance, promotes EMT in breast cancer, contributing to metastasis, while its downregulation in other cancers suggests a loss of tumor-suppressive function. Similarly, miR-143 exhibits contrasting roles, promoting metastasis in HBV-related hepatocellular carcinoma (HCC) while acting as a tumor suppressor in colorectal cancer. Such context-dependent behaviors emphasize the importance of taking a nuanced approach to miRNA-based prognosis [80]. miRNAs also serve as indicators of stem cell-like properties, which are often linked with treatment resistance and recurrence. miR-181, highly expressed in HCC and hepatic stem cells, can be a marker for cancer stem cell populations, indicating poor prognosis. The let-7 family, recognized for its role in stem cell regulation, provides insight into tumor aggressiveness and recurrence potential. Additionally, miR-200c, which modulates BMI1 expression and is an essential regulator of the self-renewal of stem cells, further highlights the link between miRNA dysregulation and stemness-related poor prognosis [81].

Invasion and metastasis-related miRNAs, encompassing miR-532-5p, miR-27b, miR-101, miR-205, miR-146a/b, and miR-122, offer significant prognostic value. For example, miR-122, suppressing angiogenesis and intrahepatic metastasis in HCC, is frequently downregulated in aggressive tumors, correlating with increased metastatic potential and reduced survival rates. Finally, miRNAs play a crucial role in predicting drug sensitivity. miR-328, for instance, regulates ABCG2 expression in breast cancer cells, influencing mitoxantrone sensitivity. This indicates that miRNA profiling can guide personalized treatment strategies and improve therapeutic outcomes [81].

In summary, miRNAs function as multifaceted prognostic biomarkers, furnishing critical insights into treatment response, metastatic risk, disease progression, and patient survival. Their diverse roles in gene regulation underscore their value in precision oncology, paving the way for improved cancer management through miRNA-based diagnostics and targeted therapeutics.

miRNA-Based Therapeutics: Current Advances and Clinical Trials

MicroRNA mimics and antagonists (anti-miRs) represent innovative strategies in cancer therapy, targeting the dysregulated expression of miRNAs associated with tumorigenesis. These approaches aim to inhibit oncogenic miRNAs or restore typical miRNA function, influencing cancer progression and treatment outcomes. Research conducted by Hongbo Qian et al. proved that miR-21 is frequently highly expressed in numerous malignancies, facilitating tumor proliferation by suppressing tumor suppressor genes. Utilizing anti-miRs to target miR-21 can reinstate the expression of these suppressor genes, resulting in reduced cancer progression as well as enhanced therapeutic outcomes. Conversely, miR-34 is often downregulated in cancer. Introducing miR-34 mimics can reinstate its function, initiating apoptosis and reduction of cell growth [82].

Further refining miRNA-based therapies, a study by Sayani Mazumder et al. found that using miRNA mimics in pairs can enhance the precision of targets while diminishing toxicity to cells. This approach minimizes off-target impacts, a common challenge in cancer therapies. Specifically, the study focused on the miRNA mimic pair let-7c-5p and miR-125b-5p, which r decreased the expression of prevalent designated genes, such as hexokinase 2 (HK2) and branched-chain amino acid transaminase 1 (BCAT1). The downregulation of these genes is essential for inhibiting cancer cell metabolism and growth. The study demonstrated that using lower doses of these miRNA pairs was more effective than a single miRNA, helping to minimize the cytotoxic effects often associated with higher doses [83]. Effective delivery of miRNA mimics and anti-miRs is paramount for successful cancer therapy. Several delivery methods have been developed, with exosomes and nanoparticles being the primary approaches, each offering unique advantages.

Exosomes, natural nanovesicles secreted by mammalian cells, possess inherent attributes that enable them to circumvent the immune system, rendering them highly efficient for miRNA-based cancer therapy. An investigation by Xin Hua et al. demonstrated that engineered exosomes loaded using miRNA-FAM through electroporation facilitated efficient miRNA entry and enabled co-delivery of chemotherapy agents such as co-load 7-coumarin. This combination approach enhanced cancer treatment efficacy by integrating miRNA-based autophagy inhibition with chemotherapy. Furthermore, integrating miRNA-FAM into an exosome-magnetic nanoparticle (MNP) system significantly improved targeted delivery and cellular uptake, ensuring therapeutic action within the tumor microenvironment [84].

Nanoparticle-based delivery offers several advantages, including protection of miRNA from degradation and targeted modification. Encapsulation of miRNA within nanoparticles shields it from enzymatic degradation by RNases in the bloodstream, ensuring its stability until it reaches target cells. Surface alterations, such as the attachment of certain ligands or molecules, further enhance nanoparticle-miRNA interactions, ensuring secure enclosure within the nanoparticle structure. Additionally, nanoparticles can be engineered for targeted delivery, directing miRNA specifically to intended cells or tissues, thereby increasing therapeutic effectiveness while reducing off-target impacts. Some nanoparticle systems are also designed for controlled release, gradually dispensing miRNA to maintain sustained therapeutic effects over time [85].

Extracellular vesicles (EVs), ranging from 30 nanometers to 10 micrometers, are natural transporters of miRNAs, making them promising for cancer therapy. These nano-sized carriers may be designed to administer therapeutic miRNAs directly to neoplastic cells, enabling a targeted approach that may inhibit tumor growth or induce cancer cell death. Techniques such as direct loading, electroporation, sonication, chemical methods, and preloading with miRNA mimics are used to load miRNAs into EVs efficiently. The ongoing research and advancements in miRNA-loaded EVs have demonstrated significant potential for clinical applications [86]. Considerable investments have accelerated the development of miRNA-based therapies, including them in clinical studies for multiple disease. However, challenges persist, particularly regarding safety concerns and adverse effects.

Promising miRNA candidates in clinical trials include TTX-MC138, an inhibitor of miR-10b, which has demonstrated efficacy in preclinical studies by inhibiting cell migration, invasion, and immune evasion mechanisms. This therapeutic approach impacts the ability of cancer cells to manipulate the TME. Another candidate, Pep-21, combines an anti-miR-21 oligonucleotide with a PD-L1-binding peptide to target tumor cells and TAMs. This therapy effectively disrupts the immunosuppressive TME by promoting a transition from an M2-like to an M1-like phenotype in TAMs, enhancing antitumor immune responses [87].

Despite these advances, clinical trials have encountered challenges and setbacks. Safety concerns and adverse immune reactions remain significant hurdles, exemplified by the withdrawal of the miR-34a mimic MRX34 due to severe immune toxicity. Limited tumor specificity, incorrect targeting, and inefficient delivery methods remain obstacles. Observed toxicities include immune-related adverse events and liver toxicity [87].

Strategies to overcome these challenges include refining delivery mechanisms through advanced nanoparticle and exosome engineering, enhancing risk assessment and preclinical validation, combining miRNA therapy with immunotherapies and other TME-modulating drugs, developing tumor-targeted vectors to improve specificity, and exploring alternative routes of administration to minimize systemic toxicity.

Case Studies and Emerging Research

MicroRNAs have become crucial biomarkers in cancer detection and prognosis owing to their regulatory roles in gene expression. Their dysregulation is linked to numerous characteristics associated with various attributes of cancer, encompassing proliferation, apoptosis, metastasis, and treatment resistance. Clinical trials investigating miRNAs have provided valuable insights into their capacity as diagnostic tools, prognostic indicators, and therapeutic goals. Several microRNAs, including miR-21, miR-155, and miR-34a, have undergone comprehensive studied across different cancer types, revealing their significance in predicting disease progression and treatment response. Table 3.2 summarizes key miRNAs currently evaluated in clinical trials, highlighting their roles in cancer prognosis and diagnosis.

Table 3.2 Key miRNAs associated with cancer diagnosis and prognosis, along with their clinical significance and trial status. Data from ClinicalTrials.gov indicate their potential use in cancer detection, progression monitoring, and therapeutic applications

miRNA	Cancer type	Clinical significance	Trial status	ClinicalTrials.gov ID
miR-21	Lung, breast, colorectal, pancreatic	Diagnostic and prognostic biomarker; associated with poor survival	Completed	NCT02192021
miR-155	breast, lung, lymphoma	Elevated in tumors; correlated with metastasis and poor prognosis	Recruiting	NCT03098099
miR-200 family	Ovarian, breast, lung	Downregulated in aggressive cancers; linked to EMT	Active	NCT01105374
miR-34a	Colorectal, prostate, NSCLC	Tumor suppressor; used in miRNA-based therapies	Terminated	NCT01829971
miR-375	Head and neck, prostate	Tumor suppressor; potential prognostic marker	Completed	NCT03636437
miR-16	Prostate, leukemia	Therapeutic miRNA mimic (MRX34)	Terminated	NCT02369198
miR-10b	Breast, glioblastoma	Linked to metastasis; potential therapeutic target	Ongoing	NCT02909610
miR-126	Lung, colorectal	Antiangiogenic miRNA; correlates with response to therapy	Completed	NCT02603224
miR-145	Bladder, breast, prostate	Tumor suppressor; associated with better prognosis	Not yet recruiting	NCT04167979
miR-221/222	Thyroid, prostate, glioblastoma	OncomiRs; associated with drug resistance	Active	NCT02716060
miR-17-92 cluster	Lung, lymphoma, colorectal	Oncogenic; regulates cell cycle and apoptosis	Recruiting	NCT04006738
miR-125b	Breast, prostate, bladder	Functions as tumor suppressor and oncogene depending on cancer type	Completed	NCT03239355

Future Directions and Challenges

MicroRNA-based therapeutics, including miRNA mimics and antagonists, have emerged as promising tools for oncological therapy by regulating the aberrant expression of oncogenic and tumor-suppressive miRNAs. For instance, anti-miRs targeting overexpressed oncogenic microRNAs such as miR-21 can restore tumor suppressor gene activity as well as inhibit tumor growth. In contrast, mimics of downregulated tumor-suppressive miRNAs like miR-34 can induce apoptosis and suppress proliferation. Recent studies have shown that combinatorial miRNA therapy, such as the co-delivery of let-7c-5p and miR-125b-5p, enhances the accuracy of targets and reduces cytotoxicity by effectively downregulating common oncogenic targets like HK2 and BCAT1 involved in tumor metabolism [88].

Targeted and stable delivery means such as exosomes and nanoparticles have been formulated to ensure therapeutic efficacy. Engineered exosomes, which are naturally biocompatible, can be loaded with therapeutic miRNAs and chemotherapeutic agents, as demonstrated by Xin Hua et al., where exosomes enhance intracellular uptake and autophagy inhibition in cancer cells [89]. On the other hand, nanoparticles offer protection against enzymatic degradation and allow surface functionalization for targeted delivery. Clinical translation remains challenging despite advances due to off-target effects, immune activation, and systemic toxicity, as seen in the discontinued MRX34 trial. Ongoing efforts focus on refining delivery strategies, integrating miRNA therapy with immunotherapy, and leveraging artificial intelligence for personalized therapeutic designs, collectively positioning miRNA therapeutics as a novel frontier in precision oncology [90].

Conclusion

Future advancements in miRNA-based therapeutics will focus on optimizing delivery systems, such as tumor-targeted nanoparticles and exosome-magnetic nanoparticle hybrids, to enhance specificity and minimize systemic toxicity. Integrating miRNA profiling with multi-omics approaches will enable personalized medicine by leveraging tumor-specific miRNA signatures like miR-155 in bladder carcinoma, and miR-451 in gastric carcinoma to predict treatment responses. Additionally, combination strategies pairing miRNA modulators with chemotherapy, radiation, or immune checkpoint inhibitors may help overcome therapy resistance driven by the tumor microenvironment. Further research into miRNA modifications, including m6A methylation, could uncover novel biomarkers and therapeutic targets. Large-scale clinical trials are essential to validate miRNA-based diagnostics in liquid biopsies like serum miR-301a-3p for pancreatic cancer and establish standardized protocols for therapeutic applications. By addressing delivery challenges and leveraging miRNAs' dual roles as oncogenes and tumor suppressors, translational research can propel miRNA-based therapies into mainstream oncology, making TME modulation a cornerstone of future cancer treatment.

References

1. Ebert MS, Sharp PA (2012) Roles for MicroRNAs in conferring robustness to biological processes. Cell 149(3):515–524. https://doi.org/10.1016/J.CELL.2012.04.005/ASSET/FB49842D-240E-41C5-A172-9E552BE34529/MAIN.ASSETS/GR5.JPG
2. Großhans H (ed) (2010) Regulation of MicroRNAs, vol 700. Springer, US. https://doi.org/10.1007/978-1-4419-7823-3
3. Lee Y (2002) MicroRNA maturation: stepwise processing and subcellular localization. EMBO J 21(17):4663–4670. https://doi.org/10.1093/emboj/cdf476
4. Burda Z, Krzywicki A, Martin OC, Zagorski M (2011) Motifs emerge from function in model gene regulatory networks. Proc Natl Acad Sci 108(42):17263–17268. https://doi.org/10.1073/pnas.1109435108
5. Braga TV, Evangelista FCG, Gomes LC, Araújo SS da S, Carvalho M das G, Sabino A de P. Evaluation of MiR-15a and MiR-16-1 as prognostic biomarkers in chronic lymphocytic leukemia. Biomed Pharmacother 2017;92:864–869. doi:https://doi.org/10.1016/j.biopha.2017.05.144
6. Shiah SG, Chou ST, Chang JY (2021) MicroRNAs: their role in metabolism, tumor microenvironment, and therapeutic implications in head and neck squamous cell carcinoma. Cancers (Basel) 13(22):5604. https://doi.org/10.3390/cancers13225604
7. Otoukesh B, Abbasi M, Gorgani H et al (2020) MicroRNAs signatures, bioinformatics analysis of miRNAs, miRNA mimics and antagonists, and miRNA therapeutics in osteosarcoma. Cancer Cell Int 20(1):254. https://doi.org/10.1186/s12935-020-01342-4
8. Wang Q, Shao X, Zhang Y et al (2023) Role of tumor microenvironment in cancer progression and therapeutic strategy. Cancer Med 12(10):11149–11165. https://doi.org/10.1002/cam4.5698
9. Pernot S, Evrard S, Khatib AM (2022) The give-and-take interaction between the tumor microenvironment and immune cells regulating tumor progression and repression. Front Immunol:13. https://doi.org/10.3389/fimmu.2022.850856
10. Andón FT, Digifico E, Maeda A et al (2017) Targeting tumor associated macrophages: the new challenge for nanomedicine. Semin Immunol 34:103–113. https://doi.org/10.1016/j.smim.2017.09.004
11. Quail DF, Joyce JA (2013) Microenvironmental regulation of tumor progression and metastasis. Nat Med 19(11):1423–1437. https://doi.org/10.1038/nm.3394
12. Josephs DH (2015) Tumour-associated macrophage polarisation and re-education with immunotherapy. Front Biosci 7(2):735. https://doi.org/10.2741/e735
13. Mira E, Carmona-Rodríguez L, Tardáguila M et al (2013) A lovastatin-elicited genetic program inhibits M2 macrophage polarization and enhances T cell infiltration into spontaneous mouse mammary tumors. Oncotarget 4(12):2288–2301. https://doi.org/10.18632/oncotarget.1376
14. Yang F (2023) The roles and challenges of immunosuppressive cells in TME for CAR-T cell therapy in GBM. Highlight Sci Eng Technol 74:1553–1558. https://doi.org/10.54097/296e0m78
15. Bilotta MT, Antignani A, Fitzgerald DJ (2022) Managing the TME to improve the efficacy of cancer therapy. Front Immunol:13. https://doi.org/10.3389/fimmu.2022.954992
16. Hafler DA, Astier AL (2015) Editorial: T cell regulation by the environment. Front Immunol:6. https://doi.org/10.3389/fimmu.2015.00229
17. Hinshaw DC, Shevde LA (2019) The tumor microenvironment innately modulates Cancer progression. Cancer Res 79(18):4557–4566. https://doi.org/10.1158/0008-5472.CAN-18-3962
18. Shimabukuro-Vornhagen A, Draube A, Liebig TM, Rothe A, Kochanek M, von Bergwelt-Baildon MS (2012) The immunosuppressive factors IL-10, TGF-β, and VEGF do not affect the antigen-presenting function of CD40-activated B cells. J Exp Clin Cancer Res 31(1):47. https://doi.org/10.1186/1756-9966-31-47
19. Yue B (2014) Biology of the extracellular matrix. J Glaucoma 23:S20–S23. https://doi.org/10.1097/IJG.0000000000000108

20. Lu P, Takai K, Weaver VM, Werb Z (2011) Extracellular matrix degradation and remodeling in development and disease. Cold Spring Harb Perspect Biol 3(12):a005058–a005058. https://doi.org/10.1101/cshperspect.a005058
21. Brassart-Pasco S, Brézillon S, Brassart B, Ramont L, Oudart JB, Monboisse JC (2020) Tumor microenvironment: extracellular matrix alterations influence tumor progression. Front Oncologia:10. https://doi.org/10.3389/fonc.2020.00397
22. Brizzi MF, Tarone G, Defilippi P (2012) Extracellular matrix, integrins, and growth factors as tailors of the stem cell niche. Curr Opin Cell Biol 24(5):645–651. https://doi.org/10.1016/j.ceb.2012.07.001
23. Eble JA, Niland S (2019) The extracellular matrix in tumor progression and metastasis. Clin Exp Metastasis 36(3):171–198. https://doi.org/10.1007/s10585-019-09966-1
24. Oudart JB, Doué M, Vautrin A et al (2016) The anti-tumor NC1 domain of collagen XIX inhibits the FAK/ PI3K/Akt/mTOR signaling pathway through αvβ3 integrin interaction. Oncotarget 7(2):1516–1528. https://doi.org/10.18632/oncotarget.6399
25. Kehlet SN, Sanz-Pamplona R, Brix S, Leeming DJ, Karsdal MA, Moreno V (2016) Excessive collagen turnover products are released during colorectal cancer progression and elevated in serum from metastatic colorectal cancer patients. Sci Rep 6(1):30599. https://doi.org/10.1038/srep30599
26. Henke E, Nandigama R, Ergün S (2020) Extracellular matrix in the tumor microenvironment and its impact on Cancer therapy. Front Mol Biosci:6. https://doi.org/10.3389/fmolb.2019.00160
27. Fischer KR, Durrans A, Lee S et al (2015) Epithelial-to-mesenchymal transition is not required for lung metastasis but contributes to chemoresistance. Nature 527(7579):472–476. https://doi.org/10.1038/nature15748
28. Reuter JA, Ortiz-Urda S, Kretz M et al (2009) Modeling inducible human tissue neoplasia identifies an extracellular matrix interaction network involved in Cancer progression. Cancer Cell 15(6):477–488. https://doi.org/10.1016/j.ccr.2009.04.002
29. Wright K, Ly T, Kriet M, Czirok A, Thomas SM (2023) Cancer-associated fibroblasts: master tumor microenvironment modifiers. Cancers (Basel) 15(6):1899. https://doi.org/10.3390/cancers15061899
30. Glabman RA, Choyke PL, Sato N (2022) Cancer-associated fibroblasts: tumorigenicity and targeting for cancer therapy. Cancers (Basel) 14(16):3906. https://doi.org/10.3390/cancers14163906
31. Kazakova AN, Lukina MM, Anufrieva KS et al (2024) Exploring the diversity of cancer-associated fibroblasts: insights into mechanisms of drug resistance. Front Cell Dev Biol:12. https://doi.org/10.3389/fcell.2024.1403122
32. De Francesco EM, Lappano R, Santolla MF, Marsico S, Caruso A, Maggiolini M (2013) HIF-1α/GPER signaling mediates the expression of VEGF induced by hypoxia in breast cancer associated fibroblasts (CAFs). Breast Cancer Res 15(4):R64. https://doi.org/10.1186/bcr3458
33. Costa A, Kieffer Y, Scholer-Dahirel A et al (2018) Fibroblast heterogeneity and immunosuppressive environment in human breast cancer. Cancer Cell 33(3):463–479.e10. https://doi.org/10.1016/j.ccell.2018.01.011
34. Shibuya M (2011) Vascular endothelial growth factor (VEGF) and its receptor (VEGFR) signaling in angiogenesis: a crucial target for anti- and pro-angiogenic therapies. Genes Cancer 2(12):1097–1105. https://doi.org/10.1177/1947601911423031
35. Johnson DE, O'Keefe RA, Grandis JR (2018) Targeting the IL-6/JAK/STAT3 signalling axis in cancer. Nat Rev Clin Oncol 15(4):234–248. https://doi.org/10.1038/nrclinonc.2018.8
36. Jorgovanovic D, Song M, Wang L, Zhang Y (2020) Roles of IFN-γ in tumor progression and regression: a review. Biomark Res 8(1):49. https://doi.org/10.1186/s40364-020-00228-x
37. Khalaf K, Hana D, Chou JTT, Singh C, Mackiewicz A, Kaczmarek M (2021) Aspects of the tumor microenvironment involved in immune resistance and drug resistance. Front Immunol:12. https://doi.org/10.3389/fimmu.2021.656364

38. Huang R, Kang T, Chen S (2024) The role of tumor-associated macrophages in tumor immune evasion. J Cancer Res Clin Oncol 150(5):238. https://doi.org/10.1007/s00432-024-05777-4
39. Wang S, Wang J, Chen Z et al (2024) Targeting M2-like tumor-associated macrophages is a potential therapeutic approach to overcome antitumor drug resistance. NPJ Precis Oncol 8(1):31. https://doi.org/10.1038/s41698-024-00522-z
40. Barrueto L, Caminero F, Cash L, Makris C, Lamichhane P, Deshmukh RR (2020) Resistance to checkpoint inhibition in Cancer immunotherapy. Transl Oncol 13(3):100738. https://doi.org/10.1016/j.tranon.2019.12.010
41. Bayat M, Sadri NJ (2024) Exosomal miRNAs: the tumor's trojan horse in selective metastasis. Mol Cancer 23(1):167. https://doi.org/10.1186/s12943-024-02081-0
42. Yang Z, Liu X, Xu H et al (2024) Integrative analysis of genomic and epigenomic regulation reveals miRNA mediated tumor heterogeneity and immune evasion in lower grade glioma. Commun Biol 7(1):824. https://doi.org/10.1038/s42003-024-06488-9
43. Grigore IA, Rajagopal A, Chow JTS, Stone TJ, Salmena L (2024) Discovery of miRNA–mRNA regulatory networks in glioblastoma reveals novel insights into tumor microenvironment remodeling. Sci Rep 14(1):27493. https://doi.org/10.1038/s41598-024-78337-y
44. Fontana L, Pelosi E, Greco P et al (2007) MicroRNAs 17-5p–20a–106a control monocytopoiesis through AML1 targeting and M-CSF receptor upregulation. Nat Cell Biol 9(7):775–787. https://doi.org/10.1038/ncb1613
45. Wang C, Wang X, Zhang D et al (2023) The macrophage polarization by miRNAs and its potential role in the treatment of tumor and inflammation (review). Oncol Rep 50(4):190. https://doi.org/10.3892/or.2023.8627
46. Cichocki F, Felices M, McCullar V et al (2011) Cutting edge: MicroRNA-181 promotes human NK cell development by regulating Notch signaling. J Immunol 187(12):6171–6175. https://doi.org/10.4049/jimmunol.1100835
47. Wang P, Gu Y, Zhang Q et al (2012) Identification of resting and type I IFN-activated human NK cell miRNomes reveals MicroRNA-378 and MicroRNA-30e as negative regulators of NK cell cytotoxicity. J Immunol 189(1):211–221. https://doi.org/10.4049/jimmunol.1200609
48. Lu LF, Thai TH, Calado DP et al (2009) Foxp3-dependent MicroRNA155 confers competitive fitness to regulatory T cells by targeting SOCS1 protein. Immunity 30(1):80–91. https://doi.org/10.1016/j.immuni.2008.11.010
49. Cortez MA, Anfossi S, Ramapriyan R et al (2019) Role of miRNAs in immune responses and immunotherapy in cancer. Genes Chromosomes Cancer 58(4):244–253. https://doi.org/10.1002/gcc.22725
50. Wang Z, Wang H, Zhou S, Mao J, Zhan Z, Duan S (2024) miRNA interplay: mechanisms and therapeutic interventions in cancer. MedComm – Oncology 3(4). https://doi.org/10.1002/mog2.93
51. Barrera LN, Ridley PM, Bermejo-Rodriguez C, Costello E, Perez-Mancera PA (2023) The role of microRNAs in the modulation of cancer-associated fibroblasts activity during pancreatic cancer pathogenesis. J Physiol Biochem 79(1):193–204. https://doi.org/10.1007/s13105-022-00899-0
52. Suzuki HI, Katsura A, Matsuyama H, Miyazono K (2015) MicroRNA regulons in tumor microenvironment. Oncogene 34(24):3085–3094. https://doi.org/10.1038/onc.2014.254
53. Muz B, de la Puente P, Azab F, Azab AK (2015) The role of hypoxia in cancer progression, angiogenesis, metastasis, and resistance to therapy. Hypoxia 83. https://doi.org/10.2147/HP.S93413
54. Alshahrani SH, Ibrahim YS, Jalil AT et al (2022) Metabolic reprogramming by miRNAs in the tumor microenvironment: focused on immunometabolism. Front Oncologia:12. https://doi.org/10.3389/fonc.2022.1042196
55. Rao G, Peng X, Tian Y, Fu X, Zhang Y (2023) Circular RNAs in hepatocellular carcinoma: biogenesis, function, and pathology. Front Genet:14. https://doi.org/10.3389/fgene.2023.1106665

56. Gordino G, Costa-Pereira S, Corredeira P et al (2022) MicroRNA-181a restricts human γδ T cell differentiation by targeting Map3k2 and Notch2. EMBO Rep 23(1). https://doi.org/10.15252/embr.202052234
57. Kuninty PR, Schnittert J, Storm G, Prakash J (2016) MicroRNA targeting to modulate tumor microenvironment. Front Oncologia:6. https://doi.org/10.3389/fonc.2016.00003
58. Kousar K, Ahmad T, Abduh MS et al (2022) miRNAs in regulation of tumor microenvironment, chemotherapy resistance, immunotherapy modulation and miRNA therapeutics in cancer. Int J Mol Sci 23(22):13822. https://doi.org/10.3390/ijms232213822
59. Liu HY, Zhang YY, Zhu BL et al (2019) miR-21 regulates the proliferation and apoptosis of ovarian cancer cells through PTEN/PI3K/AKT. Eur Rev Med Pharmacol Sci 23(10):4149–4155. https://doi.org/10.26355/eurrev_201905_17917
60. Arif KMT, Elliott EK, Haupt LM, Griffiths LR (2020) Regulatory mechanisms of epigenetic miRNA relationships in human Cancer and potential as therapeutic targets. Cancers (Basel) 12(10):2922. https://doi.org/10.3390/cancers12102922
61. Wu Y, Song Y, Xiong Y et al (2017) MicroRNA-21 (Mir-21) promotes cell growth and invasion by repressing tumor suppressor PTEN in colorectal cancer. Cell Physiol Biochem 43(3):945–958. https://doi.org/10.1159/000481648
62. Pajares MJ, Alemany-Cosme E, Goñi S, Bandres E, Palanca-Ballester C, Sandoval J (2021) Epigenetic regulation of microRNAs in Cancer: shortening the distance from bench to bedside. Int J Mol Sci 22(14):7350. https://doi.org/10.3390/ijms22147350
63. D'Angelo G, Duplan E, Boyer N, Vigne P, Frelin C (2003) Hypoxia up-regulates prolyl hydroxylase activity. J Biol Chem 278(40):38183–38187. https://doi.org/10.1074/jbc.M302244200
64. Peng X, Gao H, Xu R, Wang H, Mei J, Liu C (2020) The interplay between HIF-1α and noncoding RNAs in cancer. J Exp Clin Cancer Res 39(1):27. https://doi.org/10.1186/s13046-020-1535-y
65. Micolucci L, Matacchione G, Albertini MC et al (2023) A data-mining approach to identify NF-kB-responsive microRNAs in tissues involved in inflammatory processes: potential relevance in age-related diseases. Int J Mol Sci 24(6):5123. https://doi.org/10.3390/ijms24065123
66. Villarroya-Beltri C, Gutiérrez-Vázquez C, Sánchez-Cabo F et al (2013) Sumoylated hnRNPA2B1 controls the sorting of miRNAs into exosomes through binding to specific motifs. Nat Commun 4(1):2980. https://doi.org/10.1038/ncomms3980
67. Zhu X, Shen H, Yin X et al (2019) Macrophages derived exosomes deliver miR-223 to epithelial ovarian cancer cells to elicit a chemoresistant phenotype. J Exp Clin Cancer Res 38(1):81. https://doi.org/10.1186/s13046-019-1095-1
68. Yang D, Li Z, Gao G et al (2021) Combined analysis of surface protein profile and microRNA expression profile of exosomes derived from brain microvascular endothelial cells in early cerebral ischemia. ACS Omega 6(34):22410–22421. https://doi.org/10.1021/acsomega.1c03248
69. Chen F, Chu L, Li J et al (2020) Hypoxia induced changes in miRNAs and their target mRNAs in extracellular vesicles of esophageal squamous cancer cells. Thorac Cancer 11(3):570–580. https://doi.org/10.1111/1759-7714.13295
70. Bautista-Sánchez D, Arriaga-Canon C, Pedroza-Torres A et al (2020) The promising role of miR-21 as a Cancer biomarker and its importance in RNA-based therapeutics. Mol Ther Nucleic Acids 20:409–420. https://doi.org/10.1016/j.omtn.2020.03.003
71. Iacona JR, Lutz CS (2019) miR-146a-5p: expression, regulation, and functions in cancer. WIREs RNA 10(4). https://doi.org/10.1002/wrna.1533
72. Witten L, Slack FJ (2020) miR-155 as a novel clinical target for hematological malignancies. Carcinogenesis 41(1):2–7. https://doi.org/10.1093/carcin/bgz183
73. Yang X, Ren Y, Li X, Xia L, Wan J (2025) MiR-146a reduces inflammation in experimental pancreatitis via the TRAF6–NF-κB signaling pathway in mice. Immun Inflamm Dis 13(3). https://doi.org/10.1002/iid3.70163
74. Raue R, Frank AC, Syed SN, Brüne B (2021) Therapeutic targeting of MicroRNAs in the tumor microenvironment. Int J Mol Sci 22(4):2210. https://doi.org/10.3390/ijms22042210

75. Sarhadi VK, Armengol G (2022) Molecular biomarkers in cancer. Biomolecules 12(8):1021. https://doi.org/10.3390/biom12081021
76. Ferracin M, Veronese A, Negrini M (2010) Micromarkers: miRNAs in cancer diagnosis and prognosis. Expert Rev Mol Diagn 10(3):297–308. https://doi.org/10.1586/erm.10.11
77. Shin VY (2014) MiRNA as potential biomarkers and therapeutic targets for gastric cancer. World J Gastroenterol 20(30):10432. https://doi.org/10.3748/wjg.v20.i30.10432
78. Bertoli G, Cava C, Castiglioni I (2015) MicroRNAs: new biomarkers for diagnosis, prognosis, therapy prediction and therapeutic tools for breast cancer. Theranostics 5(10):1122–1143. https://doi.org/10.7150/thno.11543
79. Noonan EJ, Place RF, Basak S, Pookot D, Li LC (2010) miR-449a causes Rb-dependent cell cycle arrest and senescence in prostate cancer cells. Oncotarget 1(5):349–358. https://doi.org/10.18632/oncotarget.167
80. Cho WCS (2010) MicroRNAs: potential biomarkers for cancer diagnosis, prognosis and targets for therapy. Int J Biochem Cell Biol 42(8):1273–1281. https://doi.org/10.1016/j.biocel.2009.12.014
81. Yu DC, Li QG, Ding XW, Ding YT (2011) Circulating MicroRNAs: potential biomarkers for cancer. Int J Mol Sci 12(3):2055–2063. https://doi.org/10.3390/ijms12032055
82. Qian H, Maghsoudloo M, Kaboli PJ et al (2024) Decoding the promise and challenges of miRNA-based Cancer therapies: an essential update on miR-21, miR-34, and miR-155. Int J Med Sci 21(14):2781–2798. https://doi.org/10.7150/ijms.102123
83. Mazumder S, Banerjee U, Sinsinwar S, et al. Distance between miRNA responsive elements selectively enhances target specificity with reduced cellular toxicity by miRNA-mediated synergism. Published online August 4, 2024. https://doi.org/10.1101/2024.08.01.606174
84. Hua X, Liu Y, Liu X et al (2024) Engineered exosome-based drug delivery system for synergistic cancer therapy via autophagy inhibition and chemotherapy. Biomed Anal 1(2):103–110. https://doi.org/10.1016/j.bioana.2024.05.002
85. Zhou T, Qiu JM, Han XJ et al (2024) The application of nanoparticles in delivering small RNAs for cancer therapy. Discov Oncol 15(1):500. https://doi.org/10.1007/s12672-024-01341-1
86. Menjivar NG, Oropallo J, Gebremedhn S et al (2024) MicroRNA Nano-shuttles: engineering extracellular vesicles as a cutting-edge biotechnology platform for clinical use in therapeutics. Biol Proced Online 26(1):14. https://doi.org/10.1186/s12575-024-00241-6
87. St-Cyr G, Penarroya D, Daniel L, Giguère H, Alkayyal AA, Tai LH (2023) Remodeling the tumor immune microenvironment with oncolytic viruses expressing miRNAs. Front Immunol:13. https://doi.org/10.3389/fimmu.2022.1071223
88. Belderbos RA, Aerts JGJV, Vroman H (2019) Enhancing dendritic cell therapy in solid tumors with Immunomodulating conventional treatment. Mol Ther Oncolytics 13:67–81. https://doi.org/10.1016/j.omto.2019.03.007
89. Li Z, Rana TM (2014) Therapeutic targeting of microRNAs: current status and future challenges. Nat Rev Drug Discov 13(8):622–638. https://doi.org/10.1038/nrd4359
90. Vaxevanis C, Bachmann M, Seliger B (2024) Immune modulatory microRNAs in tumors, their clinical relevance in diagnosis and therapy. J Immunother Cancer 12(8):e009774. https://doi.org/10.1136/jitc-2024-009774
91. Ebi H, Sato T, Sugito N et al (2009) Counterbalance between RB inactivation and miR-17–92 overexpression in reactive oxygen species and DNA damage induction in lung cancers. Oncogene 28(38):3371–3379. https://doi.org/10.1038/onc.2009.201

Chapter 4
Epigenetic Regulation of miRNAs in Cancer

Deepa Ramasamy, Ahmad S. Kodous, Samson Mani, Arunagiri Kuha Deva Magendhra Rao, and Anandan Balakrishnan

Abstract MicroRNAs (miRNAs) are master regulators of gene expression at the posttranscriptional level. The unique tissue-specific expression of miRNA, when altered, leads to aberrant gene regulations, oncogenic signaling, and deregulated cellular pathways, transforming a normal cell into a cancer cell. The control of miRNA expression is strongly influenced by the epigenetic modifications occurring at an early stage of cellular transformation. The epigenetic changes potentially determine the chromatin accessibility and transcriptional assembly by simple base modifications. The well-known phenomenon of DNA methylation, histone methylation, and acetylation is therefore the dictator of the mighty miRNAs. The promoter regions of tumor suppressor miRNA genes are found to be hypermethylated, diminishing their transcriptional efficiency. The promoters of oncogenic microRNAs are often demethylated, releasing the tight control on miRNA transcription. The chromatin state is relaxed due to histone deacetylation or demethylation, facilitating the accessibility for an active transcription complex at the miRNA genes. Surprisingly, miRNAs are observed to be regulating the enzymes like DNMTs and HDACs that catalyze such epigenetic modifications. Therefore, DNMT, HDAC, and HAT inhibitors are being proficiently used in combinatorial cancer therapy. This chapter gives an exclusive insight into the regulatory inter-relationship of the epigenetic factors and miRNAs in cancer, with a highlight on the role of RNA methylation in miRNA biogenesis.

D. Ramasamy · Arunagiri Kuha Deva Magendhra Rao
Department of Molecular Oncology, Cancer Institute (W.I.A), Chennai, Tamil Nadu, India

A. S. Kodous
Radiation Biology Department, National Centre for Radiation Research & Technology, Egyptian Atomic-Energy Authority, Cairo, Egypt

S. Mani
Department of Research, Rajiv Gandhi Cancer Institute, and Research Centre, Delhi, India

A. Balakrishnan (✉)
Department of Genetics, Dr. ALM Postgraduate Institute of Basic Medical Sciences, University of Madras, Chennai, Tamil Nadu, India

S. Pathak, A. Banerjee (eds.), *MicroRNAs: Role in Cancer Progression, Diagnostics and Therapy*, Advances in Biochemistry in Health and Disease 34,
https://doi.org/10.1007/978-3-032-08824-6_4

Keywords MicroRNAs (miRNAs) · Epigenetic regulation · DNA methylation · Histone modification · Cancer therapy

Introduction

MicroRNAs (miRNAs) are small endogenous noncoding RNAs of 18 to 24 nucleotides in length involved in cell proliferation, differentiation, apoptosis, and immune responses. The expression of miRNAs is critically regulated at both the transcriptional and posttranscriptional levels through genetic and epigenetic mechanisms. Epigenetic regulation of miRNAs has an indispensable role in the development of cancer. Among the epigenetic alterations, DNA methylation and histone modifications largely influence miRNA expression, leading to cancer development [1, 2]. DNA methylation, particularly in the promoter regions of miRNA genes, is a well-known mechanism that can silence tumor-suppressive miRNAs in various cancers. miR-34a, found to be a critical tumor-suppressive miRNA, is often silenced by hypermethylation in colon and lung cancer [3]. miR-143/145 negatively regulates oncogenic signals that are frequently downregulated through DNA methylation in colorectal cancer [4]. The acetylation, deacetylation, and methylation of histones also play an equally significant role in regulating miRNA expression. Histone deacetylases (HDACs) can repress miRNA expression by promoting a compact chromatin structure, and histone acetylation is associated with gene activation [5]. In cancer, alterations in histone modification patterns lead to the silencing of the tumor-suppressive miR-203, which is downregulated in prostate cancer and associated with tumor progression [6]. Oncogenic miRNAs like miR-21 are upregulated in breast, liver, and lung cancers, often through mechanisms that involve epigenetic alterations [6–8]. Additionally, the chromatin remodeling complexes, SWI/SNF, are involved in regulating miRNA gene expression by altering chromatin structure around miRNA genes, thus either enhancing or silencing transcription [9, 10]. Disruption of these chromatin remodeling processes can result in the downregulation of the tumor-suppressive miRNAs, promoting cancer origin and metastasis. In addition, miRNAs themselves regulate the epigenetic landscape by targeting the genes controlling DNA methylation and histone modification. For instance, miR-29b targets DNA methyltransferases, potentially altering DNA methylation patterns and influencing the expression of other cancer-related genes [11, 12]. This reciprocal relationship between miRNAs and epigenetic regulators highlights the complexity and the dynamic regulation of gene expression (Fig. 4.1). The dysregulation of miRNAs, whether through epigenetic silencing of tumor-suppressive miRNAs or overexpression of oncogenic miRNAs, contributes significantly to cancer initiation, progression, and metastasis [2, 13, 14]. Therefore, understanding the epigenetic regulation of miRNAs can open new directions in cancer therapy. DNA methyltransferase inhibitors (DNMTis) and histone deacetylase inhibitors (HDACis), which restore the tumor-suppressive miRNAs or inhibit oncogenic miRNAs are promising epigenetic therapies [5, 15–17]. Additionally, miRNA mimics may be

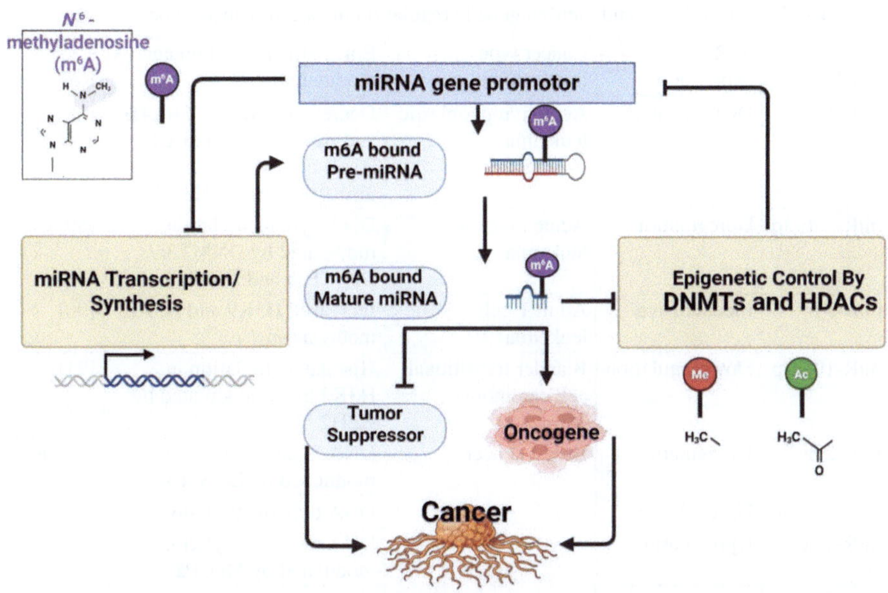

Fig. 4.1 Role of epigenetic factors in altering microRNA function. The miRNA gene promoter is controlled by epigenetic factors like DNMT and histone deacetylase, causing deregulated expression levels of miRNAs. RNA modifications like m6A are found to determine mature miRNA biosynthesis, and m6A can affect the translational control of tumor suppressor genes and oncogenes by miRNAs, contributing to cancer development

used to substitute the expression of tumor-suppressive miRNAs, while miRNA inhibitors affect oncogenic miRNAs [18, 19]. The epigenetic regulation of miRNAs is an exhaustive phenomenon that controls gene expression and cellular signaling. Harnessing the potential of epigenetic modifications and miRNA activity together should assure future RNA therapeutics for cancer.

DNA Methylation and miRNA Regulation

Epigenetic modifications control the expression of coding and noncoding RNAs. They exert myriad effects on miRNA-mediated cancer development. Epigenetic regulation of a single miRNA may imply controlling the expression levels of several coding RNA transcripts (Table 4.1). DNA methylation refers to the addition of a methyl group to the 5′ position of the cytosine in CpG dinucleotides, primarily within CpG islands located at the gene promoter regions. Notably, dysregulation of DNA methylation at miRNA promoters contributes to the aberrant expression of miRNAs. The tumor-suppressive miR-34a, which is a direct target of transcription factor p53, is frequently silenced by DNA methylation observed in lung, colon, and breast cancers. Loss of miR-34a expression allows for the uncontrolled activation of

Table 4.1 List of miRNAs and their epigenetic regulation in various cancer types

miRNA	miRNA expression	Cancer type involvement	Epigenetic regulation and its modulators	Ref
miR-124a	Downregulation	Acute lymphoblastic leukemia	Decreased levels of 3mk4H3 and AcH3 and increased levels of 2mK9H3, 3mK9H3, and 3mK27H3	[88]
miR-29b-3p	Upregulation	Acute myeloid leukemia	DNA hypomethylation: modulated by DNMT3a/ DNMT3b and DNMT1	[89]
miR-31	Inconclusive	Adult T cell leukemia	Increased H3K9 and H3K27 methylation	[90]
miR-101-3p	Downregulation	Bladder transitional cell carcinoma	Histone methylation in H3K27me3: modulated by EZH2	[91]
miR-29a	Upregulation	Breast cancer	DNA hypermethylation: modulated by DNMT3a	[92–99]
miR-29b-3p	Upregulation		DNA hypermethylation	
miR-132	Upregulation		DNA hypermethylation: modulated by MeCP2	
miR-10a	Downregulation		H3K27me3/DNA methylation	
miR-29c, miR-26b, miR-148b	Upregulation		DNA methylation: modulated by DNMT3b	
miR-148a, miR-152	Upregulation		DNA methylation: modulated by DNMT1	
let-7e, miR-1246, miR-1826, and miR-361-5p	Inconclusive		Decreased H3K4me3	
miR-31	Inconclusive		Hypermethylation	
miR-155	Inconclusive		Decreased H2A and H3 deacetylation (HDAC)	
miR-106b, miR-93, and miR-25	Inconclusive		Histone acetyltransferase modifications at EP300	
miR-22	Inconclusive	Breast cancer cell lines	Histone acetyltransferase modifications at TIP60	[100]
miR-101-3p	Upregulation	Breast/prostate	Histone methylation in H3K27me3: modulated by EZH2	[101]
miR-15a, miR-16, and miR-29b		Chronic lymphocytic leukemia	Histone deacetylation	[102]
miR-143	Downregulation	Colorectal cancer	DNA methylation: modulated by DNMT3a	[40, 41, 103]
miR-342	Upregulation		DNA methylation: modulated by DNMT1	

(continued)

Table 4.1 (continued)

miRNA	miRNA expression	Cancer type involvement	Epigenetic regulation and its modulators	Ref
miR-124	Inconclusive	Colorectal cancer cell line	Histone acetyltransferase modifications at hCLOCK	[104]
miR-29a-3p	Upregulation	Cutaneous melanoma	DNA methylation: modulated by DNMT3a/ DNMT3b	[105]
miR-152	Downregulation	Endometrial cancer	DNA methylation: modulated by DNMT1	[106]
miR-148a	Upregulation	Gastric cancer	DNA methylation: modulated by DNMT1	[107]
miR-101-3p	Downregulation	Glioblastoma	Histone methylation in H3K27me3: modulated by EZH2	[108, 109]
miR-152	Upregulation		DNA methylation: modulated by DNMT1	
miR-185	Upregulation	Glioma	DNA methylation: modulated by DNMT1	[110]
miR-101-3p	Upregulation	Hepatitis B virus (HBV)-related hepatocellular carcinoma	DNA methylation: modulated by DNMT3a	[111]
miR-200a	Upregulation	Hepatocellular carcinoma	Histone acetylation in H3ac: modulated by HDAC4	[106, 112–117]
miR-125b-5p	Upregulation		Histone methylation in H3K9me3: modulated by SUV39H1	
miR-152	Upregulation		DNA methylation: modulated by DNMT1	
miR-139-5p, miR-125b, miR-101, let-7c, and miR-200b	Inconclusive		Increased Histone methylation in H3K27me3: modulated by EZH2	
miR-449	Inconclusive		Histone deacetylation: modulated by HDAC1–3	
miR-224	Inconclusive		Histone acetylation: modulated by Ep300	
miR-373	Upregulation	Hilar cholangiocarcinoma	DNA methylation: modulated by MDB2	[118]
miR-23a	Inconclusive	Human leukemic Jurkat cells	Histone deacetylation: modulated by HDAC4	[119]
miR-148a, miR-152	Upregulation	Human malignant cholangiocytes, gastric cancer	DNA methylation: modulated by DNMT1	[107, 120]

(continued)

Table 4.1 (continued)

miRNA	miRNA expression	Cancer type involvement	Epigenetic regulation and its modulators	Ref
miR-19a, miR-25, miR-32, miR-92b, miR-96	Upregulation	Leukemia and lymphoma cells	DNA methylation: modulated by PRMT5	[120]
miR-101-1	Upregulation	Lung cancer	Histone methylation in H3K27me3: modulated by EZH2	[109, 121]
miR-31	Upregulation	Melanoma	DNA methylation: modulated by EZH2	[122, 123]
miR-29b-3p	Upregulation	Multiple myeloma	DNA methylation: modulated by DNMT3a/DNMT3b	
miR-15a and miR-16	Inconclusive	Non-Hodgkin B cell lymphomas	Histone deacetylation: modulated by HDAC3	[124]
miR-29a	Downregulation	Non-small-cell lung cancer	DNA methylation: modulated by DNMT3a/DNMT3b	[89, 125, 126]
miR-29b-3p	Upregulation			
miR-29c	Inconclusive			
miR-29b-3p	Inconclusive		DNA methylation: modulated by DNMT1/DNMT3a/DNMT3b	
miR-29b-3p	Inconclusive		DNA methylation: modulated by DNMT3b	
miR-185	Upregulation	Ovarian cancer	DNA methylation: modulated by DNMT1	[127]
miR-152	Downregulation			
miR-127	Upregulation	Primary human tumors	Increased histone H3 acetylation and H3-K4 methylation	[46]
miR-34b	Downregulation	Prostate cancer	Histone methylation in H3K4me3: modulated by HDAC1/HDAC2/HDAC4	[128–132]
miR-34b	Upregulation		DNA methylation: modulated by DNMT1	
miR-449a	Upregulation		Histone acetylation: modulated by HDAC1	
miR-101-3p	Downregulation		Histone methylation in H3K27me3: modulated by EZH2	
miR-615	Inconclusive		Increased H3K9 acetylation	
miR-17-5p	Inconclusive		Histone acetyltransferase modifications at PCAF	
miRNA-1260b	Upregulation	Prostate cancer cells	Increased H3K9-me2, H3K9me3, and H3K27me3	

(continued)

Table 4.1 (continued)

miRNA	miRNA expression	Cancer type involvement	Epigenetic regulation and its modulators	Ref
miR-138-5p	Downregulation	Squamous cell carcinoma	Histone methylation in H3K27me3: modulated by EZH2	[133]
miR-101-3p	Upregulation	T cell leukemia	Histone methylation in H3K27me3: modulated by EZH2	[46, 134, 135]
miR-128, mIR-101-3p, miR-128	Downregulation			

pro-survival pathways, through Bcl-2, Cyclin D1, and c-MYC expression [4]. The silencing of miR-34a by DNA methylation has also been linked to poor cancer treatment outcomes, making it an excellent prognostic biomarker for cancer [20]. The miR-143/145 is a cluster of tumor-suppressive miRNAs that are frequently silenced via hypermethylation in colorectal cancer. These miRNAs regulate cell proliferation through the aberrant activation of Ras and Rho pathways and promote cancer metastasis. The oncogenic miRNA, on the contrary, exhibits loss of promoter methylation with an upregulation of miRNAs targeting the tumor suppressor genes. miR-21, which is frequently upregulated in various cancers due to promoter demethylation, contributes to the suppression of tumor suppressors like PTEN, PDCD4, and TGF-βR2, leading to increased tumor growth, migration, and resistance to apoptosis [21].

miRNA promoter methylation is mediated by the DNA methyltransferase (DNMT) family of enzymes, namely DNMT1, DNMT3A, and DNMT3B. The primary function of DNMT1 is maintaining the methylation marks, while DNMT3A and DNMT3B are responsible for de novo methylation, establishing new methylation patterns [22–24]. The dysregulation of DNMTs often results in the hypermethylation of tumor-suppressive miRNA genes, leading to their silencing and contributing to the malignant transformation of normal cells into cancer cells. Particularly, higher expression of DNMT1 and DNMT3B in cancers is associated with downregulation of protein-coding genes and several noncoding RNAs, including miRNAs. In addition to direct methylation of miRNA promoters, DNMTs can also interact with chromatin-modifying enzymes, such as histone deacetylases (HDACs), to establish a condensed chromatin environment around miRNA genes [25, 26], leading to their repression.

Histone Modifications and miRNA Expression

Histone modifications are a fundamental mode of gene regulation at the chromatin state. These modifications affect chromatin accessibility, influencing the active transcription of genes, including microRNA (miRNA) genes. Histones, the protein components of chromatin, are subject to various posttranslational modifications like the covalent addition or removal of chemical groups that affect the state of chromatin. The major histone modifications are acetylation, methylation, phosphorylation, ubiquitination, and SUMOylation which are primarily found on the N-terminal end of histones. The presence or absence of specific modifications can either open or close the chromatin structure, thereby promoting or inhibiting gene expression [16, 27, 28]. The widespread histone modifications that control gene expression are histone acetylation and histone methylation. The histone acetylation process typically correlates with gene activation, as the addition of acetyl groups neutralizes the positive charge of histones, reducing their affinity for the negatively charged DNA and leading to a more open chromatin structure. In contrast, histone methylation can either activate or repress gene expression depending on the specific histone residues that are methylated and the number of methyl groups added. Methylation of histone H3 at lysine 4 (H3K4me3) is associated with gene activation, while methylation at lysine 9 (H3K9me3) and lysine 27 (H3K27me3) is typically linked to gene silencing [29–31].

The miRNA genes, transcribed by RNA polymerase II, are regulated by various transcriptional and posttranscriptional mechanisms. A pre-transcription control of miRNA is exerted by an open or closed chromatin state determined by the histone alterations. Cancer cells have characteristic patterns of histone modifications, leading to the dysregulation of miRNA expression. Histone marks often contribute to the silencing of tumor-suppressive miRNAs and the activation of oncogenic miRNAs. The aberrant histone modifications can result in the silencing of key tumor-suppressive miRNAs, such as miR-34a, miR-143, and miR-200, all of which are frequently downregulated in various cancer types [32–34]. The reactivation of these miRNAs through the modulation of histone modifications has been proposed as a promising therapeutic strategy. Drugs that target specific histone-modifying enzymes other than HDAC inhibitors are the HMT inhibitors shown to be effective in early-phase clinical trials by upregulating the tumor-suppressive miRNAs [5, 35, 36]. Understanding the role of histone modifications in miRNA regulation offers important insights into cancer biology and presents potential targets for therapeutic intervention.

Histone acetylation is often associated with transcriptional activation including miRNA genes. The acetylation of histones, particularly H3 and H4, around miRNA gene promoters facilitates the recruitment of transcriptional machinery, leading to the initiation of miRNA transcription. Studies have shown that histone acetylation is a key factor in regulating the expression of several tumor-suppressive miRNAs. The well-known tumor suppressor miR-34a is often silenced in cancers and is activated through histone acetylation in response to p53 signaling. Transcription factor Tp53, also a critical tumor suppressor, can recruit histone acetyltransferases (HATs) to the miR-34a promoter, promoting its expression and exerting its tumor-suppressive effects [18, 37, 38].

Additionally, miR-143/145, another cluster of miRNAs involved in regulating cellular processes such as cell cycle arrest and apoptosis, is activated by histone acetylation in response to stress signals [38, 39]. The histone acetylation near the miR-143/145 promoter facilitates the transcriptional activation of these miRNAs, leading to the inhibition of oncogenic signaling pathways. In the context of cancer, the use of histone deacetylase inhibitors (HDACis), which prevent the removal of acetyl groups from histones, has been shown to restore the expression of silenced miRNAs, offering potential therapeutic strategies for cancer treatment. Treatment with HDACis has been shown to increase the expression of miR-34a and miR-143/145 in various cancer models, thereby suppressing tumor growth [40, 41].

Histone methylation is often associated with gene silencing when it occurs at specific lysine residues. The methylation of histones can alter chromatin structure in a way that inhibits the binding of transcriptional activators, leading to the repression of gene expression. The tumor-suppressive miR-203 is silenced through histone methylation in various cancers, including prostate and lung cancers. In these cancers, miR-203 is silenced by trimethylation of histone H3 at lysine 9 (H3K9me3), a modification typically associated with transcriptional repression [42]. The loss of miR-203 expression contributes to the uncontrolled proliferation of cancer cells, as miR-203 targets and downregulates key oncogenes such as Bcl-2. Furthermore, the miR-200 family, which plays a key role in regulating epithelial-to-mesenchymal transition (EMT) and metastasis, is often repressed in metastatic cancers due to histone methylation. Specifically, the miR-200 promoter region is often marked by H3K27me3, a repressive histone mark, leading to the silencing of these miRNAs and promoting the epithelial-to-mesenchymal transition, a key step in cancer metastasis [43–45]. The CpG island embedded miR-127 was found upregulated in association with active transcription remarks of DNA demethylation, histone acetylation, and trimethylation (H3K4me3) [46]. miR-127 overexpression leads to downregulation of BCL6, a proto-oncogene that exerts its tumor suppressor effect.

Posttranscriptional Regulation of miRNAs

Beyond transcriptional control, miRNAs are also regulated posttranscriptionally. RNA-binding proteins (RBPs) and lncRNAs can influence miRNA stability and function. For instance, certain RBPs can bind to miRNAs, protecting them from degradation or facilitating their incorporation into the RISC complex [47, 48]. Additionally, lncRNAs can act as competitive endogenous RNAs (ceRNAs), sequestering miRNAs and preventing them from binding to their targets, compromising the miRNA activity [49]. Given their central role in cancer, miRNAs present promising targets for therapeutic intervention. Therapies that restore the tumor-suppressive miRNAs expression or inhibit the activity of oncogenic miRNAs are being explored. One such strategy involves the use of DNA methyltransferase inhibitors (DNMTis) like 5-aza-2′-deoxycytidine, which has been shown to reactivate silenced miRNAs, such as miR-34a, in cancer cells [18]. Similarly, histone deacetylase inhibitors (HDACis), such as vorinostat and romidepsin, can reverse the silencing of miRNAs regulated by histone modifications, offering an alternative therapeutic approach. Additionally, recent advancements in miRNA-based therapies, including miRNA mimics and miRNA inhibitors (antagomiRs), have shown promise in preclinical models. miRNA mimics are designed to restore the function of downregulated tumor-suppressive miRNAs, while antagomiRs are used to inhibit the function of overexpressed oncogenic miRNAs. Furthermore, antagomiR-21 has demonstrated efficacy in reducing tumor progression by targeting the oncogenic miR-21 in preclinical models [50–53]. An exciting prospect is the combination of miRNA-based therapies with existing treatments, such as chemotherapy, immunotherapy, or targeted therapies. The use of epigenetic modulators in combination with miRNA-based therapies can enhance the efficacy of conventional treatments by reversing epigenetic silencing and restoring the expression of tumor-suppressive miRNAs.

RNA Methylation and MicroRNA Regulation in Cancer

RNA methylation is an important layer of posttranscriptional regulation that influences various RNA processes, such as stability, splicing, translation, and degradation. Among the various types of RNA modifications, N6-methyladenosine (m6A) is an abundant modification found in eukaryotic mRNA and noncoding RNAs, including microRNAs (miRNAs) [54]. The regulation of miRNA biogenesis, stability, and function is now shown to highly depend on the RNA methylation mechanism.

RNA methylation is catalyzed by a family of methyltransferases and demethylases that add or remove methyl groups. The major type of RNA modification is N6-methyladenosine (m6A), which occurs at the nitrogen atom at position 6 of the adenine base. The deposition of the m6A mark is carried out by the methyltransferase complex, consisting of core proteins METTL3, METTL14, and WTAP. METTL3 is the catalytic subunit of the complex and is responsible for adding the methyl group, while METTL14 helps stabilize the complex [55–58]. The m6A modification can also be removed by m6A demethylases, including FTO (Fat Mass and Obesity-associated protein) and ALKBH5, which remove the methyl group from RNA, restoring the unmethylated form [59, 60]. In addition to methyltransferases and demethylases, RNA-binding proteins, namely YTHDF1, YTHDF2, and YTHDC1, recognize and interpret m6A modifications, mediating downstream effects such as mRNA stability and translation efficiency. These readers of the m6A mark are essential for the proper functioning of the RNA methylation machinery and help guide the fates of modified RNAs, including miRNAs.

miRNA biogenesis is a multistep process that involves the transcription of primary miRNA (pri-miRNA) by RNA polymerase II, followed by the cleavage of the pri-miRNA by the Drosha complex to produce precursor miRNA (pre-miRNA). The pre-miRNA is then exported to the cytoplasm, where it is further processed by Dicer into mature miRNA. While RNA methylation has been primarily studied in the context of mRNA processing and translation, emerging evidence indicates that m6A modifications also play a significant role in the regulation of miRNA biogenesis. The addition of m6A is found to influence the processing of pre-miRNA into mature miRNA. The m6A modification on miRNA precursors can enhance the efficiency of Dicer cleavage or affect the nuclear export of the pre-miRNA, which in turn impacts the levels of mature miRNA produced [61–63]. Additionally, m6A remarks on pri-miRNA can alter the interaction with the Drosha complex, thereby regulating the maturation of miRNA. This regulation is particularly important because the proper maturation of miRNAs ensures the generation of functional miRNA molecules, which can then execute their gene-silencing functions. miR-21, a well-known oncomiR (oncogenic miRNA) involved in promoting cancer cell proliferation, is regulated by m6A methylation. The m6A modification of pri-miR-21 enhances its processing into the precursor miRNA, thereby leading to increased levels of mature miR-21, which promotes tumorigenesis in several cancer types. On the other hand, tumor-suppressive miRNAs, such as miR-34a, can be negatively regulated by m6A methylation at the miRNA gene or precursor stages, contributing to the downregulation of these miRNAs in cancers.

In addition to influencing miRNA biogenesis, RNA methylation also determines miRNA stability. The m6A modification on mature miRNAs can modulate their degradation rates. This regulation is primarily mediated by m6A-binding proteins such as YTHDF2, which binds to m6A-modified miRNAs and promotes their decay [64–66]. In glioblastoma, m6A modification of miR-125b results in its stabilization, which contributes to its oncogenic activity by inhibiting the expression of tumor suppressor genes [67]. Similarly, miR-126, a tumor-suppressive miRNA involved in vascular remodeling and metastasis suppression, is stabilized in part through the m6A modification. The stabilization of miR-126 in cancer cells can inhibit angiogenesis and prevent metastasis (78). In cancers, altered m6A methylation patterns can result in the stabilization or destabilization of specific miRNAs, thus affecting their ability to control gene expression. The control of miRNA stability by RNA methylation can, therefore, serve as a mechanism by which cancer cells manipulate miRNA levels to promote survival, proliferation, and metastasis.

miRNAs play a critical role in regulating cancer-related pathways by targeting genes involved in cell cycle control, apoptosis, migration, and invasion. Through the regulation of miRNA biogenesis, stability, and function by RNA methylation, cancer cells differentially express key miRNAs that drive tumor progression or suppression. Oncogenic miRNAs, such as miR-21, miR-155, and miR-372, are often upregulated in cancer, and their elevated expression can be partially attributed to changes in RNA methylation patterns. The m6A methylation of miR-21 and its precursor facilitates the processing of pri-miR-21, leading to higher levels of mature miR-21 in gliomas, breast cancer, and lung cancer. The increased expression of miR-21 promotes the downregulation of tumor-suppressive genes such as PTEN, PDCD4, and TP53, contributing to cancer cell survival, invasion, and resistance to chemotherapy [68, 69]. Similarly, miR-155, another oncogenic miRNA, is overexpressed in multiple malignancies, including lymphoma, breast cancer, and colorectal cancer. The stabilization of miR-155 through m6A methylation of its precursor is shown to enhance its oncogenic activity by targeting tumor suppressors and activating inflammatory pathways [70, 71]. Conversely, tumor-suppressive miRNAs such as miR-34a, miR-143, and miR-200, which are often downregulated in cancer, can also be affected by altered RNA methylation. In these cases, reduced m6A methylation may result in decreased stability or processing efficiency of these miRNAs, which contribute to cancer development and metastasis.

The understanding of the role of RNA methylation in miRNA regulation offers new opportunities for the development of RNA-based cancer therapies. Targeting the enzymes responsible for adding or removing m6A marks from RNA, such as METTL3, FTO, and ALKBH5, could be a potential therapeutic strategy to modulate miRNA expression in cancer cells. Inhibition of METTL3 has been observed to reduce the expression of oncogenic miRNAs, such as miR-21 and miR-155, and restore the expression of tumor-suppressive miRNAs, leading to decreased cancer cell proliferation and enhanced sensitivity to chemotherapy [21, 71]. In addition, small molecules or RNA-based therapies to regulate m6A modifications on specific miRNAs could provide a more targeted approach to cancer treatment. Therapies that enhance the stability of tumor-suppressive miRNAs or inhibit the stabilization of oncogenic miRNAs could help reverse the gene dysregulation in cancer cells and restore normal cellular function.

Therapeutic Implications of miRNA and Epigenetic Modifications

Noncoding RNA (ncRNA) and miRNA-based therapy is an emerging mode of cancer therapeutics. The multidimensional role of miRNA in posttranscriptional regulation makes it an effective tool in targeting cancer cells for its promising potency and efficacy. Since miRNA is a dual sword, it can either promote tumorigenesis as oncomiRs or suppress it as antimiRs or tumor-suppressive miRNAs [72, 73]. Thus, miRNA sequence can be exploited as miRNA mimics or inhibitors based on the regulatory function that could hamper cancer progression and development. Combining the epigenetic and miRNA-based therapies would enhance the efficiency and help in targeting tumor cells. DNA methylation in regulating miRNA expression, aimed at reversing aberrant DNA methylation, is an attractive approach for cancer treatment. DNA methyltransferase inhibitors (DNMTis), such as 5-aza-2′-deoxycytidine (decitabine) and hydralazine, have shown the potential to reactivate silenced miRNAs and restore the expression levels of tumor-suppressive miRNAs. Treatment with decitabine has been shown to reactivate miR-34a expression in cancer cell lines, leading to decreased cell proliferation and enhanced sensitivity to chemotherapeutic agents [3, 4, 35]. Similarly, miR-143/145 expression can be reactivated by DNMT inhibition in colorectal cancer cells, leading to the suppression of tumor growth and metastasis. The potential activity of DNMTis as a therapeutic agent has a universal effect in decreasing promoter methylation, but in the case of miRNA genes, the downregulation of their gene targets is an additional effect. Moreover, the combination of DNMTis with histone deacetylase inhibitors (HDACis) may further enhance the reactivation of tumor-suppressive miRNAs and improve overall therapeutic efficacy [74, 75]. Vorinostat and romidepsin are known

HDACis, shown to promote histone acetylation and open chromatin structure, facilitating the transcription of silenced miRNAs [76–78]. Reversing abnormal DNA methylation patterns of miRNA genes through a combination of DNMT inhibitors offers an emerging, indirect, and cost-effective approach to noncoding RNA-based cancer therapy [79–81]. These FDA-approved drugs have been shown to suppress cellular oncogenic signals by tightly regulating the protein-coding genes. Clinical trials of miRNA in cancer therapeutics are an upfront initiative in commercializing the treatment to the public. It can be classified into two broad domains based on its applications: miRNA mimics are considered as miRNA drugs that restore the tumor-suppressive effects using miRNAs, and **miRNA inhibitors** silence the overexpressed oncogenes. Liposome-formulated MRX34 targeting miR-34a was the first mimic tested against several cancers, including lymphoma, melanoma, NSCLC, renal cell carcinoma, liver cancer, and advanced solid tumors. Although the trial has been terminated due to its immune-mediated adverse effects at an early stage with loss of life posing a major challenge in the delivery system and mimics [82, 83], it paved the way for several other miRNA-based RNA therapies in cancer. In 2014, by modifying the delivery system, bacterial minicell extracellular vesicle carriers, TARGOmiRs, were used for miR-16 targeting NSCLC and malignant pleural mesothelioma. The trial resulted in acceptable safety profiles and showed reliable tumor-suppressive effects, but it required additional support and validation with combinatorial therapies in larger cohorts [84]. Santolla et al., 2018 used an antisense oligonucleotide (ASO)-based inhibitor LNA-i-miR-221 for miR-221 that selectively intervenes A20/c-Rel/CTGF signaling. This trial provides notable insights into the formation of cancer-associated fibroblasts and the stimulatory effects of miR-221 in breast cancer patients [85]. Similarly, another ASO-based inhibitor of miR-155 named Cobomarsen hampered cellular proliferation and development in T cell lymphoma [86, 87]. Challenges that need to be addressed are optimizing a suitable delivery system and controlling the adverse immune responses would provide a significant mimic or inhibitor and reform the cancer treatment. Inculcating the combinatorial trials with epigenetic modifiers such as HDAC inhibitors, HAT inhibitors, TET modifiers, along with miRNA regulators, would enhance the potency of the results and the patient's overall survival benefit.

Conclusion and Future Perspectives

MiRNAs are intricately associated with epigenetic modifications that influence posttranscriptional gene regulation. Epigenetic control of gene expression through DNA methylation and acetylation directly affects all the transcripts. The miRNA's selective deregulation through the same DNA modifications can indirectly control the transcript level at post-maturation. Several factors like DNMTs, HDACs, and HATs determine miRNA expression and therefore are interesting candidates for gene expression or signaling pathways. Epigenetic drugs have not been used extensively in cancer therapy due to their ubiquitous activity. A target-based DNA modification using DNMTs or HDACs and selective editing will enhance the scope of personalized medicine. Controlling the expression levels of tumor suppressor miRNA through epigenetic modifiers is another interesting approach for cancer gene therapy. RNA methylation and its impact on miRNA biogenesis or maturation need further studies, as RNA modifiers can deregulate miRNA availability and activity. RNA methylation inhibitors evaluated for effective RNA modification can control oncogenic pathways through miRNA expression. A comprehensive cancer treatment combining epigenetic components and miRNAs will enhance the state-of-the-art precision medicine.

Disclosure Statement on AI Use The AI tools used in this project included Grammarly AI for grammar, paraphrasing, and proofreading.

Disclosure of Potential Conflict of Interest The authors declare no conflict of interest to disclose.

Competing Interest The authors declare they have no financial or nonfinancial interests to disclose.

References

1. Bartel DP (2004) MicroRNAs: genomics, biogenesis, mechanism, and function. Cell 116(2):281
2. Calin GA, Croce CM (2006) MicroRNA signatures in human cancers. Nat Rev Cancer 6(11):857
3. Zhang T, Hu Y, Yang N, Yu S, Pu X (2024) The microRNA-34 family and its functional role in lung cancer. Am J Clin Oncol 47(9):448
4. Lujambio A, Ropero S, Ballestar E, Fraga MF, Cerrato C, Setién F et al (2007) Genetic unmasking of an epigenetically silenced microRNA in human cancer cells. Cancer Res 67(4):1424
5. Hassell KN (2019) Histone deacetylases and their inhibitors in cancer epigenetics. Diseases 7(4):57
6. Wang F, Ma Y, Wang H, Qin H (2017) Reciprocal regulation between microRNAs and epigenetic machinery in colorectal cancer. Oncol Lett 13(3):1048
7. Saviana M, Le P, Micalo L, Del Valle-Morales D, Romano G, Acunzo M et al (2023) Crosstalk between miRNAs and DNA methylation in cancer. Genes 14(5):1075
8. Liu X, Chen X, Yu X, Tao Y, Bode AM, Dong Z et al (2013) Regulation of microRNAs by epigenetics and their interplay involved in cancer. J Exp Clin Cancer Res 32(1):96
9. Mathur R, CWM R (2018) SWI/SNF (BAF) complexes: guardians of the epigenome. Ann Rev Cancer Biol 2
10. Centore RC, Sandoval GJ, Soares LMM, Kadoch C, Chan HM (2020) Mammalian SWI/SNF chromatin remodeling complexes: emerging mechanisms and therapeutic strategies. Trends Genet 36(12):936
11. Wu F, Yang Q, Mi Y, Wang F, Cai K, Zhang Y et al (2022) miR-29b-3p inhibitor alleviates Hypomethylation-related aberrations through a feedback loop between miR-29b-3p and DNA methylation in Cardiomyocytes. Front Cell Dev Biol 10:788799
12. Yan B, Guo Q, Fu FJ, Wang Z, Yin Z, Wei YB et al (2015) The role of miR-29b in cancer: regulation, function, and signaling. Onco Targets Ther 8:539–548

13. Romano G, Veneziano D, Acunzo M, Croce CM (2017) Small non-coding RNA and cancer. Carcinogenesis 38:5
14. Calin GA, Sevignani C, Dumitru CD, Hyslop T, Noch E, Yendamuri S et al (2004) Human microRNA genes are frequently located at fragile sites and genomic regions involved in cancers. Proc Natl Acad Sci USA 101(9):2999
15. Liang B, Wang Y, Xu J, Shao Y, Xing D (2023) Unlocking the potential of targeting histone-modifying enzymes for treating IBD and CRC. Clin Epigenetics 15(1):146
16. Mandumpala JJ, Baby S, Tom AA, Godugu C, Shankaraiah N (2022) Role of histone demethylases and histone methyltransferases in triple-negative breast cancer: epigenetic mnemonics. Life Sci 292:120321
17. Lane AA, Chabner BA (2009) Histone deacetylase inhibitors in cancer therapy. J Clin Oncol 27(32):5459–5468
18. Di Martino MT, Leone E, Amodio N, Foresta U, Lionetti M, Pitari MR et al (2012) Synthetic miR-34a mimics as a novel therapeutic agent for multiple myeloma: in vitro and in vivo evidence. Clin Cancer Res 18(22):6260
19. Fu J, Imani S, Wu MY, Wu RC (2023) MicroRNA-34 family in cancers: role, mechanism, and therapeutic potential. Cancer 15(19):4723
20. He X, Yang A, McDonald DG, Riemer EC, Vanek KN, Schulte BA et al (2017) MiR-34a modulates ionizing radiation-induced senescence in lung cancer cells. Oncotarget 8(41)
21. Fang H, Xie J, Zhang M, Zhao Z, Wan Y, Yao Y (2017) MiRNA-21 promotes proliferation and invasion of triple-negative breast cancer cells through targeting PTEN. Am J Transl Res 9(3):953–961
22. Robertson J, Robertson AB, Klungland A (2011) The presence of 5-hydroxymethylcytosine at the gene promoter and not in the gene body negatively regulates gene expression. Biochem Biophys Res Commun 411(1):40–43
23. Jones PA, Takai D (2001) The role of DNA methylation in mammalian epigenetics. Science 293(5532):1068–1070
24. Malone CS, Miner MD, Doerr JR, Jackson JP, Jacobsen SE, Wall R et al (2001) CmC(A/T) GG DNA methylation in mature B cell lymphoma gene silencing. Proc Natl Acad Sci USA 98(18):10404–10409
25. Gowher H, Stockdale CJ, Goyal R, Ferreira H, Owen-Hughes T, Jeltsch A (2005) De novo methylation of nucleosomal DNA by the mammalian Dnmt1 and Dnmt3A DNA methyltransferases. Biochemistry 44(29):9899–9904
26. Tong L, Liang H, Zhuang H, Liu C, Zhang Z (2020) The relationship between hdac3 and malignant tumors: A mini review. Crit Rev Eukaryot Gene Expr 30(3):279–284
27. Sahu V, Lu C (2022) Oncohistones: hijacking the histone code. Ann Rev Cancer Biol 6:293–312
28. Ramazi S, Allahverdi A, Zahiri J (2020) Evaluation of post-translational modifications in histone proteins: A review on histone modification defects in developmental and neurological disorders. J Biosci 45(1):135
29. Wang F, Higgins JMG (2013) Histone modifications and mitosis: countermarks, landmarks, and bookmarks. Trends Cell Biol 23(4):175–184
30. Wang GG, Allis CD, Chi P (2007) Chromatin remodeling and cancer, part I: covalent histone modifications. Trends Mol Med 13(9):363
31. Sha QQ, Zhang J, Fan HY (2020) Function and regulation of histone H3 Lysine-4 methylation during oocyte meiosis and maternal-to-zygotic transition. Front Cell Develop Biol 8:597498
32. Mortazavi D, Sohrabi B, Mosallaei M, Nariman-Saleh-Fam Z, Bastami M, Mansoori Y et al (2022) Epi-miRNAs: regulators of the histone modification machinery in human cancer. J Oncol 2022:4889807
33. Toden S, Zumwalt TJ, Goel A (2021) Non-coding RNAs and potential therapeutic targeting in cancer. Biochim Biophys Acta Rev Cancer 1875(1):188491

34. Szczepanek J, Tretyn A (2023) MicroRNA-mediated regulation of histone-modifying enzymes in cancer: mechanisms and therapeutic implications. Biomolecules 13(11):1590
35. Fu J, Imani S, Wu MY, Wu RC (2023) MicroRNA-34 family in cancers: role, mechanism, and therapeutic potential. Cancers (Basel) 15(19):4723
36. Taylor-Papadimitriou J, Burchell JM (2022) Histone Methylases and demethylases regulating antagonistic methyl Marks: changes occurring in cancer. Cells 11(7):1113
37. Sykes SM, Mellert HS, Holbert MA, Li K, Marmorstein R, Lane WS et al (2006) Acetylation of the p53 DNA-binding domain regulates apoptosis induction. Mol Cell 24(6):841–851
38. Chang J, Davis-Dusenbery BN, Kashima R, Jiang X, Marathe N, Sessa R et al (2013) Acetylation of p53 stimulates miRNA processing and determines cell survival following genotoxic stress. EMBO J 32(24):3192–3205
39. Capaccia C, Diverio S, Zampini D, Guelfi G (2022) The complex interaction between P53 and miRNAs joins new awareness in physiological stress responses. Cells 11(10):1631
40. Wang R, Zhang H, Ding W, Fan Z, Ji B, Ding C et al (2020) miR-143 promotes angiogenesis and osteoblast differentiation by targeting HDAC7. Cell Death & Disease 11(3):1–15
41. Wang L, Mei SZ, Fei JC, Liu X, Dan CQ, Qian X et al (2014) MiR-143 acts as a tumor suppressor by targeting N-RAS and enhances temozolomide-induced apoptosis in glioma. Oncotarget 5(14):5416
42. Chen W, Xu XK, Li JL, Kong KK, Li H, Chen C et al (2017) MALAT1 is a prognostic factor in glioblastoma multiforme and induces chemoresistance to temozolomide through suppressing miR-203 and promoting thymidylate synthase expression. Oncotarget 8(14):22783–22799
43. Batista L, Bourachot B, Mateescu B, Reyal F, Mechta-Grigoriou F (2016) Regulation of miR-200c/141 expression by intergenic DNA-looping and transcriptional read-through. Nat Commun 7:8959
44. Neves R, Scheel C, Weinhold S, Honisch E, Iwaniuk KM, Trompeter HI et al (2010) Role of DNA methylation in miR-200c/141 cluster silencing in invasive breast cancer cells. BMC Res Notes 3:219
45. Uhlmann S, Zhang JD, Schwäger A, Mannsperger H, Riazalhosseini Y, Burmester S et al (2010) MiR-200bc/429 cluster targets PLCγ1 and differentially regulates proliferation and EGF-driven invasion than miR-200a/141 in breast cancer. Oncogene 29(30):4297–4306
46. Saito Y, Liang G, Egger G, Friedman JM, Chuang JC, Coetzee GA et al (2006) Specific activation of microRNA-127 with downregulation of the proto-oncogene BCL6 by chromatin-modifying drugs in human cancer cells. Cancer Cell 9(6):435–443
47. Hansen TB, Wiklund ED, Bramsen JB, Villadsen SB, Statham AL, Clark SJ et al (2011) miRNA-dependent gene silencing involving Ago2-mediated cleavage of a circular antisense RNA. EMBO J 30(21):4414–4422
48. Iwakawa H, oki, Tomari Y (2022) Life of RISC: formation, action, and degradation of RNA-induced silencing complex. Mol Cell 82(1):30–43
49. Salmena L, Poliseno L, Tay Y, Kats L, Pandolfi PP (2011) A ceRNA hypothesis: the rosetta stone of a hidden RNA language? Cell 146(3):353
50. Bandres E, Agirre X, Bitarte N, Ramirez N, Zarate R, Roman-Gomez J et al (2009) Epigenetic regulation of microRNA expression in colorectal cancer. Int J Cancer 125(11):2737–2743
51. Gambari R, Brognara E, Spandidos DA, Fabbri E (2016) Targeting oncomiRNAs and mimicking tumor suppressor miRNAs: new trends in the development of miRNA therapeutic strategies in oncology (review). Int J Oncol 49(1):5
52. Shah MY, Ferrajoli A, Sood AK, Lopez-Berestein G, Calin GA (2016) microRNA therapeutics in cancer — An emerging concept. EBioMedicine 12:34–42
53. Fu Z, Wang L, Li S, Chen F, Au-Yeung KKW, Shi C (2021) MicroRNA as an important target for anticancer drug development. Front Pharmacol 12:736323
54. Sun T, Wu R, Ming L (2019) The role of m6A RNA methylation in cancer. Biomed Pharmacother 112:108613
55. Boulias K, Greer EL (2023) Biological roles of adenine methylation in RNA. Nat Rev Genet 24(3):143–160

56. Chen XY, Zhang J, Zhu JS (2019) The role of m6A RNA methylation in human cancer. Mol Cancer 18(1):1–9
57. An Y, Duan H (2022) The role of m6A RNA methylation in cancer metabolism. Mol Cancer 21(1):14
58. Liu J, Yue Y, Han D, Wang X, Fu Y, Zhang L et al (2013) A METTL3–METTL14 complex mediates mammalian nuclear RNA N6-adenosine methylation. Nat Chem Biol 10(2):93–95
59. Landfors M, Nakken S, Fusser M, Dahl JA, Klungland A, Fedorcsak P (2016) Sequencing of FTO and ALKBH5 in men undergoing infertility work-up identifies an infertility-associated variant and two missense mutations. Fertil Steril 105(5):1170–1179.e5
60. Tan Z, Shi S, Xu J, Liu X, Lei Y, Zhang B et al (2022) RNA N6-methyladenosine demethylase FTO promotes pancreatic cancer progression by inducing the autocrine activity of PDGFC in an m6A-YTHDF2-dependent manner. Oncogene 41(20):2860
61. Wang H, Song X, Song C, Wang X, Cao H (2021) m6A-seq analysis of microRNAs reveals that the N6-methyladenosine modification of miR-21–5p affects its target expression. Arch Biochem Biophys 711:109023
62. Akhtar J, Lugoboni M, Junion G (2022) m6A RNA modification in transcription regulation. 12(5):266–276.
63. Zhang B, Chen Z, Tao B, Yi C, Lin Z, Li Y et al (2021) m6A target microRNAs in serum for cancer detection. Mol Cancer 20(1):1–8
64. Wang Y, Li Y, Toth JI, Petroski MD, Zhang Z, Zhao JC (2014) N6-methyladenosine modification destabilizes developmental regulators in embryonic stem cells. Nat Cell Biol 16(2):191–198
65. Huang J, Chen Z, Chen X, Chen J, Cheng Z, Wang Z (2021) The role of RNA N6-methyladenosine methyltransferase in cancers. Mol Ther Nucleic Acids 23:887–896
66. Zhao Z, Ju Q, Ji J, Li Y, Zhao Y (2022) N6-Methyladenosine methylation regulator RBM15 is a potential prognostic biomarker and promotes cell proliferation in pancreatic adenocarcinoma. Front Mol Biosci 9:109
67. Zepecki JP, Karambizi D, Eduardo Fajardo J, Snyder KM, Guetta-Terrier C, Tang OY et al (2021) miRNA-mediated loss of m6A increases nascent translation in glioblastoma. PLoS Genet 17(3):e1009086
68. Bi X, Lv X, Liu D, Guo H, Yao G, Wang L et al (2021) METTL3-mediated maturation of miR-126-5p promotes ovarian cancer progression via PTEN-mediated PI3K/Akt/mTOR pathway. Cancer Gene Ther 28(3–4):335–349
69. Chawra HS, Agarwal M, Mishra A, Chandel SS, Singh RP, Dubey G et al (2024) MicroRNA-21's role in PTEN suppression and PI3K/AKT activation: implications for cancer biology. Pathol Res Pract 254:155091
70. Bala S, Marcos M, Kodys K, Csak T, Catalano D, Mandrekar P et al (2011) Up-regulation of microRNA-155 in macrophages contributes to increased tumor necrosis factor α (TNFα) production via increased mRNA half-life in alcoholic liver disease. J Biol Chem 286(2):1436
71. Luan T, Zhang X, Wang S, Song Y, Zhou S, Lin J et al (2017) Long non-coding RNA MIAT promotes breast cancer progression and functions as ceRNA to regulate DUSP7 expression by sponging miR-155-5p. Oncotarget 8(44):76153
72. Nguyen MN, Than VT (2024) RNA therapeutics in cancer treatment. Prog Mol Biol Transl Sci 203:197–223
73. Blandino G, Dinami R, Marcia M, Anastasiadou E, Ryan BM, Palcau AC et al (2023) The new world of RNA diagnostics and therapeutics. J Exp Clin Cancer Res 42(1):189
74. Xiao W, Zhou Q, Wen X, Wang R, Liu R, Wang T et al (2021) Small-molecule inhibitors overcome epigenetic reprogramming for cancer therapy. Front Pharmacol:12
75. Nyce J, Liu L, Jones PA (1986) Variable effects of DNA-synthesis inhibitors upon DNA methylation in mammalian cells. Nucleic Acids Res 14(10):4353
76. Cheishvili D, Boureau L, Szyf M (2015) DNA demethylation and invasive cancer: implications for therapeutics. Br J Pharmacol 172(11):2705–2715

77. Yang Y, Yan Y, Chen Z, Hu J, Wang K, Tang N et al (2021) Histone deacetylase inhibitors Romidepsin and Vorinostat promote hepatitis B virus replication by inducing cell cycle arrest. J Clin Transl Hepatol 9(2):160
78. Tiffon CE, Adams JE, Van Der Fits L, Wen S, Townsend PA, Ganesan A et al (2011) The histone deacetylase inhibitors vorinostat and romidepsin downmodulate IL-10 expression in cutaneous T-cell lymphoma cells. Br J Pharmacol 162(7):1590
79. Farooqi AA, Rehman ZU, Muntane J (2014) Antisense therapeutics in oncology: current status. Onco Targets Ther 7:2035–2042
80. Cheng Y, He C, Wang M, Ma X, Mo F, Yang S et al (2019) Targeting epigenetic regulators for cancer therapy: mechanisms and advances in clinical trials. Signal Transduct Target Ther 4(1):62
81. Di Martino MT, Tagliaferri P, Tassone P (2025) MicroRNA in cancer therapy: breakthroughs and challenges in early clinical applications. J Exp Clin Cancer Res 44(1):126
82. Beg MS, Brenner AJ, Sachdev J, Borad M, Kang YK, Stoudemire J et al (2017) Phase I study of MRX34, a liposomal miR-34a mimic, administered twice weekly in patients with advanced solid tumors. Investig New Drugs 35(2):180–188
83. Hong DS, Kang YK, Borad M, Sachdev J, Ejadi S, Lim HY et al (2020) Phase 1 study of MRX34, a liposomal miR-34a mimic, in patients with advanced solid tumours. Br J Cancer 122(11):1630–1637
84. van Zandwijk N, Pavlakis N, Kao SC, Linton A, Boyer MJ, Clarke S et al (2017) Safety and activity of microRNA-loaded minicells in patients with recurrent malignant pleural mesothelioma: a first-in-man, phase 1, open-label, dose-escalation study. Lancet Oncol 18(10):1386–1396
85. Santolla MF, Lappano R, Cirillo F, Rigiracciolo DC, Sebastiani A, Abonante S et al (2018) miR-221 stimulates breast cancer cells and cancer-associated fibroblasts (CAFs) through selective interference with the A20/c-Rel/CTGF signaling. J Exp Clin Cancer Res 37(1):94
86. Seto AG, Beatty X, Lynch JM, Hermreck M, Tetzlaff M, Duvic M et al (2018) Cobomarsen, an oligonucleotide inhibitor of miR-155, co-ordinately regulates multiple survival pathways to reduce cellular proliferation and survival in cutaneous T-cell lymphoma. Br J Haematol 183(3):428–444
87. Anastasiadou E, Seto AG, Beatty X, Hermreck M, Gilles ME, Stroopinsky D et al (2021) Cobomarsen, an oligonucleotide inhibitor of miR-155, slows DLBCL tumor cell growth in vitro and in vivo. Clin Cancer Res 27(4):1139–1149
88. Agirre X, Vilas-Zornoza A, Jiménez-Velasco A, Martin-Subero JI, Cordeu L, Gárate L et al (2009) Epigenetic silencing of the tumor suppressor microRNA Hsa-miR-124a regulates CDK6 expression and confers a poor prognosis in acute lymphoblastic leukemia. Cancer Res 69(10):4443–4453
89. Fabbri M, Garzon R, Cimmino A, Liu Z, Zanesi N, Callegari E et al (2007) MicroRNA-29 family reverts aberrant methylation in lung cancer by targeting DNA methyltransferases 3A and 3B. Proc Natl Acad Sci USA 104(40):15805–15810
90. Yamagishi M, Nakano K, Miyake A, Yamochi T, Kagami Y, Tsutsumi A et al (2012) Polycomb-mediated loss of miR-31 activates nik-dependent NF-κB pathway in adult T cell leukemia and other cancers. Cancer Cell 21(1):121–135
91. Friedman JM, Liang G, Liu CC, Wolff EM, Tsai YC, Ye W et al (2009) The putative tumor suppressor microRNA-101 modulates the cancer epigenome by repressing the polycomb group protein EZH2. Cancer Res 69(6):2623–2629
92. Pogribny IP, Filkowski JN, Tryndyak VP, Golubov A, Shpyleva SI, Kovalchuk O (2010) Alterations of microRNAs and their targets are associated with acquired resistance of MCF-7 breast cancer cells to cisplatin. Int J Cancer 127(8):1785–1794
93. Tan Y, Zhang B, Wu T, Skogerbø G, Zhu X, Guo X et al (2009) Transcriptional inhibiton of Hoxd4 expression by miRNA-10a in human breast cancer cells. BMC Mol Biol 10(1):1–9
94. Sandhu R, Rivenbark AG, Coleman WB (2012) Loss of post-transcriptional regulation of DNMT3b by microRNAs: a possible molecular mechanism for the hypermethylation defect observed in a subset of breast cancer cell lines. Int J Oncol 41(2):721–732

95. Xu Q, Jiang Y, Yin Y, Li Q, He J, Jing Y et al (2013) A regulatory circuit of miR-148a/152 and DNMT1 in modulating cell transformation and tumor angiogenesis through IGF-IR and IRS1. J Mol Cell Biol 5(1):3–13
96. Mitra D, Das PM, Huynh FC, Jones FE (2011) Jumonji/ARID1 B (JARID1B) protein promotes breast tumor cell cycle progression through epigenetic repression of MicroRNA let-7e. J Biol Chem 286(47):40531–40535
97. Augoff K, McCue B, Plow EF, Sossey-Alaoui K (2012) MiR-31 and its host gene lncRNA LOC554202 are regulated by promoter hypermethylation in triple-negative breast cancer. Mol Cancer 11:5
98. Chang S, Wang RH, Akagi K, Kim KA, Martin BK, Cavallone L, et al (2011) Tumor suppressor BRCA1 epigenetically controls oncogenic microRNA-155. Nat Med 17(10):1275–1282
99. Zhou Y, Hu Y, Yang M, Jat P, Li K, Lombardo Y et al (2014) The miR-106b∼25 cluster promotes bypass of doxorubicin-induced senescence and increase in motility and invasion by targeting the E-cadherin transcriptional activator EP300. Cell Death Differ 21(3):462–474
100. Pandey AK, Zhang Y, Zhang S, Li Y, Tucker-Kellogg G, Yang H et al (2015) TIP60-miR-22 axis as a prognostic marker of breast cancer progression. Oncotarget 6(38):41290–41306
101. Ren G, Baritaki S, Marathe H, Feng J, Park S, Beach S et al (2012) Polycomb protein EZH2 regulates tumor invasion via the transcriptional repression of the metastasis suppressor RKIP in breast and prostate cancer. Cancer Res 72(12):3091–3104
102. Sampath D, Liu C, Vasan K, Sulda M, Puduvalli VK, Wierda WG et al (2012) Histone deacetylases mediate the silencing of miR-15a, miR-16, and miR-29b in chronic lymphocytic leukemia. Blood 119(5):1162–1172
103. Ng EKO, Tsang WP, Ng SSM, Jin HC, Yu J, Li JJ et al (2009) MicroRNA-143 targets DNA methyltransferases 3A in colorectal cancer. Br J Cancer 101(4):699–706
104. Yu JZ, Sun N, Bei YB, Li XB, Lu C, Hua LC (2017) Circadian gene hCLOCK contributes to progression of colorectal carcinoma and is directly regulated by tumor-suppressive microRNA-124. Mol Med Rep 16(6):7923–7930
105. Nguyen T, Kuo C, Nicholl MB, Sim MS, Turner RR, Morton DL et al (2011) Downregulation of microRNA-29c is associated with hypermethylation of tumor-related genes and disease outcome in cutaneous melanoma. Epigenetics 6(3):388–394
106. Tsuruta T, Kozaki KI, Uesugi A, Furuta M, Hirasawa A, Imoto I et al (2011) miR-152 is a tumor suppressor microRNA that is silenced by DNA hypermethylation in endometrial cancer. Cancer Res 71(20):6450–6462
107. Zhu A, Xia J, Zuo J, Jin S, Zhou H, Yao L et al (2012) MicroRNA-148a is silenced by hypermethylation and interacts with DNA methyltransferase 1 in gastric cancer. Med Oncol 29(4):2701–2709
108. Smits M, Nilsson J, Mir SE, van der Stoop PM, Hulleman E, Niers JM et al (2010) miR-101 is down-regulated in glioblastoma resulting in EZH2-induced proliferation, migration, and angiogenesis. Oncotarget 1(8):710–720
109. Smits M, Mir SE, Nilsson JRA, van der Stoop PM, Niers JM, Marquez VE et al (2011) Down-regulation of miR-101 in endothelial cells promotes blood vessel formation through reduced repression of EZH2. PLoS One 6(1):e16282
110. Zhang Z, Tang H, Wang Z, Zhang B, Liu W, Lu H et al (2011) MiR-185 targets the DNA Methyltransferases 1 and regulates global DNA methylation in human glioma. Mol Cancer 10(1):1–16
111. Cho HM, Jeon HS, Lee SY, Jeong KJ, Park SY, Lee HY et al (2011) MicroRNA-101 inhibits lung cancer invasion through the regulation of enhancer of zeste homolog 2. Exp Ther Med 2(5):963–967
112. Hang YJ, Yang F, Feng CB, Lu Z, Song HX, Ping ZW et al (2011) The histone deacetylase 4/SP1/microrna-200a regulatory network contributes to aberrant histone acetylation in hepatocellular carcinoma. Hepatology 54(6):2025–2035
113. Ngo-Yin Fan D, Ho-Ching Tsang F, Hoi-Kam Tam A, Leung-Kuen Au S, Chak-Lui Wong C, Wei L et al (2013) Histone lysine methyltransferase, suppressor of variegation 3-9

homolog 1, promotes hepatocellular carcinoma progression and is negatively regulated by microRNA-125b. Hepatology 57(2):637–647
114. Huang J, Yin P (2018) Structural insights into N6-methyladenosine (m6A) modification in the transcriptome. Genomics Proteomics Bioinformatics 16(2):85–98
115. Au SLK, Wong CCL, Lee JMF, Fan DNY, Tsang FH, Ng IOL et al (2012) Enhancer of zeste homolog 2 epigenetically silences multiple tumor suppressor microRNAs to promote liver cancer metastasis. Hepatology 56(2):622–631
116. Buurman R, Grlevik E, Schffer V, Eilers M, Sandbothe M, Kreipe H et al (2012) Histone deacetylases activate hepatocyte growth factor signaling by repressing microRNA-449 in hepatocellular carcinoma cells. Gastroenterology 143(3):811–820
117. Wang Y, Toh HC, Chow P, Chung AYF, Meyers DJ, Cole PA et al (2012) MicroRNA-224 is up-regulated in hepatocellular carcinoma through epigenetic mechanisms. FASEB J 26(7):3032–3041
118. Chen Y, Luo J, Tian R, Sun H, Zou S (2011) miR-373 negatively regulates methyl-CpG-binding domain protein 2 (MBD2) in hilar cholangiocarcinoma. Dig Dis Sci 56(6):1693–1701
119. Rathore MG, Saumet A, Rossi JF, De Bettignies C, Tempé D, Lecellier CH et al (2012) The NF-κB member p65 controls glutamine metabolism through miR-23a. Int J Biochem Cell Biol 44(9):1448–1456
120. Braconi C, Huang N, Patel T (2010) Microrna-dependent regulation of DNA methyltransferase-1 and tumor suppressor gene expression by interleukin-6 in human malignant cholangiocytes. Hepatology 51(3):881–890
121. Alajez NM, Shi W, Hui ABY, Bruce J, Lenarduzzi M, Ito E et al (2010) Enhancer of Zeste homolog 2 (EZH2) is overexpressed in recurrent nasopharyngeal carcinoma and is regulated by miR-26a, miR-101, and miR-98. Cell Death Dis 1(10):e85
122. Choe J, Lin S, Zhang W, Liu Q, Wang L, Ramirez-Moya J et al (2018) mRNA circularization by METTL3–eIF3h enhances translation and promotes oncogenesis. Nature 561(7724):556–560
123. Asangani IA, Harms PW, Dodson L, Pandhi M, Kunju LP, Maher CA et al (2012) Genetic and epigenetic loss of microRNA-31 leads to feed-forward expression of EZH2 in melanoma. Oncotarget 3(9):1011–1025
124. Zhang X, Chen X, Lin J, Lwin T, Wright G, Moscinski LC et al (2012) Myc represses miR-15a/miR-16-1 expression through recruitment of HDAC3 in mantle cell and other non-Hodgkin B-cell lymphomas. Oncogene 31(24):3002–3008
125. Amodio N, Leotta M, Bellizzi D, Di Martino MT, D'Aquila P, Lionetti M et al (2012) DNA-demethylating and anti-tumor activity of synthetic miR-29b mimics in multiple myeloma. Oncotarget 3(10):1246–1258
126. Li G, Zhao J, Peng X, Liang J, Deng X, Chen Y (2012) The mechanism involved in the loss of PTEN expression in NSCLC tumor cells. Biochem Biophys Res Commun 418(3):547–552
127. Xiang Y, Ma N, Wang D, Zhang Y, Zhou J, Wu G et al (2013) MiR-152 and miR-185 co-contribute to ovarian cancer cells cisplatin sensitivity by targeting DNMT1 directly: a novel epigenetic therapy independent of decitabine. Oncogene 33(3):378–386
128. Majid S, Dar AA, Saini S, Shahryari V, Arora S, Zaman MS et al (2013) miRNA-34b inhibits prostate cancer through demethylation, active chromatin modifications, and AKT pathways. Clin Cancer Res 19(1):73–84
129. Noonan EJ, Place RF, Pookot D, Basak S, Whitson JM, Hirata H et al (2009) MiR-449a targets HDAC-1 and induces growth arrest in prostate cancer. Oncogene 28(14):1714–1724
130. Hulf T, Sibbritt T, Wiklund ED, Bert S, Strbenac D, Statham AL et al (2011) Discovery pipeline for epigenetically deregulated miRNAs in cancer: integration of primary miRNA transcription. BMC Genomics 12:54
131. Gong AY, Eischeid AN, Xiao J, Zhao J, Chen D, Wang ZY et al (2012) MiR-17-5p targets the p300/CBP-associated factor and modulates androgen receptor transcriptional activity in cultured prostate cancer cells. BMC Cancer 12:492

132. Hirata H, Hinoda Y, Shahryari V, Deng G, Tanaka Y, Tabatabai ZL et al (2014) Genistein downregulates onco-miR-1260b and upregulates sFRP1 and Smad4 via demethylation and histone modification in prostate cancer cells. Br J Cancer 110(6):1645–1654
133. Cao P, Deng Z, Wan M, Huang W, Cramer SD, Xu J et al (2010) MicroRNA-101 negatively regulates Ezh2 and its expression is modulated by androgen receptor and HIF-1α/HIF-1β. Mol Cancer 9(1):1–12
134. Liu X, Wang C, Chen Z, Jin Y, Wang Y, Kolokythas A et al (2011) MicroRNA-138 suppresses epithelial–mesenchymal transition in squamous cell carcinoma cell lines. Biochem J 440(1):23–31
135. Sasaki D, Imaizumi Y, Hasegawa H, Osaka A, Tsukasaki K, Lim Choi Y et al (2011) Overexpression of enhancer of zeste homolog 2 with trimethylation of lysine 27 on histone H3 in adult T-cell leukemia/lymphoma as a target for epigenetic therapy. Haematologica 96(5):712–719

Chapter 5
MicroRNAs and Cancer Stem Cells: Mechanisms and Their Role in Cancer

Neeharika Sara Sennie, Sneha Pallatt, P. Ashiq Shibili, Amit Dey, Surajit Pathak, and Antara Banerjee

Abstract Cancer stem cells (CSCs) are a subset of tumour cells that possess self-renewal, differentiation, and tumour-initiating capabilities. They are responsible for tumour progression, recurrence, metastasis, and resistance to therapy. MicroRNAs (miRNAs) are short non-coding RNAs that regulate gene expression at the post-transcriptional level. They are critical modulators of CSC functions. miRNAs influence CSC behaviour and modulate key signalling pathways such as Wnt/β-catenin, Notch, Hedgehog, and PI3K/AKT, thereby affecting tumour initiation, progression, metastasis, drug resistance, and recurrence. miRNAs act as either oncomiRs or tumour suppressors, shaping the tumour microenvironment, altering responses to chemotherapy, and regulating key characteristics of cancer. They also alter the tumour microenvironment, regulate stemness, and determine therapy response. This chapter provides a comprehensive overview of miRNA-mediated regulation of CSCs across multiple cancer types, elucidates the molecular mechanisms of miRNA-CSC interactions, and provides an overview of the diagnostic and therapeutic implications.

Keywords MicroRNAs · Cancer stem cells · Tumourigenesis · miRNA regulation · Drug resistance · miRNA-based therapy

Introduction

Cancer is a diverse and complex disease characterized by the abnormal and uncontrolled growth of cells, which not only evade regulatory mechanisms but also possess the ability to spread to distant tissues through metastasis [1, 2]. Globally, cancer

N. S. Sennie · S. Pallatt · P. Ashiq Shibili · A. Dey · S. Pathak · A. Banerjee (✉)
Faculty of Allied Health Sciences, Medical Biotechnology Lab, Chettinad Academy of Research and Education (CARE), Chettinad Hospital and Research Institute, Chennai, Tamil Nadu, India
e-mail: antarabanerjee@care.edu.in

S. Pathak, A. Banerjee (eds.), *MicroRNAs: Role in Cancer Progression, Diagnostics and Therapy*, Advances in Biochemistry in Health and Disease 34,
https://doi.org/10.1007/978-3-032-08824-6_5

is ranked among the leading causes of morbidity and mortality, with over 20 million recent cases and approximately 9.7 million cancer-related deaths [3, 4]. The most common cancers include breast, lung, liver, colon, and prostate [4]. It was found that lung cancer (12% of all cases) appeared as the commonly diagnosed cancer worldwide, followed by breast cancer (11.6%), colorectal cancer (9.6%), prostate cancer (7.3%), and stomach cancer (4.9%) [3]. Over the decades, significant advancements have been made in cancer treatment, and combined with improvements in early detection, the morbidity and mortality rates of several cancers have been reduced [5]. Major cancer treatments encompass surgery, hormone therapy, gene therapy, immunotherapy, radiation therapy, laser therapy, combination therapy, and targeted therapy, with chemotherapy remaining as the most prevalent, widely used, and effective approach for cancer management [6]. Despite progress in cancer treatments, these tumour cells often develop resistance to chemotherapeutic drugs, leading to relapses and poor survival outcomes in patients [4]. A significant factor contributing to this challenge is the presence of Cancer stem cells (CSCs), a small subset of tumour cells found within these cancerous populations [6]. These cells resemble normal stem cells and possess the ability to self-renew and differentiate in response to chemotherapeutic agents. CSCs play a central role in tumour initiation and metastasis due to their ability to enter a state of quiescence through cell cycle arrest, making them resistant to conventional therapies, thereby portraying themselves as a crucial target for cancer treatment [7]. In recent years, microRNAs (miRNAs) have acted as key post-transcriptional regulators of gene expression, playing a major role in cancer progression. miRNAs are short, single-stranded, non-coding RNAs and have a length of roughly 18–25 nucleotides. They are produced from a primary miRNA transcript (pri-miRNA), which is transcribed from either the introns of protein-coding genes in the nucleus or the non-protein-coding genes [8]. These miRNAs are involved in the regulation of CSC self-renewal and differentiation [9]. They function as either oncomiRs or tumour suppressors, thereby having a significant impact on the development of cancer by regulating mRNA [10]. Several malignancies are linked to their dysregulation, including ovarian, breast, lung, and colon cancers. This chapter explores the interplay between microRNAs and cancer stem cells, highlighting their role in tumour progression, treatment resistance, and the potential for miRNA-based therapeutic strategies.

Cancer Stem Cells and Their Characteristics

CSCs exhibit stem-like characteristics that contribute to disease progression, including enhanced self-renewal, clonogenicity, proliferation, metastatic potential, homing capacity, and the ability to regenerate tumour tissue, similar to normal stem cells [11]. They possess remarkable organizational capabilities, enabling them to communicate with surrounding cells to coordinate nutrient supply and collectively evade immune responses, thereby establishing a microenvironment favourable to tumour progression. Similarly, CSCs produce diverse cell populations that often exhibit high plasticity potential and resistance to adverse environmental conditions, such as low oxygen or nutrient levels, or chemotherapeutic agents, frequently adapting through mechanisms such as induced cell death and quiescence [12]. CSCs are found in various malignancies, including leukaemia, colorectal, breast, and lung cancer, highlighting their relevance in cancer research. A hallmark of CSCs is their ability to self-renew continuously, a process modulated by key signalling pathways such as Wnt/β-catenin, Notch, and Hedgehog, which regulate stemness and tumour initiation. CSCs also possess a high tumour-initiating capacity, with even a small number capable of generating new tumours, setting themselves apart from non-stem cancer cells. The Wnt/β-catenin pathway serves as a central regulator of CSC maintenance, enhancing proliferation while suppressing differentiation to support stem-like properties [13]. Dysregulation of the pathway is associated with tumourigenesis, metastasis, and chemoresistance. The Notch signalling pathway is also central to CSC fate determination, promoting self-renewal and therapy resistance; thus, it is yet another important target for cancer therapy [14]. Likewise, the Hedgehog signalling pathway enables CSC growth and survival by controlling stemness and genes responsible for cell cycle progression, especially in lethal cancers like pancreatic and glioblastoma [15]. The TGF-β/BMP pathway serves as a tumour suppressor during early stages of cancer, or supporting CSC plasticity and immune evasion in advanced cancer [16]. The JAK/STAT signalling pathway is also central to CSC-promoted tumour progression by regulating cell proliferation, inflammation, and immune evasion, leading to therapeutic resistance and relapse of tumours [17]. CSCs are well adapted to hypoxic tumour microenvironments, where hypoxia-inducible factors (HIF-1α, HIF-2α) activate genes involved in maintaining stemness, promoting angiogenesis, and conferring resistance to therapy. Additionally, CSCs evade conventional treatments through mechanisms such as enhanced DNA repair, cellular quiescence, drug efflux mediated by ATP-binding cassette (ABC) transporters like ABCG2 and MDR1, and activation of anti-apoptotic pathways including Bcl-2 and NF-κB, ultimately promoting tumour survival and recurrence [18]. CSCs were first recognized based on their functional properties; however, their detection across various cancer types is now by specific surface markers (such as CD44, CD133, and CD24) and enzymatic activity indicators like aldehyde dehydrogenase (ALDH). The cell surface markers used for the identification of CSCs in various cancers are summarized in Table 5.1. When compared to proliferating tumour cells, quiescent CSCs are resistant to chemotherapy and radiation because they have minimal metabolic activity and no active cell replication. However, CSCs can re-enter the cell cycle after the treatment has eliminated the proliferative cancer cells [28].

miRNA-Mediated Regulation in Cancer

miRNAs are short, non-coding RNA molecules (18–25 bp), involved in the post-transcriptional regulation of genes [8]. As of the most recent reports up to 2023, approximately 2654 mature miRNAs have been identified in humans [29]. They bind their target messenger RNAs (mRNAs) at the 3′ and 5′ untranslated regions

Table 5.1 CSC surface marker profiles and their functional roles across different cancer types

Cancer type	CSC markers	Major role of markers	References
Lung cancer	CD133⁺, CD44⁺, ABCG2, ALDH, CD87⁺, SP, CD90⁺	Associated with self-renewal, drug resistance, and tumour initiation	[19]
Colon cancer	CD133⁺, CD44⁺, CD24⁺, CD166⁺, EpCAM⁺, ALDH, ESA	Promote tumourigenesis, metastasis, and therapy resistance; help identify CSC populations	[20]
Liver cancer	CD133⁺, CD44⁺, CD49f⁺, CD90⁺, ALDH, ABCG2, CD24⁺, ESA	Regulate stemness, invasion, and chemoresistance; implicated in recurrence and metastasis	[21]
Breast cancer	CD133⁺, CD44⁺, CD24⁻, EpCAM⁺, ALDH-1	Related to tumour initiation, EMT, metastasis, and resistance to therapy	[22]
Gastric cancer	CD133⁺, CD44⁺, CD24⁺	Associated with self-renewal, it contributes to metastasis and drug resistance	[23]
Leukaemia	CD34⁺, CD38⁻, CD123⁺	Self-renewing capacity and resistance to chemotherapy	[24]
Prostate cancer	CD133⁺, CD44⁺, α2β1, ABCG2, ALDH	Self-renewal, tumour progression, and resistance to androgen deprivation therapies	[25]
Pancreatic cancer	CD133⁺, CD44⁺, CD24⁺, ABCG2, ALDH, EpCAM⁺, ESA	Involved in tumour initiation, chemoresistance, and aggressive tumour behaviour	[26]
Melanoma	ABCB5⁺, CD20⁺	Associated with enhanced tumorigenicity, therapeutic resistance, and metastatic progression	[27]

(UTRs) or coding regions, resulting in either degradation of mRNA or translational repression [30] (Fig. 5.1). This regulation is crucial for many biological processes, including cell differentiation, proliferation, and apoptosis. In cancer, miRNAs function either as tumour suppressors or oncomiRs, contributing to tumour development, metastasis, and resistance to therapy [29]. Aberrant miRNA expression in diverse cancers has driven research into their potential as diagnostic markers, prognostic indicators, and therapeutic targets. Cancer-specific miRNA may serve as non-invasive biomarkers for early-stage detection and precision medicine approaches [31]. Therapeutic application of miRNAs involves either restoring tumour-suppressing miRNAs or inhibiting oncomiRs, thereby representing a promising strategy for next-generation cancer therapy.

The association of miRNAs with cancer was first identified in 2002, highlighting the decreased levels of miR-15a and miR-16-1 in patients with Chronic Lymphocytic Leukaemia (CLL) [32]. These miRNAs map to chromosome 13q14, a region frequently deleted in CLL, indicating a tumour suppressor role. Studies indicate that miRNA deletion contributes to leukaemogenesis. This finding initiated a wide range of studies showing that miRNA dysregulation is a characteristic, prevalent across a

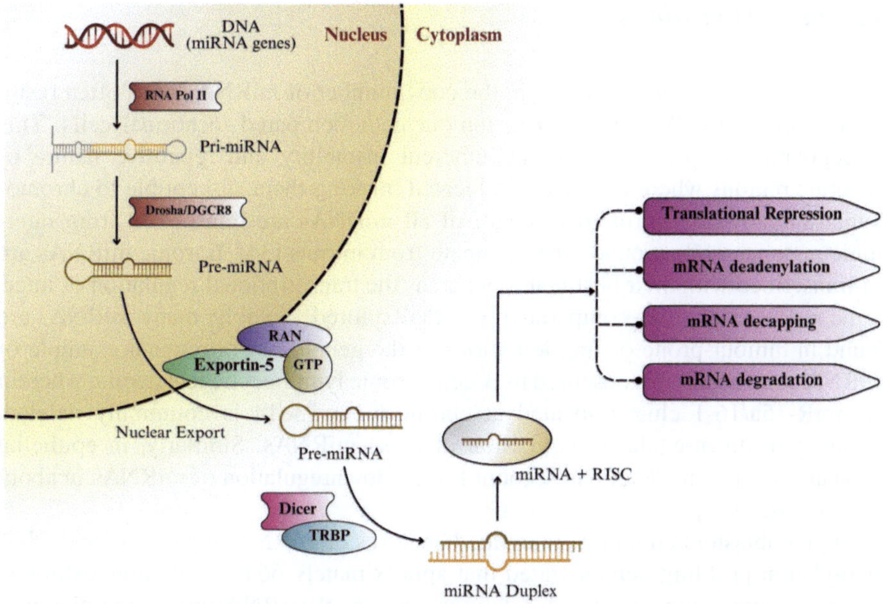

Fig. 5.1 Schematic representation of miRNA biogenesis and its role in post-transcriptional gene silencing via the RISC complex

broad range of cancers, both solid tumours and haematologic malignancies [32]. Regarding CSCs, miRNAs influence key cellular processes, including apoptosis, adhesion, migration, cell cycle regulation, and DNA repair, all of which contribute to and support CSC survival and resistance to treatment. The intricate nature of miRNA–mRNA interactions is primarily due to the ability of miRNAs to simultaneously regulate multiple genes and signalling cascades, thereby exerting broad control over various cellular processes, underscoring the necessity for precise target validation; however, only a limited number of these interactions have been experimentally confirmed [8]. miRNAs, often located within regions of genomic instability, are particularly susceptible to genetic alterations, which frequently lead to their dysregulation. Such genetic alterations may include translocations, point mutations, deletions, and gene amplifications. Additionally, epigenetic changes such as DNA hypomethylation and hypermethylation also contribute to miRNA dysregulation. Such abnormalities impact miRNA levels in both normal and tumour tissues, causing their expression to either increase or decrease in cancer [33]. miRNA dysregulation can result from multiple factors, such as genetic mutations, epigenetic changes, errors in transcriptional control, and disruptions in the miRNA biogenesis process.

Genomic Alterations

Genetic mutations and variations in the copy number of miRNA genes often result in abnormal miRNA expression in tumour cells compared to normal cells. This dysregulation is partly due to the inherent instability and repetitive nature of genomic regions where miRNAs are located, making them susceptible to chromosomal alterations. Approximately half of all miRNAs are transcribed from intergenic regions, while around 40% originate from introns [34]. Intronic miRNAs are co-transcribed with their host genes, whereas the transcriptional regulation of intergenic miRNAs remains comparatively underexplored. Notably, many miRNAs are found in tumour-prone or fragile regions of the genome. A prominent example of miRNA gene deletion is observed in B-cell chronic lymphocytic leukaemia, wherein the miR-15a/16-1 cluster located at chromosome 13q14 is commonly deleted, resulting in downregulated expression of these miRNAs. Similarly, in epithelial ovarian cancer, gene deletions account for the downregulation of miRNAs in about 15% of cases [35].

In pineoblastoma tumours, integrated analyses of miRNA sequencing and DNA methylation profiling demonstrated that approximately 66.67% of cases exhibited mutations, predominantly focal deletions, in critical miRNA-processing enzymes such as DROSHA, DGCR8, and DICER1 [36]. Moreover, conserved miRNAs like Let-7d and miR-205 are frequently mutated across various cancers. Beyond miRNAs themselves, mutations can also affect their targets and regulatory elements. For example, deletions or single-nucleotide polymorphisms (SNPs) within the 3′ untranslated region (3′UTR) of target genes may interfere with miRNA binding, thereby impairing post-transcriptional regulation. A specific example involves a mutation in the 3′UTR of the oncogene *E2F1*, which prevents miR-136-5p from binding, resulting in unchecked *E2F1* expression and the development of a malignant phenotype [37].

Epigenetic Alterations

Cancer is frequently associated with diverse epigenetic alterations, including dysregulated histone modifications, hypermethylation of tumour suppressor gene promoters, and global hypomethylation of the genomic DNA. These changes can also affect genomic regions encoding miRNAs, similar to those seen in protein-coding genes. For instance, in epithelial ovarian cancer, extensive transcriptional dysregulation correlates with reduced miRNA expression, primarily due to epigenetic silencing and genomic copy number loss. This reduction accounts for 15–36% of total miRNA downregulation [35]. A tumour-suppressive miRNA, miR-34b, regulated by p53, was found to be epigenetically silenced in ovarian cancer. Furthermore, a cluster of tumour-suppressor miRNAs located at the Dlk1-Gtl2 region on chromosome 14 was also silenced in late-stage ovarian cancer, resulting in elevated expression of their oncogenic mRNA targets [38].

Similar patterns were noted in breast cancer, where miR-193a-3p was downregulated due to promoter hypermethylation. The extent of DNA methylation correlated with the stage of cancer and grade [39]. Normally, this miRNA plays a critical role in inhibiting the migration, invasion, and proliferation of breast cancer cells. In lung cancer, downregulated expression of the tumour-suppressor miR-29 led to increased levels of DNA methyltransferases and abnormal DNA methylation. Restoration of miR-29 expression re-established normal methylation patterns and reactivated tumour-suppressor genes, thereby hindering cancer progression [40].

Transcriptional Modulation of miRNA Function

miRNA expression is intricately regulated by diverse transcriptional factors (TFs), and any alterations in these TFs can contribute to miRNA dysregulation. A dynamic regulatory relationship exists between miRNAs and TFs, where each can influence the other. Notably, a large proportion of miRNA targets are themselves transcription factors, establishing complex regulatory circuits [41]. Key TFs involved in cancer-related miRNA regulation include p53 and c-Myc. Mutations in *TP53*, the gene encoding p53, are common in many cancers. p53 influences several miRNAs involved in cell cycle arrest and apoptosis, such as miR-34a, miR-107, miR-605, and miR-1246. For example, p53 activates miR-34a by binding to its promoter, and miR-34a in turn targets SIRT1, a deacetylase that suppresses p53 activity, creating a feedback loop that influences cell fate [41].

Similarly, c-Myc is known to regulate tumour suppressor miRNAs such as let-7, miR-15a, miR-26, miR-29, and miR-30. c-Myc directly suppresses miR-122 expression through binding to gene promoter sequences, while miR-122 downregulates *E2F1* and *TFDP2*, indirectly reducing c-Myc transcription, a feedback mechanism that impacts tumour progression [42]. Additional TFs are also implicated in miRNA regulation. In gastric cancer, NME2 regulates miR-100 expression and targets genes such as *LIMS1, STARD5, and RIPK1 in gastric cancer, leading to the suppression of* apoptosis and enhanced tumor cell survival. Inhibiting miR-100 has been shown to stabilize p53 by blocking its degradation, promoting apoptosis. miR-100 directly targets the E3 ubiquitin ligase RNF144B, which under stress conditions induces apoptosis through p53 degradation [43]. In CRC, *SP1* was identified as a key TF promoting miR-146b-3p expression by binding to its promoter. Elevated *SP1* levels resulted in increased miR-146b-3p expression, thereby enhancing cancer cell proliferation, migration, and invasion. Importantly, suppressing *SP1* or miR-146b-3p reversed these effects, highlighting a potential therapeutic target [44].

Altered miRNA Biogenesis in Cancer

miRNA biogenesis is a tightly regulated multistep process occurring at both transcriptional and post-transcriptional levels, and its dysregulation has been closely associated with cancer progression [45]. miRNAs are first transcribed as primary transcripts, known as pri-miRNAs, by RNA polymerase II, generating transcripts with a 5′ cap and a 3′ poly(A) tail. However, a small subset of miRNAs is transcribed by RNA polymerase III. The canonical miRNA processing pathway involves two key ribonuclease-mediated steps: one in the nucleus and the other in the cytoplasm. Within the nucleus, the microprocessor complex, comprising the RNase III enzyme DROSHA and the double-stranded RNA-binding protein DGCR8, processes the pri-miRNA into a precursor miRNA (pre-miRNA) of approximately 70 nucleotides, characterized by a hairpin structure. DGCR8 plays a crucial role in recognizing the boundary between the single-stranded and double-stranded regions of the pri-miRNA, thereby enabling accurate cleavage by DROSHA. [46].

The pre-miRNA is subsequently transported to the cytoplasm via Exportin-5(XPO5) in a Ran-GTP-dependent manner. Upon reaching the cytoplasm, the RNase III enzyme DICER further processes the pre-miRNA into a miRNA duplex approximately 20 to 25 nucleotides in length. One strand of the miRNA duplex is preferentially incorporated into the Argonaute (AGO) protein to arrange the RNA-induced silencing complex (RISC), whereas the complementary strand is usually degraded. Through this complex, miRNAs guide the silencing of target mRNAs via translational repression, destabilization, or degradation [47]. Dysregulation in any of these steps can impair miRNA maturation and contribute to oncogenesis. For example, in cancer cells, defects in pre-miRNA lead to the retention of pre-miRNAs in the nucleus, inhibiting their further processing. Studies profiling 225 pre-miRNAs revealed higher nuclear retention in tumourigenic cells compared to normal cells, often due to reduced XPO5 expression. This reduction may result from genetic or epigenetic alterations or post-translational modifications affecting XPO5 [47].

Furthermore, the downregulation of key processing enzymes like DROSHA and DICER has been reported in several cancers. Germline mutations in miRNA biogenesis genes significantly increase cancer risk. In pleuropulmonary blastoma, heterozygous germline mutations in DICER1 truncate the protein near its functional RNase III domains, while somatic mutations in ovarian cancer alter the RNase domain of DICER1, leading to reduced enzymatic activity. Similarly, DROSHA knockdown enhances cell proliferation in lung adenocarcinoma, linking its loss to tumour progression [37].

Regulatory proteins also modulate the miRNA biogenesis machinery. For instance, the transcriptional coactivator WBP2, which promotes carcinogenesis in breast cancer, negatively regulates the microprocessor complex by binding to its components, such as DDX5 and DGCR8. This interaction suppresses pri-miRNA processing and enhances oncogenic potential by increasing cancer cell proliferation. In ovarian cancer, tumour suppressor miR-145, which targets SOX9 and

TWIST, is downregulated due to interference in its maturation [48]. The p70 S6 kinase (p70S6K), a PI3K/AKT pathway effector, targets both Dicer and tristetraprolin (TTP), thereby disrupting the maturation of miR-145. A decline in TTP expression correlates with elevated p70S6K levels, a pattern associated with poor prognosis.

Another regulatory protein, DDX17, acts as a cofactor in the microprocessor complex and influences CSC properties. Upon polyubiquitination by HectH9 (an E3 ubiquitin ligase) and activation of nuclear YAP, DDX17 dissociates from the microprocessor complex. This prevents DDX17 from recognizing specific sequence motifs at the 3′ end of pri-miRNAs, thereby impairing the maturation of tumour suppressor miRNAs. Moreover, polyubiquitinated DDX17 promotes the binding of p300 to form a YAP-DDX17-p300 complex, which enhances transcriptional activation of stemness-related genes, especially under hypoxic conditions, through acetylation of H3K56. In conclusion, disruptions in the microRNA synthesis pathway, whether caused by mutations in essential processing enzymes, impaired nuclear export, or interference from regulatory proteins, significantly aid in cancer initiation and progression by altering the expression and function of both tumour-suppressive and oncogenic miRNAs.

MicroRNA Regulation of Cancer Stem Cells Across Various Cancers

Breast Cancer

Breast cancer stem cells (BCSCs) play a major role in tumour progression, metastasis, and treatment resistance. miRNAs are key regulators of BCSCs, influencing pathways related to self-renewal, apoptosis, and survival. Based on their target genes, miRNAs can act as tumour suppressors or oncogenic regulators. For instance, the miR-200 family (miR-200a, miR-200b, miR-200c, and miR-141, miR-429) regulates EMT by targeting transcriptional repressors of E-cadherin, like ZEB1, ZEB2, and SNAIL1, preventing EMT and reducing BCSC stemness and metastasis. Perdigão-Henriques et al. [49] showed that downregulation of miR-200c correlated with the upregulation of stemness markers (*OCT4*, *SOX2*, *c-MYC*) and EMT-related genes (*SNAIL1*, *CDH2*, *TWIST1/2*). In contrast, upregulation of miR-30c was associated with similar markers, highlighting its role in controlling stemness and metastasis. miR-34a, a tumour suppressor miRNA, is frequently downregulated in cancer and CSCs. Li et al. [50] found that miR-34a suppresses key genes involved in stemness and metastasis, including *NOTCH*, *MYC*, *BCL-2*, and *CD44*, thereby inhibiting BCSC self-renewal, tumour growth, and therapy resistance. Additionally, miR-210 enhances BCSC metastasis and self-renewal under hypoxic conditions by targeting E-cadherin [51]. miR-21 promotes proliferation and invasion in triple-negative breast cancer (TNBC) by targeting tumour suppressor genes like *PTEN* [52].

Lung Cancer

CSCs in non-small cell lung cancer (NSCLC) are pivotal in driving tumour progression, therapeutic resistance, and recurrence, making them critical targets for novel treatment strategies such as miRNA-based therapies. In NSCLC, CSCs contribute significantly to therapy resistance and tumour recurrence. Researchers have identified several miRNAs that regulate CSC-like traits and drug resistance mechanisms. Koh et al. [53] targeted paclitaxel-resistant NSCLC cells, particularly $CD133^+$ CSCs, which exhibited elevated levels of MUC1-C and BMI1, key proteins in stemness maintenance. Overexpression of miR-128 reduced these markers, thereby suppressing proliferation, self-renewal, migration, and tumourigenicity while enhancing apoptosis. Similarly, miR-181b was significantly downregulated in cisplatin-resistant NSCLC cells; restoration of miR-181b expression suppressed CSC properties such as sphere formation and expression of markers like CD133, SOX2, NANOG, KLF4, and ALDH. Mechanistically, miR-181b targets Notch2, inactivating the Notch2/Hes1 pathway essential for CSC maintenance [54]. In lung adenocarcinoma (LUAD), miR-31 is overexpressed. Functional analyses revealed that miR-31 suppresses CSC traits, including adhesion, invasion, differentiation, and metastasis, by targeting the MET-PI3K-Akt signalling axis, which is crucial for CSC survival and proliferation [55]. Moreover, NSCLC tissues show elevated levels of miR-146a-5p compared to normal controls. This miRNA enhances cancer cell proliferation, migration, and apoptosis resistance by directly suppressing *TRAF6*, a vital component of the NF-κB pathway. Its suppression activates NF-κB signalling, contributing to oncogenesis [56]. Ruiz-Manriquez et al. [57] emphasized that dysregulated miRNAs target key cancer-correlated signal pathways like PI3K/Akt/mTOR, Hippo-YAP/TAZ, and Wnt/β-catenin, which either promote or inhibit tumour growth. Therefore, miRNAs are not only responsible for cancer initiation and progression but are also conceivable biomarkers for diagnostics and viable targets for therapy.

Colorectal Cancer (CRC)

Colorectal cancer stem cells (CR-CSCs) are pivotal in tumour initiation, progression, and resistance to therapy, making them key targets for more effective cancer treatments. miRNAs play critical roles in modulating the stem-like properties of CR-CSCs by regulating key pathways involved in proliferation, differentiation, and survival. A direct transcriptional target of p53, miR-34a, is frequently downregulated in colon cancer [58]. Its restoration has been shown to inhibit tumourigenesis and metastasis by targeting genes involved in stemness and proliferation. Specifically, miR-34a modulates the Notch pathway, crucial for maintaining CR-CSC self-renewal. By inhibiting Notch1 and its downstream effectors, miR-34a promotes differentiation and reduces the tumourigenic potential of CR-CSCs [50, 59].

Furthermore, miR-34a is epigenetically suppressed in CR-CSCs by long non-coding RNAs (lncRNAs) like Lnc34a, which use the epigenetic modifiers such as Dnmt3a and HDAC1 to silence miR-34a expression, thus preserving the stem-like state and enhancing the proliferative capacity of CR-CSCs [60]. miR-203 also acts as a tumour suppressor in CR-CSCs by targeting the transcription factor GATA6, which promotes stemness. Overexpression of miR-203 in HCT-116 and HT-29 cell lines decreases stemness traits, while its sequestration increases them. Stemness is further decreased by inhibiting GATA6 downstream effectors such as LRH-1 and Hes-1 [61]. Additionally, miR-142-3p, involved in haematopoietic stem cell development, has been shown to enhance colon CSC-like features. BM-MSCs produce exosomes loaded with miR-142-3p, which enhance tumourigenicity and stemness markers in CR-CSCs [62]. Jain et al. [63] highlighted that inflammatory cytokines, particularly various interleukins such as IL-1, IL-6, and IL-8, play crucial roles in activating key transcription factors like NF-κB, Smads, STAT3, SNAIL, ZEB, and Twist. These factors collectively promote EMT, a process essential for enhancing tumour cell migration, invasion, and metastasis, thereby activating CSCs. It was found that miRNAs regulate CSC characteristics by suppressing key signalling pathways required for CSC maintenance, including the Wnt/β-catenin, Notch, Hedgehog, and PI3K/Akt pathways. Tumour-suppressor miRNAs, such as miR-34a and miR-145, suppress CSC self-renewal and diminish stemness by suppressing these pathways. Conversely, oncogenic miRNAs, including miR-21 and miR-155, reinforce CSC traits and facilitate cancer development by promoting or maintaining these signalling pathways [64].

Liver Cancer

Hepatocellular carcinoma (HCC) is a highly determined liver cancer driven by CSCs that facilitate tumour initiation, progression, metastasis, and therapy resistance. Recent studies emphasize the critical role of miRNAs in regulating the stemness and differentiation of HCC stem cells by targeting key signalling molecules. miR-142-3p has been shown to influence stem cell-like features in cancer. It functions in haematopoietic stem cell development and, when delivered via exosomes from bone marrow-derived mesenchymal stem cells (BM-MSCs), enhances stem cell markers and tumourigenicity in colon CSCs [57, 65]. In HCC, tumour-associated macrophages (TAMs) can enhance the stemness and tumourigenicity of HCC cells by increasing the population of CD90+ CSCs. miR-125a/b counters this by targeting CD90, inhibiting TAM-induced CSC traits such as self-renewal, migration, and invasion, suggesting its role in modulating both intrinsic CSC properties and the CSC-TAM crosstalk in the tumour microenvironment [54]. In CSC-enriched tumour sphere cultures, miR-100 and miR-125b are downregulated, correlating with increased expression of IGF2 (Insulin-like growth factor 2), which activates the PI3K/AKT/mTOR pathway to sustain stemness. Re-expression of miR-100 and miR-125b decreases IGF2 levels, reducing CSC features and tumour growth in xenograft models, positioning these miRNAs as tumour suppressors in HCC [66]. miR-155 promotes CSC generation and maintenance by downregulating TP53INP1 (Tumour protein p53-induced nuclear protein 1), which drives tumourigenesis [67]. miR-124-3p inhibits *AKT2* and *SIRT1*, increasing the nuclear accumulation of FOXO3a, enhancing apoptosis, and sensitizing HCC cells to sorafenib treatment, overcoming drug resistance [68].

Leukaemia

In haematological malignancies such as leukaemia, CSCs, termed leukaemia stem cells (LSCs), play a pivotal role in disease initiation, progression, and resistance to therapy. Leukaemia, especially chronic myeloid leukaemia (CML) and acute myeloid leukaemia (AML), contains CSCs or LSCs, responsible for disease onset, maintenance, and therapeutic resistance. miRNAs control their self-renewal, proliferation, and drug resistance potential. miR-378 is overexpressed in CML patients. This induces proliferation and drug resistance of leukaemic cells by suppressing apoptosis. miR-378 increases CSC-like features by increasing stem cell spheres, which enhances the expression of stemness markers like OCT4 and c-Myc. This action is enabled by downregulating the tumour suppressor, FUS1 [69], which activates apoptotic pathways and suppresses oncogenic tyrosine kinases [70]. miR-185 functions as a tumour suppressor in the non-treated CML stem/progenitor cells. Its overexpression inhibits the survival of drug-resistant leukaemic cells, thereby increasing the sensitivity to tyrosine kinase inhibitors (TKIs) and killing long-term repopulating LSCs. Ghafouri-Fard et al. [70] also stated that miR-185 suppresses PAK6, thus disrupting RAS/MAPK signalling and mitochondrial function, enhancing drug sensitivity. In AML, miR-155 regulates leukaemic cell quiescence and proliferation. It has been studied that FLT3-ITD (FMS-like tyrosine kinase 3 internal tandem duplication) mutational changes activate miR-155, which suppresses SHIP1 (SH2-containing inositol phosphatase 1), thereby causing AKT activation. By this, the maturation of miR-126 gets suppressed. This results in low levels of miR-126, which sustain the leukaemic CSC pool by allowing quiescent leukaemic blasts to multiply and enter the cell cycle (Dinh Hoa [71]). The key miRNAs and their role in major cancers are summarized in Table 5.2.

The Role of miRNA Dysregulation in Modulating Cancer Stem Cell Functions

miRNAs play pivotal roles in cancer by modulating tumour growth, progression, and prognosis. Oncogenic miRNAs are often upregulated, silencing tumour suppressor genes, while tumour-suppressive miRNAs are frequently deleted or silenced

Table 5.2 Functional impact of miRNAs on cancer stem cells in different cancer types

miRNA	Target genes	Cancer type	Effect on CSCs	References
miR-34a	*BCL-2, Notch1*	Breast, colon	Suppresses self-renewal	[50]
let-7	*RAS, HMGA2*	Lung, breast	Promotes differentiation	[72]
miR-200c	*ZEB1, ZEB2*	Breast, pancreas	Inhibits EMT and metastasis	[73]
miR-21	*PTEN, PDCD4*	Colon, breast	Enhances survival and chemoresistance	[74]
miR-124	*CDK6, SNAIL2*	Glioblastoma, breast	Inhibits proliferation and invasion	[75]
miR-145	*OCT4, SOX2, KLF4*	Colon, prostate, breast	Suppresses pluripotency, stemness	[76]
miR-137	*LSD1, EZH2*	Glioblastoma	Inhibits CSC proliferation and self-renewal	[77]
miR-10b	*HOXD10, RHOC*	Breast, pancreas	Promotes invasion and metastasis	[78]
miR-155	*TP53INP1, SOCS1*	Leukaemia, breast, colon	Enhances tumourigenesis and drug resistance	[50]
miR-221/222	*PTEN, p27Kip1*	Breast, liver, glioma	Promotes proliferation and invasion	[79]
miR-378	*FUS1*	Leukaemia	Increases CSC traits and drug resistance	[70]
miR-185	*PAK6*	Leukaemia	Increases sensitivity to TKIs and suppresses LSCs	[70]

epigenetically [80]. These alterations promote drug resistance, metastasis, and survival via pathways like PI3K/AKT, MAPK, and Wnt/β-catenin [81]. miRNAs also regulate the tumour microenvironment, immune response, and angiogenesis [82]. Recent findings highlight their influence on CSC traits such as self-renewal, EMT, and metabolic reprogramming [51, 83] (Fig. 5.2).

Self-renewal Ability

Self-renewal in CSCs drives tumour growth, therapy resistance, and relapse. miRNAs regulate stemness by modulating signalling pathways and transcription factors, impacting CSC self-renewal and tumourigenic potential. The miR-34 family (miR-34a, miR-34b, and miR-34c) is a well-studied tumour-suppressive miRNA family that targets genes critical for CSC properties, including *CD44*, *Notch1*, and *BMI1*. These genes regulate stemness, self-renewal, and metastasis, and miR-34a downregulation impairs CSC self-renewal. Therapeutic delivery of miR-34a to glioblastoma models reduced stemness markers, tumourigenic potential, and therapy resistance [50]. The miR-200 family (miR-200a, miR-200b, and miR-200c) regulates pluripotency and epithelial-mesenchymal transition (EMT), targeting factors like SOX2, ZEB1, ZEB2, and BMI1. Restoring miR-200c in colon CSCs reduced sphere formation and reversed chemoresistance [84]. The let-7 family regulates

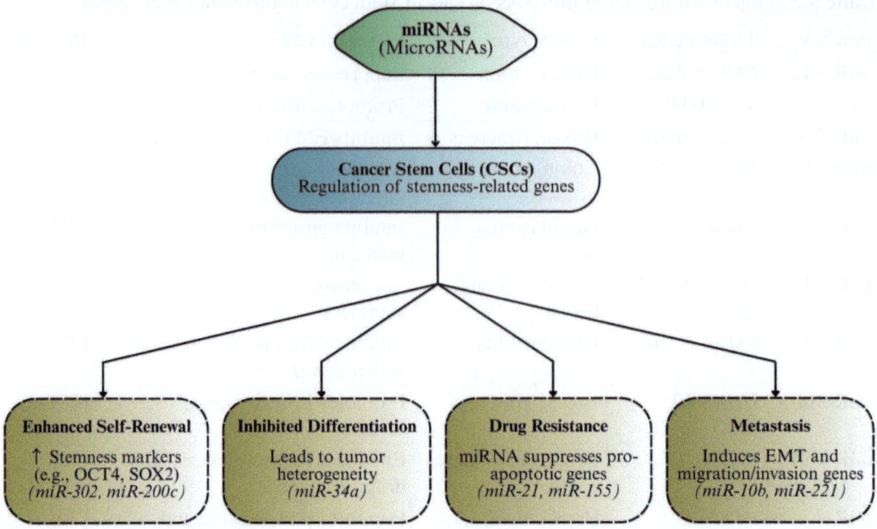

Fig. 5.2 Illustration showing the regulatory role of miRNAs in cancer stem cell functions by regulating genes involved in stemness, differentiation, drug resistance, and metastasis

self-renewal by inhibiting oncogenes such as *RAS*, *HMGA2*, and *LIN28*. Restoring let-7 in triple-negative breast cancer stem cells reduced CSC markers and tumour initiation in xenograft models [72]. Similarly, miR-145 regulates pluripotency by targeting key factors like OCT4, SOX2, and KLF4, reducing stemness in pancreatic CSCs and inhibiting tumour growth in patient-derived xenograft models [55]. In contrast, miR-21 is an oncomiR that promotes CSC self-renewal and survival in breast cancer by inhibiting negative regulators of the PI3K/AKT pathway. Inhibiting miR-21 in hepatocellular carcinoma CSCs reduced self-renewal, colony formation, and enhanced chemosensitivity [85].

Differentiation

CSCs exhibit altered differentiation, maintaining stem-like traits. miRNAs reprogram CSCs by targeting transcription factors and signalling molecules, promoting differentiation and reducing aggressiveness. miRNAs such as miR-146a, miR-203, and miR-22 play significant roles in promoting CSC differentiation, which can reduce tumourigenic potential. miR-146a inhibits the Notch1 signalling pathway, a critical regulator of stemness. Mei et al. [86] showed that delivering miR-146a into glioma stem cells enhanced differentiation and reduced tumour burden in preclinical models, suggesting that targeting Notch1 with miR-146a could eliminate the CSC population by inducing differentiation. In breast cancer, miR-203 acts as a differentiation switch by regulating ΔNp63 and BMI1, factors critical for CSC maintenance. Shen et al. [87] found that miR-203 expression is epigenetically silenced in breast CSCs, and its reintroduction decreases self-renewal while inducing luminal differentiation markers. This suggests that miR-203 could explain the differentiation seen in certain breast cancer subtypes, leading to a better prognosis. miR-22, an epigenetic regulator, inhibits TET family DNA demethylases, inducing genome-wide methylation changes associated with differentiation. Mukohyama et al. [88] found that overexpressing miR-22 in colorectal CSCs resulted in methylation changes similar to differentiated cells, suggesting that miR-22 may counteract CSC-specific methylation patterns and promote differentiation, potentially reducing tumourigenic capacity.

EMT

miRNAs regulate epithelial-to-mesenchymal transition (EMT) in CSCs by targeting transcription factors, signalling molecules, and structural proteins, influencing CSC plasticity and promoting metastasis. The miR-200 family (miR-200a, miR-200b, miR-200c) is an important regulator of EMT in CSCs, inhibiting EMT by targeting ZEB1 and ZEB2, which repress E-cadherin, a critical epithelial marker. Restoring miR-200 expression in CSCs can reverse EMT and reduce metastasis. For instance, Chen et al. [89] showed that miR-200 expression in triple-negative breast cancer stem cells reversed EMT and reduced metastatic potential. The miR-34 family also suppresses EMT by targeting SNAI1 and ZEB1, important mediators of mesenchymal transition. Li et al. [50] found that p53 upregulates miR-34 in prostate CSCs to inhibit EMT, showing the role of tumour suppressors in regulating EMT through miRNA expression. Conversely, some miRNAs promote EMT and metastasis. miR-9 induces EMT by targeting E-cadherin, resulting in loss of cell adhesion and the development of mesenchymal characteristics. Ma et al. [90] demonstrated that inhibiting miR-9 in pancreatic cancer reduced stem-like properties and metastatic potential. miR-10b also promotes CSC invasiveness by regulating HOXD10, a transcription factor that suppresses migration and invasion. Halim et al. [91] recognized miR-10b as a critical factor in breast CSC metastasis, and reducing miR-10b levels decreased metastasis in preclinical models.

Drug Resistance

miRNAs regulate drug resistance in CSCs by targeting the genes of drug efflux, DNA repair, and apoptosis, either promoting resistance or sensitizing CSCs to therapy. miRNAs play a crucial role in modulating drug resistance in CSCs. miR-451, for example, sensitizes drug-resistant CSCs by targeting MDR1, which encodes P-glycoprotein, a drug efflux pump central to multidrug resistance. Liu et al. [92] demonstrated that restoring miR-451 levels in colorectal CSCs downregulated

MDR1 expression, enhancing the intracellular accumulation of therapeutic agents and improving treatment efficacy. This suggests that miR-451 could be a promising approach in overcoming drug resistance in colorectal cancer. Similarly, miR-128 enhances chemosensitivity by targeting BMI1 and various ATP-binding cassette (ABC) transporters, which are often upregulated in CSCs to facilitate drug efflux. Hendrik Setia Budi et al. [93] found that miR-128 reduced stemness and reversed drug resistance in glioblastoma and breast cancer models, indicating its potential in promoting a drug-sensitive CSC phenotype. Conversely, miR-130a is an oncomiR that promotes drug resistance by targeting DICER1, a key enzyme in miRNA biogenesis. Huang et al. [94] showed that upregulating miR-130a in melanoma stem cells impairs DICER1 function, altering miRNA processing and increasing stemness and chemoresistance. Inhibiting miR-130a reversed *DICER1* downregulation and restored chemosensitivity, suggesting a novel regulatory link between miRNA biogenesis and drug resistance.

Tumour Microenvironment Interaction

miRNAs are central controllers of the tumour microenvironment (TME), modulating the intricate interactions between cancer stem cells (CSCs) and the surrounding stromal, immune, and endothelial cells. The miR-29 family (miR-29a, miR-29b, and miR-29c) is a tumour suppressor miRNA that can establish an anti-tumour microenvironment. In neuroblastoma models, miR-29 overexpression enhances levels of anti-tumour cytokines (e.g. CCL5, CCL11), chemokines (CXCL9, CXCL11, CXCL16), and immune-stimulatory molecules like CD40 and endostatin, an angiogenesis inhibitor [95]. These changes promote apoptosis and activate cytotoxic cells, like NK and CD8+ T cells, which release perforin and granzyme B, leading to cancer cell lysis. Besides, miR-29 suppresses pro-tumourigenic factors like osteopontin and proliferin, thereby helping in tumour suppression [95]. miR-34a influences tumour-associated macrophages (TAMs) by promoting M1 (pro-inflammatory) polarization while suppressing M2 (tumour-supportive) polarization, thereby reducing immune evasion and chemoresistance [50]. miR-15/16 regulates cancer-associated fibroblasts (CAFs), which are repressed in CAFs, and facilitates tumour growth by expressing fibroblast-secreted growth factors, such as FGF-2 and its receptor FGFR1, that aid in the survival of cancer cells. Loss of these miRNAs increases fibroblast-mediated secretion of growth factors such as FGF-2, which supports cancer cell survival. Downregulation of exosomal miR-148b in CAFs also results in reduction of DNMT1, supporting epithelial-to-mesenchymal transition and metastasis [96]. miRNAs also regulate the function of immune cells, influencing anti-tumour (e.g. CD8+ T cells, NK cells) versus immunosuppressive populations (e.g. Tregs, MDSCs), thereby controlling immune evasion and tumour growth [97, 98].

MicroRNAs as Therapeutic Agents Against Cancer Stem Cells

As a Biomarker for Cancer Stem Cell-Associated Malignancies

miRNA-based diagnostics have emerged as a promising tool for early cancer detection, classification, and monitoring, particularly in cancers driven by CSCs. Circulating miRNAs in blood and serum, found using liquid biopsies, provide a non-invasive method for real-time tumour monitoring. These circulating miRNAs are associated with both primary and metastatic tumours, as well as specific CSC populations. Chen et al. [99] highlighted the potential of circulating miRNAs to detect tumours at early stages and monitor disease progression, while miRNA panels have been developed to classify patients based on specific CSCs for prognosis and treatment response. Lopez-Bertoni et al. [100] found that miRNA panels could distinguish between tumours with high and low CSC purity, thus enhancing diagnostic accuracy across various cancers. An innovative diagnostic strategy involves exosomal miRNAs, which are released by CSCs within extracellular vesicles, protecting them from degradation. Wu et al. [101] noted that exosomal miRNAs from CSCs can serve as reliable biomarkers, offering insights into tumour aggressiveness and metastatic potential. Profiling miRNA expression has uncovered stemness-related miRNA patterns that can classify tumours and predict patient outcomes. Nakano et al. [102] emphasized the clinical application of miRNAs for prognostic modelling and guiding treatment strategies. Advances in single-cell analysis have further refined miRNA diagnostics by revealing CSC heterogeneity, which bulk testing may overlook. Prakash et al. [103] conveyed that single-cell methods can identify rare CSC subpopulations with unique miRNA profiles, enhancing diagnostic precision. Together, these miRNA-based diagnostic tools offer high specificity and sensitivity, paving the way for personalized oncology and enabling early intervention in CSC-mediated malignancies.

Ongoing Clinical Trials

miRNA-based therapeutics targeting CSCs are undergoing clinical trials, but several technical and clinical barriers need to be addressed for successful translation. Phase I studies of MRX34 (miR-34a mimic) in liver cancer and other solid tumours with CSC aspects have shown potential [104], while Cobomarsen, an oligonucleotide that inhibits miR-155, is being tested in haematological malignancies, targeting leukaemic stem cells [105]. TargomiRs (miR-16 mimic) are undergoing trials for mesothelioma, focusing on CSC populations [106]. Early-phase trials of RGLS5579, yet another oligonucleotide drug that inhibits miR-10b, are being investigated in glioblastoma targeting CSCs [107], and miR-122 regulator is used in liver cancer trials targeting hepatic CSCs [108]. However, several challenges persist. miRNAs are often unstable in vivo, limiting their therapeutic potential. Manufacturing these RNA-based drugs presents additional technical difficulties, and achieving an optimal therapeutic window is still a significant hurdle. Clinical translation is hampered by the lack of reliable biomarkers to evaluate CSC-specific efficacy, regulatory challenges with RNA-based therapies, and issues with trial design (e.g. determining the appropriate endpoints for CSC targeting). Moreover, integrating miRNA therapies with existing treatments is complex due to potential adverse drug-drug interactions or optimization challenges [104].

Next-Generation Delivery Systems for miRNA-Based Cancer Stem Cell Therapy

Effective delivery of miRNA-based therapeutics targeting CSCs remains a significant challenge due to CSCs'resistance features and their location in specific niches. To overcome these challenges, various advanced delivery systems have been developed to improve the stability, specificity, and cellular uptake of miRNA mimics and inhibitors. Lipid nanoparticles (LNPs), modified to carry CSC-targeting ligands, have shown enhanced cellular uptake and reduced off-target effects, improving therapeutic efficacy [109]. Polymeric carriers, such as PLGA and chitosan, provide sustained, controlled release, enhancing anti-tumour responses in CSC-driven cancers [110]. Aptamer-miRNA conjugates represent another promising strategy, enabling cell-specific targeting by binding to unique surface markers on CSCs. Esposito et al. [50] had shown that aptamer-guided delivery of miR-34a to $CD44^+$ CSCs inhibited tumour-initiating potential without affecting normal stem cells. Additionally, exosomes encapsulating therapeutic miRNAs can efficiently deliver their cargo to CSCs, modulating gene expression and reducing stemness traits [30]. Hydrogel-based delivery systems have also been explored for localized retention and sustained release of miRNA therapeutics. Wang et al. [23] showed that hydrogel-encapsulated miRNA mimics prolonged exposure to CSCs within tumours, enhancing anti-cancer efficacy while minimizing systemic toxicity. These evolving delivery technologies, including LNPs, aptamer-miRNA conjugates, exosomes, and hydrogels, hold significant promise in eliminating CSC populations and improving clinical outcomes in cancer patients.

Limitations and Challenges

Despite the potential of miRNA-based therapies against CSCs, clinical translation faces several challenges. These include inconsistent manufacturing, ineffective delivery to hypoxic CSC niches, and miRNA instability in biofluids [111]. Safety concerns are raised by off-target effects, like those seen in MRX34 trials [112]. Therapeutic targeting is made more difficult by CSCs's variability and plasticity [113]. Additionally, changing regulatory frameworks and unclear clinical trial endpoints for CSC-specific outcomes create further barriers, which would require new approaches to address these complex issues [114, 115].

Conclusion and Future Perspective

MicroRNAs are crucial regulators of CSCs, affecting tumour initiation, treatment resistance, and metastasis. Their ability to act as oncomiRs or tumour suppressors presents unique therapeutic opportunities. Clinical translation, however, is challenged by delivery efficiency, CSC heterogeneity, and off-target effects. Emerging delivery platforms like exosomes and nanoparticles, along with precise miRNA profiling, provide solutions to these barriers. Combining miRNA therapies with conventional cancer therapy could enhance efficacy and lower relapse rates. Further studies must emphasize optimization of delivery, the search for good biomarkers, and individualized strategies to leverage miRNAs for therapy against CSCs and revolutionize the fight against cancer. The future of miRNA-based CSC therapy lies in novel delivery and therapeutic strategies. Emerging technologies like cell-penetrating peptides, biomimetic carriers, and stimuli-responsive nanocarriers offer tumour-specific targeting with low systemic toxicity. Combinatorial miRNA approaches and artificial networks may overcome resistance. Personalized therapy is advancing through high-output profiling and computational modelling. Engineered exosomes enhance biocompatibility, reduce immunogenicity, and improve miRNA delivery to CSCs, holding promise for improved efficacy and safety in clinical applications.

Acknowledgement All authors are thankful to the Chettinad Academy of Research and Education for providing infrastructural support to complete this work.

Ethics Approval Not Applicable.

Conflict of Interest All authors declare that there are no conflicts of competing interests.

Availability of Data Not Applicable.

Author's Contribution The chapter was designed, conceptualized, and supervised by AB. NSS, SnP, ASP, and AD were involved in drafting the chapter. ASP and NSS were involved in designing the images and tables. SuP and AB critically reviewed the chapter. All authors read, reviewed, and approved the final chapter.

Funding Not Applicable.

References

1. Brown JS, Amend SR, Austin RH, Gatenby RA, Hammarlund EU, Pienta KJ (2023) Updating the definition of cancer. Mol Cancer Res 21(11):1142–1147. https://doi.org/10.1158/1541-7786.mcr-23-0411
2. Kamal N, Ilowefah MA, Hilles AR, Anua NA, Awin T, Alshwyeh HA, Aldosary SK, Jambocus NGS, Alosaimi AA, Rahman A, Mahmood S, Mediani A (2022) Genesis and mechanism of some cancer types and an overview on the role of diet and nutrition in cancer prevention. Molecules 27(6):1794. https://doi.org/10.3390/molecules27061794

3. Bray F, Laversanne M, Sung H, Ferlay J, Siegel RL, Soerjomataram I, Jemal A (2024) Global cancer statistics 2022: GLOBOCAN estimates of incidence and mortality worldwide for 36 cancers in 185 countries. CA Cancer J Clin 74(3):229–263. https://doi.org/10.3322/caac.21834
4. Ramos A, Sadeghi S, Tabatabaeian H (2021) Battling Chemoresistance in cancer: root causes and strategies to uproot them. Int J Mol Sci 22(17):9451. https://doi.org/10.3390/ijms22179451
5. Loud JT, Murphy J (2017) Cancer screening and early detection in the 21 st century. Semin Oncol Nurs 33(2):121–128. https://doi.org/10.1016/j.soncn.2017.02.002
6. Phi LTH, Sari IN, Yang Y-G, Lee S-H, Jun N, Kim KS, Lee YK, Kwon HY (2018) Cancer stem cells (CSCs) in drug resistance and their therapeutic implications in cancer treatment. Stem Cells Int 2018:1–16. https://doi.org/10.1155/2018/5416923
7. Chu X, Tian W, Ning J, Xiao G, Zhou Y, Wang Z, Zhai Z, Tanzhu G, Yang J, Zhou R (2024) Cancer stem cells: advances in knowledge and implications for cancer therapy. Signal Transduct Target Ther 9(1):170. https://doi.org/10.1038/s41392-024-01851-y
8. Ankasha SJ, Shafiee MN, Wahab NA, Ali RAR, Mokhtar NM (2018) Post-transcriptional regulation of microRNAs in cancer: from prediction to validation. Oncol Rev. https://doi.org/10.4081/oncol.2018.344
9. Yu Z, Li Y, Fan H, Liu Z, Pestell RG (2012) miRNAs regulate stem cell self-renewal and differentiation. Front Genet 3. https://doi.org/10.3389/fgene.2012.00191
10. He C, Luo B, Jiang N, Liang Y, He Y, Zeng J, Liu J, Zheng X (2018) OncomiR or antioncomiR: role of miRNAs in acute myeloid leukemia. Leuk Lymphoma 60(2):284–294. https://doi.org/10.1080/10428194.2018.1480769
11. Lathia JD, Liu H (2017) Overview of cancer stem cells and stemness for community oncologists. Target Oncol 12(4):387–399. https://doi.org/10.1007/s11523-017-0508-3
12. Aponte PM, Caicedo A (2017) Stemness in cancer: stem cells, cancer stem cells, and their microenvironment. Stem Cells Int 2017:1–17. https://doi.org/10.1155/2017/5619472
13. Zhang Y, Wang X (2020) Targeting the Wnt/β-catenin signaling pathway in cancer. J Hematol Oncol 13(1). https://doi.org/10.1186/s13045-020-00990-3
14. Yang L, Shi P, Zhao G, Xu J, Peng W, Zhang J, Zhang G, Wang X, Dong Z, Chen F, Cui H (2020) Targeting cancer stem cell pathways for cancer therapy. Signal Transduct Target Ther 5(8). https://doi.org/10.1038/s41392-020-0110-5
15. Sari IN, Phi LTH, Jun N, Wijaya YT, Lee S, Kwon HY (2018) Hedgehog signaling in cancer: a prospective therapeutic target for eradicating cancer stem cells. Cells 7(11):208. https://doi.org/10.3390/cells7110208
16. Sheikh K, Amjad M, Irfan M, Anjum S, Majeed T, Riaz M, Jassim A, Sharif E, Ibrahim W (2025) Exploring TGF-β Signaling in cancer progression: prospects and therapeutic strategies. Onco Targets Ther 18:233–262. https://doi.org/10.2147/ott.s493643
17. Sabaawy HE, Ryan BM, Khiabanian H, Pine SR (2021) JAK/STAT of all trades: linking inflammation with cancer development, tumour progression and therapy resistance. Carcinogenesis. https://doi.org/10.1093/carcin/bgab075
18. Marzagalli M, Fontana F, Raimondi M, Limonta P (2021) Cancer stem cells-key players in tumour relapse. Cancers 13(3):376
19. Zheng Y, Wang L, Yin L, Yao Z, Tong R, Xue J, Lu Y (2022) Lung cancer stem cell markers as therapeutic targets: an update on signaling pathways and therapies. Front Oncol 12:873994. https://doi.org/10.3389/fonc.2022.873994
20. Zhou Y, Xia L, Wang H, Oyang L, Su M, Liu Q, Lin J, Tan S, Tian Y, Liao Q, Cao D (2017) Cancer stem cells in progression of colorectal cancer. Oncotarget 9(70):33403–33415. https://doi.org/10.18632/oncotarget.23607
21. Xiao Y, Lin M, Jiang X, Ye J, Guo T, Shi Y, Bian X (2017) The recent advances on liver cancer stem cells: biomarkers, separation, and therapy. Anal Cell Pathol (Amst) 2017:5108653. https://doi.org/10.1155/2017/5108653
22. Ibragimova M, Tsyganov M, Litviakov N (2022) Tumour stem cells in breast cancer. Int J Mol Sci 23(9):5058. https://doi.org/10.3390/ijms23095058
23. Wang J, Qu J, Hou Q, Huo X, Zhao X, Chang L, Xu C (2024) Strategies for the isolation and identification of gastric cancer stem cells. Stem Cells Int 2024:5553852. https://doi.org/10.1155/2024/5553852

24. Vergez F, Nicolau-Travers ML, Bertoli S, Rieu JB, Tavitian S, Bories P, Luquet I, Mas V, Largeaud L, Sarry A, Huguet F, Delabesse E, Bérard E, Récher C (2020) CD34+CD38−CD123+ leukemic stem cell frequency predicts outcome in older acute myeloid leukemia patients treated by intensive chemotherapy but not hypomethylating agents. Cancer 12(5):1174. https://doi.org/10.3390/cancers12051174
25. Kushwaha PP, Verma S, Kumar S, Gupta S (2022) Role of prostate cancer stem-like cells in the development of antiandrogen resistance. Cancer Drug Resist (Alhambra, Calif) 5(2):459–471. https://doi.org/10.20517/cdr.2022.07
26. Ishiwata T, Matsuda Y, Yoshimura H, Sasaki N, Ishiwata S, Ishikawa N, Takubo K, Arai T, Aida J (2018) Pancreatic cancer stem cells: features and detection methods. Pathol Oncol Res 24(4):797–805. https://doi.org/10.1007/s12253-018-0420-x
27. Yin Q, Shi X, Lan S, Jin H, Wu D (2021) Effect of melanoma stem cells on melanoma metastasis. Oncol Lett 22(1):566. https://doi.org/10.3892/ol.2021.12827
28. Paul R, Dorsey JF, Fan Y (2022) Cell plasticity, senescence, and quiescence in cancer stem cells: biological and therapeutic implications. Pharmacol Ther 231:107985. https://doi.org/10.1016/j.pharmthera.2021.107985
29. Chakrabortty A, Patton DJ, Smith BF, Agarwal P (2023) miRNAs: potential as biomarkers and therapeutic targets for cancer. Genes 14(7):1375. https://doi.org/10.3390/genes14071375
30. O'Brien J, Hayder H, Zayed Y, Peng C (2018) Overview of MicroRNA biogenesis, mechanisms of actions, and circulation. Front Endocrinol 9(402). https://doi.org/10.3389/fendo.2018.00402
31. Metcalf GAD (2024) MicroRNAs: circulating biomarkers for the early detection of imperceptible cancers via biosensor and machine-learning advances. Oncogene:1–8. https://doi.org/10.1038/s41388-024-03076-3
32. Autore F, Ramassone A, Stirparo L, Pagotto S, Fresa A, Innocenti I, Visone R, Laurenti L (2023) Role of microRNAs in chronic lymphocytic leukemia. Int J Mol Sci 24(15):12471–12471. https://doi.org/10.3390/ijms241512471
33. Jung G, Hernández-Illán E, Moreira L, Balaguer F, Goel A (2020) Epigenetics of colorectal cancer: biomarker and therapeutic potential. Nat Rev Gastroenterol Hepatol 17(2):111–130. https://doi.org/10.1038/s41575-019-0230-y
34. Hussen BM, Hidayat HJ, Salihi A, Sabir DK, Taheri M, Ghafouri-Fard S (2021) MicroRNA: A signature for cancer progression. Biomed Pharmacother 138:111528. https://doi.org/10.1016/j.biopha.2021.111528
35. Peng Y, Croce CM (2016) The role of MicroRNAs in human cancer. Signal Transduct Target Ther 1:15004. https://doi.org/10.1038/sigtrans.2015.4
36. Pfaff E, Aichmüller C, Sill M, Stichel D, Snuderl M, Karajannis MA, Schuhmann MU, Schittenhelm J, Hasselblatt M, Thomas C, Korshunov A, Rhizova M, Wittmann A, Kaufhold A, Iskar M, Ketteler P, Lohmann D, Orr BA, Ellison DW, von Hoff K, Mynarek M, Rutkowski S, Sahm F, von Deimling A, Lichter P, Kool M, Zapatka M, Pfister SM, Jones DTW (2020) Molecular subgrouping of primary pineal parenchymal tumours reveals distinct subtypes correlated with clinical parameters and genetic alterations. Acta Neuropathol 139(2):243–257. https://doi.org/10.1007/s00401-019-02101-0
37. Dragomir MP, Knutsen E, Calin GA (2022) Classical and noncanonical functions of miRNAs in cancers. Trends Genet 38(4):379–394. https://doi.org/10.1016/j.tig.2021.10.002
38. Zhang L, Volinia S, Bonome T, Calin GA, Greshock J, Yang N, Liu C-G, Giannakakis A, Alexiou P, Hasegawa K, Johnstone CN, Megraw MS, Adams S, Lassus H, Huang J, Kaur S, Liang S, Sethupathy P, Leminen A, Simossis VA (2008) Genomic and epigenetic alterations deregulate microRNA expression in human epithelial ovarian cancer. Proc Natl Acad Sci 105(19):7004–7009. https://doi.org/10.1073/pnas.0801615105
39. Tang Y, Yang S, Wang M, Liu D, Liu Y, Zhang Y, Zhang Q (2019) Epigenetically altered miR 193a 3p promotes HER2 positive breast cancer aggressiveness by targeting GRB7. Int J Mol Med 43(6):2352–2360. https://doi.org/10.3892/ijmm.2019.4167
40. Roberti A, Valdes AF, Torrecillas R, Fraga MF, Fernandez AF (2019) Epigenetics in cancer therapy and nanomedicine. Clin Epigenetics 11(1):81. https://doi.org/10.1186/s13148-019-0675-4

41. Bracken CP, Scott HS, Goodall GJ (2016) A network-biology perspective of microRNA function and dysfunction in cancer. Nat Rev Genet 17(12):719–732. https://doi.org/10.1038/nrg.2016.134
42. Shah V, Shah J (2020) Recent trends in targeting miRNAs for cancer therapy. J Pharm Pharmacol 72(12):1732–1749. https://doi.org/10.1111/jphp.13351
43. Gong Y, Yang G, Wang Q, Wang Y, Zhang X (2020) NME2 is a master suppressor of apoptosis in gastric cancer cells via transcriptional regulation of miR-100 and other survival factors. Mol Cancer Res 18(2):287–299. https://doi.org/10.1158/1541-7786.MCR-19-0612
44. Wang D, Feng M, Ma X, Tao K, Wang G (2021) Transcription factor SP1-induced microRNA-146b-3p facilitates the progression and metastasis of colorectal cancer via regulating FAM107A. Life Sci 277:119398. https://doi.org/10.1016/j.lfs.2021.119398
45. Annese T, Tamma R, De Giorgis M, Ribatti D (2020) microRNAs biogenesis, functions and role in tumour angiogenesis. Front Oncol 10:581007. https://doi.org/10.3389/fonc.2020.581007
46. Gurtner A, Falcone E, Garibaldi F, Piaggio G (2016) Dysregulation of microRNA biogenesis in cancer: the impact of mutant p53 on Drosha complex activity. J Exp Clin Cancer Res 35:45. https://doi.org/10.1186/s13046-016-0319-x
47. Wu K, He J, Pu W, Peng Y (2018) The role of Exportin-5 in MicroRNA biogenesis and cancer. Genomics Proteomics Bioinformatics 16(2):120–126. https://doi.org/10.1016/j.gpb.2017.09.004
48. Tabatabaeian H, Lim SK, Chu T, Seah SH, Lim YP (2021) WBP2 inhibits microRNA biogenesis via interaction with the microprocessor complex. Life Sci Alliance 4(7):e202101038. https://doi.org/10.26508/lsa.202101038
49. Perdigão-Henriques R, Petrocca F, Altschuler G, Thomas M, Le MV, Tan SM, Hide W, Lieberman J (2015) miR-200 promotes the mesenchymal to epithelial transition by suppressing multiple members of the Zeb2 and Snail1 transcriptional repressor complexes. Oncogene 35(2):158–172. https://doi.org/10.1038/onc.2015.69
50. Li WJ, Wang Y, Liu R, Kasinski AL, Shen H, Slack FJ, Tang DG (2021) MicroRNA-34a: potent tumour suppressor, cancer stem cell inhibitor, and potential anticancer therapeutic. Front Cell Dev Biol 9:640587. https://doi.org/10.3389/fcell.2021.640587
51. Khan AQ, Ahmed EI, Elareer NR, Junejo K, Steinhoff M, Uddin S (2019) Role of miRNA-regulated cancer stem cells in the pathogenesis of human malignancies. Cells 8(8). https://doi.org/10.3390/cells8080840
52. Fang H, Xie J, Zhang M, Zhao Z, Wan Y, Yao Y (2017) miRNA-21 promotes proliferation and invasion of triple-negative breast cancer cells through targeting PTEN. Am J Transl Res 9(3):953–961
53. Koh H, Park H, Chandimali N, Huynh DL, Zhang JJ, Ghosh M, Gera M, Kim N, Bak Y, Yoon D-Y, Park YH, Kwon T, Jeong DK (2017) MicroRNA-128 suppresses paclitaxel-resistant lung cancer by inhibiting MUC1-C and BMI-1 in cancer stem cells. Oncotarget 8(66):110540–110551. https://doi.org/10.18632/oncotarget.22818
54. Wang Y, Wang B, Xiao S, Li Y, Chen Q (2018) miR-125a/b inhibits tumour-associated macrophages mediated in cancer stem cells of hepatocellular carcinoma by targeting CD90. J Cell Biochem 120(3):3046–3055. https://doi.org/10.1002/jcb.27436
55. Hou C, Sun B, Jiang Y, Zheng J, Yang N, Ji C, Liang Z, Shi J, Zhang R, Liu Y, Ye C, Zuo P (2016) MicroRNA-31 inhibits lung adenocarcinoma stem-like cells via down-regulation of MET-PI3K-Akt signaling pathway. Anti-Cancer Agents Med Chem 16(4):501–518. https://doi.org/10.2174/1871520615666150824152353
56. Liu X, Liu B, Li R, Wang F, Wang N, Zhang M, Bai Y, Wu J, Liu L, Han D, Li Z, Feng B, Zhou G, Wang S, Zeng L, Miao J, Yao Y, Liang B, Huang L, Wang Q (2020) miR-146a-5p plays an oncogenic role in NSCLC via suppression of TRAF6. Front Cell Dev Biol 8. https://doi.org/10.3389/fcell.2020.00847
57. Ruiz-Manriquez LM, Carrasco-Morales O, Sanchez Z EA, Osorio-Perez SM, Estrada-Meza C, Pathak S, Banerjee A, Bandyopadhyay A, Duttaroy AK, Paul S (2022) MicroRNA-mediated regulation of key signaling pathways in hepatocellular carcinoma: A mechanistic insight. Front Genet 13:910733. https://doi.org/10.3389/fgene.2022.910733

58. Jothimani G, Sriramulu S, Chabria Y, Sun XF, Banerjee A, Pathak S (2018) A review on theragnostic applications of micrornas and long non-coding RNAs in colorectal cancer. Curr Top Med Chem 18(30):2614–2629. https://doi.org/10.2174/1568026619666181221165344
59. Rupaimoole R, Slack FJ (2016) A role for miR-34 in colon cancer stem cell homeostasis. Stem Cell Investig 3:42–42. https://doi.org/10.21037/sci.2016.08.04
60. Wang L, Bu P, Ai Y, Srinivasan T, Chen HJ, Xiang K, Lipkin SM, Shen X (2016) A long non-coding RNA targets microRNA miR-34a to regulate colon cancer stem cell asymmetric division. elife 5. https://doi.org/10.7554/elife.14620
61. Lai H, Tseng W, Huang S, Chao T, Su Y (2019) MicroRNA-203 diminishes the stemness of human colon cancer cells by suppressing *GATA6* expression. J Cell Physiol 235(3):2866–2880. https://doi.org/10.1002/jcp.29192
62. Li H, Li F (2018) Exosomes from BM-MSCs increase the population of CSCs via transfer of miR-142-3p. Br J Cancer 119(6):744–755. https://doi.org/10.1038/s41416-018-0254-z
63. Jain SM, Deka D, Das A, Paul S, Pathak S, Banerjee A (2023) Role of interleukins in inflammation-mediated tumor immune microenvironment modulation in colorectal cancer pathogenesis. Dig Dis Sci 68(8):3220–3236. https://doi.org/10.1007/s10620-023-07972-8
64. Banerjee A, Deka D, Muralikumar M, Sun-Zhang A, Bisgin A, Christopher C, Zhang H, Sun XF, Pathak S (2023) A concise review on miRNAs as regulators of colon cancer stem cells and associated signalling pathways. Clin Transl Oncol 25(12):3345–3356. https://doi.org/10.1007/s12094-023-03200-x
65. Dai J, Hao Y, Chen X, Yu Q, Wang B (2023) miR-122/SENP1 axis confers stemness and chemoresistance to liver cancer through Wnt/β-catenin signaling. Oncol Lett 26(3). https://doi.org/10.3892/ol.2023.13976
66. Seol HS, Akiyama Y, Lee S-E, Shimada S, Jang SJ (2020) Loss of miR-100 and miR-125b results in cancer stem cell properties through IGF2 upregulation in hepatocellular carcinoma. Sci Rep 10(1). https://doi.org/10.1038/s41598-020-77960-9
67. Liu F, Kong X, Lv L, Gao J (2015) MiR-155 targets TP53INP1 to regulate liver cancer stem cell acquisition and self-renewal. FEBS Lett 589(4):500–506. https://doi.org/10.1016/j.febslet.2015.01.009
68. Dong Z, Wu H, He Y, Huang Z, Weng Y, Li H, Liang C, Yu W, Chen W (2022) MiRNA-124-3p.1 sensitizes hepatocellular carcinoma cells to sorafenib by regulating FOXO3a by targeting AKT2 and SIRT1. Cell Death Dis 13(1):1–11. https://doi.org/10.1038/s41419-021-04491-0
69. Ma J, Wu D, Yi J, Yi Y, Zhu X, Qiu H, Kong R, Lin J, Qian J, Deng Z (2019) MiR-378 promoted cell proliferation and inhibited apoptosis by enhanced stem cell properties in chronic myeloid leukemia K562 cells. Biomed Pharmacother 112:108623. https://doi.org/10.1016/j.biopha.2019.108623
70. Ghafouri-Fard S, Hajiesmaeili M, Shoorei H, Bahroudi Z, Taheri M, Sharifi G (2021) The impact of lncRNAs and miRNAs in regulation of function of cancer stem cells and progression of cancer. Front Cell Dev Biol 9. https://doi.org/10.3389/fcell.2021.696820
71. Hoang DH, Zhao D, Branciamore S, Maestrini D, Rodriguez IR, Kuo Y-H, Rockne R, Khaled SK, Zhang B, Xuan L, Marcucci G (2022) MicroRNA networks in FLT3-ITD acute myeloid leukemia. Proc Natl Acad Sci 119(16). https://doi.org/10.1073/pnas.2112482119
72. Ma Y, Shen N, Wicha MS, Luo M (2021) The roles of the Let-7 family of MicroRNAs in the regulation of cancer stemness. Cells 10(9):2415. https://doi.org/10.3390/cells10092415
73. Lu Y, Lu J, Li X, Zhu H, Fan X, Zhu S, Wang Y, Guo Q, Wang L, Huang Y, Zhu M, Wang Z (2014) MiR-200a inhibits epithelial-mesenchymal transition of pancreatic cancer stem cell. BMC Cancer 14:85. https://doi.org/10.1186/1471-2407-14-85
74. Liu C, Tong Z, Tan J, Xin Z, Wang Z, Tian L (2019) MicroRNA-21-5p targeting PDCD4 suppresses apoptosis via regulating the PI3K/AKT/FOXO1 signaling pathway in tongue squamous cell carcinoma. Exp Ther Med 18(5):3543–3551. https://doi.org/10.3892/etm.2019.7970
75. Xia H, Cheung WKC, Ng SS, Jiang X, Jiang S, Sze J, Leung GKK, Lu G, Chan DTM, Bian XW, Kung HF, Poon WS, Lin MC (2012) Loss of brain-enriched miR-124 microRNA enhances stem-like traits and invasiveness of glioma cells. J Biol Chem 287(13):9962–9971. https://doi.org/10.1074/jbc.M111.332627

76. Zhu Y, Wang C, Becker SA, Hurst K, Nogueira LM, Findlay VJ, Camp ER (2018) miR-145 antagonizes SNAI1-mediated Stemness and radiation resistance in colorectal cancer. Mol Ther 26(3):744–754. https://doi.org/10.1016/j.ymthe.2017.12.023
77. Wang Y, Chen R, Zhou X, Guo R, Yin J, Li Y, Ma G (2020) miR-137: a novel therapeutic target for human glioma. Mol Ther Nucleic Acids 21:614–622. https://doi.org/10.1016/j.omtn.2020.06.028
78. Ouyang H, Gore J, Deitz S, Korc M (2014) microRNA-10b enhances pancreatic cancer cell invasion by suppressing TIP30 expression and promoting EGF and TGF-β actions. Oncogene 33(38):4664–4674. https://doi.org/10.1038/onc.2013.405
79. Garofalo M, Quintavalle C, Romano G, Croce CM, Condorelli G (2012) miR221/222 in cancer: their role in tumour progression and response to therapy. Curr Mol Med 12(1):27–33. https://doi.org/10.2174/156652412798376170
80. Yang Z, Liu X, Xu H, Teschendorff AE, Xu L, Li J, Fu M, Liu J, Zhou H, Wang Y, Zhang L, He Y, Lv K, Yang H (2024) Integrative analysis of genomic and epigenomic regulation reveals miRNA mediated tumour heterogeneity and immune evasion in lower grade glioma. Commun Biol 7(1). https://doi.org/10.1038/s42003-024-06488-9
81. He Y, Sun MM, Zhang GG, Yang J, Chen KS, Xu WW, Li B (2021) Targeting PI3K/Akt signal transduction for cancer therapy. Signal Transduct Target Ther 6(1):1–17. https://doi.org/10.1038/s41392-021-00828-5
82. Anfossi S, Babayan A, Pantel K, Calin GA (2018) Clinical utility of circulating non-coding RNAs—an update. Nat Rev Clin Oncol 15(9):541–563. https://doi.org/10.1038/s41571-018-0035-x
83. Li G-M, Xiao G-Z, Qin P-F, Wan X-Y, Fu Y-J, Zheng Y-H, Luo M-Y, Ren D-L, Liu S-P, Chen H-X, Lin H-C (2022) Single-cell RNA sequencing reveals heterogeneity in the tumour microenvironment between young-onset and old-onset colorectal cancer. Biomolecules 12(12):1860. https://doi.org/10.3390/biom12121860
84. Klicka K, Grzywa TM, Mielniczuk A, Klinke A, Włodarski PK (2022) The role of miR-200 family in the regulation of hallmarks of cancer. Front Oncol 12. https://doi.org/10.3389/fonc.2022.965231
85. Rahimi Farsi N, Naghipour B, Shahabi P, Safaralizadeh R, Hajiasgharzadeh K, Dastmalchi N, Alipour MR (2023) The role of microRNAs in hepatocellular carcinoma: therapeutic targeting of tumour suppressor and oncogenic genes. Clin Exp Hepatol 9(4):307–319. https://doi.org/10.5114/ceh.2023.131669
86. Mei J, Bachoo R, Zhang C-L (2011) MicroRNA-146a inhibits glioma development by targeting Notch1. Mol Cell Biol 31(17):3584–3592. https://doi.org/10.1128/mcb.05821-11
87. Martínez-Illescas NG, Leal S, González P, Graña-Castro O, Muñoz-Oliveira JJ, Cortés-Peña A, Gómez-Gil M, Vega Z, Neva V, Romero A, Quintela-Fandino M, Ciruelos E, Sanz C, Aragón S, Sotolongo L, Jiménez S, Caleiras E, Mulero F, Sánchez C, Malumbres M (2023) miR-203 drives breast cancer cell differentiation. Breast Cancer Res 25(1). https://doi.org/10.1186/s13058-023-01690-9
88. Mukohyama J, Isobe T, Hu Q, Hayashi T, Watanabe T, Maeda M, Yanagi H, Qian X, Yamashita K, Minami H, Mimori K, Sahoo D, Kakeji Y, Suzuki A, Dalerba P, Shimono Y (2019) miR-221 targets QKI to enhance the tumourigenic capacity of human colorectal cancer. Stem Cells 79(20):5151–5158. https://doi.org/10.1158/0008-5472.can-18-3544
89. Chen H, Li Z, Zhang L, Zhang L, Zhang Y, Wang Y, Xu M, Zhong Q (2020) MicroRNA-200c inhibits the metastasis of triple-negative breast cancer by targeting ZEB2, an epithelial-mesenchymal transition regulator. Ann Clin Lab Sci 50(4):519–527. http://www.annclinlabsci.org/content/50/4/519.long
90. Ma L, Young J, Prabhala H, Pan E, Mestdagh P, Muth D, Teruya-Feldstein J, Reinhardt F, Onder TT, Valastyan S, Westermann F, Speleman F, Vandesompele J, Weinberg RA (2010) miR-9, a MYC/MYCN-activated microRNA, regulates E-cadherin and cancer metastasis. Nat Cell Biol 12(3):247–256. https://doi.org/10.1038/ncb2024
91. Halim A, Al-Qadi N, Kenyon E, Conner KN, Mondal SK, Medarova Z, Moore A (2024) Inhibition of miR-10b treats metastatic breast cancer by targeting stem cell-like properties. Oncotarget 15(1):591–606. https://doi.org/10.18632/oncotarget.28641

92. Liu R-L, Dong Y, Deng Y-Z, Wang W-J, Li W-D (2015) Tumour suppressor miR-145 reverses drug resistance by directly targeting DNA damage-related gene RAD18 in colorectal cancer. Tumour Biol 36(7):5011–5019. https://doi.org/10.1007/s13277-015-3152-5
93. Budi HS, Younus LA, Lafta MH, Parveen S, Mohammad HJ, Al-qaim ZH, Jawad MA, Mireya R, Mustafa YF, Alhachami FR, Karampoor S, Mirzaei R (2023) The role of miR-128 in cancer development, prevention, drug resistance, and immunotherapy. Front Oncol 12. https://doi.org/10.3389/fonc.2022.1067974
94. Huang J, Zhao M, Hu H, Wang J, Ang L, Zheng L (2019) MicroRNA-130a reduces drug resistance in breast cancer. Int J Clin Exp Pathol 12(7):2699. https://pmc.ncbi.nlm.nih.gov/articles/PMC6949573/
95. Pathania AS, Chava H, Chaturvedi NK, Chava S, Byrareddy SN, Coulter DW, Challagundla KB (2024) The miR-29 family facilitates the activation of NK-cell immune responses by targeting the B7-H3 immune checkpoint in neuroblastoma. Cell Death Dis 15(6). https://doi.org/10.1038/s41419-024-06791-7
96. Otmani K, Lewalle P (2021) Tumour suppressor miRNA in cancer cells and the tumour microenvironment: mechanism of deregulation and clinical implications. Front Oncol 11. https://doi.org/10.3389/fonc.2021.708765
97. Kousar K, Ahmad T, Abduh MS, Kanwal B, Shah SS, Naseer F, Anjum S (2022) miRNAs in regulation of tumour microenvironment, chemotherapy resistance, immunotherapy modulation and miRNA therapeutics in cancer. Int J Mol Sci 23(22):13822–13822. https://doi.org/10.3390/ijms232213822
98. Xing Y, Ruan G, Ni H, Qin H, Chen S, Gu X, Shang J, Zhou Y, Tao X, Zheng L (2021) Tumour immune microenvironment and its related miRNAs in tumour progression. Front Immunol 12. https://doi.org/10.3389/fimmu.2021.624725
99. Chen J, Chen S, Zhuo L, Zhu Y, Zheng H (2020) Regulation of cancer stem cell properties, angiogenesis, and vasculogenic mimicry by miR-450a-5p/SOX2 axis in colorectal cancer. Cell Death Disease 11(3). https://doi.org/10.1038/s41419-020-2361-z
100. Lopez-Bertoni H, Li Y, Laterra J (2015) Cancer stem cells: dynamic entities in an ever-evolving paradigm. Biol Med s2. https://doi.org/10.4172/0974-8369.1000s2-001
101. Wu Y, Cao Y, Chen L, Lai X, Zhang S, Wang S (2024) Role of exosomes in cancer and Aptamer-modified exosomes as a promising platform for cancer targeted therapy. Biol Proc Online 26(1):15. https://doi.org/10.1186/s12575-024-00245-2
102. Nakano Y, Isobe K, Kobayashi H, Kaburaki K, Isshiki T, Sakamoto S, Takai Y, Tochigi N, Mikami T, Iyoda A, Homma S, Kishi K (2019) Clinical importance of long non-coding RNA LINC00460 expression in EGFR-mutant lung adenocarcinoma. Int J Oncol. https://doi.org/10.3892/ijo.2019.4919
103. Pan XW, Zhang H, Xu D, Chen JX, Chen WJ, Gan SS, Qu FJ, Chu CM, Cao JW, Fan YH, Song X, Ye JQ, Zhou W, Cui XG (2020) Identification of a novel cancer stem cell subpopulation that promotes progression of human fatal renal cell carcinoma by single-cell RNA-seq analysis. Int J Biol Sci 16(16):3149–3162. https://doi.org/10.7150/ijbs.46645
104. Beg MS, Brenner AJ, Sachdev J, Borad M, Kang Y-K, Stoudemire J, Smith S, Bader AG, Kim S, Hong DS (2017) Phase I study of MRX34, a liposomal miR-34a mimic, administered twice weekly in patients with advanced solid tumors. Investig New Drugs 35(2):180–188. https://doi.org/10.1007/s10637-016-0407-y
105. Anastasiadou E, Seto AG, Beatty X, Hermreck M, Gilles M-E, Stroopinsky D, Pinter-Brown LC, Pestano L, Marchese C, Avigan D, Trivedi P, Escolar DM, Jackson AL, Slack FJ (2021) Cobomarsen, an oligonucleotide inhibitor of miR-155, slows DLBCL tumor cell growth in vitro and in vivo. Clin Cancer Res: Off J Amer Assoc Cancer Res 27(4):1139–1149. https://doi.org/10.1158/1078-0432.CCR-20-3139
106. Reid G, Kao SC, Pavlakis N, Brahmbhatt H, MacDiarmid J, Clarke S, Boyer M, van Zandwijk N (2016) Clinical development of TargomiRs, a miRNA mimic-based treatment for patients with recurrent thoracic cancer. Epigenomics 8(8):1079–1085. https://doi.org/10.2217/epi-2016-0035

107. Krichevsky AM, Uhlmann EJ (2019) Oligonucleotide therapeutics as a new class of drugs for malignant brain tumors: targeting mRNAs, regulatory RNAs, mutations, combinations, and beyond. Neurotherapeutics 16(2):319–347. https://doi.org/10.1007/s13311-018-00702-3
108. Dhungel B, Ramlogan-Steel CA, Layton CJ, Steel JC (2018) miRNA122a regulation of gene therapy vectors targeting hepatocellular cancer stem cells. Oncotarget 9(34):23577–23588. https://doi.org/10.18632/oncotarget.25280
109. Tang Y, Chen Y, Zhang Z, Tang B, Zhou Z, Chen H (2021) Nanoparticle-based RNAi therapeutics targeting cancer stem cells: update and prospective. Pharmaceutics 13(12):2116. https://doi.org/10.3390/pharmaceutics13122116
110. Kesharwani P, Halwai K, Jha SK, Al Mughram MH, Almujri SS, Almalki WH, Sahebkar A (2024) Folate-engineered chitosan nanoparticles: next-generation anticancer nanocarriers. Mol Cancer 23(1). https://doi.org/10.1186/s12943-024-02163-z
111. Chen Z, Han F, Du Y, Shi H, Zhou W (2023) Hypoxic microenvironment in cancer: molecular mechanisms and therapeutic interventions. Signal Transduct Targeted Ther 8(1). https://doi.org/10.1038/s41392-023-01332-8
112. Kim T, Croce CM (2023) MicroRNA: trends in clinical trials of cancer diagnosis and therapy strategies. Exp Mol Med 55(7):1314–1321. https://doi.org/10.1038/s12276-023-01050-9
113. Zheng X, Yu C, Xu M (2021) Linking tumour microenvironment to plasticity of cancer stem cells: mechanisms and application in cancer therapy. Front Oncol 11. https://doi.org/10.3389/fonc.2021.678333
114. Jones CH, Androsavich JR, So N, Jenkins MP, MacCormack D, Prigodich A, Welch V, True JM, Dolsten M (2024) Breaking the mold with RNA—a "RNAissance" of life science. NPJ Genom Med 9(1). https://doi.org/10.1038/s41525-023-00387-4
115. Saw PE, Song E (2024) Advancements in clinical RNA therapeutics: present developments and prospective outlooks. Cell Rep Med 5(5):101555–101555. https://doi.org/10.1016/j.xcrm.2024.101555

Chapter 6
Impending Role of Exosomal miRNA as Substantial Players in Early Cancer Diagnostics

Rajalakshmi Prakash, Akilesh Natesh, Karan Kubendiran, Swaroop Santhekadur Palakshappa, and Anbarasu Kannan

Abstract The exosomes, nano-sized cup-shaped extracellular vesicles measuring 30–150 nm in size, have extended their potential use as a tumor biomarker in recent years. These exosomes are stated to be the carbon copy of the cells of origin, mirroring the exact physiological or pathological state of their parent cell, including cancer cells. Pertaining to cancer, it is generally believed "the sooner, the better." Most of the currently available diagnostic techniques are not devoid of their side effects, in that they are invasive and hazardous. These conventional diagnostic tools are considered less beneficial as they only detect cancer in its later stages, leading to poor treatment efficiency. This is where the use of exosomes comes into play, as these are less invasive methods of diagnosis, secreted and released into circulation in the early stages of cancer, and are easily detectable. Accounting for the presence of cargo it carries, such as proteins, nucleic acids, lipids, and metabolites, exosomes have garnered considerable attention from the scientific community worldwide for their remarkable role as an efficient diagnostic tool.

Among the various nucleic acids transported by exosomes, microRNAs (miRNAs) are a class of small, noncoding RNAs that have gained significant attention due to their critical role in posttranscriptional gene regulation via interaction with target mRNAs. Notably, under pathological conditions and cellular stress, cells tend to secrete higher levels of miRNAs into the extracellular environment, often encapsulated within exosomes. These circulating, exosome-associated miRNAs have

R. Prakash · S. S. Palakshappa · A. Kannan (✉)
Cancer and Exosome Biology Laboratory, Department of Biochemistry, CSIR—Central Food Technological Research Institute, Mysuru, India

Academy of Scientific and Innovative Research (AcSIR), Ghaziabad, India
e-mail: anbarasu@cftri.res.in

A. Natesh · K. Kubendiran
Cancer and Exosome Biology Laboratory, Department of Biochemistry, CSIR—Central Food Technological Research Institute, Mysuru, India

S. Pathak, A. Banerjee (eds.), *MicroRNAs: Role in Cancer Progression, Diagnostics and Therapy*, Advances in Biochemistry in Health and Disease 34,
https://doi.org/10.1007/978-3-032-08824-6_6

emerged as promising noninvasive biomarkers for various diseases, particularly cancer. This chapter explores the biological characteristics of exosomal miRNAs, the molecular mechanisms that govern their selective packaging into the exosomal lumen, and their significant implications for early cancer detection. Special emphasis is placed on their diagnostic relevance and how their profiling may aid in enhancing the overall efficacy of cancer therapy.

Keywords Exosomes · MicroRNAs (miRNAs) · Noninvasive biomarkers · Early cancer detection · Cancer diagnostics

Introduction

Cancer: Mechanisms, Prevalence, and Evolving Therapies

Being in the state of uncontrollable cell growth or cell proliferation at a particular site or tissue of the body technically defines the term "cancer". Cancer is a multifactorial disease where the normal cells rather than shadowing the regulatory mechanism of controlled cell proliferation, differentiation, and apoptosis (programmed cell death), adapt the abnormal condition like accumulation of genetic and epigenetic alteration leading toward uncontrolled proliferation, causing them to grow indefinitely, resisting cell death, inducing angiogenesis, and invading surrounding tissues and metastasize to a distant organ [19]. All these factors combinatorially raise a normal cell to transform into the malignant ones through the acquisition of mutations in key genes of oncogenes, tumor suppressor genes, and genes regulating DNA repair. Most cancers fall into three main groups: carcinomas, sarcomas, and leukemias or lymphomas. Carcinomas include 90% of human cancers and are malignancies of epithelial cells. Sarcomas are solid tumors that occur in connective tissues; leukemias and lymphomas are malignancies that arise from the blood-forming cells [15].

As per the WHO reports, cancer remains the second leading cause of death, with around ten million deaths in 2020. The common types include lip, oral, lung, breast, prostate, colorectal, and stomach cancer. Worldwide, the incidence rate continues due to an aging population and lifestyle changes such as alcohol consumption, smoking, poor diet, and physical inactivity. Among all types of cancer, lung cancer remains the most frequently diagnosed cancer in 2022, in response for almost 2.5 million new cases found. Worldwide, breast cancer contributes 11.6%, colorectal cancer contributes 9.6%, prostate cancer contributes 7.3%, and stomach cancer contributes 4.9% of new cancer cases. Amongst all the cancer types, Lung cancer was the leading one contributing 12.4% of all new cancer cases and an estimated 1.8 million deaths. Predominantly, lung cancer and breast cancer are frequent cause of cancer related deaths in men and women respectively [7]. Approximately 400,000 children develop cancer annually worldwide. The American Cancer Society projects about 2,041,910 new cancer cases and 618,120 cancer deaths in the USA in 2025. The most common cancers in women are breast cancer (32% of cases), lung and bronchus cancer (12%), and colorectal cancer (7%). The most common cancers in men are prostate cancer (30%), lung and bronchus cancer (11%), and colorectal cancer (8%). This statistic underscores the continuing prominence of breast, prostate, lung, and colorectal cancers as the most frequently diagnosed cancers in 2025, influenced by factors such as aging populations, lifestyle, and screening practices [68].

The mechanism that drives the transformation of proto-oncogenes into oncogenes takes place due to the mutation and inactivation of tumor suppressor genes like *TP53* and *RB1*, which normally act as brakes on cell proliferation; their inactivation removes critical checkpoints in the cell cycle, leading to the transformation of a normal cell into a malignant cell. Cyclins and cyclin-dependent kinases, the two major proteins, play a role in regulating the cell cycle. These proteins act as regulators when any of these are not functioning properly, i.e., *INK4a/ARF*, which encodes for tumor suppressor proteins. A mutation in these will lead to the formation of a tumor in a specific cell type and lead to malignancy. Moreover, defects in DNA repair mechanisms contribute to genomic instability and further suit the environment of oncogenic mutations [58]. There are several advanced diagnostic tools for the detection of cancer, the detailed discussion of which will be presented in the upcoming sections of this chapter.

Cancer treatment is often tailored to the type, stage, and genetic characteristics of the tumor. Current treatment includes surgery to remove the localized tumor, using high-energy rays to destroy cancer cells, radiation therapy, and chemotherapy that uses drugs to kill which inhibit the growth of rapidly dividing cells [49]. Targeted therapy interferes with specific molecules involved in the growth of tumor cells, where an alternate option of immunotherapy enhances the body's immune responses to identify the cancer cells, but the problem is it often increases the cytokines, which act as a fuel for the growth of the tumor. Hormone therapy is often used in cancer therapy, like breast cancer and prostate cancer, with a wide range of side effects, as cancer cells often end up reprogramming metabolic pathways, suppressing immune responses and adapting to therapeutic pressure in disease complexity. In such cases, metabolic reprogramming, such as increased glycolysis (the Warburg effect), allows cancer cells to thrive even in hypoxic conditions [90].

Current advanced treatment improves the medication for cancer therapy, utilizing the progressions made in genomic and molecular biology which has revolutionized cancer care. Addressing the problem through using simple and preventive methods, like liquid biopsies, AI-driven diagnostics, and CRISPR-Cas9 editing, shows promising early detection and personalized treatment strategies. Chemotherapy often has no specific targets, and nonselectively attenuates all cell types including non cancerous cells. In order to over come this off targeting effect, and to achieve better outcome, a target-specific strategy, directing the cell surface antigens, growth factors, receptors, or signal transduction pathways that regulate cell cycle progression, cell death, metastasis, and angiogenesis, aiming to control the malignancy potential is needed. Targeting cancer cells with a combinational therapy as a new treatment option might bring a future scale to mitigate the proliferation of cancer cells.

Exosomes: Origin, Biogenesis, and Functional Relevance in Cancer

The story of exosomes begins in the early 1980s when researchers observed small vesicular structures released from the surface of maturing reticulocytes. These vesicles were initially believed to be cellular debris, unwanted materials packaged and expelled by cells as a form of waste disposal. However, in 1987, Rose M. Johnstone, who had been studying the release of transferrin receptors from maturing red blood cells, coined the term "exosomes" to describe these vesicles, marking a paradigm shift [34]. Rather than being passive waste containers, exosomes were beginning to be recognized as active participants in intercellular communication.

In the decades that followed, exosomes emerged from the shadows of being considered insignificant vesicles to assuming central roles in physiology and disease. Their significance became especially apparent in cancer biology, where tumor-derived exosomes were shown to modulate the tumor microenvironment, drive metastasis, induce angiogenesis, mediate immune suppression, and more intriguingly, carry molecular cargo particularly miRNAs that mirrored the molecular state of the tumor cells from which they originated [35, 80].

Human-derived exosomes are typically 30–150 nm in size and originate from the endosomal compartment of the cell. Their biogenesis is tightly orchestrated through a series of intracellular events starting with the formation of early endosomes via the invagination of the plasma membrane during endocytosis. These early endosomes mature into late endosomes, within which intraluminal vesicles (ILVs) are formed by inward budding of the limiting membrane. These late endosomes containing multiple ILVs are known as multivesicular bodies (MVBs). MVBs have two fates: they can either fuse with lysosomes for degradation or traffic to the plasma membrane to release ILVs into the extracellular space as exosomes [39].

The formation of these vesicles involves the Endosomal Sorting Complex Required for Transport (*ESCRT*) machinery, composed of four complexes (*ESCRT-0, -I, -II, and -III*), along with associated proteins like *TSG101* and *ALIX* [31, 74]. These components facilitate cargo sorting and membrane budding. Interestingly, exosome formation can also occur through ESCRT-independent mechanisms, primarily involving the generation of ceramide by neutral sphingomyelinase 2 (*nSMase2*), which alters membrane curvature and promotes vesicle budding [77]. Proteins like tetraspanins (*CD9, CD63, CD81*) are not only markers for exosomes but also actively participate in cargo selection and vesicle budding [4].

Once secreted, exosomes travel through extracellular fluids such as blood, saliva, urine, and cerebrospinal fluid. They carry within them a selective cargo of proteins, lipids, DNA fragments, mRNAs, long noncoding RNAs, and most notably, miRNAs. It is because of the cargo they carry and due to the fact that these exosomes are present in almost all the body fluids with a designated role and genetic makeup, they are being explored for their role as a potent diagnostic marker in the detection of several ailments, including cancer. Isolation, characterization, and quantification of exosomes serve as the basic prerequisite in exosomal biomarker development. The isolation of exosomes from various sources, including blood, serum, saliva, urine, cell culture supernatant, etc., is well documented, and the gist of the same has been provided herewith, which includes the traditional method of ultracentrifugation through a series of centrifugations with increasing gyrations [8], kit-based isolation which uses precipitation through Polyethylene Glycol (PEG) as its base, ultrafiltration using membranes of specific size, size exclusion chromatography using dedicated columns, immunomagnetic method of isolation using the principle of affinity precipitation using exosomal surface membrane proteins [38, 89], and also through the more recent size exclusion-fast protein liquid chromatography (SE-FPLC) [36]. A few of the other isolation methods include the use of Integrated Double Filtration Microfluid Device, Nano Plasmon-Enhanced Scattering (nPES), Membrane-Mediated Exosome Separation, and On-Chip isolation, etc. [89]. Followed by isolation, the exosomes are quantified and characterized using techniques such as Nanoparticle Tracking Analysis (NTA) and Dynamic Light Scattering (DLS) for quantification, and Zeta potential, Scanning Electron Microscopy, and Transmission Electron Microscopy to analyze their structure and morphological features, and Western blotting targeting surface receptors for their identification [64].

Among the exosomal cargos, miRNAs are of considerable interest because of their ability to modulate gene expression posttranscriptionally in recipient cells. Their presence in exosomes is not probabilistic but highly selective, raising the question: how are specific miRNAs loaded into exosomes? This will be discussed in the upcoming sections.

MicroRNAs: Discovery, Localization, and Biogenesis Mechanisms

MicroRNA, abbreviated as miRNA, is a class of noncoding RNA of a size ranging around 22 nucleotides. Discovered in the year 1993 by the Ambros and Ruvkun groups, *lin-4* is said to be the first miRNA, and so far, more than 2500 miRNAs have been reported in humans, which quite evidently shows their abundance [43, 50, 62] This evolutionarily conserved class of small noncoding RNAs is reported to form a hairpin-like structure. The miRNAs are known to have pan organelle localization, including the nucleus, cytoplasm, rough Endoplasmic Reticulum, Golgi complex, mitochondria, Endosomes, Multivesicular Bodies (MVB), and also the extracellular biological fluids and the exosomes [51]. Previous studies have reported that the ratio of miRNA inside the exosomes is higher than that of the parent cell [23]. Upon cellular stress and various pathological conditions, the secretion of miRNAs is reported to be elevated, characterizing such circulatory miRNAs that can be used in the development of biomarkers for the respective pathological conditions [32].

The mechanism of miRNA biogenesis includes a series of processes that occur both in and out of the nucleus and may undertake either canonical or noncanonical pathways (Fig. 6.1). To brief out, the process of miRNA biogenesis has been divided into three phases: the nuclear phase encompassing transcription and the nuclear processing of the miRNA gene, the transport of miRNA to the cytoplasm, and the cytoplasmic processing of miRNA. Like that of any other RNA, the biogenesis of miRNAs also starts in the nucleus and is often co-transcribed alongside the protein-coding gene by the action of RNA polymerase II to form a long primary miRNA (pri-miRNA of more than 1000 nucleotides long hairpin). Followed by this, owing

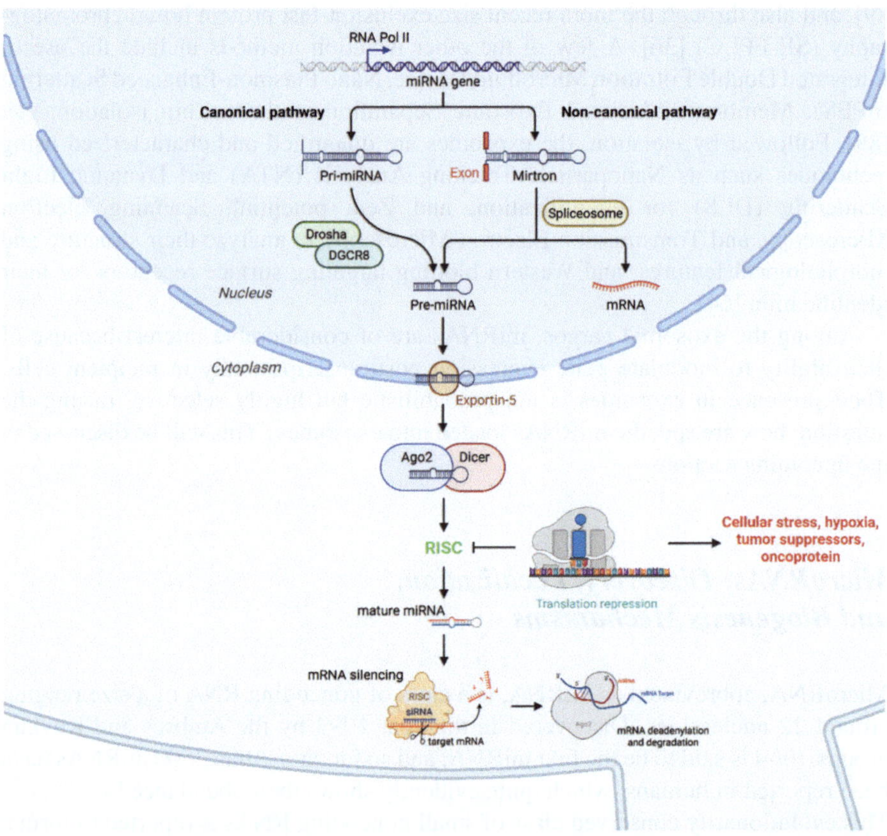

Fig. 6.1 Schematic representation of the miRNA biogenesis pathway and its regulatory mechanisms. miRNAs are transcribed in the nucleus, processed by the *Drosha-DGCR8* complex, and exported to the cytoplasm, where *Dicer* cleaves them into mature miRNA duplexes. The guide strand is incorporated into the *RISC* complex to mediate gene silencing via translational repression or mRNA degradation. This pathway is tightly regulated by oncogenes, tumor suppressors, and cellular stress responses

to the action of a microprocessor complex encompassing Drosha and DGCR8, which cleaves the pri-miRNA, a precursor miRNA hairpin or pre-miRNA of 60–70 nucleotides is formed. Further, the pre-miRNA gets transported out of the nucleus into the cytoplasmic space through the nuclear transporter protein *Exportin5-Ran-GTP* where it gets converted to a mature double-stranded miRNA duplex of 20–22 nucleotide size by the action of a *RNase III* enzyme, Dicer, where one of the strands is considered to be the functional strand and the other nonfunctional which gets degraded. Upon maturation, the functional strand of miRNA is further loaded with Argonaute 2 (*Ago2*), forming an *RISC* (RNA-induced silencing complex) to target their respective mRNA [5, 41, 46, 69].

Apart from the above canonical pathway, the biogenesis of miRNAs can also occur through noncanonical pathways where one or more steps of the canonical pathways, especially the one including *Drosha*, will be bypassed. Briefly, the non-canonical pathway of miRNA biosynthesis is a deviation from the conventional canonical pathway, which includes alternate strategies. The noncanonical pathway is broadly divided into *Drosha*-independent and *Dicer*-independent pathways, which are further divided into six and two subcategories, respectively. The *Drosha*-independent pathways include the Mirtron-derived pathway, snoRNA-derived pathway, shRNA-derived pathway, tRNA-derived pathway, tRNase Z-dependent pathway, and the one from the pre-endo-siRNA-derived pathway. Here, the processing of pri-miRNA to pre-miRNA does not require *Drosha*, but the steps after the generation of pre-miRNA will follow the processing similar to that of the canonical pathway involving *Dicer* and *Ago2*. The Dicer-independent pathways include the Ago-dependent pathway and the tRNase Z-dependent pathway, where the initial processing involves the presence of *Drosha* but bypasses *Dicer*. Finally, in both *Drosha*-independent and *Dicer*-independent pathways, the functional strand of miRNA will be loaded with *Ago2* protein to form RISC [1, 59, 84].

Conventional Cancer Diagnosis

Before the actual concept of exosomal miRNA and their influence as diagnostic biomarkers for early cancer detection, let's understand and know the conventional diagnostic markers and methods currently being practiced. Diagnosis plays a vital part in achieving an advantage in combating several pathological conditions. As far as cancer is concerned, it is widely accepted that the sooner it gets detected, the better the treatment efficiency. Cancer detected at later stages limits the therapeutic options and decreases the survival rate or life expectancy of the patient. Several previous studies reveal that the survival rates significantly increase if the cancer is detected in the early stages, when the tumor can be surgically eliminated [3, 57]. It is recently reported that the average 5–year survival rate if the cancer is detected in the early stages and treated is 91%. On the other hand, the average 5–year survival rate if the cancer is detected in a later stage is 26% [11].

At present, the field of cancer diagnostics is thriving with several efficient diagnostic tools that include both invasive and noninvasive methods. Among all the available diagnostic tools for cancer, biopsy, the more invasive diagnostic tool, is considered the gold standard [14], where a small fragment of a doubtful tissue sample will be extracted from the patient and checked for pathological intervention with reference to the morphological changes. The other invasive method of cancer diagnostics includes classic blood testing for the detection of abnormal blood count and detection of conventional cancer biomarkers *CA 125*, *CA 19–9*, *CA-15-3*, *CD117*, *CD19*, *CD 20*, *HE4*, alpha-fetoproteins (*AFP*), bladder tumor antigen (*BTA*), etc. [29, 61].

Most of the aforementioned biomarkers can also be detected with the noninvasive laboratory process through urine examination. Other noninvasive cancer diagnostics methods comprise several imaging tests, including CT scan, X-ray, mammography, ultrasound, Positron emission tomography (PET), etc. However, these diagnostic tools are often limited since most of them only detect cancer in its later stages, leading to poor treatment efficiency, and some of the aforementioned invasive techniques are said to be accompanied by adverse effects, accounting for their mode of operation [21].

Regulation of miRNA

The regulation of microRNAs (miRNAs) is a multitiered and tightly controlled process that ensures appropriate spatial and temporal gene expression. Unlike the biogenesis process, which encompasses the generation and maturation of miRNAs, regulation encompasses how miRNA levels and activity are modulated across different cellular contexts, developmental stages, and pathological conditions, especially in cancers. Understanding these regulatory nodes is important, as even subtle alterations can disrupt homeostasis and contribute to disease pathogenesis.

At the transcriptional level, miRNA genes are primarily transcribed by RNA polymerase II, and their expression is modulated by transcription factors, enhancers, and suppressors [42]. For example, tumor suppressor *p53* activates transcription of miRNAs such as the *miR-34* family, whereas oncogenic transcription factors like *MYC* and *E2F1* upregulate clusters like miR-17-92 [10], establishing a clear connection between cellular stress responses and miRNA expression. Additionally, hormones and cytokines also influence miRNA gene transcription, further integrating miRNA regulation with cellular signaling networks.

Epigenetic modifications are another major regulatory mechanism. The promoters of tumor-suppressive miRNAs are frequently silenced by DNA methylation or histone deacetylation in various cancers. For instance, *miR-124*, *miR-34a*, and *let-7* family members are often silenced via promoter hypermethylation in breast, colorectal, and lung cancers. Histone modifications such as *H3K27me3* also contribute to the repression of specific miRNA loci, making epigenetic remodeling a significant contributor to miRNA deregulation in tumorigenesis.

At the posttranscriptional level, RNA-binding proteins (RBPs) like *LIN28*, *hnRNP A1*, and *KSRP* modulate miRNA processing by interacting with primary or precursor miRNAs. *LIN28*, for example, blocks the maturation of *let-7* miRNAs by recruiting terminal uridyl transferases (TUTases) that tag *pre-let-7* for degradation [82]. This regulatory switch is frequently hijacked in cancers, where *LIN28* overexpression correlates with poor prognosis due to sustained repression of tumor-suppressive *let-7* miRNAs.

Feedback and feedforward loops between miRNAs and their targets also represent sophisticated modes of regulation. miRNAs often target the mRNAs of transcription factors that control their expression, establishing autoregulatory loops. For instance, *miR-145* represses *MYC* expression, while *MYC* in turn suppresses *miR-145*, forming a reciprocal regulatory axis relevant in multiple malignancies. Additionally, competing endogenous RNAs (ceRNAs), such as long noncoding RNAs (lncRNAs) and circular RNAs (circRNAs), can sequester miRNAs through shared binding sites, thereby modulating miRNA availability and target repression efficiency.

Finally, subcellular localization and stress-induced export influence miRNA activity. Under conditions like oxidative stress or hypoxia, cells may preferentially package specific miRNAs into exosomes or other extracellular vesicles, modulating intercellular communication and systemic signaling. These selective processes suggest an additional regulatory checkpoint linked to cellular homeostasis and the tumor microenvironment.

Collectively, these diverse layers of miRNA regulation ensure functional fidelity across physiological states. However, when deregulated, they contribute to cancer initiation, progression, immune evasion, and therapy resistance, highlighting miRNA regulators as potential diagnostic and therapeutic targets.

Exosomal miRNA and Their Sorting Mechanism

In the preceding section, we had discussed the biochemical description, biogenesis, secretion, and regulation of exosomes and miRNA, respectively. However, this section of the chapter focuses on answering the question that arose in the introduction on how the miRNAs are sorted inside the exosomes. Among the various exosomal cargoes, miRNA plays a vital role in the regulation of several mRNAs, thereby regulating protein expression. As mentioned earlier, the ratio of miRNA secreted in the exosomes is comparatively higher than that of the parent cells, which proves that the miRNAs are precisely involved in signal transduction between various cells. The miRNAs, among all other noncoding RNAs, are acknowledged as the major noncoding RNAs present in the exosomes (around 42.32%). Concerning this, the focus of this section of the chapter will be on how these miRNAs are packed or sorted inside the exosomes. Through several research articles in the past decades, it is evident that the miRNAs are not just randomly packed inside the respective exosomes; rather, they do so by following a specific selective sorting mechanism. The study by

Groot M. and Lee H. also has highlighted the variation in the selective sorting mechanisms during various pathologies, including chronic lung disease, immune response, neuroinflammation, diabetes mellitus, cancer, and heart disease, suggesting that the miRNAs are selectively sorted to address the needs of the parent cell.

The detailed schematic representation of the same has been depicted in Fig. 6.2. The sorting of miRNAs into exosomes is regulated by several sophisticated mechanisms. One of the most established models involves the recognition of specific sequence motifs within miRNAs, known as EXO motifs, such as GGAG or GGCU. These motifs are recognized by RNA-binding proteins like heterogeneous nuclear ribonucleoprotein *A2B1* (*hnRNPA2B1*), which, when sumoylated, show increased binding affinity to these miRNAs, facilitating their sorting into exosomes [81]. Another protein, Y-box protein 1 (*YBX1*), has been shown to selectively bind certain miRNAs and is critical for their exosomal packaging, especially under stress conditions such as hypoxia [72].

Another proposed mechanism involves the *Argonaute 2* (*AGO2*) protein, a core component of the RNA-induced silencing complex (*RISC*). Though *AGO2* is mainly cytoplasmic, its interaction with miRNAs destined for exosomes implies a potential role in sorting [24]. Additionally, the ceramide pathway mediated by *nSMase2* not only contributes to membrane budding but also appears to modulate the sorting of miRNAs through changes in membrane microdomains [37]. These mechanisms are not mutually exclusive and may act in concert or differentially across cell types and conditions.

Once secreted, exosomal miRNAs can be taken up by recipient cells through various modes, including endocytosis, micropinocytosis, direct fusion with the plasma membrane, or receptor-ligand interactions [60]. Upon delivery, the miRNAs can silence target genes and reprogram cellular behavior. This mechanism of action has profound implications in the context of cancer, particularly in its earliest stages.

So far, several aspects of miRNA, including their biochemical nature, their synthesis, their loading inside the exosomes, and the governing factors, etc., has been explored, but the main question remains unanswered that how these exosomes influence the regulation of cancer mRNAs and how these exosomal miRNAs are being used in early cancer detection. These questions will be addressed in the upcoming section of the chapter.

Exosomal miRNA-Based Regulation of Cancer mRNA

Exosomes provide a protective and targeted delivery mechanism for miRNA duplexes from one cell to another. This transfer often results in the posttranscriptional regulation of target mRNAs by binding to the complementary sequence, influencing cancer-related processes like proliferation, formation of the blood vessels, and developing a resistance against therapies [78] and immune suppression. Exosomal miRNAs can modulate the expression of genes involved in cell cycle regulation and apoptosis. For example, *miR-21* is at elevated levels in tumor-derived

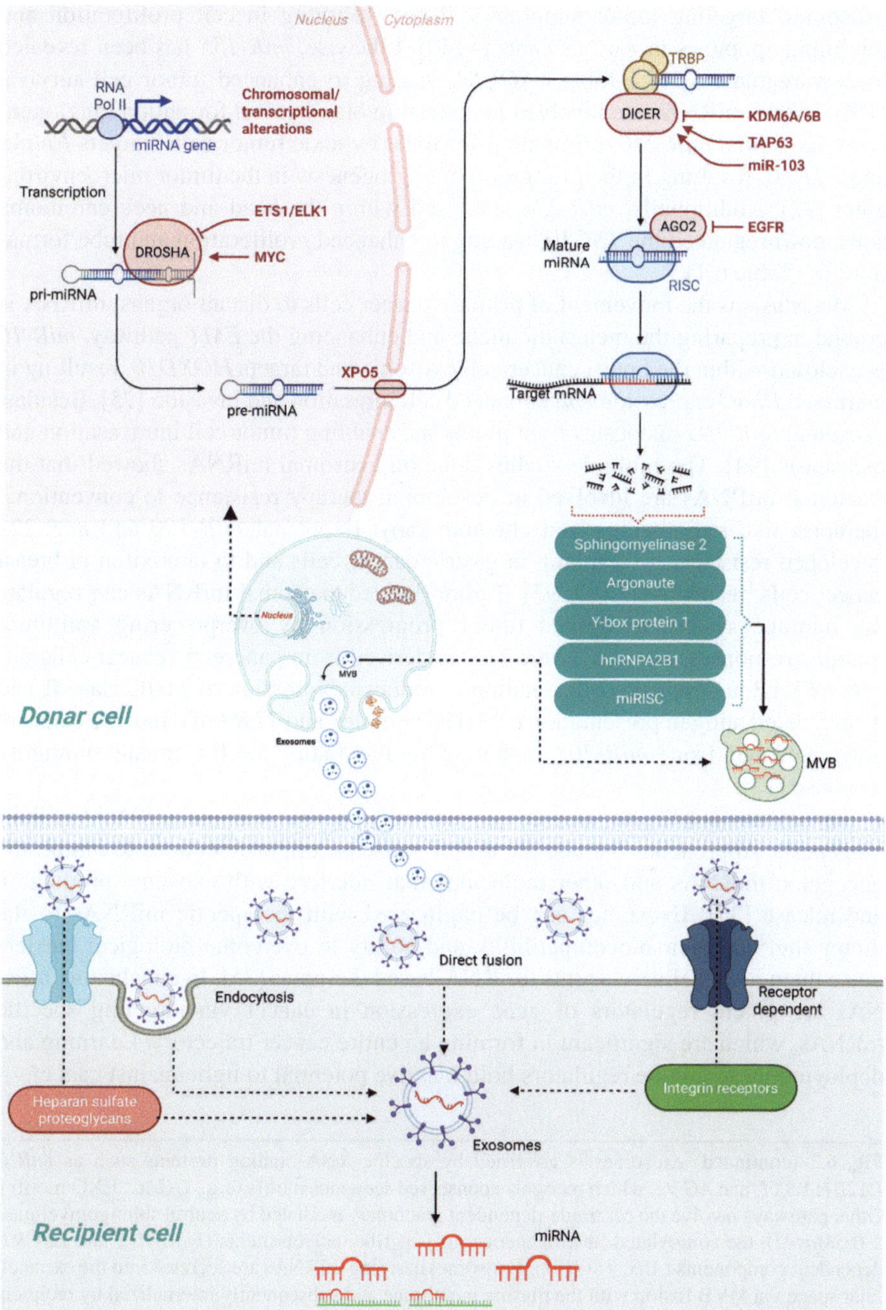

Fig. 6.2 Schematic representation of exosomal miRNA sorting and secretion mechanisms. MicroRNAs (miRNAs) are selectively packaged into intraluminal vesicles (ILVs) within multivesicular bodies (MVBs), which are later secreted as exosomes. The sorting of miRNAs into (continued)

exosomes, targeting tumor suppressor genes, resulting in cell proliferation and inhibiting apoptosis in various cancers [40]. Likewise, *miR-155* has been revealed to downregulate *STAT* inhibitor *SOCS1*, leading to enhanced tumor cell survival [16]. Certain miRNAs are involved in activating blood vessel formation (angiogenesis). Exosomal *miR-210*, originating from the hypoxic tumor cells, targets *EphA2* and *PTP1B*, resulting in the promotion of angiogenesis in the tumor microenvironment [92]. Additionally, *miR-23a*, enclosed within the head and neck carcinoma cells, downregulated the *TSG10*, leading to enhanced proliferation and tube formation [6] (Table 6.1).

Metastasis is the movement of primary cancer cells to distant organs. miRNA is crucial in preparing the metastatic niche and enhancing the *EMT* pathway. *miR-10* is enclosed within the breast cancer cell exosome and targets *HOXD10*, resulting in increased *RhoC* expression and enhanced cell migration and invasion [75]. Besides, exosomal *miR-105* dislocates tight junctions, enabling tumor cell intravasation and metastasis [94]. The multiple studies done on exosomal miRNAs showed that the exosomal miRNAs are involved in developing therapy resistance to conventional therapies like radiotherapy and chemotherapy. Exosomal *miR-100* and *miR-222* developed resistance to cisplatin in gastric cancer cells and to tamoxifen in breast cancer cells, respectively [83, 93]. Tumor-derived exosomal miRNAs can regulate the immune system to uphold tumor progression by overpowering antitumor immune responses, exosomal *miR-212-3p* derived from pancreatic cancer cells targets *RFXAP* in dendritic cells, leading to reduced expression of MHC class II and dysregulated antigen presentation on MHC complex and *TLR4* of dendritic cells are downregulated from *miR-203*, resulting in the fading of the innate immunity response [17].

Therapeutic implications are employed to disrupt the exosomal miRNAs or their biogenesis. Approaches include the use of antisense oligonucleotides to inhibit the oncogenic miRNAs and other molecules that interfere with exosome production and release [37]. Exosomes can be engineered with therapeutic miRNAs to the tumor site, and their biocompatibility and ability to overcome biological barriers make them ideal delivery agents for RNA-based therapies [35]. In conclusion, miRNAs are potent regulators of gene expression in cancer via targeting specific mRNAs, which are significant in forming an entire cancer trajectory. Learning and deploying these minute regulators hold massive potential to fight against cancer.

Fig. 6.2 (continued) exosomes is governed by specific RNA-binding proteins such as *hnRNPA2B1*, *YBX1*, and *AGO2*, which recognize conserved sequence motifs (e.g., *GGAG*, EXO motifs). Other pathways involve the ceramide-dependent machinery mediated by neutral sphingomyelinase 2 (*nSMase2*), the sumoylated heterogeneous nuclear ribonucleoproteins (*hnRNPs*), and *ESCRT*-dependent components (*Alix*, *TSG101*). Exosomes carrying miRNAs are secreted into the extracellular space via MVB fusion with the plasma membrane and subsequently internalized by recipient cells through endocytosis, phagocytosis, or direct fusion. Upon entry, exosomal miRNAs modulate gene expression in recipient cells by targeting complementary mRNAs, influencing numerous cellular processes including immune modulation, tumor progression, and therapy resistance. This figure highlights the precise regulatory checkpoints of miRNA packaging, transport, and uptake, illustrating the complexity and therapeutic potential of exosomal miRNA communication

Table 6.1 Exosomal miRNA and their role in cancer

miRNA	Targets	Biological role/ pathway	Associated disease	References
miR-21	Tumor microenvironment	Promotes proliferation, inhibits apoptosis (via *PTEN/AKT*)	Breast, lung, and colorectal cancer	[71]
miR-155	Endothelial, immune cells	Induces inflammation (via *NF-κB*)	Breast, prostate, lung, and colorectal cancer	[56]
miR-1246	Immune cells	Immune evasion, EMT regulation	Pancreatic ductal adenocarcinoma, oral squamous cell carcinoma, and hepatocellular carcinoma	[66]
miR-223	Cancer cells	Regulates invasion, proliferation, cell-to-cell communication	Hepatocellular carcinoma and breast cancer	[20]
miR-200a	Tumor microenvironment	EMT inhibition, associated with *CA-125*	Ovarian cancer	[70]
miR-200b	Tumor microenvironment	EMT inhibition	Associated with various cancers	[67]
miR-200c	Tumor microenvironment	EMT inhibition	Ovarian cancer	[65]
miR-373	Tumor microenvironment	Promotes tumor progression	Breast, medullary thyroid, gastric, and bladder cancer	[18]
miR-222	Breast cancer cells	Drug resistance via *PTEN* downregulation	Bladder cancer, breast cancer, and ovarian cancer	[12]
miR-145-3p	Colorectal cancer cells	5-FU resistance modulation	Osteosarcoma, breast cancer, and lung cancer	[25]
miR-4644	Endothelial cells	Promotes angiogenesis	Bladder cancer, hepatocellular carcinoma, colorectal cancer, and pancreatic cancer	[86]
miR-4772-3p	Tumor microenvironment	Prognostic biomarker for recurrence	Colon cancer	[47]
miR-1290	Tumor microenvironment	Prognostic marker promotes resistance	Colon, pancreatic, lung, and liver cancer	[85]
miR-320a-3p	Immune cells	Prognostic marker	Gastric cancer, breast cancer, colon cancer, lung cancer, and ovarian cancer	[28]
miR-1229	Tumor microenvironment	Promotes metastasis	Breast cancer, colorectal cancer, and gastric cancer	[30]
miR-150	Tumor microenvironment	Enhances tumor growth	Melanoma and papillary thyroid carcinoma	[13]
miR-146a	Immune cells	Promotes immune suppression	Cervical cancer	[45]
miR-1307-5p	Tumor microenvironment	Lymph node metastasis, tumor staging	Lung adenocarcinoma	[13]

Exosomal miRNA Biomarker Panels in Clinical Use

Exosomal microRNAs (exo-miRNAs) have emerged as promising biomarkers in oncology due to their stability and noninvasive nature. These small, noncoding RNAs are encapsulated within lipid bilayered extracellular vesicles called exosomes, which protect them from degradation by RNase. Their release into biological fluids like blood and urine makes them ideal candidates for liquid biopsies in cancer diagnostics. Several exo-miRNA panels have reached clinical validation phases, underlining their significance.

For instance, the ExoDx Prostate IntelliScore (EPI) is a noninvasive urine test that uses exosomal RNA signatures, including PCA3 and ERG, to evaluate the risk of high-grade prostate cancer. This assay has shown considerable improvement over PSA testing alone in identifying patients requiring biopsy [53].

Another notable development is the miRview mets test by Rosetta Genomics, which helps identify the tissue of origin in metastatic cancers. Though it primarily uses total miRNA profiling, research is pushing toward exosome-derived miRNA refinement for greater specificity.

Moreover, exosomal miRNA panels like *miR-21*, *miR-1246*, and *miR-1229* have been proposed as part of multi-marker panels for colorectal and breast cancer screening [26, 63]. These miRNAs regulate oncogenic pathways such as *PI3K/AKT* and *Wnt* signaling, thereby contributing to tumor progression (Table 6.2).

The ExoDx Lung (*ALK*) test, in development, utilizes plasma-derived exosomes to detect *ALK* rearrangements in non-small cell lung cancer (NSCLC), helping in therapy selection without requiring a tumor biopsy.

Table 6.2 Examples of exosomal miRNA-based diagnostic trials and applications

Cancer type	Diagnostic marker(s)	Sample type	Clinical utility/trial status
Prostate cancer	*ERG, PCA3* (ExoDx prostate, EPI)	Urine	FDA-authorized risk stratification tool; validated in multiple prospective studies [52, 79]
Thyroid carcinoma (PTC)	*miR-130a-3p, miR-29a*	Serum exosomes	Differentiates PTC from benign nodules; levels correlate with recurrence risk [22]
Colorectal cancer	*miR-92a*	Plasma exosomes	High sensitivity (~89%) and specificity (~70%) in distinguishing CRC vs controls [48]
Breast cancer	*miR-21, miR-155, miR-1246*	Plasma exosomes	Elevated in early- and late-stage disease; potential for monitoring response (sensitivity ≥70%) [88]
Clear cell renal carcinoma	*miR-146a & miR-126*	Plasma exosomes	Predictive of checkpoint immunotherapy response (AUC ~0.75) [33]

All listed examples are based on peer-reviewed studies or FDA-authorized assays published up to 2024. The diagnostic markers are exosomal miRNAs measured in noninvasive biofluids, demonstrating clinical performance in early detection, differential diagnosis, or therapeutic monitoring

Despite their promise, only a few exosomal miRNA panels have gained widespread clinical adoption, primarily due to technical and regulatory hurdles. Standardization in exosome isolation, RNA quantification, and data interpretation remains a significant challenge. However, ongoing advancements in microfluidics and high-throughput sequencing are addressing these limitations, paving the way for the integration of exosomal miRNA panels into routine oncology diagnostics [76].

Exosomal miRNA-Based Early Detection of Specific Cancer

The ability of exosomal miRNAs to reflect the tumor-specific molecular landscape makes them potent candidates for early cancer detection. Several miRNAs packaged in exosomes have shown high sensitivity and specificity in discriminating cancer patients from healthy individuals or those with benign conditions, across various cancer types.

In lung cancer, exo-*miR-21*, *miR-126*, and *miR-210* have demonstrated diagnostic potential in early-stage NSCLC. A study by Cazzoli et al. [9] showed that a panel of these exosomal miRNAs could distinguish early NSCLC patients from healthy controls with over 80% accuracy.

In breast cancer, exosomal *miR-1246* and *miR-21* are overexpressed in patients' serum and are linked with tumor aggressiveness. Notably, Hannafon et al. [27] reported that *miR-1246* levels in exosomes could discriminate early-stage cancer with a diagnostic accuracy exceeding 90%.

For colorectal cancer (CRC), Ogata-Kawata et al. [63] identified a panel of seven exo-miRNAs, including *miR-21*, *miR-23a*, and *miR-1246*, which could distinguish CRC patients from healthy individuals. These biomarkers correlate with tumor stage and metastasis, supporting their clinical relevance in early detection (Table 6.3).

In pancreatic cancer, a notoriously silent and aggressive malignancy, exo-*miR-196a* and *miR-1246* have emerged as potential early biomarkers. Melo et al. [54] demonstrated that these miRNAs could be detected in serum exosomes from patients months before radiological detection, suggesting their value in presymptomatic screening.

In hepatocellular carcinoma (HCC), exosomal *miR-21*, *miR-122*, and *miR-224* have shown promise. Sohn et al. [73] reported elevated levels of exo-*miR-21* in HCC patient plasma, with diagnostic potential comparable to or exceeding alpha-fetoprotein (AFP).

In ovarian cancer, exosomal *miR-200* family members and *miR-21* have demonstrated strong diagnostic utility. Meng et al. [55] found that exo-*miR-200a* and *miR-200c* levels were significantly elevated in early-stage ovarian cancer patients compared to healthy women.

Table 6.3 Exosomal miRNAs in early cancer detection

Cancer type	Exosomal miRNAs involved	Study type	Source (Biofluid)	References
Lung cancer	miR-21, miR-126, miR-210	Clinical/*in vitro*	Plasma	[9]
Breast cancer	miR-1246, miR-21	Clinical/*in vitro*	Serum	[27]
Colorectal cancer	miR-1246, miR-23a, miR-1246	Clinical	Serum	[63]
Pancreatic cancer	miR-196a, miR-1246	Clinical/*in vitro*	Serum	[54]
Hepatocellular carcinoma	miR-21, miR-122, miR-224	Clinical	Plasma	[73]
Ovarian cancer	miR-200a, miR-200c, miR-21	Clinical/*in vitro*	Serum	[55]

Collectively, these findings suggest that tumor-derived exosomal miRNAs are not only detectable in early cancer stages but also provide cancer-type specificity, making them indispensable in precision diagnostics.

Emerging Liquid Biopsy Platforms

Liquid biopsy is revolutionizing cancer diagnostics by offering noninvasive, real-time insights into tumor biology. Among the various analytes used in liquid biopsies are circulating tumor DNA (ctDNA), circulating tumor cells (CTCs), and extracellular vesicles (EVs). Exosomes have gained significant traction due to their high abundance, stability, and cell-specific molecular cargo, especially exosomal miRNAs.

Emerging platforms are focusing on optimizing exosome isolation and analysis to enhance diagnostic yield. One such approach involves microfluidics-based exosome capture using devices like ExoChip or ExoSearch, which use antibodies against tetraspanins (*CD63*, *CD9*, *CD81*) to selectively capture exosomes from biofluids [91]. These platforms enable rapid processing with minimal sample volumes, making them ideal for clinical applications.

Another advancement is a nano-plasmonic sensor integrated with surface-enhanced Raman scattering (SERS) to detect exosomal miRNAs directly without the need for RNA extraction. For example, the SERS-based EXoRNA platform allows for multiplexed detection of miRNAs with high sensitivity [87].

High-throughput sequencing and digital PCR have also been integrated with exosomal RNA analysis to quantify miRNAs from individual vesicles, increasing diagnostic specificity. Droplet digital PCR (ddPCR) is valuable in distinguishing single-nucleotide variants and miRNA expression changes in cancer.

Artificial intelligence (AI) and machine learning algorithms are being incorporated into these platforms for biomarker discovery. By analyzing large-scale miRNA expression profiles, these tools can identify miRNA panels most predictive of early-stage cancers, improving the accuracy of diagnosis.

Emerging companies like Exosome Diagnostics, Bio-Techne, and NanoView Biosciences are at the forefront of commercializing such platforms. Exosome Diagnostics' ExoDx platform is already used for prostate and lung cancer risk assessment, while Bio-Techne has developed exosome capture kits optimized for various biofluids.

Despite these advances, challenges such as inter-patient variability, standardization of isolation protocol, and regulatory hurdles remain. The development of integrated systems combining exosome capture, analysis, and data interpretation into one device is critical for mainstream adoption.

Conclusion

When it comes to cancer, the sooner it gets detected or diagnosed, the better the treatment outcome and the chances of survival. In the aforementioned sections, we have highlighted the conventional diagnostic tools currently in practice, their pros and cons, which might serve in shedding light on what needs to be done to fill the gap and to improve the existing strategies, either aiding in early cancer detection or in being failproof, accurate, and specific. Here, one such plausible diagnostic option has been discussed where the circulatory exosomal miRNA can be used as a biomarker for early cancer detection. The miRNA content of exosomes not only provides a fingerprint of the tumor's molecular landscape but also holds functional relevance. The enrichment of specific miRNAs in exosomes is dynamic and can change in response to therapy, making them excellent biomarkers for treatment monitoring and early relapse detection.

The field has also ventured into utilizing engineered exosomes as therapeutic delivery vehicles, where miRNAs can be artificially loaded into exosomes and delivered to target cells. These holds promise for precision medicine, especially for tumors located in anatomically challenging sites. Exosomes' intrinsic biocompatibility, low immunogenicity, and natural ability to cross biological barriers make them ideal vectors [2]. Clinical trials are already underway exploring the utility of exosomal miRNAs as both diagnostic tools and therapeutic agents.

Despite these advancements, several challenges remain. The standardization of exosome isolation methods, whether by ultracentrifugation, size exclusion chromatography, immunoaffinity, or microfluidic devices, remains a bottleneck [44]. Moreover, the sensitivity and reproducibility of miRNA quantification require optimization across platforms such as qRT-PCR, digital droplet PCR, and next-generation sequencing.

References

1. Abdelfattah AM, Park C, Choi MY (2014) Update on non-canonical microRNAs. Biomol Concepts 5(4):275–287. https://doi.org/10.1515/bmc-2014-0012
2. Alvarez-Erviti L, Seow Y, Yin H, Betts C, Lakhal S, Wood MJ (2011) Delivery of siRNA to the mouse brain by systemic injection of targeted exosomes. Nat Biotechnol 29(4):341–345. https://doi.org/10.1038/nbt.1807. Epub 2011 Mar 20
3. Anand U, Dey A, Chandel AKS, Sanyal R, Mishra A, Pandey DK, De Falco V, Upadhyay A, Kandimalla R, Chaudhary A, Dhanjal JK, Dewanjee S, Vallamkondu J, Pérez de la Lastra JM (2023) Cancer chemotherapy and beyond: current status, drug candidates, associated risks and progress in targeted therapeutics. Genes Dis 10(4):1367–1401. https://doi.org/10.1016/j.gendis.2022.02.007
4. Andreu Z, Yáñez-Mó M (2014) Tetraspanins in extracellular vesicle formation and function. Front Immunol 5:442. https://doi.org/10.3389/fimmu.2014.00442
5. Annese T, Tamma R, De Giorgis M, Ribatti D (2020) microRNAs biogenesis, functions and role in tumor angiogenesis. Front Oncol 10. https://doi.org/10.3389/fonc.2020.581007
6. Bao L, You B, Shi S, Shan Y, Zhang Q, Yue H, Zhang J, Zhang W, Shi Y, Liu Y, Wang X, Liu D, You Y (2018) Metastasis-associated miR-23a from nasopharyngeal carcinoma-derived exosomes mediates angiogenesis by repressing a novel target gene TSGA10. Oncogene 37(21):2873–2889. https://doi.org/10.1038/s41388-018-0183-6
7. Bray F, Laversanne M, Sung H, Ferlay J, Siegel RL, Soerjomataram I, Jemal A (2024) Global cancer statistics 2022: GLOBOCAN estimates of incidence and mortality worldwide for 36 cancers in 185 countries. CA Cancer J Clin 74(3):229–263. https://doi.org/10.3322/caac.21834
8. Bunggulawa EJ, Wang W, Yin T, Wang N, Durkan C, Wang Y, Wang G (2018) Recent advancements in the use of exosomes as drug delivery systems. J Nanobiotechnol 16(1):81. https://doi.org/10.1186/s12951-018-0403-9
9. Cazzoli R, Buttitta F, Di Nicola M, Malatesta S, Marchetti A, Rom WN, Pass HI (2013) microRNAs derived from circulating exosomes as noninvasive biomarkers for screening and diagnosing lung cancer. J Thorac Oncol 8(9):1156–1162. https://doi.org/10.1097/JTO.0b013e318299ac32
10. Chang T-C, Yu D, Lee Y-S, Wentzel EA, Arking DE, West KM, Dang CV, Thomas-Tikhonenko A, Mendell JT (2008) Widespread microRNA repression by Myc contributes to tumorigenesis. Nat Genet 40(1):43–50. https://doi.org/10.1038/ng.2007.30
11. Chen X, Gole J, Gore A, He Q, Lu M, Min J, Yuan Z, Yang X, Jiang Y, Zhang T, Suo C, Li X, Cheng L, Zhang Z, Niu H, Li Z, Xie Z, Shi H, Zhang X et al (2020) Non-invasive early detection of cancer four years before conventional diagnosis using a blood test. Nat Commun 11(1):3475. https://doi.org/10.1038/s41467-020-17316-z
12. Chen W-X, Wang D-D, Zhu B, Zhu Y-Z, Zheng L, Feng Z-Q, Qin X-H (2021) Exosomal miR-222 from adriamycin-resistant MCF-7 breast cancer cells promote macrophages M2 polarization via PTEN/Akt to induce tumor progression. Aging (Albany NY) 13(7):10415–10430. https://doi.org/10.18632/aging.202802
13. Cho O (2025) Plasma exosomal miR-150-3p, NMT2, and PRDM1 as predictive biomarkers of acute tumor response in patients with cervical cancer undergoing chemoradiotherapy. Am J Cancer Res 15(2):546–558. https://doi.org/10.62347/SPQY5709
14. Connal S, Cameron JM, Sala A, Brennan PM, Palmer DS, Palmer JD, Perlow H, Baker MJ (2023) Liquid biopsies: the future of cancer early detection. J Transl Med 21(1):118. https://doi.org/10.1186/s12967-023-03960-8
15. Dakal TC, Dhabhai B, Pant A, Moar K, Chaudhary K, Yadav V, Ranga V, Sharma NK, Kumar A, Maurya PK, Maciaczyk J, Schmidt-Wolf IGH, Sharma A (2020) Oncogenes and tumor suppressor genes: functions and roles in cancers. MedComm. 2024 May 31 5(6):e582. https://doi.org/10.1002/mco2.582
16. Deng T, Zhang H, Yang H, Wang H, Bai M, Sun W, Wang X, Si Y, Ning T, Zhang L, Li H, Ge S, Liu R, Lin D, Li S, Ying G, Ba Y (2020) Exosome miR-155 derived from gastric carcinoma

promotes angiogenesis by targeting the c-MYB/VEGF Axis of endothelial cells. Mol Ther Nucleic Acids 19:1449–1459. https://doi.org/10.1016/j.omtn.2020.01.024

17. Ding G, Zhou L, Qian Y, Fu M, Chen J, Chen J, Xiang J, Wu Z, Jiang G, Cao L (2015) Pancreatic cancer-derived exosomes transfer miRNAs to dendritic cells and inhibit RFXAP expression via miR-212-3p. Oncotarget 6(30):29877–29888
18. Eichelser C, Stückrath I, Müller V, Milde-Langosch K, Wikman H, Pantel K, Schwarzenbach H (2014) Increased serum levels of circulating exosomal microRNA-373 in receptor-negative breast cancer patients. Oncotarget 5(20):9650–9663. https://doi.org/10.18632/oncotarget.2520
19. Fares J, Fares MY, Khachfe HH, Salhab HA, Fares Y (2020) Molecular principles of metastasis: a hallmark of cancer revisited. Signal Transduct Target Ther 5(1):1–17. https://doi.org/10.1038/s41392-020-0134-x
20. Favero A, Segatto I, Perin T, Belletti B (2021) The many facets of miR-223 in cancer: Oncosuppressor, oncogenic driver, therapeutic target, and biomarker of response. Wiley Interdiscipl Rev RNA 12(6):e1659. https://doi.org/10.1002/wrna.1659
21. Ferguson JL, Turner SP (2018) Bone cancer: diagnosis and treatment principles. Am Fam Physician 98(4):205–213
22. Ghosh S, Rajendran RL, Mahajan AA, Chowdhury A, Bera A, Guha S, Chakraborty K, Chowdhury R, Paul A, Jha S, Dey A, Dubey A, Gorai S, Das P, Hong CM, Krishnan A, Gangadaran P, Ahn B-C (2024) Harnessing exosomes as cancer biomarkers in clinical oncology. Cancer Cell Int 24(1):278. https://doi.org/10.1186/s12935-024-03464-5
23. Goldie BJ, Dun MD, Lin M, Smith ND, Verrills NM, Dayas CV, Cairns MJ (2014) Activity-associated miRNA are packaged in Map 1b-enriched exosomes released from depolarized neurons. Nucleic Acids Res 42(14):9195–9208. https://doi.org/10.1093/nar/gku594
24. Guduric-Fuchs J, O'Connor A, Camp B, O'Neill CL, Medina RJ, Simpson DA (2012) Selective extracellular vesiclemediated export of an overlapping set of microRNAs from multiple cell types. BMC Genomics 13:357. https://doi.org/10.1186/1471-2164-13-357
25. Hang W, Feng Y, Sang Z, Yang Y, Zhu Y, Huang Q, Xi X (2019) Downregulation of miR-145-5p in cancer cells and their derived exosomes may contribute to the development of ovarian cancer by targeting CT. Int J Mol Med 43(1):256–266. https://doi.org/10.3892/ijmm.2018.3958
26. Hannafon BN, Ding W-Q (2013) Intercellular communication by exosome-derived microRNAs in cancer. Int J Mol Sci 14(7):14240–14269. https://doi.org/10.3390/ijms140714240
27. Hannafon BN, Trigoso YD, Calloway CL, Zhao YD, Lum DH, Welm AL, Zhao ZJ, Blick KE, Dooley WC, Ding WQ (2016) Plasma exosome microRNAs are indicative of breast cancer. Breast Cancer Res 18(1):90. https://doi.org/10.1186/s13058-016-0753-x
28. Hao X, Xin R, Dong W (2020) Decreased serum exosomal miR-320a expression is an unfavorable prognostic factor in patients with hepatocellular carcinoma. J Int Med Res 48(4):0300060519896144. https://doi.org/10.1177/0300060519896144
29. Henry NL, Hayes DF (2012) Cancer biomarkers. Mol Oncol 6(2):140–146. https://doi.org/10.1016/j.molonc.2012.01.010
30. Hu H-Y, Yu C-H, Zhang H-H, Zhang S-Z, Yu W-Y, Yang Y, Chen Q (2019) Exosomal miR-1229 derived from colorectal cancer cells promotes angiogenesis by targeting HIPK2. Int J Biol Macromol 132:470–477. https://doi.org/10.1016/j.ijbiomac.2019.03.221
31. Hurley JH (2008) ESCRT complexes and the biogenesis of multivesicular bodies. Curr Opin Cell Biol 20(1):4–11. https://doi.org/10.1016/j.ceb.2007.12.002. Epub 2008 Jan 28
32. Iftikhar, H, Carney GE (2016) Evidence and potential in vivo functions for biofluid miRNAs: from expression profiling to functional testing—Iftikhar—2016—bioessays—Wiley Online Library, March 2. https://onlinelibrary.wiley.com/doi/10.1002/bies.201500130
33. Ivanova E, Asadullina D, Gilyazova G, Rakhimov R, Izmailov A, Pavlov V, Khusnutdinova E, Gilyazova I (2023) Exosomal MicroRNA levels associated with immune checkpoint inhibitor therapy in clear cell renal cell carcinoma. Biomedicine 11(3):801. https://doi.org/10.3390/biomedicines11030801

34. Johnstone RM, Adam M, Hammond JR, Orr L, Turbide C (1987) Vesicle formation during reticulocyte maturation. Association of plasma membrane activities with released vesicles (exosomes). J Biol Chem 262(19):9412–9420
35. Kalluri R, LeBleu VS (2020) The biology, function, and biomedical applications of exosomes. Science 367(6478):eaau6977. https://doi.org/10.1126/science.aau6977
36. Kapoor KS, Harris K, Arian KA, Ma L, Schueng Zancanela B, Church KA, McAndrews KM, Kalluri R (2024) High throughput and rapid isolation of extracellular vesicles and exosomes with purity using size exclusion liquid chromatography. Bioact Mater 40:683–695. https://doi.org/10.1016/j.bioactmat.2024.08.002
37. Kosaka N, Iguchi H, Yoshioka Y, Takeshita F, Matsuki Y, Ochiya T (2010) Secretory mechanisms and intercellular transfer of microRNAs in living cells. J Biol Chem 285(23):17442–17452. https://doi.org/10.1074/jbc.M110.107821
38. Kotmakçı M, Akbaba GE, Kotmakçı M, Akbaba GE (2017) Exosome isolation: is there an optimal method with regard to diagnosis or treatment? In: Novel implications of exosomes in diagnosis and treatment of cancer and infectious diseases. IntechOpen. https://doi.org/10.5772/intechopen.69407
39. Kowal J, Tkach M, Théry C (2014) Biogenesis and secretion of exosomes. Curr Opin Cell Biol 29:116–125. https://doi.org/10.1016/j.ceb.2014.05.004. Epub 2014 Jun 22
40. Kumarswamy R, Thum T (2013) Non-coding RNAs in cardiac remodeling and heart failure. Circ Res 113(6):676–689. https://doi.org/10.1161/CIRCRESAHA.113.300226
41. Lao TD, Le TAH (2020) MicroRNAs: biogenesis, functions and potential biomarkers for early screening, prognosis and therapeutic molecular monitoring of nasopharyngeal carcinoma. Processes 8(8):Article 8. https://doi.org/10.3390/pr8080966
42. Lee Y, Kim M, Han J, Yeom K-H, Lee S, Baek SH, Kim VN (2004) MicroRNA genes are transcribed by RNA polymerase II. EMBO J 23(20):4051–4060. https://doi.org/10.1038/sj.emboj.7600385
43. Li J, Zhang Y (2019) Current experimental strategies for intracellular target identification of microRNA. ExRNA 1(1):6. https://doi.org/10.1186/s41544-018-0002-9
44. Li P, Kaslan M, Lee SH, Yao J, Gao Z (2017) Progress in exosome isolation techniques. Theranostics 7(3):789–804. https://doi.org/10.7150/thno.18133
45. Li L-J, Gu Y-J, Wang L-Q, Wan W, Wang H-W, Yang X-N, Ma L-L, Yang L-H, Meng Z-H (2021) Serum exosomal microRNA-146a as a novel diagnostic biomarker for acute coronary syndrome. J Thorac Dis 13(5):3105–3114. https://doi.org/10.21037/jtd-21-609
46. Lin S, Gregory RI (2015) MicroRNA biogenesis pathways in cancer. Nat Rev Cancer 15(6):321–333. https://doi.org/10.1038/nrc3932
47. Liu C, Eng C, Shen J, Lu Y, Takata Y, Mehdizadeh A, Chang GJ, Rodriguez-Bigas MA, Li Y, Chang P, Mao Y, Hassan MM, Wang F, Li D (2016) Serum exosomal miR-4772-3p is a predictor of tumor recurrence in stage II and III colon cancer. Oncotarget 7(46):76250–76260. https://doi.org/10.18632/oncotarget.12841
48. Liu J, Yoo J, Ho JY, Jung Y, Lee S, Hur SY, Choi YJ (2021) Plasma-derived exosomal miR-4732-5p is a promising noninvasive diagnostic biomarker for epithelial ovarian cancer. J Ovarian Res 14(1):59. https://doi.org/10.1186/s13048-021-00814-z
49. Liu B, Zhou H, Tan L, Siu KTH, Guan X-Y (2024) Exploring treatment options in cancer: tumor treatment strategies. Signal Transduct Target Ther 9(1):175. https://doi.org/10.1038/s41392-024-01856-7
50. Macfarlane L-A, Murphy PR (2010) MicroRNA: biogenesis, function and role in cancer. Curr Genomics 11(7):537–561. https://doi.org/10.2174/138920210793175895
51. Makarova JA, Shkurnikov MU, Wicklein D, Lange T, Samatov TR, Turchinovich AA, Tonevitsky AG (2016) Intracellular and extracellular microRNA: an update on localization and biological role. Prog Histochem Cytochem 51(3–4):33–49. https://doi.org/10.1016/j.proghi.2016.06.001
52. McKiernan J, Donovan MJ, O'Neill V, Bentink S, Noerholm M, Belzer S, Skog J, Kattan MW, Partin A, Andriole G, Brown G, Wei JT, Thompson IM, Carroll P (2016) A novel urine

exosome gene expression assay to predict high-grade prostate cancer at initial biopsy. JAMA Oncol 2(7):882–889. https://doi.org/10.1001/jamaoncol.2016.0097
53. McKiernan J, Donovan MJ, Margolis E, Partin A, Carter B, Brown G, Torkler P, Noerholm M, Skog J, Shore N, Andriole G, Thompson I, Carroll P (2018) A prospective adaptive utility trial to validate performance of a novel urine exosome gene expression assay to predict high-grade prostate cancer in patients with prostate-specific antigen 2-10ng/ml at initial biopsy. Eur Urol 74(6):731–738. https://doi.org/10.1016/j.eururo.2018.08.019
54. Melo SA, Luecke LB, Kahlert C, Fernandez AF, Gammon ST, Kaye J, LeBleu VS, Mittendorf EA, Weitz J, Rahbari N, Reissfelder C, Pilarsky C, Fraga MF, Piwnica-Worms D, Kalluri R (2015) Glypican-1 identifies cancer exosomes and detects early pancreatic cancer. Nature 523(7559):177–182. https://doi.org/10.1038/nature14581
55. Meng X, Müller V, Milde-Langosch K, Trillsch F, Pantel K, Schwarzenbach H (2016) Diagnostic and prognostic relevance of circulating exosomal miR-373, miR-200a, miR-200b and miR-200c in patients with epithelial ovarian cancer. Oncotarget 7(13):16923–16935. https://doi.org/10.18632/oncotarget.7850
56. Mikamori M, Yamada D, Eguchi H, Hasegawa S, Kishimoto T, Tomimaru Y, Asaoka T, Noda T, Wada H, Kawamoto K, Gotoh K, Takeda Y, Tanemura M, Mori M, Doki Y (2017) MicroRNA-155 controls exosome synthesis and promotes gemcitabine resistance in pancreatic ductal adenocarcinoma. Sci Rep 7(1):42339. https://doi.org/10.1038/srep42339
57. Miller KD, Nogueira L, Devasia T, Mariotto AB, Yabroff KR, Jemal A, Kramer J, Siegel RL (2022) Cancer treatment and survivorship statistics, 2022. CA Cancer J Clin 72(5):409–436. https://doi.org/10.3322/caac.21731
58. Ming Z, Lim SY, Rizos H (2020) Genetic alterations in the INK4a/ARF locus: effects on melanoma development and progression. Biomolecules 10(10):1447. https://doi.org/10.3390/biom10101447
59. Miyoshi K, Miyoshi T, Siomi H (2010) Many ways to generate microRNA-like small RNAs: non-canonical pathways for microRNA production. Mol Gen Genomics 284(2):95–103. https://doi.org/10.1007/s00438-010-0556-1
60. Mulcahy LA, Pink RC, Carter DR (2014) Routes and mechanisms of extracellular vesicle uptake. J Extracell Vesicles 3. https://doi.org/10.3402/jev.v3.24641
61. Nagpal M, Singh S, Singh P, Chauhan P, Zaidi MA (2016) Tumor markers: a diagnostic tool. Natl J Maxillofac Surg 7(1):17–20. https://doi.org/10.4103/0975-5950.196135
62. O'Brien J, Hayder H, Zayed Y, Peng C (2018) Overview of MicroRNA biogenesis, mechanisms of actions, and circulation. Front Endocrinol 9:402. https://doi.org/10.3389/fendo.2018.00402
63. Ogata-Kawata H, Izumiya M, Kurioka D, Honma Y, Yamada Y, Furuta K, Gunji T, Ohta H, Okamoto H, Sonoda H, Watanabe M, Nakagama H, Yokota J, Kohno T, Tsuchiya N (2014) Circulating exosomal microRNAs as biomarkers of colon cancer. PLoS One 9(4):e92921. https://doi.org/10.1371/journal.pone.0092921
64. Patel GK, Khan MA, Zubair H, Srivastava SK, Khushman M, Singh S, Singh AP (2019) Comparative analysis of exosome isolation methods using culture supernatant for optimum yield, purity and downstream applications. Sci Rep 9:5335. https://doi.org/10.1038/s41598-019-41800-2
65. Qiao P, Du H, Guo X, Yu M, Zhang C, Shi Y (2024) Serum exosomal miR-200c is a potential diagnostic biomarker for breast cancer. Biomarkers 29(7):419–426. https://doi.org/10.1080/1354750X.2024.2406520
66. Qiu W, Guo X, Li B, Wang J, Qi Y, Chen Z, Zhao R, Deng L, Qian M, Wang S, Zhang Z, Guo Q, Zhang S, Pan Z, Zhao S, Xue H, Li G (2021) Exosomal miR-1246 from glioma patient body fluids drives the differentiation and activation of myeloid-derived suppressor cells. Mol Ther 29(12):3449–3464. https://doi.org/10.1016/j.ymthe.2021.06.023
67. Reese M, Flammang I, Yang Z, Dhayat SA (2020) Potential of exosomal microRNA-200b as liquid biopsy marker in pancreatic ductal adenocarcinoma. Cancer 12(1):197. https://doi.org/10.3390/cancers12010197

68. Sathishkumar K, Chaturvedi M, Das P, Stephen S, Mathur P (2022) Cancer incidence estimates for 2022 & projection for 2025: result from National Cancer Registry Programme, India. Indian J Med Res 156(4–5):598–607. https://doi.org/10.4103/ijmr.ijmr_1821_22
69. Shang R, Lee S, Senavirathne G, Lai EC (2023) microRNAs in action: biogenesis, function and regulation. Nat Rev Genet 24(12):816–833. https://doi.org/10.1038/s41576-023-00611-y
70. Shen L, Chen G, Xia Q, Shao S, Fang H (2019) Exosomal miR-200 family as serum biomarkers for early detection and prognostic prediction of cholangiocarcinoma. Int J Clin Exp Pathol 12(10):3870–3876
71. Shi R, Wang P-Y, Li X-Y, Chen J-X, Li Y, Zhang X-Z, Zhang C-G, Jiang T, Li W-B, Ding W, Cheng S-J (2015) Exosomal levels of miRNA-21 from cerebrospinal fluids associated with poor prognosis and tumor recurrence of glioma patients. Oncotarget 6(29):26971–26981. https://doi.org/10.18632/oncotarget.4699
72. Shurtleff MJ, Temoche-Diaz MM, Karfilis KV, Ri S, Schekman R (2016) Y-box protein 1 is required to sort microRNAs into exosomes in cells and in a cell-free reaction. elife 5:e19276. https://doi.org/10.7554/eLife.19276
73. Sohn W, Kim J, Kang SH, Yang SR, Cho J-Y, Cho HC, Shim SG, Paik Y-H (2015) Serum exosomal microRNAs as novel biomarkers for hepatocellular carcinoma. Exp Mol Med 47(9):e184. https://doi.org/10.1038/emm.2015.68
74. Stuffers S, Sem Wegner C, Stenmark H, Brech A (2009) Multivesicular endosome biogenesis in the absence of ESCRTs. Traffic 10(7):925–937. https://doi.org/10.1111/j.1600-0854.2009.00920.x. Epub 2009 Apr 10
75. Swarbrick A (2008) Review of: tumour invasion and metastasis initiated by microRNA-10b in breast cancer. Breast Cancer Online 11(1):e2. https://doi.org/10.1017/S1470903108006688
76. Théry C, Witwer KW, Aikawa E, Alcaraz MJ, Anderson JD, Andriantsitohaina R, Antoniou A, Arab T, Archer F, Atkin-Smith GK, Ayre DC, Bach J-M, Bachurski D, Baharvand H, Balaj L, Baldacchino S, Bauer NN, Baxter AA, Bebawy M et al (2018) Minimal information for studies of extracellular vesicles 2018 (MISEV2018): a position statement of the International Society for Extracellular Vesicles and update of the MISEV2014 guidelines. J Extracell Vesicles 7(1):1535750. https://doi.org/10.1080/20013078.2018.1535750
77. Trajkovic K, Hsu C, Chiantia S, Rajendran L, Wenzel D, Wieland F, Schwille P, Brügger B, Simons M (2008) Ceramide triggers budding of exosome vesicles into multivesicular endosomes. Science 319(5867):1244–1247. https://doi.org/10.1126/science.1153124
78. Turchinovich A, Drapkina O, Tonevitsky A (2019) Transcriptome of extracellular vesicles: state-of-the-art. Front Immunol 10. https://doi.org/10.3389/fimmu.2019.00202
79. Tutrone R, Donovan MJ, Torkler P, Tadigotla V, McLain T, Noerholm M, Skog J, McKiernan J (2020) Clinical utility of the exosome based ExoDx prostate(IntelliScore) EPI test in men presenting for initial biopsy with a PSA 2-10 ng/mL. Prostate Cancer Prostatic Dis 23(4):607–614. https://doi.org/10.1038/s41391-020-0237-z
80. Valadi H, Ekström K, Bossios A, Sjöstrand M, Lee JJ, Lötvall JO (2007) Exosome-mediated transfer of mRNAs and microRNAs is a novel mechanism of genetic exchange between cells. Nat Cell Biol 9(6):654–659. https://doi.org/10.1038/ncb1596
81. Villarroya-Beltri C, Gutiérrez-Vázquez C, Sánchez-Cabo F, Pérez-Hernández D, Vázquez J, Martin-Cofreces N, Martinez-Herrera DJ, Pascual-Montano A, Mittelbrunn M, Sánchez-Madrid F (2013) Sumoylated hnRNPA2B1 controls the sorting of miRNAs into exosomes through binding to specific motifs. Nat Commun 4:2980. https://doi.org/10.1038/ncomms3980
82. Viswanathan SR, Daley GQ, Gregory RI (2008) Selective blockade of microRNA processing by Lin28. Science (New York, N Y) 320(5872):97–100. https://doi.org/10.1126/science.1154040
83. Wei Y, Lai X, Yu S, Chen S, Ma Y, Zhang Y, Li H, Zhu X, Yao L, Zhang J (2014) Exosomal miR-221/222 enhances tamoxifen resistance in recipient ER-positive breast cancer cells. Breast Cancer Res Treat 147(2):423–431. https://doi.org/10.1007/s10549-014-3037-0
84. Winter J, Jung S, Keller S, Gregory RI, Diederichs S (2009) Many roads to maturity: microRNA biogenesis pathways and their regulation. Nat Cell Biol 11(3):228–234. https://doi.org/10.1038/ncb0309-228

85. Wu Y, Wei J, Zhang W, Xie M, Wang X, Xu J (2020) Serum Exosomal miR-1290 is a potential biomarker for lung adenocarcinoma. Onco Targets Ther 13:7809–7818. https://doi.org/10.2147/OTT.S263934
86. Yan L, Li Q, Sun K, Jiang F (2020) MiR-4644 is upregulated in plasma exosomes of bladder cancer patients and promotes bladder cancer progression by targeting UBIAD1. Am J Transl Res 12(10):6277–6289
87. Yoshioka Y, Kosaka N, Konishi Y, Ohta H, Okamoto H, Sonoda H, Nonaka R, Yamamoto H, Ishii H, Mori M, Furuta K, Nakajima T, Hayashi H, Sugisaki H, Higashimoto H, Kato T, Takeshita F, Ochiya T (2014) Ultra-sensitive liquid biopsy of circulating extracellular vesicles using ExoScreen. Nat Commun 5(1):3591. https://doi.org/10.1038/ncomms4591
88. Youssef E, Palmer D, Fletcher B, Vaughn R (2025) Exosomes in precision oncology and beyond: from bench to bedside in diagnostics and therapeutics. Cancer 17(6):940. https://doi.org/10.3390/cancers17060940
89. Yu L-L, Zhu J, Liu J-X, Jiang F, Ni W-K, Qu L-S, Ni R-Z, Lu C-H, Xiao M-B (2018) A comparison of traditional and novel methods for the separation of exosomes from human samples. Biomed Res Int 2018(1):3634563. https://doi.org/10.1155/2018/3634563
90. Zafar A, Khatoon S, Khan MJ, Abu J, Naeem A (2025) Advancements and limitations in traditional anti-cancer therapies: a comprehensive review of surgery, chemotherapy, radiation therapy, and hormonal therapy. Discov Oncol 16(1):607. https://doi.org/10.1007/s12672-025-02198-8
91. Zhang P, He M, Zeng Y (2016) Ultrasensitive microfluidic analysis of circulating exosomes using a nanostructured graphene oxide/polydopamine coating. Lab Chip 16(16):3033–3042. https://doi.org/10.1039/c6lc00279j
92. Zhang S, Chu WC, Lai RC, Lim SK, Hui JHP, Toh WS (2016) Exosomes derived from human embryonic mesenchymal stem cells promote osteochondral regeneration. Osteoarthr Cartil 24(12):2135–2140. https://doi.org/10.1016/j.joca.2016.06.022
93. Zheng P, Chen L, Yuan X, Luo Q, Liu Y, Xie G, Ma Y, Shen L (2017) Exosomal transfer of tumor-associated macrophage-derived miR-21 confers cisplatin resistance in gastric cancer cells. J Exp Clin Cancer Res 36(1):53. https://doi.org/10.1186/s13046-017-0528-y
94. Zhou W, Fong MY, Min Y, Somlo G, Liu L, Palomares MR, Yu Y, Chow A, O'Connor STF, Chin AR, Yen Y, Wang Y, Marcusson EG, Chu P, Wu J, Wu X, Li AX, Li Z, Gao H et al (2014) Cancer-secreted miR-105 destroys vascular endothelial barriers to promote metastasis. Cancer Cell 25(4):501–515. https://doi.org/10.1016/j.ccr.2014.03.007

Chapter 7
Advanced Bioinformatics and Machine Learning Methods in the Prediction of miRNAs and their Targets in Cancer

Mursaleen Baba and Habeeb Shaik Mohideen

Abstract MicroRNAs (miRNAs) are non-coding genes that are known to control gene expression by binding with messenger RNAs (mRNAs), mostly at the post-transcriptional level. The ability to fine-tune the regulation of gene expression has placed miRNAs at the centre of cancer research, where they have been reported to affect processes like tumour growth, metastasis, apoptosis, and genome stability. Due to this, understanding how miRNAs behave in cancer has gained great importance for identifying potential biomarkers and developing more precise therapies. In this chapter, we take a closer look at how various computational methods, especially those involving machine learning (ML), are enhancing our ability to identify miRNAs and predict their target genes. It starts by describing the biology of the miRNAs and how variations in expression levels contribute to cancers. Traditionally used tools like TargetScan, miRanda, and RNAhybrid are covered, along with their limitations of accuracy and consistency.

Further, we focus on ML, which offers a more versatile approach by integrating large biological data, including sequence patterns, structural features, and gene expression profiles. Different ML models, ranging from supervised to deep learning, have been discussed for their strengths and use cases in bioinformatics, genomics, and cancer research. Recent studies applying ML to real-world cancer datasets further demonstrate its translational potential in identifying clinically relevant miRNAs and their gene targets. Though promising, ML-based methods are prone to perennial problems, such as dealing with high-dimensional data, missing values, and, most importantly, overfitting, which is associated with small sample sizes. This chapter highlights the importance of robust feature choice, stable model assessment, and interdisciplinary collaboration towards creating models that are not only predictive but also biologically relevant.

M. Baba · H. S. Mohideen (✉)
Bioinformatics and Integrative Omics Lab, Department of Genetic Engineering, School of Bioengineering, College of Engineering and Technology, SRM Institute of Science and Technology, Kattankulathur, Chengalpattu, Tamil Nadu, India

S. Pathak, A. Banerjee (eds.), *MicroRNAs: Role in Cancer Progression, Diagnostics and Therapy*, Advances in Biochemistry in Health and Disease 34,
https://doi.org/10.1007/978-3-032-08824-6_7

Keywords MicroRNAs (miRNAs) · Cancer biomarkers · Machine learning (ML) · Target prediction · Computational biology

Biological Significance of miRNAs in Cancer

miRNAs are responsible for controlling gene expression. Translational control of target messenger RNAs (mRNAs) is achieved by miRNAs binding to complementary sequences in their 3′ untranslated regions (3′ UTRs) [1]. miRNA biogenesis is a highly controlled multi-step process that begins with RNA polymerase II/III transcription to create primary miRNAs (pri-miRNAs). The nuclear RNase III enzyme Drosha and its partner DGCR8 subsequently cleave the pri-miRNAs to produce precursor miRNAs (pre-miRNAs) [2]. Pre-miRNAs are transferred from the nucleus to the cytoplasm by Exportin-5 and then cleaved by Dicer to form mature miRNA duplexes. One strand of the duplex is integrated into the RNA-induced silencing complex (RISC), which targets the miRNA at its target mRNA [2].

In cancer biology, miRNAs are key regulators of oncogenes and tumour suppressor genes. Dysregulation of miRNAs can induce tumourigenesis by multiple mechanisms, such as stimulating invasion, metastasis, suppressing apoptosis and genome stability, and cell proliferation [3]. For example, oncogenic miRNAs (oncomiRs) like the miR-17 ~ 92 cluster are overexpressed in a subset of cancers and promote the progression of cancer. In contrast, tumour-inhibitory miRNAs such as the let-7 family repress tumour-promoting oncogenes such as MYC and RAS family members to suppress tumour development [4]. Moreover, global downregulation of miRNA biogenesis machinery components, including DROSHA and DICER1, has also been found to correlate with mutations in rare cancers and unfavourable clinical outcomes [5]. This underlines the two-way contribution of miRNAs to cancer development and their promise as diagnostic markers and therapeutic targets.

Traditional Bioinformatics Approaches for miRNA Prediction

Sequence-Based Homology Analysis

Homology analysis of the sequences is one of the most basic methods for predicting miRNAs as well as their associated genes or targets. This approach uses the miRNA sequence conservation across species. It postulates that functional miRNAs are retained within evolutionary lineages because of their fundamental roles in the gene expression machinery. Researchers contend that by comparing genomic sequences of different organisms, it is possible to find regions adjacent to already known miRNAs and identify them as putative novel miRNA candidates. This method is especially useful for designing miRNAs that are conserved in coding sequences of various species because those are bound to have some biological importance [6].

Homology-based methods tend to rely on databases like miRBase [7], which stores known sequences of miRNAs. In these databases, researchers align the query sequences to find regions with a high level of sequence conservation to the already known miRNA sequences. On the other hand, such a method is likely to overlook some of the more advanced and less conserved or specific to the species considered miRNAs, which may be equally important for gene-regulating devices. Moreover, the lack of innovative approaches in terms of sequence divergence leading to novel discoveries of miRNAs hinders these methods [7].

Thermodynamic Stability and Secondary Structure Prediction

Predicting miRNA requires assessing the thermodynamic stability and the secondary structure of pre-miRNAs. Pre-miRNAs are processed further into mature miRNAs, and generally, they exist within a stem-loop structure. The stem loops need to be stable for the miRNA biogenesis machinery to recognise and process them. RNAfold and Mfold are tools that predict the secondary structure of RNA sequences and estimate their thermodynamic stability, which can help delineate potential candidates for pre-miRNAs [8, 9].

The underlying explanation supporting this method is that true pre-miRNAs are likely to have specific characteristics in their stem-loop structures, like a remarkable degree of base pairing and exceptionally low free energy. These represent specific structural features that researchers evaluate to separate real from pseudo pre-miRNA candidates. Even so, this approach can be misled by the sequences forming stable structures that are not processed into miRNAs. Often, a combination of structural analysis with other predictive approaches yields higher precision.

Feature Selection for miRNA Prediction

Sequence-Based Features

Sequence-based features are primary elements in predicting miRNAs, reflecting intrinsic aspects of miRNAs and their target mRNAs. They can include nucleotide composition, dinucleotide counts or frequencies, and sequence motifs. The most relevant motifs for miRNA specificity are found in their seed region, which includes nucleotides 2 to 8 at the 5′ end of the miRNA and is often involved in target recognition and binding. Therefore, predictive algorithms would normally examine the level of complementary base pairing in the seed region and an mRNA target's 3′ untranslated region (3′ UTR), which could be impacted by the methods used to set output. A common feature across predicted sites is evolutionary conservation of both seed region/mRNA 3′ UTR interaction and binding site conservation across the taxonomy of other organisms. Sequences that are conserved across species are more reliable sequences, as they tend to represent broader functional classes of miRNA targets.

More sophisticated approaches, like miPIE, leverage sequence-based features as well as expressions for obtaining better predictions [10]. In the methods of miPIE, it combines different metrices, which include dinucleotide composition, conservation, and transcriptomic patterns based on datasets obtained from next-generation sequencing (NGS) [10]. In the case of MBSTAR and other similar methods, hundreds of features representing different sequence factors, inherently like accumulation of patterns of nucleotide conversions, representative of minor changes to the sequence, may help add more sophistication to the output of the prediction [11]. There is no question that most of the predictions will rely primarily on sequence-based features to help identify plausible miRNAs and miRNA targets; however, care must be taken to select the optimal features to avoid redundancy or overfitting.

Structural Descriptors

Structural descriptors evaluate the secondary structure and thermodynamic stability of precursor miRNAs (pre-miRNAs) that are essential during maturation into mature miRNAs. "True" pre-miRNAs have well-defined, stable stem-loop structures that will become modules used by the biogenic machinery. For example, from algorithms such as RNAfold and Mfold, we can compute free energy values and conclude that the lower the free energy values, the more thermodynamically favourable the structure is. The structural descriptors can also involve the size of loops, the number and length of bulges, and the length of stem portions of the hairpin structure.

In addition, the accessibility of binding sites on target mRNAs is an important structural feature. Although accessibility does not have structural properties per se, because of the local secondary structures in mRNAs, it may prove if it is exposed or buried within the folded structures. PITA, for example, considers accessibility based on energy differences between binding and the costs of accessibility [12]. Additionally, shape analysis methods were proven to be a complement to structural feature selection that was not limited by structural definitions because it was performed from sequence data on a multidimensional RNA-seq platform [13].

Although structural descriptors provide useful information on miRNA structure and function, prediction of complex structures can also be computationally expensive and error-prone. Including structural information in conjunction with features based on sequence has been shown to be helpful in improving prediction accuracy reliably.

Challenges in Computational miRNA Identification and Target Prediction

Prediction of miRNAs through computational methods and their targets presents many challenges based on the intricacy of miRNA-mRNA interactions. The specificity of the interaction depends on factors like seed region complementarity, binding site accessibility, and stability of secondary structure [14, 15]. Tools such as TargetScan, miRanda, RNAhybrid, and PITA take these parameters into consideration but tend to produce different results based on algorithmic assumptions [14]. For instance:

1. TargetScan goes through conserved binding sites and computes cumulative scores from seed-pairing stability and site prevalence.
2. miRanda measures thermodynamic stability together with evolutionary conservation.
3. PITA emphasises site accessibility by calculating energy differences between binding formation and accessibility costs.

Even with improvements in prediction tools, problems like false positives are still common. ML-based methodologies have been highlighted as potential solutions by combining diverse experimental datasets for improving predictions. Nevertheless, the efficacy of these ML models relies extensively on the quality and diversity of training datasets [14].

Limitations of Non-Machine Learning-Based Methods

Analysing the sequences and structures of miRNAs using traditional bioinformatics methods precedes automating the process; however, these techniques offer limited capabilities in comparison to modern computational approaches. Most of these methods use some sort of set parameters, which excludes them from considering the intricacies of miRNA biogenesis. For example, the homology approach will fail to identify novel miRNAs whose sequences do not have some conserved counterparts, while the structural analysis approach will pseudo-dynamically classify structures based on rigid thermodynamic assumptions.

These techniques are considered inefficient because they require manual oversight to curate data. The handcrafted approach may introduce inconsistency and variability and is highly inefficient when sifting through large genomic datasets. Reliance on expert judgement adds inconsistencies, making results too subjective. These biases make methodologies vulnerable as the volume of available genomic data expands. This poses a new problem, as traditional methods become increasingly deficient, unadaptable, complex, and diverse, especially across species.

The issues presented can be solved seamlessly with the implementation of various ML techniques. The main advantage stems from the ability to automate processes with the aid of big datasets, enabling adjustments and improvements in accuracy over time. ML models stand to benefit a lot due to algorithmic incorporation of different features such as sequence, structure, and evolutionary conservation aiding in the identification of miRNAs and their targets.

Fundamentals of Machine Learning in Bioinformatics

Core Concepts of Machine Learning

The ML method infringes on the field of artificial intelligence because it tries to construct algorithms that can learn from data and generate predictions without the need for programming. ML has three main learning paradigms: supervised learning, unsupervised learning, and reinforcement learning [16]. Models to predict outcomes are trained using labelled datasets in supervised learning [17], while unsupervised learning deals with the detection of hidden patterns or structures in data lacking labels [18]. Reinforcement learning, in which agents are trained to make a series of decisions, is aided by interaction with the environment and feedback in the form of incentives and penalties.

Major components of ML are feature extraction, model construction, and model evaluation. Feature extraction is the selection of relevant variables from a primary data source. The process of minimising the error of an algorithm within set parameters to enhance predictive accuracy is referred to as model training [19]. Evaluation, on the other hand, uses metrics like accuracy, precision, F1-score, and recall examining a model's assessment at specific dataset parameters. [19]. The ability to process complex datasets such as genomic sequences and proteomic structures has been further enhanced using cheaper complex multi-layered neural networks known as deep learning [20, 21].

Role of Machine Learning in Biology

The rapid growth of biology's incorporation with ML is largely due to the contemporary biological experiments that produce high-dimensional datasets that need to be interpreted [22], also known as twenty-first-century biology. In genomics analysis, ML algorithms are integrated in gene prediction through the identification of coding regions in each DNA sequence alongside rearrangement of genomes [23]. In the field of proteomics, ML algorithms are applied in the accurate prediction of protein structures and function annotation of proteins through databases and biomedical literature [24]. In systems biology, ML aids in the reconstruction of gene regulatory networks (GRNs), cellular process modelling, and discovery of disease biomarkers [25, 26].

In cancer studies, ML can be used in the process of cancer screening and cancer detection [27]. ML models can also be constructed to forecast the outcome of certain drugs based on a patient's profile [28].

Challenges in Applying Machine Learning to Omics Data

The application of ML to bioinformatics has transformed biological research towards feature extraction, predictive analytics, and data mining within intricate datasets. This has resulted in significant advancements in understanding the pathophysiology of diseases and in designing effective treatment methods. Regardless of the potential advantages, the use of ML on the omics datasets has noticeable challenges. The main factors affecting omics data are their high-dimensional nature, which includes thousands of features alongside a restricted number of samples. The limited sample size per feature leads to overfitting, where the model can accurately predict the training data but is unable to provide any reasonable output with data that was not used to train it. There is also the problem of incomplete information, which arises because of several reasons. In the case of multi-omics datasets, for instance, the absence of values may be due to certain technological constraints or other variations in the experiments being done. Missing data requires advanced algorithms capable of estimating or reconstructing the absent values without needing complete datasets, along with sophisticated mechanisms that deal with incomplete data [29–31].

Furthermore, the inter-platform and inter-individual heterogeneity of omics data can serve as an additional hurdle. Variations in experimental procedures can affect datasets with unwanted noise or biased information, impeding integration and subsequent analyses. Some feature selection techniques do exist, for instance, ridge regression or PCA, but they tend to not only overlook but actively ignore biologically relevant information [29]. Additionally, the lack of standardised pipelines for preprocessing omics data can hinder reproducibility.

Biological complexity presents challenges for ML models as well. To precisely capture underlying patterns in non-linear interactions between biomolecules, sophisticated techniques such as ensemble approaches or deep neural networks are frequently needed. Nevertheless, the implementation of these models necessitates significant skill and is computationally demanding [21, 29]. To overcome these obstacles, statisticians, computer scientists, and biologists must work together across disciplinary boundaries to create solid analytical frameworks that strike a compromise between biological relevance and computing efficiency.

Advanced Machine Learning Methodologies in miRNA Prediction

Sequence Classification

The process of identifying and classifying miRNAs is essential for understanding how genes are regulated, primarily during the initiation stages of cancer formation. Typically, the process of classifying miRNAs involves numerous stages of feature extraction driven by human interpretation. Not only is this process time-consuming, but also the patterns exhibited by miRNA sequences can be too intricate for humans to process accurately [31].

To address these challenges, ML has been an effective method for classifying miRNAs while also automating the process. One of these methods, miRFam [32], combines n-gram feature characteristics with a multiclass classifier support vector machine (SVM) to distinguish miRNAs into "families" or signs of evolutionary relationships among miRNAs. Using this method, miRFam shows 98% pairwise classification accuracy on datasets containing upwards of 300 families, each with five members, compared to the alternative of examining and separating individuals or groupings of miRNA traits on a family tree of unique characteristics (Fig. 7.1) [32].

Along with supervised learning methods, semi-supervised learning systems have also been used to simultaneously utilise labelled and unlabelled data. A dual semi-supervised ML method that combines a multi-view co-training and active-learning system has shown improved accuracy when compared with conventional supervised classification methods such as miRDeep and miPIE. A significant benefit of the dual-stage ML framework is that it lessens the dependence on annotated and classified datasets containing miRNA sample data without compromising the performance [33]. This is useful in studying previously sequenced genomes or for studies focusing on non-model species that lack any annotations or classifications [34].

Structure and Sequence Analysis Using Deep Learning

By allowing for models to learn complicated patterns directly from raw data, deep learning is fundamentally changing how biological sequences are analysed. For miRNA prediction, deep learning models can incorporate both sequence and structural information, which is beneficial [11].

The CL-PMI model, for example, combines CNNs and long short-term memory networks (LSTM), which allows for the modelling of linear sequences along with the spatial properties of RNA secondary structure [35]. Using predicted secondary structures from tools such as RNAfold and taking advantage of the recurrent nature of the LSTM, CL-PMI has shown good performance using cross-species datasets. This indicates that using both sequence and structure properties is key to approaching miRNA identification. The hybrid CNN-LSTM model demonstrated high

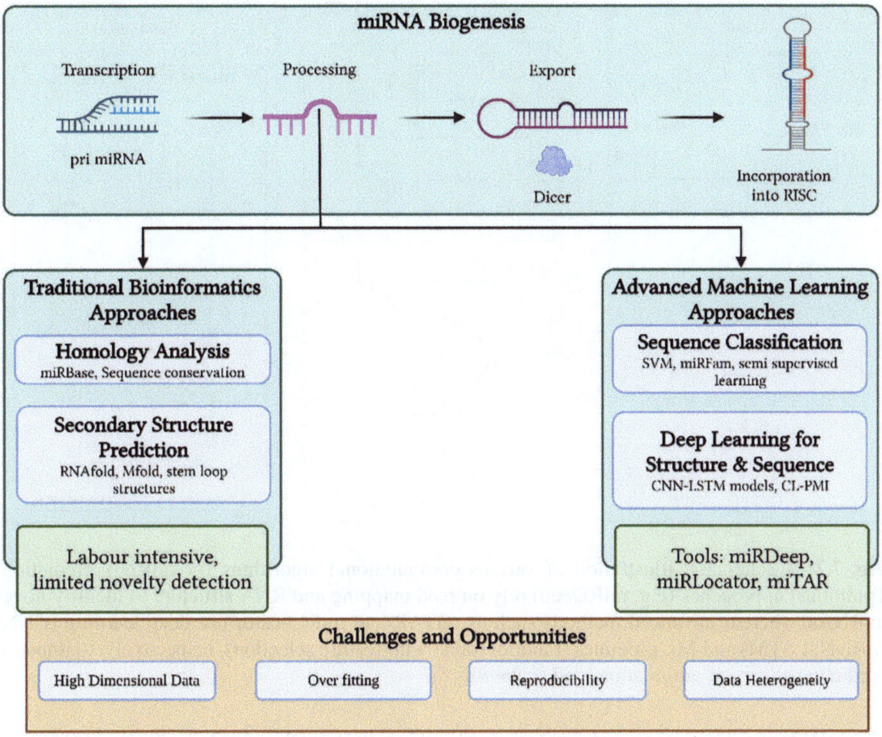

Fig. 7.1 A comparison of classic bioinformatics and advanced machine learning approaches to miRNA prediction in the context of miRNA biogenesis. Traditional approaches, like homology analysis and secondary structure prediction, rely on sequence conservation and thermodynamic stability, but are ineffective at detecting new miRNAs. Machine learning algorithms, such as SVM, CNN-LSTM, and CL-PMI, allow for sequence classification and structure-integrated prediction, resulting in increased scalability and discovery potential. Key computational tools include miRDeep, miRLocator, and miTAR. Despite progress, issues such as high-dimensional data, overfitting, repeatability, and data heterogeneity remain

accuracy and F1 scores for predicting pre-miRNAs. The hybrid approach provides an understanding of local patterns with CNNs and long-range patterns with LSTMs of the same pre-miRNA representation [31].

Comparative Analysis of miRNA Prediction Tools

A variety of computational tools for miRNA prediction are now available and have been developed using a range of approaches. In this section, we review three major tools: miRDeep, miRLocator, and miTAR (Fig. 7.2).

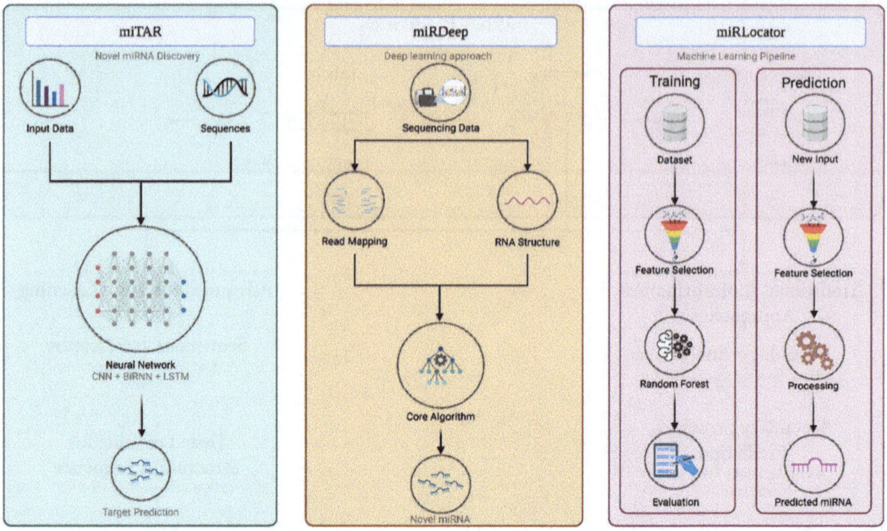

Fig. 7.2 A schematic illustration of various computational algorithms for miRNA prediction. Traditional approaches (e.g. miRDeep) rely on read mapping and RNA structure to identify novel miRNAs, whereas advanced methods such as miTAR and miRLocator use deep learning (CNN, BiRNN, LSTM) and ML pipelines (random forest with feature selection), respectively, to improve prediction accuracy, automation, and scalability

1. miRDeep: This is an integrated application that finds miRNAs from RNA-Seq data, resulting in a comprehensive report with built-in modules for read alignment and secondary structure prediction and providing a user-friendly interface for miRNA discovery [36].
2. miRLocator: This is a tool for the identification of miRNA precursors that applies ML methods and is focused on the localisation or identification of the gene of miRNAs in the genome for the discovery of putative novel miRNAs [37].
3. miTAR: This method combines recurrent neural networks (RNNs), which recognise sequential patterns, with convolutional neural networks (CNNs), which are excellent at learning spatial data. Thus, this method offers the benefit of learning miRNA:target's intrinsic spatial and sequential properties [38].

Applications of Machine Learning-Based miRNA Prediction in Cancer

Breast Cancer

Breast cancer represents an extensively studied malignancies in the context of miRNA-based machine learning applications. Recent advances have demonstrated the remarkable potential of ML algorithms in identifying clinically relevant miRNA signatures that can guide therapeutic decisions. A study by Juntao Li et al. (2022)

employed a new method where they combined ensemble regularised multinomial logistic regression and Cox regression for identifying miRNA biomarkers in breast cancer; they identified 22 miRNAs as oncogenes or inhibitors of breast cancer [39].

The clinical relevance of these miRNAs was further validated by various methods such as expression analysis, KEGG pathway analysis, regulatory network analysis, K-M survival analysis, PPI analysis, miRNA-protein-drug interaction network and GO enrichment analysis, enhancing the reasonableness of the predicted miRNAs.

Furthermore, there is one more study where researchers used ML ensemble of feature selection methods to identify the breast cancer subtype-specific miRNA biomarkers; they were able to identify 27 miRNAs as potential biomarkers [40].

Lung Cancer

A great example of the successful use of machine learning (ML) in miRNA research is lung cancer and its various subtypes. Using sophisticated machine learning algorithms, miRNA expression data has been integrated with genomic, proteomic, and clinical data to provide previously unheard-of insights into disease mechanisms and treatment vulnerabilities.

A study in 2024 established a deep learning algorithm to combine extracellular miRNA (exmiRs) and mRNA for cancer studies. [41]. In this study, the authors aimed to identify exmiRs and mRNAs as potential biomarkers for lung cancer. They created CrossPred, a deep learning model that generates a shared representation space for mRNAs and exmiRs using contrastive learning and multiple encoders. This model supports cross-prediction, which means it can forecast mRNA expression using exmiR data and vice versa. Their method also helped to discover key mRNAs and exmiRs linked to cancer.

Another groundbreaking study was done at the John Hopkins University, USA, where a group of scientists created a machine learning-based liquid biopsy tool that detects lung cancer risk by analysing cell-free DNA (cfDNA) fragments in blood samples [42]. According to this study, DNA fragments are released into the bloodstream by dying cells. Different fragmentation patterns are seen in lung cancer cells. With a 99.8% negative predictive meaning, it is very unlikely to miss real cancer cases—the machine learning model was trained on data from 576 participants and validated on data from 382 more. The test may increase screening accessibility because it is easy to use, affordable, and suitable for use in a typical doctor's office.

Colorectal Cancer

A 2022 study sought to find serum microRNA (miRNA) expression patterns that could act as non-invasive biomarkers for early identification and prognosis of colorectal cancer (CRC) [43]. This study examines the use of machine learning to identify early indicators of colorectal cancer by examining patterns in blood serum microRNA (miRNA) expression. Before using statistical and computational methods to determine which miRNAs were significantly different between the two groups, researchers profiled a broad range of circulating miRNAs from cancer patients and healthy people. By combining these discoveries with machine learning, they created predictive models that used miRNA expression profiles to categorise samples. The objective was to investigate whether these models could serve as a basis for blood-based, non-invasive screening instruments that are easier to use and less invasive than conventional diagnostic processes.

Another study from 2024 developed a pre-trained nucleotide-to-graph neural network system to predict miRNA-mRNA interactions in CRC [44]. This tool converts RNA sequences into de Bruijn graphs and learns features using graph neural networks (GNNs). CNNs, GRUs, and BiGRUs are examples of traditional deep learning models that are outperformed by the Gra-CRC-miRTar framework. In terms of accuracy, AUROC, and F1-score, the GIN-based architecture performed the best. It identified 172 of the 201 miRNA-mRNA interactions in CRC that have been experimentally verified.

Our understanding of the regulatory complexity of cancer has improved thanks to the incorporation of machine learning into miRNA research, which has also expedited the conversion of these discoveries into insights that can be used in clinical settings. ML-driven tools have shown remarkable accuracy in identifying miRNA biomarkers and clarifying miRNA-mRNA interaction networks with practical clinical relevance, ranging from colorectal to lung and breast cancers. By handling high-dimensional data, identifying new patterns, and enabling personalised predictions via ensemble learning, deep architectures, and graph-based models, these techniques surpass conventional bioinformatics techniques. Precision oncology will be further fuelled by the convergence of machine learning (ML) with multi-omics data, strong validation pipelines, and domain-specific expertise as computational biology develops.

Conclusion

Using ML in miRNA research has significantly expanded what we can explore in cancer biology. miRNAs, though small, have a major influence on how genes are regulated. When their normal function is disrupted, it can contribute to the development of cancer. Because of this, researchers are now looking at miRNAs not only as potential markers for diagnosing and predicting cancer outcomes but also as promising targets for developing new treatments.

While conventional bioinformatics software has set the stage, it tends to have difficulty with the intricate nature of miRNA-mRNA interactions. ML presents a more adaptive and scalable solution. Leveraging large, heterogeneous datasets and learning hard-to-define patterns by machine, ML models can enhance the accuracy and complexity of miRNA target prediction.

But difficulties persist. Most omics data are high-dimensional but accompanied by few samples, which then causes overfitting issues. Data may also be noisy or missing, particularly when obtained from disparate platforms or experimental setups. To truly enable ML in this field, we require improved algorithms, more effective validation methods, and more intensive cross-disciplinary partnerships.

Ultimately, the marriage of domain expertise with computational capability has genuine potential for pushing forward cancer research. Done judiciously, ML might help transform miRNA biology's complexity into salient insights that serve both patients and researchers alike. The growing application of ML in real-world cancer studies highlights its practicality and effectiveness in identifying diagnostic and therapeutic miRNAs, providing a valuable link between computational research and clinical utility.

References

1. Lin S, Gregory RI (2015) MicroRNA biogenesis pathways in cancer. Nat Rev Cancer 15:321–333. https://doi.org/10.1038/nrc3932
2. Santhekadur PK, Kumar DP (2020) RISC assembly and post-transcriptional gene regulation in hepatocellular carcinoma. Genes Dis 7:199–204. https://doi.org/10.1016/j.gendis.2019.09.009
3. Frixa T, Donzelli S, Blandino G (2015) Oncogenic MicroRNAs: key players in malignant transformation. Cancers (Basel) 7:2466–2485. https://doi.org/10.3390/cancers7040904
4. Ma Y, Shen N, Wicha MS, Luo M (2021) The roles of the Let-7 family of MicroRNAs in the regulation of cancer stemness. Cells 10. https://doi.org/10.3390/cells10092415
5. Robertson J, Jorcyk C, Oxford J (2018) DICER1 syndrome: DICER1 mutations in rare cancers. Cancers (Basel) 10:143. https://doi.org/10.3390/cancers10050143
6. Potla P, Ali SA, Kapoor M (2021) A bioinformatics approach to microRNA-sequencing analysis. Osteoarthr Cartil Open 3:100131. https://doi.org/10.1016/j.ocarto.2020.100131
7. Kozomara A, Birgaoanu M, Griffiths-Jones S (2019) miRBase: from microRNA sequences to function. Nucleic Acids Res 47:D155–D162. https://doi.org/10.1093/nar/gky1141
8. Nguyen TL, Nguyen TD, Bao S et al (2020) The internal loops in the lower stem of primary microRNA transcripts facilitate single cleavage of human microprocessor. Nucleic Acids Res 48:2579–2593. https://doi.org/10.1093/nar/gkaa018
9. Han D, Zhang J, Tang G (2012) MicroRNAfold: pre-microRNA secondary structure prediction based on modified NCM model with thermodynamics-based scoring strategy. Int J Data Min Bioinform 6:272. https://doi.org/10.1504/IJDMB.2012.049246
10. Peace RJ, Sheikh Hassani M, Green JR (2019) miPIE: NGS-based prediction of miRNA using integrated evidence. Sci Rep 9:1548. https://doi.org/10.1038/s41598-018-38107-z
11. Parveen A, Mustafa SH, Yadav P, Kumar A (2019) Applications of machine learning in miRNA discovery and target prediction. Curr Genomics 20:537–544. https://doi.org/10.2174/1389202921666200106111813
12. Zheng H, Fu R, Wang J-T et al (2013) Advances in the techniques for the prediction of microRNA targets. Int J Mol Sci 14:8179–8187. https://doi.org/10.3390/ijms14048179
13. Mégret L, Nair SS, Dancourt J et al (2020) Combining feature selection and shape analysis uncovers precise rules for miRNA regulation in Huntington's disease mice. BMC Bioinformatics 21:75. https://doi.org/10.1186/s12859-020-3418-9
14. Riolo G, Cantara S, Marzocchi C, Ricci C (2020) miRNA targets: from prediction tools to experimental validation. Methods Protoc 4. https://doi.org/10.3390/mps4010001

15. Rojo Arias JE, Busskamp V (2019) Challenges in microRNAs' targetome prediction and validation. Neural Regen Res 14:1672–1677. https://doi.org/10.4103/1673-5374.257514
16. Sarker IH (2021) Machine learning: algorithms, real-world applications and research directions. SN Comput Sci 2:160. https://doi.org/10.1007/s42979-021-00592-x
17. Wang Y, Cui Z, Ke R (2023) Machine learning basics. In: Machine learning for transportation research and applications. Elsevier, Amsterdam, pp 25–40
18. Yazici İ, Shayea I, Din J (2023) A survey of applications of artificial intelligence and machine learning in future mobile networks-enabled systems. Eng Sci Technol Int J 44:101455. https://doi.org/10.1016/j.jestch.2023.101455
19. Nogales RE, Benalcázar ME (2023) Analysis and evaluation of feature selection and feature extraction methods. Int J Comput Intell Syst 16:153. https://doi.org/10.1007/s44196-023-00319-1
20. Inza I, Calvo B, Armañanzas R et al (2010) Machine learning: an indispensable tool in bioinformatics. Methods Mol Biol 593:25–48. https://doi.org/10.1007/978-1-60327-194-3_2
21. Auslander N, Gussow AB, Koonin EV (2021) Incorporating machine learning into established bioinformatics frameworks. Int J Mol Sci 22. https://doi.org/10.3390/ijms22062903
22. Chakravarthi GP, Babu VR, Ramamurty DSVNM et al (2024) AI and machine learning in biotechnology: a paradigm shift in biochemical innovation. Int J Plant Anim Environ Sci 14. https://doi.org/10.26502/ijpaes.4490166
23. Le D-H (2020) Machine learning-based approaches for disease gene prediction. Brief Funct Genomics 19:350–363. https://doi.org/10.1093/bfgp/elaa013
24. Meng Y, Zhang Z, Zhou C et al (2025) Protein structure prediction via deep learning: an in-depth review. Front Pharmacol 16:1498662. https://doi.org/10.3389/fphar.2025.1498662
25. Noordijk B, Garcia Gomez ML, ten Tusscher KHWJ et al (2024) The rise of scientific machine learning: a perspective on combining mechanistic modelling with machine learning for systems biology. Front Syst Biol 4. https://doi.org/10.3389/fsysb.2024.1407994
26. Greener JG, Kandathil SM, Moffat L, Jones DT (2022) A guide to machine learning for biologists. Nat Rev Mol Cell Biol 23:40–55. https://doi.org/10.1038/s41580-021-00407-0
27. Koh D-M, Papanikolaou N, Bick U et al (2022) Artificial intelligence and machine learning in cancer imaging. Commun Med 2:133. https://doi.org/10.1038/s43856-022-00199-0
28. Badwan BA, Liaropoulos G, Kyrodimos E et al (2023) Machine learning approaches to predict drug efficacy and toxicity in oncology. Cell Rep Methods 3:100413. https://doi.org/10.1016/j.crmeth.2023.100413
29. Flores JE, Claborne DM, Weller ZD et al (2023) Missing data in multi-omics integration: recent advances through artificial intelligence. Front Artif Intell 6. https://doi.org/10.3389/frai.2023.1098308
30. Torres-Martos Á, Bustos-Aibar M, Ramírez-Mena A et al (2023) Omics data preprocessing for machine learning: a case study in childhood obesity. Genes (Basel) 14. https://doi.org/10.3390/genes14020248
31. Tasdelen A, Sen B (2021) A hybrid CNN-LSTM model for pre-miRNA classification. Sci Rep 11:14125. https://doi.org/10.1038/s41598-021-93656-0
32. Ding J, Zhou S, Guan J (2011) miRFam: an effective automatic miRNA classification method based on n-grams and a multiclass SVM. BMC Bioinformatics 12:216. https://doi.org/10.1186/1471-2105-12-216
33. Sheikh Hassani M, Green JR (2019) Multi-view co-training for microRNA prediction. Sci Rep 9:10931. https://doi.org/10.1038/s41598-019-47399-8
34. Sheikh Hassani M, Green JR (2019) A semi-supervised machine learning framework for microRNA classification. Hum Genomics 13:43. https://doi.org/10.1186/s40246-019-0221-7
35. Wang H, Ma Y, Dong C et al (2019) CL-PMI: a precursor MicroRNA identification method based on convolutional and long short-term memory networks. Front Genet 10. https://doi.org/10.3389/fgene.2019.00967

36. An J, Lai J, Lehman ML, Nelson CC (2013) miRDeep*: an integrated application tool for miRNA identification from RNA sequencing data. Nucleic Acids Res 41:727–737. https://doi.org/10.1093/nar/gks1187
37. Zhang T, Ju L, Zhai J et al (2019) miRLocator: a python implementation and web server for predicting miRNAs from pre-miRNA sequences. Methods Mol Biol 1932:89–97
38. Gu T, Zhao X, Barbazuk WB, Lee J-H (2021) miTAR: a hybrid deep learning-based approach for predicting miRNA targets. BMC Bioinformatics 22:96. https://doi.org/10.1186/s12859-021-04026-6
39. Li J, Zhang H, Gao F (2022) Identification of miRNA biomarkers for breast cancer by combining ensemble regularized multinomial logistic regression and Cox regression. BMC Bioinformatics 23:434. https://doi.org/10.1186/s12859-022-04982-7
40. Sarkar JP, Saha I, Sarkar A, Maulik U (2021) Machine learning integrated ensemble of feature selection methods followed by survival analysis for predicting breast cancer subtype specific miRNA biomarkers. Comput Biol Med 131:104244. https://doi.org/10.1016/j.compbiomed.2021.104244
41. Athaya T, Li X, Hu H (2024) A deep learning method to integrate extracellular miRNA with mRNA for cancer studies. Bioinformatics 40. https://doi.org/10.1093/bioinformatics/btae653
42. Mazzone PJ, Bach PB, Carey J et al (2024) Clinical validation of a cell-free DNA fragmentome assay for augmentation of lung cancer early detection. Cancer Discov 14:2224–2242. https://doi.org/10.1158/2159-8290.CD-24-0519
43. Pawelka D, Laczmanska I, Karpinski P et al (2022) Machine-learning-based analysis identifies miRNA expression profile for diagnosis and prediction of colorectal cancer: a preliminary study. Cancer Genomics Proteomics 19:503–511. https://doi.org/10.21873/cgp.20336
44. Yin R, Zhao H, Li L et al (2024) Gra-CRC-miRTar: the pre-trained nucleotide-to-graph neural networks to identify potential miRNA targets in colorectal cancer. Comput Struct Biotechnol J 23:3020–3029. https://doi.org/10.1016/j.csbj.2024.07.014

Chapter 8
MicroRNA-Mediated Cell Cycle Regulation in Cancer

Sangeetha Gopalakrishnan, Premkumar Jayapal, Venkataramanan Swaminathan, and James John

Abstract MicroRNAs (miRNAs) are small noncoding RNAs that regulate gene expression at the posttranscriptional level and play a critical role in maintaining cellular homeostasis. Dysregulation of miRNAs is a hallmark of many cancers, where they function either as oncogenes (oncomiRs) or tumor suppressors. This review highlights the intricate relationship between miRNAs and cell cycle regulation in cancer progression. OncomiRs such as miR-21, miR-155, and the miR-17-92 cluster are frequently overexpressed in cancers and promote tumor growth by targeting key tumor-suppressor genes. In contrast, tumor-suppressor miRNAs like let-7, miR-34a, and the miR-200 family are often downregulated, leading to uncontrolled

S. Gopalakrishnan
Department of Medical Lab Technology, School of Allied Health Science, Sathyabama Institute of Science and Technology, Chennai, India

Division of Laboratories, Biochemistry & Hematology Section, Central Leprosy Teaching and Research Institute, Chengalpattu, India

P. Jayapal
Department of Medical Lab Technology, School of Allied Health Science, Sathyabama Institute of Science and Technology, Chennai, India

Department of Biomedical Engineering, School of Bio & Chemical Engineering, Sathyabama Institute of Science and Technology, Chennai, India

V. Swaminathan
Department of Diagnostic & Allied Health Sciences, Management & Science University, University Drive, Shah Alam, Selangor, Malaysia

J. John (✉)
Department of Medical Lab Technology, School of Allied Health Science, Sathyabama Institute of Science and Technology, Chennai, India

Department of Biomedical Engineering, School of Bio & Chemical Engineering, Sathyabama Institute of Science and Technology, Chennai, India

Department of Diagnostic & Allied Health Sciences, Management & Science University, University Drive, Shah Alam, Selangor, Malaysia
e-mail: jamesjohn.ahs@sathyabama.ac.in

S. Pathak, A. Banerjee (eds.), *MicroRNAs: Role in Cancer Progression, Diagnostics and Therapy*, Advances in Biochemistry in Health and Disease 34,
https://doi.org/10.1007/978-3-032-08824-6_8

proliferation and metastasis. Moreover, aberrant miRNA expression contributes to chemoresistance by modulating apoptosis and drug response pathways. The dual role of miRNAs in cancer pathogenesis underscores their potential as diagnostic biomarkers and therapeutic targets. Advances in miRNA-based therapies, including mimics and inhibitors, offer promising avenues for cancer treatment. This review provides a comprehensive overview of miRNA-mediated cell cycle control and its implications in oncogenesis, paving the way for future research and clinical applications in cancer management.

Keywords MicroRNAs (miRNAs) · Cell cycle regulation · OncomiRs · Tumor suppressor miRNAs · Cancer therapeutics

Introduction

Overview of MicroRNAs (miRNAs)

MicroRNAs, also known as miRNAs, are small endogenous RNAs which regulate various genes at the posttranscriptional level. With increasing research on miRNAs, they are now recognized as key to gene expression and potential biomarkers. Several miRNA are currently undergoing preclinical testing as inhibitors and have shown promise as innovative therapeutic agents [34]. Generally, miRNAs reduce the expression of messenger RNAs (mRNAs) that contain complementary sequences; however, their function is more complex than a simple "on-off" mechanism. Additionally, many miRNA targets are noncoding RNAs themselves [40]. Thus, miRNAs are broadly defined as small noncoding RNAs that have essential roles in posttranscriptional gene regulation. They miRNA regulate their targets through destabilizing mRNA and via inhibiting the translational [7].

Importance of Cell Cycle Regulation in Cancer

Over the past decade, microRNAs (miRNAs) have been widely recognized for their role in both cancer-promoting and tumor-suppressing pathways [15]. The proto-oncogene *c-MYC* is strongly linked to miRNAs and seems to have a potential role in cancer development. It is activated in nearly half of all cancers and produces a transcription factor that controls various cell functions such as metastasis, stem cell-like properties, cell cycle progression, epithelial-mesenchymal transition (EMT), and blood vessel formation, all of which contribute to tumor growth and spread [13, 38]. The cancer-promoting effects of c-MYC are driven by its target genes, which include a complex mix of protein-coding and noncoding RNAs, including many miRNAs [13]. Notably, c-MYC directly influences the expression of miR-17-92, miR-34a, miR-15a/16-1, and miR-9, all of which are involved in different tumor-promoting and tumor-suppressing pathways [35, 44]. Moreover, miRNAs regulate both the expression and activity of c-MYC, creating a regulatory feedback loop [14]. Understanding how these networks influence and control c-MYC functions is essential for progress in cancer research. In the future, miRNAs linked to c-MYC could be used for diagnosis and treatment, providing new approaches to cancer therapy [26].

How Dysregulated miRNAs Contribute to Tumorigenesis

miRNAs regulate gene expression after transcription by binding to the 3′ untranslated regions 3′ untranslated region of target mRNAs, causing their degradation or preventing their translation [2]. Disruptions in miRNA regulation are a key feature of cancer, as they can interfere with essential cellular functions like cell growth, programmed cell death, differentiation, and metastasis [47].

Oncogenic miRNAs (OncomiRs) and Their Role in Cancer Progression

Oncogenic miRNAs (oncomiRs) are overexpressed in cancer and contribute to tumor development by suppressing tumor suppressor genes. These miRNAs enhance proliferation, inhibit apoptosis, and promote metastasis. miR-21 is the most studied oncomiRs, frequently overexpressed in various cancers, including lung, breast, and glioblastoma. It targets PTEN, a tumor suppressor involved in the PI3K/AKT pathway, leading to uncontrolled cell survival and proliferation [57]. miR-155 is another well-known oncomiR that plays a role in hematological malignancies by inhibiting genes such as SOCS1, which negatively regulates JAK/STAT signaling [56]. miR-17-92 cluster, often amplified in cancers like lymphoma and lung cancer, targets key regulators of apoptosis such as *BIM* and *PTEN*, promoting cell cycle progression and survival [39].

Tumor Suppressor miRNAs and Their Downregulation in Cancer

Tumor suppressor miRNAs are often reduced in cancer unlike oncomiRs which shows overexpression, allowing uncontrolled tumor growth and resistance to cell death; let-7 family of miRNAs is known for its tumor-suppressive role by targeting Ras and MYC oncogenes. Downregulation of let-7 is observed in lung and breast cancer, leading to increased cell proliferation [30]. miR-34a, which is reported to be a direct target of *p53*, is frequently silenced in cancers through promoter hypermethylation. It regulates genes involved in apoptosis (*BCL-2*), cell cycle arrest (*CDK4, CDK6*), and metastasis (*SNAIL*) [24, 25]. miR-200 family is crucial for suppressing EMT by targeting transcription factors, i.e., *ZEB1* and *ZEB2*, that promote metastasis. Downregulation of miR-200 leads to enhanced tumor invasion and drug resistance [5] (Table 8.3).

miRNA-Mediated Resistance to Apoptosis and Chemotherapy

Aberrant miRNA expression also contributes to chemotherapy resistance by modulating apoptosis and drug metabolism pathways. miR-221/222 overexpression introduces the resistance to TRAIL-dependent apoptosis in breast cancer by targeting p27Kip1 and PTEN, leading to enhanced survival signaling [18]. miR-125b has been implicated in doxorubicin resistance by inhibiting p53-dependent apoptosis, reducing chemotherapy efficacy [32]. Loss of miR-31 in colorectal cancer leads to resistance to 5-FU treatment, as it regulates pathways involved in drug response .

miRNAs as Biomarkers and Therapeutic Targets in Cancer

Because of their stability in bodily fluids and unique expression in diseases, miRNAs are being considered potential biomarkers for cancer diagnosis and prognosis. Circulating miR-21 levels, for example, have been studied as a biomarker for detecting lung and breast cancer [27]. miRNA-based therapies, including miRNA mimics to restore tumor suppressor miRNAs and anti-miRs to inhibit oncomiRs, are being investigated as innovative cancer treatments. In addition to their intracellular roles, miRNAs are also secreted in extracellular vesicles such as exosomes, facilitating intercellular communication within the tumor microenvironment. Exosomal miRNAs can modulate immune responses, angiogenesis, and metastasis by transferring oncogenic or tumor-suppressive signals to recipient cells, thus playing a critical role in cancer progression and therapy resistance.

Basics of Cell Cycle Regulation

The cell cycle is a tightly regulated process responsible for accurate DNA replication and cell division. It is governed by an intricate network of molecular signals, including cyclins, cyclin-dependent kinases (CDKs), and checkpoint proteins. Dysregulation of the cell cycle is a hallmark of cancer, leading to unchecked cell proliferation and genomic instability [23].

Phases of the Cell Cycle (G1, S, G2, M)

The eukaryotic cell cycle is a highly regulated process consisting of four sequential phases, ensuring proper growth of a cell, DNA replication, and division. The G1 phase (Gap 1) is a preparatory stage where the cell grows, synthesizes essential proteins, and assesses environmental conditions before committing to DNA replication. This phase serves as a critical checkpoint, determining whether a cell will proceed through the cycle or enter a quiescent state (G0 phase). The S phase (synthesis) is when DNA replication occurs, ensuring two identical copies of the genome are produced, followed by the G2 phase (Gap 2) enabling the cell to prepare for mitosis by synthesizing proteins required for chromosomal condensation and spindle formation. Finally, the M phase (mitosis) facilitates the equal segregation of chromosomes into two daughter cells through mitotic spindle machinery, comprising prophase, metaphase, anaphase, and telophase, followed by cytokinesis, which completes cell division. Cells can transiently or permanently exit the cycle into G0 phase, a resting state where they remain quiescent until receiving appropriate signals to re-enter the cycle [1].

Role of Cyclins, CDKs, and Checkpoint Regulators: The Cell Cycle Is Controlled by a Complex Interplay of Proteins That Ensure Proper Progression and Prevent Genomic Instability

Cyclins and Cyclin-Dependent Kinases (CDKs)

Cyclins and cyclin-dependent kinases (CDKs) serve as the primary regulators of the cell cycle, ensuring precise control over cell cycle transitions. These are synthesized and degraded in a phase-specific manner, while CDKs require activation by their respective cyclins to drive cell cycle progression. Cyclin D/CDK4 and CDK6 govern the G1 phase, facilitating the transition to the S phase, whereas Cyclin E/CDK2 plays a crucial role during DNA replication and G1/S transition. During the S phase, Cyclin A/CDK2 ensures proper replication, and in the G2/M transition, Cyclin B/CDK1 (Cdc2) regulates mitotic entry. Dysregulation of cyclins and CDKs is commonly observed in various cancers, leading to unscheduled proliferation and loss of cell cycle control [36].

B. Cell Cycle Checkpoints

Checkpoints act as surveillance mechanisms that monitor DNA integrity and cellular conditions. G1/S checkpoint determines whether the cell proceeds to DNA replication. It is regulated by the retinoblastoma protein (pRB) and E2F transcription factors. DNA damage activates *p53*, which induces *p21*, an inhibitor of CDKs, leading to cell cycle arrest [50]. G2/M checkpoint ensures that DNA replication is complete and free from damage before entering mitosis. Key players include Chk1 and Chk2, which inhibit Cdc25, preventing activation of CDK1 [55]. SAC (the spindle assembly checkpoint) avoids early chromosome separation by blocking the anaphase-promoting complex/cyclosome (APC/C) until proper attachment of all kinetochores to the mitotic spindle is ensured [42].

Key Tumor Suppressors (p53, RB, p21) in Cell Cycle Control: The Balance Between Tumor Suppressors and Oncogenes Ensures Normal Cell Cycle Progression

Tumor suppressors and oncogenes play crucial roles in maintaining cell cycle regulation and preventing malignant transformation. p53 functions as a key guardian of the genome by activating DNA repair mechanisms and inducing apoptosis in response to severe DNA damage [52]. Similarly, pRB inhibits E2F transcription factors and, at the same time, prevents early entry to S phase [6]. Additionally, p21 and p27 act as CDK inhibitors, enforcing checkpoint control to halt cell cycle progression when necessary [50]. Conversely, oncogenes such as MYC promote uncontrolled proliferation by enhancing cyclin/CDK activity, while Ras drives downstream signaling pathways leading to increased cyclin D expression and mitogenic signaling [23, 38].

MicroRNAs and Their Role in Cancer Biology

Biogenesis of miRNAs

miRNAs regulate gene expression posttranscriptionally by binding to messenger RNAs (mRNAs), leading to their degradation or translational repression. Initially, RNA polymerase II transcribes them as primary miRNAs (pri-miRNAs), which are then processed into precursor miRNAs (pre-miRNAs) by the Drosha-DGCR8 complex within the nucleus. These pre-miRNAs are exported to the cytoplasm via Exportin-5, where Dicer further processes them into mature miRNA duplexes. The functional strand is integrated into the RNA-induced silencing complex (RISC), which guides the complex to target mRNAs, ultimately suppressing translation or promoting mRNA degradation [3] (Fig. 8.1).

Primary miRNAs (pri-miRNAs) are transcribed in the nucleus by RNA polymerase II and processed by the microprocessor complex, comprising Drosha and DGCR8, into precursor miRNAs (pre-miRNAs). These pre-miRNAs are exported to the cytoplasm via Exportin-5, where they are further cleaved by the enzyme Dicer to produce a miRNA duplex. One strand of the duplex (guide strand) is then loaded into the RISC complex, while the other (passenger strand) is degraded. The mature RISC-miRNA complex binds to target mRNAs, resulting in gene silencing through translational repression or mRNA degradation.

Mechanism of miRNA-Mediated Gene Silencing

miRNAs interfere with gene expression mainly by binding to 3′ untranslated region (3′UTR) of target mRNAs, either suppressing translation or promoting mRNA degradation based on complementarity. Perfect complementarity triggers mRNA

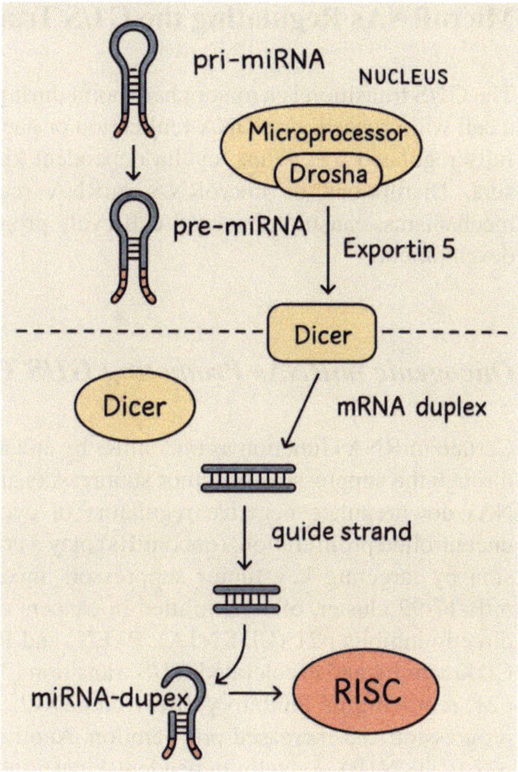

Fig. 8.1 Biogenesis of microRNAs (miRNAs) and their incorporation into the RNA-Induced Silencing Complex (RISC)

cleavage, while partial complementarity blocks translation by interfering with ribosomal activity. The RISC complex, particularly Argonaute (AGO) proteins, plays a key role in facilitating gene silencing. Additionally, miRNAs can modulate gene expression indirectly by influencing chromatin structure and epigenetic modifications [22].

General Impact of miRNAs in Cancer (OncomiRs vs. Tumor Suppressor miRNAs)

miRNAs can act as either oncogenic miRNAs (oncomiRs) or tumor suppressor miRNAs, depending on their target genes and cellular environment. OncomiRs, such as miR-21 and miR-155, drive tumor development by suppressing tumor suppressor genes, leading to increased cell growth, survival, and metastasis. In contrast, tumor suppressor miRNAs like let-7 and miR-34 inhibit oncogenic pathways by targeting cell cycle regulators and anti-apoptotic genes [8]. miRNA dysregulation plays a key role in cancer progression, contributing to uncontrolled cell growth, resistance to cell death, blood vessel formation, and metastasis. Abnormal miRNA expression is common in many cancers, making them valuable biomarkers and potential therapeutic targets [16].

MicroRNAs Regulating the G1/S Transition

The G1/S transition is a major checkpoint during the cell cycle that decides whether a cell will proceed with DNA replication or pause for growth. This process is carefully regulated by cyclins, cyclin-dependent kinases (CDKs), and tumor suppressors. Disruptions in microRNA miRNA regulation can interfere with these mechanisms, causing abnormal cell cycle progression and contributing to cancer development.

Oncogenic miRNAs Promoting G1/S Transition

Certain miRNAs function as oncomiRs by enhancing the G1/S transition, primarily through the suppression of tumor suppressors and cell cycle inhibitors. These miRNAs downregulate negative regulators of cyclin/CDK activity, thus facilitating uncontrolled proliferation. (oncomiRs) play a critical role in driving cancer progression by targeting key tumor suppressors involved in cell cycle regulation. The miR-17-92 cluster, often amplified in cancers such as lymphoma and lung cancer, directly inhibits p21 (CDKN1A), PTEN, and RB1, which cause increased cyclin/CDK activity and accelerated G1/S transition [37]. Similarly, miR-106b suppresses p21, removing its inhibitory effect on CDK2, thereby promoting rapid cell cycle progression and increased proliferation. Another key player, miR-221/222, targets p27 (CDKN1B), a cyclin-dependent kinase inhibitor, leading to enhanced CDK2 activity and facilitating S-phase entry. The overexpression of these miRNAs in various cancers disrupts normal checkpoint control, leading to uncontrolled proliferation and tumor development.

Example: miR-17-92 Cluster Targeting Tumor Suppressors (e.g., p21, PTEN)

The miR-17-92 cluster, often referred to as "oncomiR-1," comprises six miRNAs: miR-17, miR-18a, miR-19a, miR-19b, miR-20a, and miR-92a, which promote tumorigenesis by targeting multiple tumor suppressors. It is frequently overexpressed in cancers, particularly lymphoma, lung, and breast cancer [37]. The dysregulation of key tumor suppressor targets, such as *p21 (CDKN1A)* and *PTEN*, plays a significant role in driving uncontrolled cell cycle progression and tumorigenesis. p21 functions as a CDK inhibitor, preventing the activation of CDK2 and enforcing the G1/S checkpoint; however, its inhibition by oncogenic miRNAs removes this restraint, allowing unchecked entry into the S phase and promoting proliferation. Similarly, PTEN plays a critical role in suppressing uncontrollable cell growth that downregulates the PI3K/AKT signaling pathway, which controls cell growth and survival. The loss or suppression of PTEN leads to hyperactivation of AKT signaling, enhancing cell proliferation, survival, and resistance to apoptosis [45]. The downregulation of these key regulators by oncogenic miRNAs significantly contributes to the loss of cell cycle control, a hallmark of cancer.

Tumor Suppressor miRNAs Inhibiting G1/S Transition

Certain miRNAs function as tumor suppressors by regulating oncogenic pathways and enforcing G1 arrest. The let-7 family controls cell proliferation by targeting key oncogenes like Ras and MYC, preventing activation of MAPK and PI3K/AKT signaling essential for G1/S progression [49, 54]. Loss of let-7 is associated with aggressive cancers, including lung, breast, and colorectal carcinoma [30]. Similarly, miR-34a, a p53 target, inhibits cyclin D1 and CDK4, blocking the G1/S transition and promoting apoptosis [24, 25]. Its downregulation is linked to colorectal, breast, and lung cancers. The miR-15a/16 cluster, often deleted in chronic lymphocytic leukemia (CLL), suppresses tumor growth by targeting cyclin D1 for G1 arrest and BCL-2 for apoptosis [9–12]. Loss of miR-15a/16 contributes to malignancies like CLL, prostate cancer, and neuroblastoma. Restoring these miRNAs has shown promise in re-establishing cell cycle control and inducing apoptosis, emphasizing their therapeutic potential [9–12, 24, 25].

MicroRNAs in S Phase and DNA Replication

miRNAs play a pivotal role in regulating DNA replication and repair, ensuring genomic stability and preventing uncontrolled cell proliferation. These small non-coding RNAs modulate the expression of key genes involved in the DNA damage response (DDR), including those regulating cell cycle checkpoints, homologous recombination (HR), and non-homologous end joining (NHEJ) [2]. Dysregulation of miRNAs can impair DDR pathways, leading to genomic instability, a hallmark of cancer, by promoting mutations, chromosomal aberrations, and defective cell cycle arrest mechanisms [29].

Several oncomiRs and tumor suppressor miRNAs are implicated in DDR regulation. For instance, miR-155 has been shown to suppress RAD51, a critical protein in homologous recombination, thereby compromising DNA repair efficiency and increasing mutation rates in tumor cells [19–21]. Similarly, miR-182 downregulates BRCA1, impairing DNA double-strand break repair and sensitizing cells to DNA-damaging agents [41]. In contrast, tumor suppressor miRNAs such as miR-34a, a direct transcriptional target of p53, play a protective role by enhancing DDR. miR-34a downregulates CDK4 and cyclin D1, enforcing cell cycle arrest at the G1/S checkpoint and allowing DNA repair before replication [24, 25]. Additionally, miR-125b regulates ATM, a crucial DDR kinase, thereby modulating the activation of p53 and checkpoint responses following DNA damage [33].

The deregulation of these miRNAs can have profound consequences in cancer development and treatment response. Loss of tumor suppressor miRNAs leads to ineffective DNA repair and increased susceptibility to mutagenic events, while overexpression of oncogenic miRNAs accelerates proliferation despite genomic defects [23]. Understanding these miRNA-mediated mechanisms in DDR not only provides insight into cancer progression but also opens new avenues for miRNA-based therapeutic interventions aimed at restoring genomic integrity. Dysregulation of DNA repair-associated miRNAs can impair genomic stability, leading to accumulation of mutations that promote immune evasion. Additionally, altered expression of these miRNAs can downregulate antigen presentation pathways, reduce neoantigen generation, and suppress immune recognition, thereby facilitating cancer immunoevasion.

miRNAs Modulating DNA Replication and Repair

Several miRNAs influence DNA replication and repair by targeting key regulators of the cell cycle, checkpoint pathways, and DNA repair machinery. These miRNAs can either promote or inhibit the gene expression related to DNA damage response (DDR), thereby affecting how cells respond to replicative stress and genomic instability.

miR-155, miR-21, and miR-182 (oncogenic miRNAs) play a role in impairing DNA repair pathways, thereby promoting tumorigenesis. For example, miR-155 suppresses RAD51, a critical protein involved in homologous recombination (HR), leading to defective DNA repair and increased mutation rates [19–21]. Similarly, miR-182 targets BRCA1, compromising double-strand break (DSB) repair and sensitizing cancer cells to DNA-damaging agents [41]. Additionally, miR-21 negatively regulates GADD45, a key gene involved in genome maintenance, thereby reducing the efficiency of DDR mechanisms and enhancing genomic instability [51].

Conversely, tumor suppressor miRNAs, such as miR-34a, miR-16, and miR-125b, reinforce DDR and maintain genomic integrity. miR-34a, a direct transcriptional target of p53, inhibits CDK4 and cyclin D1, enforcing G1/S checkpoint arrest and allowing time for DNA repair before replication proceeds [24, 25]. miR-16 targets CHK1, preventing premature mitotic entry in response to DNA damage and ensuring proper cell cycle progression [46]. Additionally, miR-125b regulates ATM, a key DDR kinase, thereby modulating p53 activation and checkpoint responses following DNA damage [33]. The dysregulation of these miRNAs is frequently observed in cancer and contributes to replication stress, defective DNA repair, and increased mutation rates, ultimately driving tumor progression. Targeting these miRNAs through miRNA-based therapies, such as miRNA mimics or anti-miRNA oligonucleotides, holds promise for restoring DDR function and enhancing cancer treatment efficacy.

Table 8.1 miRNAs involved in DNA replication and repair

miRNA	Target genes/ pathways	Function in DDR	Implications in cancer	References
miR-34a	*p53, Cyclin D1, CDK4*	Reinforces G1 checkpoint, induces apoptosis	Tumor suppressor; loss leads to chemoresistance	[24, 25]
miR-21	*CHK1, CHK2*	Overrides G2/M checkpoint, suppresses DNA repair	OncomiR; promotes tumor progression	[17]
miR-16	*Cyclin E1*	Delays G1/S transition	Loss linked to leukemia and aggressive tumors	[9–12]
miR-24	*H2AX*	Inhibits DNA double-strand break repair	Reduces checkpoint activation	[31]
miR-200	*Wee1, PLK1*	Delays mitotic entry, prevents EMT	Tumor suppressor in breast, lung, and ovarian cancer	[28]

miR-155, an oncogenic miRNA, has been implicated in disrupting DDR by targeting RAD51, a crucial protein reported in the homologous recombination (HR) repair pathway. Downregulation of RAD51 by miR-155 leads to defective DNA double-strand break repair, making cells more susceptible to instability and tumor aggression [19–21]. Additionally, miR-155 modulates the ATR/CHK1 checkpoint pathway, impairing cell cycle arrest in response to DNA damage, thereby increasing the risk of mutations. As a direct target of p53, miR-34a functions as a critical tumor suppressor miRNA by enhancing DDR mechanisms. It regulates G1/S checkpoint control by inhibiting cyclin E2 and CDK6, ensuring that DNA replication occurs accurately [24, 25]. Additionally, miR-34a SIRT1 expression, a downregulator of p53, thereby reinforces p53-mediated cell cycle arrest and apoptosis in response to DNA damage [53]. Loss of miR-34a leads to a defective DDR, allowing cancer cells to proliferate despite genomic abnormalities (Table 8.1).

MicroRNAs Regulating the G2/M Transition

The G2/M transition is a critical checkpoint where cells assess DNA integrity before committing to mitosis. This phase is tightly regulated by key proteins such as checkpoint kinases (CHK1, CHK2), cyclin B/CDK1, and polo-like kinases (PLKs), ensuring that cells only proceed to mitosis if their DNA is intact. Dysregulation of specific oncogenic and tumor suppressor miRNAs can either promote uncontrolled mitotic entry or delay mitotic progression, influencing cancer development and progression.

Oncogenic miRNAs Facilitating Mitotic Entry

Several oncogenic miRNAs drive an unchecked G2/M transition by suppressing checkpoint kinases and regulators of cell cycle arrest following response to DNA damage. These miRNAs contribute to genomic instability and tumorigenesis. miR-21 Targeting Checkpoint Kinases (CHK1, CHK2): One of the most well-characterized oncogenic miRNAs, miR-21, facilitates mitotic entry by targeting CHK1 and CHK2, key checkpoint kinases responsible for enforcing G2/M arrest in response to DNA damage [17]. By downregulating these kinases, miR-21 weakens the G2/M checkpoint, allowing cells to progress into mitosis although they have damaged DNA. This leads to chromosomal instability, aneuploidy, and increased tumor progression. Additionally, miR-21 enhances the expression of cyclin B1 and CDK1, further driving mitotic progression and contributing to aggressive tumor phenotypes, such as in breast and colorectal cancers [51].

Tumor Suppressor miRNAs Delaying Mitotic Progression

Certain tumor suppressor miRNAs act as a protective mechanism by reinforcing checkpoint control, delaying mitotic entry, and inhibiting EMT is a key process in cancer metastasis. miR-200 Family Inhibiting EMT and Mitotic Regulators: The miR-200 family, best known for suppressing EMT and metastasis, also plays a role in regulating the G2/M transition by targeting mitotic regulators such as Wee1 kinase and PLK1 (polo-like kinase 1) [28]. Wee1 kinase inhibits CDK1, preventing premature mitotic entry, while PLK1 is essential for mitotic progression. Loss of miR-200 expression leads to unchecked mitotic entry, metastasis, and risk of increased tumor progression and cancers such as breast, lung, and ovarian cancer [5]. Conversely, overexpression of miR-200 enhances checkpoint control, ensuring proper DNA repair before mitosis, thereby preventing aneuploidy and tumor development.

These findings highlight the dual role of miRNAs in regulating the G2/M transition, where oncogenic miRNAs drive mitotic entry and genomic instability, while tumor suppressor miRNAs reinforce checkpoint mechanisms to prevent uncontrolled proliferation (Table 8.2). Targeting these miRNAs through therapeutic interventions such as miRNA mimics and inhibitors offers potential strategies to restore cell cycle control and improve cancer treatment outcomes.

MicroRNAs and Cell Cycle Checkpoints

Cell cycle checkpoints serve as essential surveillance systems that halt cell cycle progression in response to DNA damage or mitotic errors, preventing genomic instability. miRNAs play a key role in modulating these checkpoints by targeting

Table 8.2 miRNAs regulating the G2/M checkpoint

miRNA	Target genes	Effect on G2/M transition	Impact on cancer progression	References
miR-21	CHK1, CHK2	Promotes unchecked mitotic entry	Increases genomic instability	[17]
miR-200	Weel, PLK1	Enforces checkpoint control	Suppresses metastasis	[28]
miR-155	WEE1, TP53BP1	Overrides G2 arrest	Enhances tumor proliferation	
miR-17-92	RB1, E2F	Accelerates mitotic progression	Promotes high-grade malignancies	[37]

key regulatory proteins involved in DNA damage repair, checkpoint activation, and spindle assembly checkpoint (SAC) control. Dysregulation of these miRNAs can lead to checkpoint failure, allowing damaged cells to continue dividing, thereby promoting tumorigenesis.

miRNAs Influencing the G1 and G2 DNA Damage Checkpoints

The G1 and G2 checkpoints play a crucial role in maintaining genome integrity by ensuring that damaged DNA is either repaired before DNA replication (G1 checkpoint) or before mitotic entry (G2 checkpoint). These checkpoints are tightly regulated by a network of tumor suppressors, checkpoint kinases, and DNA repair factors, preventing the propagation of mutations that could lead to genomic instability and cancer. Several miRNAs modulate these checkpoints by targeting key regulators of cell cycle arrest, apoptosis, and DNA repair mechanisms, thereby influencing cell fate decisions in response to genotoxic stress.

miR-34a: A Key p53-Dependent Checkpoint Regulator

One of the most well-characterized tumor suppressor miRNAs, miR-34a, acts as a key downstream effector of the p53 tumor suppressor pathway. p53 induces miR-34a expression in response to DNA damage, reinforcing checkpoint control by suppressing cyclin D1 and CDK4, thereby preventing aberrant cell cycle progression at the G1 checkpoint [24, 25]. By inhibiting cell cycle regulators and promoting pro-apoptotic signaling, miR-34a serves as a critical barrier against tumorigenesis, particularly in p53-mutant cancers, where its downregulation leads to unchecked proliferation and resistance to apoptosis.

miR-16: Reinforcing the G1 Checkpoint

miR-16 is another tumor suppressor miRNA that plays a key role in regulating the G1/S transition by targeting cyclin E1, a crucial driver of S-phase entry. By down-regulating cyclin E1, miR-16 delays the G1-to-S phase transition, thereby reinforcing the G1 checkpoint [9–12]. Loss of miR-16 expression has been associated with chronic lymphocytic leukemia (CLL) and other malignancies, where its depletion leads to uncontrolled cell cycle progression and enhanced survival of cancer cells.

miR-24: Regulating DNA Damage Response and Checkpoint Activation

miR-24 plays a dual role in cell cycle regulation and DNA repair, primarily by targeting H2AX, a histone variant essential for double-strand break (DSB) repair and checkpoint activation. By inhibiting H2AX, miR-24 weakens the DNA damage response (DDR), affecting checkpoint signaling and repair efficiency [31]. While miR-24-mediated suppression of H2AX may facilitate rapid cell cycle progression under normal conditions, its overexpression in cancer cells can compromise DNA repair, leading to accumulation of mutations and genomic instability.

In the G2 checkpoint, miRNAs regulate the kinases required for cell cycle arrest: miR-21 downregulates CHK1/CHK2, preventing proper activation of the G2 checkpoint and allowing mitotic entry despite unresolved DNA damage [17]; miR-155 suppresses RAD51, a critical DNA repair protein, leading to defective homologous recombination and increased genomic instability [19–21].

Role of miRNAs in the Spindle Assembly Checkpoint (SAC)

The spindle assembly checkpoint (SAC) ensures that chromosomes are correctly attached to spindle microtubules before the cell enters anaphase. miRNAs can either reinforce SAC activation by targeting mitotic regulators or promote mitotic slippage by inhibiting SAC components. miR-125b regulating p53 and ATM pathways: miR-125b plays a dual role in checkpoint regulation by targeting both p53 and ATM (Ataxia Telangiectasia Mutated), two critical regulators of DNA damage response and SAC control [32]. By downregulating p53, miR-125b suppresses the G1 checkpoint, allowing cells to bypass DNA damage-induced arrest. Additionally, by inhibiting ATM, it compromises SAC activation, leading to premature mitotic exit and chromosomal instability hallmarks of cancer progression. miR-200c inhibits BubR1, a crucial component of the SAC, leading to defective chromosome segregation and aneuploidy [28].

Therapeutic Implications of Targeting MicroRNAs in Cancer

miRNAs are gaining attention as potential diagnostic, prognostic, and therapeutic tools in cancer due to their key role in regulating the cell cycle, apoptosis, and tumor progression. The abnormal expression of certain miRNAs in different cancers has led to the development of miRNA-based biomarkers and therapies, including miRNA mimics and inhibitors. However, challenges like efficient delivery and precise target specificity remain significant obstacles to their clinical application. Targeting DNA repair-associated miRNA dysregulation to overcome cancer immunoevasion is being explored using various delivery platforms, including lipid nanoparticles, viral vectors, and exosome-based systems. These delivery methods enable precise modulation of miRNA expression within tumor cells, aiming to restore immune surveillance by enhancing antigen presentation and reducing immune escape mechanisms (Table 8.3).

miRNA-Based Biomarkers for Cancer Prognosis

Dysregulated miRNAs have become promising non-invasive biomarkers for early cancer detection, prognosis, and tracking treatment responses. Unlike traditional biopsy-based diagnostics, circulating miRNAs found in blood, urine, and other biofluids offer a stable and accessible alternative for detecting tumor-specific molecular changes. Their stability in biofluids is attributed to their encapsulation within exosomes, microvesicles, and protein complexcs, protecting them from degradation. Advanced techniques such as quantitative real-time PCR (qRT-PCR), microarrays, and next-generation sequencing (NGS) enable precise miRNA profiling, facilitating their use as diagnostic and prognostic tools in oncology (Table 8.4).

Table 8.3 OncomiRs and tumor-suppressor miRNAs as prognostic biomarkers

miRNA type	Examples	Cancer types	Clinical implications	References
OncomiRs	miR-21, miR-155, miR-17-92	Lung, breast, colorectal, leukemia	Poor prognosis, chemoresistance	
Tumor suppressors	miR-34a, let-7, miR-15a/16	Ovarian, pancreatic, lung, leukemia	Loss linked to aggressive tumors	[24, 25]

Table 8.4 Techniques for detecting circulating miRNAs as cancer biomarkers

Method	Principle	Advantages	Limitations	References
qRT-PCR	Amplifies and quantifies miRNA	High sensitivity, specificity	Requires prior knowledge of targets	
Microarrays	Hybridization-based miRNA profiling	High throughput, cost-effective	Less sensitive than qRT-PCR	[37]
Next-generation sequencing (NGS)	Unbiased miRNA discovery	High sensitivity, identifies novel miRNAs	Expensive, complex data analysis	Zhang et al. (2010)

OncomiRs as Prognostic Markers

Certain oncomiRs, which are overexpressed miRNAs in cancer, have been linked to tumor progression, metastasis, and therapy resistance, making them valuable prognostic indicators. Notably, miR-21 is frequently overexpressed in lung, breast, colorectal, and gastric cancers; miR-21 promotes cell survival, angiogenesis, and immune evasion, correlating with poor prognosis and chemoresistance. miR-155 is reported to be elevated in hematological malignancies, breast cancer, and pancreatic cancer; miR-155 modulates inflammatory pathways and oncogenic signaling, contributing to tumor aggressiveness and immune evasion (Zhang et al. 2010). miR-17-92 cluster is frequently upregulated in lymphomas, lung, and colorectal cancers, promoting cell proliferation and inhibition of apoptotic pathways, leading to high metastatic potential [37].

Tumor Suppressor miRNAs as Predictive Markers

The loss or reduced expression of tumor suppressor miRNAs is linked to aggressive tumor growth and poor survival outcomes. miR-34a, a key downstream target of p53, inhibits tumor growth by promoting cell cycle arrest and apoptosis. Its depletion is associated with chemoresistance and metastasis in cancers such as pancreatic, colorectal, and ovarian cancer [24, 25]; let-7, one of the first identified tumor suppressor miRNAs, suppresses oncogenes like RAS and MYC. Lower let-7 levels are linked to lung cancer progression and poor treatment responses [30]. Similarly, miR-15a/16 acts as tumor suppressors by targeting BCL-2, an anti-apoptotic protein. Their deletion or downregulation is commonly found in chronic lymphocytic leukemia (CLL) and other cancers, promoting tumor cell survival [9–12].

miRNA Mimics and Inhibitors as Therapeutic Strategies

Targeting miRNAs therapeutically involves either restoring tumor-suppressive miRNAs (miRNA mimics) or inhibiting oncogenic miRNAs (antagomiRs/miRNA inhibitors). miRNA Mimics: Synthetic miRNAs are designed to replace lost tumor suppressor miRNAs. Example: MRX34, a miR-34a mimic, was tested in clinical trials for solid tumors [4]. AntagomiRs (miRNA Inhibitors): Chemically modified oligonucleotides are designed to suppress oncomiRs. Example: AntagomiR-21 targets miR-21 to restore PTEN expression and suppress tumor growth. Other promising approaches include the following: small molecule inhibitors compounds that modulate miRNA expression by affecting transcription factors or regulatory pathways; miRNA sponges decoy molecules that bind to and sequester multiple copies of an oncomiR, thereby reducing its activity.

Challenges in miRNA-Based Therapies (Delivery, Specificity)

Despite their potential, miRNA-based therapies face significant delivery and specificity challenges: stability and -degradation—miRNAs are susceptible to nuclease degradation in circulation, requiring chemical modifications (e.g., 2′-O-methylation) for enhanced stability. Efficient delivery systems: Nanoparticles, liposomes, and exosome-based delivery platforms are being explored to improve miRNA uptake in tumor cells. Off-target effects: Since miRNAs regulate multiple genes, unintended off-target effects can lead to undesired toxicity. Strategies such as tissue-specific promoters and modified delivery vehicles are under investigation to enhance specificity.

Future Perspectives and Conclusion

miRNAs have emerged as pivotal regulators of cancer biology, influencing cell cycle control, tumor progression, and therapeutic resistance. The expanding research in miRNA-cancer interactions offers novel insights into tumor biology and targeted treatments, paving the way for personalized medicine approaches. However, challenges related to delivery, specificity, and safety must be addressed for their effective clinical application. miRNA-based strategies offer novel opportunities for cancer diagnosis and therapy, with ongoing clinical trials and advancements in delivery technologies. However, overcoming challenges in specificity, stability, and tumor targeting will be crucial for their successful clinical translation.

Emerging Research on miRNA-Cancer Interactions

Recent advancements in next-generation sequencing (NGS), single-cell RNA sequencing, and AI-driven miRNA analysis have greatly enhanced our understanding of miRNA regulation in cancer. Key emerging areas include:

- **miRNA Crosstalk with Other Noncoding RNAs:** Interactions between miRNAs, long noncoding RNAs (lncRNAs), and circular RNAs (circRNAs) in cancer progression.
- **Exosomal miRNAs in Cancer Metastasis:** Tumor-derived exosomal miRNAs play a vital role in cell-to-cell communication, immune evasion, and metastasis.
- **CRISPR-Based miRNA Editing:** Genome-editing tools like CRISPR/Cas13 enable precise miRNA modifications, offering potential therapeutic applications.

Potential for Personalized Medicine Using miRNA Profiling

The unique miRNA expression signatures observed across different cancer types have positioned miRNAs as valuable tools in personalized medicine, aiding in cancer diagnosis, prognosis, and treatment optimization. One of the most promising applications is early cancer detection and prognosis, where circulating miRNAs serve as non-invasive liquid biopsy biomarkers. Unlike traditional tissue biopsies, which are invasive and may not capture tumor heterogeneity, miRNA-based liquid biopsies enable real-time monitoring of tumor dynamics, improving early detection and disease surveillance. Additionally, miRNA profiling has been instrumental in predicting therapy responses, allowing for patient stratification based on drug sensitivity. Certain miRNAs are known to mediate chemotherapy resistance or sensitivity, making them valuable indicators for tailoring chemotherapy and immunotherapy regimens to enhance treatment efficacy. Moreover, miRNA-based strategies are being explored in combination therapies, where miRNA mimics or inhibitors are used alongside targeted inhibitors, chemotherapy, or immune checkpoint inhibitors. By modulating the expression of key oncogenic or tumor suppressor pathways, these approaches hold the potential to overcome drug resistance, reduce side effects, and improve patient outcomes. The integration of miRNA-based diagnostics and therapeutics into personalized medicine frameworks represents a significant step toward more precise, individualized cancer treatment strategies.

Summary of Key Points and Final Remarks

miRNAs function as critical regulators of cell cycle checkpoints, influencing tumorigenesis through their role as oncomiRs or tumor suppressor miRNAs. The dysregulation of miRNAs in cancer leads to uncontrolled proliferation, metastasis, and resistance to therapy. miRNA-based therapies (mimics, inhibitors, and nanoparticles) hold promise, but challenges such as target specificity and efficient delivery must be addressed. Future research should focus on miRNA-based diagnostics, combination therapies, and precision medicine approaches for improved cancer treatment.

In conclusion, while miRNA-based therapeutics are still in their early stages, their integration into personalized cancer medicine could revolutionize cancer diagnosis, prognosis, and therapy. Continued advancements in biotechnology, RNA therapeutics, and artificial intelligence-driven miRNA analysis will be instrumental in translating these promising findings into clinical applications.

References

1. Alberts B, Johnson A, Lewis J, Raff M, Roberts K, Walter P (2015) Molecular biology of the cell, 6th edn. Garland Science
2. Bartel DP (2009) MicroRNAs: target recognition and regulatory functions. Cell 136(2):215–233. https://doi.org/10.1016/j.cell.2009.01.002
3. Bartel DP (2018) Metazoan microRNAs. Cell 173(1):20–51. https://doi.org/10.1016/j.cell.2018.03.006
4. Beg MS, Brenner AJ, Sachdev J, Borad M, Kang YK, Stoudemire J et al (2017) Phase I study of MRX34, a liposomal miR-34a mimic, in patients with advanced solid tumors. Mol Cancer Ther 16(5):1106–1115. https://doi.org/10.1158/1535-7163.MCT-16-0530
5. Bracken CP, Gregory PA, Kolesnikoff N, Bert AG, Wang J, Shannon MF, Goodall GJ (2008) A double-negative feedback loop between ZEB1–SIP1 and the microRNA-200 family regulates epithelial-mesenchymal transition. Cancer Res 68(19):7846–7854. https://doi.org/10.1158/0008-5472.CAN-08-1942
6. Burkhart DL, Sage J (2008) Cellular mechanisms of tumour suppression by the retinoblastoma gene. Nat Rev Cancer 8(9):671–682. https://doi.org/10.1038/nrc2399
7. Bushati N, Cohen SM (2007) microRNA functions. Annu Rev Cell Dev Biol 23:175–205. https://doi.org/10.1146/annurev.cellbio.23.090506.123406. PMID: 17506695
8. Calin GA, Croce CM (2006) MicroRNA signatures in human cancers. Nat Rev Cancer 6(11):857–866. https://doi.org/10.1038/nrc1997
9. Calin GA, Cimmino A, Fabbri M, Ferracin M, Wojcik SE, Shimizu M et al (2008a) MiR-15a and miR-16 induce apoptosis by targeting BCL2. Proc Natl Acad Sci 105(32):5166–5171. https://doi.org/10.1073/pnas.0801239105
10. Calin GA, Cimmino A, Fabbri M, Ferracin M, Wojcik SE, Shimizu M et al (2008b) MiR-15a and miR-16-1 cluster functions in human leukemia. Proc Natl Acad Sci 105(13):5166–5171. https://doi.org/10.1073/pnas.0800121105
11. Calin GA, Cimmino A, Fabbri M, Ferracin M, Wojcik SE, Shimizu M et al (2008c) MiR-15a and miR-16-1 cluster functions in human leukemia. Proc Natl Acad Sci 105(13):5166–5171. https://doi.org/10.1073/pnas.0800294105
12. Calin GA, Cimmino A, Fabbri M, Ferracin M, Wojcik SE, Shimizu M et al (2008d) miR-16 modulates cell growth through targeting CCNE1. Oncogene 27(48):6748–6757. https://doi.org/10.1038/onc.2008.27
13. Dang CV (2012) MYC on the path to cancer. Cell 149(1):22–35. https://doi.org/10.1016/j.cell.2012.03.003
14. Deng X, Wu X, Bauer NA, Hu X (2016) MicroRNA regulation of c-MYC in cancer. Tumour Biol 37(9):11667–11677. https://doi.org/10.1007/s13277-016-5046-7
15. Di Leva G, Croce CM (2013) miRNA profiling of cancer. Curr Opin Genet Dev 23(1):3–11. https://doi.org/10.1016/j.gde.2013.01.002
16. Esquela-Kerscher A, Slack FJ (2006) Oncomirs – microRNAs with a role in cancer. Nat Rev Cancer 6(4):259–269. https://doi.org/10.1038/nrc1840
17. Feng Y, Zhang X (2012) miR-21 promotes tumorigenesis and genomic instability by targeting checkpoint kinases. Oncogene 31(29):3655–3664. https://doi.org/10.1038/onc.2011.530
18. Garofalo M, Quintavalle C, Romano G, Croce CM, Condorelli G (2009) miR-221/222 in cancer: their role in tumor progression and response to therapy. Curr Mol Med 9(6):739–745. https://doi.org/10.2174/156652409788848725
19. Gasparini P, Lovat F, Fassan M (2014a) Protective role of miR-155 in breast cancer through RAD51 targeting impairs homologous recombination after irradiation. Proc Natl Acad Sci 111(12):4536–4541. https://doi.org/10.1073/pnas.1319855111
20. Gasparini P, Lovat F, Caserini P et al (2014b) miR-155 modulates the DNA damage response through RAD51 targeting. Oncogene 33(4):480–488. https://doi.org/10.1038/onc.2012.620

21. Gasparini P, Lovat F, Fassan M (2014) Protective role of miR-155 in breast cancer through RAD51 targeting impairs DNA repair. Cancer Res 74(2):532–541. https://doi.org/10.1158/0008-5472.CAN-13-1624
22. Ha M, Kim VN (2014) Regulation of microRNA biogenesis. Nat Rev Mol Cell Biol 15(8):509–524. https://doi.org/10.1038/nrm3838
23. Hanahan D, Weinberg RA (2011) Hallmarks of cancer: the next generation. Cell 144(5):646–674. https://doi.org/10.1016/j.cell.2011.02.013
24. He L, He X, Lowe SW, Hannon GJ (2007) microRNAs join the p53 network – another piece in the tumour-suppression puzzle. Nat Rev Cancer 7(11):819–822. https://doi.org/10.1038/nrc2232
25. He L, He X, Lim LP, de Stanchina E, Xuan Z, Liang Y et al (2007) A microRNA component of the p53 tumor suppressor network. Nature 447(7148):1130–1134. https://doi.org/10.1038/nature05939
26. Huang T, Alvarez AA, Pangeni RP, Horbinski C (2014) MicroRNAs and c-MYC interplay in cancer. J Hematol Oncol 7(1):65. https://doi.org/10.1186/s13045-014-0065-5
27. Huang X, Liang M, Tang W (2017) Circulating microRNAs as biomarkers in the diagnosis and prognosis of cancer. Front Genet 8:150. https://doi.org/10.3389/fgene.2017.00150
28. Humphries B, Wang Z, Yang C, Yang C (2014) MicroRNA regulation of cell cycle progression in tumor transformation and metastasis. Cancer Lett 354(1):1–7. https://doi.org/10.1016/j.canlet.2014.07.039
29. Jeggo PA, Pearl LH, Carr AM (2011) DNA repair, genome stability, and cancer: a historical perspective. Nat Rev Cancer 11(10):692–699. https://doi.org/10.1038/nrc3085
30. Johnson SM, Grosshans H, Shingara J, Byrom M, Jarvis R, Cheng A et al (2005) RAS is regulated by the let-7 microRNA family. Cell 120(5):635–647. https://doi.org/10.1016/j.cell.2005.01.014
31. Lal A, Pan Y, Navarro F, Dykxhoorn DM, Moreau L, Meire E et al (2009) miR-24 inhibits cell proliferation by suppressing DNA repair pathways in human cells. Cancer Res 69(13):5233–5240. https://doi.org/10.1158/0008-5472.CAN-08-3633
32. Le MT, Teh C, Shyh-Chang N et al (2009) MicroRNA-125b is a novel negative regulator of p53. Genes Dev 23(7):862–876. https://doi.org/10.1101/gad.1767609
33. Le MTN, Teh C, Shyh-Chang N et al (2011) MicroRNA-125b is a novel negative regulator of p53. Genes Dev 25(2):199–205. https://doi.org/10.1101/gad.1975911
34. Lu TX, Rothenberg ME (2018) MicroRNA. J Allergy Clin Immunol 141(4):1202–1207. https://doi.org/10.1016/j.jaci.2017.08.034. Epub 2017 Oct 23. PMID: 29074454; PMCID: PMC5889965
35. Lujambio A, Ropero S, Ballestar E, Fraga MF, Cerrato C, Setien F et al (2007) Genetic unmasking of an epigenetically silenced microRNA in human cancer cells. Cancer Res 67(4):1424–1429. https://doi.org/10.1158/0008-5472.CAN-06-4218
36. Malumbres M, Barbacid M (2009) Cell cycle, CDKs, and cancer: a changing paradigm. Nat Rev Cancer 9(3):153–166. https://doi.org/10.1038/nrc2602
37. Mendell JT (2008) miRiad roles for the miR-17-92 cluster in development and disease. Cell 133(2):217–222. https://doi.org/10.1016/j.cell.2008.04.001
38. Meyer N, Penn LZ (2008) Reflecting on 25 years with MYC. Nat Rev Cancer 8(12):976–990. https://doi.org/10.1038/nrc2231
39. Mogilyansky E, Rigoutsos I (2013) The miR-17/92 cluster: a comprehensive update on its genomics, genetics, functions and increasingly important and numerous roles in health and disease. Cell Death Differ 20(12):1603–1614. https://doi.org/10.1038/cdd.2013.125
40. Mohr AM, Mott JL (2015) Overview of microRNA biology. Semin Liver Dis 35(1):3–11. https://doi.org/10.1055/s-0034-1397344. Epub 2015 Jan 29. PMID: 25632930; PMCID: PMC4797991
41. Moskwa P, Buffa FM, Pan Y et al (2011) miR-182-mediated downregulation of BRCA1 impacts DNA repair and sensitivity to PARP inhibitors. Mol Cell 41(2):210–220. https://doi.org/10.1016/j.molcel.2010.12.005

42. Musacchio A, Salmon ED (2007) The spindle-assembly checkpoint in space and time. Nat Rev Mol Cell Biol 8(5):379–393. https://doi.org/10.1038/nrm2163
43. Nana-Sinkam SP, Croce CM (2011) MicroRNAs as therapeutic targets in cancer. Transl Res 157(4):216–225. https://doi.org/10.1016/j.trsl.2011.01.009
44. O'Donnell KA, Wentzel EA, Zeller KI, Dang CV, Mendell JT (2005) c-MYC-regulated microRNAs modulate E2F1 expression. Nature 435(7043):839–843. https://doi.org/10.1038/nature03677
45. Olive V, Bennett MJ, Walker JC, Ma C, Jiang I, Cordon-Cardo C et al (2013) miR-19 is a key oncogenic component of the miR-17-92 cluster. Genes Dev 23(24):2839–2849. https://doi.org/10.1101/gad.186827.113
46. Palumbo T, Faucz FR, Azevedo M et al (2014) miR-16 targets CHK1 and influences cell survival in adrenocortical tumors. Endocr Relat Cancer 21(4):611–623. https://doi.org/10.1530/ERC-14-0094
47. Peng Y, Croce CM (2016) The role of microRNAs in human cancer. Signal Transduct Target Ther 1(1):15004. https://doi.org/10.1038/sigtrans.2015.4
48. Rupaimoole R, Slack FJ (2017) MicroRNA therapeutics: towards a new era for the management of cancer and other diseases. Nat Rev Drug Discov 16(3):203–222. https://doi.org/10.1038/nrd.2016.246
49. Samsonova A, Aravin AA, Kulakovskiy IV (2022) Let-7 family miRNAs in development and cancer: insights from network analysis and multi-omics data. Mol Oncol 16(12):2154–2171. https://doi.org/10.1002/1878-0261.13185
50. Sherr CJ, Roberts JM (1999) CDK inhibitors: positive and negative regulators of G1-phase progression. Genes Dev 13(12):1501–1512. https://doi.org/10.1101/gad.13.12.1501
51. Si ML, Zhu S, Wu H, Lu Z, Wu F, Mo YY (2007) miR-21-mediated tumor growth. Oncogene 26(19):2799–2803. https://doi.org/10.1038/sj.onc.1210083
52. Vousden KH, Prives C (2009) Blinded by the light: the growing complexity of p53. Cell 137(3):413–431. https://doi.org/10.1016/j.cell.2009.04.037
53. Yamakuchi M, Ferlito M, Lowenstein CJ (2008) miR-34a repression of SIRT1 regulates apoptosis. Proc Natl Acad Sci 105(36):13421–13426. https://doi.org/10.1073/pnas.0812308105
54. Yu F, Yao H, Zhu P, Zhang X, Pan Q, Gong C et al (2007) Let-7 regulates self-renewal and tumorigenicity of breast cancer cells. Cell 131(6):1109–1123. https://doi.org/10.1016/j.cell.2007.10.054
55. Zhou BB, Elledge SJ (2000) The DNA damage response: putting checkpoints in perspective. Nature 408(6811):433–439. https://doi.org/10.1038/35044005
56. Tili E, Michaille JJ, Cimino A, Costinean S, Dumitru CD, Adair B, Fabbri M, Alder H, Liu CG, Calin GA, Croce CM (2007) Modulation of miR-155 and miR-125b levels following lipopolysaccharide/TNF-alpha stimulation and their possible roles in regulating the response to endotoxin shock. J Immunol 179(8):5082–9. https://doi.org/10.4049/jimmunol.179.8.5082
57. Meng F, Henson R, Wehbe-Janek H, Ghoshal K, Jacob ST, Patel T (2007) MicroRNA-21 regulates expression of the PTEN tumor suppressor gene in human hepatocellular cancer. Gastroenterology. 133(2):647–58. https://doi.org/10.1053/j.gastro.2007.05.022

Chapter 9
MicroRNAs in Breast Cancer: Diagnostic and Therapeutic Potential

Serdar Karakurt, Irem Mukaddes Bilgiseven, Kemal Bas, and Cigdem Gokcek Sarac

Abstract MicroRNAs (miRNAs) have become key players in the pathogenesis, development, and response to treatment of breast cancer. This chapter discusses the clinical and preclinical implications of miRNAs, with special emphasis on their function in drug resistance, tumor suppression, and therapeutic responsiveness. The miR-34a-mediated tumor-suppressive effects of combination treatments, the prognostic value of circulating miRNA by miR-34a and miR-195, and miRNA polymorphism (e.g., miR-23b and miR-107) effects on post-operative pain treatment are discussed in detail. Further, miRNAs are investigated as diagnostic and prognostic biomarkers and, therefore, as possible targets for translation in personalized medicine. Although it is challenging to target miRNAs therapeutically, their multi-targeting efficacy is superior to that of single-gene therapies. The findings underscore the growing promise of miRNA-based therapies and their standing in clinical oncology for the better management of breast cancer.

Keywords MicroRNAs (miRNAs) · Breast cancer · Drug resistance · Biomarkers · Personalized medicine

Introduction

Breast cancer is known worldwide as one of the most common and challenging malignancies. Breast cancer accounts for about 30% of the types of cancer observed in women, with a mortality-to-incidence rate of 15% [1]. Despite the advances in conventional treatments, it is associated with significant morbidity and mortality

S. Karakurt (✉) · I. M. Bilgiseven · K. Bas
Faculty of Science, Department of Biochemistry, Selcuk University, Konya, Türkiye

C. Gokcek Sarac
Faculty of Engineering, Department of Biomedical Engineering, Akdeniz University, Antalya, Türkiye

S. Pathak, A. Banerjee (eds.), *MicroRNAs: Role in Cancer Progression, Diagnostics and Therapy*, Advances in Biochemistry in Health and Disease 34,
https://doi.org/10.1007/978-3-032-08824-6_9

rates. A family history of breast cancer is still the best-known risk factor for breast cancer. Women with mutations in breast cancer susceptibility genes (BRCA1 or BRCA2) have a 70% or greater risk of developing breast cancer [2]. Conditions such as menstrual age, pregnancy and breastfeeding, or the age of menopause have an impact on a woman's risk of breast cancer with a BRCA mutation. Invasive breast cancer is classified according to the presence or absence of human epidermal growth factor type 2 (HER2) and hormone receptor (HR) expressions [3]. In the treatment of invasive breast cancer, it is important to make the classification correctly in terms of prognosis and drug sensitivity. Although the main treatment for breast cancer is surgery or radiation therapy, endocrine therapy is another option to suppress breast cancer recurrence [4]. Furthermore, HER2, which is overexpressed in 20–25% of breast cancers, is the most well-known therapeutic target in breast cancer [5]. Breast cancer treatment can be performed by administering chemotherapy with HER2-directed agents with neoadjuvant therapy. With the development of technology, identifying new molecular determinants of prognostic and predictive factors is a current research area.

MicroRNAs (miRNAs) are small noncoding RNA molecules 18–25 nucleotides long. The stability and translation efficiency of mRNAs are regulated by miRNAs. miRNAs specifically show altered expression levels in some types of cancer. miRNAs are associated with tumor progression, tumor suppressor, or metastasis [6]. The tumor's endogenous level of tumor suppressor miRNAs is low. Dysregulation of high endogenous levels of oncogenic miRNAs in the tumor causes tumor formation [7]. The tumor is formed due to an unlimited proliferation of cells, and the level of apoptosis in the cells decreases. miRNAs act as regulators of apoptosis in the cell. miR-21 inhibits the tumor suppressor factor PDCD4, leading to delayed cell proliferation [8]. miR-21 regulates a wide pathway associated with specific features of cancer, such as invasion and metastasis. Epithelial-mesenchymal transmission is crucial in breast cancer metastasis (Fig. 9.1). Decreased expression of E-cadherin affects the Wnt/β-catenin signaling pathway [9]. WNT-1 is one of the ligands of the Wnt/β-catenin pathway and is a direct target of miR-148a. Changes in the expression of miR-148a affect the concentration of WNT-1 and suppress the migration of breast cancer cells [10]. miR-30c affects the Nf-κB signaling pathway. Low expression of miR-30c indicates an advanced tumor stage and the presence of metastasis, while high expression of miR-30c is associated with decreased cell viability and increased apoptosis [11, 12]. miR-10b, miR-21, miR-31, miR-125b, miR-130a-3p, miR-155, miR-181a, miR-181b, miR-183, miR-195, and miR-451a are biomarkers that can be used to distinguish breast cancer from healthy breast tissue [13]. miRNA17-5p, miRNA-18a-5p, miRNA-20b-eta5p, miRNA-29b-3p, miRNA-93-5p, miRNA-124-3p, miRNA-195-5p, miRNA-338-3p, miRNA-340-5p, and miRNA-486-5p were associated with breast cancer recurrence rather than expression in tumor tissues [14].

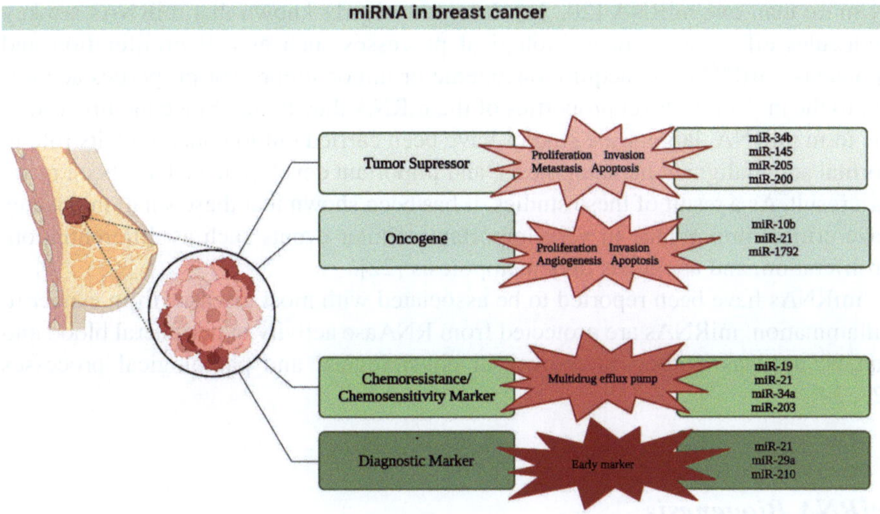

Fig. 9.1 Functional characterization of different miRNAs in breast cancer

miRNA Biogenesis and Function

miRNA is a single-stranded RNA molecule that is approximately 22 nucleotides (usually 20–24 nucleotides) long and regulates gene expression. There are two well-defined types of small RNAs: miRNAs and short interfering RNAs (siRNAs). The first discovered small RNA is miRNA [15]. miRNAs, which regulate gene expression by mRNA cleavage or translation inhibition, have a biomarker potential in terms of investigating their role in normal development and diseases. miRNAs are functional RNA molecules transcribed from RNA genes in protein-coding introns or exons and nonprotein-coding regions on the genome, but not translated into protein. The human genome has been discovered to have hundreds of highly conserved gene regions encoding miRNAs. To date, over 1000 miRNAs have been identified in the human genome [16]. Lee and colleagues in Victor Ambros' laboratory discovered the first microRNA in Caenorhabditis elegans, a roundworm, in 1993. The gene called lin-4 was reported to transcribe a small RNA of 22 nucleotides in length, although it does not encode any protein. The miRNA discovered as lin-4 decreased the expression of the target gene, and no protein product was found in DNA sequence analysis [17].

miRNAs are formed as a result of three successive steps. In the first step, transcription of primary miRNAs (pri-miRNAs) from miRNA genes occurs. In the second step, pri-miRNAs are converted into precursor miRNAs (pre-miRNAs) in the nucleus. In the third and final step, mature miRNAs are formed in the cytoplasm [18, 19]. miRNAs perform their functions by recognizing their own nucleotide sequences and complementary target genes. It is also known that each miRNA can regulate the expression of more than one mRNA, and each mRNA can be targeted by more than one miRNA [20, 21]. It is now clearly known that miRNAs are key molecules effective in many biological processes such as cell proliferation and apoptosis. miRNAs can acquire oncogenic or tumor suppressor properties according to the molecular-level properties of the mRNA they target. Since the first use of the term miRNA, large-scale studies have been carried out to understand its role in normal and malignant hematopoiesis, and important developments have been made as a result. As a result of these studies, it has been shown that these small molecules have critical importance in very important cellular events such as differentiation, proliferation, and apoptosis in hematopoiesis [22].

miRNAs have been reported to be associated with most diseases from cancer to inflammation. miRNAs are protected from RNAase activity in peripheral blood and can be used as a good biomarker in physiological and pathological processes [23, 24].

miRNA Biogenesis

miRNA biogenesis occurs in two steps: in the nucleus and the cytoplasm. Specialized molecules are involved in these steps (Fig. 9.2). miRNAs are transcribed from genomic DNA, usually by RNA polymerase II. In the first step, a long RNA called pri-miRNA is formed, which has the structure of a hairpin and contains the mature miRNA sequence [25]. The hairpin structure is cut by a microprocessor consisting of Drosha (RNAase III enzyme) and its cofactor DiGeorge critical syndrome region 8 (DGCR8) (Pasha). After this cut, a 60–70 nucleotide-long RNA molecule called pre-miRNA is formed. This pre-miRNA is transported from the nucleus to the cytoplasm via Exportin-5 (XPO5) [26]. The pre-miRNA that passes into the cytoplasm is catalyzed by Dicer-TRBP [transactivation-responsible (TAR) RNA-binding protein] and cut by Dicer1 (Dicer), also an enzyme from the RNase III family, to form a duplex of approximately 22 nucleotides. This duplex molecule contains both mature miRNA and antisense miRNA. A chain called a guide miRNA is loaded into the RISC (RNA-induced silencing complex) and called miRISC or miRNPs (miRibonucleoprotein). RISC also contains the Argonaute2 (Ago2) protein, a catalytic component that ensures chain selection for the more stable 5′ end. Following miRNA-mRNA interaction, protein expression of the target mRNA may occur as a result of translational repression, mRNA degradation, or mRNA decay accelerated by uncapping and deadenylation [27].

miRNA Expression in Breast Cancer

Alterations in miRNA expression can cause many diseases [28]. miR-15a and miR-16-1 may be located in genome regions where miRNAs may be associated with cancer [29, 30]. Inherited mutations and single-nucleotide polymorphisms in

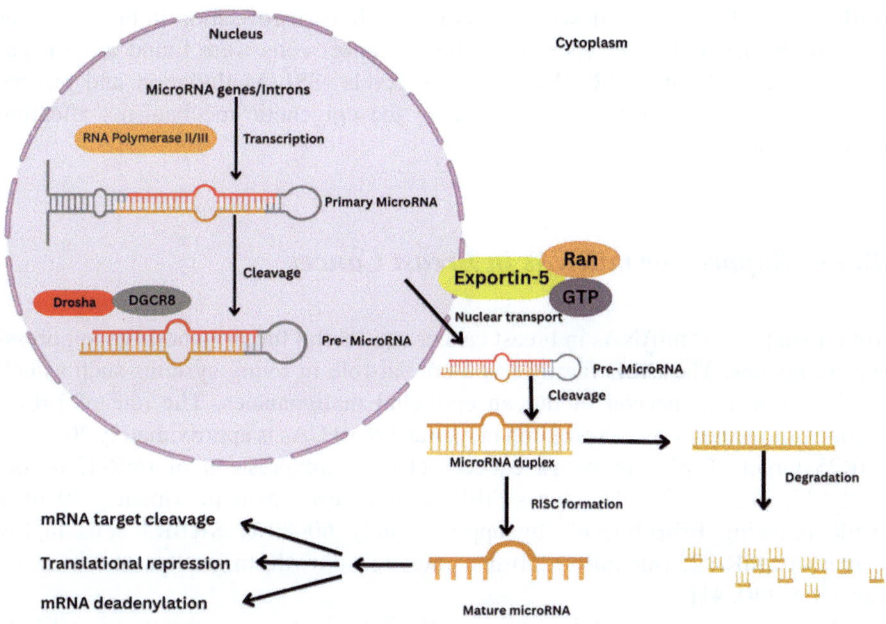

Fig. 9.2 Overview of miRNA biogenesis

the primary transcripts of miR-15a and miR-16-1 have been shown to cause decreased expression of these miRNAs in chronic lymphocytic leukemia [31]. Adenosine deaminase (ADARs) has been found to affect the expression level of miRNAs that recognize adenosine residues. In the case of miR-142, after ADARs regulate pri-miR-142, there is a decrease in its processing by Drosha, and it is degraded within RISC. As a result, it causes lower levels of miR-142 [32]. Alterations in the activity of Drosha and Dicer have been detected at the miRNA level in different tumor types. For example, silencing of Dicer or Drosha is associated with cellular transformation or tumorigenesis. Such as, down-expression of Dicer in a breast cancer cell line associated with tumor metastasis [33]. Further miRNAs themselves can influence the processing of other miRNAs in a process described as "complex, interplay regulation." Tang et al. have shown that miR-709 is present in the nucleus of various cells and interacts with pri-miR-15a/16-1 in mice. Consequently, the interaction between miR-709 and pri-miR-15a/16-1 promotes the maturation of pri-miR-15a/16-1 [34]. Tumor growth and disease progression in breast cancer are affected by the interaction between miR-790 and p53 pathway [35]. Changes in miRNA expressions in cancer may be related to epigenetic changes similar to changes in DNA methylation. Studies have shown that miRNA genes are associated with CpG islands [36]. Saito et al. suggested that changes in methylation level may lead to silencing of tumor suppressor miRNAs. As a result of this inhibition, up-expression of miR-127 in bladder cancer cell lines and human fibroblasts was observed. miR-127, silenced in different cancers, targets BCL-6. In addition, miR-9-1 was found to be down-expressed after hypermethylation in breast cancer [37]. High methylation levels in basal breast cancer cells were found to be negatively correlated with miR-29c expression levels [38]. Methylation and histone deacetylase (HDAC) inhibition are among the epigenetic mechanisms affecting miRNA expression [39].

Tumor Suppressor miRNAs in Breast Cancer

Tumor suppressor miRNAs in breast cancer can inhibit tumorigenesis by suppressing oncogenes. The ErbB family has a crucial role in living systems such as cell proliferation and survival of human epithelial malignancies. The rate of ErbB2 amplification and/or overexpression in human breast CAs is approximately 20–30%. miR25 targets ErbB2 in breast cancer cells. Overexpression of miR-125a and miR-125b in SKBR3 cells reduces ErbB2 protein levels by approximately 40–65% while reducing ErbB3 levels by approximately 60–80%. SKBR3 cells highly expressed miR-125a or miR-125b that impaired growth, migration, and invasion capacities [40, 41].

The genomic location of the human miR-17/20 cluster is on chromosome 13q31 and is associated with loss of heterozygosity in various cancers, including breast cancer. The miR-17/20 cluster has been found to reduce the amount of cyclin D1 and suppress breast cancer cell proliferation. Reduced miR-17/20 expression in breast cancer cell lines is inversely associated with high cyclin D1 levels. Decreased miR-17/20 expression in human breast cancer samples is associated with higher cyclin D1 levels compared to normal breast tissue. Decreased expression of miR-17/20 is associated with increased levels of cyclin D1. A gene study found that mice lacking miR-17/20 did not live very long. Therefore, more tissue-specific studies are needed [42].

miR-27b

Cytochrome P450 (CYP450s) contain essential enzymes involved in drug metabolism and bioactivation and are responsible for ~75% of drug metabolism in the body. CYP1B1 is an important cancer-associated molecule that catalyzes procarcinogens and 17β-estradiol and overexpressed in various human cancers, including breast cancer. There is a near-perfect match in CYP1B1 mRNA with the miR-27b sequence. Furthermore, analyses of breast tumor tissue and normal tissue showed an inverse relationship between miR-27b and CYP1B1 expression levels in tumor tissue. Most breast cancer patients have decreased miR-27b expression levels and increased CYP1B1 expression levels [16, 41].

miR-125a and miR-125b

One of the epidermal growth factor receptor (EGFR) family of tyrosine kinases is *HER2 (ERBB2). HER2* does not possess a ligand-binding domain of its own and therefore cannot bind growth factor. Hence, it binds tightly with another ligand-binding member of this family to form a heterodimer, thereby stabilizing its ligand-binding activity and increasing kinase-mediated downstream signaling pathways such as mitogen-activated protein kinase (MAPK) and phosphatidyl inositol-3 kinase. Amplification of this gene has been reported in many cancer types including breast tumors [40]. Both miRNAs are potential tumor suppressors, and bioinformatic studies have shown that the target seed sequences for miR-125a and miR-125b are in the 3′UTR region of HER2 and HER3. Overexpression of miR-125a and miR-125b in SKBR3 (ErbB2(HER2)-dependent human breast cancer cell line) cells has been shown to suppress transcript (mRNA) and protein levels of HER2 and HER3, leading to decreased anchorage-dependent growth, cell motility, and invasive capacity associated with low levels of p-ERK1/2 and p-AKT [43, 44].

miR-17-5p

miR-17-5p, also known as miR-91, is located on chromosome 13q31, and this genomic region has undergone loss of heterozygosity in various types of cancer, including breast cancer. miR-17-5p is complementary to the mRNA of the AIB1 (amplified in breast cancer 1) gene, a steroid receptor coactivator that increases the transcriptional activity of both ER-α and E2F. Therefore, AIB1 is a direct target of miR-17-5p. miR-17-5p inhibits E2F1 and ERα functions by suppressing AIB1 mRNA translation. miR-17-5p-mediated AIB1 downregulation also causes a decrease in estrogen receptor-mediated and estrogen receptor-independent breast cancer cell proliferation. Recently, it has been shown that the cyclin D1 gene (*CCND1*), which is overexpressed in approximately 50% of human cancers, is also a direct target of miR17-5p in breast cancer cells. miR-17-5p prevents the proliferation of breast cancer cells by suppressing the synthesis of cyclin D1 protein, and this effect is eliminated by CCND1 siRNA in breast cancer cells lacking cyclin D1 [45].

miR-34a

miR-34a is downregulated in many types of cancer and is transcriptionally regulated by p53. A study comparing normal epithelial cell lines and HER-2+ cell lines reported that miR-34a levels were reduced in triple-negative and mesenchymal breast cancer cell lines. The researchers suggested that p53 mutations in these subtypes of breast cancer contribute to the low miR-34a expression [46].

Oncogenic miRNAs in Breast Cancer

miR-27a

miR-27a has been reported as an oncomiR in breast cancer and regulates the transcriptional cofactor ZBTB10 (Zinc Finger and BTB Domain Containing 10) gene. This zinc finger ZBTB10 gene, a potential target of miR-27a, is a recognized repressor of oncogene-specific proteins (Sp), thought to be a transcription factor that enables the transition of the cell cycle from G0-G1 to S phase. Inactivation of miR-27a in MDA-MB-231 cells has been reported to increase ZBTB10 expression and decrease the amount of Sp genes [47]. However, miR-27a suppresses Myt-1 expression in MDA-MB-231 cells, increases cdc2/cyclin B activity, and promotes breast cancer cell proliferation. However, miR-27a suppresses Myt-1 expression in MDA-MB-231 cells, increases cdc2/cyclin B activity, and promotes breast cancer cell proliferation. In therapeutic studies, downregulation of miR-27a was reported in SKBR3 cell lines after treatment with LAQ824, a histone deacetylase inhibitor, using a proapoptotic dose [16].

miR-21

miR-21 is also overexpressed in most tumors including breast cancer. It is highly upregulated in breast tumors compared to normal breast tissues and shows oncogenic function [48]. Studies investigating the oncogenic role of miR-21 in cell cultures were also performed in mouse xenografts, and it was reported that miR-21 knockdown suppressed breast tumor development by inhibiting cell growth [48]. Tropomyosin 1 (TMP1), a tumor suppressor in MCF-7 cells, is one of the targets of miR-21. The fact that miR-21 downregulates the tumor suppressor gene programmed cell death reveals the effect of miR-21 on cell proliferation. In pharmacological studies, researchers have shown that miR-21 antagonism in MCF-7 cells causes apoptosis, and this is associated with low Bcl-2 protein expression in cells transfected with anti-miR-21 and tumors originating from these cells [49]. They suggested that Bcl-2 may be an indirect target of miR-21. Another important target of miR-21 is PTEN, a phosphatase and tensin homologue tumor suppressor gene. Maspin is also a direct target of miR-21 [50]. These 3 genes are responsible for decreases in invasive capacity in metastatic breast cancer cell lines.

In conclusion, both the importance of miRNAs in cell cycle regulation and their oncogenic, tumor suppressor, or metastatic properties reveal the importance of miRNAs in breast cancer diagnosis and treatment. In recent studies, considering that miRNAs play an important role in the development of neurodegenerative diseases, autoimmune disorders, cancer, and viral infections, the importance of revealing RNA mechanisms and therapeutic use in diagnosis and treatment is getting stronger daily [51]. Therefore, the use of miRNAs in the treatment of these diseases is expected to break new ground, and the therapeutic approaches to be developed are thought to be promising for both patients and clinicians.

Diagnostic Potential of miRNAs in Breast Cancer

Breast cancer is one of the most prevalent malignant diseases affecting women worldwide. The high incidence of the disease has prompted comprehensive research toward elucidating its molecular basis, identifying potential biomarkers to improve the diagnostic and prognostic procedures, along with the treatment response [52]. The necessity for effective biomarkers is increasingly critical as healthcare providers attempt to seek more personalized and targeted cancer treatment strategies [53]. Among the candidates, microRNAs (miRNAs) are attracting considerable interest [54]. These small noncoding RNA molecules play a very important role in the regulation of gene expression and are involved in serious biological processes, such as cell proliferation, differentiation, and apoptosis, which are essential in maintaining cellular homeostasis and determining whether cells grow or die [55]. The dysregulation of miRNAs is documented to be involved in several cancers, including the initiation and progression of breast cancer, making them novel candidates for biomarker development [54]. miRNAs have the capability of allowing clinicians to gain deeper insight into the disease course and, if possible, enabling personalized therapy approaches for a certain patient [56].

Breast cancer is not a single entity; it comprises a broad spectrum of subtypes that differ in molecular profiles, clinical behaviors, and treatments. Understanding how various miRNAs are associated with these subtypes will lead to better diagnostic accuracy [54]. This chapter will discuss the prognostic and predictive capabilities of miRNAs, explaining how they may provide insight into disease outcomes and response to therapies to inform treatment decisions. Although miRNAs are seen as promising indicators for breast cancer, their widespread clinical application must overcome numerous difficulties. Stability within biological samples, an actual need for uniformity of assays, and complexities involved in interpreting their expression-just some of the huge obstacles [23]. Besides, intrinsic heterogeneity in breast cancer displays further complexities to these processes that define the identification and validation of miRNA biomarkers. Here, the aim is to make these challenges clear, while discussing possible strategies for confronting them, in order to provide an in-depth insight into the current state and the possible way forward for further research.

miRNAs in Subtypes of Breast Cancer

Breast cancer is an extremely complicated and varied illness that appears in many forms. It consists of distinct molecular subtypes that exhibit different biological features, clinical presentations, and degrees of responsiveness to various treatment modalities. Intrinsic heterogeneity of the disease poses serious challenges in effective management and involves critical implications for choices pertaining to strategies of treatment and prognostic assessment [57]. These two concepts, therefore, are translated into the classification of subtypes of breast cancer into: estrogen receptor-positive (ER+), HER2-positive (HER2+), and triple-negative breast cancer (TNBC), serving as a huge determinant to the development of personalized therapies [58]. Incidentally, classification plays a very important part in decisions regarding treatment as, through that, the individual care provider is able to fit the therapeutic approach to uniquely portrayed biological features of each subset. These molecular subtypes are important in improving patient care and increasing overall outcomes [59].

The role of miRNAs in the pathophysiology has increasingly been the focus of attention by scientists and researchers in studying the complex molecular mechanisms underlying the different subtypes of breast cancer. Indeed, recent investigations have provided an explanation for the involvement of certain miRNAs concerning various subtypes of breast cancer and hence suggested their role not only as diagnostic markers but also as therapeutic targets [60]. Therefore, as an example, miR-21 was defined as a critical biomarker, almost uniquely associated with poor prognosis in patients affected by ER+ breast cancer [61]. Overexpression of miR-21 has been linked to some conditions promoting tumor growth and acquiring resistance to endocrine therapies, which are usually used as a first-line treatment in patients with hormone receptor-positive tumors [62]. The presence and levels of expression of miR-21 may turn out to be a powerful biomarker that enhances the ability of clinicians to determine in each patient the likelihood of response to treatment and the overall prognosis. This knowledge could potentially guide more aggressive intervention strategies in those patients identified as being at higher risk for poor outcomes. Furthermore, miRNAs regulated by estrogen receptor activity, such as miR-342-3p, have also been demonstrated in ER+ breast cancer [63].

In the case of another subtype, that is, HER2+ breast cancer, the upregulation of miR-125b has been observed [64]. This is coupled with the enhancement in the cellular proliferation and invasive capacity of the cancerous cells. These types of observations indicate the involvement of miR-125b in disease progression [65] and provide evidence that it is a critical player in the HER2+ subtype of breast cancer which opens up great possibilities in applying this miRNA concomitantly with current HER2-directed therapies and hopes to improve treatment efficacy and patients' outcomes in this aggressive subtype of cancer.

In the study of TNBC, which is characterized by very aggressive features and a lack of targeted therapeutic options, miR-200c was identified as a key modulator regulating the epithelial-mesenchymal transition (EMT) process, a crucial stage in the progression and metastasis of TNBC [66]. EMT is one of the critical biologic processes associated with metastasis and tumor spread. Thus, dysregulation of miR-200c in TNBC explains their aggressive clinical course and underlines the importance of deciphering miRNAs within their biological perspective of breast cancer [67]. In TNCB, miR-18a-5p, miR-9-5p, miR-135b-5p, miR-522-3p are upregulated, while miR-449a and miR-190b are downregulated. Furthermore, miR-21 is overexpressed in HER2+ and triple-negative types, which is linked to aggressive tumor behaviors and poor prognosis [68], miR-155 increases the metastatic capacity of breast cancer cells by modulating the epithelial-mesenchymal transition process [69], and certain miRNAs in the miR-17-92 cluster are upregulated in HER2+ breast cancer [70]. Furthermore, circulating miRNAs such as miR-503, miR-1307, miR-212, and miR-592 have been reported to have potential in predicting breast cancer progression and response to treatment [71]. Key miRNAs, including miR-21, miR-23a, and miR-34a, are associated with biomolecular processes, such as apoptosis, angiogenesis, and metastasis, and have an important role in determining prognosis and molecular subtypes [55].

Taken together, these observations depict how miRNAs could be greatly helpful in understanding unique biological features and complexities related to the different subtypes of breast cancer [60]. The increasing amount of research on miRNAs continuously improves our understanding of the complexities in breast cancer but simultaneously opens up new possibilities for therapeutic intervention [71]. The identification of specific miRNAs could therefore provide investigators and clinicians with opportunities to develop innovative treatment modalities aimed at targeting the underlying molecular drivers of each subtype [56]. As our knowledge grows and research moves forward, the possibility of new paradigms in treatment will look ever brighter.

Prognostic and Predictive miRNA Biomarkers in Breast Cancer

The role of miRNAs as prognostic and predictive biomarkers in relation to breast cancer has been one of the most dynamically elaborated fields of both basic and clinical research for many years. This intense area of study has revealed numerous key miRNAs functioning as biomarkers that were found to be significantly correlated with patient outcome, tumor characteristics, and general survival rates. Metastasis is one of the leading causes of death from breast cancer, and miRNAs, like miR-520c, miR-10b, miR-373, and miR-155, may accelerate this process by increasing migration and invasion of cancer cells [72]. Among these, miR-155 is one of the most studied miRNAs, as it has gained much attention for its association with high metastatic levels. A tremendous number of studies have found that high levels of miR-155 are associated with a higher chance of disease progression among breast cancer patients [60]. The strong correlation of miR-155 expression with disease progression emphasizes the urgent need for further investigation into the expression patterns of different miRNAs in the context of breast cancer. These molecular signatures are of more than academic interest; they could be critical markers of the disease's aggressiveness and a potential course of a patient's illness. For example, miR-182 has been associated with poor clinical outcomes and identified as a potential prognostic biomarker [73], while researches have indicated that high levels of miR-10b are associated with increased risk of metastasis and poor prognosis [74], and low expression of miR-335 increases metastatic potential and leads to decreased survival rates [75]. Besides, it has been reported that miR-210 and miR-215 are significant in the early detection of metastasis up to 2 years prior to clinical diagnosis [76]. The miR-200 family (miR-200a, miR-200b, and miR-200c) and miR-203 have been reported to be effective in predicting metastasis and overall survival and to be of clinical benefit in metastatic breast cancer (MBC) patients [67]. Moreover, miRNAs, such as miR-21, miR-34a, and miR-23a, are linked to apoptosis, angiogenesis, and metastasis in disease progression [55]. Unlike miR-155, the function of miR-34a has emerged clearly in the realm of breast cancer research as a well-defined tumor suppressor. Most of the evidence has established that higher expression levels of miR-34a are linked to a good prognosis for patients. This is not only a significant finding but also highlights the potential of miR-34a as a good biomarker. Furthermore, it also offers intriguing possibilities for its potential use as a therapeutic agent in ongoing efforts to combat this common cancer [77].

The dual functions of miRNAs, serving both as prognosis biomarkers and potential therapeutic targets, hold great promise to significantly impact and improve breast cancer treatments toward personalized therapy. Beyond their prognostic value, miRNAs also hold immense promise as predictive biomarkers, which can be used to estimate the likelihood of treatment responses in breast cancer patients. One such example is miR-27a, which has become a very prominent molecule in the discussion of resistance to tamoxifen, a common and widely used option for treating ER+ breast cancer [78]. Expression levels of miR-27a may act as important markers for assessing appropriate treatment approaches. Certainly, more personalized therapies rely on specific predispositions in resistance to treatment. Besides, miR-205 is one of the most important miRNAs studied in recent years, associated with sensitivity to HER2-targeted therapy. This link is significant as it may indicate that miR-205 expression levels could influence the treatment strategy for patients with HER2+ breast cancer. Profiling of miRNA expression to determine which patients will elicit the most appropriate response to particular treatments represents a major step toward the development of precision medicine. It sets the scene for developing more effective personalized strategies for the management of breast cancer and ultimately for improving patient care [79].

In summary, the knowledge presently available on miRNAs in relation to breast cancer points out two faces of the same coin: as prognostic and predictive biomarkers, the avenues toward personalized medicine. The use of miRNA profiling will help the health professional make better choices on the treatment options for ultimately improved patient outcomes and survival rates in the face of the complex challenges breast cancer poses. The ongoing research efforts in this field are critical to enabling the creation of more innovative and effective strategies for fighting this widespread and multifaceted disease that plagues many people and families across the globe. A deeper comprehension of the various roles that miRNAs play in the context of breast cancer will result from this, allowing for the customization of treatments and giving patients new hope. Findings from studies thus carry potential not only regarding novel approaches to therapeutic decision-making but also in adding to the basic knowledge to facilitate changes that are much needed to bring about massive changes in the treatment arena and patient care in this important area of oncology.

Challenges in Clinical Application

Despite an enormous potential of miRNAs as biomarkers of the pathophysiology of breast cancer, there are still massive challenges to their clinical utilization. Some of the big technical challenges involve the precise identification and quantification of miRNAs. For example, the currently used Quantitative Real-Time-PCR (qRT-PCR) and Next-Generation Sequencing (NGS) are two of the few powerful analytic tools in the laboratory that often need to compromise one or several of their best features, including sensitivity, specificity, and reproducibility. Indeed, the technical limitations were noted in the study of Schwarzenbach et al. [71] that one of the major challenges is the consistency and reliability of the measurement of miRNAs across different biological samples. The relatively low quantity of miRNAs in various biological samples a crucial element in making their assessment more difficult-further complicates this. These difficulties are made worse by RNA contamination and degradation, which makes it more difficult to get accurate and reliable data. Mestdagh et al. showed how the problems in this respect then result in biased data that finally hinder the reliable use of miRNAs as biomarkers in clinical diagnostics [80].

Apart from the technical challenges, the standardization of miRNA detection methodologies is very important to ensure the reliability and reproducibility of the results for different laboratories and clinical settings. Indeed, variations in sample handling, processing techniques, and analytical methodologies can easily result in inconsistent and conflicting reports that seriously hamper the translation of miRNA biomarkers into everyday clinical practice. Through the overcoming of such challenges, the efforts made have resulted in a concerted development of standardized protocols in miRNA research. One of the significant efforts in this field is the guidelines entitled "Minimum Information for the Publication of Quantitative Real-Time PCR Experiments (MIQE)," which were created to increase the credibility of miRNA studies [81]. The main goal of these guidelines is to reduce variability but increase the quality of the overall miRNA research for its possible clinical application. One big hindrance to the practical application of miRNA biomarkers in clinical practice is the requirement for a strong validation in large independent cohorts. Although many studies reported associations between specific miRNAs and breast cancer outcomes, it has to be taken into consideration that the findings often come from either small sample sizes or validation in diverse populations is very limited [82]. Hence, comprehensive validation studies involving larger and diverse patient groups are an inevitable necessity for confirming the clinical utility of miRNA biomarkers and establishing their roles in everyday clinical practice. Without such validation, it becomes very difficult to establish the reliability of these biomarkers in guiding diagnostic procedures or treatment decisions.

Another major obstacle to the effective use of miRNAs as diagnostic or prognostic tools is the inherent biological complexity associated with them. miRNAs can show tissue-specific expression and may target several genes in different biological pathways, making the interpretation of their exact roles in the context of breast cancer quite complex. As Ebert and Sharp discussed, it is important to have a more in-depth comprehension of the functional implications of miRNA dysregulation, especially when considering the different subtypes of breast cancer [83]. This is not only critical for the developing targeted therapeutic approaches but also for the precise prediction of patient outcomes and the design of personalized treatment plans. In summary, these are the challenges that will need to be addressed in order to explore miRNAs as useful biomarkers for breast cancer in a clinical setting.

In conclusion, microRNAs, which are more commonly referred to as miRNAs, are one of the most promising and fascinating frontiers in the ongoing search of reliable biomarkers in the areas of diagnosis, prognosis, and treatment response in breast cancer. These small, noncoding RNA molecules are intricately associated with a variety of distinct subtypes of breast cancer and play a critical regulatory function in the expression of genes. This will improve comprehension of the fundamental principles of malignancies and facilitate the identification of potential therapeutic targets, thereby enabling the development of successful and personalized treatment regimens for patients. Although progress has been made in comprehending the function of miRNAs in breast cancer, there are still numerous significant obstacles in the technical and clinical aspects of their application. These challenges include the further standardization of miRNA detection methods, the validation of miRNAs as reliable biomarkers in diverse patient populations, and a deeper comprehension of the biological function that these molecules play in the complex process of initiation and progression of tumors. Overcoming the issues highlighted above is critical to ensuring the successful translation of miRNAs into the clinic. This, therefore, calls for further studies geared toward overcoming the existing challenges and translating the research from the laboratory to the clinic. The scientific community should shift the focus of studies toward determining the functional role of miRNAs in breast cancer and develop ways that ensure they are reliable and accurate in the clinics.

Therapeutic Potential of miRNAs in Breast Cancer

miRNA-Based Therapeutics: Overview

miRNA mimics are small, synthetic, and double-stranded structures that can precisely match endogenous miRNA. Since miRNA mimics can be modified, they provide more advantageous use than their naturally occurring forms. AntagomiRs are antisense oligonucleotides that bind directly to miRNA and block its activity. While miRNA antagonists are produced to inactivate endogenous miRNAs that function in diseased tissues, miRNA mimics are used to compensate for the lack of function. In other words, miRNA mimics are used when miRNA expression decreases and causes disease. miRNA mimics work by suppressing or restoring the expression of these miRNAs. In cases where miRNA expression is increased, antagomiRs are used to inhibit miRNA activity [84].

miRNAs in Targeting Specific Pathways in Breast Cancer

About 50% of the genes that code for human miRNA are located in sensitive chromosomal regions or cancer-related regions [85]. MiRNAs act as oncogenes or tumor suppressors in various tumor types and regulate cytostatic drug response. Irregularity of cell proliferation in breast cancer is the main cause of breast tumor formation. The cell division capabilities of healthy cells are limited by a certain number of cell divisions. Healthy cells that have reached a certain density stop multiplying, stopping in the G0 phase of the cell cycle. When healthy cells stop proliferating, it is a response to growth-inhibiting influences in their environment. However, this mechanism does not work normally in cancer cells. miRNAs regulate the abnormal cell cycle of breast cancer by interacting with protein kinases, protein kinase inhibitors, cyclin proteins, and similar factors. Cyclin E1 is known to be an important cell cycle regulator. It is also an important target for miR-16, miR-497, miR-483-3p, and miR-30c-2-3p, whose expression is reduced in breast cancer. Overexpression of miR-483-3p significantly reduced cell proliferation in breast cancer and inhibited G1-S crossover [86]. miR-34a suppresses the cell cycle through induction of the G1 cell cycle. miR-34c affects the cell cycle by causing G2/M arrest in breast cancer [87]. The MAPK/ERK signaling pathway is linked to cell growth and cell proliferation and is inhibited via miR-543, thereby inhibiting tumor development. In the EGF-treated miR-543 mimic group, cell proliferation and cell turnover increased, and cell apoptosis decreased. Thus, the inhibition of miR-543 on breast cancer occurred by suppressing the MAPK/ERK signaling pathway [88]. In ER-positive breast cancer cells, anti-miR-203 targeted the tumor suppressor protein SOCS3, suppressed cell proliferation by inhibiting cyclin D1 and pStat3, and inhibited tumor growth in breast cancer in vivo [89].

When a normal cell is genetically disrupted, the apoptosis signaling pathway is induced. But cancer cells are resistant to this response, and a disruption in the apoptosis pathway can turn breast cells into malignant cells [90]. Overexpression of miR-148a in breast cancer cells increased the rate of apoptosis by reducing cell viability and chemoresistance [91]. Upregulation of miR-7-5p decreases cell growth in breast cancer. Proteasome activator subunit 3 (REGγ) regulates tumor growth, and miR-7-5p inhibits REGγ and induces apoptosis [92]. miR-205 regulates PI2K/AKT signaling by targeting VEGF-A and FGF3 and induces apoptosis, especially in resistant breast cancer cells [93]. miR-370-3p, miR-7974, miR-4675, miR-150-5p, miR-106a, miR-330, miR-188-5p, and miR-331 inhibit apoptosis, promoting cell proliferation and migration in breast cancer cells.

In breast cancer metastasis, tumor cells detach from the breast area and spread to other parts of the body through the lymphatic system or bloodstream. In particular, bone is the most common site of breast cancer metastasis. Circulating miR-24-2-5p has been shown to reduce tumor progression and inhibit osteogenic differentiation in bone in breast cancer cells with high metastatic potential [94]. The *TAC1* gene is activated in connection with stress and supports the easy entry of breast cancer cells into the bone marrow and the proliferation of cancer cells. miR-130a and miR-206 suppress tumor cells by preventing the entry of breast cancer cells into the bone marrow stroma [95]. miR-338-5p is known to have a tumor-suppressive role in various malignancies such as glioma, prostate, lung, and stomach. As a result of miR-338-5p suppressing ETS1 expression in breast cancer, cell proliferation and invasion are reduced [96].

Delivery Systems for miRNA Therapeutics

Various nanotherapeutic approaches are used in the treatment of breast cancer with variations such as different shapes, sizes, and surface charges. These approaches include liposomes, nanoparticles, dendrimers, quantum dots, or polymer micelles. In addition, natural compounds such as flavonoids can inhibit breast cancer proliferation and metastasis by targeting specific miRNAs. miRNA-based treatment options are preferred because miRNAs can target proteins involved in tumor progression, drug resistance, and tumor metastasis. However, although the stability of miRNAs is variable, as it is prone to degradation by nucleases present in the blood, chemical modification of anti-miR oligonucleotides can solve this problem [97]. In addition, the involvement of a single miRNA in more than one signaling pathway can be a challenging process as a result of off-target effects. In breast cancer, miRNA-based drug delivery systems can deliver miRNAs to tumor cells, thereby increasing the expression of tumor suppressor miRNAs or silencing oncogenic miRNAs. Lipid-based carriers such as liposomes and nanoparticles promote cellular uptake by encapsulating miRNA molecules. Quercetin activated caspase-3 by upregulating miR-146a in breast cancer cells and stimulated apoptosis. It also inhibited cell proliferation by reducing epidermal growth factor receptor (EGFR) expression [98]. MiR-34a is known to regulate important tumorigenic factors such as Bcl-2 and P53, which are effective in cisplatin resistance in breast cancer. The developed miR34a-Cis-Lip-RGD liposome lowered the threshold of resistance of breast cancer cells to apoptosis and suppressed tumor growth in the in vivo mouse model [99]. Exosomes are known to play a crucial role in cell-to-cell communication. Magnetic nanobeads and SERS nanoprobes were used to detect exosomes originating from the tumor. The SERS nanoprobe is designed in the form of a core-shell with gold core-silver shell nanorods, and a protective layer of silica shell. Exosome-specific antibodies are placed on the outermost surface of the SERS nanoprobe. The presence or absence of target exosomes in breast cancer could be determined qualitatively and quantitatively [100]. The miRNA/C-PEG-Nio-AuNP combination was created for the effective distribution of miRNA-33a mimic and AuNPs in breast cancer cells. This combination facilitated cellular uptake and suppressed cell growth by regulating Bax/Bcl-2 expression levels [101].

Clinical Trials and Preclinical Studies

The critical role of miRNAs in deadly diseases, including cancer, has been debated for many years. miRNAs are used in the detection of drug-resistant cells and in the upstream and downstream regulation of gene expression for the effective treatment of breast cancer. However, the development of miRNA drugs to be used in the treatment of breast cancer still carries some challenges. miRNA-34a is a tumor suppressor and is administered as a combination therapy with chemotherapeutic drugs. The miRNA-34a-docetaxel-nanotransporter combination suppressed tumor growth in the in vivo mouse model. In clinical trials, circulating miRNA-34a was elevated in the blood of patients with metastases of breast cancer. In addition, the increased level of miRNA-34a represents a poor prognosis and is considered a more aggressive type of cancer [102]. Chemotherapy-induced neutropenia is a toxicity that has been very important in clinical trials. It is difficult to identify patients with this risk. The development of neutropenia was high in patients with treatment-related decreased circulating expression of miRNA-195. The use of miRNA-195 as a biomarker in breast cancer is already known; in addition to this, it has been shown to play a regulatory role in circulating neutrophil proliferation [103]. Tramadol is an analgesic therapeutic that acts by binding to the μ-opioid receptor (MOR), used in the treatment of pain after breast cancer surgery. MORs are encoded by the OPRM1 gene, and OPRM1 expression is regulated by miRNA-23b and miR-107. As a result of tramadol treatment of miR-23b and miR-107 polymorphisms, pain management has improved, and the risk of side effects has decreased [104].

miRNAs are one of the main causes of cancer formation. Therefore, a large number of miRNAs have been classified as oncogenic or tumor suppressors. Studies describing miRNA profiles have identified specific miRNAs involved in breast cancer and focused on their potential to be used as biomarkers. In addition to diagnostic miRNAs, miRNAs that explain the effectiveness of therapeutic drugs play a crucial role in the improvement of cancer treatment. For this reason, miRNAs have also been included in the game in clinical oncology. miRNAs can be difficult to target, and side effects may be encountered during treatment. However, it should not be forgotten that cancer is not a simple disease caused by a single genetic mutation or epigenetic changes. Therefore, it is not possible to treat cancer based on a single target. Thanks to the unique aspects of miRNAs, treatment results based on a single miRNA may be equivalent to or superior to treatment using more than one drug. Therapeutic combinations, including miRNA mimics and antisense miRNA inhibitors, may show high selectivity and efficacy in treatment. Thus, it is an inevitable fact that miRNA-based therapies will be technological therapeutics that can be used in cancer treatment.

Conclusion

It has been determined that increased expression levels of miR-125a and miR-125b in SKBR3 cells reduce mRNA levels in both HER2 and HER3. As a result, it was observed that the proliferation of SKBR3 cells decreased. The molecular mechanisms of this disease are quite complex, and microRNAs (miRNAs) provide important data on the progression of cancer or the reduction of its effects in this process. MiRNAs are small RNA molecules that are involved in the regulation of gene expression and can have important effects on the development, progression, and response to treatment of cancer. In breast cancer, changes in the expression levels of certain miRNAs can increase the growth, invasion, and metastasis potential of tumor cells. For example, miR-21 is frequently expressed in breast cancer cells, and this can support the proliferation of cancer cells. On the other hand, miRNAs such as miR-34a and miR-146a have tumor suppressor properties; a decrease in the levels of these miRNAs can pave the way for tumor development. Such miRNAs can prevent the spread of tumors by targeting genes that control the growth of cancer cells. MiRNAs also stand out as biomarkers that can be used in early diagnosis and prognosis of breast cancer. Detection of miRNAs in cancerous tissues or the bloodstream can help diagnose the disease in its early stages. In addition, miRNAs are being investigated as a potential tool for monitoring response to treatment and developing personalized treatment strategies during the treatment process. In conclusion, miRNAs have the potential to cure breast cancer pathogenesis and treatment. Targeting miRNAs has the potential to develop new therapeutic strategies in cancer treatment, and future clinical studies will more clearly reveal the role of these molecules in breast cancer treatment.

References

1. Siegel RL et al (2021) Cancer statistics, 2021. CA Cancer J Clin 71(1):7–33
2. Semple J et al (2015) Does the age of breast cancer diagnosis in first-degree relatives impact on the risk of breast cancer in BRCA1 and BRCA2 mutation carriers? Breast Cancer Res Treat 154(1):163–169
3. Yamada D et al (2022) Relationship between breast density, breast cancer subtypes, and prognosis. Clin Breast Cancer 22(6):560–566
4. Yamamoto Y et al (2024) The Japanese breast cancer society clinical practice guidelines for breast cancer, 2022 edition: changes from the 2018 edition and general statements on breast cancer treatment. Breast Cancer 31(3):340–346
5. Loibl S et al (2021) Breast cancer. Lancet 397(10286):1750–1769
6. Abd-Aziz N, Kamaruzman NI, Poh CL (2020) Development of microRNAs as potential therapeutics against cancer. J Oncol 2020(1):8029721
7. Li L et al (2012) Regulation of breast cancer tumorigenesis and metastasis by miRNAs. Expert Rev Proteomics 9(6):615–625
8. Qi L et al (2009) Expression of miR-21 and its targets (PTEN, PDCD4, TM1) in flat epithelial atypia of the breast in relation to ductal carcinoma in situ and invasive carcinoma. BMC Cancer 9:1–8

9. Loh H-Y et al (2019) The regulatory role of microRNAs in breast cancer. Int J Mol Sci 20(19):4940
10. Xue J et al (2016) MicroRNA-148a inhibits migration of breast cancer cells by targeting MMP-13. Tumour Biol 37:1581–1590
11. Shukla K et al (2015) MicroRNA-30c-2-3p negatively regulates NF-κB signaling and cell cycle progression through downregulation of TRADD and CCNE1 in breast cancer. Mol Oncol 9(6):1106–1119
12. Elhelbawy NG et al (2021) miRNA-148a and miRNA-30c expressions as potential biomarkers in breast cancer patients. Biochem Biophys Rep 27:101060
13. Niedźwiecki S et al (2018) Serum levels of circulating miRNA-21, miRNA-10b and miRNA-200c in triple-negative breast cancer patients. Ginekol Pol 89(8):415–420
14. Giordano C et al (2023) miRNAs in the box: potential diagnostic role for extracellular vesicle-packaged miRNA-27a and miRNA-128 in breast cancer. Int J Mol Sci 24(21):15695
15. Kim VN (2005) Small RNAs: classification, biogenesis, and function. Mol Cells 19(1):1–15
16. Shenouda SK, Alahari SK (2009) MicroRNA function in cancer: oncogene or a tumor suppressor? Cancer Metastasis Rev 28:369–378
17. Lee RC, Feinbaum RL, Ambros V (1993) The C. elegans heterochronic gene lin-4 encodes small RNAs with antisense complementarity to lin-14. Cell 75(5):843–854
18. Kwak PB, Iwasaki S, Tomari Y (2010) The microRNA pathway and cancer. Cancer Sci 101(11):2309–2315
19. Saydam F, Değirmenci İ, Güneş HV (2011) MikroRNA\'lar ve kanser. Dicle Tıp Dergisi 38(1)
20. Pillai RS (2005) MicroRNA function: multiple mechanisms for a tiny RNA? RNA 11(12):1753–1761
21. Sun W et al (2010) microRNA: a master regulator of cellular processes for bioengineering systems. Annu Rev Biomed Eng 12(1):1–27
22. Vasilatou D et al (2010) The role of microRNAs in normal and malignant hematopoiesis. Eur J Haematol 84(1):1–16
23. Mitchell PS et al (2008) Circulating microRNAs as stable blood-based markers for cancer detection. Proc Natl Acad Sci 105(30):10513–10518
24. Wang K et al (2009) Circulating microRNAs, potential biomarkers for drug-induced liver injury. Proc Natl Acad Sci 106(11):4402–4407
25. Winter J et al (2009) Many roads to maturity: microRNA biogenesis pathways and their regulation. Nat Cell Biol 11(3):228–234
26. Yi R et al (2003) Exportin-5 mediates the nuclear export of pre-microRNAs and short hairpin RNAs. Genes Dev 17(24):3011–3016
27. Jafri I et al (2019) Erythrocyte miRNA 144 and miRNA 451 as cell aging biomarkers in African American adults. Open Biochem J 13(1):81
28. Iorio MV, Croce CM (2012) Causes and consequences of microRNA dysregulation. Cancer J 18(3):215–222
29. Calin GA et al (2002) Frequent deletions and down-regulation of micro-RNA genes miR15 and miR16 at 13q14 in chronic lymphocytic leukemia. Proc Natl Acad Sci U S A 99(24):15524–15529
30. Calin GA et al (2004) Human microRNA genes are frequently located at fragile sites and genomic regions involved in cancers. Proc Natl Acad Sci 101(9):2999–3004
31. Hu Z et al (2008) Genetic variants of miRNA sequences and non–small cell lung cancer survival. J Clin Invest 118(7):2600–2608
32. Yang W et al (2006) Modulation of microRNA processing and expression through RNA editing by ADAR deaminases. Nat Struct Mol Biol 13(1):13–21
33. Grelier G et al (2009) Prognostic value of dicer expression in human breast cancers and association with the mesenchymal phenotype. Br J Cancer 101(4):673–683
34. Tang R et al (2012) Mouse miRNA-709 directly regulates miRNA-15a/16-1 biogenesis at the posttranscriptional level in the nucleus: evidence for a microRNA hierarchy system. Cell Res 22(3):504–515

35. Odame E et al (2021) Enhancer RNAs: transcriptional regulators and workmates of NamiRNAs in myogenesis. Cell Mol Biol Lett 26:1–20
36. Weber B et al (2007) Methylation of human microRNA genes in normal and neoplastic cells. Cell Cycle 6(9):1001–1005
37. Lehmann U et al (2008) Epigenetic inactivation of microRNA gene hsa-mir-9-1 in human breast cancer. J Pathol 214(1):17–24
38. Poli E et al (2015) Molecular subtype-specific expression of microRNA-29c in breast cancer is associated with CpG dinucleotide methylation of the promoter. PLoS One 10(11):e0142224
39. Roccaro AM et al (2010) microRNA-dependent modulation of histone acetylation in Waldenström macroglobulinemia. Blood J Am Soc Hematol 116(9):1506–1514
40. O'Day E, Lal A (2010) MicroRNAs and their target gene networks in breast cancer. Breast Cancer Res 12:1–10
41. Yu Z et al (2010) microRNA, cell cycle, and human breast cancer. Am J Pathol 176(3):1058–1064
42. Eiriksdottir G et al (1998) Mapping loss of heterozygosity at chromosome 13q: loss at 13q12-q13 is associated with breast tumour progression and poor prognosis. Eur J Cancer 34(13):2076–2081
43. Wang S et al (2013) Functional cooperation of miR-125a, miR-125b, and miR-205 in entinostat-induced downregulation of erbB2/erbB3 and apoptosis in breast cancer cells. Cell Death Dis 4(3):e556–e556
44. Lehmann TP et al (2015) Relative levels of let-7a, miR-17, miR-27b, miR-125a, miR-125b and miR-206 as potential molecular markers to evaluate grade, receptor status and molecular type in breast cancer. Mol Med Rep 12(3):4692–4702
45. Calvano Filho CMC et al (2014) Triple-negative and luminal a breast tumors: differential expression of miR-18a-5p, miR-17-5p, and miR-20a-5p. Tumour Biol 35:7733–7741
46. Sun F et al (2008) Downregulation of CCND1 and CDK6 by miR-34a induces cell cycle arrest. FEBS Lett 582(10):1564–1568
47. Chen H et al (2020) MiR-27a facilitates breast cancer progression via GSK-3β. Technol Cancer Res Treat 19:1533033820965576
48. Song N, Liang B, Wang D (2016) The function of MiR-21 expression differences and pathogenesis on familial and triple negative breast cancer serum. Pak J Pharm Sci 29
49. Zhang M-Y et al (2019) Mechanism of propofol in inhibiting emt of breast cancer by controlling mir-21 expression. ACTA Medica Mediterranea 35(6):3139–3145
50. Huang S-X et al (2020) Maspin inhibits MCF-7 cell invasion and proliferation by downregulating miR-21 and increasing the expression of its target genes. Oncol Lett 19(4):2621–2628
51. Syed RU et al (2024) MicroRNA-21 (miR-21) in breast cancer: from apoptosis dysregulation to therapeutic opportunities. Pathol Res Pract 262:155572
52. Yadav BS, Chanana P, Jhamb S (2015) Biomarkers in triple negative breast cancer: a review. World J Clin Oncol 6(6):252
53. Sung H et al (2021) Global cancer statistics 2020: GLOBOCAN estimates of incidence and mortality worldwide for 36 cancers in 185 countries. CA Cancer J Clin 71(3):209–249
54. Iorio MV, Croce CM (2012) MicroRNA dysregulation in cancer: diagnostics, monitoring and therapeutics. A comprehensive review. EMBO Mol Med 4(3):143–159
55. Bartel DP (2004) MicroRNAs: genomics, biogenesis, mechanism, and function. Cell 116(2):281–297
56. Garzon R et al (2008) Distinctive microRNA signature of acute myeloid leukemia bearing cytoplasmic mutated nucleophosmin. Proc Natl Acad Sci 105(10):3945–3950
57. Polyak K (2011) Heterogeneity in breast cancer. J Clin Invest 121(10):3786–3788
58. Waks AG, Winer EP (2019) Breast cancer treatment: a review. JAMA 321(3):288–300
59. Perou CM et al (2000) Molecular portraits of human breast tumours. Nature 406(6797):747–752

60. Iorio MV, Croce CM (2012) microRNA involvement in human cancer. Carcinogenesis 33(6):1126–1133
61. Najdi R et al (2009) A Wnt kinase network alters nuclear localization of TCF-1 in colon cancer. Oncogene 28(47):4133–4146
62. Chen J, Wang X (2014) MicroRNA-21 in breast cancer: diagnostic and prognostic potential. Clin Transl Oncol 16(3):225–233
63. Oh W et al (2010) Hdm2 negatively regulates telomerase activity by functioning as an E3 ligase of hTERT. Oncogene 29(28):4101–4112
64. Jiang P et al (2006) The bad guy cooperates with good cop p53: bad is transcriptionally up-regulated by p53 and forms a bad/p53 complex at the mitochondria to induce apoptosis. Mol Cell Biol 26:9071
65. Lai X, Schmitz U, Vera J (2022) The role of MicroRNAs in cancer biology and therapy from a systems biology perspective. Adv Exp Med Biol 1385:1–22
66. Gregory PA et al (2008) The miR-200 family and miR-205 regulate epithelial to mesenchymal transition by targeting ZEB1 and SIP1. Nat Cell Biol 10(5):593–601
67. Humphries B, Yang C (2015) The microRNA-200 family: small molecules with novel roles in cancer development, progression and therapy. Oncotarget 6(9):6472
68. Zhu S et al (2008) MicroRNA-21 targets tumor suppressor genes in invasion and metastasis. Cell Res 18(3):350–359
69. Jiang S et al (2010) MicroRNA-155 functions as an OncomiR in breast cancer by targeting the suppressor of cytokine signaling 1 gene. Cancer Res 70(8):3119–3127
70. Grillari J, Hackl M, Grillari-Voglauer R (2010) miR-17–92 cluster: ups and downs in cancer and aging. Biogerontology 11:501–506
71. Giuliano M et al (2011) Circulating tumor cells as prognostic and predictive markers in metastatic breast cancer patients receiving first-line systemic treatment. Breast Cancer Res 13:1–9
72. Sherer NM et al (2007) Retroviruses can establish filopodial bridges for efficient cell-to-cell transmission. Nat Cell Biol 9(3):310–315
73. Bajaj R et al (2020) Prognostic role of microRNA 182 and microRNA 18a in locally advanced triple negative breast cancer. PLoS One 15(11):e0242190
74. Ma L, Teruya-Feldstein J, Weinberg RA (2007) Tumour invasion and metastasis initiated by microRNA-10b in breast cancer. Nature 449(7163):682–688
75. Ozawa PM et al (2020) Identification of miRNAs enriched in extracellular vesicles derived from serum samples of breast cancer patients. Biomolecules 10(1):150
76. Fasanaro P et al (2008) MicroRNA-210 modulates endothelial cell response to hypoxia and inhibits the receptor tyrosine kinase ligand Ephrin-A3. J Biol Chem 283(23):15878–15883
77. Geoghegan JC et al (2014) Chondroitin sulfate is the primary receptor for a peptide-modified AAV that targets brain vascular endothelium in vivo. Mol Ther Nucl Acids:3
78. Ljepoja B et al (2019) MiRNA-27a sensitizes breast cancer cells to treatment with selective estrogen receptor modulators. Breast 43:31–38
79. Garzon R, Calin GA, Croce CM (2009) MicroRNAs in cancer. Annu Rev Med 60(1):167–179
80. Mestdagh P et al (2014) Evaluation of quantitative miRNA expression platforms in the microRNA quality control (miRQC) study. Nat Methods 11(8):809–815
81. Bustin SA et al (2009) The MIQE guidelines: M inimum I nformation for publication of Q uantitative real-time PCR E xperiments. Oxford University Press
82. Turchinovich A et al (2011) Characterization of extracellular circulating microRNA. Nucleic Acids Res 39(16):7223–7233
83. Ebert MS, Sharp PA (2012) Roles for microRNAs in conferring robustness to biological processes. Cell 149(3):515–524
84. Dasgupta I, Chatterjee A (2021) Recent advances in miRNA delivery systems. Methods Protocols 4(1):10
85. Melo SA, Esteller M (2011) Dysregulation of microRNAs in cancer: playing with fire. FEBS Lett 585(13):2087–2099

86. Huang X, Lyu J (2018) Tumor suppressor function of miR-483-3p on breast cancer via targeting of the cyclin E1 gene. Exp Ther Med 16(3):2615–2620
87. Achari C et al (2014) Expression of miR-34c induces G2/M cell cycle arrest in breast cancer cells. BMC Cancer 14(1):538
88. Chen P et al (2017) MicroRNA 543 suppresses breast cancer cell proliferation, blocks cell cycle and induces cell apoptosis via direct targeting of ERK/MAPK. Onco Targets Ther 10(Null):1423–1431
89. Muhammad N et al (2016) Anti-miR-203 suppresses ER-positive breast cancer growth and stemness by targeting SOCS3. Oncotarget 7(36):58595–58605
90. Wong RS (2011) Apoptosis in cancer: from pathogenesis to treatment. J Exp Clin Cancer Res 30:1–14
91. Li Q et al (2017) MicroRNA-148a promotes apoptosis and suppresses growth of breast cancer cells by targeting B-cell lymphoma 2. Anti-Cancer Drugs 28(6):588–595
92. Mirzaei Z et al (2024) The role of mir-7-5p in cancer: function, prognosis, diagnosis, and therapeutic implications. Mol Biol Rep 52(1):12
93. Hu Y et al (2016) miRNA-205 targets VEGFA and FGF2 and regulates resistance to chemotherapeutics in breast cancer. Cell Death Dis 7(6):e2291–e2291
94. Puppo M et al (2024) Protective effects of miR-24-2-5p in early stages of breast cancer bone metastasis. Breast Cancer Res 26(1):186
95. Akshaya RL, Saranya I, Selvamurugan N (2023) MicroRNAs mediated interaction of tumor microenvironment cells with breast cancer cells during bone metastasis. Breast Cancer 30(6):910–925
96. Chen WJ et al (2024) MiR-338-5p, a novel metastasis-related miRNA, inhibits triple-negative breast cancer progression by targeting the ETS1/NOTCH1 axis. Heliyon 10(15):e34949
97. Pagoni M et al (2023) miRNA-based technologies in cancer therapy. J Pers Med 13(11):1586
98. Tao S-F, He H-F, Chen Q (2015) Quercetin inhibits proliferation and invasion acts by up-regulating miR-146a in human breast cancer cells. Mol Cell Biochem 402:93–100
99. Bardania H et al (2024) Synergistic breast cancer therapy with RGD-decorated liposomes co-delivering mir-34a and cisplatin. Cancer Nanotechnol 15(1):60
100. Zong S et al (2016) Facile detection of tumor-derived exosomes using magnetic nanobeads and SERS nanoprobes. Anal Methods 8(25):5001–5008
101. Ahmadi SM et al (2024) Efficient delivery of gold nanoparticles and miRNA-33a via cationic PEGylated Niosomal formulation to MCF-7 breast cancer cells. AAPS PharmSciTech 25(7):213
102. Roth C et al (2010) Circulating microRNAs as blood-based markers for patients with primary and metastatic breast cancer. Breast Cancer Res 12:1–8
103. Larsen MT et al (2013) MicroRNA profiling in human neutrophils during bone marrow granulopoiesis and in vivo exudation. PLoS One 8(3):e58454
104. Vidic Z et al (2023) Association of OPRM1, MIR23B, and MIR107 genetic variability with acute pain, chronic pain and adverse effects after postoperative tramadol and paracetamol treatment in breast cancer. Radiol Oncol 57(1):111–120

Chapter 10
miR-200c as a Master Regulator for Breast Cancer Therapeutics and Diagnosis

Sunny Sharma and Chandi C. Mandal

Abstract Cancer remains a major global health issue, with researchers striving to develop more effective targeted therapies. Among all cancer types, breast cancer is a leading cause of death in women worldwide. The high mortality rate of breast cancer is attributed not only to the aggressive subtypes and tumor heterogeneity but also to the complex molecular mechanisms underlying its metastasis and therapy resistance. Epithelial-to-mesenchymal transition (EMT) is instrumental in contributing towards the aggressiveness and metastatic potential. MicroRNAs (miRNAs) are small, noncoding RNA molecules that regulate gene expression posttranscriptionally. One key miRNA, miR-200c, plays a vital role in suppressing EMT by targeting the transcriptional repressor of E-cadherin such as ZEB1 and ZEB2, thereby preserving epithelial characteristics. In addition to its role in EMT, miR-200c also affects cancer stem ness, immune response, and resistance to treatment, making it a strong candidate for both diagnostic and therapeutic applications. The downregulation of miR200c is associated with poor patient survival and high-grade tumors like TNBC. Its presence in blood and exosomes provides a noninvasive method for early detection and disease monitoring. Experimental therapies involving miR-200c like nanoparticle-based delivery systems or drugs that reactivate its expression have shown encouraging results in reducing tumor growth and improving drug response. miR200c can be a compelling target for diagnostic and therapeutic purpose in breast cancer, due to its ability to modulate cancer hallmarks. Further research in the direction of optimized delivery system and clinical validation is crucial to harness the potential of miR200c against breast cancer.

Keywords miR-200c · Breast Cancer · Epithelial to mesenchymal transition · Diagnosis and therapeutics

S. Sharma · C. C. Mandal (✉)
Department of Biochemistry, School of Life Sciences, Central University of Rajasthan, Ajmer, Rajasthan, India

S. Pathak, A. Banerjee (eds.), *MicroRNAs: Role in Cancer Progression, Diagnostics and Therapy*, Advances in Biochemistry in Health and Disease 34,
https://doi.org/10.1007/978-3-032-08824-6_10

Introduction

Cancer represents one of the greatest health challenges of the modern era, with researchers worldwide continuing their tireless efforts to develop effective treatments and cures. This multifaceted disease constitutes a major global health concern, with recent statistics indicating approximately 19.3 million new diagnoses annually. Recent epidemiological shifts have positioned breast cancer as the most frequently diagnosed malignancy, accounting for about 11.7% of cases (approximately 2.3 million diagnoses), lung cancer accounts for 11.4%, but is now exceeded by other types. Colorectal cancer follows at 10%, with prostate and stomach cancers contributing 7.3% and 5.6% respectively [1, 2]. Breast cancer specifically presents as a diverse disease with distinct subtypes and remains the primary cause of cancer mortality among women globally. Small noncoding RNA molecules called microRNAs play crucial regulatory roles in gene expression at the posttranscriptional level and significantly influence cancer development, advancement, and spread. The miR-200 family, particularly miR-200c, serves a fundamental role in maintaining epithelial characteristics in carcinomas. Within healthy mammary tissue, miR-200c preserves epithelial integrity by inhibiting epithelial-to-mesenchymal transition (EMT) ZEB1 and ZEB2 are transcription factors that typically downregulate E-cadherin and other molecules involved in cell adhesion [3, 4]. Notably, miR-200c deficiency occurs frequently in breast cancer, particularly in aggressive or basal-like tumors, and correlates with increased cell migration capability, stem-cell-like properties, and poorer clinical outcomes [4, 5]. Comprehending miR-200c's complex function as both a tumor suppressor and a potential biomarker in circulation is essential for enhancing breast cancer treatment approaches. MicroRNAs represent a revolutionary discovery in molecular biology that has transformed our understanding of gene regulation and disease processes. Among the thousands of identified microRNAs, the miR-200 family has emerged as a critical regulator of cellular identity and behavior, particularly in epithelial tissues and their associated malignancies [6]. This chapter explores the therapeutic and diagnostic significance of miR-200c in breast cancer, with special emphasis on its role in EMT and regulatory mechanisms.

Historical Discovery and Biological Mechanism of MicroRNAs

MicroRNAs (miRNAs) are short, noncoding RNA molecules around 22 nucleotides in length that regulate various biological functions [6]. Horvitz's lab discovered that the lin-4 gene, although it does not code for a protein, plays a key role in regulating larval development in *C. elegans* by blocking the production of the LIN-14 protein

[7], where they found an interesting fact was a small transcript of 22 nucleotides long. Both Ambros and Ruvkun continued to study lin-4 and LIN-14 after leaving Horvitz's lab. Ambros and Ruvkun observed that lin-4 RNAs exhibited antisense complementarity to multiple sites within the 3′ untranslated region (UTR) of the LIN-14 gene [8]. These complementary regions were located in a section of the 3′ UTR previously implicated in mediating LIN-14 repression by the LIN-4 gene product [9]. The Ruvkun lab further demonstrated that these complementary sites are critical for LIN-14 regulation by LIN-4, showing that this regulation significantly reduces LIN-14 protein levels without altering LIN-14 mRNA levels. Collectively, these findings supported a model in which LIN-4 RNAs bind to the LIN-14 3′ UTR to drive translational repression of LIN-14, facilitating the transition from first to second larval stage cell divisions [6, 8]. The shorter LIN-4 RNA is now acknowledged as the first discovered member of a widespread group of miRNAs [10]. Victor Ambros and Gary Ruvkun received the 2024 Nobel Prize for Physiology or Medicine for this discovery. Most miRNAs are first made as primary transcripts (pri-miRNAs) from DNA, then processed into precursor forms (pre-miRNAs), and finally into mature miRNAs. These mature miRNAs usually attach to the 3′ untranslated region (3′ UTR) of target mRNAs to block their expression [11]. The process closely resembles standard transcription, with RNA polymerase II producing the initial pri-miRNA, which is then modified by adding a 5′ cap and a 3′ poly-A tail.

Short RNA molecules that do not code for proteins, typically 19 to 25 nucleotides long, miRNAs, constitute an extensive and continually expanding class of regulatory elements with increasingly diverse identified functions. In the three decades since their initial identification, scientific research has documented an impressive 38,589 miRNA genetic loci distributed across 271 species, yielding a total of 48,860 mature miRNA sequences. Within the human genome specifically, scientists have characterized 1917 hairpin precursors that give rise to 2654 mature miRNA sequences, highlighting these molecules' significant regulatory importance across various biological systems [12]. These tiny RNA regulators participate in controlling sophisticated physiological mechanisms including embryonic development, tissue and organ formation, and cancer progression [13–15]. Despite significant research progress, the complete functional repertoire of most miRNAs remains incompletely characterized. Research has revealed a complex relationship between miRNAs and cancer development. While certain specific miRNAs can promote tumor formation, comprehensive expression analyses have consistently demonstrated that overall miRNA levels typically decrease in cancerous tissues compared to healthy counterparts [16]. This reduction particularly affects key regulatory miRNA families, including the miR-200 group.

miR-200 Family: Structure, Function, and Role in EMT Regulation in Breast Cancer

1. **Structure and Evolution**

The miR-200 family includes five well-conserved members, grouped into two genomic clusters according to their chromosomal positions and sequence similarity. The first cluster, located on chromosome 1p36, includes miR-200a, miR-200b, and miR-429, while the second cluster on chromosome 12p13 comprises miR-200c and miR-141 [17, 18]. This genomic organization is evolutionarily preserved across species, suggesting fundamental biological importance. These miRNAs share similar seed sequences of the critical nucleotide region that determines target recognition with only a single nucleotide difference distinguishing the two clusters (Fig. 10.1) [19]. This slight variation results in partly overlapping yet distinct target repertoires, enabling fine-tuned regulation of complex cellular processes. The remarkable conservation of the miR-200 family across vertebrate evolution underscores its essential role in development and tissue homeostasis [20, 21].

2. **Biogenesis and Molecular Functions of miR-200c**

MiR-200c, like other miRNAs, undergoes a sophisticated biogenesis process beginning with transcription in the nucleus and culminating in the formation of mature miRNA in the cytoplasm [22]. Following transcription by RNA polymerase II, the primary miR-200c transcript undergoes sequential processing by the microprocessor complex containing Drosha and DGCR8, followed by Exportin-5-mediated nuclear export and cytoplasmic cleavage by Dicer [23, 24]. The resulting mature miR-200c incorporates into the RNA-induced silencing complex (RISC), which guides it to target messenger RNAs. MiR-200c primarily functions through posttranscriptional gene silencing by binding to complementary sequences in the 3′ untranslated regions (UTRs) of target mRNAs [25]. This interaction typically results in translational repression or mRNA degradation, effectively silencing gene expression. Through this mechanism, miR-200c regulates numerous cellular processes

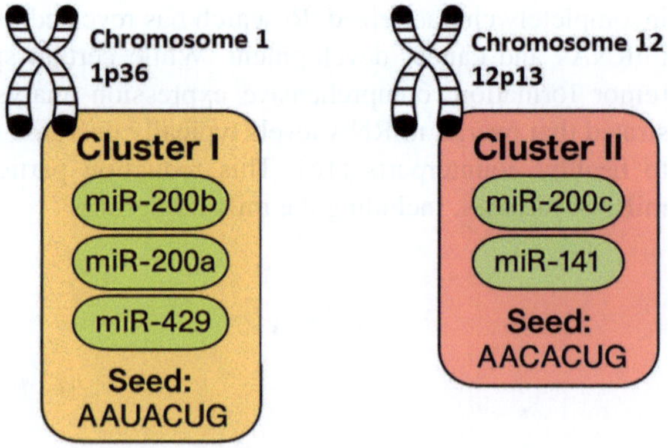

Fig. 10.1 Genomic organization and seed sequence comparison of the miR-200 family

including differentiation, proliferation, migration, and programmed cell death [26, 27].

3. **miR-200c as a Master Regulator of Epithelial Phenotype**

A defining characteristic of miR-200c is its role as a guardian of epithelial identity. This function is largely mediated through its reciprocal regulatory relationship with epithelial-mesenchymal transition (EMT) transcription factors, particularly ZEB1 and ZEB2 [28, 29]. By directly targeting and suppressing these factors, miR-200c maintains expression of epithelial markers like E-cadherin while inhibiting mesenchymal characteristics. In normal breast epithelium, miR-200c maintains tissue architecture and cellular adhesion, crucial for epithelial barrier function and homeostasis [30]. The loss of miR-200c expression represents a critical step in EMT initiation, allowing increased expression of ZEB1/2 and subsequent repression of epithelial genes [31]. This molecular switch facilitates cellular plasticity essential for both normal development and, unfortunately, cancer progression (Fig. 10.2) [32].

4. **Dysregulation of miR-200c in Breast Cancer**

Breast cancer, a heterogeneous disease with multiple molecular subtypes, exhibits distinct patterns of miR-200c expression across different classifications. Generally, miR-200c levels are reduced in more aggressive subtypes, particularly in triple-negative and basal-like breast cancers, compared to luminal subtypes (Fig. 10.3) [33]. This differential expression correlates with clinical outcomes, with lower miR-200c levels associated with poorer prognosis [34]. The mechanisms underlying miR-200c dysregulation in breast cancer are multifaceted, involving genetic, epigenetic, and transcriptional alterations. Promoter hypermethylation

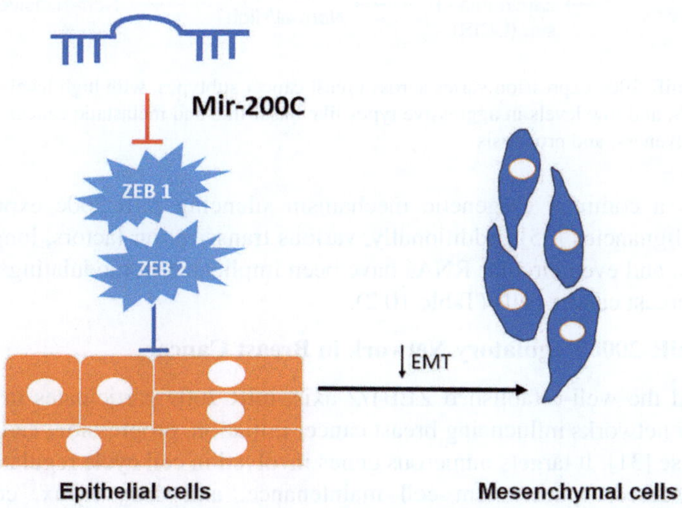

Fig. 10.2 miR-200c suppresses EMT in breast cancer by targeting ZEB1/2

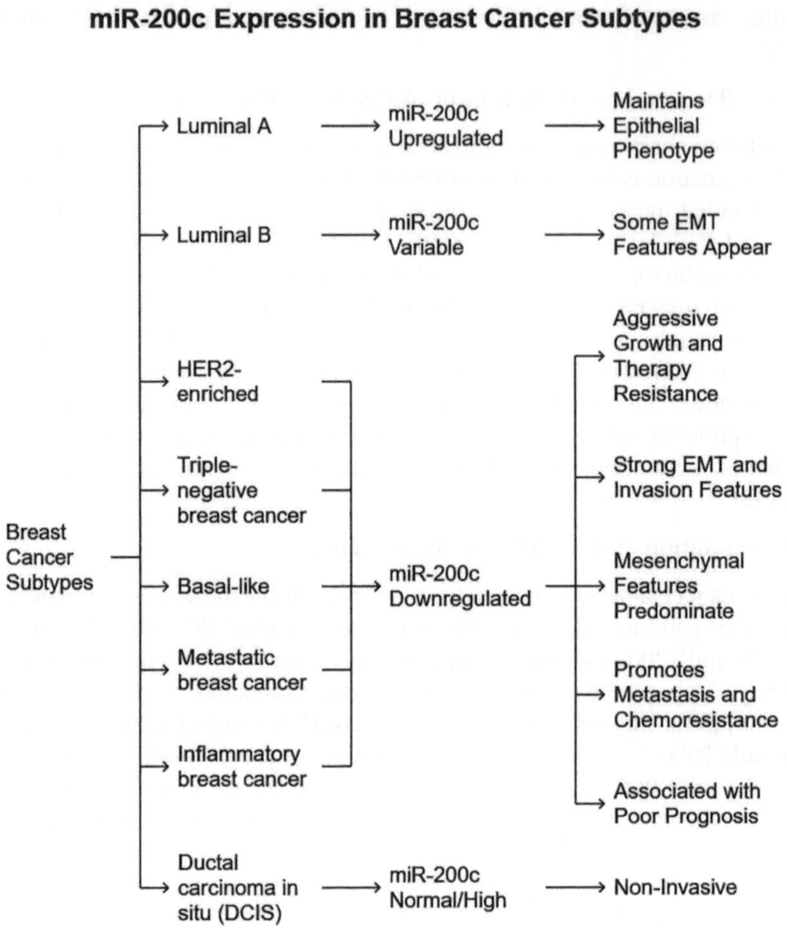

Fig. 10.3 miR-200c expression varies across breast cancer subtypes, with high levels in Luminal A and DCIS, and low levels in aggressive types like basal-like and metastatic cancer, influencing EMT, invasiveness, and prognosis

represents a common epigenetic mechanism silencing miR-200c expression in breast malignancies [35]. Additionally, various transcription factors, long noncoding RNAs, and even circular RNAs have been implicated in modulating miR-200c levels in breast cancer cells (Table 10.2).

5. The miR-200c Regulatory Network in Breast Cancer

Beyond the well-established ZEB1/2 axis, miR-200c participates in extensive regulatory networks influencing breast cancer initiation, progression, and therapeutic response [31]. It targets numerous genes involved in cell cycle regulation, apoptosis resistance, cancer stem cell maintenance, and drug efflux, collectively contributing to its tumor-suppressive function (Fig. 10.4) [36, 37]. Recent research has revealed that miR-200c also regulates metabolic reprogramming, immune

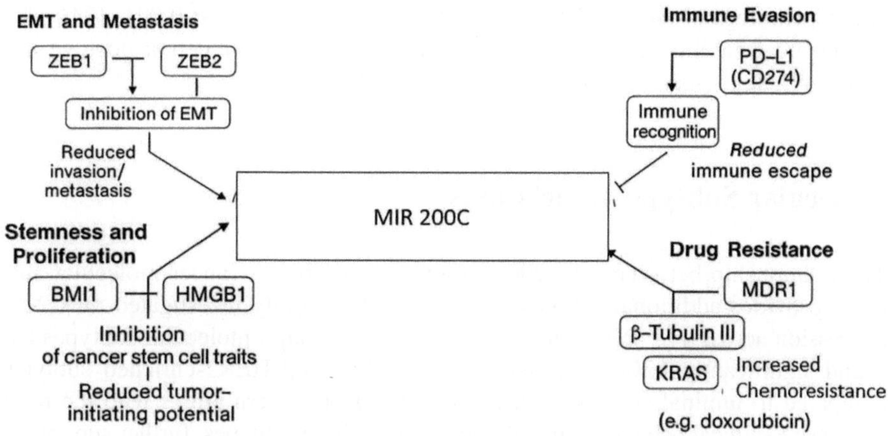

Fig. 10.4 Functional roles of miR-200c in cancer. miR-200c suppresses EMT, metastasis, stemness, immune evasion, and drug resistance by targeting ZEB1/2, BMI1, HMGB1, PD-L1, MDR1, and KRAS-related pathways

evasion mechanisms, and tumor microenvironment interactions in breast cancer [38, 39]. This expanding functional repertoire positions miR-200c as a central node in breast cancer regulatory networks, with far-reaching implications for cancer biology and clinical management.

miR-200c maintains epithelial traits by inhibiting ZEB1 and ZEB2, which repress E-cadherin. Its downregulation promotes EMT, marked by reduced E-cadherin, and increased N-cadherin and vimentin, leading to metastasis.

Introduction to miR-200c as a Clinical Biomarker

The potential utility of miR-200c as a biomarker in breast cancer stems from its fundamental role in maintaining epithelial integrity and suppressing epithelial-mesenchymal transition (EMT). The characteristic dysregulation of miR-200c across breast cancer subtypes and disease stages presents a compelling opportunity for clinical application in diagnostics, prognostication, and therapeutic monitoring. This section explores the extensive evidence supporting miR-200c's diagnostic significance, addressing both tissue-based and liquid biopsy approaches.

I. **Tissue-Based miR-200c Expression Patterns**

Prognostic Value in Primary Breast Tumors

Madhavan et al. conducted a prospective study of 173 breast cancer patients, demonstrating that diminished miR-200c expression in primary tumors independently predicted shorter recurrence-free survival after adjusting for traditional clinicopathological factors (HR = 2.31, 95% CI: 1.52–3.49, $p < 0.001$) [34]. This prognostic capability appeared particularly robust in estrogen receptor-positive subtypes [47], suggesting subtype-specific utility [48].

Molecular Subtype Associations

The relationship between miR-200c expression and breast cancer molecular subtypes provides additional diagnostic nuance [49]. A study investigated miR-200c expression across 412 breast tumors representing all major molecular subtypes and found significantly lower expression in basal-like and HER2-enriched subtypes compared to luminal subtypes [50]. This expression pattern aligns with the more aggressive clinical behavior typically observed in these subtypes, further supporting miR-200c's role as a marker of tumor biology [51].

In a multi-institutional study examining 526 breast cancer samples, a study demonstrated that miR-200c expression followed a gradient across molecular subtypes, with highest levels in luminal A tumors and progressively decreasing expression in luminal B, HER2-enriched, and triple-negative tumors (Table 10.1) [52]. Importantly, this expression pattern remained consistent across independent validation cohorts, highlighting the robustness of this molecular relationship.

II. **Circulating miR-200c as a Noninvasive Biomarker**

Detection in Blood-Based Liquid Biopsies

The presence of miR-200c in circulation offers opportunities for noninvasive monitoring through liquid biopsy approaches. Contrary to expression patterns observed in primary tumors, several studies have reported elevated levels of circulating miR-200c in breast cancer patients compared to healthy controls, particularly in advanced diseases [53].

In a landmark study by Antolín et al. (2015), whole-blood miR-200c levels were significantly higher in metastatic breast cancer patients compared to early-stage patients and healthy controls (fold change = 4.92) [54]. Receiver operating characteristic (ROC) analysis demonstrated excellent discriminatory ability between metastatic patients and healthy controls (area under curve [AUC] = 0.882, 95% CI: 0.813–0.951). This suggests that circulating miR-200c may serve as a valuable indicator of metastatic disease burden [55].

The apparent discrepancy between decreased tissue expression and increased circulating levels likely reflects tumor biology rather than contradictory findings [56]. As explained by a study, active EMT processes in advancing tumors may result in enhanced miR-200c release into circulation, potentially as a compensatory mechanism or as a consequence of altered cellular trafficking [39].

Table 10.1 Diagnostic significance of miR-200c in breast cancer: biomarker applications in detection and prognosis

Breast cancer subtypes	miR-200c expression	Summary of roles and impact	Key References
Luminal A	Elevated	Supports epithelial characteristics and linked to favorable outcomes	[40]
Luminal B	Mixed (up regulated or down regulated)	Expression fluctuates depending on tumor grade; some mesenchymal traits may emerge	[41, 42]
HER2-enriched	Reduced	Lower levels associated with aggressive behavior and resistance to targeted therapies	[43]
Triple-negative breast cancer	Strongly reduced	Marked by high invasiveness and EMT activity; miR-200c is frequently suppressed	[41]
Basal-like	Decreased	Characterized by mesenchymal traits; reduced miR-200c enhances invasion potential	[44]
Metastatic breast cancer	Downregulated	Lower expression contributes to metastasis and resistance to chemotherapy	[45]
Inflammatory breast cancer	Suppressed	Correlated with poor prognosis and aggressive disease behavior	[40]
Ductal carcinoma in situ (DCIS)	Normal or elevated	Typically preserved in early-stage, noninvasive lesions	[46]

Table 10.2 Representative studies of miR-200c targets and pathways in Breast Cancer

Study (year)	Model/system	miR-200c targets/pathways	Key findings
Cochrane et al. (2011)	TNBC and endometrial cancer cells (in vitro)	ZEB1, ZEB2	Reintroduction of miR-200c suppressed EMT by upregulating E-cadherin and reducing cell migration, invasion, and resistance to anoikis
Song et al. (2015)	Breast cancer cell lines and mouse xenograft models	KRAS, AKT/ERK signaling	miR-200c expression was reduced in tumors; it directly targeted KRAS, thereby inhibiting the AKT/ERK pathway and decreasing cell proliferation and survival
Zhang et al. (2020)	TNBC tissues and cell lines	ZEB2 (novel target)	miR-200c was found to be downregulated in TNBC; it directly targeted ZEB2, thereby reversing EMT and limiting invasion and metastasis
Williams et al. (2021)	Mouse mammary tumor model (claudin-low breast cancer)	Immune checkpoints (PD-L1), EMT factors	Restoring miR-200c levels reduced EMT and immune suppression, resulting in tumor growth inhibition and cytokine environment modulation
Eroles et al. (2023)	Human breast cancer cells (in vitro) and in vivo nanoparticle delivery	ZEB1, ZEB2	Nanoparticles carrying miR-200c-3p effectively downregulated ZEB1/2, suppressing tumor growth and metastasis

Exosomal miR-200c for Enhanced Stability and Specificity

Exosomes are small extracellular vesicles that protect and transport microRNAs in circulation have emerged as particularly promising carriers of diagnostic information (Fig. 10.5) [57]. The study by Qiao research team analyzed serum samples from 51 breast cancer patients and 47 healthy controls. The expression levels of serum exosomal miR-200c were significantly lower in BC patients compared to healthy controls. Receiver operating characteristic (ROC) curve analysis demonstrated that serum exosomal miR-200c could effectively distinguish BC patients from healthy individuals, with an area under the curve (AUC) of 0.854 (95% confidence interval [CI]: 0.777–0.930). This indicates a high diagnostic accuracy [58]. These findings suggest that serum exosomal miR-200c holds promise as a noninvasive biomarker for the early detection of breast cancer, potentially offering greater diagnostic accuracy than traditional tumor markers [59].

Similarly, Jin and colleagues (2019) reported that serum exosomal miR-200c levels were significantly altered even in stage I breast cancer patients compared to controls, suggesting utility for early detection [60]. Furthermore, combining exosomal miR-200c with conventional serum tumor markers (CA15–3, CEA) improved diagnostic accuracy, highlighting the potential for integrative biomarker approaches [61].

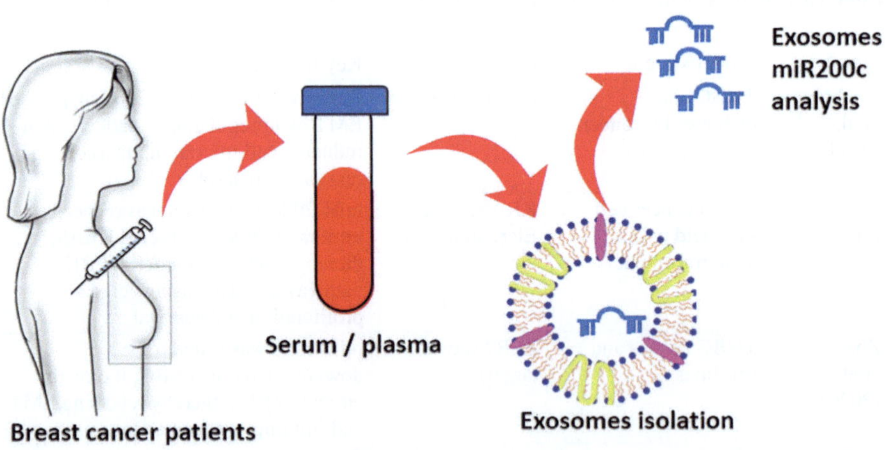

Fig. 10.5 Schematic representation of exosomal miR-200c analysis in breast cancer patients. Blood samples are collected from breast cancer patients, and serum or plasma is obtained. Exosomes are subsequently isolated from these samples, and miR-200c expression is analyzed to explore its potential as a diagnostic or prognostic biomarker

III. **Predictive Applications in Treatment Response**

Neoadjuvant Therapy Response Assessment

Beyond diagnosis and prognosis, miR-200c expression patterns may predict response to neoadjuvant therapy [65].

A study demonstrated that patients achieving pCR after neoadjuvant chemotherapy had significantly higher pretreatment miR-200c levels than nonresponders [66]. Logistic regression analysis identified high miR-200c expression as an independent predictor of pathological complete response, even after adjusting for tumor characteristics [67].

Monitoring Treatment Efficacy

Dynamic changes in circulating miR-200c during treatment may provide valuable information about therapeutic efficacy. Shao and colleagues (2019) monitored plasma miR-200c levels in 146 breast cancer patients before and after surgical intervention, observing significant decreases post-surgery in patients without detectable residual disease [68]. Persistent elevation or rapid re-elevation of miR-200c after therapy predicted earlier disease recurrence, suggesting utility for surveillance.

IV. **Multi-marker Panels and Integrated Approaches**

Combinatorial Strategies for Enhanced Accuracy

While miR-200c alone demonstrates significant diagnostic potential, its integration into multi-marker panels substantially enhances clinical utility. Zhang et al. (2022) developed a six-microRNA panel including miR-200c that achieved remarkable diagnostic accuracy for breast cancer detection [69]. This panel maintained high sensitivity across all molecular subtypes, addressing a common limitation of single-marker approaches [70].

Similarly, Madhavan and colleagues (2016) demonstrated that combining tissue miR-200c with other EMT-related microRNAs (miR-200b, miR-429) significantly improved prognostic accuracy compared to any individual marker (Table 10.3) [34]. This integrated approach better captured the complex biology underlying breast cancer progression.

Table 10.3 Circulating miR-200c as a noninvasive biomarker

Observation	Tissue miR-200c	Plasma miR-200c	Reference(s)
Early/localized tumor	High or normal	Low or undetectable	[62]
EMT/invasive/metastatic tumor	Downregulated	Elevated	[63]
Post-effective treatment	May rebound	Decreases (if tumor burden lowers)	[64]

Integration with Conventional Clinical Parameters

Perhaps most promising is the integration of miR-200c assessment with conventional clinical parameters. Wang et al. (2019) incorporated miR-200c expression into a prognostic model alongside traditional clinicopathological factors, demonstrating superior prognostic accuracy compared to standard methods (integrated AUC = 0.891 vs. 0.782 for conventional factors alone [50]. This integrated approach represents a practical pathway toward clinical implementation.

A pioneering study developed a clinical decision support algorithm incorporating circulating miR-200c alongside conventional biomarkers and imaging findings [71]. This integrated approach demonstrated superior accuracy in distinguishing malignant from benign breast lesions (AUC = 0.924) compared to conventional methods alone, highlighting the translational potential of miR-200c in clinical practice.

V. **Technological Considerations in miR-200c Detection**

Standardization and Methodological Challenges

Despite promising results, several methodological challenges must be addressed before widespread clinical implementation. Detection platforms for miR-200c vary considerably across studies, from quantitative PCR to microarray and next-generation sequencing approaches, each with distinct sensitivity and dynamic range characteristics [72].

VI. **Emerging Detection Technologies**

Conclusions

Clinical Perspective and Future Directions

The extensive body of evidence supports miR-200c's significant potential as a biomarker in breast cancer detection, prognostication, and treatment monitoring [73].

Both tissue-based and circulating miR-200c assessments offer complementary insights into tumor biology and disease progression. While technical challenges remain, ongoing standardization efforts and technological innovations continue addressing implementation barriers.

Future research directions should focus on large-scale prospective validation studies with standardized methodologies, development of integrated decision support tools incorporating miR-200c alongside conventional parameters, and exploration of therapeutic strategies targeting the miR-200c regulatory network. With continued investigation and technological refinement, miR-200c assessment could become an invaluable component of precision oncology approaches in breast cancer management.

Therapeutic Potential

The tumor-suppressive role of miR-200c suggests it could be harnessed therapeutically. Strategies include restoring miR-200c expression in tumors and targeting its regulatory network. Preclinical studies demonstrate that miR-200c replenishment inhibits metastasis and sensitizes cancer cells to therapy [28]. For example, vector-mediated delivery of miR-200c into TNBC cell lines led to reduced EMT, invasion, and enhanced chemotherapy response [5]. In vivo, inducible miR-200c expression in a mouse mammary carcinoma model reversed EMT, decreased metastasis, and slowed tumor growth [3]. miR-200c also reprograms the tumor microenvironment: its restoration in mesenchymal tumors promoted macrophage- mediated anti-tumor immunity [3], suggesting combination with immunotherapy. Nanotechnology approaches aim to deliver miR-200c mimics to tumors. Eroles et al. developed tumor-targeted mesoporous silica nanoparticles carrying miR-200c-3p mimics; treatment in breast cancer models downregulated ZEB1/2, inhibited growth, and was well-tolerated [74]. Other vehicles (liposomes, polymers) (Table 10.4) have been explored in preclinical breast cancer models, often with positive results in reducing metastasis [29]. Additionally, small molecules that epigenetically upregulate miR-200c (e.g., DNA-demethylating agents) are under investigation to reactivate miR-200c in tumors [75].

Table 10.4 Therapeutic potential platforms

Platform	Advantages	Limitations
Liposomes	Biocompatible; clinically approved; protects miRNA from degradation; and can be targeted [78]	Short circulation time; moderate tumor delivery; and encapsulation issues [79]
MSNs	High loading capacity; controlled; and targeted release [80]	Safety concerns and complex scale-up [81]
Polymeric NPs (e.g., PLGA)	Biodegradable; tunable release; and easy to modify [78]	Blood instability; possible immune response; and low tumor targeting without ligands [82]

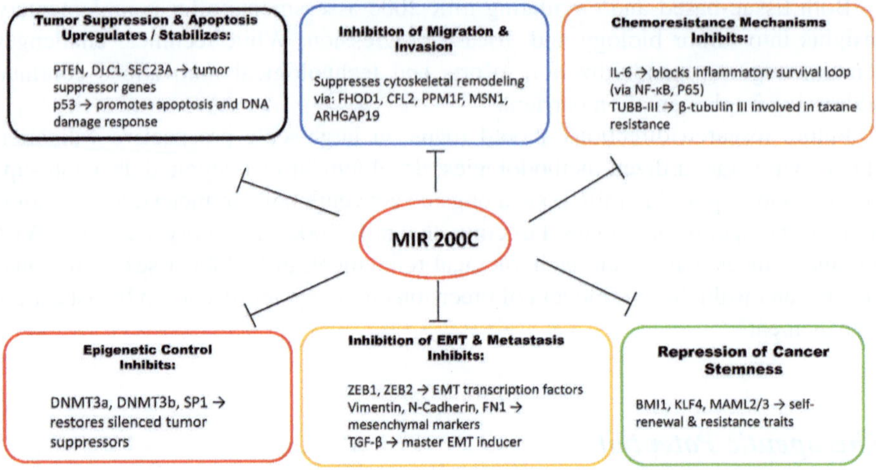

Fig. 10.6 Mechanistic map of miR-200c-mediated regulation in cancer therapy resistance

Beyond direct delivery, miR-200c can enhance chemotherapy and targeted therapy efficacy. It downregulates drug-resistance genes (e.g., MDR1, β-tubulin III) and stemness factors, rendering cells more chemosensitive [4, 76]. For instance, transfection of miR-200c into doxorubicin-resistant breast cancer cells restored drug sensitivity. Furthermore, network studies suggest combining miR-200c modulation with blockade of complementary pathways (e.g., targeting PD-L1 or TGF-β) may yield synergistic effects, given miR-200c's role in regulating immune checkpoints [3].

Despite promise, no miR-200c–based therapy has entered clinical trials yet. Challenges include efficient and specific delivery, stability of miRNA mimics, and potential off-target effects. Nonetheless, miRNA therapeutics are under active clinical investigation (e.g., miR-34 mimics in liver cancer), highlighting feasibility. In the near future, miR-200c could be combined with conventional treatments to improve outcomes in aggressive breast cancers (Fig. 10.6) [77].

Comparison of miR-200c Delivery Platforms: Translational Considerations

Delivering miR-200c effectively to tumor tissues is one of the main challenges in translating laboratory findings into clinical therapies. Several delivery technologies have been explored to improve stability, targeting, and therapeutic efficiency. Below is a comparison of major delivery systems.

Current Clinical Studies

While direct clinical trials of miR-200c therapies are not yet reported, several translational studies assess miR-200c as a biomarker in clinical cohorts. As noted above, large-cohort analyses have quantified plasma and tumor miR-200c and correlated levels with treatment response and survival [53, 83]. For example, in locally advanced breast cancer patients undergoing neoadjuvant chemotherapy, higher tumor miR-200c-3p expression predicted worse overall survival independently of pathological response [53, 83]. Such findings suggest that tumor miR-200c could refine risk stratification and identify patients needing more aggressive therapy.

On the diagnostic front, multicenter studies are validating circulating miRNA panels including miR-200c for early detection. One recent Chinese study confirmed that serum exosomal miR-200c levels distinguished early breast cancer from benign lesions and controls, with high sensitivity and specificity [59]. These efforts are paving the way for clinical assays. In metastatic disease, dynamic monitoring of miR-200c (and other miR-200 family members) during therapy has prognostic value for progression-free survival [84].

To date, no registered clinical trial is specifically evaluating miR-200c replacement therapy in breast cancer. However, lessons from miRNA trials in other cancers inform this field [85]. The MRX34 trial (miR-34 mimic in hepatocellular carcinoma) demonstrated that miRNA can be administered safely to patients, albeit with immunological side effects. Future breast cancer trials may adopt similar nanoparticle or viral delivery strategies for miR-200c mimics once robust preclinical efficacy is established.

Limitations and Future Direction

Despite promising data, several limitations temper enthusiasm. First, miR-200c's role is context-dependent: it is highly expressed in epithelial-like luminal tumors and repressed in basal-like cancers, leading to variable patterns in tissue and blood samples. This heterogeneity complicates biomarker development. Additionally, different studies report opposite trends (e.g., plasma miR-200c low in early-stage BC but high in metastatic settings [53, 83], reflecting the dynamic regulation during progression. Standardization of assay techniques (tissue vs. plasma vs. exosomes), normalization controls, and multi-center validation are needed [85].

Therapeutically, delivery remains the main challenge. Systemic administration of naked miRNA mimics leads to rapid degradation. Nanocarriers show promise, but must achieve efficient tumor targeting and avoid unintended effects on normal tissues. Off-target suppression of other genes and immune reactions are concerns. More research into tissue-specific promoters, targeted conjugates, and controlled-release systems is needed.

Future work should also explore the network of miR-200c regulators and interactors. For example, long noncoding RNAs (lncRNAs) and circular RNAs (circRNAs) that "sponge" miR-200c can dampen its activity; targeting these sponges might indirectly raise miR-200c function [48]. Epigenetic therapies that demethylate the miR-200c promoter (e.g., histone deacetylase inhibitors) are another avenue. Combination strategies for instance, using miR-200c mimics alongside immune checkpoint inhibitors which represent an exciting frontier given miR-200c's role in modulating PD-L1 expression [3].

Conclusions

miR-200c is a pivotal microRNA in breast cancer that suppresses malignant progression by targeting key EMT drivers and oncogenic pathways. Its dysregulation correlates with metastasis and poor outcome, making it a candidate biomarker. Preclinical models support miR-200c replenishment as a therapeutic approach to inhibit metastasis and overcome drug resistance. While clinical translation is still emerging, ongoing research into miR-200c's mechanisms, delivery methods, and integration into biomarker panels holds promise. Continued efforts in large-cohort validation and advanced drug delivery technologies are required to realize the full diagnostic and therapeutic potential of miR-200c in breast cancer.

References

1. Sung H et al (2021) Global cancer statistics 2020: GLOBOCAN estimates of incidence and mortality worldwide for 36 cancers in 185 countries. CA Cancer J Clin 71(3):209–249
2. Lia T et al (2022) Development and validation of pyroptosis-related lncRNAs prediction model for bladder cancer. Biosci Rep 42(1)
3. Williams MM et al (2021) MicroRNA-200c restoration reveals a cytokine profile to enhance M1 macrophage polarization in breast cancer. NPJ Breast Cancer 7(1):64
4. Howe EN, Cochrane DR, Richer JK (2011) Targets of miR-200c mediate suppression of cell motility and anoikis resistance. Breast Cancer Res 13(2):R45
5. Chen H et al (2020) MicroRNA-200c inhibits the metastasis of triple-negative breast cancer by targeting ZEB2, an epithelial-mesenchymal transition regulator. Ann Clin Lab Sci 50(4):519–527
6. Bartel DP (2004) MicroRNAs: genomics, biogenesis, mechanism, and function. Cell 116(2):281–297
7. Kumar S, Nag A, Mandal CC (2015) A comprehensive review on miR-200c, a promising cancer biomarker with therapeutic potential. Curr Drug Targets 16(12):1381–1403
8. Lee RC, Feinbaum RL, Ambros V (1993) The C. elegans heterochronic gene lin-4 encodes small RNAs with antisense complementarity to lin-14. Cell 75(5):843–854
9. Wightman B et al (1991) Negative regulatory sequences in the lin-14 3′-untranslated region are necessary to generate a temporal switch during Caenorhabditis elegans development. Genes Dev 5(10):1813–1824

10. Lagos-Quintana M et al (2001) Identification of novel genes coding for small expressed RNAs. Science 294(5543):853–858
11. Treiber T, Treiber N, Meister G (2019) Regulation of microRNA biogenesis and its crosstalk with other cellular pathways. Nat Rev Mol Cell Biol 20(1):5–20
12. Kozomara A, Birgaoanu M, Griffiths-Jones S (2019) miRBase: from microRNA sequences to function. Nucleic Acids Res 47(D1):D155–d162
13. Wienholds E et al (2005) MicroRNA expression in zebrafish embryonic development. Science 309(5732):310–311
14. Yi R et al (2006) Morphogenesis in skin is governed by discrete sets of differentially expressed microRNAs. Nat Genet 38(3):356–362
15. Esquela-Kerscher A, Slack FJ (2006) Oncomirs - microRNAs with a role in cancer. Nat Rev Cancer 6(4):259–269
16. Lee EJ et al (2008) Systematic evaluation of microRNA processing patterns in tissues, cell lines, and tumors. RNA 14(1):35–42
17. Humphries B, Yang C (2015) The microRNA-200 family: small molecules with novel roles in cancer development, progression and therapy. Oncotarget 6(9):6472
18. Feng X et al (2014) MiR-200, a new star miRNA in human cancer. Cancer Lett 344(2):166–173
19. Korpal M, Kang Y (2008) The emerging role of miR-200 family of microRNAs in epithelial-mesenchymal transition and cancer metastasis. RNA Biol 5(3):115–119
20. Klicka K et al (2022) The role of miR-200 family in the regulation of hallmarks of cancer. Front Oncol 12:965231
21. Dykxhoorn DM (2010) MicroRNAs and metastasis: little RNAs go a long way. Cancer Res 70(16):6401–6406
22. Ha M, Kim VN (2014) Regulation of microRNA biogenesis. Nat Rev Mol Cell Biol 15(8):509–524
23. Lee Y, et al (2002) MicroRNA maturation: stepwise processing and subcellular localization
24. Winter J et al (2009) Many roads to maturity: microRNA biogenesis pathways and their regulation. Nat Cell Biol 11(3):228–234
25. Filipowicz W, Bhattacharyya SN, Sonenberg N (2008) Mechanisms of post-transcriptional regulation by microRNAs: are the answers in sight? Nat Rev Genet 9(2):102–114
26. Lu J, Clark AG (2012) Impact of microRNA regulation on variation in human gene expression. Genome Res 22(7):1243–1254
27. Burk U et al (2008) A reciprocal repression between ZEB1 and members of the miR-200 family promotes EMT and invasion in cancer cells. EMBO Rep 9(6):582–589
28. Gregory PA et al (2008) The miR-200 family and miR-205 regulate epithelial to mesenchymal transition by targeting ZEB1 and SIP1. Nat Cell Biol 10(5):593–601
29. Park S-M et al (2008) The miR-200 family determines the epithelial phenotype of cancer cells by targeting the E-cadherin repressors ZEB1 and ZEB2. Genes Dev 22(7):894–907
30. Górecki I, Rak B (2021) The role of microRNAs in epithelial to mesenchymal transition and cancers; focusing on mir-200 family. Cancer Treat Res Commun 28:100385
31. Bracken CP et al (2008) A double-negative feedback loop between ZEB1-SIP1 and the microRNA-200 family regulates epithelial-mesenchymal transition. Cancer Res 68(19):7846–7854
32. Brabletz S, Brabletz T (2010) The ZEB/miR-200 feedback loop—a motor of cellular plasticity in development and cancer? EMBO Rep 11(9):670–677
33. Castilla MÁ et al (2012) MicroRNA-200 family modulation in distinct breast cancer phenotypes. PLoS One 7(10):e47709
34. Madhavan D et al (2016) Circulating miRNAs with prognostic value in metastatic breast cancer and for early detection of metastasis. Carcinogenesis 37(5):461–470
35. Jones R et al (2017) Re-expression of miR-200c suppresses proliferation, colony formation and in vivo tumor growth of murine claudin-low mammary tumor cells. Oncotarget 8(14):23727
36. Shimono Y et al (2009) Downregulation of miRNA-200c links breast cancer stem cells with normal stem cells. Cell 138(3):592–603

37. Chang B-P et al (2014) miR-200c inhibits metastasis of breast cancer cells by targeting HMGB1. J Huazhong Univ Sci Technolog Med Sci 34:201–206
38. Chen D et al (2013) MicroRNA-200c overexpression inhibits tumorigenicity and metastasis of CD117+ CD44+ ovarian cancer stem cells by regulating epithelial-mesenchymal transition. J Ovarian Res 6:1–11
39. Guo H et al (2025) MicroRNA-200c in cancer generation. Invasion Metastasis 26(2):710
40. Lesurf R et al (2016) Subtype-specific patterns of genomic instability and miRNA expression in breast cancer. J Cancer Res 76(17):5016–5026
41. Gregory PA et al (2008) The miR-200 family and miR-205 regulate epithelial to mesenchymal transition by targeting ZEB1 and SIP1. J Nat Cell Biol 10(5):593–601
42. Shimono Y et al (2009) Downregulation of miRNA-200c links breast cancer stem cells with normal stem cells. J Cell 138(3):592–603
43. Neve RM et al (2006) A collection of breast cancer cell lines for the study of functionally distinct cancer subtypes. J Cancer Cell 10(6):515–527
44. Park SM et al (2008) The miR-200 family determines the epithelial phenotype of cancer cells by targeting the E-cadherin repressors ZEB1 and ZEB2. J Genes Dev 22(7):894–907
45. Korpal M et al (2011) Direct targeting of Sec23a by miR-200s influences cancer cell secretome and promotes metastatic colonization. J Nat Med 17(9):1101–1108
46. Neves R et al (2010) MicroRNA expression in cell lines and in clinical samples of triple-negative breast cancer. J PLOS ONE 5(4):e12343
47. Hur K et al (2013) MicroRNA-200c modulates epithelial-to-mesenchymal transition (EMT) in human colorectal cancer metastasis. Gut 62(9):1315–1326
48. Wellner U et al (2009) The EMT-activator ZEB1 promotes tumorigenicity by repressing stemness-inhibiting microRNAs. Nat Cell Biol 11(12):1487–1495
49. Wang J et al (2014) Downregulation of miR-486-5p contributes to tumor progression and metastasis by targeting protumorigenic ARHGAP5 in lung cancer. Oncogene 33(9):1181–1189
50. Wang H et al (2018) Circulating microRNAs as potential cancer biomarkers: the advantage and disadvantage. Clin Epigenetics 10:1–10
51. Bertucci F et al (2006) Gene expression profiling shows medullary breast cancer is a subgroup of basal breast cancers. Cancer Res 66(9):4636–4644
52. Bertucci F et al (2019) Genomic characterization of metastatic breast cancers. Nature 569(7757):560–564
53. Antolín S et al (2015) Circulating miR-200c and miR-141 and outcomes in patients with breast cancer. BMC Cancer 15:297
54. Antolín S et al (2015) Circulating miR-200c and miR-141 and outcomes in patients with breast cancer. BMC Cancer 15:1–15
55. Schwarzenbach H et al (2014) Clinical relevance of circulating cell-free microRNAs in cancer. Nat Rev Clin Oncol 11(3):145–156
56. Guo J et al (2009) Differential expression of microRNA species in human gastric cancer versus non-tumorous tissues. J Gastroenterol Hepatol 24(4):652–657
57. Qiao L et al (2015) Advanced research on vasculogenic mimicry in cancer. J Cell Mol Med 19(2):315–326
58. Jin W et al (2011) DNA methylation-dependent regulation of TrkA, TrkB, and TrkC genes in human hepatocellular carcinoma. Biochem Biophys Res Commun 406(1):89–95
59. Qiao P et al (2024) Serum exosomal miR-200c is a potential diagnostic biomarker for breast cancer. Biomarkers 29(7):419–426
60. Jin Y et al (2019) Circulating microRNAs as potential diagnostic and prognostic biomarkers in hepatocellular carcinoma. Sci Rep 9(1):10464
61. Yoshioka Y et al (2012) Micromanaging iron homeostasis: hypoxia-inducible micro-RNA-210 suppresses iron homeostasis-related proteins. J Biol Chem 287(41):34110–34119
62. Widman AJ et al (2024) Ultrasensitive plasma-based monitoring of tumor burden using machine-learning-guided signal enrichment. Nat Med 30(6):1655–1666

63. Semba T, Ishimoto T (2024) Spatial analysis by current multiplexed imaging technologies for the molecular characterisation of cancer tissues. Br J Cancer 131(11):1737–1747
64. Zhang KC et al (2015) Prognostic role of miR-200c in various malignancies: a systematic review and meta-analysis. Int J Clin Exp Med 8(2):1931–1943
65. Teng Y et al (2016) miRNA-200a/c as potential biomarker in epithelial ovarian cancer (EOC): evidence based on miRNA meta-signature and clinical investigations. Oncotarget 7(49):81621–81633
66. Liu Y et al (2017) MiRNAs predict the prognosis of patients with triple negative breast cancer: a meta-analysis. PLoS One 12(1):e0170088
67. Chang QQ et al (2019) LncRNA PVT1 promotes proliferation and invasion through enhancing Smad3 expression by sponging miR-140-5p in cervical cancer. Radiol Oncol 53(4):443–452
68. Shao B et al (2019) Plasma microRNAs predict chemoresistance in patients with metastatic breast cancer. Technol Cancer Res Treat 18:1533033819828709
69. Zhang L et al (2015) A circulating miRNA signature as a diagnostic biomarker for non-invasive early detection of breast cancer. Breast Cancer Res Treat 154:423–434
70. Thi Chung Duong T et al (2022) Diagnostic and prognostic value of miR-200 family in breast cancer: a meta-analysis and systematic review. Cancer Epidemiol 77:102097
71. Rodríguez-Martínez A et al (2019) Exosomal miRNA profile as complementary tool in the diagnostic and prediction of treatment response in localized breast cancer under neoadjuvant chemotherapy. Breast Cancer Res 21:1–9
72. Mestdagh P et al (2014) Evaluation of quantitative miRNA expression platforms in the microRNA quality control (miRQC) study. Nat Methods 11(8):809–815
73. Kroh EM et al (2010) Analysis of circulating microRNA biomarkers in plasma and serum using quantitative reverse transcription-PCR (qRT-PCR). Methods 50(4):298–301
74. Garrido-Cano I et al (2023) Delivery of miR-200c-3p using tumor-targeted mesoporous silica nanoparticles for breast cancer therapy. ACS Appl Mater Interfaces 15(32):38323–38334
75. Pecot CV et al (2011) RNA interference in the clinic: challenges and future directions. Nat Rev Cancer 11(1):59–67
76. Liu S et al (2015) miR-200c/Bmi1 axis and epithelial-mesenchymal transition contribute to acquired resistance to BRAF inhibitor treatment. Pigment Cell Melanoma Res 28(4):431–441
77. Monroig-Bosque Pdel C, Rivera CA, Calin GA (2015) MicroRNAs in cancer therapeutics: "from the bench to the bedside". Expert Opin Biol Ther 15(10):1381–1385
78. Suedee R et al (2010) Development of a pH-responsive drug delivery system for enantioselective-controlled delivery of racemic drugs. J Control Release 142(1):122–131
79. Rupaimoole R, Slack FJ (2017) MicroRNA therapeutics: towards a new era for the management of cancer and other diseases. Nat Rev Drug Discov 16(3):203–222
80. Mukhopadhyay TK, Morstein J, Trauner D (2022) Photopharmacological control of cell signaling with photoswitchable lipids. Curr Opin Pharmacol 63:102202
81. El Andaloussi S et al (2013) Extracellular vesicles: biology and emerging therapeutic opportunities. Nat Rev Drug Discov 12(5):347–357
82. Athauda D, Foltynie T (2016) Insulin resistance and Parkinson's disease: a new target for disease modification? Prog Neurobiol 145-146:98–120
83. Navarro-Manzano E et al (2022) Prognostic and predictive effects of tumor and plasma miR-200c-3p in locally advanced and metastatic breast cancer. Cancers (Basel) 14(10)
84. Fischer C et al (2022) Circulating miR-200 family as predictive markers during systemic therapy of metastatic breast cancer. Arch Gynecol Obstet 306(3):875–885
85. Cookson VJ et al (2012) Circulating microRNA profiles reflect the presence of breast tumours but not the profiles of microRNAs within the tumours. Cell Oncol (Dordr) 35(4):301–308

Chapter 11
Oral Cancer, MiRNAs, and Cannabioids: A Complicated *Ménage à Trois*

Andrea De Laurentiis, Manuel Luis Wolfson, Ganna Dmytrenko, Mariana Farina, and Fernando Correa

Abstract Oral and oropharyngeal cancers remain significant global health concerns, with well-established risk factors including tobacco and alcohol use, along with others such as areca nut chewing, HPV infection, and poor oral hygiene.

The association between cannabis consumption and oral carcinogenesis is still under investigation. However, cannabinoids are increasingly used by cancer patients for symptom relief, and recent studies suggest they may modulate cancer-related pathways through alterations in microRNA (miRNA) expression. miRNAs, in turn, can influence the endocannabinoid system.

Saliva, a readily accessible and noninvasive biofluid, offers a promising medium for detecting changes in miRNA profiles linked to oral cancer and premalignant conditions. These salivary miRNAs may serve as early diagnostic biomarkers and help predict disease outcomes.

This chapter explores the current evidence on the relationship between cannabis use, salivary miRNA expression, and the development of oral cancer, underscoring the potential of salivary miRNAs in early detection and prognosis.

A. De Laurentiis · M. Farina · F. Correa (✉)
Universidad de Buenos Aires (UBA), Facultad de Odontología, Cátedra de Fisiología, Ciudad Autónoma de Buenos Aires, Argentina

Centro de Estudios Farmacológicos y Botánicos, Universidad de Buenos Aires-Consejo Nacional de Investigaciones Científicas y Técnicas, Ciudad Autónoma de Buenos Aires, Argentina
e-mail: fernando.correa@odontologia.uba.ar

M. L. Wolfson
Centro de Estudios Farmacológicos y Botánicos, Universidad de Buenos Aires-Consejo Nacional de Investigaciones Científicas y Técnicas, Ciudad Autónoma de Buenos Aires, Argentina

G. Dmytrenko
Universidad de Buenos Aires (UBA), Facultad de Odontología, Cátedra de Fisiología, Ciudad Autónoma de Buenos Aires, Argentina

S. Pathak, A. Banerjee (eds.), *MicroRNAs: Role in Cancer Progression, Diagnostics and Therapy*, Advances in Biochemistry in Health and Disease 34,
https://doi.org/10.1007/978-3-032-08824-6_11

Keywords Oral cancer · Oropharyngeal cancer · Cannabinoids · MicroRNAs · Salivary biomarkers · Early detection · CBD · THC · Saliva · Oral premalignant lesions

Introduction

Oral Cancer: Epidemiology and Etiology

The International Classification of Diseases for Oncology (ICD-O) delineates oral cavity cancer as including malignancies of the inner lip, dorsal tongue, gingiva, hard and soft palates, buccal mucosa, floor of the mouth, and "other specified parts of the mouth." In contrast, oropharyngeal cancer encompasses tumors arising in the base of the tongue, tonsillar region, anterior epiglottis, lateral oropharyngeal wall, and pharyngeal areas such as Waldeyer's ring [23, 127]. Data from the 2022 GLOBOCAN report by the World Health Organization's (WHO) International Association of Cancer Registries place oral cancer as the 16th most commonly diagnosed cancer globally, with approximately 389,485 new cases annually and an age-standardized incidence rate (ASR) of 4.0 per 100,000 population [18]. Oropharyngeal cancer ranks 24th, with 106,316 new cases reported each year and a global ASR of 1.9 per 100,000 individuals [18]. The prognosis for oral cancer is closely tied to the stage at diagnosis: early-stage cases have a 5-year survival rate of near 90%, while late-stage diagnoses correspond to survival rates of approximately 30% [90]. Worldwide mortality for oral cancer stands at 188,230 deaths annually, with a global ASR of 1.9 per 100,000, whereas oropharyngeal cancer accounts for 52,268 deaths and an ASR of 0.5 per 100,000 [18]. Although oral cancer is more commonly observed in developing countries, recent trends suggest an increase in incidence within high-income nations [110]. Regional distribution highlights Asia as the most affected continent, accounting for 64.2% of cases, followed by Europe (17.4%), North America (7.6%), Latin America and the Caribbean (5.6%), Africa (3.8%), and Oceania (1.3%) [110]. Men are disproportionately affected, with a global ASR of 5.8 per 100,000 compared to 2.3 per 100,000 in women [18], and the disease is most frequently diagnosed after the fifth decade of life [84]. Despite higher incidence rates in men, a notable increase in tongue cancer among younger women without a history of tobacco or alcohol use has been reported, suggesting possible alternative etiologic mechanisms or emerging disease subtypes in this population [23, 97]. In particular, certain oropharyngeal cancers—including those of the tonsils, base of the tongue, and pharynx—have been increasingly linked to human papillomavirus (HPV) infection in nonsmoking, nondrinking individuals [12].

Moreover, the high mortality rates associated with oral cancers are largely attributed to delayed diagnoses, aggressive tumor behavior, frequent recurrences, and the potential for metastasis [27]. In light of these challenges, the American Dental Association has emphasized the importance of routine visual and tactile examinations for both oral and oropharyngeal cancers, underscoring the pivotal role of dental professionals in facilitating early detection through regular screenings [64]. Despite such professional guidelines, there remains a critical need for the advancement of more accurate and sensitive molecular diagnostic tools to enable earlier and more reliable identification of these malignancies.

Oral cancer can arise either independently or through the malignant transformation of oral premalignant dysplastic lesions (OPMLs), also known as oral potentially malignant disorders (OPMDs). These include conditions such as leukoplakia, proliferative verrucous leukoplakia, erythroplakia, melanoplakia, oral lichen planus, oral lichenoid lesions, submucous fibrosis, or a combination thereof [62]. Leukoplakia, one of the most prevalent premalignant lesions, is described as a "white plaque or patch in the oral mucosa that cannot be characterized as any other detectable lesion" [81]. The risk of progression to oral cancer over five years ranges from 2.2% to 32%, depending on the histopathological grade of dysplasia identified in biopsy specimens [21]. A particularly concerning subtype, proliferative verrucous leukoplakia, tends to present as multifocal lesions with aggressive clinical behavior, frequent recurrences, and a high likelihood of malignant transformation [8, 107]. Erythroplakia is defined as a distinct, velvety red macule or plaque typically greater than 1.5 cm in diameter, commonly found on the floor of the mouth, soft palate, or buccal mucosa, and similarly "cannot be characterized as any other recognizable condition" [81]. Owing to its elevated malignant potential, surgical removal is often recommended [84]. Oral lichen planus is another potentially malignant condition that can appear in papular, atrophic, or erosive forms, the latter of which are particularly painful and prone to secondary infections. Malignant transformation rates for this lesion have been reported to range between 0.4% and 5.6% [8, 81]. Oral submucous fibrosis, primarily affecting populations in South Asia, is a chronic disorder associated with the use of betel quid and is characterized by progressive fibrosis of the oral mucosa, leading to irreversible soft tissue changes [44].

Oral squamous cell carcinoma (OSCC) accounts for approximately 90% of all oral cancer cases, making it by far the most prevalent histological subtype. The remaining 10% comprise a range of non-squamous malignancies, including tumors of the minor salivary glands (such as mucoepidermoid and adenoid cystic carcinomas), basal cell carcinomas, mesenchymal sarcomas, hematological neoplasms, and oral melanomas [28, 88]. OSCCs are histopathologically graded as well-, moderately-, or poorly differentiated based on cellular atypia and tissue architecture. The disease often begins without significant pain or specific symptoms, which contributes to delays in diagnosis. As a result, a significant proportion of cases approximately 50% are identified at stage IV, when prognosis is considerably worse [125]. Survival outcomes vary dramatically depending on the stage at diagnosis: early-stage OSCC (stage I) is associated with a 5-year survival rate of around 80%, while this rate drops to 20–40% for cases diagnosed at stages III or IV [3]. OSCC typically progresses by invading adjacent tissues, spreading to cervical lymph nodes, and potentially metastasizing to distant organs [88]. Clinical signs indicative of advanced disease and poorer prognosis include trismus, restricted movement of the tongue, and cutaneous infiltration with dermal lymphatic spread [84]. Although surgical resection remains the cornerstone of treatment, adjuvant therapies particularly in high-risk patients are often required to reduce the likelihood of locoregional recurrence. External beam radiotherapy is the standard postoperative modality, and various chemotherapeutic agents such as cisplatin, 5-fluorouracil, and docetaxel have been combined with radiotherapy, though studies have yielded mixed results regarding efficacy [96]. Despite comprehensive treatment, up to one-third of patients experience relapse, with locoregional recurrence being the predominant mode of treatment failure [16].

Tobacco consumption, whether by smoking or chewing, alongside alcohol intake, remains the most well-established risk factors for OSCC [78]. Other notable contributors include the chewing of areca nuts and betel leaves, advanced age, inadequate oral hygiene, exposure to wood dust, chronic mechanical trauma, oral colonization by *Candida albicans*, and infection with human papillomavirus (HPV) [1, 73, 84]. Tobacco products harbor numerous carcinogenic compounds such as polycyclic hydrocarbons, acetaldehyde, urethane, and nitrosamines. The risk of developing OSCC correlates positively with the cumulative tobacco exposure measured in pack-years, and although cessation lowers this risk, it does not completely eliminate it [29]. Moreover, the combined use of tobacco and alcohol exerts a synergistic effect, markedly increasing oral cancer risk [78]. Novel tobacco delivery systems, including smokeless tobacco (snus), hookahs, water pipes, electronic cigarettes, and oral dissolvable products, are frequently promoted as less harmful alternatives; however, ongoing research continues to examine their carcinogenic potential [21]. Betel quid, a preparation consisting of betel leaves, sliced areca nuts, slaked lime, and often tobacco is extensively used in South Asia and the Pacific regions, where the burden of oral cancer is disproportionately high [8, 45, 110]. The principal carcinogenic agents within betel quid are polyphenols, classified as Group I carcinogens by the International Agency for Research on Cancer [131]. While tobacco smoking has declined in many developed countries, contributing to a reduction in oral cancer incidence, there has been a notable rise in HPV-positive oropharyngeal cancers in recent years [21]. This increase is largely attributed to changing sexual behaviors, specifically greater acceptance and practice of oral sex, resulting in higher oral HPV exposure among younger populations and females under 45 years of age [40, 124].

Although biopsy remains the gold standard for the definitive diagnosis of oral cancer, it is an invasive procedure that can cause significant discomfort to patients. Therefore, there is an urgent need to develop noninvasive diagnostic methods capable of detecting cancer at early stages. microRNAs (miRNAs), particularly those detectable in saliva, represent a novel and promising avenue for noninvasive biomarker discovery in OSCC. While protein and DNA methylation markers in cancerous cells also offer significant diagnostic and prognostic value, salivary miRNAs present distinct advantages due to their accessibility and the minimally invasive nature of their collection. Protein biomarkers such as altered p53, overexpression of the epidermal growth factor receptor (EGFR), and cyclin D1 are often found to be overexpressed in OSCC and are commonly associated with increased lesion aggressiveness [98, 99]. Beyond proteins, DNA methylation markers have also shown potential in OSCC diagnostics. Genes such as p16 and DAPK1 frequently exhibit aberrant methylation patterns, with hypermethylation of the p16 promoter region and increased expression of DAPK1 both correlating with poor prognosis [99]. A growing body of evidence has revealed distinct differences in the miRNA expression profiles between oral cancer tissues and their normal counterparts, highlighting miRNAs as important molecular tools for understanding disease progression. Accordingly, multiple studies have examined plasma and salivary miRNAs as potential noninvasive biomarkers for early detection and ongoing surveillance of oral cancer.

Plasma and Salivary miRNAs as Biomarkers in Oral Cancer

miRNAs are small noncoding RNA molecules, typically 18 to 25 nucleotides in length, which play critical roles in regulating cell growth, proliferation, and apoptosis through modulation of gene expression and/or transcriptional repression [1]. These miRNAs are commonly secreted into bodily fluids encapsulated within membrane-bound vesicles known as exosomes. Additionally, freely circulating miRNAs in serum or plasma are protected from degradation by RNases through their association with members of the Argonaute protein family [74, 89]. Notably, miRNAs possess the unique capability to simultaneously target multiple protein-coding genes, either within the same signaling pathway or across different pathways, often exerting synergistic regulatory effects. Consequently, even minor alterations in miRNA expression can induce substantial changes in the expression profiles of numerous protein-coding genes, thereby affecting diverse cellular phenotypes.

In dental tissues, miRNAs have been implicated in several key physiological and pathological processes, including periodontal disease, tooth eruption and movement, dental pulp health, differentiation of dental cells, enamel mineralization, and tumorigenesis. Alterations in specific miRNA expression profiles contribute to cancer development and progression by functioning as oncogenes or tumor suppressors. Distinct miRNA signatures have been proposed as biomarkers for prognosis, early detection, and accurate diagnosis of oral and oropharyngeal cancers [34]. Recent investigations have identified changes in circulating miRNA levels in the plasma or serum of oral cancer patients. Yan et al. [137] compared plasma miRNA profiles from Danish and Chinese patients, revealing elevated levels of miR-26a-5p, miR-148a-3p, and miR-21-5p alongside decreased miR-486-5p in the Danish cohort. Conversely, the Chinese cohort showed reduced miR-375, miR-92b-3p, and miR-486-5p without significant increases in the miRNAs elevated in the Danish group. A Canadian study reported increased plasma concentrations of a broad panel of miRNAs, including miR-16, miR-338, miR-let-7b, Let-7d, miR-92a, miR-30e, miR-320, miR-7, miR-25, miR-195, miR-624, miR-142, miR-192-5p, miR-150-5p, miR-451, miR-17-5p, miR-26a, and miR-486-5p in patients presenting with high-risk oral lesions [72]. In a Turkish population, Ayaz et al. [7] found a completely different set of miRNAs in the plasma of laryngeal squamous cell carcinoma patients. The inconsistent findings across studies may be attributed to differences in sample handling, including centrifugation speed and duration, which can influence the yield of microvesicles. Furthermore, coagulation processes can affect cell-free miRNA concentrations, contributing to variability between plasma and serum measurements. The lack of standardized endogenous controls for miRNA expression normalization further complicates comparisons and reproducibility among studies [146].

Saliva is a complex biofluid secreted by the three major salivary glands parotid, submandibular, and sublingual as well as numerous minor salivary glands distributed throughout the oral mucosa [120]. The composition of saliva is further influenced by gingival crevicular fluid, particularly in inflammatory states such as gingivitis and periodontitis [55]. This fluid comprises a diverse array of components, including enzymes, antibodies, hormones, antimicrobial agents, and cytokines. In addition, saliva contains desquamated epithelial cells from the oral mucosa and, in cases of neoplasia, may harbor leukocytes [112, 117, 120]. Compared to blood sampling, saliva collection offers a less invasive, safer, and more cost-efficient method to obtain a "liquid biopsy," especially for diagnosing diseases of the oral cavity [112]. Salivary miRNAs are derived from salivary gland secretions, gingival crevicular fluid, and exfoliated oral epithelial cells, and exist either complexed with protective macromolecules that shield them from degradation or encapsulated within exosomes [48, 94]. To date, over 458 distinct miRNA species have been identified in saliva, with miR-335-3p being the most prevalent [48, 135].

Numerous studies have documented alterations in salivary miRNA expression profiles among patients with oral cancer and/or premalignant lesions. Early detection of such changes in salivary miRNAs is therefore critical for identifying OPMLs with potential for malignant progression [8]. Specific miRNAs have been implicated in promoting cell proliferation, growth, and cell cycle progression, while others are associated with anchorage-independent colony formation and enhanced vascular invasion [141]. Notably, overexpression of miR-21, miR-181b, and miR-345 correlates with leukoplakia severity and its malignant transformation, as these miRNAs are significantly elevated in lesions exhibiting increased mitotic activity, altered nucleus-to-cytoplasm ratios, and abnormal superficial mitosis [19]. Similarly, elevated miR-31 expression has been identified as a risk factor for OPML progression and may serve as a predictor of recurrence and malignant conversion [56]. In contrast, miR-26a, miR-29a, and miR-375 are consistently downregulated in leukoplakia lesions [108, 116]. The salivary upregulation of miR-27b can effectively discriminate oral cancer patients from healthy controls [134]. Supporting this, a pilot study by Di Stasio et al. [30] reported overexpression of both miR-27b and miR-181b in saliva from patients with high-grade oral dysplasia, suggesting their utility as potential biomarkers for disease severity. Collectively, these findings underscore the promise of specific salivary miRNAs as early biomarkers for monitoring malignant transformation in patients with OPML [141].

The sensitivity and specificity of miRNAs for the diagnosis of oral cancer varies depending on the specific miRNA and the study [8, 79, 134]. A recent study identified a panel of eight miRNAs in saliva that demonstrated high accuracy in discriminating between patients with oral cancer and healthy controls [9]. This panel includes: miR-7-5p, miR-10b-5p, miR-182-5p, miR-215-5p, miR-431-5p, miR-486-3p, miR-3614-5p, and miR-4707-3p [9]. The efficiency of this panel to discriminate between oral cancer and healthy controls was AUC: 0.954, sensitivity: 86%, specificity: 90%; whereas the efficiency to discriminate between oral cancer and premalignant lesions was AUC: 0.911, sensitivity: 90%, specificity: 82.7% [9]. The same authors identified a panel of four salivary microRNAs such as miR-105p, miR-182-5p, miR-215-5p, and miR-3614-5p that effectively discriminates between patients with oral premalignant lesions and healthy individuals, yielding an AUC of 0.807. The panel demonstrated a sensitivity of 59.6%, specificity of 90%, positive predictive value of 83.3%, and negative predictive value of 72% [9]. A noteworthy aspect of their methodology was the inclusion of both upregulated and downregulated miRNAs, diverging from traditional approaches that typically focus only on overexpressed markers. Despite the encouraging performance metrics, these findings require further validation in studies involving larger and more diverse cohorts.

Several other miRNAs have been investigated as potential biomarkers for oral lesions. He et al. [50] reported elevated levels of miR-24-3p, while Momen-Heravi et al. [83] observed increased miR-27b and decreased miR-136 in saliva samples from OSCC patients. Zahran et al. [142] found heightened salivary miR-21 and miR-184 alongside reduced miR-145 expression in oral cancer cases. High expression of miR-135b-5p, miR-18a-5p, and miR-30a-5p has been correlated with poorer survival outcomes, whereas a miRNA panel including miR-31-3p, miR-139-5p, and miR-30a-5p has been proposed for distinguishing oral cancer from noncancerous tissues [121].

In saliva-derived extracellular vesicles, Gai et al. [37] identified miR-412-3p, miR-512-3p, miR-302b-3p, and miR-517b-3p as significantly overexpressed in OSCC patients, suggesting their utility as biomarkers. Similarly, Liu et al. [66] reported increased miR-31 in both saliva and plasma of oral cancer patients, associating it with enhanced tumor proliferation, invasion, and metastasis. Other miRNAs such as miR-146a-5p, miR-9, miR-127, miR-191, and miR-222 have also been shown to be upregulated in the saliva from OSCC patients compared to controls, while miR-200 and miR-125a are downregulated [77, 80, 95, 104]. Notably, Duz et al. [31] demonstrated that salivary miR-139-5p levels were reduced in tongue squamous cell carcinoma patients but returned to baseline after tumor resection, indicating its potential role in diagnosis and follow-up. Analysis from oral cytology samples revealed miR-21 upregulation and miR-375 downregulation in oral tongue cancer patients [49]. Plasma miR-21 levels decreased following tumor removal in patients with favorable prognosis, but remained elevated in those with poor outcomes, highlighting its prognostic value [54].

Yap et al. [140] successfully isolated and analyzed miRNAs from oral swabs, detecting dysregulation of miR-24-3p, miR-21-5p, let-7c-5p, miR-99a-5p, and miR-100-5p in OSCC patients with 81.5% specificity and 86.8% sensitivity. Additional miRNAs suggested as biomarkers for tongue and head and neck squamous cell carcinomas (HNSCCs) include miR-139-5p, miR-let-7a-5p, and miR-3928 [33, 133]. miR-181 and miR-296 have been associated with metastatic OSCC, while miR-31 and miR-130b are linked to nonmetastatic tumors [113, 138]. Other miRNAs implicated in OSCC pathogenesis include miR-140-5p, miR-125b-5p, miR-155-3p, miR-214-5p, and miR-146a [42, 59, 93].

In a pilot study, Vageli et al. [130] evaluated a novel panel comprising miR-21, miR-136, miR-3928, and miR-29b in saliva and serum, demonstrating significant overexpression in OSCC patients compared to healthy controls. Additionally, miR-21 levels were notably higher among smokers, underscoring its potential as a biomarker in this subgroup [130].

Uma Maheswari et al. [129] examined salivary miR-21 and miR-31 expression in patients with OPML, including leukoplakia, oral submucous fibrosis (OSMF), oral lichen planus, and OSMF combined with leukoplakia. Both miRNAs were significantly upregulated in severe dysplasia compared to controls. Receiver operating characteristic analysis revealed that miR-21 exhibited superior diagnostic performance for OPML, with 66% specificity, 69% sensitivity, and an area under the curve of 0.820. Furthermore, miR-31 expression correlated positively with smoking duration. These results were confirmed by Liu et al. [70], who linked miR-21 and miR-31 to premalignant lesion initiation and progression, while high miR-31 expression was associated with poor prognosis and reduced overall survival [57]. Tu et al. [128] reported markedly decreased salivary miR-375 levels in OPML patients, with low miR-375 expression in oral cancer correlating with worse survival outcomes [144].

The multifaceted roles of these miRNAs in oral carcinogenesis and progression have been comprehensively reviewed by Rishabh et al. [105]. For an in-depth analysis of each miRNA's function and clinical relevance, readers are referred to this detailed review.

It is important to highlight that the sensitivity and specificity of miRNAs as biomarkers for oral cancer vary considerably across studies, due to differences in methodologies, sample sizes, and study populations. Consequently, further research is necessary to validate these findings and to identify the optimal combination of miRNAs for accurate oral cancer diagnosis [8, 79, 134]. Notably, Liu et al. [69] performed a systematic review and meta-analysis encompassing twenty studies on the diagnostic accuracy of blood- and saliva-derived miRNAs in detecting OSCC. They reported a pooled sensitivity and specificity for salivary miRNAs of 0.70 (95% CI: 0.63–0.77) and 0.82 (95% CI: 0.72–0.90), respectively, with an AUC of 0.80 (95% CI: 0.77–0.84). In contrast, blood-based miRNAs demonstrated slightly higher sensitivity at 0.79 (95% CI: 0.73–0.84), similar specificity of 0.82 (95% CI: 0.77–0.86), and an AUC of 0.88 (95% CI: 0.84–0.90). More recently, Nguyen and Nonaka [87] conducted a meta-analysis evaluating diagnostic accuracy of blood and salivary miRNAs for oropharyngeal squamous cell carcinoma, concluding that combined analysis of both sources enhances diagnostic performance, with pooled sensitivity of 0.70, specificity of 0.75, a diagnostic odds ratio of 7, and an AUC of 0.78.

In summary, saliva has emerged as a promising biofluid for oral cancer diagnosis due to its noninvasive collection and direct contact with oral lesions. Advantages of saliva include its ease of collection through low-cost, noninvasive methods, its ability to reflect the tumor microenvironment, and the potential of salivary miRNAs to reveal early alterations indicative of cancer, facilitating earlier diagnosis and improved prognosis. Moreover, serial saliva sampling enables monitoring of disease progression and recurrence through miRNA profile changes. Despite these benefits, several limitations exist for using saliva as a liquid biopsy in oral cancer [120, 123]. The lack of standardized protocols for sample processing and miRNA detection leads to variability and complicates cross-study comparisons. Saliva composition is influenced by individual factors such as diet, oral hygiene, and comorbidities, potentially affecting diagnostic accuracy. Additionally, miRNA profile interpretation is complex, requiring specialized expertise and advanced bioinformatics tools. While multiple miRNAs show promise as biomarkers, large-scale validation studies are imperative to confirm their clinical utility. Furthermore, specificity remains a concern, as some miRNAs are deregulated under diverse pathological conditions beyond oral cancer, which may cause false-positive results [27, 74, 102, 120, 123].

Therefore, establishing standardized protocols for saliva collection, storage, and processing is crucial to preserve miRNA integrity and ensure reproducibility. Appropriate normalization strategies are also required to mitigate variability in RNA quantity and quality. Large multicenter cohorts are necessary to validate candidate miRNA biomarkers and develop robust diagnostic algorithms. Integrating salivary miRNA analysis with clinical examination and biopsy could enhance early detection accuracy and sensitivity of oral cancer [27, 74, 102]. Within this framework, miRNA molecular signatures hold significant promise for assessing OSCC risk, guiding therapeutic decisions, and predicting treatment outcomes. The discovery of reliable, noninvasive biomarkers would greatly improve OSCC clinical management. Current research is actively exploring novel biomarkers across multiple biological domains including genomics, transcriptomics, proteomics, metabolomics, immunomics, and microbiomics of saliva [101].

Cannabinoids and Oral Cancers: Friends or Foes?

Cannabis remains the most commonly used recreational drug, typically consumed through smoking, vaporization, or ingestion. Smoking is often carried out using joints, blunts, or bongs [122]. The link between cannabis consumption and oral cancer continues to be debated. Cannabis smoke has been shown to contain carcinogenic compounds similar to those in tobacco smoke [43]. Joints arguably the most prevalent form of cannabis delivery are generally loosely packed and lack filters, resulting in higher concentrations of inhaled smoke than conventional tobacco cigarettes [114]. Cannabis use has been associated with various oral health alterations, including xerostomia, leukoedema, and elevated colonization by *Candida albicans* [132]. Furthermore, chronic exposure to cannabis smoke may detrimentally affect oral homeostasis, contributing to dental caries, periodontal disease, mucosal conditions, and suboptimal oral hygiene [22, 60, 132]. Several reports have described specific oral mucosal changes in cannabis users, such as "cannabis stomatitis" and the presence of lesions like leukoedema, leukoplakia, and hyperkeratosis [118], as well as premalignant alterations particularly leukoplakia and erythroplakia are in high-risk anatomical sites like the floor of the mouth and the tongue [2, 10]. Gallagher et al. [38] noted a 3.5–5-fold increased risk of head and neck cancer in individuals diagnosed with cannabis use disorder. Other studies have also suggested a positive correlation between cannabis use and malignancies of the oral or oropharyngeal regions [36, 73, 143]. However, a number of epidemiological studies have not confirmed such associations [4, 11, 39, 71, 91, 106]. Rosenblatt et al. [106], after adjusting for major confounders, found no association between OSCC and cannabis exposure. Similarly, no link has been reported between smoking cannabis with head and neck cancer but instead an increased risk for other malignancies such as prostate, cervical, and brain cancers was noted [46]. In summary, while some evidence points toward a potential relationship between cannabis and oral or oropharyngeal cancers, these findings remain inconclusive and are complicated by confounding factors such as concurrent tobacco and alcohol use.

Conversely, emerging evidence suggests that cannabinoids may exert therapeutic effects in the context of oral pathologies [10]. *Cannabis sativa* produces over 500 biologically active compounds, including approximately 100 cannabinoids primarily Δ^9-tetrahydrocannabinol (THC) and cannabidiol (CBD) [76]. These phytocannabinoids engage the CB1 and CB2 cannabinoid receptors that are widely distributed across human tissues [53]. Expression of both receptors has been confirmed in several oral tissues, including salivary glands [35, 100], dental pulp, periodontal ligament [61, 67, 92], oral mucosa, the tongue [52], and taste papillae [17]. Studies have also reported elevated cannabinoid receptor expression in mobile tongue squamous cell carcinoma [126], although Alhouayek et al. [6] found no significant difference in CB1 or CB2 mRNA levels between tumor and healthy tissues, highlighting the potential discrepancy between transcriptional and protein-level expression. Liu et al. [68] demonstrated increased CB1/CB2 mRNA levels in HPV-positive head and neck squamous cell carcinoma (HNSCC) samples, and THC exposure promoted cell proliferation in corresponding cell lines. In contrast, Blal et al. [15] observed that purified CBD induced significant apoptosis in HNSCC cells, with enhanced efficacy when co-administered with THC, supporting an entourage effect hypothesis. These contrasting findings point to a complex role of cannabinoid signaling in oncogenesis. Further, Shewale et al. [115] found no association between marijuana use and HPV-positive OSCC, and even a negative correlation with HPV-negative OSCC. Liang et al. [63] similarly identified a reduced risk of HNSCC in individuals with long-term cannabis use. Semlali et al. [111] reported that THC exposure in Ca9–22 gingival carcinoma cells reduced proliferation and activated autophagic and apoptotic pathways. Whyte et al. [136] demonstrated cannabinoid-independent cytotoxicity via THC-mediated inhibition of mitochondrial respiration in Tu183 cells. CBD has also been shown to induce cell cycle arrest and apoptosis in various HNSCC cell lines [41], and higher concentrations (>25 μM) of CBD were found to promote DNA damage and cytotoxicity in HSC-3 oral carcinoma cells [14]. Similarly, Blal et al. [15] reported CBD-induced apoptosis in Scc4, Scc9, Scc25, and Cal27 cell lines. Notably, Go et al. [41] identified a synergistic cytotoxic effect when combining CBD with cisplatin in an in vivo HNSCC model. Such synergistic interactions have also been observed in other malignancies treated with combinations of cannabinoids and chemotherapeutic agents, including vincristine, doxorubicin, cytarabine, and bortezomib [51, 65, 103]. The adjuvant use of cannabinoids alongside anticancer therapies has been reviewed in depth by Nahler [85]. Overall, these findings reinforce the need for further research to delineate whether cannabinoids act predominantly as antineoplastic or tumor-promoting agents in oral cancers.

Phytocannabinoids also exhibit clinically recognized benefits in supportive cancer care. They have been approved as antiemetics for patients undergoing chemotherapy for oral cancer [25, 119] and are used to manage radiotherapy-induced pain in head and neck cancer [86]. Additionally, cannabinoids have been reported to enhance appetite and improve mood, thereby contributing to better quality of life in newly diagnosed HNSCC patients [145]. Nonetheless, not all studies support these findings, and some have failed to demonstrate improvements in patient well-being from THC or synthetic cannabinoids during radiotherapy [24, 47].

In addition to their potential in cancer treatment and symptom management, CBD has shown promise as a therapeutic agent for oral mucositis due to its anti-inflammatory and antioxidant properties [26]. Oral mucositis is a debilitating side effect frequently associated with chemoradiotherapy in head and neck cancer patients. Emerging data suggest that the anti-inflammatory mechanisms of phytocannabinoids may involve epigenetic pathways, including modulation of microRNA expression.

Interplay Between Cannabinoids and microRNAs in Cancer

Cannabinoids are increasingly utilized by cancer patients to alleviate symptoms such as pain, nausea, vomiting, and cancer-associated weight loss. Concurrently, numerous studies have established the involvement of miRNAs in key processes of cancer pathogenesis, including sustained proliferative signaling, evasion of apoptosis, autonomous replication, angiogenesis, invasion, and metastatic potential [109]. A bidirectional relationship exists between cannabinoids and miRNAs: cannabinoids can modulate the expression of various miRNAs, while miRNAs can, in turn, regulate components of the endocannabinoid system (Fig. 11.1). It has been proposed that some of the anticancer effects attributed to cannabinoids may be mediated through alterations in miRNA expression profiles. For example, both THC and CBD were shown to suppress lipopolysaccharide (LPS)-induced expression of miR-146a in BV2 microglial cells, while CBD also significantly upregulated miR-34a in both resting and LPS-stimulated cells [58]. Similarly, Alharris et al. [5] demonstrated that CBD induces apoptosis in the human neuroblastoma cell lines SH-SY5Y and IMR-32 through the activation of serotonin and vanilloid receptors.

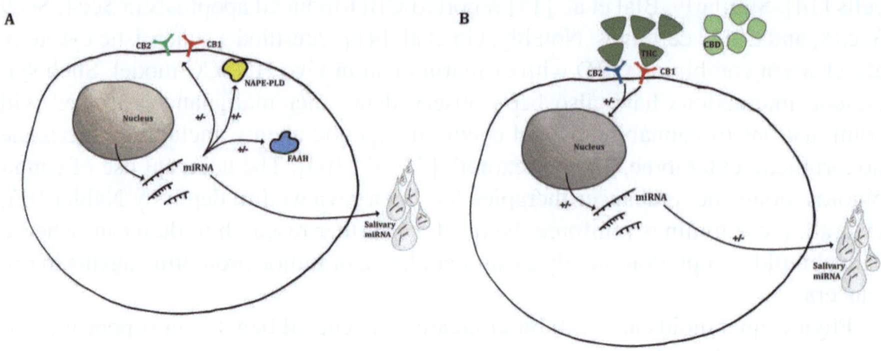

Fig. 11.1 (**a**) Oral cancer cells produce miRNAs that influence the expression of various components of the endocannabinoid system. (**b**) Conversely, activation of the endocannabinoid system by phytocannabinoids modulates miRNA production in oral cancer cells

This receptor targeting by CBD led to downregulation of hsa-let-7a, subsequently increasing the expression of caspase-3 and GAS-7, both of which are involved in apoptotic pathways. Additionally, CBD increased levels of hsa-miR-1972, which correlated with decreased expression of *BCL2L1* and *SIRT2* genes implicated in cancer cell survival and proliferation. The suppression of SIRT2 by hsa-miR-1972 may contribute to CBD-induced cytotoxicity [5].

Immune cells are central to the recognition and elimination of cancer cells. Emerging evidence suggests that the interplay between miRNAs and the endocannabinoid system influences immune responses within the tumor microenvironment, with potential implications for cancer progression and therapeutic response. It has been shown that THC treatment of normal lymph node cells resulted in lower expression of miR-17, miR-92, miR-421, and miR-374b [139], which are involved in tumor growth and promotion [13]. THC also downregulated miR-34-5p and let-7a-5p in mononuclear cells infiltrating the lungs in a murine model of acute respiratory distress syndrome [82]. Conversely, THC was shown to upregulate a set of anti-inflammatory miRNAs, including miR-10a, miR-24, miR-99b, miR-145, miR-149, and miR-197, in the intestinal tissues of simian immunodeficiency virus (SIV)-infected macaques [20]. Similarly, McDew-White et al. [75] reported that chronic THC treatment exerted anti-inflammatory effects in the gingival tissues of SIV-infected rhesus monkeys, partially through downregulation of miR-142-3p, miR-21, and miR-223.

Exogenous cannabinoids have been reported to display both oncogenic and tumor-suppressive properties. A principal anticancer mechanism of cannabinoids and endocannabinoids involves the induction of apoptosis and autophagy, inhibition of cancer cell migration and invasion, and the promotion of cell cycle arrest, primarily via activation of CB1 and CB2 receptors. These pathways are susceptible to interference by miRNAs, which may undermine the tumor-suppressive effects of the endocannabinoid system. For instance, miR-139-5p functions as an oncogenic miRNA by inhibiting endocannabinoid signaling, thereby promoting cellular proliferation and preventing cell cycle arrest. Similarly, miR-27a has been implicated in enhancing inflammation, while miR-30b-5p has been shown to suppress inflammasome activation downstream of CB2 signaling [32]. Moreover, the role of tumor-derived exosomes and their embedded miRNAs in modulating the endocannabinoid system is increasingly recognized as a critical factor in tumor progression and a potential target for therapeutic intervention.

Despite advances in characterizing the interactions between cannabinoids and miRNAs in various disease models, research specifically examining their roles in oral cancers remains limited. Therefore, future studies are urgently needed to explore how cannabinoids such as THC and CBD may influence miRNA expression and endocannabinoid signaling in OSCC and other malignancies of the oral cavity.

Acknowledgments The authors would like to thank Anabel Cecilia Rodríguez, Maximiliano Cella, María Sol Oricchio Gagliardo, Melany Elizabeth Lombardi, Eduardo Nieves, Daniel Guerrero, Vanesa Fernández, Alcira Mazziotti, Rosana Leiva, Alejandra Verón, Patricia Fernández, Cristina Lincon, and Daniela Néspola for their excellent technical and administrative support.

References

1. Aghiorghiesei O, Zanoaga O, Nutu A, Braicu C, Campian RS, Lucaciu O, Neagoe IB (2022) The world of oral cancer and its risk factors viewed from the aspect of MicroRNA expression patterns. Genes (Basel) 13. https://doi.org/10.3390/genes13040594
2. Ahrens AGMS, Bressi T (2007) Marijuana as promoter for oral cancer? More than a suspect. Addict Disord Their Treat 6:117–119. https://doi.org/10.1097/01.adt.0000210727.30885.88
3. Al Rawi N, Elmabrouk N, Abu Kou R, Mkadmi S, Rizvi Z, Hamdoon Z (2021) The role of differentially expressed salivary microRNA in oral squamous cell carcinoma. A systematic review. Arch Oral Biol 125:105108. https://doi.org/10.1016/j.archoralbio.2021.105108
4. Aldington S, Harwood M, Cox B, Weatherall M, Beckert L, Hansell A, Pritchard A, Robinson G, Beasley R, Cannabis and Respiratory Disease Research Group (2008) Cannabis use and cancer of the head and neck: case-control study. Otolaryngol Head Neck Surg 138:374–380. https://doi.org/10.1016/j.otohns.2007.12.002
5. Alharris E, Singh NP, Nagarkatti PS, Nagarkatti M (2019) Role of miRNA in the regulation of cannabidiol-mediated apoptosis in neuroblastoma cells. Oncotarget 10:45–59. https://doi.org/10.18632/oncotarget.26534
6. Alhouayek M, Boldrup L, Fowler CJ (2019) Altered mRNA expression of genes involved in endocannabinoid signalling in squamous cell carcinoma of the oral tongue. Cancer Investig 37:327–338. https://doi.org/10.1080/07357907.2019.1638394
7. Ayaz L, Görür A, Yaroğlu HY, Ozcan C, Tamer L (2013) Differential expression of microRNAs in plasma of patients with laryngeal squamous cell carcinoma: potential early-detection markers for laryngeal squamous cell carcinoma. J Cancer Res Clin Oncol 139:1499–1506. https://doi.org/10.1007/s00432-013-1469-2
8. Balakittnen J, Weeramange CE, Wallace DF, Duijf PHG, Cristino AS, Kenny L, Vasani S, Punyadeera C (2023) Noncoding RNAs in oral cancer. Wiley Interdiscip Rev RNA 14:1–45. https://doi.org/10.1002/wrna.1754
9. Balakittnen J, Ekanayake Weeramange C, Wallace DF, Duijf PHG, Cristino AS, Hartel G, Barrero RA, Taheri T, Kenny L, Vasani S, Batstone M, Breik O, Punyadeera C (2024) A novel saliva-based miRNA profile to diagnose and predict oral cancer. Int J Oral Sci 16:1–13. https://doi.org/10.1038/s41368-023-00273-w

10. Bellocchio L, Inchingolo AD, Inchingolo AM, Lorusso F, Malcangi G, Santacroce L, Scarano A, Bordea IR, Hazballa D, D'Oria MT, Gargiulo Isacco C, Nucci L, Serpico R, Tartaglia GM, Giovanniello D, Contaldo M, Farronato M, Dipalma G, Inchingolo F (2021) Cannabinoids drugs and oral health – from recreational side-effects to medicinal purposes: a systematic review. Int J Mol Sci 22. https://doi.org/10.3390/ijms22158329
11. Berthiller J, Lee Y-CA, Boffetta P, Wei Q, Sturgis EM, Greenland S, Morgenstern H, Zhang Z-F, Lazarus P, Muscat J, Chen C, Schwartz SM, Eluf Neto J, Wünsch Filho V, Koifman S, Curado MP, Matos E, Fernandez L, Menezes A, Daudt AW, Ferro G, Brennan P, Hashibe M (2009) Marijuana smoking and the risk of head and neck cancer: pooled analysis in the INHANCE consortium. Cancer Epidemiol Biomarkers Prev 18:1544–1551. https://doi.org/10.1158/1055-9965.EPI-08-0845
12. Bhat GR, Hyole RG, Li J (2021) Head and neck cancer: current challenges and future perspectives. Adv Cancer Res 152:67–102. https://doi.org/10.1016/bs.acr.2021.05.002
13. Bian H, Zhou Y, Zhou D, Zhang Y, Shang D, Qi J (2019) The latest progress on miR-374 and its functional implications in physiological and pathological processes. J Cell Mol Med 23:3063–3076. https://doi.org/10.1111/jcmm.14219
14. Billi M, Pagano S, Pancrazi GL, Valenti C, Bruscoli S, Di Michele A, Febo M, Grignani F, Marinucci L (2025) DNA damage and cell death in human oral squamous cell carcinoma cells: the potential biological effects of cannabidiol. Arch Oral Biol 169:106110. https://doi.org/10.1016/j.archoralbio.2024.106110
15. Blal K, Besser E, Procaccia S, Schwob O, Lerenthal Y, Abu Tair J, Meiri D, Benny O (2023) The effect of cannabis plant extracts on head and neck squamous cell carcinoma and the quest for cannabis-based personalized therapy. Cancers (Basel) 15. https://doi.org/10.3390/cancers15020497
16. Blatt S, Krüger M, Sagheb K, Barth M, Kämmerer PW, Al-Nawas B, Sagheb K (2022) Tumor recurrence and follow-up intervals in oral squamous cell carcinoma. J Clin Med 11. https://doi.org/10.3390/jcm11237061
17. Borsani E, Majorana A, Cocchi MA, Conti G, Bonadeo S, Padovani A, Lauria G, Bardellini E, Rezzani R, Rodella LF (2014) Epithelial expression of vanilloid and cannabinoid receptors: a potential role in burning mouth syndrome pathogenesis. Histol Histopathol 29:523–533. https://doi.org/10.14670/HH-29.10.523
18. Bray F, Laversanne M, Sung H, Ferlay J, Siegel RL, Soerjomataram I, Jemal A (2024) Global cancer statistics 2022: GLOBOCAN estimates of incidence and mortality worldwide for 36 cancers in 185 countries. CA Cancer J Clin 74:229–263. https://doi.org/10.3322/caac.21834
19. Brito JAR, Gomes CC, Guimarães ALS, Campos K, Gomez RS (2014) Relationship between microRNA expression levels and histopathological features of dysplasia in oral leukoplakia. J Oral Pathol Med 43:211–216. https://doi.org/10.1111/jop.12112
20. Chandra LC, Kumar V, Torben W, Vande Stouwe C, Winsauer P, Amedee A, Molina PE, Mohan M (2015) Chronic administration of Δ9-tetrahydrocannabinol induces intestinal anti-inflammatory microRNA expression during acute simian immunodeficiency virus infection of rhesus macaques. J Virol 89:1168–1181. https://doi.org/10.1128/JVI.01754-14
21. Chaturvedi AK, Freedman ND, Abnet CC (2022) The evolving epidemiology of oral cavity and oropharyngeal cancers. Cancer Res 82:2821–2823. https://doi.org/10.1158/0008-5472.CAN-22-2124
22. Chisini LA, Cademartori MG, Francia A, Mederos M, Grazioli G, Conde MCM, Correa MB (2019) Is the use of cannabis associated with periodontitis? A systematic review and meta-analysis. J Periodontal Res 54:311–317. https://doi.org/10.1111/jre.12639
23. Conway DI, Purkayastha M, Chestnutt IG (2018) The changing epidemiology of oral cancer: definitions, trends, and risk factors. Br Dent J 225:867–873. https://doi.org/10.1038/sj.bdj.2018.922
24. Côté M, Trudel M, Wang C, Fortin A (2016) Improving quality of life with nabilone during radiotherapy treatments for head and neck cancers: a randomized double-blind placebo-controlled trial. Ann Otol Rhinol Laryngol 125:317–324. https://doi.org/10.1177/0003489415612801

25. Cretu B, Zamfir A, Bucurica S, Scheau AE, Savulescu Fiedler I, Caruntu C, Caruntu A, Scheau C (2024) Role of cannabinoids in oral cancer. Int J Mol Sci 25:1–20. https://doi.org/10.3390/ijms25020969
26. Cuba LF, Salum FG, Cherubini K, Figueiredo MAZ (2017) Cannabidiol: an alternative therapeutic agent for oral mucositis? J Clin Pharm Ther 42:245–250. https://doi.org/10.1111/jcpt.12504
27. Dar GM, Agarwal S, Kumar A, Nimisha A, Sharma AK, Verma R, Sattar RSA, Ahmad E, Ali A, Mahajan B, Saluja SS, Meher R (2022) A non-invasive miRNA-based approach in early diagnosis and therapeutics of oral cancer. Crit Rev Oncol Hematol 180:103850. https://doi.org/10.1016/j.critrevonc.2022.103850
28. Dhanuthai K, Rojanawatsirivej S, Thosaporn W, Kintarak S, Subarnbhesaj A, Darling M, Kryshtalskyj E, Chiang C-P, Shin H-I, Choi S-Y, Lee S-S, Aminishakib P (2018) Oral cancer: a multicenter study. Med Oral Patol Oral Cir Bucal 23:e23–e29. https://doi.org/10.4317/medoral.21999
29. Di Credico G, Edefonti V, Polesel J, Pauli F, Torelli N, Serraino D, Negri E, Luce D, Stucker I, Matsuo K, Brennan P, Vilensky M, Fernandez L, Curado MP, Menezes A, Daudt AW, Koifman R, Wunsch-Filho V, Holcatova I, Ahrens W, Lagiou P, Simonato L, Richiardi L, Healy C, Kjaerheim K, Conway DI, Macfarlane TV, Thomson P, Agudo A, Znaor A, Boaventura Rios LF, Toporcov TN, Franceschi S, Herrero R, Muscat J, Olshan AF, Zevallos JP, La Vecchia C, Winn DM, Sturgis EM, Li G, Fabianova E, Lissowska J, Mates D, Rudnai P, Shangina O, Swiatkowska B, Moysich K, Zhang Z-F, Morgenstern H, Levi F, Smith E, Lazarus P, Bosetti C, Garavello W, Kelsey K, McClean M, Ramroth H, Chen C, Schwartz SM, Vaughan TL, Zheng T, Menvielle G, Boccia S, Cadoni G, Hayes RB, Purdue M, Gillison M, Schantz S, Yu G-P, Brenner H, D'Souza G, Gross ND, Chuang S-C, Boffetta P, Hashibe M, Lee Y-CA, Dal Maso L (2019) Joint effects of intensity and duration of cigarette smoking on the risk of head and neck cancer: a bivariate spline model approach. Oral Oncol 94:47–57. https://doi.org/10.1016/j.oraloncology.2019.05.006
30. Di Stasio D, Romano A, Boschetti CE, Montella M, Mosca L, Lucchese A (2022) Salivary miRNAs expression in potentially malignant disorders of the oral mucosa and oral squamous cell carcinoma: a pilot study on miR-21, miR-27b, and miR-181b. Cancers (Basel) 15. https://doi.org/10.3390/cancers15010291
31. Duz MB, Karatas OF, Guzel E, Turgut NF, Yilmaz M, Creighton CJ, Ozen M (2016) Identification of miR-139-5p as a saliva biomarker for tongue squamous cell carcinoma: a pilot study. Cell Oncol 39:187–193. https://doi.org/10.1007/s13402-015-0259-z
32. Ebrahimi N, Far NP, Fakhr SS, Faghihkhorasani F, Miraghel SA, Chaleshtori SR, Rezaei-Tazangi F, Beiranvand S, Baziyar P, Manavi MS, Zarrabi A, Nabavi N, Ren J, Aref AR (2023) The endocannabinoid system, a new gatekeeper in the pharmacology of human hepatocellular carcinoma. Environ Res 228:115914. https://doi.org/10.1016/j.envres.2023.115914
33. Fadhil RS, Wei MQ, Nikolarakos D, Good D, Nair RG (2020) Salivary microRNA miR-let-7a-5p and miR-3928 could be used as potential diagnostic bio-markers for head and neck squamous cell carcinoma. PLoS One 15:e0221779. https://doi.org/10.1371/journal.pone.0221779
34. Farshbaf A, Mohajertehran F, Sahebkar A, Garmei Y, Sabbagh P, Mohtasham N (2022) The role of altered microRNA expression in premalignant and malignant head and neck lesions with epithelial origin. Health Sci Rep 5:e921. https://doi.org/10.1002/hsr2.921
35. Fernandez-Solari J, Prestifilippo JP, Ossola CA, Rettori V, Elverdin JC (2010) Participation of the endocannabinoid system in lipopolysaccharide-induced inhibition of salivary secretion. Arch Oral Biol 55:583–590. https://doi.org/10.1016/j.archoralbio.2010.05.006
36. Firth NA (1997) Marijuana use and oral cancer: a review. Oral Oncol 33:398–401. https://doi.org/10.1016/S0964-1955(97)00022-5
37. Gai C, Camussi F, Broccoletti R, Gambino A, Cabras M, Molinaro L, Carossa S, Camussi G, Arduino PG (2018) Salivary extracellular vesicle-associated miRNAs as potential biomarkers in oral squamous cell carcinoma. BMC Cancer 18:1–11. https://doi.org/10.1186/s12885-018-4364-z

38. Gallagher TJ, Chung RS, Lin ME, Kim I, Kokot NC (2024) Cannabis use and head and neck cancer. JAMA Otolaryngol Head Neck Surg 1:1–8. https://doi.org/10.1001/jamaoto.2024.2419
39. Ghasemiesfe M, Barrow B, Leonard S, Keyhani S, Korenstein D (2019) Association between marijuana use and risk of cancer: a systematic review and meta-analysis. JAMA Netw Open 2:e1916318. https://doi.org/10.1001/jamanetworkopen.2019.16318
40. Gillison ML, Chaturvedi AK, Anderson WF, Fakhry C (2015) Epidemiology of human papillomavirus-positive head and neck squamous cell carcinoma. J Clin Oncol 33:3235–3242. https://doi.org/10.1200/JCO.2015.61.6995
41. Go YY, Kim SR, Kim DY, Chae SW, Song JJ (2020) Cannabidiol enhances cytotoxicity of anti-cancer drugs in human head and neck squamous cell carcinoma. Sci Rep 10:1–11. https://doi.org/10.1038/s41598-020-77674-y
42. Gołąbek K, Hudy D, Świętek A, Gaździcka J, Dąbrowska N, Miśkiewicz-Orczyk K, Zięba N, Misiołek M, Strzelczyk JK (2023) miR-125b-5p, miR-155-3p, and miR-214-5p and target E2F2 gene in oral squamous cell carcinoma. Int J Mol Sci 24. https://doi.org/10.3390/ijms24076320
43. Graves BM, Johnson TJ, Nishida RT, Dias RP, Savareear B, Harynuk JJ, Kazemimanesh M, Olfert JS, Boies AM (2020) Comprehensive characterization of mainstream marijuana and tobacco smoke. Sci Rep 10:7160. https://doi.org/10.1038/s41598-020-63120-6
44. Gupta S, Jawanda MK (2021) Oral submucous fibrosis: an overview of a challenging entity. Indian J Dermatol Venereol Leprol 87:768–777. https://doi.org/10.25259/IJDVL_371_20
45. Gupta PC, Warnakulasuriya S (2002) Global epidemiology of areca nut usage. Addict Biol 7:77–83. https://doi.org/10.1080/13556210020091437
46. Hashibe M, Straif K, Tashkin DP, Morgenstern H, Greenland S, Zhang Z-F (2005) Epidemiologic review of marijuana use and cancer risk. Alcohol 35:265–275. https://doi.org/10.1016/j.alcohol.2005.04.008
47. Häuser W, Welsch P, Radbruch L, Fisher E, Bell RF, Moore RA (2023) Cannabis-based medicines and medical cannabis for adults with cancer pain. Cochrane Database Syst Rev 6:CD014915. https://doi.org/10.1002/14651858.CD014915.pub2
48. Hayashi T, Koyama N, Mizukoshi K, Azuma Y, Kashimata M (2014) Extracellular microRNAs in saliva. J Oral Biosci 56:91–96. https://doi.org/10.1016/j.job.2014.05.003
49. He Q, Chen Z, Cabay RJ, Zhang L, Luan X, Chen D, Yu T, Wang A, Zhou X (2016) microRNA-21 and microRNA-375 from oral cytology as biomarkers for oral tongue cancer detection. Oral Oncol 57:15–20. https://doi.org/10.1016/j.oraloncology.2016.03.017
50. He L, Ping F, Fan Z, Zhang C, Deng M, Cheng B, Xia J (2020) Salivary exosomal miR-24-3p serves as a potential detective biomarker for oral squamous cell carcinoma screening. Biomed Pharmacother 121:109553. https://doi.org/10.1016/j.biopha.2019.109553
51. Holland ML, Panetta JA, Hoskins JM, Bebawy M, Roufogalis BD, Allen JD, Arnold JC (2006) The effects of cannabinoids on P-glycoprotein transport and expression in multidrug resistant cells. Biochem Pharmacol 71:1146–1154. https://doi.org/10.1016/j.bcp.2005.12.033
52. Hossain MZ, Ando H, Unno S, Kitagawa J (2020) Targeting peripherally restricted cannabinoid receptor 1, cannabinoid receptor 2, and endocannabinoid-degrading enzymes for the treatment of neuropathic pain including neuropathic orofacial pain. Int J Mol Sci 21. https://doi.org/10.3390/ijms21041423
53. Howlett AC, Abood ME (2017) CB1 and CB2 receptor pharmacology. Adv Pharmacol 80:169–206. https://doi.org/10.1016/bs.apha.2017.03.007
54. Hsu C-M, Lin P-M, Wang Y-M, Chen Z-J, Lin S-F, Yang M-Y (2012) Circulating miRNA is a novel marker for head and neck squamous cell carcinoma. Tumour Biol 33:1933–1942. https://doi.org/10.1007/s13277-012-0454-8
55. Humphrey SP, Williamson RT (2001) A review of saliva: normal composition, flow, and function. J Prosthet Dent 85:162–169. https://doi.org/10.1067/mpr.2001.113778

56. Hung K-F, Liu C-J, Chiu P-C, Lin J-S, Chang K-W, Shih W-Y, Kao S-Y, Tu H-F (2016) MicroRNA-31 upregulation predicts increased risk of progression of oral potentially malignant disorder. Oral Oncol 53:42–47. https://doi.org/10.1016/j.oraloncology.2015.11.017
57. Jakob M, Mattes LM, Küffer S, Unger K, Hess J, Bertlich M, Haubner F, Ihler F, Canis M, Weiss BG, Kitz J (2019) MicroRNA expression patterns in oral squamous cell carcinoma: hsa-mir-99b-3p and hsa-mir-100-5p as novel prognostic markers for oral cancer. Head Neck 41:3499–3515. https://doi.org/10.1002/hed.25866
58. Juknat A, Gao F, Coppola G, Vogel Z, Kozela E (2019) MiRNA expression profiles and molecular networks in resting and LPS-activated BV-2 microglia-effect of cannabinoids. PLoS One 14:1–25. https://doi.org/10.1371/journal.pone.0212039
59. Kai Y, Peng W, Ling W, Jiebing H, Zhuan B (2014) Reciprocal effects between microRNA-140-5p and ADAM10 suppress migration and invasion of human tongue cancer cells. Biochem Biophys Res Commun 448:308–314. https://doi.org/10.1016/j.bbrc.2014.02.032
60. Keboa MT, Enriquez N, Martel M, Nicolau B, Macdonald ME (2020) Oral health implications of cannabis smoking: a rapid evidence review. J Can Dent Assoc 86:k2
61. Konermann A, Jäger A, Held SAE, Brossart P, Schmöle A (2017) In vivo and in vitro identification of endocannabinoid signaling in periodontal tissues and their potential role in local pathophysiology. Cell Mol Neurobiol 37:1511–1520. https://doi.org/10.1007/s10571-017-0482-4
62. Kumari P, Debta P, Dixit A (2022) Oral potentially malignant disorders: etiology, pathogenesis, and transformation into oral cancer. Front Pharmacol 13:825266. https://doi.org/10.3389/fphar.2022.825266
63. Liang C, McClean MD, Marsit C, Christensen B, Peters E, Nelson HH, Kelsey KT (2009) A population-based case-control study of marijuana use and head and neck squamous cell carcinoma. Cancer Prev Res (Phila) 2:759–768. https://doi.org/10.1158/1940-6207.CAPR-09-0048
64. Lingen MW, Abt E, Agrawal N, Chaturvedi AK, Cohen E, D'Souza G, Gurenlian JA, Kalmar JR, Kerr AR, Lambert PM, Patton LL, Sollecito TP, Truelove E, Tampi MP, Urquhart O, Banfield L, Carrasco-Labra A, D'Souza G, Gurenlian JA, Kalmar JR, Kerr AR, Lambert PM, Patton LL, Sollecito TP, Truelove E, Tampi MP, Urquhart O, Banfield L, Carrasco-Labra A (2017) Evidence-based clinical practice guideline for the evaluation of potentially malignant disorders in the oral cavity. J Am Dent Assoc 148:712–727.e10. https://doi.org/10.1016/j.adaj.2017.07.032
65. Liu WM, Scott KA, Shamash J, Joel S, Powles TB (2008) Enhancing the in vitro cytotoxic activity of Delta9-tetrahydrocannabinol in leukemic cells through a combinatorial approach. Leuk Lymphoma 49:1800–1809. https://doi.org/10.1080/10428190802239188
66. Liu C-J, Lin S-C, Yang C-C, Cheng H-W, Chang K-W (2012) Exploiting salivary miR-31 as a clinical biomarker of oral squamous cell carcinoma. Head Neck 34:219–224. https://doi.org/10.1002/hed.21713
67. Liu C, Qi X, Alhabeil J, Lu H, Zhou Z (2019) Activation of cannabinoid receptors promote periodontal cell adhesion and migration. J Clin Periodontol 46:1264–1272. https://doi.org/10.1111/jcpe.13190
68. Liu C, Sadat SH, Ebisumoto K, Sakai A, Panuganti BA, Ren S, Goto Y, Haft S, Fukusumi T, Ando M, Saito Y, Guo T, Tamayo P, Yeerna H, Kim W, Hubbard J, Sharabi AB, Gutkind JS, Califano JA (2020) Cannabinoids promote progression of HPV-positive head and neck squamous cell carcinoma via p38 MAPK activation. Clin Cancer Res 26:2693–2703. https://doi.org/10.1158/1078-0432.CCR-18-3301
69. Liu D, Xin Z, Guo S, Li S, Cheng J, Jiang H (2021) Blood and salivary MicroRNAs for diagnosis of oral squamous cell carcinoma: a systematic review and meta-analysis. J Oral Maxillofac Surg 79:1082.e1–1082.e13. https://doi.org/10.1016/j.joms.2020.12.043

70. Liu W, Shi H, Zhou Z, Zhang C, Shen X (2023) Association of salivary miRNAs with onset and progression of oral potentially malignant disorders: searching for noninvasive biomarkers. J Dent Sci 18:432–436. https://doi.org/10.1016/j.jds.2022.08.002
71. Llewellyn CD, Linklater K, Bell J, Johnson NW, Warnakulasuriya S (2004) An analysis of risk factors for oral cancer in young people: a case-control study. Oral Oncol 40:304–313. https://doi.org/10.1016/j.oraloncology.2003.08.015
72. Maclellan SA, Lawson J, Baik J, Guillaud M, Poh CF-Y, Garnis C (2012) Differential expression of miRNAs in the serum of patients with high-risk oral lesions. Cancer Med 1:268–274. https://doi.org/10.1002/cam4.17
73. Marks MA, Chaturvedi AK, Kelsey K, Straif K, Berthiller J, Schwartz SM, Smith E, Wyss A, Brennan P, Olshan AF, Wei Q, Sturgis EM, Zhang ZF, Morgenstern H, Muscat J, Lazarus P, McClean M, Chen C, Vaughan TL, Wunsch-Filho V, Curado MP, Koifman S, Matos E, Menezes A, Daudt AW, Fernandez L, Posner M, Boffetta P, Amy Lee YC, Hashibe M, D'Souza G (2014) Association of marijuana smoking with oropharyngeal and oral tongue cancers: pooled analysis from the INHANCE consortium. Cancer Epidemiol Biomarkers Prev 23:160–171. https://doi.org/10.1158/1055-9965.EPI-13-0181
74. Mazumder S, Datta S, Ray JG, Chaudhuri K, Chatterjee R (2019) Liquid biopsy: miRNA as a potential biomarker in oral cancer. Cancer Epidemiol 58:137–145. https://doi.org/10.1016/j.canep.2018.12.008
75. McDew-White M, Lee E, Alvarez X, Sestak K, Ling BJ, Byrareddy SN, Okeoma CM, Mohan M (2022) Cannabinoid control of gingival immune activation in chronically SIV-infected rhesus macaques involves modulation of the indoleamine-2,3-dioxygenase-1 pathway and salivary microbiome. EBioMedicine 75:103769. https://doi.org/10.1016/j.ebiom.2021.103769
76. Mechoulam R, Hanuš LO, Pertwee R, Howlett AC (2014) Early phytocannabinoid chemistry to endocannabinoids and beyond. Nat Rev Neurosci 15:757–764. https://doi.org/10.1038/nrn3811
77. Mehdipour M, Shahidi M, Manifar S, Jafari S, Mashhadi Abbas F, Barati M, Mortazavi H, Shirkhoda M, Farzanegan A, Elmi Rankohi Z (2018) Diagnostic and prognostic relevance of salivary microRNA-21, -125a, -31 and -200a levels in patients with oral lichen planus – a short report. Cell Oncol 41:329–334. https://doi.org/10.1007/s13402-018-0372-x
78. Mello FW, Melo G, Pasetto JJ, Silva CAB, Warnakulasuriya S, Rivero ERC (2019) The synergistic effect of tobacco and alcohol consumption on oral squamous cell carcinoma: a systematic review and meta-analysis. Clin Oral Investig 23:2849–2859. https://doi.org/10.1007/s00784-019-02958-1
79. Menini M, De Giovanni E, Bagnasco F, Delucchi F, Pera F, Baldi D, Pesce P (2021) Salivary micro-RNA and oral squamous cell carcinoma: a systematic review. J Pers Med 11. https://doi.org/10.3390/jpm11020101
80. Min S-K, Jung SY, Kang HK, Park S-A, Lee JH, Kim M-J, Min B-M (2017) Functional diversity of miR-146a-5p and TRAF6 in normal and oral cancer cells. Int J Oncol 51:1541–1552. https://doi.org/10.3892/ijo.2017.4124
81. Mithani SK, Mydlarz WK, Grumbine FL, Smith IM, Califano JA (2007) Molecular genetics of premalignant oral lesions. Oral Dis 13:126–133. https://doi.org/10.1111/j.1601-0825.2006.01349.x
82. Mohammed A, Alghetaa H, Sultan M, Singh NP, Nagarkatti P, Nagarkatti M (2020) Administration of Δ9-tetrahydrocannabinol (THC) post-staphylococcal enterotoxin B exposure protects mice from acute respiratory distress syndrome and toxicity. Front Pharmacol 11:893. https://doi.org/10.3389/fphar.2020.00893
83. Momen-Heravi F, Trachtenberg AJ, Kuo WP, Cheng YS (2014) Genomewide study of salivary MicroRNAs for detection of oral cancer. J Dent Res 93:86S–93S. https://doi.org/10.1177/0022034514531018
84. Montero PH, Patel SG (2015) Cancer of the oral cavity. Surg Oncol Clin N Am 24:491–508. https://doi.org/10.1016/j.soc.2015.03.006

85. Nahler G (2024) Phytocannabinoids as chemotherapy adjuncts – a review for users. Onco 4:287–321. https://doi.org/10.3390/onco4040021
86. Neben-Wittich MA (2021) Medical cannabis to treat symptoms from head and neck radiation therapy. Pract Radiat Oncol 11:313–316. https://doi.org/10.1016/j.prro.2021.03.007
87. Nguyen H, Nonaka T (2024) Salivary miRNAs as auxiliary liquid biopsy biomarkers for diagnosis in patients with oropharyngeal squamous cell carcinoma: a systematic review and meta-analysis. Front Genet 15:1–10. https://doi.org/10.3389/fgene.2024.1352838
88. Nocini R, Lippi G, Mattiuzzi C (2020) Biological and epidemiologic updates on lip and oral cavity cancers. Ann Cancer Epidemiol 4:1. https://doi.org/10.21037/ace.2020.01.01
89. O'Brien J, Hayder H, Zayed Y, Peng C (2018) Overview of MicroRNA biogenesis, mechanisms of actions, and circulation. Front Endocrinol (Lausanne) 9:402. https://doi.org/10.3389/fendo.2018.00402
90. Omar E (2015) Current concepts and future of noninvasive procedures for diagnosing oral squamous cell carcinoma – a systematic review. Head Face Med 11. https://doi.org/10.1186/s13005-015-0063-z
91. Osazuwa-Peters N, Adjei-Boakye E, Loux TM, Varvares MA, Schootman M (2016) Insufficient evidence to support or refute the association between head and neck cancer and marijuana use. J Evid Based Dent Pract 16:127–129. https://doi.org/10.1016/j.jebdp.2016.06.004
92. Ossola CA, Balcarcel NB, Astrauskas JI, Bozzini C, Elverdin JC, Fernández-Solari J (2019) A new target to ameliorate the damage of periodontal disease: the role of transient receptor potential vanilloid type-1 in contrast to that of specific cannabinoid receptors in rats. J Periodontol 90:1325–1335. https://doi.org/10.1002/JPER.18-0766
93. Palmieri A, Carinci F, Martinelli M, Pezzetti F, Girardi A, Cura F, Rubini C, Scapoli L (2014) Role of the MIR146A polymorphism in the origin and progression of oral squamous cell carcinoma. Eur J Oral Sci 122:198–201. https://doi.org/10.1111/eos.12121
94. Park NJ, Zhou X, Yu T, Brinkman BMN, Zimmermann BG, Palanisamy V, Wong DT (2007) Characterization of salivary RNA by cDNA library analysis. Arch Oral Biol 52:30–35. https://doi.org/10.1016/j.archoralbio.2006.08.014
95. Park NJ, Zhou H, Elashoff D, Henson BS, Kastratovic DA, Abemayor E, Wong DT (2009) Salivary microRNA: discovery, characterization, and clinical utility for oral cancer detection. Clin Cancer Res 15:5473–5477. https://doi.org/10.1158/1078-0432.CCR-09-0736
96. Parmar A, Macluskey M, Mc Goldrick N, Conway DI, Glenny A-M, Clarkson JE, Worthington HV, Chan KK (2021) Interventions for the treatment of oral cavity and oropharyngeal cancer: chemotherapy. Cochrane Database Syst Rev 12:CD006386. https://doi.org/10.1002/14651858.CD006386.pub4
97. Patel SC, Carpenter WR, Tyree S, Couch ME, Weissler M, Hackman T, Hayes DN, Shores C, Chera BS (2011) Increasing incidence of oral tongue squamous cell carcinoma in young white women, age 18 to 44 years. J Clin Oncol 29:1488–1494. https://doi.org/10.1200/JCO.2010.31.7883
98. Pillai J, Chincholkar T, Dixit R, Pandey M (2021) A systematic review of proteomic biomarkers in oral squamous cell cancer. World J Surg Oncol 19:1–28. https://doi.org/10.1186/s12957-021-02423-y
99. Prasad M, Sekar R, Priya MDL, Varma SR, Karobari MI (2024) A new perspective on diagnostic strategies concerning the potential of saliva-based miRNA signatures in oral cancer. Diagn Pathol 19:1–14. https://doi.org/10.1186/s13000-024-01575-1
100. Prestifilippo JP, Fernández-Solari J, de la Cal C, Iribarne M, Suburo AM, Rettori V, McCann SM, Elverdin JC (2006) Inhibition of salivary secretion by activation of cannabinoid receptors. Exp Biol Med (Maywood) 231:1421–1429. https://doi.org/10.1177/153537020623100816
101. Radaic A, Kamarajan P, Cho A, Wang S, Hung G, Najarzadegan F, Wong DT, Ton-That H, Wang C, Kapila YL (2024) Biological biomarkers of oral cancer. Periodontol 96:250–280. https://doi.org/10.1111/prd.12542

102. Rajan C, Roshan VGD, Khan I, Manasa VG, Himal I, Kattoor J, Thomas S, Kondaiah P, Kannan S (2021) MiRNA expression profiling and emergence of new prognostic signature for oral squamous cell carcinoma. Sci Rep 11:1–12. https://doi.org/10.1038/s41598-021-86316-w
103. Ramer R, Hinz B (2017) Cannabinoids as anticancer drugs. Adv Pharmacol 80:397–436. https://doi.org/10.1016/bs.apha.2017.04.002
104. Rapado-González Ó, Majem B, Muinelo-Romay L, Álvarez-Castro A, Santamaría A, Gil-Moreno A, López-López R, Suárez-Cunqueiro MM (2018) Human salivary microRNAs in cancer. J Cancer 9:638–649. https://doi.org/10.7150/jca.21180
105. Rishabh K, Khadilkar S, Kumar A, Kalra I, Kumar AP, Kunnumakkara AB (2021) Review micrornas as modulators of oral tumorigenesis – a focused review. Int J Mol Sci 22:1–32. https://doi.org/10.3390/ijms22052561
106. Rosenblatt KA, Daling JR, Chen C, Sherman KJ, Schwartz SM (2004) Marijuana use and risk of oral squamous cell carcinoma. Cancer Res 64:4049–4054. https://doi.org/10.1158/0008-5472.CAN-03-3425
107. Rotaru D, Chisnoiu R, Picos A, Picos A, Chisnoiu A (2020) Treatment trends in oral lichen planus and oral lichenoid lesions (Review). Exp Ther Med 20:1–1. https://doi.org/10.3892/etm.2020.9328
108. Roy R, Singh R, Chattopadhyay E, Ray A, De Sarkar N, Aich R, Paul RR, Pal M, Roy B (2016) MicroRNA and target gene expression based clustering of oral cancer, precancer and normal tissues. Gene 593:58–63. https://doi.org/10.1016/j.gene.2016.08.011
109. Salamat JM, Abbott KL, Flannery PC, Ledbetter EL, Pondugula SR (2022) Interplay between the cannabinoid system and microRNAs in cancer. ACS Omega 7:9995–10000. https://doi.org/10.1021/acsomega.2c00635
110. Sarode G, Maniyar N, Sarode SC, Jafer M, Patil S, Awan KH (2020) Epidemiologic aspects of oral cancer. Dis Mon 66:100988. https://doi.org/10.1016/j.disamonth.2020.100988
111. Semlali A, Beji S, Ajala I, Rouabhia M (2021) Effects of tetrahydrocannabinols on human oral cancer cell proliferation, apoptosis, autophagy, oxidative stress, and DNA damage. Arch Oral Biol 129:105200. https://doi.org/10.1016/j.archoralbio.2021.105200
112. Setti G, Pezzi ME, Viani MV, Pertinhez TA, Cassi D, Magnoni C, Bellini P, Musolino A, Vescovi P, Meleti M (2020) Salivary MicroRNA for diagnosis of cancer and systemic diseases: a systematic review. Int J Mol Sci 21. https://doi.org/10.3390/ijms21030907
113. Severino P, Oliveira LS, Andreghetto FM, Torres N, Curioni O, Cury PM, Toporcov TN, Paschoal AR, Durham AM (2015) Small RNAs in metastatic and non-metastatic oral squamous cell carcinoma. BMC Med Genet 8:31. https://doi.org/10.1186/s12920-015-0102-4
114. Sheehan TJ, Hamnett HJ, Beasley R, Fitzmaurice PS (2019) Chemical and physical variations of cannabis smoke from a variety of cannabis samples in New Zealand. Forensic Sci Res 4:168–178. https://doi.org/10.1080/20961790.2018.1445937
115. Shewale JB, Pickard RKL, Xiao W, Jiang B, Gillison ML (2021) Independent association of marijuana use and poor oral hygiene with HPV-negative but not HPV-positive head and neck squamous cell carcinomas. Cancer 127:2099–2110. https://doi.org/10.1002/cncr.33440
116. Shi W, Yang J, Li S, Shan X, Liu X, Hua H, Zhao C, Feng Z, Cai Z, Zhang L, Zhou D (2015) Potential involvement of miR-375 in the premalignant progression of oral squamous cell carcinoma mediated via transcription factor KLF5. Oncotarget 6:40172–40185. https://doi.org/10.18632/oncotarget.5502
117. Shpitzer T, Hamzany Y, Bahar G, Feinmesser R, Savulescu D, Borovoi I, Gavish M, Nagler RM (2009) Salivary analysis of oral cancer biomarkers. Br J Cancer 101:1194–1198. https://doi.org/10.1038/sj.bjc.6605290
118. Singh H, Katz J, Saleh W, Cha S (2019) Impact of cannabis on the port of entry-oral tissues: an overview. Int J Oral Dent Health 5:10–14. https://doi.org/10.23937/2469-5734/1510098
119. Smith LA, Azariah F, Lavender VTC, Stoner NS, Bettiol S (2015) Cannabinoids for nausea and vomiting in adults with cancer receiving chemotherapy. Cochrane Database Syst Rev 2015:CD009464. https://doi.org/10.1002/14651858.CD009464.pub2

120. Song M, Bai H, Zhang P, Zhou X, Ying B (2023) Promising applications of human-derived saliva biomarker testing in clinical diagnostics. Int J Oral Sci 15:1–17. https://doi.org/10.1038/s41368-022-00209-w
121. Stojkovic G, Jovanovic I, Dimitrijevic M, Jovanovic J, Tomanovic N, Stankovic A, Arsovic N, Boricic I, Zeljic K (2023) Meta-signature guided investigation of miRNA candidates as potential biomarkers of oral cancer. Oral Dis 29:1550–1564. https://doi.org/10.1111/odi.14185
122. Streck JM, Hughes JR, Klemperer EM, Howard AB, Budney AJ (2019) Modes of cannabis use: a secondary analysis of an intensive longitudinal natural history study. Addict Behav 98:106033. https://doi.org/10.1016/j.addbeh.2019.106033
123. Sullivan R, Montgomery A, Scipioni A, Jhaveri P, Schmidt AT, Hicks SD (2022) Confounding factors impacting microRNA expression in human saliva: methodological and biological considerations. Genes (Basel) 13. https://doi.org/10.3390/genes13101874
124. Sun R, Dou W, Liu W, Li J, Han X, Li S, Wu X, Wang F, Xu X, Li J (2023) Global, regional, and national burden of oral cancer and its attributable risk factors from 1990 to 2019. Cancer Med 12:13811–13820. https://doi.org/10.1002/cam4.6025
125. Sundermann BV, Uhlmann L, Hoffmann J, Freier K, Thiele OC (2018) The localization and risk factors of squamous cell carcinoma in the oral cavity: a retrospective study of 1501 cases. J Craniomaxillofac Surg 46:177–182. https://doi.org/10.1016/j.jcms.2017.10.019
126. Theocharis S, Giaginis C, Alexandrou P, Rodriguez J, Tasoulas J, Danas E, Patsouris E, Klijanienko J (2016) Evaluation of cannabinoid CB1 and CB2 receptors expression in mobile tongue squamous cell carcinoma: associations with clinicopathological parameters and patients' survival. Tumour Biol 37:3647–3656. https://doi.org/10.1007/s13277-015-4182-8
127. Tranby EP, Heaton LJ, Tomar SL, Kelly AL, Fager GL, Backley M, Frantsve-Hawley J (2022) Oral cancer prevalence, mortality, and costs in medicaid and commercial insurance claims data. Cancer Epidemiol Biomarkers Prev 31:1849–1857. https://doi.org/10.1158/1055-9965.EPI-22-0114
128. Tu H-F, Lin L-H, Chang K-W, Cheng H-W, Liu C-J (2022) Exploiting salivary miR-375 as a clinical biomarker of oral potentially malignant disorder. J Dent Sci 17:659–665. https://doi.org/10.1016/j.jds.2021.09.020
129. Uma Maheswari TN, Nivedhitha MS, Ramani P (2020) Expression profile of salivary micro RNA-21 and 31 in oral potentially malignant disorders. Braz Oral Res 34:1–10. https://doi.org/10.1590/1807-3107BOR-2020.VOL34.0002
130. Vageli DP, Doukas PG, Shah R, Boyi T, Liu C, Judson BL (2023) A novel saliva and serum miRNA panel as a potential useful index for oral cancer and the association of miR-21 with smoking history: a pilot study. Cancer Prev Res (Phila) 16:653–659. https://doi.org/10.1158/1940-6207.CAPR-23-0219
131. Verma TP, Adhikari S, Kandpal M, Jha HC (2024) Differential effects of areca nut, tobacco, and lime on oral squamous cell carcinoma progression. Front Cancer Control Soc 2. https://doi.org/10.3389/fcacs.2024.1414881
132. Versteeg PA, Slot DE, van der Velden U, van der Weijden GA (2008) Effect of cannabis usage on the oral environment: a review. Int J Dent Hyg 6:315–320. https://doi.org/10.1111/j.1601-5037.2008.00301.x
133. Wang K, Jin J, Ma T, Zhai H (2017) MiR-139-5p inhibits the tumorigenesis and progression of oral squamous carcinoma cells by targeting HOXA9. J Cell Mol Med 21:3730–3740. https://doi.org/10.1111/jcmm.13282
134. Wang J, Lv N, Lu X, Yuan R, Chen Z, Yu J (2021) Diagnostic and therapeutic role of microRNAs in oral cancer (Review). Oncol Rep 45:58–64. https://doi.org/10.3892/or.2020.7854
135. Weber JA, Baxter DH, Zhang S, Huang DY, How Huang K, Jen Lee M, Galas DJ, Wang K (2010) The MicroRNA spectrum in 12 body fluids. Clin Chem 56:1733–1741. https://doi.org/10.1373/clinchem.2010.147405

136. Whyte DA, Al-Hammadi S, Balhaj G, Brown OM, Penefsky HS, Souid AK (2010) Cannabinoids inhibit cellular respiration of human oral cancer cells. Pharmacology 85:328–335. https://doi.org/10.1159/000312686
137. Yan Y, Wang X, Venø MT, Bakholdt V, Sørensen JA, Krogdahl A, Sun Z, Gao S, Kjems J (2017) Circulating miRNAs as biomarkers for oral squamous cell carcinoma recurrence in operated patients. Oncotarget 8:8206–8214. https://doi.org/10.18632/oncotarget.14143
138. Yang CC, Hung PS, Wang PW, Liu CJ, Chu TH, Cheng HW, Lin SC (2011) miR-181 as a putative biomarker for lymph-node metastasis of oral squamous cell carcinoma. J Oral Pathol Med 40:397–404. https://doi.org/10.1111/j.1600-0714.2010.01003.x
139. Yang X, Bam M, Nagarkatti PS, Nagarkatti M (2016) RNA-seq analysis of δ9-Tetrahydrocannabinol-treated T cells reveals altered gene expression profiles that regulate immune response and cell proliferation. J Biol Chem 291:15460–15472. https://doi.org/10.1074/jbc.M116.719179
140. Yap T, Seers C, Koo K, Cheng L, Vella LJ, Hill AF, Reynolds E, Nastri A, Cirillo N, McCullough M (2019) Non-invasive screening of a microRNA-based dysregulation signature in oral cancer and oral potentially malignant disorders. Oral Oncol 96:113–120. https://doi.org/10.1016/j.oraloncology.2019.07.013
141. Yete S, Saranath D (2020) MicroRNAs in oral cancer: biomarkers with clinical potential. Oral Oncol 110:105002. https://doi.org/10.1016/j.oraloncology.2020.105002
142. Zahran F, Ghalwash D, Shaker O, Al-Johani K, Scully C (2015) Salivary microRNAs in oral cancer. Oral Dis 21:739–747. https://doi.org/10.1111/odi.12340
143. Zhang ZF, Morgenstern H, Spitz MR, Tashkin DP, Yu GP, Marshall JR, Hsu TC, Schantz SP (1999) Marijuana use and increased risk of squamous cell carcinoma of the head and neck. Cancer Epidemiol Biomarkers Prev 8:1071–1078
144. Zhang B, Li Y, Hou D, Shi Q, Yang S, Li Q (2017) MicroRNA-375 inhibits growth and enhances radiosensitivity in oral squamous cell carcinoma by targeting insulin like growth factor 1 receptor. Cell Physiol Biochem 42:2105–2117. https://doi.org/10.1159/000479913
145. Zhang H, Xie M, Archibald SD, Jackson BS, Gupta MK (2018) Association of marijuana use with psychosocial and quality of life outcomes among patients with head and neck cancer. JAMA Otolaryngol Head Neck Surg 144:1017–1022. https://doi.org/10.1001/jamaoto.2018.0486
146. Zhang Z, Liu T, Dong M, Ahamed MA, Guan W (2024) Sample-to-answer salivary miRNA testing: new frontiers in point-of-care diagnostic technologies. Wiley Interdiscip Rev Nanomed Nanobiotechnol 16:1–31. https://doi.org/10.1002/wnan.1969

Chapter 12
MicroRNAs in Lung Cancer

Venkataramanan Swaminathan and James John

Abstract MicroRNAs are a class of small, noncoding RNAs that function as important regulators of gene expression at the posttranscriptional level. These molecules act by binding to complementary sequences on target messenger RNAs, leading to either degradation or translational silencing of these targets. In the context of lung cancer, one of the most common cancers and also one of its leading causes of morbidity and mortality in the world, microRNAs are key functionally relevant players not only in initiation and progression but also in therapeutic response. A review of circulating microRNAs in lung cancer pathobiology: potential role of tumor markers as data sources and extraction. Conclusion: this global review summarizes what we currently know about microRNAs in lung cancer pathobiology, including their patterns of differential expression and their complex biological function, whether as oncogenes or tumor suppressor genes. This range is also considered for its effect on several important signaling pathways controlling important cellular behaviors such as epithelial-mesenchymal transition (EMT), apoptosis, and metastatic dissemination and following the complex interactions in the tumor ecosystem. Furthermore, this review reflects on the immense potential of microRNAs in the clinic as biomarkers for diagnosis and prognosis and their potential as molecular targets in lung cancer therapy. It explores current technological limitations in microRNA research and therapeutic development, and provides insights into emerging strategies and future research directions in this dynamic field. By delineating the multifaceted functions of microRNAs in lung cancer,

V. Swaminathan (✉)
Faculty of Health & Life Sciences, Department of Diagnostic and Allied Health Science, Management and Science University (MSU), Shah Alam, Malaysia

Department of Medical laboratory Technology, School of Allied Health Science, Sathyabama Institute of Science and Technology, Chennai, India
e-mail: s_venkataramanan@msu.edu.my

J. John
Department of Medical laboratory Technology, School of Allied Health Science, Sathyabama Institute of Science and Technology, Chennai, India

S. Pathak, A. Banerjee (eds.), *MicroRNAs: Role in Cancer Progression, Diagnostics and Therapy*, Advances in Biochemistry in Health and Disease 34,
https://doi.org/10.1007/978-3-032-08824-6_12

investigators have laid the groundwork for novel therapeutic approaches, better stratification of patients, and improved prognostic markers.

Keywords MicroRNA · Lung cancer · Biomarkers · Oncogene · Tumor

Introduction to MicroRNAs

MicroRNAs (miRNAs) are short (<25 nucleotide) noncoding RNA molecules that act as posttranscriptional regulators (Czech et al. 2021). Discovered in *Caenorhabditis elegans* in 1993, miRNAs have since evolved into major regulators of various biological processes ranging from cell division to differentiation and apoptosis [1]. These small RNAs usually act by hybridization to the 3′ untranslated region (3′UTR) of specific messenger RNAs (mRNAs), resulting in translational repression or degradation of the mRNA. More than 2000 miRNAs are encoded in the human genome, and miRNAs together regulate about 60% of protein-coding genes [11].

miRNAs are produced through a complex multistep process that takes place in both the nucleus and cytoplasm. Initially transcribed by RNA polymerase II as primary miRNAs (pri-miRNAs), these molecules form one or multiple hairpin structures. The microprocessor complex, composed of Drosha and DGCR8, cleaves these pri-miRNAs into 70-nucleotide-long precursor miRNAs (pre-miRNAs). Exportin-5 facilitates the export of pre-miRNAs to the cytoplasm in a Ran-GTP-dependent manner. Once in the cytoplasm, Dicer cleaves the pre-miRNA hairpin to generate a miRNA duplex. The mature miRNA strand is then incorporated into the RNA-induced silencing complex (RISC), while the complementary strand is typically degraded. Guided by the mature miRNA, the RISC complex recognizes target mRNAs and binds to them through complementary base pairing to induce posttranscriptional gene silencing [29].

Aberrantly expressed miRNAs [(miRNA dysregulation)] have been associated with initiation and progression in human malignancies, including lung cancer. Copy number alterations (CNAs), epigenetic alterations, and defects in miRNA biogenesis machinery can lead to aberrant miRNA expression [32]. Cancer-associated miRNAs can be classified as either tumor suppressor or oncogenic (oncomiRs) based on their targets and functions. Tumor suppressor miRNAs usually target oncogenes and are commonly downregulated in cancer, whereas oncogenic miRNAs usually target tumor suppressor genes and are generally upregulated in malignant tissues.

Lung cancer, an eminent cause of cancer-related mortality worldwide, is defined by widespread genomic and epigenomic changes with severely altered miRNA expression profiles. miRNAs have been studied extensively in lung cancer, yielding information on the pathogenesis of lung cancer that has been important for understanding disease mechanisms and developing new classes of diagnostics and therapeutics (Fig. 12.1) [15].

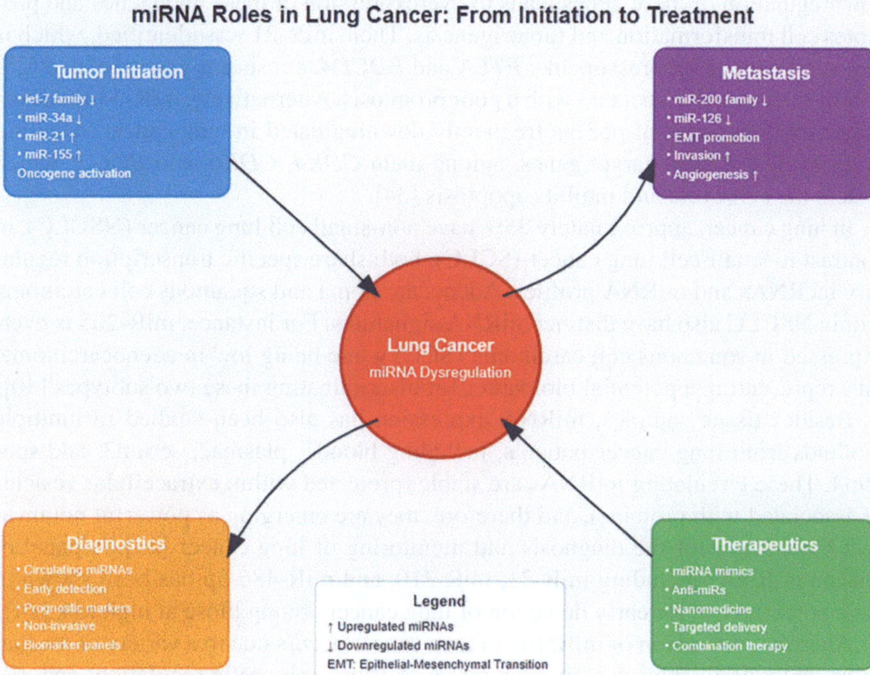

Fig. 12.1 Conceptual overview of miRNA roles in lung cancer

Figure 12.1 illustrates a comprehensive overview that ties together all the major themes tumor initiation, metastasis, diagnostics, and therapeutics.

MicroRNA Expression and Lung Cancer

miRNA expression profiling in lung cancer has revealed widespread downregulation of distinctive miR-17/92 clusters, Dicer activity leading to increased levels of specific miRNA in advanced lung cancers, as well as numerous miRNA–mRNA interaction networks that are involved in the molecular pathological processes of lung cancers. The advent of high-throughput technologies, including microarray analysis, next-generation sequencing (NGS), and quantitative reverse transcription-polymerase chain reaction (qRT-PCR), has enabled the comprehensive profiling of miRNA expression in lung tumors and has uncovered unique miRNA signatures associated with histological subtypes, lung cancer stages, and molecular features [4].

Reduced expression of let-7 family miRNAs in lung cancer tissues relative to adjacent normal tissue was reported by one of the first studies to show altered miRNA expression in lung cancer [35]. Members of the let-7 family act as tumor suppressors, targeting oncogenes like RAS, MYC, and HMGA2. The downregulation of these genes leads to overexpression of these oncogenes and promotes cell transformation and tumorigenesis. Then, miR-21 was identified, which is targeted to tumor suppressors like *PTEN* and *PDCD4*, consistently upregulated [25] in lung cancer and associated with a poor prognosis. Alternatively, miR-34a, a direct transcriptional target of p53, is frequently downregulated in lung cancer, resulting in upregulation of its target genes, among them *CDK4*, *CDK6*, and *BCL-2*, which fosters the cell cycle and inhibits apoptosis [34].

In lung cancer, approximately 85% have non-small cell lung cancer (NSCLC), in contrast to small-cell lung cancer (SCLC), both share specific transcription regulatory lncRNAs and miRNA profiles. Adenocarcinoma and squamous cell carcinoma within NSCLC also have distinct miRNA signatures. For instance, miR-205 is overexpressed in squamous cell carcinoma (SCC) while being low in adenocarcinoma, thus representing a potential biomarker for discriminating these two subtypes [40].

Besides tissue samples, miRNA expression has also been studied in multiple biofluids from lung cancer patients, including blood1, plasma2, serum3, and sputum4. These circulating miRNAs are stable (protected within extracellular vesicles or associated with proteins), and therefore, they are emerging as powerful noninvasive biomarkers for the diagnosis and monitoring of lung cancer [23]. A panel of plasma miRNAs including miR-21, miR-210, and miR-486-5p has been shown to have potential use for early detection of lung cancer among those at high risk [14].

Aberrant expression of miRNAs in lung cancer occurs due to a variety of mechanisms including genomic amplification or deletion, epigenetic regulation, and disruption of the miRNA biogenesis machinery. Recent developments in single-cell sequencing technologies have broken down our understanding of miRNA expression in lung cancer into more granular details that reveal intratumoral heterogeneity in miRNA profiles that may underlie treatment resistance and disease progression [6].

The Role of miRNAs in Lung Cancer Pathogenesis

MicroRNAs play a crucial role in the development of lung cancer by controlling multiple cancer characteristics, such as continuous proliferative signaling, overcoming growth inhibitors, resistance to cell death, angiogenesis, and initiation of invasion and metastasis. These small RNAs drive the initiation, advancement, and spread of lung tumors by targeting various essential signaling pathways and cellular mechanisms.

A few important signaling pathways in lung cancer are regulated by miRNAs. MiR-486-5p targets *PTEN*, a negative regulator of the PI3K/AKT pathway. The upregulation of miR-486-5p promotes the pathway activation and tumor growth. MiR-34a, a negative regulator of the MAPK pathway, targets MAPK1 thereby inhibiting cell proliferation [20]. miR-140-5p downregulated in advanced lung cancer targets and negatively regulates components of the TGF-β pathway (TGFBR1 and SMAD3), thus promoting TGF-β signaling and EMT [44].

Epithelial-mesenchymal transition (EMT) is an essential process in cancer progression, allowing epithelial cells to gain mesenchymal characteristics to promote invasion and metastasis. The transcription factor's E-cadherin repressor activity is also inhibited by the miR-200 family through ZEB1 and ZEB2 target genes, promoting an epithelial phenotype (43). Thus, the loss of miR-200 family expression promotes increased EMT and metastatic capability [21]. On the other hand, miR-21 induces EMT by targeting inhibitors of these mechanisms like *PTEN* and *PDCD4* [25].

MicroRNAs in the regulation of cell cycle progression and apoptosis in lung cancer. Furthermore, miR-15a and miR-16 target an antiapoptotic protein, BCL2, and their downregulation leads to reduced apoptosis [22]. For example, miR-34a regulates cell cycle and apoptosis by targeting CDK4, CDK6, and BCL2, and it is commonly downregulated in lung cancer, particularly in p53 mutated cases [34].

It has been well-recognized that the tumor microenvironment (TME) plays a critical role in tumor growth and plasticity, and miRNAs are involved in this interplay between tumor cells and TME. MiR-210 is activated under hypoxic conditions and targets EFNA3, inhibiting endothelial cells from migrating and angiogenesis. MiR-126 is considered to be downregulated in lung cancer and inhibits angiogenesis by targeting VEGF and is associated with reduced metastatic potential [38].

Recent research has elucidated the pivotal role of miRNAs in orchestrating immune responses in the lung cancer microenvironment. As an example, natural killer cells are known to upregulate miR-34a that is also cooperatively upregulated by cytotoxic lymphoid cells used to enhance natural killer cell activity against lung cancer cells through UL16-binding protein rooting [17]. MiR-200c regulates PD-L1 expression in lung cancer cells to modulate sensitivity to immune checkpoint blockade therapy.

miRNAs as Tumor Suppressors and Oncogenes in Lung Cancer

MicroRNAs function either as tumor suppressors or oncogenes in lung cancer depending on their targets. Several key miRNAs exemplify these roles, with significant clinical and therapeutic implications.

Tumor Suppressor miRNAs

As shown in Table 12.1, several miRNAs function as tumor suppressors in lung cancer. The let-7 family comprises 12 members in humans and targets multiple oncogenes, including RAS, MYC, and HMGA2. In lung cancer, let-7 expression is frequently reduced, leading to enhanced expression of these oncogenes and

Table 12.1 Key miRNAs in lung cancer pathogenesis

miRNA	Expression in lung cancer	Target genes	Functions	Clinical relevance
let-7 family	Downregulated	*RAS, MYC, HMGA2*	Cell cycle regulation, proliferation	Associated with poor prognosis, potential therapeutic target
miR-21	Upregulated	*PTEN, PDCD4, SPRY2*	Inhibits apoptosis, promotes invasion	Associated with advanced stage, lymph node metastasis
miR-34a	Downregulated	*CDK4, CDK6, BCL2, MET*	Induces cell cycle arrest, apoptosis	Regulated by p53, enhances chemosensitivity
miR-155	Upregulated	*SOCS1, FOXO3a, TP53INP1*	Promotes proliferation, inflammation	Associated with poor prognosis
miR-200 family	Downregulated	*ZEB1, ZEB2*	Inhibits EMT, maintains epithelial phenotype	Associated with metastasis, chemoresistance
miR-126	Downregulated	*VEGF, PIK3R2, KRAS*	Inhibits angiogenesis, proliferation	Associated with poor survival, potential biomarker

miR-126 downregulated VEGF, inhibition associated with poor PIK3R2, angiogenesis, survival, and potential KRAS proliferation biomarker

promoting tumor growth [19]. A single nucleotide polymorphism in the let-7 complementary site of the KRAS 3'UTR has been associated with increased risk of NSCLC, highlighting the importance of this interaction [33].

MiR-34a targets several genes involved in cell cycle progression, apoptosis, and invasion, such as *CDK4, CDK6, BCL2*, and *MET*, and is directly regulated by *p53* [34]. Many tumors have decreased miR-34a expression, especially in lung cancer with mutated p53. Reconstitution of miR-34a expression inhibits proliferation, induces apoptosis, and increases sensitivity to common chemotherapeutic agents [20].

MiR-143 and miR-145, usually expressed as a cluster, target several oncogenes like *KRAS, NRAS, MYC, EGFR*, and *OCT4*. Both of these miRNAs display frequent downregulation in lung cancer, resulting in increased expression of their corresponding targeted oncogenes [42]. Whether the expression of these miRNAs is reduced in lung and other cancers has been linked to deletion of the miR-143/145 cluster at chromosome 5q31, a region commonly lost in lung cancer, providing genetic evidence for the reduced expression of these miRNAs in cancer.

The miR-200 family is essential for the inhibition of EMT by targeting the ZEB1 and ZEB2 transcription factors. The downregulation of miR-200 family members facilitates their EMT, which facilitates invasion and metastasis [21]. Restoring miR-200 expression reverses EMT and reduces invasive and metastatic potential.

Indeed, miR-126 targets various genes critical to oncogenesis such as *VEGF, PIK3R2*, and *KRAS*. Loss of its expression is associated with advanced stage, lymph node metastasis, and poor prognosis [38]. Mechanistically, miR-126 suppresses lung cancer cells angiogenesis, proliferation, and metastasis.

Oncogenic miRNAs

MiR-21 stands out as one of the most commonly elevated miRNAs in different human tumor types, particularly in lung cancer. Noteworthy targets of this microRNA include PTEN, PDCD4, and SPRY2 [25]. In the context of lung cancer, heightened miR-21 expression correlates with later disease stages, lymph node metastasis, and unfavorable prognoses. Inhibiting miR-21 drives down cell proliferation, migration, and invasion in lung cancer cells while enhancing their responsiveness to chemotherapy and radiotherapy [31].

The expression of miR-155 is dysregulated in response to inflammatory stimuli and other signals, suggesting an important contribution to multiple cancers by suppressing the expression of tumor suppressor genes, such as *SOCS1, FOXO3a*, and *TP53INP1* [19, 20]. Their levels were associated with advanced disease stage, lymph node metastasis, and poor prognosis in lung cancer. Inhibition of miR-155 activity was found to inhibit proliferation of lung cancer cells, and enhance their sensitivity to chemotherapy.

The miR-17-92 cluster was reported to target *PTEN, BIM*, and *TGFBR2* that are some of the tumor suppressor genes [30]. The cluster is usually overexpressed and amplificated in lung cancer promoting cell proliferation, inhibiting cell apoptosis and stimulating angiogenesis [5]. Silencing of some members or the whole branch represses lung cancer cell proliferation and tumorigenesis.

miR-221 and miR-222 are known to target several tumor suppressor genes, including *PTEN, TIMP3*, and *PUMA*. High expression of miR-221/222 correlates with a poor prognosis in lung cancer and with advanced stage and lymph node metastasis [13]. Inhibition of microRNA-221/222 decreases lung cancer cell proliferation, migration, and invasion, and sensitizes lung cancer cells to TRAIL-mediated apoptosis.

Numerous therapeutic approaches aimed at these miRNAs are being generated, such as miRNA mimics to restore tumor suppressor miRNAs and anti-miRs or miRNA sponges intended to inhibit oncogenic miRNAs. An example is MRX34, a liposomal formulation of miR-34a mimic that was advanced into a phase I clinical trial for multiple cancers (including lung cancer), the first miRNA-based drug to enter the clinic [2].

miRNA Biomarkers for Early Detection and Prognosis in Lung Cancer

Early identification of lung cancer greatly enhances survival rates, emphasizing the requirement for precise and specific biomarkers. MicroRNAs have surfaced as prospective contenders because of their endurance in biological fluids, expression specific to tissues, and reliable quantification techniques.

Diagnostic Biomarkers

As shown in Table 12.2, numerous studies have investigated circulating miRNAs as biomarkers for early detection of lung cancer. Chen et al. [5] identified a panel of serum miRNAs, including miR-21, miR-210, and miR-486-5p, that distinguished lung cancer patients from healthy controls with high sensitivity and specificity. Wozniak et al. [39] identified a 24-miRNA signature in plasma that detected early-stage lung cancer with 91.5% sensitivity and 97.5% specificity.

Specific miRNA signatures have been associated with different histological subtypes. Nadal et al. [28] identified a plasma miRNA signature comprising miR-205, miR-19a, and miR-30b that distinguished squamous cell carcinoma from adenocarcinoma with high accuracy.

Beyond blood-based biomarkers, miRNAs have been detected in sputum, bronchoalveolar lavage fluid, and exhaled breath condensate. Xing et al. [41] identified a sputum miRNA panel comprising miR-21, miR-31, and miR-210 that detected lung cancer with 82% sensitivity and 86% specificity.

Exosomal miRNAs offer additional advantages as biomarkers. Jin et al. [18] reported an exosomal miRNA signature that detected early-stage lung cancer with 94% sensitivity and 89% specificity.

Despite promising results, standardization of protocols for sample collection, processing, and miRNA quantification is essential for clinical implementation. Recent technological advances, such as digital PCR and NGS, have enhanced the sensitivity and specificity of miRNA detection, potentially addressing some challenges.

Prognostic and Predictive Biomarkers

MicroRNA expression profiles have been associated with various clinical parameters, offering insights into disease progression and treatment response. Lin et al. [23] identified a 7-miRNA signature predicting disease-free survival in stage I NSCLC, potentially identifying high-risk patients for adjuvant therapy.

Table 12.2 Circulating miRNAs as biomarkers for lung cancer

miRNA biomarker	Sample type	Diagnostic performance	Clinical application	References
miR-21, miR-210, miR-486-5p	Serum	Sensitivity: 80%, specificity: 85%	Early detection	[5]
miR-205, miR-19a, miR-30b	Plasma	Accuracy: 88%	Distinguishing SCC from ADC	[28]
miR-21, miR-31, miR-210	Sputum	Sensitivity: 82%, specificity: 86%	Early detection	[41]
24-miRNA signature	Plasma	Sensitivity: 91.5%, specificity: 97.5%	Early-stage detection	[39]

Exosomal miRNA plasma sensitivity: 94%, early detection Jin et al., panel, specificity: 89% 2017

Individual miRNAs also correlate with prognosis. High expression of oncogenic miRNAs (miR-21, miR-155, and miR-210) associates with poor prognosis, while high expression of tumor suppressor miRNAs (let-7, miR-34a, and miR-200 family) indicates better outcomes [15].

MicroRNAs modulate treatment response in lung cancer by targeting genes involved in drug metabolism, cell survival, DNA repair, and immune responses. miR-21 targets PTEN and PDCD4, involved in cisplatin-induced apoptosis, and its high expression associates with platinum-based chemotherapy resistance [25]. MiR-34a enhances cisplatin sensitivity by targeting SIRT1 and increasing p53 acetylation [34]. MiR-200 family members enhance docetaxel sensitivity by targeting ZEB1 and ZEB2, inhibiting EMT and stemness [21].

In targeted therapies, miR-30c and miR-221/222 target BIM, involved in EGFR-TKI-induced apoptosis, and their high expression associates with EGFR-TKI resistance in EGFR-mutant NSCLC [13]. In immunotherapy, miR-200 family members target PD-L1, and low miR-200 expression associates with high PD-L1 expression and poor response to PD-1/PD-L1 inhibitors.

Several miRNA-based predictive models have been developed. Yang et al. [43] developed a 3-miRNA signature (miR-21, miR-24, and miR-29a) predicting platinum-based chemotherapy response in NSCLC with 87% accuracy. Wu et al. [40] reported a 4-miRNA signature predicting EGFR-TKI response in EGFR-mutant NSCLC with 92% accuracy.

Therapeutic Strategies Targeting miRNAs in Lung Cancer

Figure 12.2 illustrates the therapeutic development pipeline and the key challenges faced at each stage, making it perfect for the therapeutics section.

The dysregulation of miRNAs in lung cancer provides rationale for therapeutic strategies aimed at normalizing miRNA expression. Two main approaches have been developed: miRNA replacement therapy and miRNA inhibition therapy.

miRNA Replacement Therapy

miRNA replacement therapy involves delivering synthetic miRNA mimics or expression vectors encoding tumor suppressor miRNAs to cancer cells. let-7 restoration inhibits lung tumor growth in mouse models by targeting RAS, MYC, and HMGA2 [19]. miR-34a delivery inhibits lung tumor growth and enhances treatment sensitivity by targeting genes involved in cell cycle progression, apoptosis, and stemness [20].

MRX34, a liposomal formulation of miR-34a mimic, entered clinical testing in a phase I trial (NCT01829971) for advanced solid tumors, including lung cancer.

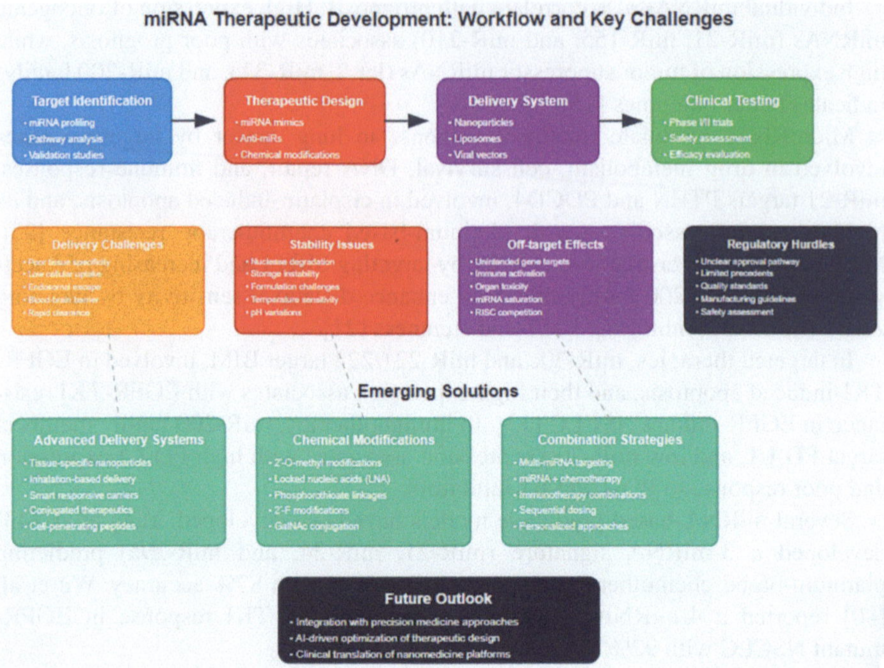

Fig. 12.2 miRNA-based therapeutic workflow and challenges

Although terminated early due to immune-related adverse events, it demonstrated feasibility of miRNA mimic delivery [2].

Other miRNA-based therapies in development include mesomiR-1, a miR-16 mimic delivered using EnGeneIC Dream Vectors, showing promising results in preclinical models of mesothelioma and NSCLC and being evaluated in a phase I trial (NCT02369198) [37]. A let-7 mimic delivered using lipid nanoparticles has shown efficacy in preclinical models of KRAS-mutant NSCLC and is being developed for clinical testing [36].

miRNA Inhibition Therapy

miRNA inhibition therapy involves the use of antisense oligonucleotides (anti-miRs) or miRNA sponges to inhibit oncogenic miRNAs. Anti-miRs use base pairing to bind target miRNAs, preventing these targets from interacting with mRNAs. By endowing oligonucleotides with chemical modifications like 2′-O-methyl, 2′-O-methoxyethyl, and locked nucleic acid (LNA), their stability and binding affinities can be improved [44]. miRNA sponges have more than one individual binding site for specific types of miRNAs, competing with targets from the endogenous process.

miR-21 inhibition diminishes lung tumor growth and improves treatment sensitivity by derepressing PTEN and PDCD4 [31]. Knockdown of cytochrome b561 reduces cecramine-induced tumor growth and promotes chemosensitivity by derepressing SOCS1 and FOXO3a, similar to miR-155 inhibition.

Challenges await in translating promising preclinical results into clinical development. Targeting tumor cells specifically requires efficient delivery of these molecules while minimizing the uptake in normal tissues. Over the years, several delivery systems have been reported, encompassing lipid-based, polymer-based, and inorganic nanoparticles with targeting ligands for improved tumor specificity [12].

Local delivery to the lungs via inhalation is a potentially promising method, promising a higher concentration of the drug in the lungs while reducing systemic exposure. At least three inhalable formulations have been identified to perform well in preclinical models [12].

Other challenges are related to off-target effects, as miRNAs generally are able to target more than one gene, and development of resistance (through mutations in the miRNA processing, modifications in the binding sites, or activation of compensatory pathways). Combination strategies, either targeting multiple miRNAs or coadministering miRNA-based therapies with conventional treatments, may offer solutions to such challenges [40].

Future Perspectives in miRNA-Based Nanomedicine for Lung Cancer

The emerging field of nanomedicine offers revolutionary approaches to overcome traditional challenges in miRNA-based therapy for lung cancer. Nanoparticle-based delivery systems represent a promising frontier with significant advantages over conventional delivery methods, particularly for targeting the distinctive physiology of the lungs.

Inhalation-based nanodelivery systems deserve special consideration for lung cancer applications. Pulmonary administration enables direct targeting of lung tissue while minimizing systemic exposure, potentially addressing both efficacy and toxicity concerns. Recent developments in inhalable nanoformulations include lipid-polymer hybrid nanoparticles that demonstrate superior stability during nebulization. These platforms can protect miRNA cargo from degradation while facilitating cellular uptake in lung tissues. Preliminary studies with inhalable miR-34a-loaded nanoparticles in orthotopic lung cancer models demonstrate enhanced tumor penetration and sustained release profiles compared to systemic administration.

Smart nanocarriers represent another innovative direction. These systems respond to tumor-specific microenvironmental cues like acidic pH, elevated reactive oxygen species, or specific enzyme activities prevalent in lung cancer. For example, pH-responsive polymeric nanoparticles that release miRNA cargo selectively in the acidic tumor microenvironment have shown promising results in preclinical lung cancer models. Similarly, redox-responsive nanocarriers exploit elevated glutathione levels in cancer cells to trigger miRNA release specifically within tumor tissue.

Multimodal nanotherapeutics that combine miRNA delivery with complementary treatment modalities may yield synergistic benefits. Nanoparticles co-loaded with miRNA and chemotherapeutic agents have demonstrated enhanced efficacy in resistant lung cancer models. For instance, co-delivery of miR-34a with paclitaxel results in significantly greater tumor suppression than either treatment alone, potentially through miR-34a-mediated sensitization of cancer cells to chemotherapy.

Theranostic nanoplatforms incorporating imaging capabilities with miRNA delivery represent another promising direction. These systems enable real-time tracking of biodistribution, tumor accumulation, and therapeutic efficacy. Superparamagnetic iron oxide nanoparticles conjugated with miRNA have demonstrated utility for both MRI imaging and therapeutic delivery in lung cancer models.

Yet, there are still many challenges. The manufacturing of nanoformulations at scales with reproducible quality poses a challenge. Also, the significant interdependent relationship between nanoparticle characteristics and biological systems requires strategic optimization. Future investigations must therefore concentrate on increasing tumor specific delivery, enhancing circulation time, and decreasing immunogenicity.

Another challenge comes from regulatory considerations. miRNA-loaded nanoparticles come with unique challenges in quality control, safety assessment, and standardization due to their complex nature. It will be necessary to define suitable and specific regulatory pathways for clinically translating nanomedicine-based miRNA therapeutics.

Overall, although substantial progress has been made in developing nanomedicine strategies for miRNA delivery in lung cancer, further research involving significant interdisciplinary collaboration will be required to translate these advances into clinical application. The intersection of nanotechnology, RNA biology, and cancer medicine may ultimately give rise to transformative therapies for lung cancer patients.

miRNAs in Lung Cancer Metastasis and Tumor Microenvironment

miRNAs in Metastasis

Metastasis is a process comprised of several steps, including epithelial–mesenchymal transition (EMT), invasion, intravasation, survival in circulation, extravasation, and colonization. MicroRNAs are negative regulators targeting genes implicated in cell adhesion, migration, invasion, and microenvironment interactions, oscillating the diverse aspects of this cascade.

The miR-200 microRNA family regulates epithelial mesenchymal transition (EMT) by directly targeting ZEB1 and ZEB2. Indeed, miR-200 family downregulation promotes the epithelial-mesenchymal transition (EMT) and increases migration and invasion [21] Likewise, miR-205 targets ZEB1, ZEB2, and PRKCE and is downregulated in metastatic lung cancer [40].

In contrast, miR-21 is known to induce EMT and promote invasion by targeting PTEN, PDCD4, and SPRY2. High level of miR-21 expressions correlates with node metastasis and poor prognosis [25]. Example: miR-10b, a downstream target of HOXD10, is overexpressed in metastatic lung cancer and induces migration and invasive capabilities [26].

MicroRNAs also regulate survival in circulation and colonization of distant sites. miR-126 inhibits metastasis by targeting VEGF and factors involved in angiogenesis and endothelial recruitment [38].

Metastatic organotropism is partly regulated by miRNAs. miR-378 promotes brain metastasis in NSCLC by targeting MYT1, a transcription factor regulating neural development [45].

MicroRNAs also regulate interactions between tumor cells and the metastatic niche. miR-192 secreted by lung cancer cells modulates endothelial barrier function by targeting ALCAM, facilitating extravasation and colonization. miR-105 targets ZO-1 in endothelial cells, enhancing vascular permeability and metastatic potential [46].

Recent studies highlight miRNAs in regulating the pre-metastatic niche. miR-21 and miR-29a released by lung cancer cells bind to toll-like receptors on immune cells, triggering inflammatory responses that create a favorable environment for metastatic colonization [10]. miR-23a modifies the bone marrow microenvironment to support metastatic growth [44].

miRNAs and Tumor Microenvironment

The tumor microenvironment (TME) plays crucial roles in cancer development, progression, and treatment response. MicroRNAs mediate interactions between tumor cells and the TME, influencing angiogenesis, immune evasion, and fibroblast activation.

Several miRNAs regulate angiogenesis in the lung cancer microenvironment. miR-126 inhibits angiogenesis by targeting VEGF, and its reduced expression contributes to enhanced tumor angiogenesis [38].

Cancer-associated fibroblasts (CAFs) contribute to tumor growth, invasion, and therapy resistance. miR-21, upregulated in lung cancer cells, transfers to fibroblasts through exosomes, promoting CAF phenotype by targeting PTEN and activating AKT [25]. Conversely, CAF-derived exosomal miR-196a promotes lung cancer cell proliferation and metastasis by targeting CDKN1B and activating the PI3K/AKT/mTOR pathway [16].

MicroRNAs regulate various aspects of tumor-immune interactions. miR-34a enhances natural killer cell activity against tumor cells by targeting UL16 binding protein (ULBP2), a ligand for the NK cell activating receptor NKG2D [17]. miR-146a modulates tumor-associated macrophage activity by targeting TRAF6 and IRAK1, key components of the NF-κB pathway [3].

MicroRNAs also regulate immune checkpoint molecules. miR-200c targets PD-L1, and its downregulation contributes to enhanced PD-L1 expression and immune evasion. miR-34a targets PD-L1 and enhances T cell activation, and its reduced expression contributes to immune suppression [7].

MicroRNAs modulate extracellular matrix (ECM) composition and remodeling. MiR-29 family members, downregulated in lung cancer, target multiple ECM proteins, including collagens, elastin, and fibrillin, contributing to enhanced ECM deposition and fibrosis [8]. miR-200b targets VEGF and Ets-1, regulators of matrix metalloproteinase expression, and its downregulation contributes to enhanced MMP expression and ECM degradation [21].

Technological Barriers in miRNA-Based Diagnostics

While miRNAs show promise as diagnostic biomarkers of lung cancer, translation into clinical practice is hampered by several technological obstacles. Overcoming these obstacles is critical in order to translate miRNA research through the preclinical and clinical stages and ultimately into the clinic to aid patients.

To the best of our knowledge, there is no standardization involving miRNA extraction protocols, which poses a major obstacle for miRNA-based diagnostics. Substantial variability in miRNA measurements can arise from differences in collection methods, storage conditions, and RNA extraction techniques. Such as when blood samples are hemolysised during collection, releasing cellular miRNAs into plasma or serum, thereby biasing the extracellular miRNA profile. Similarly, sample storage time and temperature prior to analysis/processing also greatly influence miRNA stability and subsequent recovery rates. Variability in metabolomic data continues to make establishing consistent reference ranges and diagnostic cutoffs across laboratories and clinical settings challenging (Melander et al. 2023).

Another major difficulty is the quantification of miRNAs. Current technologies (e.g., quantitative PCR/qPCR, microarrays, next-generation sequencing/NGS) all suffer from specific limitations. qPCR is sensitive but needs normalisation to reference genes that may also have variation in abundance within a sample and/or between disease states. Microarrays have high throughput but cross-hybridization problems and small dynamic ranges. NGS offers extensive profiling, but it comes at a cost with respect to data analysis and interpretation. Moreover, there is still a technically challenging issue with detecting miRNAs at low abundance, especially in biofluids in which concentrations of miRNAs may even be at picomolar concentrations.

miRNA diagnostics are also hampered by analytical challenges. The choice of suitable endogenous controls for normalization is a point of contention, conventional reference genes displaying unstable expression in cancer settings. Because of the short length and high sequence similarity among miRNA family members, specific detection of miRNA is challenging, and there is a potential for cross-reactivity; and isomiRs (i.e., variations of the same miRNA differing by a few nucleotides in sequence length) constitute an analytical challenge, or may fulfill different biological functions that current diagnostic practices do not differentiate.

miRNA is also influenced by pre-analytical variables significantly. Even disease-independent factors like patient age, gender, body mass index, medication status, and circadian rhythms may affect circulating miRNA levels. Research studies often fail to control for such variables adequately and the findings are therefore inconsistent and poorly reproducible in different cohorts.

miRNA diagnostics, however, may have substantial technological development barriers for miniaturization and point-of-care applications. Despite the promising use of microfluidic and nanomaterial-based detection systems, these systems have had shortcomings such as sensitivity limits, lack of multiplexing capabilities, as well as difficulty being translated to real-world clinical use. The complexity of these systems results in systems that can be expensive and require specialized technical expertise to implement, limiting widespread adoption.

Another challenge is computational limits. Typically, the generation of miRNA data especially from high-throughput platforms requires advanced bioinformatic methods to process and interpret the resulting datasets. This inconsistency arises from the differences in feature selection methods, machine learning algorithms, and statistical techniques employed by most studies to derive diagnostic miRNA signatures. Moreover, integrating miRNA data with other molecular biomarkers and clinical parameters needs sophisticated computational frameworks that are currently under development.

There are more difficult limitations in clinical implementation with regulatory considerations. Little is known about the path to FDA approval (or other regulatory approval) for miRNA-based diagnostics, and few precedents exist to inform development. Analytical validity, clinical validity, and clinical utility will have to be rigorously addressed before widespread clinical adoption can commence.

Cost effectiveness poses a final big hurdle. Clinically viable miRNA diagnostics will need to consider not only performance but also commercial aspects in order to be affordable and integrated into existing healthcare systems. The existing techniques for global miRNA profiling are still quite costly compared to standard diagnostic methods creating challenges in their practical utility in resource-limited contexts.

Overcoming these technological barriers will take a multidisciplinary approach, spanning the areas of molecular biology, bioengineering, computational science, and clinical medicine. Addressing such barriers will require strategic investments, including efforts to develop new technologies, enact standardization initiatives, and conduct large-scale validation studies to ensure the broader utility of miRNAs as diagnostic tools for lung cancer.

Challenges and Future Perspectives in miRNA Research for Lung Cancer

Despite significant advances in understanding miRNAs in lung cancer, several challenges remain in translating this knowledge into clinical applications. This section discusses these challenges and outlines future directions.

Technical and Biological Challenges

A significant obstacle to the routine use of miRNA as a biomarker is the technical variability of miRNA quantification between different platforms and laboratories. Ultimately, however, standardized methodologies for the collection, processing, and quantification of miRNAs will be indispensable to enable reproducible results and facilitate clinical translation (Fig. 12.2) [9].

Another challenge is the biological complexity of miRNA function. MicroRNAs can have multiple targets, while one gene can also be regulated by multiple miRNAs to form a complex regulatory network. Moreover, some miRNAs exhibit context-dependent roles, such that miRNA activity in one cell type or circumstance may differ from that in another cell type [32], posing a challenge in assessing the net effect of miRNA modulation in the clinic.

miRNA research is challenging to some extent due to tumor heterogeneity. Lung tumors contain heterogeneous cellular populations with different miRNA expression profiles. This heterogeneity can contribute to variable responses to miRNA-based therapies and resistance to treatment. This critical challenge may be resolved using single-cell sequencing technologies, which can characterize miRNA expression at the single-cell level, allowing for investigations into intratumoral heterogeneity and, consequently, more personalized therapeutic assignments [6].

Efficiently delivering the miRNAs to target cells in a specific manner remains a key hurdle in miRNA-based therapies. Although numerous delivery systems have been implemented, each has restrictions in terms of efficiency, specificity, and possible toxicity. The clinical success of the therapeutic is dependent on development of systems that are more effective in delivering them [12].

There are concerns of off-target effects and toxicity in miRNA-based therapies. MicroRNAs target multiple genes so that modulation of a particular miRNA might influence multiple cellular processes, which could occasionally lead to undesired effects. qSo delivery systems can trigger immune responses or undesired effects. These concerns may be mitigated with chemical modifications to reduce side effects, target site blockers that mitigate specific mRNAs, and refining delivery systems to limit their immunogenicity and toxicity [27].

Future Directions

There may be difficulties, but this field is still one full of potential. Their functional validation is made possible by advanced technologies, including CRISPR-Cas9 genome editing. If the binding sites can be mutated in such a way that they are incapable of binding to the miRNA, compared to a CRISPR-Cas9-mediated knockout of the miRNA itself, then a direct picture of the effects and interactions can be made [44].

Machine learning-based integration of miRNA data with other molecular and clinical data to enhance the predictive performance of miRNAs, artificial intelligence (AI), and machine learning (ML) methods will likely be the way forward. Chen et al. [6] trained deep learning on miRNA expression data combined with mRNA expression, DNA methylation, and clinical data to achieve comprehensive survival prediction in NSCLC, which significantly outperformed models trained on any single type of data.

Liquid biopsies open new possibilities for noninvasive detection and monitoring of miRNAs. Profiling exosomal miRNAs that are protected from degradation, originating from their cells of origin, may provide more sensitive and specific biomarkers for detection and monitoring [18].

The knowledge of lung cancer pathology gained from multi-omics integration can lead to a better understanding of the involved complex biology and the identification of new treatment targets and biomarkers. Integrating miRNA data with genomic, transcriptomic, proteomic, and metabolomic data can reveal functionally relevant miRNA–mRNA interactions and validate impacts on protein expression [17].

miRNA-based approaches are increasingly used as precision oncology, involving the treatment of patients based on their molecular characteristics. miRNA expression profiles may eventually help identify which patients would best respond to miRNA-based therapies or other treatment regimens. Changes in miRNA expression over the course of treatment may also indicate resistance mechanisms that could be exploited to adjust treatment [24].

Limitations of Current miRNA Therapeutics

Current miRNA-based therapeutic approaches face several critical limitations that hinder their clinical translation and effectiveness in lung cancer treatment:

- **Delivery Efficiency**: Achieving targeted delivery of miRNA therapeutics to specific lung tissues remains challenging, with current delivery systems showing limited tissue specificity and potential off-target accumulation in healthy organs.
- **Stability Issues**: Naked miRNAs are rapidly degraded by nucleases in biological fluids, requiring complex formulation strategies that often compromise bioavailability and increase manufacturing costs significantly.

- **Dosing Optimization**: Establishing optimal therapeutic doses is complicated by variable pharmacokinetic profiles, individual patient differences, and the narrow therapeutic window between efficacy and toxicity.
- **Immunogenicity Concerns**: Foreign miRNA sequences and delivery vectors can trigger unwanted immune responses, potentially reducing therapeutic efficacy and causing adverse reactions in treated patients.
- **Manufacturing Scalability**: Large-scale production of high-quality miRNA therapeutics remains technically challenging and expensive, limiting their potential for widespread clinical application and accessibility.
- **Resistance Mechanisms**: Cancer cells can develop resistance to miRNA-based therapies through mutations in target sites, altered miRNA processing machinery, or activation of compensatory pathways.
- **Regulatory Complexity**: The regulatory pathway for miRNA therapeutics approval remains unclear, with limited precedents and evolving guidelines that create uncertainty for clinical development programs.

Conclusions

MicroRNAs represent a novel form of posttranscriptional gene regulation with a key role in the pathophysiology of cancer. In lung cancer, miRNAs emerge as essential regulators of the pathogenesis of disease, from the initiation and progression of the tumor, to metastasis and response to therapy. Their dysregulation is involved in the complicated molecular pathogenesis of SLE and gives potential for diagnostic, prognostic, and therapeutic applications.

Thus, miRNAs can serve as valuable diagnostic markers and may well be useful for early lung cancer diagnosis, especially in a noninvasive manner with circulating miRNAs in different biofluids. They are appealing candidates for early detection approaches that could lead to greatly improved survival rates due to their stability in circulation, tissue-specific expression patterns, and sensitive detection methods.

Several studies have shown the prognostic and predictive role of miRNA with specific signatures or single miRNA related with disease outcomes and treatment responses. These biomarkers may help inform treatment decisions, determining patients who would benefit from targeted measures or more aggressive treatment.

Therapeutically, the capacity of miRNAs to coordinately play roles in different cancer-related pathways provides the basis for formulating strategies with broad anticancer actions. miRNA-based therapies are a promising strategy to target the complex networks of molecular alterations in lung cancer, either by restoring tumor suppressor miRNAs or blocking oncogenic miRNAs.

While there are several advances, there are still challenges in translating miRNA research into clinical application, including technical variability, biological complexity, tumor heterogeneity, delivery challenges, and potential off-target effects. The solution to these challenges involves an integrated approach across disciplines from basic sciences, clinical research to technology development.

Emerging technologies and concepts, including CRISPR-Cas9 genome editing, AI, liquid biopsies, multi-omics integration, and precision oncology will continue to better define the role of miRNAs in lung cancer and enable the formulation of improved strategies for early detection, prognosis, and treatment in the clinic.

MicroRNAs are an interesting and promising field in lung cancer, and researchers are exploring them as a mechanism to create new clinical applications that can improve patients. It is a worthy and attainable goal that miRNA discoveries can be translated into cost-effective benefits for lung cancer patients as research continues to evolve.

References

1. Bartel DP (2018) Metazoan MicroRNAs. Cell 173(1):20–51. https://doi.org/10.1016/j.cell.2018.03.006
2. Beg MS, Brenner AJ, Sachdev J, Borad M, Kang YK, Stoudemire J, Smith S, Bader AG, Kim S, Hong DS (2017) Phase I study of MRX34, a liposomal miR-34a mimic, administered twice weekly in patients with advanced solid tumors. Investig New Drugs 35(2):180–188. https://doi.org/10.1007/s10637-016-0407-y
3. Boldin MP, Taganov KD, Rao DS, Yang L, Zhao JL, Kalwani M, Garcia-Flores Y, Luong M, Devrekanli A, Xu J, Sun G, Tay J, Linsley PS, Baltimore D (2011) miR-146a is a significant brake on autoimmunity, myeloproliferation, and cancer in mice. J Exp Med 208(6):1189–1201. https://doi.org/10.1084/jem.20101823
4. Calin GA, Croce CM (2006) MicroRNA-cancer connection: the beginning of a new tale. Cancer Res 66(15):7390–7394. https://doi.org/10.1158/0008-5472.CAN-06-0800
5. Chen X, Ba Y, Ma L, Cai X, Yin Y, Wang K, Guo J, Zhang Y, Chen J, Guo X, Li Q, Li X, Wang W, Zhang Y, Wang J, Jiang X, Xiang Y, Xu C, Zheng P, Zhang J, Zhang CY (2008) Characterization of microRNAs in serum: a novel class of biomarkers for diagnosis of cancer and other diseases. Cell Res 18(10):997–1006. https://doi.org/10.1038/cr.2008.282
6. Chen L, Gibbons DL, Goswami S, Cortez MA, Ahn YH, Byers LA, Zhang X, Yi X, Dwyer D, Lin W, Diao L, Wang J, Roybal J, Patel M, Ungewiss C, Peng D, Antonia S, Mediavilla-Varela M, Robertson G, Suraokar M, Qin FX (2014) Metastasis is regulated via microRNA-200/ZEB1 axis control of tumour cell PD-L1 expression and intratumoral immunosuppression. Nat Commun 5:5241. https://doi.org/10.1038/ncomms6241
7. Cortez MA, Ivan C, Valdecanas D, Wang X, Peltier HJ, Ye Y, Araujo L, Carbone DP, Shilo K, Giri DK, Kelnar K, Martin D, Komaki R, Gomez DR, Krishnan S, Calin GA, Bader AG, Welsh JW (2019) PDL1 regulation by p53 via miR-34. J Natl Cancer Inst 111(1):38–44. https://www.researchgate.net/publication/341288362_PD-L1_an_Important_Immune_Checkpoint_Regulator_Is_Suppressed_by_miR-34a_in_Head_and_Neck_Squamous_Cell_Carcinoma/fulltext/5eb9443d299bf1287f7cc301/PD-L1-an-Important-Immune-Checkpoint-Regulator-Is-Suppressed-by-miR-34a-in-Head-and-Neck-Squamous-Cell-Carcinoma.pdf?origin=scientificContributions
8. Cushing L, Kuang PP, Qian J, Shao F, Wu J, Little F, Thannickal VJ, Cardoso WV, Lü J (2011) miR-29 is a major regulator of genes associated with pulmonary fibrosis. Am J Respir Cell Mol Biol 45(2):287–294. https://doi.org/10.1165/rcmb.2010-0323OC
9. Du L, Schageman JJ, Irnov G, L., Hammond, S. M., Minna, J. D., Gazdar, A. F., & Pertsemlidis, A. (2010) MicroRNA expression distinguishes SCLC from NSCLC lung tumor cells and suggests a possible pathological relationship between SCLCs and NSCLCs. J Exp Clin Cancer Res 29(1):75. https://doi.org/10.1186/1756-9966-29-75

10. Fabbri M, Paone A, Calore F, Galli R, Gaudio E, Santhanam R, Lovat F, Fadda P, Mao C, Nuovo GJ, Zanesi N, Crawford M, Ozer GH, Wernicke D, Alder H, Caligiuri MA, Nana-Sinkam P, Perrotti D, Croce CM (2012) MicroRNAs bind to toll-like receptors to induce pro-metastatic inflammatory response. Proc Natl Acad Sci USA 109(31):E2110–E2116. https://doi.org/10.1073/pnas.1209414109
11. Friedman RC, Farh KK, Burge CB, Bartel DP (2009) Most mammalian mRNAs are conserved targets of microRNAs. Genome Res 19(1):92–105. https://doi.org/10.1101/gr.082701.108
12. Fujita Y, Krause G, Scheffner M, Zechner D, Leddy HE, Behrens J, Sommer T, Birchmeier W (2002) Hakai, a c-Cbl-like protein, ubiquitinates and induces endocytosis of the E-cadherin complex. Nat Cell Biol 4(3):222–231. https://doi.org/10.1038/ncb758
13. Garofalo M, Di Leva G, Romano G, Nuovo G, Suh SS, Ngankeu A, Taccioli C, Pichiorri F, Alder H, Secchiero P, Gasparini P, Gonelli A, Costinean S, Acunzo M, Condorelli G, Croce CM (2009) RETRACTED: miR-221&222 regulate TRAIL resistance and enhance tumorigenicity through PTEN and TIMP3 downregulation. Cancer Cell 16(6):498–509. https://doi.org/10.1016/j.ccr.2009.10.014. (Retraction published Cancer Cell. 2022 Nov 14;40(11):1440. doi: 10.1016/j.ccell.2022.10.005)
14. Halvorsen AR, Silwal-Pandit L, Meza-Zepeda LA, Vodak D, Vu P, Sagerup C, Hovig E, Myklebost O, Børresen-Dale AL, Brustugun OT, Helland Å (2016) TP53 mutation spectrum in smokers and never smoking lung cancer patients. Front Genet 7:85. https://doi.org/10.3389/fgene.2016.00085
15. Hayes J, Peruzzi PP, Lawler S (2014) MicroRNAs in cancer: biomarkers, functions and therapy. Trends Mol Med 20(8):460–469. https://doi.org/10.1016/j.molmed.2014.06.005
16. Hu JL, Wang W, Lan XL, Zeng ZC, Liang YS, Yan YR, Song FY, Wang FF, Zhu XH, Liao WJ, Liao WT, Ding YQ, Liang L (2019) CAFs secreted exosomes promote metastasis and chemotherapy resistance by enhancing cell stemness and epithelial-mesenchymal transition in colorectal cancer. Mol Cancer 18(1):91. https://doi.org/10.1186/s12943-019-1019-x
17. Huang W, Chen Z, Zhang L, Tian D, Wang D, Fan D, Wu K, Xia L (2015) Interleukin-8 induces expression of FOXC1 to promote transactivation of CXCR1 and CCL2 in hepatocellular carcinoma cell lines and formation of metastases in mice. Gastroenterology 149(4):1053–67.e14. https://doi.org/10.1053/j.gastro.2015.05.058
18. Jin X, Chen Y, Chen H, Fei S, Chen D, Cai X, Liu L, Lin B, Su H, Zhao L, Su M, Pan H, Shen L, Xie D, Xie C (2017) Evaluation of tumor-derived exosomal miRNA as potential diagnostic biomarkers for early-stage non-small cell lung cancer using next-generation sequencing. Clin Cancer Res 23(17):5311–5319. https://doi.org/10.1158/1078-0432.CCR-17-0577
19. Johnson SM, Grosshans H, Shingara J, Byrom M, Jarvis R, Cheng A, Labourier E, Reinert KL, Brown D, Slack FJ (2005) RAS is regulated by the let-7 microRNA family. Cell 120(5):635–647. https://doi.org/10.1016/j.cell.2005.01.014
20. Kasinski AL, Slack FJ (2012) miRNA-34 prevents cancer initiation and progression in a therapeutically resistant K-ras and p53-induced mouse model of lung adenocarcinoma. Cancer Res 72(21):5576–5587. https://doi.org/10.1158/0008-5472.CAN-12-2001
21. Kim J, Yao F, Xiao Z, Sun Y, Ma L (2018) MicroRNAs and metastasis: small RNAs play big roles. Cancer Metastasis Rev 37(1):5–15. https://doi.org/10.1007/s10555-017-9712-y
22. Li L, Yuan L, Luo J, Gao J, Guo J, Xie X (2013) MiR-34a inhibits proliferation and migration of breast cancer through down-regulation of Bcl-2 and SIRT1. Clin Exp Med 13(2):109–117. https://doi.org/10.1007/s10238-012-0186-5
23. Lin PY, Yu SL, Yang PC (2010) MicroRNA in lung cancer. Br J Cancer 103(8):1144–1148. https://doi.org/10.1038/sj.bjc.6605901
24. Liu X, Sempere LF, Ouyang H, Memoli VA, Andrew AS, Luo Y, Demidenko E, Korc M, Shi W, Preis M, Dragnev KH, Li H, Direnzo J, Bak M, Freemantle SJ, Kauppinen S, Dmitrovsky E (2010) MicroRNA-31 functions as an oncogenic microRNA in mouse and human lung cancer cells by repressing specific tumor suppressors. J Clin Invest 120(4):1298–1309. https://doi.org/10.1172/JCI39566

25. Liu S, Song L, Zhang L, Zeng S, Gao F (2015) miR-21 modulates resistance of HR-HPV positive cervical cancer cells to radiation through targeting LATS1. Biochem Biophys Res Commun 459(4):679–685. https://doi.org/10.1016/j.bbrc.2015.03.004
26. Ma L, Reinhardt F, Pan E, Soutschek J, Bhat B, Marcusson EG, Teruya-Feldstein J, Bell GW, Weinberg RA (2010) Therapeutic silencing of miR-10b inhibits metastasis in a mouse mammary tumor model. Nat Biotechnol 28(4):341–347. https://doi.org/10.1038/nbt.1618
27. Monroig PDC, Chen L, Zhang S, Calin GA (2015) Small molecule compounds targeting miRNAs for cancer therapy. Adv Drug Deliv Rev 81:104–116. https://doi.org/10.1016/j.addr.2014.09.002
28. Nadal E, Truini A, Nakata A, Lin J, Reddy RM, Chang AC, Ramnath N, Gotoh N, Beer DG, Chen G (2015) A novel Serum 4-microRNA signature for lung cancer detection. Sci Rep 5:12464. https://doi.org/10.1038/srep12464
29. O'Brien J, Hayder H, Zayed Y, Peng C (2018) Overview of MicroRNA biogenesis, mechanisms of actions, and circulation. Front Endocrinol 9:402. https://doi.org/10.3389/fendo.2018.00402
30. Oliveto S, Mancino M, Manfrini N, Biffo S (2017) Role of microRNAs in translation regulation and cancer. World J Biol Chem 8(1):45–56. https://doi.org/10.4331/wjbc.v8.i1.45
31. Pan X, Wang ZX, Wang R (2010) MicroRNA-21: a novel therapeutic target in human cancer. Cancer Biol Ther 10(12):1224–1232. https://doi.org/10.4161/cbt.10.12.14252
32. Rupaimoole R, Slack FJ (2017) MicroRNA therapeutics: towards a new era for the management of cancer and other diseases. Nat Rev Drug Discov 16(3):203–222. https://doi.org/10.1038/nrd.2016.246
33. Ryan BM, Robles AI, Harris CC (2010) Genetic variation in microRNA networks: the implications for cancer research. Nat Rev Cancer 10(6):389–402. https://doi.org/10.1038/nrc2867
34. Slabáková E, Culig Z, Remšík J, Souček K (2017) Alternative mechanisms of miR-34a regulation in cancer. Cell Death Dis 8(10):e3100. https://doi.org/10.1038/cddis.2017.495
35. Takamizawa J, Konishi H, Yanagisawa K, Tomida S, Osada H, Endoh H, Harano T, Yatabe Y, Nagino M, Nimura Y, Mitsudomi T, Takahashi T (2004) Reduced expression of the let-7 microRNAs in human lung cancers in association with shortened postoperative survival. Cancer Res 64(11):3753–3756. https://doi.org/10.1158/0008-5472.CAN-04-0637
36. Trang P, Wiggins JF, Daige CL, Cho C, Omotola M, Brown D, Weidhaas JB, Bader AG, Slack FJ (2011) Systemic delivery of tumor suppressor microRNA mimics using a neutral lipid emulsion inhibits lung tumors in mice. Mol Ther 19(6):1116–1122. https://doi.org/10.1038/mt.2011.48
37. van Zandwijk N, Pavlakis N, Kao SC, Linton A, Boyer MJ, Clarke S, Huynh Y, Chrzanowska A, Fulham MJ, Bailey DL, Cooper WA, Kritharides L, Ridley L, Pattison ST, MacDiarmid J, Brahmbhatt H, Reid G (2017) Safety and activity of microRNA-loaded minicells in patients with recurrent malignant pleural mesothelioma: a first-in-man, phase 1, open-label, dose-escalation study. Lancet Oncol 18(10):1386–1396. https://doi.org/10.1016/S1470-2045(17)30621-6
38. Wang J, Chen J, Sen S (2016) MicroRNA as biomarkers and diagnostics. J Cell Physiol 231(1):25–30. https://doi.org/10.1002/jcp.25056
39. Wozniak MB, Scelo G, Muller DC, Mukeria A, Zaridze D, Brennan P (2015) Circulating MicroRNAs as non-invasive biomarkers for early detection of non-small-cell lung cancer. PLoS One 10(5):e0125026. https://doi.org/10.1371/journal.pone.0125026
40. Wu K, Zou J, Lin C, Jie ZG (2019) MicroRNA-140-5p inhibits cell proliferation, migration and promotes cell apoptosis in gastric cancer through the negative regulation of THY1-mediated Notch signaling. Biosci Rep 39(7):BSR20181434. https://doi.org/10.1042/BSR20181434. Retraction published Biosci Rep. 2023 Jan 31;43(1):BSR-2018-1434_RET. doi: 10.1042/BSR-2018-1434_RET
41. Xing L, Todd NW, Yu L, Fang H, Jiang F (2010) Early detection of squamous cell lung cancer in sputum by a panel of microRNA markers. Mod Pathol 23(8):1157–1164. https://doi.org/10.1038/modpathol.2010.111

42. Xu Q, Liu LZ, Qian X, Chen Q, Jiang Y, Li D, Lai L, Jiang BH (2012) MiR-145 directly targets p70S6K1 in cancer cells to inhibit tumor growth and angiogenesis. Nucleic Acids Res 40(2):761–774. https://doi.org/10.1093/nar/gkr730
43. Yang M, Shen H, Qiu C, Ni Y, Wang L, Dong W, Liao Y, Du J (2013) High expression of miR-21 and miR-155 predicts recurrence and unfavourable survival in non-small cell lung cancer. Eur J Cancer 49(3):604–615. https://doi.org/10.1016/j.ejca.2012.09.031
44. Zhang J, Li S, Li L, Li M, Guo C, Yao J, Mi S (2015) Exosome and exosomal microRNA: trafficking, sorting, and function. Genomics Proteomics Bioinformatics 13(1):17–24. https://doi.org/10.1016/j.gpb.2015.02.001
45. Zhao X, Wang Y, Deng R, Zhang H, Dou J, Yuan H, Hou G, Du Y, Chen Q, Yu J (2016) miR186 suppresses prostate cancer progression by targeting Twist1. Oncotarget 7(22):33136–33151. https://doi.org/10.18632/oncotarget.8887
46. Zhou W, Fong MY, Min Y, Somlo G, Liu L, Palomares MR, Yu Y, Chow A, O'Connor ST, Chin AR, Yen Y, Wang Y, Marcusson EG, Chu P, Wu J, Wu X, Li AX, Li Z, Gao H, Ren X, Wang SE (2014) Cancer-secreted miR-105 destroys vascular endothelial barriers to promote metastasis. Cancer Cell 25(4):501–515. https://doi.org/10.1016/j.ccr.2014.03.007

Chapter 13
MicroRNAs and Their Roles in Epigenetic Regulation of Tumour Microenvironment in Prostate Cancer

Saikanth Varma and Sanjay Basak

Abstract Prostate cancer (PCa) is the second most common malignancy in men worldwide. Despite advances in treatment, PCa often progresses to castration-resistant and chemo-resistant stages, underscoring the need for a deeper understanding of the tumour microenvironment (TME) and its regulatory mechanisms. The dynamic TME plays a pivotal role in PCa initiation, progression, and therapeutic response. In recent years, microRNAs (miRNAs) have emerged as critical regulators of gene expression within the TME, influencing cellular processes such as proliferation, angiogenesis, immune modulation, and metastasis. These small non-coding RNAs act either as oncogenic miRNAs (oncomiRs) promoting tumour growth or tumour suppressor miRNAs inhibiting tumour progression. This chapter explores the complex role of miRNAs in modulating various components of the PCa microenvironment, including cancer-associated fibroblasts (CAFs), tumour-associated macrophages (TAM), immune cells, extracellular matrix (ECM), angiogenesis, and hypoxia responses. Additionally, we discuss the epigenetic regulation of miRNA expression through DNA methylation and histone modifications, which further contribute to PCa pathogenesis. Finally, we evaluate the therapeutic potential of miRNA-based interventions, such as diagnostic biomarkers, mimics, and inhibitors, and outline current challenges in miRNA delivery, specificity, and clinical translation.

Keywords Prostate cancer (PCa) · Tumour microenvironment (TME) · MicroRNAs (miRNAs) · oncomiRs · Epigenetic regulation

S. Varma · S. Basak (✉)
Molecular Biology Division, National Institute of Nutrition, Indian Council of Medical Research, Hyderabad, India
e-mail: basak.sanjay@icmr.gov.in

S. Pathak, A. Banerjee (eds.), *MicroRNAs: Role in Cancer Progression, Diagnostics and Therapy*, Advances in Biochemistry in Health and Disease 34,
https://doi.org/10.1007/978-3-032-08824-6_13

Abbreviations

α-SMA	α-Smooth muscle actin
AGO	Argonaute
AR	Androgen receptor
BFSS	Biochemical failure-free survival
BHP	Benign prostate hyperplasia
CAF	Cancer-associated fibroblasts
CEMIP	Cell migration-inducing protein
CFSS	Clinical failure-free survival
CRPC	Castration resistant prostate cancer
ECM	Extracellular matrix
ECM-CAFs	Extracellular-associated CAFs
EGFR	Epidermal growth factor receptor
EMT	Epithelial mesenchymal transition
FGF	Fibroblast growth factor
FIH-1	Factor inhibiting HIF-1
FSP	Fibroblast specific protein
GDF15	Growth differentiation factor 15
GEM	Genetically engineered mouse
HIFs	Hypoxia inducible factors
HSF1	Heat shock transcription factor 1
IL-6	Interleukin 6
LRP1B	Low-density lipoprotein receptor-related 1B
m^7G	7-methylguanosine
MDSCs	Myeloid-derived suppressor cells
MIF	Migratory inhibitory factor
miRISC	miRNA-induced silencing complex
miRNA	MicroRNA
MMPs	Matrix metalloproteinases
MREs	MiRNA response elements
mRNA	Messenger RNA
NCAM	Neutral cell adhesion molecule
NF-Exo	Normal fibroblast exosomes
PCa	Prostate cancer
PDGF	Platelet derived growth factor
PHD	Prolyl-hydroxylase domain
PIN	Prostate intraepithelial neoplasia
Pre-miRNAs	Precursor miRNAs
Pri-miRNAs	Primary miRNAs
PSA	Prostate specific antigen
RHOB	Ras homolog family member B
ROS	Reactive oxygen species
SOX4	SRY-Box transcription factor 4.

TAMs	Tumour-associated macrophages
TGF-β	Transforming growth factor β
TME	Tumour microenvironment
Tregs	Regulatory T cells
TSP-2	Thrombospondin 2
UGSM	Urogenital sinus mesenchyme
UTR	Untranslated region
VEGF	Vascular endothelial growth factor

Introduction

Nature, Origin, and Models of Prostate Cancer

The prostate, a small, walnut-sized gland found in men, is a crucial part of the male reproductive system. Situated just below the bladder and surrounding the upper part of the urethra, it produces seminal fluid, a key component for sperm nourishment and transport. The adult prostate consists of three main types of epithelial cells: luminal, basal, and neuroendocrine. Luminal cells, the most abundant, are responsible for the gland's secretory function. They form a single layer around the lumen and express specific markers such as androgen receptor (AR), prostate-specific antigen (PSA), cytokeratin CK8/18, and the homeobox gene Nkx3.1 [1]. Basal cells, with their cuboidal structure, are located between the luminal cells and the basement membrane. They express markers like cytokeratin 5 and 14, as well as p63 [2], and play a key role in safeguarding luminal cells from oncogenic transformation [3]. Both luminal and basal cells originate from urogenital sinus epithelial cells, which are derived from the endoderm [4]. Neuroendocrine cells, though less abundant, are distinguished by their expression of markers such as synaptophysin, chromogranin A, and somatostatin.

Prostate cancer (PCa) is the second most common malignancy in men, a disease that significantly impacts the lives of patients and their families. As per the American Cancer Society, the estimates for PCa in the United States are about 313,780 new cases and 35,770 deaths from PCa in 2025 [5]. Moreover, the prevalence of PCa was higher in developed countries [6]. Patients suffering from PCa are usually diagnosed with localized or advanced disease. During the progression of PCa, a condition known as prostatic intraepithelial neoplasia (PIN) emerges, which is characterized by a reduction in basal cell numbers. In contrast, prostate adenocarcinoma predominantly exhibits a luminal phenotype, with basal cells notably absent.

The "cell of origin" for cancer refers to the specific cell type that undergoes a transformation and is responsible for tumour development. Since PCa is largely composed of luminal-like cells, luminal cells were initially believed to be the main cells of origin. The absence of basal cell markers such as cytokeratin and p63 is a diagnostic feature in prostate needle biopsies to confirm adenocarcinoma [7]. Despite anti-hormonal therapy, PCa often recurs and progresses into a castration-resistant state. However, a new perspective is emerging: unlike luminal cells, which rely on androgens, basal cells are androgen-independent and are now considered a potential cellular origin for PCa.

Tissue recombination techniques have been employed to explore the origins of PCa. Research demonstrated that basal cells derived from benign human prostate tissue could initiate PCa when implanted into immunodeficient mice. In a study, primary luminal or basal prostate cells were genetically modified using a lentiviral vector carrying activated/myristoylated Akt, ERG, and a fluorescent marker [8]. These modified cells were combined with murine urogenital sinus mesenchyme (UGSM) cells in Matrigel and injected subcutaneously into mice. Only basal cells generated tumour outgrowths resembling human PCa in histological and molecular characteristics [8]. Further studies using oncogenes like *Myc* and myristoylated *Akt* also confirmed that tumours originating from basal cells exhibited features of both adenocarcinoma and squamous cell carcinoma [9].

Xenograft models have been widely used for in vivo research. In this approach, human PCa cells are implanted into immunodeficient mice. For instance, the LNCaP cell line has generated several genetically related variants with different metastatic behaviours [10]. However, it is important to note that xenograft models have limitations, including a mismatched microenvironment and a functional immune system, which should be considered when interpreting results.

In recent years, genetically engineered mouse (GEM) models have emerged as a comprehensive tool for studying PCa initiation and progression. These models provide a unique opportunity to delve into the molecular and cellular mechanisms driving tumour development within the natural tissue environment while maintaining an intact immune system. GEM models can be created by overexpressing viral oncogenes or introducing gene knockouts, such as *PTEN* and *Nkx3*, to study their role in PCa [11]. This approach offers a more comprehensive and physiologically relevant platform for investigating new therapeutic strategies, providing a solid foundation for future PCa research.

Tumour Microenvironment in Prostate Cancer

The tumour microenvironment (TME) is a critical factor in cancer progression, a complex ecosystem that surrounds a tumour within the body. It comprises immune and non-immune cells, crucial in disease progression and tumour development. This environment consists of a network of mesenchymal/stromal cells, fibroblasts, endothelial cells, immune cells, neural crest cells, and fat cells. These components produce signalling molecules, including cytokines, chemokines, extracellular matrices, and matrix-degrading enzymes. As a tumour develops, the microenvironment changes significantly due to the continuous interaction between cancer cells and surrounding cells. These interactions have a profound impact on tumour cell survival, proliferation, resistance to therapy, angiogenesis, immune evasion, and metastasis, underscoring the urgency of understanding and targeting the tumour microenvironment in cancer research.

Non-cancerous cells within the tumour microenvironment, known as the stroma, differ significantly in PCa compared to normal prostate. In PCa, the global gene expression profile of the reactive stroma is altered [12]. A healthy prostate typically contains a stroma composed primarily of smooth muscle cells. However, during cancer progression, these smooth muscle cells are replaced by specialized fibroblasts called cancer-associated fibroblasts (CAF). CAFs contribute to tumour growth by secreting key factors involved in angiogenesis and metastasis, including fibroblast growth factor (FGF), transforming growth factor β (TGF-β), growth differentiation factor 15 (GDF15), platelet-derived growth factor (PDGF), and vascular endothelial growth factor (VEGF) [13, 14]. Additionally, markers such as fibroblast-specific protein (FSP), vimentin, and alpha-smooth muscle actin (α-SMA), which are specific to fibroblasts, are found at higher levels in cancer stroma [15]. The extracellular matrix (ECM) within the prostate is also significantly altered in PCa. Normally, fibroblasts contribute to ECM formation by producing collagen types I and III. However, in the presence of CAFs, this process is disrupted, reducing ECM components. CAFs also secrete matrix metalloproteinases (MMPs), which break down ECM, creating a pathway for cancer cells to spread and metastasize.

The immune system, a key player in the tumour microenvironment, is responsible for protecting the prostate from infections and abnormal cell growth. However, the development of PCa is associated with increased infiltration of immune cells, including lymphocytes, macrophages, and mast cells, into the stroma [16]. These infiltrating immune cells release factors that can either support or inhibit tumour growth and CAF activity [17]. While PCa is generally considered an "immunologically cold" tumour, meaning it has a low presence of tumour-fighting T cells, some cases exhibit immune infiltration. Cold tumours are characterized by reduced T cell activity, impaired dendritic cell function, and the presence of immunosuppressive cells such as regulatory T cells (Tregs), tumour-associated macrophages (TAMs), and myeloid-derived suppressor cells (MDSCs), [18]. The role of immune cells in the prostate tumour microenvironment is a crucial area of research, as it is not conducive to the survival and function of tumour-infiltrating immune cells, making it difficult for the body to mount an effective immune response against the cancer.

Introduction to MicroRNAs (miRNAs) and Their Regulatory Functions

miRNA Biogenesis

The first microRNA (miRNA) was discovered in 1993 in *Caenorhabditis elegans* by Ambros and Ruvkun groups [19]. The discovery of miRNA has revolutionized the field of molecular biology. miRNAs play an important role in various biological processes and development. They belong to a family of small non-coding RNAs with an average length of 22 nucleotides. Most miRNAs are derived from non-coding transcript-intronic regions and a few from protein-coding regions-exons. Moreover, the miRNAs are sometimes transcribed as long transcripts called clusters with similar seed regions [20]. The biogenesis of miRNA starts with processing RNA polymerase II/III transcripts. miRNA biogenesis can be classified into canonical and non-canonical pathways.

Canonical Pathway

Most of the miRNAs are processed from the canonical pathways, the dominant miRNA biogenesis pathway. Primary miRNAs (pri-miRNAs) are transcribed from their genes and are processed into precursor miRNAs (pre-miRNAs) in the nucleus by the microprocessor complex. The microprocessor complex consists of DGCR8 and Drosha, a ribonuclease III enzyme [21]. The DGCR8 recognizes N6 methylated GGAC region and other motifs in the pri-miRNA [22]. Meanwhile, the enzyme Drosha cleaves the pri-miRNA at the base of the hairpin structure of pri-miRNA generating a two nucleotide 3′ overhang pre-miRNA. The pre-miRNA is then exported out of the nucleus into the cytoplasm with the help of the exportin 5/RanGTP complex. Dicer, RNase III endonuclease, processes the pre-miRNA into mature miRNA duplex, a double-stranded RNA, by removing the terminal loop [23]. The mature miRNA is loaded onto the Argonaute multifamily protein to form a miRNA-induced silencing complex (miRISC). The strand selection depends on the thermodynamic stability at the 5′ end of the miRNA duplex or a 5′ U at the first nucleotide position. The strand loaded onto the AGO proteins has either low 5′ stability or 5′ uracil and is considered a guide strand. The other unloaded strand is called the passenger strand and undergoes degradation by cellular machinery. The names 5p and 3p of the miRNA arise from the directionality of the miRNA strand, where 5p and 3p strands originate from the 5′ and 3′ ends of the pre-miRNA hairpin, respectively [24].

Non-canonical Pathway

The non-canonical miRNA biogenesis pathway, a complex system that diverges from the conventional canonical miRNA processing pathway involves alternative mechanisms for generating functional miRNAs. Unlike the canonical pathway, which relies on Drosha, Dicer, exportin 5, and AGO2 for sequential cleavage and miRNA processing, the non-canonical pathway bypasses one or both of these steps through various processing routes. The non-canonical miRNA biogenesis can be categorized into Drosha/DGCR8-independent and Dicer-independent pathways. In Drosha-independent pathways, one such miRNA is mirtrons, which originate from intronic sequences and are processed independent of Drosha. Instead of Drosha-mediated cleavage, mirtrons are spliced out by the spliceosome and subsequently undergo lariat debranching before entering the miRNA maturation pathway. Once processed, they are exported from the nucleus by Exportin-5 and cleaved by Dicer in the cytoplasm to produce mature miRNAs [25]. The other such miRNA is 7-methylguanosine (m^7G)-capped pre-miRNA. In this, the RNA is exported directly into the cytoplasm with the help of exportin 1 without any processing by the Drosha enzyme. The presence of m^7G cap prevents the binding and loading of 5p strand onto the Argonaute proteins and promotes 3p strand formation [26]. In the case of Dicer-independent pathways, miRNA is processed by Drosha to form endogenous short hairpin RNA transcripts. For maturation, the miRNAs require AGO2 within the cytoplasm [27].

Function of miRNAs

miRNAs, the key players in gene regulation, exert their influence through post-transcriptional silencing, primarily mediated by the miRISC [28]. miRISC, a multi-protein complex consisting of AGO protein and guide strand, plays a critical role in targeting specific messenger RNAs (mRNAs) for degradation or translational repression, based on the degree of complementarity between the miRNA and its target sequence on mRNA, called miRNA response elements (MREs).

Among the four different AGO proteins in humans (AGO1–4), AGO2 possesses endonuclease activity, allowing it to cleave target mRNAs when fully complementary miRNA and MRE base pairing occurs directly [29]. This type of interaction causes disturbances in the AGO and the guide strand (3′ end) complex, leading to the degradation of the miRISC [30]. However, most miRNA-mRNA interactions in animal cells involve partial complementarity, leading to translational repression rather than direct cleavage [31]. Once loaded into miRISC, the miRNA guides the complex to its target mRNA by recognizing complementary sequences, primarily in the 3′ untranslated region (3′ UTR), which are not fully complementary, preventing the mRNA cleavage by the AGO2. The binding of miRISC to the target mRNA recruits additional factors that inhibit translation initiation, promote ribosome drop-off, or accelerate mRNA deadenylation and decay. Deadenylation is facilitated by CCR4-NOT and PAN2-PAN3 complexes, which shorten the poly(A) tail, marking the mRNA for degradation by exonucleases like exoribonuclease 1 (XRN1) [31, 32]. Moreover, removing the 5′ cap of the mRNA by the decapping enzyme DCP2 promotes rapid mRNA degradation by XRN1 in the 5′ to 3′ direction [32].

Overall, miRISC-mediated gene silencing is a highly dynamic and finely tuned process that plays important roles in maintaining cellular homeostasis, regulating developmental pathways, and influencing disease progression. The ability of miRNAs to regulate gene expression by targeting multiple genes highlights their significance in all biological processes.

MicroRNA in Prostate Cancer: An Overview

miRNAs, the master regulators of gene expression at the post-transcriptional level, have the potential to be a beacon of hope in cancer prevention. They bind to complementary sequences in the target mRNA, initiating either mRNA degradation or cleavage and decay. Alterations in the expression of several miRNAs are linked with cancer development, but miRNAs can also play a dual role in promoting tumour development or cancer prevention. Some miRNAs, known as oncomiRs, promote tumour development and formation, while others, known as tumour suppressor miRNAs, inhibit tumour initiation. The potential of these miRNAs in cancer prevention is a promising area of research. The regulation of gene expression by miRNAs is complex, with each miRNA capable of binding and post-transcriptionally modulating the expression of several mRNAs across different pathways. Moreover, a single mRNA is targeted by multiple miRNAs to regulate its expression [33, 34]. Any dysregulations in the miRNA levels in cancer can occur through epigenetic modifications such as genetic alterations and CpG island hypermethylation in the promoter region [35].

Classification of miRNAs in Cancer: oncomiRs vs Tumour Suppressor miRNAs

oncomiRs

The aberrant expression of miRNAs, small non-coding RNA molecules, is reported to be causatively linked with the pathogenesis of several cancers. oncomiRs, or oncogenic miRNAs, are a subset of miRNAs that are reported to promote tumour development by downregulating the expression of tumour suppressor genes, which are genes that normally inhibit cell growth and division. Various miRNAs have been reported to act as oncomiRs in PCa. Some of the most common miRNAs overexpressed in the PCa are miR-21, miR-32, miR-221, miR-222, miR-18a, and miR-429.

Table 13.1 Oncogenic miRNAs (oncomiRs), their targets, and functions in prostate cancer

miRNA	Model	Target genes	Functions	References
miR-429	IF11 and IA8 PCa cells	*P27Kip1*	Cell proliferation	[36]
miR-483-5p	LNCaP, DU-145, and PC-3 cells	*RBM5*	Cell proliferation and invasion	[37]
miR-155	PCa tissue and cells	*ANX7*	Cell proliferation	[38]
miR-153	PCa cells	*PTEN* and *FOXO1*	Cell proliferation	[39]
miR-186-5p	PCa and metastatic PCa tissues and PC-3, MDA-PCa-2b, and LNCaP	*AKAP12*	Cell proliferation and invasion	[40]
miR-345-5p	PCa tissue specimens of metastatic CRPC, hormone naïve PCa	*CDKN1A*	Cell growth and migration	[41]
miR-193a-3p	LNCaP cells and LNCaP-derived CRPC C4-2B cells	*AJUBA*	Cell migration	[42]
miR-145	PC-3 cells	*P53*	Migration, invasion, EMT, and stemness	[43]
miR-129-5p	PC-3 cells	*CAMK2N1*	Chemoresistance to docetaxel	[44]
miR-21	AR-positive PCa cells	–	Cell proliferation, tumour growth, and cancer recurrence	[45, 46]
miR-96	PCa patients and PCa cells	*FOXO1*	Growth and cellular proliferation	[47]
miR-133b	PC-3 and LNCaP cells	*RB1CC1*	Cell cycle progression and proliferation	[48]
miR-106b	DU145 and LNCaP cells	*CASP7*	Cancer recurrence and disease-specific survival	[49]
miR-182-5p	PCa tissues	*ARRDC3*	Cell proliferation, invasion, and migration	[50]
miR-148	LNCaP cells	*CAND1*	Cell proliferation and tumour growth	[51]
miR-133b	Tumour specimens and PC-3 cells	*FAIM*	Proliferation and metabolic activity	[52]
miR-183	PCa cohorts	*PSA*	Cell growth and clinical progression	[53]
miR-221/222	LNCaP cells	*P27Kip1*	Clonogenicity and proliferation	[54]
miR-221/222	LNCaP and PC-3	*SIRT1*	Proliferation and migration	[55]

Oncogenic miRNAs (oncomiRs), their targets, and functions in prostate cancer are presented in Table 13.1.

miRNA can bind with mRNA and modulate the expression of the genes, leading to alterations in the signalling pathways. For instance, in PCa, the loss of miR-125b and the upregulation of miR-22 promote prostate tumourigenesis. Studies in the parental p69 cells and metastatic subline M12 have shown that tumour suppressor miR-125b and oncomiR miR-22 can modulate the PI3K/AKT and MAPK/ERK signalling pathways in the prostate, which are key pathways in the PCa progression contributing to the prostate tumourigenesis [56]. Some miRNAs are known to act as either oncomiR or tumour suppressors, depending on the cancer type. For example, miR-9 has been characterized as either an oncomiR or a tumour suppressor in a variety of solid tumours, such as breast and ovarian carcinomas. Meanwhile, miR-9 in PCa acts as an oncomiR, and their levels were elevated in the prostate tumour tissue compared to the benign glandular epithelium in radical prostate biopsies. Moreover, in the M12 cell line, the inhibition of the miR-9 significantly halted the migration capacity and the invasive potential. Also, it reduced tumour growth and metastasis in the male athymic nude mice. Further analysis showed that miR-9 targets suppressor of cytokine signalling 5 (SOCS5) and e-cadherin mRNA [57]. Similarly, miR-301a is known to be associated with PCa recurrence. miR-301a overexpression in the PCa cell lines PC3 and LNCaP showed enhanced tumour growth in both in vitro and xenograft models. miR-301a acted as an oncomiR by targeting tumour suppressor p63 and inhibiting their expression, leading to the loss of e-cadherins and inducing epithelial to mesenchymal transition (EMT) [58]. miRNA microarray analysis performed on the radical prostatectomy with high-grade prostate intraepithelial neoplasia and PCa showed that miR-155, miR-210, miR-153, and miR-200c were downregulated in both compared to non-neoplastic prostate tissue. The putative targets of these miRNAs were STAT3, ZEB1, and BACH1 genes. In the non-cancer cell line of the prostate (RWPE-1), treatment with miR-155 antagomir elevated the mRNA expression of the STAT3 gene. In contrast, the STAT3 mRNA expression decreased upon treatment with miR-155 mimetics in PC-3 cells [59]. The metastatic potential of the PCa can cause secondary malignancies and is the primary cause of mortality. In metastatic PCa, the miR-195 was poorly expressed, and their low expression levels exhibited shorter relapse-free survival times. Overexpression of miR-195 inhibited cell migration, invasion and EMT. FGF2 was identified as a target of miR-195, and the knockdown of FGF2 suppressed EMT in PCa cell lines [60]. Similarly, miR-573 expression is significantly lower in metastatic cancer than in primary PCa. miR-573 inhibits cell migration, invasion, and TGF-β-induced EMT in vitro and in vivo models of lung metastasis. FGFR1 gene is the direct target of miR-573, where the knockdown of FGFR1 and overexpression of miR-573 reduces EMT and invasion of PCa cells [61]. In prostate tumours, miR-15 and miR-16 were downregulated in CAFs, promoting tumour growth and progression by repressing the transcript levels of FGF-2 and FGFR1 [62]. The miR-100 is downregulated in the human PCa tissue compared to the normal prostate. Decreased expression of miR-100 is much more pronounced in the bone metastatic PCa samples than primary PCa. Research shows that loss of miR-100 promotes invasion, migration, EMT, colony and spheroid formation in PCa cells PC-3 and DU145 cells along with stemness factors like Oct4, Klf4, and c-Myc. The loss of miR-100 is negatively associated with bone metastasis in PCa patients. Luciferase reporter assay shows that Argonaute 2 (AGO2) is the target of miR-100, and its loss promotes metastasis of PCa through upregulation of AGO2 expression [63].

The PCa tumour microenvironment supports tumour progression and metastasis by promoting EMT and cell invasion. miRNAs regulate migration and invasion of the PCa by modulating the expression of genes involved in the EMT. Studies have shown that in cancer-associated fibroblasts, the miR-1247 expression levels are decreased in PCa cells. The downregulation of miR-1274 promotes EMT, the invasiveness of the cancer cell and stemness treatment, thereby enhancing lung metastasis of PCa cells in vivo experiments. Neuropilin-1, a coreceptor in the epidermal growth factor receptor signalling, is a target for miR-1247 [64]. Another tumour suppressor miRNA, miR-145, targets the ZEB2 gene, an EMT activator and promoter of metastasis. Suppression of ZEB2 by overexpressing miR-145 triggers a reduction in the migration, invasion, EMT and stemness of the PCa cell lines. Moreover, the expression of ZEB2 is negatively associated with miR-145 and positively associated with serum total and free prostate-specific antigen (PSA), Gleason score, and bone metastasis in specimens obtained from primary PCa patients [65]. miR-145 overexpression inhibited cell viability and colony formation in PC-3 cells from bone metastasis of PCa and cancer stem cell properties [66]. Additionally, miR-145 is a direct target of the p53 gene, and overexpression of p53 inhibits invasion and migration and suppresses the expression of mesenchymal markers in PC-3 cells by modulating miR-145 [43]. MiR-145 has also been reported to target the HEF1 gene, whose upregulation has been linked to promoting tumour invasion, migration, and EMT. Inhibition of HEF1 expression repressed bone metastasis of the PC-3 cells in vivo [67]. Another tumour promoter miRNA is miR-21, which regulates metastasis and progression of PCa to advanced stages like castration-resistant PCa. Moreover, high levels of miR-21 are linked to poor therapeutic response in PCa patients [68], highlighting its potential as a prognostic marker in PCa. Many other miRNAs, such as miR-2055, miR-92b, miR-3195, and miR-3687, were significantly altered in the castration-resistant PCa in comparison to primary PCa tissues [69].

Tumour Suppressor MiRs

Tumour suppressor miRNAs, their targets, and functions in prostate cancer are consolidated in Table 13.2. In human prostate sublines derived from the parental p69 cell line, vimentin expression was elevated in the metastatic derivative M12 subline. Further analysis showed that miR-17-3p was reduced and positively correlated with vimentin expression and tumourigenicity. Similar results were observed in in vivo male athymic nude mice and clinical human prostatectomy specimens [88]. Studies have shown that alterations in miR-34c in the PCa cell lines promote reduced cell proliferation and apoptosis. Overexpression of miR-34c decreased cell growth by decreasing cellular proliferation rate and increasing apoptosis. The oncogenes involved in cell proliferation (E2F3) and apoptosis (BCL-2) are inversely regulated by miR-34c, suggesting its potential in suppressing tumour [89]. In VCaP and LNCaP cells, overexpression of the same gene cluster miR-132/212 inhibited TGF-beta-induced EMT at both transcript and protein levels. Luciferase reporter assay

Table 13.2 Tumour suppressor miRNAs, their targets, and functions in prostate cancer

miRNA	Model	Target gene	Functions	References
miR-99a-3p	PCa cells	*NCAPG*	Inhibited cell proliferation, migration, and invasion	[70]
miR-221-5p	C4-2 and PC-3 M-Pro4 cells	–	Migration, extravasation, and metastasis	[71]
miR-539	PCa cells	*DLX1*	Proliferation, migration, invasion, and EMT	[72]
miR-100	PCa tissue	*AGO2*	Invasion, migration, stemness, and EMT	[63]
miR-26a-5p	PCa tissues and a panel of PCa cell lines	*SERBP1*	Proliferation and motility	[73]
miR-193a	PC-3, DU145, and xenograft models	*HOTAIR* and *EZH2*	Tumour growth	[74]
miR-1247	PCa cells	*NRP1*	EMT and cell invasion	[64]
miR-573	PCa metastatic tissues and cells (Vcap, PC3, LNCaP, 22RV1, RWPE)	*FGFR1*	Cell migration, invasion, and TGF-β1-induced EMT	[61]
miR-130a	Primary tumours and PCa cell lines	*DEPD1C* and *SEC23B*	Cell viability and invasion	[75]
miR-1 and miR-200	Pten- and TP53-null prostate adenocarcinoma	*SLUG*	Mesenchymal differentiation and EMT	[76]
miR-1	PCa tissue specimens and nude mice	*TWIST1*	Bone metastasis	[77]
miR-125b	PCa cells	*PHF8*	Hyperproliferation of PCa cells	[78]
miR-205	Benign PCa WPE1-NA22 and a highly malignant WPE1-NB26 cell line	*BCL2L2*	Chemotherapeutic agent-induced apoptosis	[79]
miR-31	Benign PCa WPE1-NA22 and a highly malignant WPE1-NB26 cell line	*E2F6*	Chemotherapeutic agent-induced apoptosis	[79]
miR-31	Primary and metastatic PCa tissues, benign prostate epithelial cell line, RWPE-1, and PCA cell lines, LNCaP, PC3, DU145, VCaP, 22Rv1	*AR*	Cellular homeostasis, evolution, and progression	[80]
miR-34b/c	PC-3 and DU145	*CTNNB1*	Cell migration and EMT	[81]
miR-101	LNCaP, 22Rv1, DU145, and PC-3	*AR*	Autophagy	[82]
miR-101	DU145 and PC-3 cells	*RLIP76*	Apoptosis	[83]

(continued)

Table 13.2 (continued)

miRNA	Model	Target gene	Functions	References
miR-30a	DU145 and PC-3 cells	*TP53INP1*	Radiosensitivity	[84]
miR-331-3p	LNCaP-FGC, CWR-22RV1, and DU145 and PCa tissues	*ERBB-2*	PI3K/AKT signalling	[85]
miR-494-3p	PC-3 and DU145	*CXCR4*	Proliferation, invasion, and migration	[86]
miR-497	PCa tissues and PC-3 AR cells	*IKKβ*	Proliferation, migration, and invasion	[87]

identified the SRY-Box Transcription Factor 4 (SOX4) gene as the target of miR-132/212. Moreover, the expression levels of miR-132/212 were inversely associated with the Gleason score and SOX4 expression in PCa tissues [90]. Tumour suppressor miR-203 overexpression in the bone metastatic model of PCa suppressed cancer metastasis by inhibiting the prometastatic genes involved in EMT, motility, and invasion [91]. Several signalling pathways, including TGF-beta, Wnt and beta-catenin signalling, can mediate the metastasis of the cancer. RUNX2 is a bone-related transcription factor that regulates the expression of osteopontin, osteocalcin, and MMP11. In PCa tissues, miR-466 is under-expressed compared to healthy normal tissues. miR-466 directly regulates the expression of osteogenic transcription factor RUNX2, and overexpression of this miRNA caused a notable decrease in tumour growth and metastasis [92]. Several miRNAs can regulate the expression of genes by binding to the UTR regions of the mRNA.

Roles of miRNAs in Regulating the Tumour Microenvironment

miRNA-Mediated Regulation of Cancer-Associated Fibroblasts

The tumour microenvironment, a dynamic entity that evolves alongside cancer, actively supports its progression. Within the stroma of PCa, cancer-associated fibroblasts (CAFs) reign as the dominant cell type. Their pivotal role in tumour development and therapeutic outcomes cannot be overstated. In PCa, CAFs orchestrate tumour growth and metastasis through their dynamic interactions with cancer cells. Single-cell transcriptomics have identified 9 CAF-related miRNAs, forming the basis of a prognostic model to predict the recurrence of PCa in primary PCa tissues. In vitro studies on fibroblast cell lines have shown that the downregulation of miR-106b-5p leads to decreased cell proliferation, migration, and colony formation. Furthermore, in vivo xenograft models have demonstrated reduced tumour growth upon the downregulation of miR-106b-5p [93]. Co-culture of CAFs and PCa cell lines (C4-2 and LNCaP) has significantly increased cell proliferation and migration by upregulating the expression of miR-18b-5p. Treatment with an inhibitor of miR-18b-5p has attenuated the proliferation and migration of PCa cells supported by the CAFs. Moreover, in the xenograft mice model of PCa, the knockdown of miR-18b-5p has greatly enhanced the survival time of the mice. Additionally, the luciferase assay has confirmed that FBXL3 is the direct target of miR-18b-5p [94].

A significant hurdle in treating PCa is the cancer's chemoresistance, or its insensitivity to chemotherapy. CAF exosomes have been found to shield PCa cells from erastin-induced ferroptosis by reducing lipid reactive oxygen species (ROS) accumulation and mitochondrial damage. Exosomal CAFs miR-432-5p has targeted the CHAC1 mRNA and depleted glutathione levels in the PCa cells, thereby protecting cells from ferroptosis and inducing docetaxel resistance in PCa [95]. Furthermore, the CAF-secreted exosomal miR-423-5p has decreased the sensitivity and increased taxane drug resistance in the PCa cells by possibly binding and inhibiting the GRME2 through the transforming growth factor-beta signalling. Blocking the TGF-β pathways has decreased the drug resistance induced by the CAF exosomes in the PCa cells and in vivo [96]. Another study has isolated the CAFs and normal fibroblast from the PCa and adjacent normal tissues and found that the exosomes derived from the CAFs have dramatically increased the migration, invasion, EMT, and metastasis of the PCa cell compared to normal fibroblast exosomes (NF-Exo). The miR-1290 levels were elevated in the CAF exosomes and targeted the GSK3β mRNA inhibiting GSK3β/β-catenin signalling promoting tumour growth and metastasis [97]. Another CAF-derived exosomal miRNA, miR-146-5p, has increased the migration potential and invasiveness of the PCa cells under androgen-deprived conditions. Reduced miR-146-5p levels have promoted metastasis by possibly modulating the epidermal growth factor receptor (EGFR)/ERK pathway [98]. Studies have shown that the PCa-derived CAF transcriptome is similar to that of normal fibroblasts treated with cytokine interleukin-6 (IL-6) [99].

Influence on Immune Cells (T Cells, Macrophages, Natural Killer [NK] Cells)

Natural Killer Cells

Androgen deprivation therapy is a well-known therapy for most advanced PCa patients. However, in most cases, it develops into castration-resistant PCa (CRPC). High doses of androgen-dihydrotestosterone can weaken natural killer (NK) cell-mediated suppression of CRPC cells. Mechanistically, androgen receptor (AR) activation increases circularRNA-FKBP5 (circFKBP5) expression, which sponges miR-513a-5p, leading to the upregulation of PD-L1 and reduced immune surveillance. Preclinical models indicate that combining high-dose DHT with PD-L1 inhibition enhances CRPC suppression, highlighting a potential therapeutic strategy for advanced PCa [100]. ICAM1 is the primary target of miR-296-3p, whose ectopic expression promotes tolerance to NK cells and vice versa. The miR-296-3p expression is lower in P69 cells than in M12 cells. Similar outcomes were observed in athymic nude mice models, which are immunocompromised mice often used in cancer research, where pulmonary tumour extravasation decreased upon knockdown of miR-296-3p in the M12 cell line [101]. NK cells extensively regulate cancer metastasis. In lymphocytes purified from PCa tissues, miR-224 was downregulated. MiR-224 overexpression evaded PCa from NK cells mediated through NCR1/NKp46 signalling by HIF-1α, with NCR1 being a target of miR-224 [102].

Tumour-Associated Macrophages

Tumour-associated macrophages (TAMs) play a key role in the resistance of tumour therapeutics. TAMs also release extracellular vesicles containing miRNAs of regulatory effects. The onco-vesicles of TAM have higher miR-95 levels, indicating a direct possible interaction of miR-96 with PCa cells. The tumour promoter miR-96 targets and binds with JunB mRNA to enhance proliferation, invasion capacity and EMT of PCa cells [103]. miRNAs are released by the PCa cells and cells in the tumour environment. These miRNAs can cross-exert their effect and dramatically alter the tumour outcomes like formation, progression, and therapeutic response. Exosomes collected from LNCaP cells and urine samples of PCa patients and overexpression of miR-203 favoured macrophage polarization towards M1 type and apoptosis in treated cells over untreated cells. Moreover, in animal models, miR-203, with its potential to significantly suppress PCa tumour growth in nude mice, offers hope for its therapeutic implications. The introduction of miR-203 antagomiR caused a reversal of the changes, suggesting the anti-tumour role of miRNA [104]. Contrarily, miRNA Let-7b also modulates macrophage polarization of TAMs and augments cell motility and angiogenesis of the PCa [105, 106]. Moreover, studies have also shown the involvement of SOCS1/STAT pathway in PCa by anti-tumourigenic M2 macrophage polarization upon inhibition of Let-7b-5p [107]. Chemokine (C-X-C motif) ligand 14 (CXCL14) plays a role in maintaining inflammation in the microenvironment and cancer progression, and their levels are reported to be elevated in PCa tissues. Suppression of CXCL14 inhibits tumour growth in the xenograft model of mice and promotes M2 macrophage infiltration through NF-κB signalling [108]. miR-451 levels are reported to be decreased in clinical metastatic PCa, and macrophage migration inhibitory factor (MIF) is the direct target. Ectopic expression of miR-451 suppresses tumour growth, migration, and metastasis by targeting MIF in PCa cells [109].

T Cells

T cells play a pivotal role in PCa by influencing tumour progression and immune response. Their ability to target and destroy cancer cells, albeit often diminished by the immunosuppressive tumour microenvironment, underscores their importance. Regulatory T cells (Tregs), in contrast, contribute to immune evasion by suppressing anti-tumour immunity, promoting cancer growth. PCa employs miRNA to modulate the T cell activity and function to evade immunorecognition. PCa cells secrete chemokines like CXCL9, which helps recruit more CD4 (+) cells than normal prostate cells. The heightened recruitment of CD4 (+) cells promotes prostate tumour invasion by modulating FGF11-miR-541 signals to androgen receptor signals, which alter the MMP9 for PCa invasion [110]. mRNA and protein expression analysis of the UBASH3B gene in activated T lymphocytes, which negatively regulates T cell receptor signalling in the PCa and BPH, was performed, and it was found that UBASH3B was upregulated in PCa samples. Moreover, their expression correlated with a poor prognosis of PCa. Bioinformatic analysis showed this gene also correlated with 11 different types of tumour infiltrating immune cells from naïve B cells, macrophages, neutrophils, and dendritic cells [111].

The Warburg effect, a classic feature of cancers, is also observed in PCa. PCa cells exhibit a unique metabolic behaviour, prioritizing glucose uptake and lactate production over oxidative phosphorylation, even in the presence of sufficient oxygen. The lactate produced in the tumour cell and/or its microenvironment plays a role in immunosuppression. The glycolytic CAFs-produced lactate alters the polarization of the T cell, promoting a pro-invasive effect. CAFs can cause a reduction in the percentage of antitumour Th1 subset via SIRT1-mediated degradation of transcription factor T-bet and also increase Tregs cells by driving naïve T cell polarization through Nf-kB and FoxP3 activation [112].

In advanced PCa, radiotherapy is a standard treatment option. However, as the treatment progresses, cancer often develops resistance to radiation therapy. Studies have shown that certain miRNAs, such as miR-195 and miR-16, can significantly influence the outcome of the therapy. The biochemical recurrence-free survival of PCa patients was positively associated with miR-195 and miR-16 levels. These miRNAs target PD-L1, an immune checkpoint expression, and restoring these miRNAs improved radiotherapy by activating T cells in the TME by inhibiting PD-L1 expression. Furthermore, they promoted the proliferation of cytotoxic CD8+ T cells and inhibited Tregs and myeloid-derived suppressor cells [113]. Various oncomiRs and tumour suppressor miRNAs in prostate cancer and its tumour microenvironment are presented in Fig. 13.1.

miRNA Involvement in Angiogenesis and Hypoxia Response

Rapid cancer progression creates a low-oxygen environment in most solid tumours, including PCa. This sustained low oxygen environment, or hypoxic conditions, in cancers leads to dysfunctional vascularization and promotes the acquisition of the EMT phenotype, leading to cell migration and metastasis [114]. PCa metastasis to other tissues, such as bone and the vertebral column, drastically reduces the survival rate of PCa patients to one-third. A hypoxic environment activates hypoxia-inducible factors (HIFs), which in turn activate a cascade of other host signalling pathways in the cell. This hypoxia can accelerate proliferation, migration, EMT, and tumour

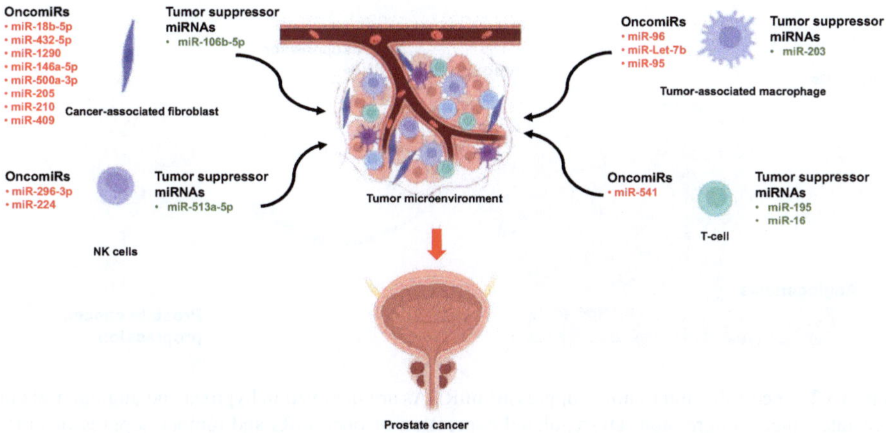

Fig. 13.1 oncomiRs and tumour suppressor miRNAs in prostate cancer and its tumour microenvironment. The expression of microRNAs (miRNAs) and their association with tumour promotion or suppression within prostate cancer and its surrounding tumour microenvironment are depicted. These miRNAs, crucial in regulating the cellular functions of key stromal and immune components, including cancer-associated fibroblasts (CAFs), tumour-associated macrophages (TAMs), natural killer (NK) cells, and T lymphocytes, play a significant role in prostate cancer biology

metastasis in PCa. The role of miRNAs in this process is significant, as they are involved in the regulation of HIF-1α and other factors, making them potential therapeutic targets.

The major transcription factor activated and stabilized in hypoxia is hypoxia-inducible factor-1α (HIF-1α), which controls the transcription of genes related to angiogenesis. Notably, HIF-1α levels were significantly increased in PCa tissues compared to normal prostate tissue, with a concomitant decrease in miR-199b expression. The overexpression of miR-199b in PCa cell lines such as DU-145 and PC-3 led to a decrease in HIF-1α levels, indicating that miR-199b binds to the 3′ UTR region and negatively regulates HIF-1α expression, as confirmed by bioinformatics analysis [115]. Similarly, another hypoxia-inducible miRNA, miR-182, is also reported to modulate HIF-1α signalling by targeting the prolyl hydroxylase domain enzymes (PHD) and factor inhibiting HIF-1 (FIH-1). Overexpression of miR-182 in both normoxia and hypoxia led to decreased PHD and FIH-1 expression, thereby increasing angiogenesis by elevating HIF-1α levels [116]. miRNAs implicated in hypoxia and angiogenesis in prostate cancer progression are depicted in Fig. 13.2.

In vitro studies suggest that hypoxia promotes the growth, viability, and invasiveness of the PC-3 cell lines by decreasing the expression of miR-132 [117]. MiR-210, a micromanager of hypoxia, is directly linked to hypoxia. Researchers investigated the prognostic effect of miR-210 in tumour cells and fibroblasts in PCa specimens obtained from prostatectomy. They found that the overexpression of this miRNA has not been linked with biochemical failure-free survival (BFSS) and clinical failure-free survival (CFSS) in PCa. However, in fibroblast, significant associations

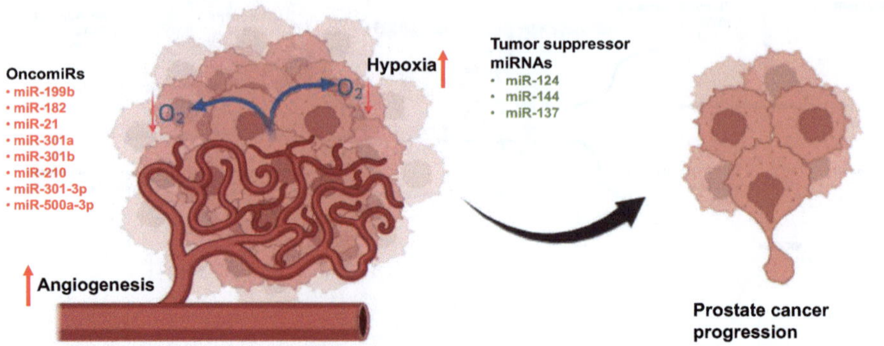

Fig. 13.2 oncomiRs and tumour suppressor miRNAs are involved in hypoxia and angiogenesis in prostate cancer progression. Dysregulated expression of oncomiRs and tumour suppressor miRNAs are associated with cellular hypoxia. It leads to increased angiogenesis, accelerating its growth, proliferation, migration, epithelial-to-mesenchymal transition, and tumour metastasis in prostate cancer

were reported with poor CFFS but not BFSS [118]. Studies have shown that the upregulation of miR-21 is significantly associated with disease progression. Exposure to hypoxic conditions of LNCaP cells caused upregulation of miR-21 in both in vitro and in vivo xenograft models. Moreover, the overexpression of miR-21 in normal prostate cells RWPE-1 enhanced the colony formation and migration potential of the cells by targeting the tumour suppressor gene Ras Homolog Family Member B (RHOB) [119]. Another miRNA induced by hypoxia in PCa is miR-210, which is correlated with Gleason grade and cancer progression [120]. Neural cell adhesion molecule (NCAM) is the direct target of miR-120 [121]. Hypoxia-inducible miR-301b-3p binds and regulates low-density lipoprotein receptor-related protein 1B (LRP1B). Downregulation of LRP1B expression by miR-301-3p accelerated tumour growth, proliferation, migration, and metastasis of the PCa cells [120]. Furthermore, oncomiRs miR-301a and miR-301b are hypoxia-responsive miRNAs whose expression is enhanced under hypoxia in PCa cells. MiR-301a/b contributes to increased cell viability, autophagy, and radioresistance with reduced cell apoptosis in the LNCaP cells by targeting the NDRG2 [122, 123]. In contrast, overexpression of hypoxia-responsive miR-124 and miR-144 in PCa cells (PC-3 and DU145) inhibits autophagy and enhances radiosensitivity by downregulation of PIM1 [124]. Similarly, enforced expression of miR-205 and miR-30a reduced irradiation-induced autophagy and enhanced radiosensitivity in PCa cells by inhibiting TP53INP1 [84, 125].

Hypoxia causes the PCa to proliferate aggressively, invade and metastasize the tumour. Researchers studied the exosomal miRNA profiling of the human LNCaP cells exposed to hypoxic and normoxic conditions. They found that 11 and 9 miRNAs were significantly higher and lower in hypoxic condition exosomes than in normoxic exosomes. Moreover, exosome miRNA analysis isolated from the serum of PCa patients exhibited increased miRNA-855 and decreased miR-521 compared to healthy individuals [126]. Under hypoxic conditions, the exosomes released by the CAFs promoted PCa metastasis in the PCa cells and in vivo models. Transcriptome analysis showed that miR-500a-3p levels were increased in the exosomes secreted by CAFs under a hypoxic microenvironment. FBXW7 is the direct target of miR-500a-3p, and overexpression of FBXW7 in PCa cells reduced the metastatic ability and decreased heat shock transcription factor 1 (HSF1) expression [127]. Hypoxia reduces miR-137 expression, inhibiting cell proliferation and migration and promoting apoptosis in human PCa cells by targeting LGR4 via the EGFR/ERK signalling pathway [128]. This finding suggests the significant role of miR-500a-3p in promoting PCa metastasis, raising concerns about its implications for cancer progression.

Effects on Extracellular Matrix Remodelling

An extracellular matrix (ECM) is essential for maintaining tissue architecture and is important in many physiological and pathological conditions. It is a non-cellular network of various components that provides support and structure to cells and tissues. ECM is highly dynamic and a reservoir of many bioactive molecules. ECM comprise proteins and polysaccharides like collagens, fibronectins, laminins, tenascins, etc. Its composition is markedly heterogeneous and tissue-specific. ECM regulates differentiation, proliferation, adhesion, polarity, apoptosis, and migration [129, 130].

In cancer, the ECM undergoes remodelling, a process that involves significant compositional changes and stiffening. These alterations in the ECM play a crucial role in tumour development and progression. The transformation of the ECM in tumours, making it distinct from normal tissue, is a key hallmark of solid tumours [131]. Collagen is the most abundant ECM protein in tissues. In PCa, the deposition of collagen increases and the collagen fibres are reorganized and straightened out [132]. ECM-associated CAFs (ECM-CAF) promote cancer progression and collagen deposition. In contrast, lymphocyte-associated CAFs (Lym-CAF) showed antitumour phenotype, and YAP1 inhibition promoted switching from pro-tumourigenic to tumour suppressor [133]. Advanced high-grade PCa expresses enzymes like matrix metalloproteinases (MMPs) for tumour invasion and metastasis. PCa research on in vitro models primarily employs monolayer cultures where the tumour's interaction with the ECM is missing. Studies have shown that MMPs are not expressed in PCa cell lines under normal cell culture conditions. However, in the same cell lines, the addition of fibronectin induces high levels of pro and active MMP2 forms [134]. Additionally, fibronectin exposure to LNCaP and PC-3 cells modulated the expression of five miRNAs miR 29b, miR-21, miR-221, miR-222, and miR-29b. These miRNAs were involved in cell proliferation, division, and migration and regulated PI3K/AKT signalling [135].

Thrombospondin 2 (TSP-2), a glycoprotein, mediates the cell's attachment to ECM. As PCa progresses to advanced metastatic forms, TSP-2 expression increases.

In turn, TSP-2 increases migration and MMP2 expression and negatively regulates the miR-376c. Animal study suggests that the knockdown of TSP-2 in the PCa cells abolished bone metastasis [136]. Asporin, a non-collagenous part of the ECM, has diverse functions, and many studies have found elevated expression of asporin in many cancer types, including PCa [137]. TGF-β can be released from ECM degradation, and in PCa, TGF-β can cause trans-differentiation of the fibroblast to myofibroblast regulating CAFs [138]. Activin A, a member of the TGF-β superfamily, can also modulate the expression of several miRNAs in LNCaP cells [139]. CAFs and stromal cells also express Tenascin-C in PCa, which is associated with proliferation, migration, and metastasis of PCa and reduced overall survival of PCa patients [140]. The silencing of cell migration-inducing protein (CEMIP) in ECM-detached cancer cells promotes ferroptosis by inhibiting cystine uptake, ultimately hindering the NRF2 phosphorylation [141]. When the cells detach from their neighbouring cells or ECM, they undergo programmed cell death called anoikis. Cancer cells are resistant to anoikis, which helps them to metastasize to bone. A key prognostic gene, TUBB3, is involved in anoikis-resistant PCa cells, and the downregulation of this gene promotes anoikis and hampers the invasion and migration capabilities of ECM-detached PCa cells [142]. Other ECM proteins, such as Cyr61 and spondin-2, are linked to PCa development [143, 144]. TMPRSS2 promotes PCa growth and metastasis by activating matriptase, which enhances the breakdown of ECM components, including nidogen-1 and laminin β1, in tumour xenografts [145].

Modulation of miRNA by Epigenetics

Epigenetic modifications such as DNA methylations and histone modifications lead to chromatin remodelling and regulate the expression of the genes involved in the development and diseases, including cancer. Like the transcriptome, the genetic material encoding the miRNAs are also susceptible to epigenetic modifications, regulating miRNA expression in diseases. Like various types of cancer, aberrant activation and inactivation of miRNA expression have been linked to PCa development, including cell proliferation, survival, and metastasis. For instance, DNA methylation in the CpG islands of the promoter region of the miR-200b-429 was found in different PCa cell lines, such as LNCaP, Du145, and C42B cell lines. Moreover, miR-200b-429 cluster was also found to be methylated in the primary prostate and metastatic tumours [146]. DNA methyltransferases (DNMTs), which cause the addition of methyl groups to the cytosine, are responsible for gene silencing. miR-615 is an epigenetically activated miRNA by DNA demethylation and H3K9 acetylation [147]. In normal healthy cells, miR-152 targets the 3′ UTR of the DNMT1 mRNA, thereby repressing the expression of DNMT1. Whereas in the PCa, the miR-152 expression is downregulated, leading to upregulated DNMT1 expression associated with PCa aggressiveness [148]. Similarly, DNMT3b mRNA expression is directly regulated by miR-145 by binding to the 3′ UTR of its mRNA. In PCa, miR-145 levels are upregulated, which downregulates DNMT3b expression leading to miR-145 promoter hypomethylation in a feedback loop mechanism [149].

Similarly, transcription repression can also be regulated by histone deacetylases (HDACs) and EZH2, which cause the removal of the acetyl group from histone proteins and the addition of methyl groups in histone proteins, respectively. In PCa, EZH2 suppresses the expression of miR-181a, miR181b, miR-203, let-7, and miR-200c [150]. HDACs are transcriptional repressors as they lead to compact packaging of DNA around the histone, making DNA inaccessible to the enzymatic machinery. Aberrant expression of the HDACs has been reported in cancer tumourigenesis and influences clinical outcomes. Researchers have found a possible link between the HDAC4 nuclear accumulation and the development of tumour type with androgen insensitivity in human PCa tissues. Moreover, the overexpression of HDAC1 and HDAC2 directly correlated with the Gleason score and tumour proliferation rate in PCa tissues [151]. The reduced expression of miR-17-5p, miR-449a, and miR-10 leads to enhanced PCa cell growth and proliferation by upregulating the p300/CBP-associated factor, HDAC-1, and EZH2 protein, respectively [152, 153].

Therapeutic Potential of Targeting miRNAs in Prostate Cancer

miRNA-Based Diagnostic and Prognostic Biomarkers

Due to its small size, stability, and presence in biological fluids, miRNA holds significant promise as a non-invasive diagnostic tool. The expression levels of miRNA are valuable in understanding tumour development, aggressiveness, metastasis, treatment response, and recurrence. Many miRNAs are differentially expressed in the tumour of different stages of development. For instance, miR-301a is differentially expressed in benign prostate hyperplasia (BPH) and adjacent benign samples in PCa samples and serves as a potential marker for metastasis [154]. Moreover, serum levels of miR-301a can be used as a prognosis marker [155]. Studies have shown that serum levels of miR-15a and miR-16-1, along with PSA levels, can be used as specific non-invasive biomarkers, rather than PSA alone, in the prognosis and diagnosis of patients with PCa [156].

Compared to healthy control, the expression of miR-139-5p expression in the peripheral blood of PCa patients with BPH was significantly increased. Moreover, associations of Gleason score, PSA levels, and pathological tumour stage with high miR-139-5p were established, suggesting miR-139-5p as a potential non-invasive biomarker for PCa screening [157]. The elevated expression of miR-374b-5p in the whole peripheral blood correlated with the Gleason score and PSA levels and can be useful in distinguishing BPH patients [158]. The diagnostic accuracy of the circulating miR-375 was assessed by the QUADAS-2 tool, which found that miR-375 can be used as a powerful biomarker for PCa diagnostics [159]. Some studies have even suggested a panel of miRNA signatures as a diagnostic tool to detect PCa—a panel of miRNAs like miR-152-3p, miR-4289, miR-326, and miR-98-5p in plasma are upregulated in PCa compared to healthy controls [160]. In another study, circulatory serum levels of miR-222-3p and miR-223-5p were downregulated in PCa in comparison to BPH and this miRNA and their combination can be used as biomarkers for PCa diagnosis [161].

In addition to circulatory miRNA, urinary miRNA could also be used as a non-invasive diagnostic marker for PCa. For instance, the miR-1913/miR-3659 expression ratio in the urine of the PCa patients was upregulated compared to BPH. The urinary miRNA ratio could be used in addition to PSA levels for diagnostic purposes [162]. All these panels of miRNAs present an intriguing possibility for the future of PCa diagnosis.

miRNA Mimics and Inhibitors in PCa Therapy

miRNA mimics and inhibitors are not just emerging, but they are also promising therapeutic tools for treating PCa. These synthetic RNA molecules, designed to replicate the function of endogenous miRNAs, hold the potential to enhance their expression and regulatory effects. In contrast, miRNA inhibitors (antimiRs) function by binding to endogenous miRNAs, preventing their interaction with target mRNAs. This inhibition restores the expression of genes previously suppressed by miRNAs, offering a potential approach for therapeutic intervention [30].

The primary focus of miRNA-based therapy is to overcome chemoresistance, regulate tumour progression, and modify treatment responses in PCa. Studies indicate that more than 50% of miRNA genes are located within cancer-associated genomic regions, highlighting their critical role in oncogenesis. Abnormal miRNA expression is a hallmark of cancer, with tumour-suppressive miRNAs often downregulated or deleted, while oncomiRs are frequently overexpressed. miRNA mimics play a crucial role in enhancing the efficacy of taxane-based chemotherapy, particularly by targeting pathways involved in cell proliferation and focal adhesion. These mimics restore tumour-suppressive miRNAs, counteracting their loss in cancer cells, whereas miRNA inhibitors suppress oncogenic miRNAs to prevent their tumour-promoting effects. A key advantage of miRNA mimics as therapeutic agents is their ability to regulate the expression of multiple genes simultaneously, making them a powerful tool in precision oncology [163]. Among the most extensively studied miRNA mimics, miR-217 and miR-181b-5p have demonstrated significant potential in enhancing the sensitivity of Docetaxel and cabazitaxel in PC3 cells. These mimics primarily function by downregulating cell proliferation and focal adhesion, thereby improving the therapeutic efficacy of taxane-based treatments [163]. Taxanes exert their anti-cancer effects by binding directly to microtubules, inhibiting chromosomal segregation, and ultimately inducing cell cycle arrest at the G2-M phase of mitosis [164]. This role of miRNA mimics in enhancing the efficacy of taxane-based chemotherapy provides reassurance about the potential of these therapies in PCa treatment.

Similarly, miR-128 has been identified as a key regulator in suppressing tumour regeneration across PCa models. This miRNA plays a critical role in negatively regulating prostate CSCs. Its overexpression in $CD44^+$ PCa cells inhibits tumour regeneration, whereas in $CD44^-$ cells, it exhibits the opposite effect [165]. Additionally, miR-205, a well-established tumour suppressor, is frequently downregulated in PCa. Studies suggest that miR-205 enhances radiation and chemotherapy's effectiveness when combined therapies are used [166]. Another promising miRNA, let-7c, is often downregulated in PCa and targets key oncogenic pathways by regulating HMGA2, MYC, TNSF10, and CASP3. Unlike other miRNAs that promote cell death, let-7c-based therapy focuses on restoring cellular functionality and normal gene regulation [167].

miR-145 is recognized as a key tumour suppressor, regulating critical oncogenes such as MYC and RAS. Animal studies have demonstrated that miR-145 treatment significantly reduces tumour growth, while cessation of treatment leads to tumour recurrence. Notably, MYC and RAS expression levels were reduced by 56% in LNCaP and PC cell lines following miR-145 therapy [168]. Additionally, miR-145-5p mimics hold therapeutic potential for treating PCa with bone metastases by inhibiting EMT [169].

miR-214-5p targets SOX4 (SRY-Box Transcription Factor 4) and inhibits PCa cell proliferation by suppressing key growth factor pathways [170]. Similarly, miR-187 has emerged as a promising therapeutic target, with overexpression shown to suppress malignant phenotypes in PCa [171]. MiR-18a plays an oncogenic role in PCa by downregulating STK4 protein expression. In vivo studies have shown that the transfection of 22Rv-1 cells with antagomiR-18a suppressed tumour growth, supporting a negative association between miR-18a expression and STK4 protein levels in PCa [165].

Another tumour-suppressive miRNA, miR-34a, has demonstrated the ability to induce apoptosis. Its therapeutic potential has been explored through the development of MRX34, a liposomal nanoparticle-based miRNA mimic, which has shown considerable efficacy in preclinical studies across multiple cancer types. This liposomal nanoparticle delivery system enhances therapeutic outcomes, particularly when used in combination with chemotherapy or immune checkpoint inhibitors. The success of MRX34 in preclinical studies across multiple cancer types underscores the potential of miRNA-based therapies in combination with other treatment modalities [172].

Despite the therapeutic promise of miRNA mimics and inhibitors, challenges remain in improving in vivo efficacy. These challenges include issues related to delivery systems, off-target effects, and the need for better understanding of miRNA biology in the context of cancer. Recent research prioritizes miRNA combination therapies to enhance specificity and effectiveness. Notably, miRNA-based therapies targeting miR-15/16, miR-34a, and miR-205 are currently in clinical trials, but none have received FDA approval to date [173, 174].

Challenges in miRNA-Based Therapeutics

Despite the promising potential of miRNA-based therapeutics in revolutionizing cancer treatment, several challenges hinder their clinical application. One of the primary obstacles is the instability of miRNAs in biological environments, as they are highly susceptible to degradation by nucleases. To overcome this, researchers have explored various delivery mechanisms, including liposomal nanoparticles, viral vectors, and chemical modifications, yet achieving optimal stability and targeted delivery remains a challenge.

Another major concern is the risk of off-target effects, as miRNAs can regulate multiple mRNAs simultaneously. This broad activity raises the possibility of unintended gene silencing, which could lead to toxicity in healthy tissues. Additionally, synthetic miRNA mimics or inhibitors may trigger immune responses, potentially causing inflammation or other adverse effects such as cytokine release syndrome. Ensuring these therapies do not provoke harmful immune reactions is not just important, but essential for their safe clinical use and the well-being of patients.

The transition from preclinical research to clinical application also presents difficulties. While animal models and cell culture studies provide valuable insights, the differences in miRNA expression patterns between humans and experimental models can lead to unpredictable therapeutic responses. Furthermore, miRNAs operate within complex gene regulatory networks, and a comprehensive understanding of their interactions is still lacking. Identifying the most effective miRNA targets without interfering with essential cellular processes is a crucial challenge in developing these therapies.

Conclusion

Prostate cancer (PCa) is the second most common malignancy in men, significantly devastating individuals and their families. Extensive research uses cell lines, xenografts and genetically engineered models to understand its pathogenesis and therapeutic targets controlling tumour microenvironment. The immune system is the key target of the tumour microenvironment, where host T cells continue to target and destroy cancer cells, albeit often diminished by the immunosuppressive tumour microenvironment. The specific cell types that transform and are responsible for the cellular origin of prostate cancer are the luminal cells, which line the inner surface of the prostate gland, and the basal cells, which are located in the outer layer of the prostate gland.

The aberrant expression of small non-coding RNA, miRNAs, is causally linked with the pathogenesis of prostate cancer. A subset of miRNAs, oncomiRs, promotes tumour development by downregulating the expression of tumour suppressor genes, which generally inhibit cell growth and division. Due to their small size, stability, and presence in biological fluids, miRNAs promise to be a noninvasive tool in the prognosis and diagnosis of patients with PCa.

From a treatment perspective, androgen deprivation therapy often develops into castration-resistant PCa and is limited to overcoming cancer's chemoresistance in advanced PCa patients. However, the potential of miRNA mimics and inhibitors is immense, promising miRNA-based therapy to overcome chemoresistance, regulate tumour progression, and modify treatment responses in PCa. These miRNA mimics enhance the efficacy of taxane-based chemotherapy and provide reassurance about the potential of these therapies in PCa treatment. Despite the promise of miRNA mimics and inhibitors, issues related to delivery systems, off-target effects, and associated factors are the major challenges for treatment regimens. Improving delivery mechanisms, including liposomal nanoparticles, viral vectors, and chemical modifications, remains a challenge, as does achieving optimal stability and targeted delivery.

Several miRNA-based therapies are currently in clinical trials for prostate cancer, but none have received FDA approval to date. Regulatory challenges include proving long-term safety, consistency, and efficacy in large patient populations. To move forward, advancements in nanotechnology-based delivery systems are needed, and this is a call to action. Improved target specificity and combination therapies involving chemotherapy, immunotherapy, or hormonal treatments may enhance the effectiveness of miRNA therapeutics. Addressing these barriers is not just important, it is critical in unlocking the full potential of miRNA-based approaches for prostate cancer treatment.

Acknowledgements Saikanth Varma received an ICMR JRF fellowship from the Government of India.

Conflict of Interest The authors declare no conflicts of interest, financial or otherwise.

References

1. Bhatia-Gaur R, Donjacour AA, Sciavolino PJ, Kim M, Desai N, Young P, Norton CR, Gridley T, Cardiff RD, Cunha GR, Abate-Shen C, Shen MM (1999) Roles for Nkx3.1 in prostate development and cancer. Genes Dev 13(8):966–977. https://doi.org/10.1101/gad.13.8.966
2. Signoretti S, Waltregny D, Dilks J, Isaac B, Lin D, Garraway L, Yang A, Montironi R, McKeon F, Loda M (2000) p63 is a prostate basal cell marker and is required for prostate development. Am J Pathol 157(6):1769–1775. https://doi.org/10.1016/s0002-9440(10)64814-6
3. El-Alfy M, Pelletier G, Hermo LS, Labrie F (2000) Unique features of the basal cells of human prostate epithelium. Microsc Res Tech 51(5):436–446. https://doi.org/10.1002/1097-0029(20001201)51:5<436::aid-jemt6>3.0.co;2-t
4. Staack A, Donjacour AA, Brody J, Cunha GR, Carroll P (2003) Mouse urogenital development: a practical approach. Differentiation 71(7):402–413. https://doi.org/10.1046/j.1432-0436.2003.7107004.x
5. Siegel RL, Kratzer TB, Giaquinto AN, Sung H, Jemal A (2025) Cancer statistics, 2025. CA Cancer J Clin 75(1):10–45. https://doi.org/10.3322/caac.21871

6. Rawla P (2019) Epidemiology of prostate cancer. World J Oncol 10(2):63–89. https://doi.org/10.14740/wjon1191
7. Humphrey PA (2007) Diagnosis of adenocarcinoma in prostate needle biopsy tissue. J Clin Pathol 60(1):35–42. https://doi.org/10.1136/jcp.2005.036442
8. Goldstein AS, Huang J, Guo C, Garraway IP, Witte ON (2010) Identification of a cell of origin for human prostate cancer. Science (New York, NY) 329(5991):568–571. https://doi.org/10.1126/science.1189992
9. Stoyanova T, Cooper AR, Drake JM, Liu X, Armstrong AJ, Pienta KJ, Zhang H, Kohn DB, Huang J, Witte ON, Goldstein AS (2013) Prostate cancer originating in basal cells progresses to adenocarcinoma propagated by luminal-like cells. Proc Natl Acad Sci 110(50):20111–20116. https://doi.org/10.1073/pnas.1320565110
10. Thalmann GN, Sikes RA, Wu TT, Degeorges A, Chang SM, Ozen M, Pathak S, Chung LW (2000) LNCaP progression model of human prostate cancer: androgen-independence and osseous metastasis. Prostate 44(2):91–103. https://doi.org/10.1002/1097-0045(20000701)44:2<91::aid-pros1>3.0.co;2-l
11. Parisotto M, Metzger D (2013) Genetically engineered mouse models of prostate cancer. Mol Oncol 7(2):190–205. https://doi.org/10.1016/j.molonc.2013.02.005
12. Dakhova O, Ozen M, Creighton CJ, Li R, Ayala G, Rowley D, Ittmann M (2009) Global gene expression analysis of reactive stroma in prostate cancer. Clin Cancer Res 15(12):3979–3989. https://doi.org/10.1158/1078-0432.ccr-08-1899
13. Chaffer CL, Weinberg RA (2011) A perspective on cancer cell metastasis. Science (New York, NY) 331(6024):1559–1564. https://doi.org/10.1126/science.1203543
14. Levesque C, Nelson PS (2018) Cellular constituents of the prostate stroma: key contributors to prostate cancer progression and therapy resistance. Cold Spring Harb Perspect Med 8(8). https://doi.org/10.1101/cshperspect.a030510
15. Marsh T, Pietras K, McAllister SS (2013) Fibroblasts as architects of cancer pathogenesis. Biochim Biophys Acta 1832(7):1070–1078. https://doi.org/10.1016/j.bbadis.2012.10.013
16. Gurel B, Lucia MS, Thompson IM Jr, Goodman PJ, Tangen CM, Kristal AR, Parnes HL, Hoque A, Lippman SM, Sutcliffe S, Peskoe SB, Drake CG, Nelson WG, De Marzo AM, Platz EA (2014) Chronic inflammation in benign prostate tissue is associated with high-grade prostate cancer in the placebo arm of the prostate cancer prevention trial. Cancer Epidemiol Biomarkers Prev 23(5):847–856. https://doi.org/10.1158/1055-9965.epi-13-1126
17. Gueron G, De Siervi A, Vazquez E (2012) Advanced prostate cancer: reinforcing the strings between inflammation and the metastatic behavior. Prostate Cancer Prostatic Dis 15(3):213–221
18. Chen Daniel S, Mellman I (2013) Oncology meets immunology: the cancer-immunity cycle. Immunity 39(1):1–10. https://doi.org/10.1016/j.immuni.2013.07.012
19. Lee RC, Feinbaum RL, Ambros V (1993) The C. elegans heterochronic gene lin-4 encodes small RNAs with antisense complementarity to lin-14. Cell 75(5):843–854
20. Tanzer A, Stadler PF (2004) Molecular evolution of a microRNA cluster. J Mol Biol 339(2):327–335
21. Denli AM, Tops BB, Plasterk RH, Ketting RF, Hannon GJ (2004) Processing of primary microRNAs by the microprocessor complex. Nature 432(7014):231–235
22. Alarcón CR, Lee H, Goodarzi H, Halberg N, Tavazoie SF (2015) N 6-methyladenosine marks primary microRNAs for processing. Nature 519(7544):482–485
23. Zhang H, Kolb FA, Jaskiewicz L, Westhof E, Filipowicz W (2004) Single processing center models for human Dicer and bacterial RNase III. Cell 118(1):57–68
24. O'Brien J, Hayder H, Zayed Y, Peng C (2018) Overview of MicroRNA biogenesis, mechanisms of actions, and circulation. Front Endocrinol 9. https://doi.org/10.3389/fendo.2018.00402
25. Babiarz JE, Ruby JG, Wang Y, Bartel DP, Blelloch R (2008) Mouse ES cells express endogenous shRNAs, siRNAs, and other microprocessor-independent, Dicer-dependent small RNAs. Genes Dev 22(20):2773–2785

26. Xie M, Li M, Vilborg A, Lee N, Shu M-D, Yartseva V, Šestan N, Steitz JA (2013) Mammalian 5′-capped microRNA precursors that generate a single microRNA. Cell 155(7):1568–1580
27. Yang J-S, Maurin T, Robine N, Rasmussen KD, Jeffrey KL, Chandwani R, Papapetrou EP, Sadelain M, O'Carroll D, Lai EC (2010) Conserved vertebrate mir-451 provides a platform for Dicer-independent, Ago2-mediated microRNA biogenesis. Proc Natl Acad Sci 107(34):15163–15168
28. Kawamata T, Tomari Y (2010) Making risc. Trends Biochem Sci 35(7):368–376
29. Jo MH, Shin S, Jung S-R, Kim E, Song J-J, Hohng S (2015) Human Argonaute 2 has diverse reaction pathways on target RNAs. Mol Cell 59(1):117–124
30. Krützfeldt J, Rajewsky N, Braich R, Rajeev KG, Tuschl T, Manoharan M, Stoffel M (2005) Silencing of microRNAs in vivo with 'antagomirs'. Nature 438(7068):685–689. https://doi.org/10.1038/nature04303
31. Jonas S, Izaurralde E (2015) Towards a molecular understanding of microRNA-mediated gene silencing. Nat Rev Genet 16(7):421–433
32. Braun JE, Truffault V, Boland A, Huntzinger E, Chang C-T, Haas G, Weichenrieder O, Coles M, Izaurralde E (2012) A direct interaction between DCP1 and XRN1 couples mRNA decapping to 5′ exonucleolytic degradation. Nat Struct Mol Biol 19(12):1324–1331
33. Lewis BP, Shih IH, Jones-Rhoades MW, Bartel DP, Burge CB (2003) Prediction of mammalian microRNA targets. Cell 115(7):787–798. https://doi.org/10.1016/s0092-8674(03)01018-3
34. Krek A, Grün D, Poy MN, Wolf R, Rosenberg L, Epstein EJ, MacMenamin P, da Piedade I, Gunsalus KC, Stoffel M, Rajewsky N (2005) Combinatorial microRNA target predictions. Nat Genet 37(5):495–500. https://doi.org/10.1038/ng1536
35. Lin S, Gregory RI (2015) MicroRNA biogenesis pathways in cancer. Nat Rev Cancer 15(6):321–333. https://doi.org/10.1038/nrc3932
36. Ouyang Y, Gao P, Zhu B, Chen X, Lin F, Wang X, Wei J, Zhang H (2015) Downregulation of microRNA-429 inhibits cell proliferation by targeting p27Kip1 in human prostate cancer cells. Mol Med Rep 11(2):1435–1441. https://doi.org/10.3892/mmr.2014.2782
37. Yang ZG, Ma XD, He ZH, Guo YX (2017) miR-483-5p promotes prostate cancer cell proliferation and invasion by targeting RBM5. Int Braz J Urol 43(6):1060–1067. https://doi.org/10.1590/s1677-5538.ibju.2016.0595
38. Cai ZK, Chen Q, Chen YB, Gu M, Zheng DC, Zhou J, Wang Z (2015) microRNA-155 promotes the proliferation of prostate cancer cells by targeting annexin 7. Mol Med Rep 11(1):533–538. https://doi.org/10.3892/mmr.2014.2744
39. Wu Z, He B, He J, Mao X (2013) Upregulation of miR-153 promotes cell proliferation via downregulation of the PTEN tumor suppressor gene in human prostate cancer. Prostate 73(6):596–604. https://doi.org/10.1002/pros.22600
40. Jones DZ, Schmidt ML, Suman S, Hobbing KR, Barve SS, Gobejishvili L, Brock G, Klinge CM, Rai SN, Park J, Clark GJ, Agarwal R, Kidd LR (2018) Micro-RNA-186-5p inhibition attenuates proliferation, anchorage independent growth and invasion in metastatic prostate cancer cells. BMC Cancer 18(1):421. https://doi.org/10.1186/s12885-018-4258-0
41. Tinay I, Tan M, Gui B, Werner L, Kibel AS, Jia L (2018) Functional roles and potential clinical application of miRNA-345-5p in prostate cancer. Prostate 78(12):927–937. https://doi.org/10.1002/pros.23650
42. Jia L, Gui B, Zheng D, Decker KF, Tinay I, Tan M, Wang X, Kibel AS (2017) Androgen receptor-regulated miRNA-193a-3p targets AJUBA to promote prostate cancer cell migration. Prostate 77(9):1000–1011. https://doi.org/10.1002/pros.23356
43. Ren D, Wang M, Guo W, Zhao X, Tu X, Huang S, Zou X, Peng X (2013) Wild-type p53 suppresses the epithelial-mesenchymal transition and stemness in PC-3 prostate cancer cells by modulating miR-145. Int J Oncol 42(4):1473–1481. https://doi.org/10.3892/ijo.2013.1825
44. Wu C, Miao C, Tang Q, Zhou X, Xi P, Chang P, Hua L, Ni H (2020) MiR-129-5p promotes docetaxel resistance in prostate cancer by down-regulating CAMK2N1 expression. J Cell Mol Med 24(3):2098–2108. https://doi.org/10.1111/jcmm.14050

45. Amankwah EK, Anegbe E, Park H, Pow-Sang J, Hakam A, Park JY (2013) miR-21, miR-221 and miR-222 expression and prostate cancer recurrence among obese and non-obese cases. Asian J Androl 15(2):226–230. https://doi.org/10.1038/aja.2012.160
46. Ribas J, Ni X, Haffner M, Wentzel EA, Salmasi AH, Chowdhury WH, Kudrolli TA, Yegnasubramanian S, Luo J, Rodriguez R, Mendell JT, Lupold SE (2009) miR-21: an androgen receptor-regulated microRNA that promotes hormone-dependent and hormone-independent prostate cancer growth. Cancer Res 69(18):7165–7169. https://doi.org/10.1158/0008-5472.can-09-1448
47. Haflidadóttir BS, Larne O, Martin M, Persson M, Edsjö A, Bjartell A, Ceder Y (2013) Upregulation of miR-96 enhances cellular proliferation of prostate cancer cells through FOXO1. PLoS One 8(8):e72400. https://doi.org/10.1371/journal.pone.0072400
48. Li X, Wan X, Chen H, Yang S, Liu Y, Mo W, Meng D, Du W, Huang Y, Wu H, Wang J, Li T, Li Y (2014) Identification of miR-133b and RB1CC1 as independent predictors for biochemical recurrence and potential therapeutic targets for prostate cancer. Clin Cancer Res 20(9):2312–2325. https://doi.org/10.1158/1078-0432.ccr-13-1588
49. Hudson RS, Yi M, Esposito D, Glynn SA, Starks AM, Yang Y, Schetter AJ, Watkins SK, Hurwitz AA, Dorsey TH, Stephens RM, Croce CM, Ambs S (2013) MicroRNA-106b-25 cluster expression is associated with early disease recurrence and targets caspase-7 and focal adhesion in human prostate cancer. Oncogene 32(35):4139–4147. https://doi.org/10.1038/onc.2012.424
50. Yao J, Xu C, Fang Z, Li Y, Liu H, Wang Y, Xu C, Sun Y (2016) Androgen receptor regulated microRNA miR-182-5p promotes prostate cancer progression by targeting the ARRDC3/ITGB4 pathway. Biochem Biophys Res Commun 474(1):213–219. https://doi.org/10.1016/j.bbrc.2016.04.107
51. Murata T, Takayama K, Katayama S, Urano T, Horie-Inoue K, Ikeda K, Takahashi S, Kawazu C, Hasegawa A, Ouchi Y, Homma Y, Hayashizaki Y, Inoue S (2010) miR-148a is an androgen-responsive microRNA that promotes LNCaP prostate cell growth by repressing its target CAND1 expression. Prostate Cancer Prostatic Dis 13:356–361. https://doi.org/10.1038/pcan.2010.32
52. Patron JP, Fendler A, Bild M, Jung U, Müller H, Arntzen M, Piso C, Stephan C, Thiede B, Mollenkopf HJ, Jung K, Kaufmann SH, Schreiber J (2012) MiR-133b targets antiapoptotic genes and enhances death receptor-induced apoptosis. PLoS One 7(4):e35345. https://doi.org/10.1371/journal.pone.0035345
53. Larne O, Östling P, Haflidadóttir BS, Hagman Z, Aakula A, Kohonen P, Kallioniemi O, Edsjö A, Bjartell A, Lilja H, Lundwall Å, Ceder Y (2015) miR-183 in prostate cancer cells positively regulates synthesis and serum levels of prostate-specific antigen. Eur Urol 68(4):581–588. https://doi.org/10.1016/j.eururo.2014.12.025
54. Galardi S, Mercatelli N, Giorda E, Massalini S, Frajese GV, Ciafrè SA, Farace MG (2007) miR-221 and miR-222 expression affects the proliferation potential of human prostate carcinoma cell lines by targeting p27Kip1. J Biol Chem 282(32):23716–23724. https://doi.org/10.1074/jbc.M701805200
55. Yang X, Yang Y, Gan R, Zhao L, Li W, Zhou H, Wang X, Lu J, Meng QH (2014) Down-regulation of mir-221 and mir-222 restrain prostate cancer cell proliferation and migration that is partly mediated by activation of SIRT1. PLoS One 9(6):e98833. https://doi.org/10.1371/journal.pone.0098833
56. Budd WT, Seashols-Williams SJ, Clark GC, Weaver D, Calvert V, Petricoin E, Dragoescu EA, O'Hanlon K, Zehner ZE (2015) Dual action of miR-125b as a tumor suppressor and OncomiR-22 promotes prostate cancer tumorigenesis. PLoS One 10(11):e0142373. https://doi.org/10.1371/journal.pone.0142373
57. Seashols-Williams SJ, Budd W, Clark GC, Wu Q, Daniel R, Dragoescu E, Zehner ZE (2016) miR-9 acts as an OncomiR in prostate cancer through multiple pathways that drive tumour progression and metastasis. PLoS One 11(7):e0159601. https://doi.org/10.1371/journal.pone.0159601

58. Nam RK, Benatar T, Wallis CJ, Amemiya Y, Yang W, Garbens A, Naeim M, Sherman C, Sugar L, Seth A (2016) MiR-301a regulates E-cadherin expression and is predictive of prostate cancer recurrence. Prostate 76(10):869–884. https://doi.org/10.1002/pros.23177
59. Cha YJ, Lee JH, Han HH, Kim BG, Kang S, Choi YD, Cho NH (2016) MicroRNA alteration and putative target genes in high-grade prostatic intraepithelial neoplasia and prostate cancer: STAT3 and ZEB1 are upregulated during prostate carcinogenesis. Prostate 76(10):937–947. https://doi.org/10.1002/pros.23183
60. Liu C, Guan H, Wang Y, Chen M, Xu B, Zhang L, Lu K, Tao T, Zhang X, Huang Y (2015) miR-195 inhibits EMT by targeting FGF2 in prostate cancer cells. PLoS One 10(12):e0144073. https://doi.org/10.1371/journal.pone.0144073
61. Wang L, Song G, Tan W, Qi M, Zhang L, Chan J, Yu J, Han J, Han B (2015) MiR-573 inhibits prostate cancer metastasis by regulating epithelial-mesenchymal transition. Oncotarget 6(34):35978–35990. https://doi.org/10.18632/oncotarget.5427
62. Musumeci M, Coppola V, Addario A, Patrizii M, Maugeri-Saccà M, Memeo L, Colarossi C, Francescangeli F, Biffoni M, Collura D, Giacobbe A, D'Urso L, Falchi M, Venneri MA, Muto G, De Maria R, Bonci D (2011) Control of tumor and microenvironment cross-talk by miR-15a and miR-16 in prostate cancer. Oncogene 30(41):4231–4242. https://doi.org/10.1038/onc.2011.140
63. Wang M, Ren D, Guo W, Wang Z, Huang S, Du H, Song L, Peng X (2014) Loss of miR-100 enhances migration, invasion, epithelial-mesenchymal transition and stemness properties in prostate cancer cells through targeting Argonaute 2. Int J Oncol 45(1):362–372. https://doi.org/10.3892/ijo.2014.2413
64. Taddei ML, Cavallini L, Ramazzotti M, Comito G, Pietrovito L, Morandi A, Giannoni E, Raugei G, Chiarugi P (2019) Stromal-induced downregulation of miR-1247 promotes prostate cancer malignancy. J Cell Physiol 234(6):8274–8285. https://doi.org/10.1002/jcp.27679
65. Ren D, Wang M, Guo W, Huang S, Wang Z, Zhao X, Du H, Song L, Peng X (2014) Double-negative feedback loop between ZEB2 and miR-145 regulates epithelial-mesenchymal transition and stem cell properties in prostate cancer cells. Cell Tissue Res 358(3):763–778. https://doi.org/10.1007/s00441-014-2001-y
66. Huang S, Guo W, Tang Y, Ren D, Zou X, Peng X (2012) miR-143 and miR-145 inhibit stem cell characteristics of PC-3 prostate cancer cells. Oncol Rep 28(5):1831–1837. https://doi.org/10.3892/or.2012.2015
67. Guo W, Ren D, Chen X, Tu X, Huang S, Wang M, Song L, Zou X, Peng X (2013) HEF1 promotes epithelial mesenchymal transition and bone invasion in prostate cancer under the regulation of microRNA-145. J Cell Biochem 114(7):1606–1615. https://doi.org/10.1002/jcb.24502
68. Singh VK, Rajak N, Singh Y, Singh AK, Giri R, Garg N (2024) Role of MicroRNA-21 in prostate cancer progression and metastasis: molecular mechanisms to therapeutic targets. Ann Surg Oncol 31(7):4795–4808. https://doi.org/10.1245/s10434-024-15453-z
69. Rönnau CGH, Fussek S, Smit FP, Aalders TW, van Hooij O, Pinto PMC, Burchardt M, Schalken JA, Verhaegh GW (2021) Upregulation of miR-3195, miR-3687 and miR-4417 is associated with castration-resistant prostate cancer. World J Urol 39(10):3789–3797. https://doi.org/10.1007/s00345-021-03723-4
70. Arai T, Okato A, Yamada Y, Sugawara S, Kurozumi A, Kojima S, Yamazaki K, Naya Y, Ichikawa T, Seki N (2018) Regulation of NCAPG by miR-99a-3p (passenger strand) inhibits cancer cell aggressiveness and is involved in CRPC. Cancer Med 7(5):1988–2002. https://doi.org/10.1002/cam4.1455
71. Kiener M, Chen L, Krebs M, Grosjean J, Klima I, Kalogirou C, Riedmiller H, Kneitz B, Thalmann GN, Snaar-Jagalska E, Spahn M, Kruithof-de Julio M, Zoni E (2019) miR-221-5p regulates proliferation and migration in human prostate cancer cells and reduces tumor growth in vivo. BMC Cancer 19(1):627. https://doi.org/10.1186/s12885-019-5819-6

72. Sun B, Fan Y, Yang A, Liang L, Cao J (2019) MicroRNA-539 functions as a tumour suppressor in prostate cancer via the TGF-β/Smad4 signalling pathway by down-regulating DLX1. J Cell Mol Med 23(9):5934–5948. https://doi.org/10.1111/jcmm.14402
73. Guo K, Zheng S, Xu Y, Xu A, Chen B, Wen Y (2016) Loss of miR-26a-5p promotes proliferation, migration, and invasion in prostate cancer through negatively regulating SERBP1. Tumour Biol 37(9):12843–12854. https://doi.org/10.1007/s13277-016-5158-z
74. Ling Z, Wang X, Tao T, Zhang L, Guan H, You Z, Lu K, Zhang G, Chen S, Wu J, Qian J, Liu H, Xu B, Chen M (2017) Involvement of aberrantly activated HOTAIR/EZH2/miR-193a feedback loop in progression of prostate cancer. J Exp Clin Cancer Res 36(1):159. https://doi.org/10.1186/s13046-017-0629-7
75. Ramalho-Carvalho J, Martins JB, Cekaite L, Sveen A, Torres-Ferreira J, Graça I, Costa-Pinheiro P, Eilertsen IA, Antunes L, Oliveira J, Lothe RA, Henrique R, Jerónimo C (2017) Epigenetic disruption of miR-130a promotes prostate cancer by targeting SEC23B and DEPDC1. Cancer Lett 385:150–159. https://doi.org/10.1016/j.canlet.2016.10.028
76. Liu YN, Yin JJ, Abou-Kheir W, Hynes PG, Casey OM, Fang L, Yi M, Stephens RM, Seng V, Sheppard-Tillman H, Martin P, Kelly K (2013) MiR-1 and miR-200 inhibit EMT via slug-dependent and tumorigenesis via slug-independent mechanisms. Oncogene 32(3):296–306. https://doi.org/10.1038/onc.2012.58
77. Chang YS, Chen WY, Yin JJ, Sheppard-Tillman H, Huang J, Liu YN (2015) EGF receptor promotes prostate cancer bone metastasis by downregulating miR-1 and activating TWIST1. Cancer Res 75(15):3077–3086. https://doi.org/10.1158/0008-5472.can-14-3380
78. Ma Q, Chen Z, Jia G, Lu X, Xie X, Jin W (2015) The histone demethylase PHF8 promotes prostate cancer cell growth by activating the oncomiR miR-125b. Onco Targets Ther 8:1979–1988. https://doi.org/10.2147/ott.s85443
79. Bhatnagar N, Li X, Padi SKR, Zhang Q, Tang MS, Guo B (2010) Downregulation of miR-205 and miR-31 confers resistance to chemotherapy-induced apoptosis in prostate cancer cells. Cell Death Dis 1(12):e105–e105. https://doi.org/10.1038/cddis.2010.85
80. Lin PC, Chiu YL, Banerjee S, Park K, Mosquera JM, Giannopoulou E, Alves P, Tewari AK, Gerstein MB, Beltran H, Melnick AM, Elemento O, Demichelis F, Rubin MA (2013) Epigenetic repression of miR-31 disrupts androgen receptor homeostasis and contributes to prostate cancer progression. Cancer Res 73(3):1232–1244. https://doi.org/10.1158/0008-5472.can-12-2968
81. Liu H, Yin J, Wang H, Jiang G, Deng M, Zhang G, Bu X, Cai S, Du J, He Z (2015) FOXO3a modulates WNT/β-catenin signaling and suppresses epithelial-to-mesenchymal transition in prostate cancer cells. Cell Signal 27(3):510–518. https://doi.org/10.1016/j.cellsig.2015.01.001
82. Guo J, Huang X, Wang H, Yang H (2015) Celastrol induces autophagy by targeting AR/miR-101 in prostate cancer cells. PLoS One 10(10):e0140745. https://doi.org/10.1371/journal.pone.0140745
83. Yang J, Song Q, Cai Y, Wang P, Wang M, Zhang D (2015) RLIP76-dependent suppression of PI3K/AKT/Bcl-2 pathway by miR-101 induces apoptosis in prostate cancer. Biochem Biophys Res Commun 463(4):900–906. https://doi.org/10.1016/j.bbrc.2015.06.032
84. Xu CG, Yang MF, Fan JX, Wang W (2016) MiR-30a and miR-205 are downregulated in hypoxia and modulate radiosensitivity of prostate cancer cells by inhibiting autophagy via TP53INP1. Eur Rev Med Pharmacol Sci 20(8):1501–1508
85. Epis MR, Giles KM, Barker A, Kendrick TS, Leedman PJ (2009) miR-331-3p regulates ERBB-2 expression and androgen receptor signaling in prostate cancer. J Biol Chem 284(37):24696–24704. https://doi.org/10.1074/jbc.M109.030098
86. Shen PF, Chen XQ, Liao YC, Chen N, Zhou Q, Wei Q, Li X, Wang J, Zeng H (2014) MicroRNA-494-3p targets CXCR4 to suppress the proliferation, invasion, and migration of prostate cancer. Prostate 74(7):756–767. https://doi.org/10.1002/pros.22795
87. Kong XJ, Duan LJ, Qian XQ, Xu D, Liu HL, Zhu YJ, Qi J (2015) Tumor-suppressive microRNA-497 targets IKKβ to regulate NF-κB signaling pathway in human prostate cancer cells. Am J Cancer Res 5(5):1795–1804

88. Zhang X, Ladd A, Dragoescu E, Budd WT, Ware JL, Zehner ZE (2009) MicroRNA-17-3p is a prostate tumor suppressor in vitro and in vivo, and is decreased in high grade prostate tumors analyzed by laser capture microdissection. Clin Exp Metastasis 26(8):965–979. https://doi.org/10.1007/s10585-009-9287-2
89. Hagman Z, Larne O, Edsjö A, Bjartell A, Ehrnström RA, Ulmert D, Lilja H, Ceder Y (2010) miR-34c is downregulated in prostate cancer and exerts tumor suppressive functions. Int J Cancer 127(12):2768–2776. https://doi.org/10.1002/ijc.25269
90. Fu W, Tao T, Qi M, Wang L, Hu J, Li X, Xing N, Du R, Han B (2016) MicroRNA-132/212 upregulation inhibits TGF-β-mediated epithelial-mesenchymal transition of prostate cancer cells by targeting SOX4. Prostate 76(16):1560–1570. https://doi.org/10.1002/pros.23241
91. Saini S, Majid S, Yamamura S, Tabatabai L, Suh SO, Shahryari V, Chen Y, Deng G, Tanaka Y, Dahiya R (2011) Regulatory role of mir-203 in prostate cancer progression and metastasis. Clin Cancer Res 17(16):5287–5298. https://doi.org/10.1158/1078-0432.ccr-10-2619
92. Colden M, Dar AA, Saini S, Dahiya PV, Shahryari V, Yamamura S, Tanaka Y, Stein G, Dahiya R, Majid S (2017) MicroRNA-466 inhibits tumor growth and bone metastasis in prostate cancer by direct regulation of osteogenic transcription factor RUNX2. Cell Death Dis 8(1):e2572. https://doi.org/10.1038/cddis.2017.15
93. Xu W, Liu S, Ma L, Cheng L, Li Q, Qing L, Yang Y, Dong Z (2024) Identification of miRNA signature in cancer-associated fibroblast to predict recurrent prostate cancer. Comput Biol Med 180:108989. https://doi.org/10.1016/j.compbiomed.2024.108989
94. Luo J, Tao H, Wen Z, Chen L, Hu H, Guan H (2024) Tumor-associated fibroblasts promotes proliferation and migration of prostate cancer cells by suppressing FBXL3 via upregulating hsa-miR-18b-5p. Nan fang yi ke da xue xue bao = J South Med Univ 44(7):1284–1296. https://doi.org/10.12122/j.issn.1673-4254.2024.07.08
95. Zhao J, Shen J, Mao L, Yang T, Liu J, Hongbin S (2024) Cancer associated fibroblast secreted miR-432-5p targets CHAC1 to inhibit ferroptosis and promote acquired chemoresistance in prostate cancer. Oncogene 43(27):2104–2114. https://doi.org/10.1038/s41388-024-03057-6
96. Shan G, Gu J, Zhou D, Li L, Cheng W, Wang Y, Tang T, Wang X (2020) Cancer-associated fibroblast-secreted exosomal miR-423-5p promotes chemotherapy resistance in prostate cancer by targeting GREM2 through the TGF-β signaling pathway. Exp Mol Med 52(11):1809–1822. https://doi.org/10.1038/s12276-020-0431-z
97. Wang S, Du P, Cao Y, Ma J, Yang X, Yu Z, Yang Y (2022) Cancer associated fibroblasts secreted exosomal miR-1290 contributes to prostate cancer cell growth and metastasis via targeting GSK3β. Cell Death Discov 8(1):371. https://doi.org/10.1038/s41420-022-01163-6
98. Zhang Y, Zhao J, Ding M, Su Y, Cui D, Jiang C, Zhao S, Jia G, Wang X, Ruan Y, Jing Y, Xia S, Han B (2020) Loss of exosomal miR-146a-5p from cancer-associated fibroblasts after androgen deprivation therapy contributes to prostate cancer metastasis. J Exp Clin Cancer Res 39(1):282. https://doi.org/10.1186/s13046-020-01761-1
99. Doldi V, Callari M, Giannoni E, D'Aiuto F, Maffezzini M, Valdagni R, Chiarugi P, Gandellini P, Zaffaroni N (2015) Integrated gene and miRNA expression analysis of prostate cancer associated fibroblasts supports a prominent role for interleukin-6 in fibroblast activation. Oncotarget 6(31):31441–31460. https://doi.org/10.18632/oncotarget.5056
100. Tang M, Sun Y, Huang CP, Chen L, Liu B, You B, Wang Z, Chang C (2022) High dose androgen suppresses natural killer cytotoxicity of castration-resistant prostate cancer cells via altering AR/circFKBP5/miRNA-513a-5p/PD-L1 signals. Cell Death Dis 13(8):746. https://doi.org/10.1038/s41419-022-04956-w
101. Liu X, Chen Q, Yan J, Wang Y, Zhu C, Chen C, Zhao X, Xu M, Sun Q, Deng R, Zhang H, Qu Y, Huang J, Jiang B, Yu J (2013) MiRNA-296-3p-ICAM-1 axis promotes metastasis of prostate cancer by possible enhancing survival of natural killer cell-resistant circulating tumour cells. Cell Death Dis 4(11):e928. https://doi.org/10.1038/cddis.2013.458
102. Chen CH, Li SX, Xiang LX, Mu HQ, Wang SB, Yu KY (2018) HIF-1α induces immune escape of prostate cancer by regulating NCR1/NKp46 signaling through miR-224. Biochem Biophys Res Commun 503(1):228–234. https://doi.org/10.1016/j.bbrc.2018.06.007

103. Guan H, Peng R, Fang F, Mao L, Chen Z, Yang S, Dai C, Wu H, Wang C, Feng N, Xu B, Chen M (2020) Tumor-associated macrophages promote prostate cancer progression via exosome-mediated miR-95 transfer. J Cell Physiol 235(12):9729–9742. https://doi.org/10.1002/jcp.29784
104. Zhang LS, Chen QC, Zong HT, Xia Q (2024) Exosome miRNA-203 promotes M1 macrophage polarization and inhibits prostate cancer tumor progression. Mol Cell Biochem 479(9):2459–2470. https://doi.org/10.1007/s11010-023-04854-5
105. Wang Z, Xu L, Hu Y, Huang Y, Zhang Y, Zheng X, Wang S, Wang Y, Yu Y, Zhang M, Yuan K, Min W (2016) miRNA let-7b modulates macrophage polarization and enhances tumor-associated macrophages to promote angiogenesis and mobility in prostate cancer. Sci Rep 6:25602. https://doi.org/10.1038/srep25602
106. Costanzi E, Romani R, Scarpelli P, Bellezza I (2020) Extracellular vesicles-mediated transfer of miRNA let-7b from PC3 cells to macrophages. Genes 11(12). https://doi.org/10.3390/genes11121495
107. Rong J, Xu L, Hu Y, Liu F, Yu Y, Guo H, Ni X, Huang Y, Zhao L, Wang Z (2020) Inhibition of let-7b-5p contributes to an anti-tumorigenic macrophage phenotype through the SOCS1/STAT pathway in prostate cancer. Cancer Cell Int 20:470. https://doi.org/10.1186/s12935-020-01563-7
108. Tian HY, Liang Q, Shi Z, Zhao H (2022) Exosomal CXCL14 contributes to M2 macrophage polarization through NF-κB signaling in prostate cancer. Oxidative Med Cell Longev 2022:7616696. https://doi.org/10.1155/2022/7616696
109. Wang G, Yao L, Yang T, Guo L, Gu S, Liu J, Yang K (2019) MiR-451 suppresses the growth, migration, and invasion of prostate cancer cells by targeting macrophage migration inhibitory factor. Transl Cancer Res 8(2):647–654. https://doi.org/10.21037/tcr.2019.03.28
110. Hu S, Li L, Yeh S, Cui Y, Li X, Chang HC, Jin J, Chang C (2015) Infiltrating T cells promote prostate cancer metastasis via modulation of FGF11→miRNA-541→androgen receptor (AR)→MMP9 signaling. Mol Oncol 9(1):44–57. https://doi.org/10.1016/j.molonc.2014.07.013
111. Wang Z, Wang Y, Peng M, Yi L (2019) UBASH3B is a novel prognostic biomarker and correlated with immune infiltrates in prostate cancer. Front Oncol 9:1517. https://doi.org/10.3389/fonc.2019.01517
112. Comito G, Iscaro A, Bacci M, Morandi A, Ippolito L, Parri M, Montagnani I, Raspollini MR, Serni S, Simeoni L, Giannoni E, Chiarugi P (2019) Lactate modulates CD4(+) T-cell polarization and induces an immunosuppressive environment, which sustains prostate carcinoma progression via TLR8/miR21 axis. Oncogene 38(19):3681–3695. https://doi.org/10.1038/s41388-019-0688-7
113. Tao Z, Xu S, Ruan H, Wang T, Song W, Qian L, Chen K (2018) MiR-195/−16 family enhances radiotherapy via T cell activation in the tumor microenvironment by blocking the PD-L1 immune checkpoint. Cell Physiol Biochem 48(2):801–814. https://doi.org/10.1159/000491909
114. Muz B, De La Puente P, Azab F, Kareem Azab A (2015) The role of hypoxia in cancer progression, angiogenesis, metastasis, and resistance to therapy. Hypoxia (Auckl) 11:83–92
115. Shang W, Chen X, Nie L, Xu M, Chen N, Zeng H, Zhou Q (2013) MiR199b suppresses expression of hypoxia-inducible factor 1α (HIF-1α) in prostate cancer cells. Int J Mol Sci 14(4):8422–8436. https://doi.org/10.3390/ijms14048422
116. Li Y, Zhang D, Wang X, Yao X, Ye C, Zhang S, Wang H, Chang C, Xia H, Wang YC, Fang J, Yan J, Ying H (2015) Hypoxia-inducible miR-182 enhances HIF1α signaling via targeting PHD2 and FIH1 in prostate cancer. Sci Rep 5:12495. https://doi.org/10.1038/srep12495
117. Zha JM, Xu XF, Li DP, Gui Q, Liang RR, Zhou XM (2016) Hypoxia promotes the growth and invasiveness of prostate cancer cells by down-regulating miR-132 in vitro. Zhonghua nan ke xue = Natl J Androl 22(12):1071–1076
118. Andersen S, Richardsen E, Moi L, Donnem T, Nordby Y, Ness N, Holman ME, Bremnes RM, Busund LT (2016) Fibroblast miR-210 overexpression is independently associated with clinical failure in prostate cancer – a multicenter (in situ hybridization) study. Sci Rep 6:36573. https://doi.org/10.1038/srep36573

119. Angel CZ, Stafford MYC, McNally CJ, Nesbitt H, McKenna DJ (2023) MiR-21 is induced by hypoxia and down-regulates RHOB in prostate cancer. Cancer 15(4). https://doi.org/10.3390/cancers15041291
120. Zheng H, Bai L (2019) Hypoxia induced microRNA-301b-3p overexpression promotes proliferation, migration and invasion of prostate cancer cells by targeting LRP1B. Exp Mol Pathol 111:104301. https://doi.org/10.1016/j.yexmp.2019.104301
121. Angel CZ, Lynch SM, Nesbitt H, McKenna MM, Walsh CP, McKenna DJ (2020) miR-210 is induced by hypoxia and regulates neural cell adhesion molecule in prostate cells. J Cell Physiol 235(9):6194–6203. https://doi.org/10.1002/jcp.29548
122. Wang W, Liu M, Guan Y, Wu Q (2016) Hypoxia-responsive Mir-301a and Mir-301b promote Radioresistance of prostate cancer cells via downregulating NDRG2. Med Sci Monit 22:2126–2132. https://doi.org/10.12659/msm.896832
123. Guo YJ, Liu JX, Guan YW (2016) Hypoxia induced upregulation of miR-301a/b contributes to increased cell autophagy and viability of prostate cancer cells by targeting NDRG2. Eur Rev Med Pharmacol Sci 20(1):101–108
124. Gu H, Liu M, Ding C, Wang X, Wang R, Wu X, Fan R (2016) Hypoxia-responsive miR-124 and miR-144 reduce hypoxia-induced autophagy and enhance radiosensitivity of prostate cancer cells via suppressing PIM1. Cancer Med 5(6):1174–1182. https://doi.org/10.1002/cam4.664
125. Wang W, Liu J, Wu Q (2016) MiR-205 suppresses autophagy and enhances radiosensitivity of prostate cancer cells by targeting TP53INP1. Eur Rev Med Pharmacol Sci 20(1):92–100
126. Panigrahi GK, Ramteke A, Birks D, Abouzeid Ali HE, Venkataraman S, Agarwal C, Vibhakar R, Miller LD, Agarwal R, Abd Elmageed ZY, Deep G (2018) Exosomal microRNA profiling to identify hypoxia-related biomarkers in prostate cancer. Oncotarget 9(17):13894–13910. https://doi.org/10.18632/oncotarget.24532
127. Liu Z, Lin Z, Jiang M, Zhu G, Xiong T, Cao F, Cui Y, Niu YN (2024) Cancer-associated fibroblast exosomes promote prostate cancer metastasis through miR-500a-3p/FBXW7/HSF1 axis under hypoxic microenvironment. Cancer Gene Ther 31(5):698–709. https://doi.org/10.1038/s41417-024-00742-2
128. Zhang H, Liang F, Yue J, Liu P, Wang J, Wang Z, Li H, Cheng D, Du J, Zhang K, Du P (2020) MicroRNA-137 regulates hypoxia-mediated migration and epithelial-mesenchymal transition in prostate cancer by targeting LGR4 via the EGFR/ERK signaling pathway. Int J Oncol 57(2):540–549. https://doi.org/10.3892/ijo.2020.5064
129. Yue B (2014) Biology of the extracellular matrix: an overview. J Glaucoma 23(8 Suppl 1):S20–S23. https://doi.org/10.1097/ijg.0000000000000108
130. Frantz C, Stewart KM, Weaver VM (2010) The extracellular matrix at a glance. J Cell Sci 123(Pt 24):4195–4200. https://doi.org/10.1242/jcs.023820
131. Luthold C, Hallal T, Labbé DP, Bordeleau F (2022) The extracellular matrix stiffening: a trigger of prostate cancer progression and castration resistance? Cancer 14(12). https://doi.org/10.3390/cancers14122887
132. Provenzano PP, Eliceiri KW, Campbell JM, Inman DR, White JG, Keely PJ (2006) Collagen reorganization at the tumor-stromal interface facilitates local invasion. BMC Med 4(1):38. https://doi.org/10.1186/1741-7015-4-38
133. Song H, Lu T, Han D, Zhang J, Gan L, Xu C, Liu S, Li P, Zhang K, Hu Z, Li H, Li Y, Zhao X, Zhang J, Xing N, Shi C, Wen W, Yang F, Qin W (2024) YAP1 inhibition induces phenotype switching of cancer-associated fibroblasts to tumor suppressive in prostate cancer. Cancer Res 84(22):3728–3742. https://doi.org/10.1158/0008-5472.can-24-0932
134. Moroz A, Delella FK, Lacorte LM, Deffune E, Felisbino SL (2013) Fibronectin induces MMP2 expression in human prostate cancer cells. Biochem Biophys Res Commun 430(4):1319–1321. https://doi.org/10.1016/j.bbrc.2012.12.031
135. Martinucci B, Cucielo MS, Minatel BC, Cury SS, Caxali GH, Aal MCE, Felisbino SL, Pinhal D, Carvalho RF, Delella FK (2022) Fibronectin modulates the expression of miRNAs in prostate cancer cell lines. Front Vet Sci 9:879997. https://doi.org/10.3389/fvets.2022.879997

136. Chen PC, Tang CH, Lin LW, Tsai CH, Chu CY, Lin TH, Huang YL (2017) Thrombospondin-2 promotes prostate cancer bone metastasis by the up-regulation of matrix metalloproteinase-2 through down-regulating miR-376c expression. J Hematol Oncol 10(1):33. https://doi.org/10.1186/s13045-017-0390-6
137. Simkova D, Kharaishvili G, Slabakova E, Murray PG, Bouchal J (2016) Glycoprotein asporin as a novel player in tumour microenvironment and cancer progression. Biomed Pap Med Fac Univ Palacky Olomouc Czech Repub 160(4):467–473. https://doi.org/10.5507/bp.2016.037
138. Untergasser G, Gander R, Lilg C, Lepperdinger G, Plas E, Berger P (2005) Profiling molecular targets of TGF-beta1 in prostate fibroblast-to-myofibroblast transdifferentiation. Mech Ageing Dev 126(1):59–69. https://doi.org/10.1016/j.mad.2004.09.023
139. Ottley EC, Nicholson HD, Gold EJ (2016) Activin a regulates microRNAs and gene expression in LNCaP cells. Prostate 76(11):951–963. https://doi.org/10.1002/pros.23184
140. Ni WD, Yang ZT, Cui CA, Cui Y, Fang LY, Xuan YH (2017) Tenascin-C is a potential cancer-associated fibroblasts marker and predicts poor prognosis in prostate cancer. Biochem Biophys Res Commun 486(3):607–612. https://doi.org/10.1016/j.bbrc.2017.03.021
141. Liu B, Li X, Wang D, Yu Y, Lu D, Chen L, Lv F, Li Y, Cheng L, Song Y, Xing Y (2022) CEMIP promotes extracellular matrix-detached prostate cancer cell survival by inhibiting ferroptosis. Cancer Sci 113(6):2056–2070. https://doi.org/10.1111/cas.15356
142. Dong B, Gu Y, Sun X, Wang X, Zhou Y, Rong Z, Zhang J, Shi X, Zhang Z, He X, Chen L, Xiong Q, Pang X, Cui Y (2024) Targeting TUBB3 suppresses Anoikis resistance and bone metastasis in prostate cancer. Adv Healthc Mater 13(28):e2400673. https://doi.org/10.1002/adhm.202400673
143. D'Antonio KB, Toubaji A, Albadine R, Mondul AM, Platz EA, Netto GJ, Getzenberg RH (2010) Extracellular matrix associated protein CYR61 is linked to prostate cancer development. J Urol 183(4):1604–1610. https://doi.org/10.1016/j.juro.2009.12.006
144. Lucarelli G, Rutigliano M, Bettocchi C, Palazzo S, Vavallo A, Galleggiante V, Trabucco S, Di Clemente D, Selvaggi FP, Battaglia M, Ditonno P (2013) Spondin-2, a secreted extracellular matrix protein, is a novel diagnostic biomarker for prostate cancer. J Urol 190(6):2271–2277. https://doi.org/10.1016/j.juro.2013.05.004
145. Ko CJ, Huang CC, Lin HY, Juan CP, Lan SW, Shyu HY, Wu SR, Hsiao PW, Huang HP, Shun CT, Lee MS (2015) Androgen-induced TMPRSS2 activates matriptase and promotes extracellular matrix degradation, prostate cancer cell invasion, tumor growth, and metastasis. Cancer Res 75(14):2949–2960. https://doi.org/10.1158/0008-5472.can-14-3297
146. Cornblatt G, Yegnasubramanian S, Lee B, Nelson W (2008) Epigenetic regulation of microRNAs in prostate cancer. Cancer Res 68(9_Supplement):52–52
147. Hulf T, Sibbritt T, Wiklund ED, Bert S, Strbenac D, Statham AL, Robinson MD, Clark SJ (2011) Discovery pipeline for epigenetically deregulated miRNAs in cancer: integration of primary miRNA transcription. BMC Genomics 12(1):54. https://doi.org/10.1186/1471-2164-12-54
148. Theodore SC, Davis M, Zhao F, Wang H, Chen D, Rhim J, Dean-Colomb W, Turner T, Ji W, Zeng G, Grizzle W, Yates C (2014) MicroRNA profiling of novel African American and Caucasian prostate cancer cell lines reveals a reciprocal regulatory relationship of miR-152 and DNA methyltranferase 1. Oncotarget 5(11):3512–3525. https://doi.org/10.18632/oncotarget.1953
149. Xue G, Ren Z, Chen Y, Zhu J, Du Y, Pan D, Li X, Hu B (2015) A feedback regulation between miR-145 and DNA methyltransferase 3b in prostate cancer cell and their responses to irradiation. Cancer Lett 361(1):121–127. https://doi.org/10.1016/j.canlet.2015.02.046
150. Cao Q, Mani RS, Ateeq B, Dhanasekaran SM, Asangani IA, Prensner JR, Kim JH, Brenner JC, Jing X, Cao X, Wang R, Li Y, Dahiya A, Wang L, Pandhi M, Lonigro RJ, Wu YM, Tomlins SA, Palanisamy N, Qin Z, Yu J, Maher CA, Varambally S, Chinnaiyan AM (2011) Coordinated regulation of polycomb group complexes through microRNAs in cancer. Cancer Cell 20(2):187–199. https://doi.org/10.1016/j.ccr.2011.06.016

151. Halkidou K, Cook S, Leung HY, Neal DE, Robson CN (2004) Nuclear accumulation of histone deacetylase 4 (HDAC4) coincides with the loss of androgen sensitivity in hormone refractory cancer of the prostate. Eur Urol 45(3):382–389; author reply 389. https://doi.org/10.1016/j.eururo.2003.10.005
152. Noonan EJ, Place RF, Pookot D, Basak S, Whitson JM, Hirata H, Giardina C, Dahiya R (2009) miR-449a targets HDAC-1 and induces growth arrest in prostate cancer. Oncogene 28(14):1714–1724. https://doi.org/10.1038/onc.2009.19
153. Singh PK, Campbell MJ (2013) The interactions of microRNA and epigenetic modifications in prostate cancer. Cancer 5(3):998–1019. https://doi.org/10.3390/cancers5030998
154. Damodaran C, Das TP, Papu John AM, Suman S, Kolluru V, Morris TJ, Faber EN, Rai SN, Messer JC, Alatassi H, Ankem MK (2016) miR-301a expression: a prognostic marker for prostate cancer. Urol Oncol 34(8):336.e313–336.e320. https://doi.org/10.1016/j.urolonc.2016.03.009
155. Kolluru V, Chandrasekaran B, Tyagi A, Dervishi A, Ankem M, Yan X, Maiying K, Alatassi H, Shaheen SP, Messer JC, Edwards A, Haddad A, Damodaran C (2018) miR-301a expression: diagnostic and prognostic marker for prostate cancer. Urol Oncol 36(11):503.e509–503.e515. https://doi.org/10.1016/j.urolonc.2018.07.014
156. Zidan HE, Abdul-Maksoud RS, Elsayed WSH, Desoky EAM (2018) Diagnostic and prognostic value of serum miR-15a and miR-16-1 expression among egyptian patients with prostate cancer. IUBMB Life 70(5):437–444. https://doi.org/10.1002/iub.1733
157. Pang C, Liu M, Fang W, Guo J, Zhang Z, Wu P, Zhang Y, Wang J (2016) MiR-139-5p is increased in the peripheral blood of patients with prostate cancer. Cell Physiol Biochem 39(3):1111–1117. https://doi.org/10.1159/000447819
158. Pang C, Song X, Fu C, Zhang Y, Zhang Y, Liu M (2020) Diagnostic value of peripheral blood miR-374b-5p in patients with prostate cancer. Clin Lab 66(1). https://doi.org/10.7754/Clin.Lab.2019.190620
159. Nitusca D, Marcu A, Seclaman E, Bardan R, Sirbu IO, Balacescu O, Bucur AI, Ursoniu S, Marian C (2022) Diagnostic value of microRNA-375 as future biomarker for prostate cancer detection: a meta-analysis. Medicina (Kaunas, Lithuania) 58(4). https://doi.org/10.3390/medicina58040529
160. Moya L, Meijer J, Schubert S, Matin F, Batra J (2019) Assessment of miR-98-5p, miR-152-3p, miR-326 and miR-4289 expression as biomarker for prostate cancer diagnosis. Int J Mol Sci 20(5). https://doi.org/10.3390/ijms20051154
161. Dülgeroğlu Y, Eroğlu O (2020) Serum levels of miR-223-3p and miR-223-5p in prostate diseases. MicroRNA (Shariqah, United Arab Emirates) 9(4):303–309. https://doi.org/10.2174/2211536609666201106090458
162. Byun YJ, Piao XM, Jeong P, Kang HW, Seo SP, Moon SK, Lee JY, Choi YH, Lee HY, Kim WT, Lee SC, Cha EJ, Yun SJ, Kim WJ (2021) Urinary microRNA-1913 to microRNA-3659 expression ratio as a non-invasive diagnostic biomarker for prostate cancer. Investig Clin Urol 62(3):340–348. https://doi.org/10.4111/icu.20200488
163. Lin HM, Nikolic I, Yang J, Castillo L, Deng N, Chan CL, Yeung NK, Dodson E, Elsworth B, Spielman C, Lee BY, Boyer Z, Simpson KJ, Daly RJ, Horvath LG, Swarbrick A (2018) MicroRNAs as potential therapeutics to enhance chemosensitivity in advanced prostate cancer. Sci Rep 8(1):7820. https://doi.org/10.1038/s41598-018-26050-y
164. Fitzpatrick JM, de Wit R (2014) Taxane mechanisms of action: potential implications for treatment sequencing in metastatic castration-resistant prostate cancer. Eur Urol 65(6):1198–1204. https://doi.org/10.1016/j.eururo.2013.07.022
165. Jin M, Zhang T, Liu C, Badeaux MA, Liu B, Liu R, Jeter C, Chen X, Vlassov AV, Tang DG (2014) miRNA-128 suppresses prostate cancer by inhibiting BMI-1 to inhibit tumor-initiating cells. Cancer Res 74(15):4183–4195. https://doi.org/10.1158/0008-5472.can-14-0404
166. Chauhan N, Manojkumar A, Jaggi M, Chauhan SC, Yallapu MM (2022) microRNA-205 in prostate cancer: overview to clinical translation. Biochim Biophys Acta Rev Cancer 1877(6):188809. https://doi.org/10.1016/j.bbcan.2022.188809

167. Mulholland EJ, Green WP, Buckley NE, McCarthy HO (2019) Exploring the potential of MicroRNA let-7c as a therapeutic for prostate cancer. Mol Ther Nucleic Acids 18:927–937. https://doi.org/10.1016/j.omtn.2019.09.031
168. Iscaife A, Reis ST, Morais DR, Viana NI, da Silva IA, Pimenta R, Bordini A, Dip N, Srougi M, Leite KRM (2018) Treating metastatic prostate cancer with microRNA-145. Apoptosis 23(7-8):388–395. https://doi.org/10.1007/s10495-018-1461-z
169. Luo B, Yuan Y, Zhu Y, Liang S, Dong R, Hou J, Li P, Xing Y, Lu Z, Lo R, Kuang GM (2022) microRNA-145-5p inhibits prostate cancer bone metastatic by modulating the epithelial-mesenchymal transition. Front Oncol 12:988794. https://doi.org/10.3389/fonc.2022.988794
170. Xu G, Meng Y, Wang L, Dong B, Peng F, Liu S, Li S, Liu T (2021) miRNA-214-5p inhibits prostate cancer cell proliferation by targeting SOX4. World J Surg Oncol 19(1):338. https://doi.org/10.1186/s12957-021-02449-2
171. Li C, Sun Z, Song Y, Zhang Y (2022) Suppressive function of bone marrow-derived mesenchymal stem cell-derived exosomal microRNA-187 in prostate cancer. Cancer Biol Ther 23(1):1–14. https://doi.org/10.1080/15384047.2022.2123675
172. Hong DS, Kang YK, Borad M, Sachdev J, Ejadi S, Lim HY, Brenner AJ, Park K, Lee JL, Kim TY, Shin S, Becerra CR, Falchook G, Stoudemire J, Martin D, Kelnar K, Peltier H, Bonato V, Bader AG, Smith S, Kim S, O'Neill V, Beg MS (2020) Phase 1 study of MRX34, a liposomal miR-34a mimic, in patients with advanced solid tumours. Br J Cancer 122(11):1630–1637. https://doi.org/10.1038/s41416-020-0802-1
173. Ghamlouche F, Yehya A, Zeid Y, Fakhereddine H, Fawaz J, Liu YN, Al-Sayegh M, Abou-Kheir W (2023) MicroRNAs as clinical tools for diagnosis, prognosis, and therapy in prostate cancer. Transl Oncol 28:101613. https://doi.org/10.1016/j.tranon.2022.101613
174. Hussein MA, Munirathinam G (2023) MicroRNAs in prostate cancer: implications for treatment response and therapeutic targets. Cancer 15(20). https://doi.org/10.3390/cancers15205023

Chapter 14
Role of miRNAs in Bone Metastasis of Prostate Cancer: A Comprehensive Update

Shilajit Khan, Bouddhaditya Roy, and Amlan Ghosh

Abstract Prostate cancer (PCa) is a malignancy unique to males. Bone metastasis represents the most common adverse outcome in PCa, involving homing of malignant cells from primary tumor into the bone followed by its (tumor cell) final settlement and colonization to disrupt the normal osteoblast-osteoclast remodeling cycle. Signaling pathways like TGFβ, NFκB, Wnt/β-catenin, BMP, and PI3K/Akt have been found to play important role in PCa bone metastasis. Researches have elucidated the impact of microRNAs (miRNAs) in different signaling pathways associated with pathological conditions like cancer. miRNAs involve in silencing of target mRNAs by binding to their 3′ UTRs. Based on the target mRNAs (tumor suppressive/oncogenic), miRNAs may be oncogenic or tumor-suppressive, respectively, thus establishing their role in tumorigenesis as well as associated adverse outcomes. In this chapter, we discuss on various events of bone metastasis, the associated signaling pathways, and, the impact of miRNAs on regulation of said pathways to assess their (miRNAs) influence on the disease outcome for scientific and clinical interest.

Keywords Prostate cancer · Bone metastasis · miRNA · mRNA · Clinical importance

S. Khan · A. Ghosh (✉)
Genetics of Non-communicable Diseases, Department of Life Sciences, Presidency University, Kolkata, West Bengal, India
e-mail: amlan.dbs@presiuniv.ac.in

B. Roy
College of Medicine and JNM Hospital, WBUHS, Kalyani, West Bengal, India

S. Pathak, A. Banerjee (eds.), *MicroRNAs: Role in Cancer Progression, Diagnostics and Therapy*, Advances in Biochemistry in Health and Disease 34,
https://doi.org/10.1007/978-3-032-08824-6_14

Introduction

Prostate cancer (PCa) is a malignancy of male reproductive organ affecting mainly aged individuals, and accounts for significant cause of mortality [50]. In PCa, bone metastasis (BM) or spreading of malignant cells from primary tumor to bone followed by their colonization and formation of secondary tumor, is considered as the most common adverse outcome affecting the patient survival significantly [20]. In bone metastasis of PCa, axial skeleton like pelvis, ribs, and spine are most commonly/preferentially affected [36]; apart from these, skull and proximal part of long bones like humerus and femur are also included among the affected sites (American Cancer Society 2020) (https://www.cancer.org/content/dam/CRC/PDF/Public/6759.00.pdf). Interestingly and alarmingly, in about 10% of patients, the bone metastasis is detected at the time of disease diagnosis [20]. In a prospective study with Indian patient population, bone metastasis was diagnosed in about 61% of the diseased individuals [9]. The treatment modalities for bone metastasis in PCa mainly focus on management of symptoms like relieve of pain and also improvement of bone health to slow down the disease progression [12]. However, bone metastasis severely affects the life span of PCa patients as evident from very low rate of 5-year survival (33%) of affected individuals [6]. Therefore, the conventional treatment plan for PCa of advanced stage needs substantial improvement for better management of the disease, and, for this, molecular markers associated with disease pathogenesis that can be targeted by novel therapeutics, need to be identified. Thus, the situation warranted a detailed review work on molecular signaling pathways frequently dysregulated during bone metastasis in PCa along with identification of the key modulators of said pathways, so that, novel therapeutics can be designed targeting them in future research. Among key signaling pathways, dysregulation of Wnt/β-catenin, TGFβ, NFκB, BMP, and PI3K/AKT pathways was frequently found in bone metastasis of PCa. miRNAs represent a class of small noncoding RNAs that regulate the expression of protein coding genes at posttranscriptional level [52]. By regulating the expression of key modulators, miRNAs can influence the dysregulation of a signaling pathway and thus affect associated disease phenotype. This review work aimed to summarize candidate miRNAs from the published literature, associated with the dysregulation of bone metastasis related key signaling pathways in PCa, so that future research can be directed to design targeted therapeutics against them.

Bone Metastasis in Prostate Cancer

Bone Metastasis: An Overview

In PCa, bone represents the most common site of metastasis, compared to other body parts. Actually, according to seed and soil hypothesis, metastatic PCa cells are considered as seeds that find a suitable environment for growth in bone tissue which thus serves as soil. Actually, PCa cells, after reaching the bone, can successfully interact with residential bone cells including osteoclast and osteoblast to establish a microenvironment successfully that actually contribute to the development of secondary or metastatic tumor [20]. The entire process occurs in a stepwise manner involving a continuous interaction between tumor cells and osteoblast/osteoclast through a series of signaling molecules belonging to different molecular pathways.

In PCa, bone metastasis initiates with acquisition of mesenchymal phenotype by malignant cells through epithelial-to-mesenchymal transition (EMT), characterized by decreased expression level of E-cadherin followed by a rise in N-cadherin protein level, commonly known as Cadherin switch [21]. Tumor cells after acquisition of mesenchymal phenotype, get detached from primary tumor, invade the surrounding extra-cellular matrix (ECM) with the aid of enzymes like matrix metalloproteinases (MMPs) and enter the blood vessels (intravastion). However, around 85% of the intravasating PCa cells undergo intravascular cell death [42]. The surviving tumor cells through their surface receptor CD44 interact with endothelial adhesion molecule VCAM-1, to get attached to vascular endothelium first, then, traverse through the wall of blood vessels (extravasation) to reach the target site (bone) [31]. Research work suggested a role of chemotactic factors in bone metastasis of PCa; SDF-1/CXCR4 axis represents one of such chemotaxis pathways in which bone marrow secretes SDF-1 chemokine to guide the homing of CXCR4-positive PCa cells to reach the target site [73]. In the bone marrow, for initial survival, PCa cells interact with human bone marrow endothelial (HBME) cells, and in this process, cell surface adhesion molecule like p-selectin of HBME, and integrin molecules ($\alpha v\beta 3$, $\alpha 5\beta 1$, $\alpha 3\beta 1$) of PCa cells play a crucial role [33]. Next PCa cells expressing CXCL16 start to interact with CXCR6 receptor on bone marrow mesenchymal stem cells (BMSCs) leading to transformation of the latter into tumor-associated fibroblasts that release additional amounts of SDF-1 to reinforce the SDF-1/CXCR4 axis between bone and metastatic cells [31]. After reaching the bone marrow, for acclimatization to the surrounding environment, metastatic PCa cells enter into a dormant state and may remain there for several years; the phenomenon is thought to be achieved by a complex interaction of PCa cells with bone marrow endothelial cells, mesenchymal stem cells, and bone marrow stromal cells through a series of growth factors, chemokines, cytokines, and adhesion molecules; among signaling molecules, growth-arrest specific protein 6 (GAS6) secreted by PCa cells and AXL, TGF-β2, BMP7 produced by component bone marrow cells are important [31, 56, 69]. Meanwhile, to improve their adhesion and survival in the bone microenvironment, PCa cells gradually adopt the characteristic features of mainly osteoblast cells; the phenomenon is called osteo-mimicry in which bone stromal cell regulated expression of osteocalcin and bone sialoprotein by PCa cells are thought to play a crucial role [27]. Now within bone microenvironment, PCa cells by some autocrine/paracrine signaling re-initiate their proliferation to form metastatic bone lesions that may be of osteoblastic, osteolytic, or mixed type [8]. In PCa, majority of bone lesions are of osteoblastic/osteosclerotic type in which tumor cells secrete a series of signaling molecules (TGFβ, endothelin-1, IGFs, BMPs, etc.) to interact with osteoblasts leading to their preferential activation [20]. Now, due to excessive osteoblastic activity and downregulated activity of osteoclasts, an imbalance of normal bone remodeling process occurs that ultimately results in the formation of under-mineralized, irregular, and fragile bone tissue (woven bone) prone to fracture [20]. Interestingly, osteoblasts after being activated by PCa cells, also secrete growth factors like IGF-1, IL-6, IL-8, etc., that in turn activate tumor cells and promote their proliferation adjacent to osteosclerotic lesions [1, 20]. Osteolytic bone lesions are comparatively less frequent in PCa, and occur due to enhanced differentiation and activity of osteoclast stimulated by tumor cell derived signaling factors like parathyroid hormone-related peptide (PTHrP), TNF-α, cytokines like IL-1 and IL-6; tumor cell derived PTHrP mediates reduced expression of osteoprotegerin (OPG) by osteoblast, and thereby facilitates the interaction between RANKL and RANK (receptor that activates NF-kB) expressed on the surface of osteoblasts and pre-osteoclasts respectively leading to differentiation of the latter into mature osteoclast and its enhanced activity. Higher resorption activity of osteoclast and decreased activity of osteoblast ultimately results into excessive degradation of bone, and during this process, osteoclasts also secrete some factors that serve as signals for survival and further growth of PCa cells adjacent to the lesion [20]. The mixed type of bone lesion is characterized by the presence of both bone destroying (osteolytic) and forming (osteosclerotic) lesions within the same metastatic deposit of bone tissue. Mixed bone lesions are mainly found in PCa having advanced clinicopathological features like higher Gleason score, lymph node metastasis, etc. [35]. Various events associated with PCa bone metastasis has been depicted schematically in Fig. 14.1.

Diagnosis and Treatment

In PCa, early and accurate detection of bone metastasis is crucial for effective management of its short- and long-term medical consequences. To achieve this, advanced imaging techniques and biomarkers are employed followed by selection of suitable therapeutic strategies.

Diagnosis by Imaging and Biomarkers

Imaging techniques commonly used to diagnose BM in PCa include computed tomography (CT), magnetic resonance imaging (MRI), and positron emission tomography/computed tomography (PET/CT). In CT, large metastatic bone lesions can be identified by visualization of parameters like cortical destruction, and sclerotic bone growth; however, it does not seem to be a suitable tool in diagnosing bone marrow metastasis due to low sensitivity [24, 41]. The implementation of dual energy CT (DECT) and quantitative CT (QCT) can achieve superior lesion detection and structural examination capabilities, respectively [3, 19]. MRI scans like diffusion-weighted imaging (DWI) and whole-body MRI (WB-MRI) have excellent

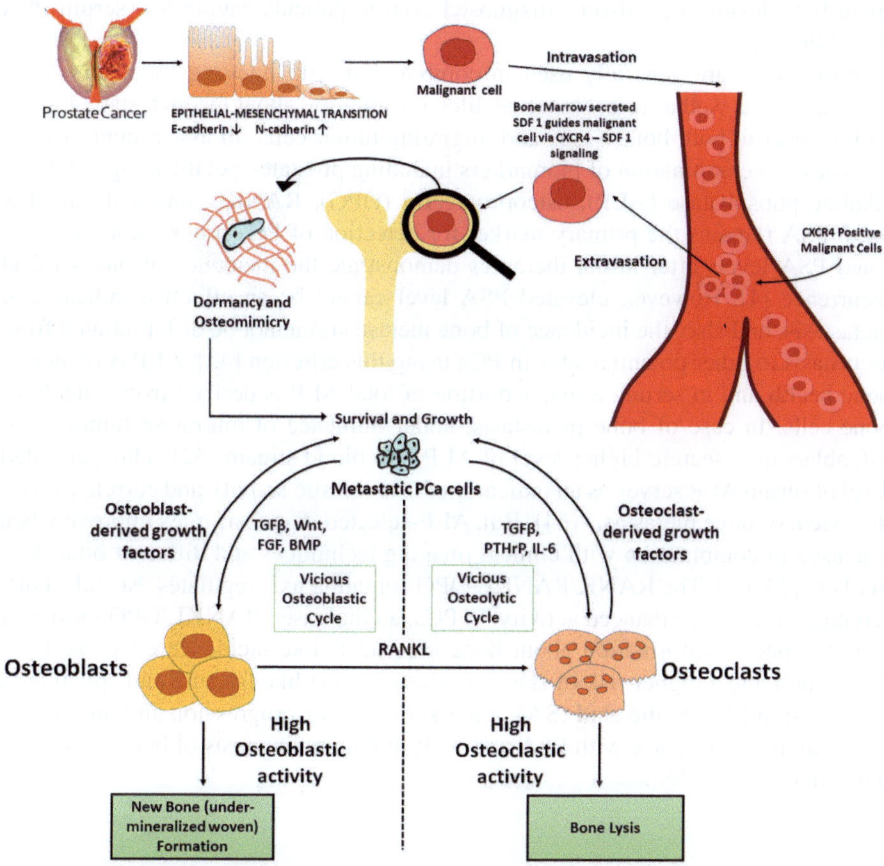

Fig. 14.1 A schematic diagram showing various stepwise events associated with bone metastasis in prostate cancer. ↑, upregulation and ↓, downregulation

soft-tissue detection capabilities; this type of imaging is suitable for visualization of bone marrow and it enables physicians to spot early-stage bone metastases before cortical destruction [44, 46]. PET/CT becomes the most superior method to detect bone metastasis in prostate cancer. In ^{18}F-NaF PET/CT, sodium fluoride with fluorine-18, a radioactive tracer usually accumulates at the site of very high bone turn over due to pathological conditions like cancer, is used [7]. Although ^{18}F-NaF PET/CT has been shown to have very high sensitivity in diagnosing bone metastasis, its specificity does not seem to be sufficient, and hence careful interpretation of the data by trained person is mandatory to identify and eliminate the false positive cases [38]. In ^{68}Ga-PSMA-11 PET/CT, a radioactive tracer ^{68}Ga-PSMA-11 is used that enables detection of prostate cancer cells by imaging on virtue of its binding specificity with prostate-specific membrane antigen (PSMA), a membrane bound glycoprotein over-expressed in PCa cells. In this imaging technique, the presence of very small bone lesions can also be diagnosed even in patients having low serum PSA level [49, 54].

Biomarkers are generally used in combination with imaging for diagnosis of bone lesions as well as to get more detailed information about the metastatic lesions with respect to both bone tissue and migrating tumor cells. In assessment of bone metastasis, a combination of biomarkers including prostate-specific antigen (PSA), alkaline phosphatase (ALP), osteoprotegerin (OPG), RANKL, and sialic acid is used. PSA remains the primary marker for detection of PCa progression, and elevated PSA levels after initial therapies demonstrate the incidence of biochemical recurrence [4]. However, elevated PSA level cannot be an effective indicator of metastasis, and also, the incidence of bone metastasis cannot be differentiated from metastasis to other potential sites in PCa using this criterion [45]. ALP is related to bone health and in serum, a major portion of total ALP is derived from osteoblast bone cells. In case of bone metastasis, under influence of migrating tumor cells, osteoblast may secrete higher level of ALP into blood stream [32]. Thus, elevated level of serum ALP serves as an indicator of osteoblastic activity and correlates with the extent of bone metastasis [64]. But, ALP-selected diagnosis rates improve when it is used in combination with clinical imaging techniques and different other biomarkers [23, 65]. The RANK/RANKL/OPG signaling axis regulates osteoclast differentiation and its enhanced activity. In PCa, an increased RANKL/OPG ratio may provide specific information about bone metastasis like incidence of active bone destruction and a higher risk of skeletal-related events like fracture and spinal cord compression [71]. Sialic acid (SA), a marker of tumor progression and metastasis, shows high concordance with PSA and ALP, aiding in diagnosis of bone metastasis in prostate cancer [58].

Treatment

Patients with bone metastatic PCa develop bone related complications like bone pain, spinal cord compression, and fragile/under-mineralized bone liable for a fracture (skeletal related events/SRE). Therefore, a therapeutic strategy is designed accordingly including conventional systemic therapy for slowing cancer progression, and additional therapies using bone modifying agents for pain management and prevention of bone related complications [43]. In systemic therapy, generally androgen synthesis inhibitor (abiraterone) or/and androgen receptor inhibitors (enzalutamide, apalutamide, darolutamide) along with anti-mitotic chemotherapeutic drugs (docetaxel, cabazitaxel) are used [43]. Among bone modifying agents, bisphosphonates (Zoledronate, clodronate) that inhibit osteoclast maturation leading to its apoptotic death, are commonly used to treat osteolytic bone lesions in metastatic PCa [20]; however, usage of the drug has been found to be associated with some adverse and pathological conditions like renal failure, osteonecrosis of the jaw, etc. [17]. As an alternative to bisphosphonate, denosumab, a monoclonal antibody targeting RANKL was developed. Although better outcome with respect to delayed skeletal related events was obtained, however, associated complications like hypocalcemia, osteonecrosis of the jaw was seen to be more severe in denosumab treated group than bisphosphonate treated one [20]. For the management of bone metastasis related pain, radiopharmaceuticals like external beam radiation therapy (EBRT), radium-223 (a calcium memetic radio-nucleotide that accumulates in portions of bone tissue with high turnover and induce apoptosis of metastatic PCa cells) are recommended. In addition, Zoledronate, a bisphosphonate, was also found to be effective in pain relief [43].

However, despite having these treatment modalities, the 5-year survival rate of metastatic (bone) PCa patients still remains very low (33%) [6]. Hence, for better management, molecular events along with their modulators associated with this advanced stage of the disease should be explored in details so that novel targets can be identified for designing effective therapeutic strategies. In this chapter, microRNAs modulating the key signaling pathways associated with BM in PCa were discussed, and their potency to serve as potential targets by natural compounds was explored in quest to develop alternate and effective strategies for better patient outcome.

Role of miRNA in Molecular Pathogenesis of PCa Bone Metastasis

miRNA: Biogenesis and Mechanism of Action

MicroRNAs (miRNAs) are a class of small noncoding RNAs (~19 to 24 nucleotides in length), endogenously derived and evolutionarily conserved. miRNAs cover about 1–5% of the human genome, and regulate the expression of about 30% of protein coding genes by binding to respective mRNAs to disrupt their translation [10, 52,]. miRNAs are transcribed by RNA polymerase II from intronic/exonic regions of protein coding gene as large primary miRNAs (pri-miRNAs) of several hundred nucleotides in length containing a stem loop-like structure; pri-miRNAs undergo several steps of processing to generate mature miRNAs following canonical or noncanonical biogenesis pathway [10, 52, 53]. In the canonical pathway, the pri-miRNA is cleaved into approximately 70 to 120 nucleotides long precursor miRNA (pre-miRNA) by a microprocessor complex containing ~160 kDa nuclear RNase III enzyme called Drosha and a double-stranded RNA binding protein Pasha (DiGeorge syndrome critical region gene 8/DGCR8) [10, 52]. pre-miRNA is then exported to cytoplasm via Exportin 5 (XPO5) and Ras-related nuclear protein (RAN), and subsequently gets cleaved near the loop into small dsRNA by another RNase III enzyme, Dicer. Next, the miRNA duplex is loaded onto an Argonaute (AGO) protein, and its one strand becomes the passenger that gets degraded, while the other remains as guide to form the mature miRNA, existing in a ribonucleoprotein complex, called RISC (RNA induced silencing complex) [52]. The noncanonical pathway, also called Mirtron Pathway is microprocessor-independent, and here the pre-miRNA is formed by processing of introns of a pre-mRNA transcript into a lariat mirtron and its subsequent debranching by the lariat debranching enzyme. Then, the pre-miRNA is exported into the cytoplasm and further processed to generate mature miRNA in a similar fashion to the canonical pathway [10, 52].

The miRNA-mRNA interaction is mediated by complementary base pairing between a 7-nucleotide seed sequence of miRNA with the microRNA response element (MRE) at 3′ UTR region of its target mRNA [74]. Now, argonaute recruits an adapter protein, trinucleotide repeat containing 6 (TNRC6) that binds to poly-A binding protein (PABPC) at 3′ end of mRNA leading to its destabilization and inhibition of translation [52].

miRNAs in Cancer Pathogenesis

miRNAs have been found to play pivotal role in the pathogenesis of various disease conditions including cancer [10, 52].

miRNAs may play oncogenic or tumor-suppressive role depending upon downstream target mRNAs. Oncogenic miRNAs promote cancer progression by attenuating tumor suppressor genes and in contrast, tumor-suppressive miRNAs repress cancer progression by targeting the mRNAs of oncogenic function [5, 60]. Like other malignancies, numerous miRNAs were found to be involved in the pathogenesis of prostate cancer onset, progression, as well as associated complications like bone metastasis. In this chapter, we highlight key oncogenic and tumor-suppressive miRNAs involved in modulation of various signaling pathways associated with PCa bone metastasis (Fig. 14.2).

miRNAs Regulating Bone Metastasis in Prostate Cancer and Associated Signaling Pathways

miR-210-3p, miR-218-5p, miR-532-3p, and miR-141-3p

Among these miRNAs, miR-210-3p was found to have an oncogenic role, and tumor suppressive function of the other three were reported in prostate tumorigenesis. An elevated expression of miR-210-3p in PCa compared to adjacent normal, and association of expression level with higher serum PSA level and Gleason score suggested its potential oncogenic role in disease pathogenesis; this seemed to be validated from positive impact of the miRNA on EMT, migration, and invasion in PCa cells (VCaP and C4-2B) [51]. Interestingly, miR-210-3p was seen to be significantly over-expressed in metastatic PCa lesions compared to nonmetastatic one, suggesting its influence on the incidence of bone metastasis. Silencing of miR-210-3p was seen to suppress bone metastasis in vivo as evident from fewer number of bone lesions induced by PC3 cells with silenced miRNA compared to that induced by

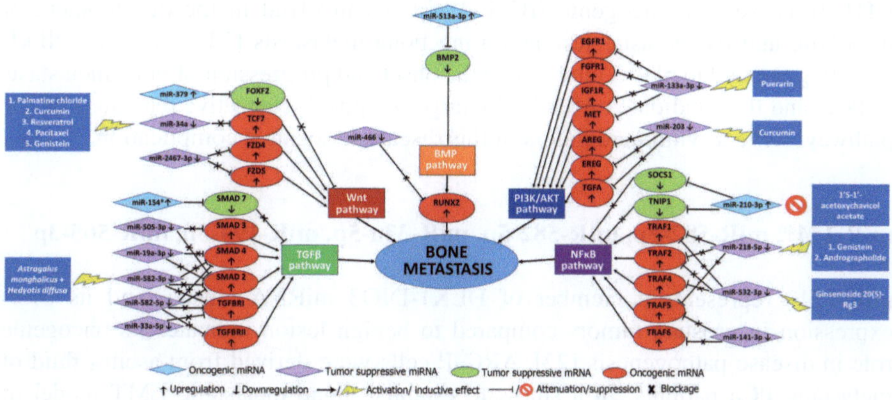

Fig. 14.2 A schematic diagram showing the miRNA network modulating key signaling pathways associated with bone metastasis of prostate cancer along with various natural products mediating up/downregulation of the miRNAs

vector treated cells in BALB/c-nu mice; better bone metastasis free survival was also recorded in animals receiving miR-210-3p silenced PC3 cells than those receiving control cells [51]. On the other hand, an inverse association of the expression level of miR-532-3p, miR-218-5p, and miR-141-3p in prostatic tumor with higher serum PSA level and Gleason score suggested their potential tumor suppressive role in PCa; this was further confirmed from their (miRNAs) ability to suppress aggressive tumor phenotypes like proliferation, migration, invasion, and EMT in vitro [28, 47, 67]. The association of these three miRNAs with PCa bone metastasis was evident from their significant reduced expression in bone metastatic tumor (PCa) compared to non-bone metastatic one. Low expression level of miR-532-3p and miR-218-5p (serum level) was seen to be associated with poor bone metastasis free survival of PCa patients, and upregulation of miR-532-3p and miR-141-3p by ectopic expression resulted suppression of bone metastatic potential of PC3 cells in vivo [28, 47, 67]. To identify the downstream target mRNAs through which these miRNAs exert their impact on bone metastasis, luciferase reporter assays were conducted. SOCS1 and TNIP1, two negative regulators of NFκB, were confirmed as the target mRNAs of miR-210-3p [51]. The tumor-suppressive miRNAs, namely, miR-218-5p, miR-532-3p, and miR-141-3p were shown to attenuate tumor necrosis factor receptor-associated factors (TRAF1, TRAF2, TRAF4, TRAF5, and TRAF6), which serve as adaptor proteins of NFκB signaling [28, 47, 67]. Thus, in PCa, upregulation of miR-210-3p and downregulation of miR-218-5p, miR-532-3p, and miR-141-3p seemed to contribute to the aberrant NFκB signaling leading to over-expression of downstream genes (TWIST1, MMP13, IL11) associated with aggressive tumor phenotypes like EMT, invasion, and migration [28, 47, 51, 67]. Actually, in PCa bone metastasis, EMT is considered as the most important and initiating event in which involvement of NFκB signaling is well established [34]. But, additionally, in metastatic PCa cells, upregulation of NFκB signaling was also reported, and this molecular event was found to induce over-expression of osteolytic (RANKL,

PTHrP) as well as osteogenic (BMP-2) factors involved in the development of osteolytic and osteoblastic lesions during bone metastasis [34, 72]. Thus, NFκB signaling seemed to play a vital role in the onset and progression of bone metastasis in PCa, and the candidate miRNAs, by targeting positive/negative regulators of the pathway, serve as vital modulators of this disease-associated complication.

miR-154*, miR-582-3p, miR-582-5p, miR-33a-5p, miR-19a-3p, miR-505-3p

miR-154* represents a member of DLK1-DIO3 miRNA cluster, and its over-expression in prostate tumors compared to benign lesion indicates its oncogenic role in disease pathogenesis [22]. ARCaP cells were derived from ascites fluid of metastatic PCa patients. In a study, to establish bone metastatic EMT model of human prostate cancer, intracardiac inoculation or orthotopic implantation of ARCaP cells of mesenchymal subtype ($ARCaP_M$) was done in SCID/Beige mice, and higher capability of bone metastasis incidence was found in comparison to that of epithelial subtype ($ARCaP_E$). In this study, significant over-expression of miR-154* in $ARCaP_M$ cells compared to its epithelial counterpart ($ARCaP_E$) was found; also, the bone metastasis incidence of $ARCaP_{M,}$ was seen to be suppressed in vivo following inhibition of miR-154* compared to control cells. By luciferase assay, SMAD7, a negative regulator of TGFβ signaling, was established as the downstream target of miR-154* [22]. On the contrary, expressions of miR-582-3p, miR-582-5p, miR-33a-5p, miR-19a-3p, and miR-505-3p were found to be reduced in PCa compared to adjacent normal as well as in bone-metastatic PCa tissues compared to non-bone-metastatic ones within malignant tumors [16, 29, 63, 66]; in majority of the studies, TCGA dataset was initially used to obtain the finding followed by its validation in independent set of clinical samples. Low expression level of miR-582-3p, miR-582-5p, miR-33a-5p, and miR-505-3p was found to be associated with poor bone metastasis free survival of PCa patients. Upregulation of miR-505-3p by miRNA mimics was found to suppress the migration and invasion of PCa cells in vitro; similar finding was obtained after the upregulation of other four miRNAs (miR-582-3p, miR-582-5p, miR-33a-5p, and miR-19a-3p), and additionally here, the bone metastasis of PCa cells was found to be suppressed in vivo. In quest of downstream target mRNAs of these miRNAs, component genes of TGFβ pathway positively regulating the signaling were identified. SMAD2, SMAD4, and TGFBRI were found to be silenced by miR-582-3p, whereas the attenuation of SMAD2, SMAD4, TGFBRI, and TGFBRII was dealt with overexpression of miR-582-5p [29]. miR-33a-5p was found to target TGFBRI [16], while miR-19a-3p and miR-505-3p were strictly involved in attenuating SMAD-2&4 and SMAD-2&3, respectively [63, 66]. The binding of all these miRNAs to 3′ UTR of their respective target mRNAs was validated by Luciferase assays. Now, among target mRNAs of TGFβ pathway, TGFBRI and TGFBRII are the receptor proteins, whereas SMAD2, SMAD3, SMAD4, and SMAD7 are intracellular signaling proteins. Therefore, upregulation and downregulation of mentioned five candidate miRNAs in PCa would result in aberrant suppression of negative (SMAD7) and activation of positive regulators (TGFBRs and other SMADs) of TGFβ signaling, contribution of which in disease-associated bone metastasis is well established.

miR-379, miR-34a, and miR-2467-3p

Like miR-154*, miR-379 is also a member of DLK1-DIO3 mRNA cluster found to be over-expressed in clinical PCa samples; miR-379 was also upregulated in mesenchymal cells compared to epithelial counterpart of bone metastatic EMT model ($ARCaP_M$ vs $ARCaP_E$) of human PCa, and its inhibition led to the reversal of EMT [22]. Forkhead box F2 (FOXF2) was predicted as the putative target mRNA of miR-379. Although the exact downstream target of FOXF2 for modulating bone metastasis in PCa has not been elucidated, the candidate has been shown to inhibit Wnt pathway in colon cancer [22]. Actually, FOXF2 encodes a member of forkhead box family of transcription factor found to play a tumor suppressive role in PCa by imposing inhibitory effect on PCa progression and metastasis [30]. FOXF2 was seen to downregulate matrix metalloproteinase-1 (MMP1), Wnt-5a, and upregulate tissue inhibitor of metalloproteinase-3 (TIMP3) [25]; therefore, miR-379 overexpression in PCa as seen by Gururajan et al. [22] might cause upregulation of matrix metalloproteinase(s) and down regulation of its inhibitor(s) leading to increased degradation of extra-cellular matrix which is a pre-requisite for development of BM. In this study, FOXF2 regulated transcriptional modulation of MMP(s) and its inhibitor(s) was suggested to also occur via noncanonical wnt signaling, although the underlying mechanism remained unexplored.

While the connection of miR-379 with Wnt/β-catenin pathway in PCa is still under the scope of investigation, miR-34a and miR-2467-3p have been found to regulate the aforesaid pathway directly. Downregulation of miR-2467-3p in primary prostate tumors compared to adjacent normal, and gradual decrease in miR-34a along advanced clinical stage and adverse outcome suggested their tumor suppressive role in PCa development [13, 40]. Both miRNAs were downregulated in PC3 cell (derived from a bone metastatic prostate adenocarcinoma), indicating their negative impact on bone metastasis of malignant cells. Using luciferase reporter assays, TCF7 was confirmed as the target mRNA of miR-34a, and FZD4 and FZD5 as those of miR-2467-3p. FZD4 and FZD5 encode "Frizzled" receptor for binding of Wnt ligand whereas, TCF7 remains in the nucleus to initiate transcription of wnt target genes following interaction with β-catenin. Thus, in PCa, miR-2467-3p and miR-34a downregulation seemed to cause aberrant wnt signaling via over-expression of frizzled receptors and TCF7, respectively, and a significant negative correlation between TCF7 and miR-34a in clinical PCa samples seemed to validate the hypothesis in part [13, 40]. The contribution of Wnt/β-catenin pathway (canonical) in PCa bone metastasis is well established. In PCa cell line C4-2B, a derivative of LNCaP, Wnt-3a, and Wnt-5a induced the over-expression of BMP-4 and BMP-6, respectively; PCa cell (C4-2B) derived BMPs were found to promote osteoblastic phenotypes (enhanced alkaline phosphatase activity and increased osteocalcin level) in pre-osteoblast cells (MC3T3-E1) [15]. In another study, a prostate cancer xenograft tumor PCa-118b, derived from metastatic bone lesion of PCa patients, was found to induce ectopic bone formation after transplantation to SCID mice, and in the associated osteoblast differentiation process, the contributing role of BMP-4 derived from tumor xenograft was reported [37].

Therefore, miR-379, miR-2467-3p, and miR-34a, by differential expression, seemed to promote aberrant Wnt signaling in PCa leading to over-expression of downstream target genes contributing to the development of bone metastasis.

miR-513a-3p and miR-466

The role of miR-513a-3p in PCa pathogenesis was explored in one study so far. The positive impact of miR-513a-3p on proliferation, migration, and invasion in metastatic PCa cell lines (PC3 and DU145) indicated a potential role of the miRNA in acquisition of aggressive tumor phenotypes leading to the development of bone metastasis[18]; however, the predicted impact of the miRNA on PCa bone metastasis was not validated in vivo. Also, in this study, bone-morphogenetic protein-2 (BMP-2) was found as a downstream target of miR-513a-3p, and their interaction was validated by luciferase reporter assay. BMP-2 was shown to play a dual role in PCa pathogenesis. Decreased expression of BMP-2 was seen frequently in PCa, and significant association of the low expression level with higher Gleason score, and incidence of biochemical recurrence indicated the importance of the protein in disease progression and prognosis [26, 61]. In the study conducted by Feng et al. [18], miR-513a-3p was found to attenuate the expression of BMP2 in metastatic PCa cell line [18], and this might be highlighted as the underlying mechanism for the potential/predicted influence of the miRNA on the development of disease-associated bone metastasis.

miR-466 plays a tumor suppressive role in PCa as evident from its reduced expression in clinical tumor (prostate) tissue compared to adjacent normal and in PCa cell lines (PC-3, DU145, LNCaP, MDaPCa2b, LAPC4) compared to nonmalignant prostate epithelial cell (RWPE-1) [14]. Among PCa cell lines, reduced expression of miR-466 in metastatic PC-3 and DU145, and reversal of EMT phenotypes of these cells following ectopic expression of the miRNA suggested its (miR-466) potential negative impact on bone metastasis; this was further validated in vivo using athymic nude mice in which significant decrease in bone metastasis was evident following orthotopic incorporation of bioluminescent PC3M-Luc-C6 cells constitutively expressing miR-466 compared to that expressing cont-miR [14]. Also, in this study, RUNX2, was predicted as the downstream target of miR-466. The interaction of miR-466 to 3′ UTR of RUNX2 was confirmed by Luciferase reporter assays, and the functional impact of miRNA mediated attenuation of RUNX2 was confirmed by knockdown experiment. RUNX2, a downstream target of BMP pathway, is an osteogenic transcription factor. RUNX2 target genes like osteopontin, osteocalcin, ANGPTs, MMP11, etc., contributed significantly to disease progression and bone metastasis. Thus, miR-466 by attenuating RUNX2, mediated the downregulation of its (RUNX2) target genes leading to suppression of bone metastasis in PCa [14].

miR-133a-3p and miR-203

Significant downregulation of miR-133a-3p in malignant prostate compared to adjacent normal as well as benign lesions, and within malignant tumors, a significant decrease in expression along with higher Gleason score, advanced clinical stage, and metastasis indicated its suppressive role in PCa development and progression [62]. In addition, a significantly lower expression of miR-133a-3p in metastatic PCa compared to nonmetastatic one, and a significant association of low expression level with poor metastasis free survival suggested a potential negative impact of the miRNA on bone metastasis. Similarly, the association of miR-203 with PCa bone metastasis was also evident from its significantly lower expression in metastatic tumors compared to both nonmetastatic tumor and normal prostate tissue [55]. The over-expression (by agomir/miR precursor) of miR-133a-3p and miR-203 in PC-3 and Ras-mutated bone metastatic prostate cancer cells also resulted in the decrease in bone metastatic potential of the cells in vivo [55, 62]. Both these tumor-suppressive miRNAs were reported to have multiple target mRNAs. miR-133a-3p attenuated EGFR, FGFR1, IGF1R, and MET, and miR-203 targeted AREG, EREG, and TGFA by directly binding to 3′ UTR, as observed in the luciferase reporter assays [55, 62]. All the target genes of miR-133a-3p encode the receptors of PI3K-AKT pathway, and those of miR-203 encode the ligands for EGFR. Actually, PI3K-AKT pathway can modulate several other signaling pathways through phosphorylation of AKT. AKT phosphorylation can activate NFκB pathway which may trigger the expression of osteolytic factors RANKL, PTHrP, and osteogenic factor BMP-2, thus facilitating the progression of bone metastasis [72]. AKT phosphorylation also inactivates GSK-3β and this inactivation leads to activated Wnt/β-catenin signaling in PCa cells leading to prostate cancer progression [68].

Clinical Importance

miRNAs can serve as clinical bio-markers for diagnosis and prognosis of malignant conditions including PCa. As suggested by existing published literatures, the serum level of oncogenic miRNAs, miR-154*, miR-379, and miR-210-3p was elevated in patients with bone metastatic PCa [22, 51]; this information may be implemented to develop less invasive miRNA-based biomarker for diagnosis of this adverse condition (bone metastasis) as well as to assess prognosis of patients in response to different therapeutics. Similarly, the potency of some tumor suppressive miRNAs like miR-582-3p and 5p [29], miR-218-5p [47], miR-203 [62], and miR-133a-3p [55] to serve as clinical biomarker in PCa and associated bone metastasis was suggested.

Apart from diagnostic and prognostic importance, the miRNAs also hold strong potential in therapeutic avenue. In order to prevent PCa progression to bone, while the expression of oncogenic miRNAs need to be reduced by targeting them therapeutically, the tumor-suppressive miRNAs should be restored to their normal levels. A wide range of natural products/its derivatives have been found to modulate the expression of the aforementioned miRNAs in different malignancies (Table 14.1). For example, tumor-suppressive miRNA miR-34a has been shown to be upregulated by Palmatine chloride [11], curcumin and resveratrol in breast cancer [70], by diallyl disulfide (DADS, a natural garlic disulfide) [11] and paclitaxel in gastric cancer [70] and by genistein in PCa [48]. Similarly, curcumin has been found to induce miR-203 expression by partially demethylating its promoter in bladder cancer [70]. 1′S-1′-acetoxychavicol acetate (ACA) is a natural compound extracted from ginger; in cervical cancer cells, treatment with ACA was found to downregulate oncogenic miR-210 leading to apoptotic cell death [70]. Genistein and andrographolide have been shown to upregulate miR-218 in gliosarcoma and ovarian cancer, respectively [2]. Ginsenoside 20(S)-Rg3 is a steroidal saponin that has been isolated from red ginseng and it triggered the upregulation of miR-532-3p in ovarian cancer cells [57]. miR-582-3p, a tumor suppressive miRNA, was found to be upregulated by combined actions of *Astragalus mongholicus* and *Hedyotis diffusa* in lung adenocarcinoma [59]. Finally, Puerarin extracted from the root of *Pueraria lobata* has been shown to upregulate the expression of miR-133a-3p in breast cancer cells [39].

Conclusion and Future Perspectives

In PCa bone metastasis, the existing treatment protocol mainly includes relieving pain, prevention of bone fracture, and improvement of bone health that do not seem to be sufficient for effective management of this advanced and complicated disease condition. To seek a way-out from this problem, the present review work focused on miRNAs, as their importance in PCa bone metastasis was evident from different in vitro and in vivo experiments. miRNAs exert positive or negative impact on bone metastasis through modulating associated molecular pathways; hence, aberrant changes in their expression level can serve as diagnostic/prognostic markers, and they can also be targeted for development of novel and effective therapeutic measures. Elevated serum level of some miRNAs (miR-210-3p, miR-154*, miR-379) in PCa patients indicated potency of these candidates to serve as less invasive disease biomarkers; the finding needs to be validated in independent studies. Moreover, all the miRNAs up/downregulated in metastatic PCa should be tested in serum or other body fluids to identify potential disease biomarkers.

For effective management of PCa bone metastasis, novel therapeutics need to be identified in order to suppress oncogenic miRNAs and restore the normal levels of tumor-suppressive ones. A number of natural products (curcumin, resveratrol, etc.) were seen to modulate some of the aforesaid miRNAs. But, majority of these studies were conducted in different other malignancies, and hence the effectiveness of

Table 14.1 Summary of the various natural products/their derivatives (along with sources) modulating various candidate miRNAs in different malignancies

Natural products/its derivatives	Source	Target miRNA	Cancer type	References
Palmatine chloride	Palmatine extracted from roots of *Coptis japonica*	miR-34a (↑)	Breast cancer	[11]
Curcumin	*Curcuma longa*	miR-34a (↑)	Breast cancer	[70]
		miR-203 (↑)	Bladder cancer	[70]
Resveratrol	Grapes, peanuts, pines, etc.	miR-34a (↑)	Breast cancer	[70]
Diallyl sulfide (DADS)	Garlic	miR-34a (↑)	Gastric cancer	[11]
Paclitaxel	Stem bark of western yew	miR-34a (↑)	Gastric cancer	[70]
1′ S-1′-acetoxy chavicol	Ginger	miR-210 (↓)	Cervical cancer	[70]
Genistein	Dietary soybeans, fava beans, lupins	miR-34a (↑)	Prostate cancer	[48]
		miR-218 (↑)	Gliosarcoma	[2]
Andrographolide	*Andrographis paniculata*	miR-218 (↑)	Ovarian cancer	[2]
Ginsenoside 20(S)-Rg3	Red ginseng	miR-532-3p (↑)	Ovarian cancer	[57]
Astragalus mongholicus + *Hedyotis diffusa*	Mongolian/Chinese plants	miR-582-3p (↑)	Lung adenocarcinoma	[59]
Puerarin	Roots of *Pueraria lobata*	miR-133a-3p (↑)	Breast cancer	[39]

Impact of natural product on miRNA; upregulation (↑), downregulation (↓)

these products on miRNAs as well as on associated molecular pathways needs to be tested in metastatic PCa. Moreover, for a considerable number of miRNAs associated with PCa bone metastasis, no natural products modulating their expression has still been reported, and hence further research in this directive is warranted.

Acknowledgments We are thankful to the Hon'ble Vice Chancellor, Presidency University, Kolkata, India. Financial support for this work was provided by the Department of Biotechnology, Govt. of India, New Delhi, through DBT-BUILDER program, Presidency University, Kolkata (No. BT/INF/22/SP45088/2022 dated 17.02.2022).

References

1. Abrahamsson PA (2004) Pathophysiology of bone metastases in prostate cancer. Eur Urol Suppl 3(5 Spec. Iss):3–9. https://doi.org/10.1016/j.eursup.2004.09.001
2. Alnuqaydan AM (2020) Targeting micro-RNAs by natural products: a novel future therapeutic strategy to combat cancer. Am J Transl Res 12(7):3531–3556
3. Altwalah J, Alsalman H, Sheikh A (2024) Diagnostic role of dual-energy computed tomography in the assessment of musculoskeletal oncology: a literature review. Curr Radiol Rep 12(3):19–29. https://doi.org/10.1007/s40134-023-00423-2

4. Awiwi MO, Gjoni M, Vikram R, Altinmakas E, Dogan H, Bathala TK, Naik S, Ravizzini G, Kandemirli SG, Elsayes KM, Salem UI (2023) MRI and PSMA PET/CT of biochemical recurrence of prostate cancer. Radiographics 43(12):e230112. https://doi.org/10.1148/rg.230112
5. Babashah S, Soleimani M (2011) The oncogenic and tumour suppressive roles of microRNAs in cancer and apoptosis. Eur J Cancer 47(8):1127–1137. https://doi.org/10.1016/j.ejca.2011.02.008
6. Baldessari C, Pipitone S, Molinaro E, Cerma K, Fanelli M, Nasso C, Oltrecolli M, Pirola M, D'Agostino E, Pugliese G, Cerri S, Vitale MG, Madeo B, Dominici M, Sabbatini R (2023) Bone metastases and health in prostate cancer: from pathophysiology to clinical implications. Cancer 15(5):1518. https://doi.org/10.3390/cancers15051518
7. Bastawrous S, Bhargava P, Behnia F, Djang DSW, Haseley DR (2014) Newer PET application with an old tracer: role of 18F-NaF skeletal PET/CT in oncologic practice. Radiographics 34(5):1295–1316. https://doi.org/10.1148/rg.345130061
8. Ben-Ghedalia-Peled N, Vago R (2022) Wnt signaling in the development of bone metastasis. Cells 11(23):3934. https://doi.org/10.3390/cells11233934
9. Bhargava Reddy KV, Muni Prasad M (2018) A prospective study on incidence of bone metastases at various prostate specific antigen levels in carcinoma prostate in south Indian population. Int J Contemp Med Res 5(1):1–4
10. Bhaskaran M, Mohan M (2014) MicroRNAs: history, biogenesis, and their evolving role in animal development and disease. Vet Pathol 51(4):759–774. https://doi.org/10.1177/0300985813502820
11. Biersack B (2016) Non-coding RNA/microRNA-modulatory dietary factors and natural products for improved cancer therapy and prevention: alkaloids, organosulfur compounds, aliphatic carboxylic acids and water-soluble vitamins. Non-coding RNA Res 1(1):51–63. https://doi.org/10.1016/j.ncrna.2016.09.001
12. Boopathi E, Birbe R, Shoyele SA, Den RB, Thangavel C (2022) Bone health management in the continuum of prostate cancer disease. Cancer 14(17):4305. https://doi.org/10.3390/cancers14174305
13. Chen WY, Liu SY, Chang YS, Yin JJ, Yeh HL, Mouhieddine TH, Hadadeh O, Abou-Kheir W, Liu YN (2015) MicroRNA-34a regulates WNT/TCF7 signaling and inhibits bone metastasis in Ras-activated prostate cancer. Oncotarget 6(1):441–457. https://doi.org/10.18632/oncotarget.2690
14. Colden M, Dar AA, Saini S, Dahiya PV, Shahryari V, Yamamura S, Tanaka Y, Stein G, Dahiya R, Majid S (2017) MicroRNA-466 inhibits tumor growth and bone metastasis in prostate cancer by direct regulation of osteogenic transcription factor RUNX2. Cell Death Dis 8(1). https://doi.org/10.1038/cddis.2017.15
15. Dai J, Hall CL, Escara-Wilke J, Mizokami A, Keller JM, Keller ET (2008) Prostate cancer induces bone metastasis through Wnt-induced bone morphogenetic protein-dependent and independent mechanisms. Cancer Res 68(14):5785–5794. https://doi.org/10.1158/0008-5472.CAN-07-6541
16. Dai Y, Wu Z, Lang C, Zhang X, He S, Yang Q, Guo W, Lai Y, Du H, Peng X, Ren D (2019) Copy number gain of ZEB1 mediates a double-negative feedback loop with miR-33a-5p that regulates EMT and bone metastasis of prostate cancer dependent on TGF-β signaling. Theranostics 9(21):6063–6079. https://doi.org/10.7150/thno.36735
17. Drake MT, Clarke BL, Khosla S (2008) Bisphosphonates: mechanism of action and role in clinical practice. Mayo Clin Proc 83(9):1032–1045. https://doi.org/10.4065/83.9.1032
18. Feng H, Deng Z, Peng W, Wei X, Liu J, Wang T (2023) Circular RNA EPHA3 suppresses progression and metastasis in prostate cancer through the miR-513a-3p/BMP2 axis. J Transl Med 21(1). https://doi.org/10.1186/s12967-023-04132-4
19. Genant HK, Engelke K, Prevrhal S (2008) Advanced CT bone imaging in osteoporosis. Rheumatology (Oxford, England) 47 Suppl 4(Suppl 4):iv9–iv16. https://doi.org/10.1093/rheumatology/ken180

20. Goode AE, Wang N, Munkley J (2023) Prostate cancer bone metastases biology and clinical management (review). Oncol Lett 25(4):163. https://doi.org/10.3892/ol.2023.13749
21. Gravdal K, Halvorsen OJ, Haukaas SA, Akslen LA (2007) A switch from E-cadherin to N-cadherin expression indicates epithelial to mesenchymal transition and is of strong and independent importance for the progress of prostate cancer. Clin Cancer Res 13(23):7003–7011. https://doi.org/10.1158/1078-0432.CCR-07-1263
22. Gururajan M, Josson S, Chu GCY, Lu CL, Lu YT, Haga CL, Zhau HE, Liu C, Lichterman J, Duan P, Posadas EM, Chung LWK (2014) MiR-154∗ and miR-379 in the DLK1-DIO3 MicroRNA mega-cluster regulate epithelial to mesenchymal transition and bone metastasis of prostate cancer. Clin Cancer Res 20(24):6559–6569. https://doi.org/10.1158/1078-0432.CCR-14-1784
23. Jiang T, Zeng Q, He J (2023) Do alkaline phosphatases have great potential in the diagnosis, prognosis, and treatment of tumors?. Transl Cancer Res 12(10):2932–2945. https://doi.org/10.21037/tcr-23-1190
24. Heindel W, Gübitz R, Vieth V, Weckesser M, Schober O, Schäfers M (2014) The diagnostic imaging of bone metastases. Dtsch Arztebl Int 111(44):741–747. https://doi.org/10.3238/arztebl.2014.0741
25. Hirata H, Ueno K, Shahryari V, Deng G, Tanaka Y, Tabatabai ZL, Hinoda Y, Dahiya R (2013) MicroRNA-182-5p promotes cell invasion and proliferation by down regulating FOXF2, RECK and MTSS1 genes in human prostate cancer. PLoS One 8(1). https://doi.org/10.1371/journal.pone.0055502
26. Horvath LG, Henshall SM, Kench JG, Turner JJ, Golovsky D, Brenner PC, O'Neill GF, Kooner R, Stricker PD, Grygiel JJ, Sutherland RL (2004) Loss of BMP2, Smad8, and Smad4 expression in prostate cancer progression. Prostate 59(3):234–242. https://doi.org/10.1002/pros.10361
27. Huang WC, Xie Z, Konaka H, Sodek J, Zhau HE, Chung LW (2005) Human osteocalcin and bone sialoprotein mediating osteomimicry of prostate cancer cells: role of cAMP-dependent protein kinase A signaling pathway. Cancer Res 65(6):2303–2313. https://doi.org/10.1158/0008-5472.CAN-04-3448
28. Huang S, Wa Q, Pan J, Peng X, Ren D, Huang Y, Chen X, Tang Y (2017) Downregulation of miR-141-3p promotes bone metastasis via activating NF-κB signaling in prostate cancer. J Exp Clin Cancer Res 36(1). https://doi.org/10.1186/s13046-017-0645-7
29. Huang S, Zou C, Tang Y, Wa Q, Peng X, Chen X, Yang C, Ren D, Huang Y, Liao Z, Huang S, Zou X, Pan J (2019) miR-582-3p and miR-582-5p suppress prostate cancer metastasis to bone by repressing TGF-β signaling. Mol Ther Nucl Acids 16:91–104. https://doi.org/10.1016/j.omtn.2019.01.004
30. Jia D, Zhou Z, Kwon OJ, Zhang L, Wei X, Zhang Y, Yi M, Roudier MP, Regier MC, Dumpit R, Nelson PS, Headley M, True L, Lin DW, Morrissey C, Creighton CJ, Xin L (2022) Stromal FOXF2 suppresses prostate cancer progression and metastasis by enhancing antitumor immunity. Nat Commun 13(1). https://doi.org/10.1038/s41467-022-34665-z
31. Jiang H (2025) Prostate cancer bone metastasis: molecular mechanisms of tumor and bone microenvironment. Cancer Manag Res 17:219–237. https://doi.org/10.2147/CMAR.S495169
32. Jiang T, Zeng Q, He J (2023) Do alkaline phosphatases have great potential in the diagnosis, prognosis, and treatment of tumors? Transl Cancer Res 12(10):2932–2945. https://doi.org/10.21037/tcr-23-1190
33. Jin JK, Dayyani F, Gallick GE (2011) Steps in prostate cancer progression that lead to bone metastasis. Int J Cancer 128(11):2545–2561. https://doi.org/10.1002/ijc.26024
34. Jin R, Sterling JA, Edwards JR, DeGraff DJ, Lee C, Park SI, Matusik RJ (2013) Activation of NF-kappa B signaling promotes growth of prostate cancer cells in bone. PLoS One 8(4). https://doi.org/10.1371/journal.pone.0060983
35. Kurokawa R, Kato S, Koyama H, Ishida M, Kurokawa M, Kuroda R, Ushiku T, Kume H, Abe O (2022) Osteolytic or mixed bone metastasis is not uncommon in patients with high-risk prostate cancer. Eur J Radiol 157:110595. https://doi.org/10.1016/j.ejrad.2022.110595

36. la Manna F, Karkampouna S, Zoni E, de Menna M, Hensel J, Thalmann GN, Kruithof-De Julio M (2019) Metastases in prostate cancer. Cold Spring Harb Perspect Med 9(3). https://doi.org/10.1101/cshperspect.a033688
37. Lee YC, Cheng CJ, Bilen MA, Lu JF, Satcher RL, Yu-Lee LY, Gallick GE, Maity SN, Lin SH (2011) BMP4 promotes prostate tumor growth in bone through osteogenesis. Cancer Res 71(15):5194–5203. https://doi.org/10.1158/0008-5472.CAN-10-4374
38. Lee JW, Park YJ, Jeon YS, Kim KH, Lee JE, Hong SH, Lee SM, Jang SJ (2020) Clinical value of dual-phase F-18 sodium fluoride PET/CT for diagnosing bone metastasis in cancer patients with solitary bone lesion. Quant Imaging Med Surg 10(11):2098–2111. https://doi.org/10.21037/QIMS-20-607
39. Li Z, Xu W, Ren X, Xu J, Chen J (2019) Puerarin promotes DUSP1 expression by regulating miR-133a-3p in breast cancer. Mol Med Rep 19(1):205–212. https://doi.org/10.3892/mmr.2018.9682
40. Liao S, Fang X, Zhou K, Zhao T, Ji L, Zhang W, Zhong X, Feng F, Lv J, Kang Y, Zhu D (2023) LINC00482 sponged miR-2467-3p to promote bone metastasis of prostate cancer through activating Wnt/β-catenin signaling pathway. J Bone Oncol 41. https://doi.org/10.1016/j.jbo.2023.100494
41. Macedo F, Ladeira K, Pinho F, Saraiva N, Bonito N, Pinto L, Gonçalves F (2017) Bone metastases: an overview. Oncol Rev 11(1). https://doi.org/10.4081/oncol.2017.321
42. Msaouel P, Pissimissis N, Halapas A, Koutsilieris M (2008) Mechanisms of bone metastasis in prostate cancer: clinical implications. Best Pract Res Clin Endocrinol Metab 22(2):341–355. https://doi.org/10.1016/j.beem.2008.01.011
43. Nishimura K (2023) Management of bone metastasis in prostate cancer. J Bone Miner Metab 41(3):317–326. https://doi.org/10.1007/s00774-023-01435-w
44. O'Sullivan GJ (2015) Imaging of bone metastasis: an update. World J Radiol 7(8):202. https://doi.org/10.4329/wjr.v7.i8.202
45. Ojuka D, Majicha S, Ngugi P (2018) Can PSA predict bone metastasis amongst newly diagnosed prostate cancer patients? Ann Afr Surg 15(1). https://doi.org/10.4314/aas.v15i1.2
46. Oprea-Lager DE, Cysouw MCF, Boellaard R, Deroose CM, de Geus-Oei LF, Lopci E, Bidaut L, Herrmann K, Fournier LS, Bäuerle T, deSouza NM, Lecouvet FE (2021) Bone metastases are measurable: the role of whole-body MRI and positron emission tomography. Front Oncol 11:772530. https://doi.org/10.3389/fonc.2021.772530
47. Peng P, Chen T, Wang Q, Zhang Y, Zheng F, Huang S, Tang Y, Yang C, Ding W, Ren D, Huang Z, Guo Y (2019) Decreased miR-218-5p levels as a serum biomarker in bone metastasis of prostate cancer. Oncol Res Treat 42(4):165–180. https://doi.org/10.1159/000495473
48. Phuah NH, Nagoor NH (2014) Regulation of microRNAs by natural agents: new strategies in cancer therapies. Biomed Res Int 2014:804510. https://doi.org/10.1155/2014/804510
49. Pomykala KL, Czernin J, Grogan TR, Armstrong WR, Williams J, Calais J (2020) Total-Body ^{68}Ga-PSMA-11 PET/CT for bone metastasis detection in prostate cancer patients: potential impact on bone scan guidelines. J Nuclear Med 61(3):405–411. https://doi.org/10.2967/jnumed.119.230318
50. Rawla P (2019) Epidemiology of prostate cancer. World J Oncol 10(2):63–89. https://doi.org/10.14740/wjon1191
51. Ren D, Yang Q, Dai Y, Guo W, Du H, Song L, Peng X (2017) Oncogenic miR-210-3p promotes prostate cancer cell EMT and bone metastasis via NF-KB signaling pathway. Mol Cancer 16(1). https://doi.org/10.1186/s12943-017-0688-6
52. Saliminejad K, Khorram Khorshid HR, Soleymani Fard S, Ghaffari SH (2019) An overview of microRNAs: biology, functions, therapeutics, and analysis methods. J Cell Physiol 234(5):5451–5465. https://doi.org/10.1002/jcp.27486
53. Sargolzaei J, Etemadi T, Alyasin A (2020) The P53/microRNA network: a potential tumor suppressor with a role in anticancer therapy. Pharmacol Res 160:105179. https://doi.org/10.1016/j.phrs.2020.105179

54. Sasikumar A (2017) Specificity of ^{68}Ga-PSMA PET/CT for prostate cancer – myths and reality. Indian J Nuclear Med 32(1):11–12. https://doi.org/10.4103/0972-3919.198449
55. Siu MK, Abou-Kheir W, Yin JJ, Chang YS, Barrett B, Suau F, Casey O, Chen WY, Fang L, Hynes P, Hsieh YY, Liu YN, Huang J, Kelly K (2014) Loss of EGFR signaling regulated miR-203 promotes prostate cancer bone metastasis and tyrosine kinase inhibitors resistance. Oncotarget 5(11):3770–3784. https://doi.org/10.18632/oncotarget.1994
56. Sosa MS, Bragado P, Aguirre-Ghiso JA (2014) Mechanisms of disseminated cancer cell dormancy: an awakening field. Nat Rev Cancer 14(9):611–622. https://doi.org/10.1038/nrc3793
57. Sumaira S, Vijayarathna S, Hemagirri M, Adnan M, Hassan MI, Patel M, Gupta R, Shanmugapriya, Chen Y, Gopinath SCB, Kanwar JR, Sasidharan S (2024) Plant bioactive compounds driven microRNAs (miRNAs): a potential source and novel strategy targeting gene and cancer therapeutics. Non-Coding RNA Res 9(4):1140–1158. https://doi.org/10.1016/j.ncrna.2024.06.003
58. Sun J, Tian T, Wang N, Jing X, Qiu L, Cui H, Liu Z, Liu J, Yan L, Li D (2024) Pretreatment level of serum sialic acid predicts both qualitative and quantitative bone metastases of prostate cancer. Front Endocrinol 15:1338420. https://doi.org/10.3389/fendo.2024.1338420
59. Sun H, Xu L, Liu S, Ma T (2025) Astragalus mongholicus and Hedyotis diffusa willd inhibit cell proliferation by attenuating the miR-582-3p-p27 signaling pathway in LUAD. Sci Rep 15(1). https://doi.org/10.1038/s41598-025-97996-z
60. Svoronos AA, Engelman DM, Slack FJ (2016) OncomiR or tumor suppressor? The duplicity of MicroRNAs in cancer. Cancer Res 76(13):3666–3670. https://doi.org/10.1158/0008-5472.CAN-16-0359
61. Tae BS, Cho S, Kim HC, Kim CH, Kang SH, Lee JG, Kim JJ, Park HS, Cheon J, Oh MM, Kang SG (2018) Decreased expression of bone morphogenetic protein-2 is correlated with biochemical recurrence in prostate cancer: immunohistochemical analysis. Sci Rep 8(1). https://doi.org/10.1038/s41598-018-28566-9
62. Tang Y, Pan J, Huang S, Peng X, Zou X, Luo Y, Ren D, Zhang X, Li R, He P, Wa Q (2018) Downregulation of miR-133a-3p promotes prostate cancer bone metastasis via activating PI3K/AKT signaling. J Exp Clin Cancer Res 37(1). https://doi.org/10.1186/s13046-018-0813-4
63. Tang Y, Wu B, Huang S, Peng X, Li X, Huang X, Zhou W, Xie P, He P (2019) Downregulation of miR-505-3p predicts poor bone metastasis-free survival in prostate cancer. Oncol Rep 41(1):57–66. https://doi.org/10.3892/or.2018.6826
64. Thio QCBS, Karhade AV, Notman E, Raskin KA, Calderón SL, Ferrone ML, Bramer JAM, Schwab JH (2020) Serum alkaline phosphatase is a prognostic marker in bone metastatic disease of the extremity. J Orthop 22:346–351. https://doi.org/10.1016/j.jor.2020.08.008
65. van der Doelen MJ, Stockhaus A, Ma Y, Mehra N, Yachnin J, Gerritsen WR, Nilsson S, van Oort IM, Ullén A (2021) Early alkaline phosphatase dynamics as biomarker of survival in metastatic castration-resistant prostate cancer patients treated with radium-223. Eur J Nucl Med Mol Imaging 48(10):3325–3334. https://doi.org/10.1007/s00259-021-05283-6
66. Wa Q, Li L, Lin H, Peng X, Ren D, Huang Y, He P, Huang S (2018) Downregulation of MIR-19a-3p promotes invasion, migration and bone metastasis via activating TGF-β signaling in prostate cancer. Oncol Rep 39(1):81–90. https://doi.org/10.3892/or.2017.6096
67. Wa Q, Zou C, Lin Z, Huang S, Peng X, Yang C, Ren D, Xu D, Guo Y, Liao Z, Wang B, Hu H, Huang S, He P (2020) Ectopic expression of miR-532-3p suppresses bone metastasis of prostate cancer cells via inactivating NF-κB signaling. Mol Ther Oncolytics 17:267–277. https://doi.org/10.1016/j.omto.2020.03.024
68. Wei A, Fan B, Zhao Y, Zhang H, Wang L, Yu X, Yuan Q, Yang D, Wang S (2016) ST6Gal-I overexpression facilitates prostate cancer progression via the PI3K/Akt/GSK-3β/β-catenin signaling pathway. Oncotarget 7(40):65374–65388. https://doi.org/10.18632/oncotarget.11699
69. Yumoto K, Eber MR, Wang J, Cackowski FC, Decker AM, Lee E, Nobre AR, Aguirre-Ghiso JA, Jung Y, Taichman RS (2016) Axl is required for TGF-β2-induced dormancy of prostate cancer cells in the bone marrow. Sci Rep 6. https://doi.org/10.1038/srep36520

70. Zhang B, Tian L, Xie J, Chen G, Wang F (2020) Targeting miRNAs by natural products: A new way for cancer therapy. Biomed Pharmacother 130:110546. https://doi.org/10.1016/j.biopha.2020.110546
71. Zhang Y, Liang J, Liu P, Wang Q, Liu L, Zhao H (2022) The RANK/RANKL/OPG system and tumor bone metastasis: potential mechanisms and therapeutic strategies. Front Endocrinol 13:1063815. https://doi.org/10.3389/fendo.2022.1063815
72. Zhu W, Hu X, Xu J, Cheng Y, Shao Y, Peng Y (2015) Effect of PI3K/Akt signaling pathway on the process of prostate cancer metastasis to bone. Cell Biochem Biophys 72(1):171–177. https://doi.org/10.1007/s12013-014-0433-3
73. Taichman RS, Cooper C, Keller ET, Pienta KJ, Taichman NS, McCauley LK (2002) Use of the stromal cell-derived factor-1/CXCR4 pathway in prostate cancer metastasis to bone. Cancer Res 62(6):1832–1837.
74. Dykes IM, Emanueli C (2017) Transcriptional and post-transcriptional gene regulation by long non-coding RNA. Genomics Proteomics Bioinformatics 15(3):177–186. https://doi.org/10.1016/j.gpb.2016.12.005

Chapter 15
MicroRNAs for the Treatment and Diagnosis of Cancer in Bone Metastasis

Prashansha Goel and Chandi C. Mandal

Abstract Bone metastasis represents one of the most critical challenges in cancer management, often remaining undetected until advanced stages due to its asymptomatic progression. Recent insights reveal that microRNAs (miRNAs), small non-coding RNA molecules, are emerging as pivotal regulators in the metastatic spread of cancer to bone. Unlike traditional protein-coding gene profiles, miRNA expression signatures offer enhanced sensitivity and specificity in early tumor and metastasis detection. Certain miRNAs such as miR-21, miR-10b, and miR-218-5p have been identified as pro-metastatic agents in breast cancer, promoting osteoclastogenesis, extracellular matrix modeling, and tumor microenvironment modulation. Notably, miRNAs also influence osteoblast and osteoclast activity, thereby contributing to bone remodeling imbalances and facilitating metastatic colonization. The dual role of miRNAs as both oncogenic drivers and tumor suppressors positions them as promising diagnostic and therapeutic tools. For instance, miR-34a, currently under clinical trial as liposomal mimic, demonstrates potential in halting bone metastasis by suppressing Notch signaling and osteoclast differentiation. Furthermore, the unique presence and stability of circulating miRNAs in blood make them viable candidates for noninvasive biomarkers. Collectively, these findings underscore the transformative role of miRNAs in understanding and combating bone metastasis, offering new hope for early detection and personalized therapy in metastatic cancers.

Keywords Bone metastasis · MicroRNAs · miR-34a · miR-21 · Diagnosis · Therapeutic targets · Cancer

P. Goel · C. C. Mandal (✉)
Department of Biochemistry, School of Life Sciences, Central University of Rajasthan, Ajmer, Rajasthan, India

S. Pathak, A. Banerjee (eds.), *MicroRNAs: Role in Cancer Progression, Diagnostics and Therapy*, Advances in Biochemistry in Health and Disease 34,
https://doi.org/10.1007/978-3-032-08824-6_15

Introduction

Cancer is characterized by the uncontrolled growth and spread of abnormal cells, and remains a leading cause of death globally. While primary tumors can often be managed, the true clinical challenge arises when cancer metastasizes, particularly to distant organs such as bone. This severely complicates treatment and significantly reduces survival rates [1, 2].

Bone metastasis occurs when cancer cells from a primary tumor site, such as the breast, prostate, or lungs, migrate and colonize the bone. Unlike primary bone cancers, these secondary tumors disrupt normal bone homeostasis and commonly affect the spine, pelvis, ribcage, and limbs. Patients with bone metastases often experience severe complications such as persistent pain, fractures, nerve compression, and skeletal-related events, all of which impact their quality of life and therapeutic outcomes [3].

Recent advances in molecular oncology have identified microRNAs (miRNAs) as crucial players in the regulation of cancer progression and metastasis. miRNAs are short (21–25 nucleotides), noncoding RNA molecules that post-transcriptionally regulate gene expression by binding to the 3′ untranslated regions of target mRNAs. They are involved in diverse cellular processes including cell differentiation, proliferation, apoptosis, and immune regulation. Aberrant miRNA expression has been linked to various cancers and is known to influence both tumor initiation and metastatic behavior [4]. In the context of bone metastasis, specific miRNAs modulate osteoclast and osteoblast activity, reshape the bone microenvironment, and enable cancer cells to survive and proliferate at secondary bone sites. Given their disease-specific expression and stability in circulation, miRNAs hold immense potential as both diagnostic biomarkers and therapeutic agents in metastatic cancers, especially those targeting bone [5, 6].

Bone Metastasis

Bone metastasis is a multistep process involving the spread of cancer cells from a primary tumor to bone, where they establish secondary growths. This progression includes cellular mechanisms such as epithelial-mesenchymal transition (EMT), cancer cell dormancy, and interaction with the bone microenvironment. Bone metastasis is most frequently observed in cancers of the breast, prostate, and multiple myeloma, often resulting in skeletal complications like bone pain, fractures, hypercalcemia, and spinal cord compression collectively referred to as skeletal-related events (SREs), or more recently, skeletal symptomatic events (SSEs), a clinical endpoint for evaluating patient quality of life.

Bones are maintained by three types of cells: osteocytes (mature bone cells), osteoblast (bone-forming cells), and osteoclast (bone-resorbing cells). In bone metastasis, this balance is disrupted, leading to increased bone formation or destruction. MicroRNAs (miRNAs) are emerging as key regulators in this disruption, influencing both tumor behavior and bone cell activity [3, 7, 8].

Types of Bone Metastasis

Bone metastasis can present as:

- Osteoblastic metastasis: Characterized by excessive bone formation and denser bone structure, commonly seen in prostate cancer.
- Osteolytic metastasis: Involves increased bone resorption, often observed in breast and lung cancers.
- Mixed metastasis: Shows features of both types.

These subtypes reflect distinct cellular and molecular pathways often regulated by tumor-secreted miRNAs, which alter osteoblast and osteoclast activity [9]. Understanding these biological processes is essential to appreciate how miRNAs regulate bone metastatic progression, which will be discussed in the next section.

Bone Marrow Colonization and an Overview of Generic Mechanism of Bone Metastasis

Once cancer cells escape the primary tumor, they must survive in circulation and find a conducive environment to settle commonly the bone marrow. According to Paget's "seed and soil" theory, metastatic cancer cells (seeds) grow only in supportive microenvironments (soil), such as the bone marrow, which is rich in signaling molecules and structural components [10]. As per this theory it was stated that, metastatic cells migrate to bone marrow where they endure, later on either they will grow or stay in dormant state up to several years. Studies have shown that around 30% of breast cancer (BRCA) patients show the presence of bone marrow micro metastases during diagnosing primary cancer, whereas half of them will never be shown up with establishment of clinical metastases [11].

EMT and Cancer Stem Cells (CSCs)

Epithelial-mesenchymal transition (EMT) is a cellular reprograming event that enables cancer cells to lose adhesion and gain motility, essential for invasion and metastasis. EMT is closely tied to cancer stem cells (CSCs), which have self-renewal capabilities and high metastatic potential. miRNAs such as miR-200 family and miR-205 inhibit EMT by targeting transcription factors like ZEB1 and ZEB2

[12]. Conversely, miR-10b and miR-21 promote EMT by downregulating tumor suppressors [13, 14].

Dormancy and Disseminated Tumor Cells (DTCs)

Disseminated tumor cells (DTCs) often lodge in the bone marrow and enter a dormant, nondividing state. This dormancy allows them to evade therapy and remain undetected for years. Dormancy is regulated by the bone niche and immune cues. miRNAs such as miR-23b and miR-126 have been shown to promote dormancy by inhibiting proliferation and inflammatory pathways, offering potential biomarkers for early-stage metastasis [15, 16].

Bone Marrow Niches and Microenvironmental Cues

Within the bone marrow, DTCs interact with specific microenvironments called niches that regulate their fate. The endosteal niche maintains dormancy, while the vascular niche supports reactivation and proliferation. These niches are rich in factors such as SDF-1, TGF-B, and RANKL. miRNAs like miR-221/222 and miR-31 regulate DTC anchorage, migration, and survival through pathways like CXCR4 and integrin signaling [7, 17, 18].

This colonization phase creates the foundation for more advance adaptations, such as osteomimicry and bone remodeling disruptions-as illustrated in Fig. 15.1.

Osteomimicry and Deregulation of Osteoblast/ Osteoclast Activity

Cancer cells that survive in the bone niche often undergo osteomimicry, a process by which they adopt bone cell-like gene expression patterns (e.g., *Runx2*, *Cadherin-11*, *Cathepsin K*). This enhances their ability to integrate into the bone microenvironment and evade immune detection [19, 20].

Osteoblast and Osteoclast Interactions

Bone remodeling is driven by a delicate balance between osteoblasts and osteoclasts. Cancer cells disrupt this balance by secreting cytokines like IL-6, IL-1, M-CSF, and PTHrP, which promote osteoclastogenesis and inhibit osteoblast

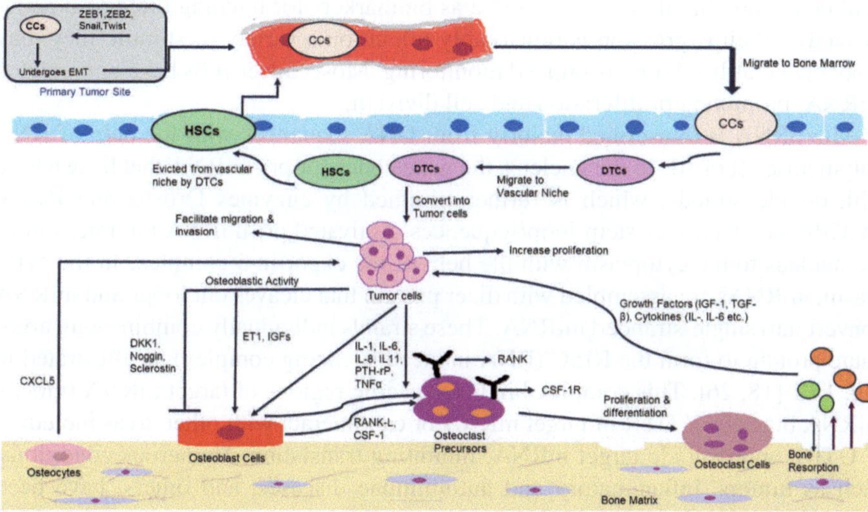

Fig. 15.1 Schematic representation of bone metastasis mechanisms. Cancer cells (CCs) migrate from the primary tumor site and enter the bloodstream. Upon reaching the bone marrow, they become disseminated tumor cells (DTCs) and occupy vascular niches shared with hematopoietic stem cells (HSCs). DTCs may then transition into active tumor cells, initiating signaling cascades that disrupt osteoblast/osteoclast balance. Tumor-secreted cytokines enhance bone resorption and create a favorable microenvironment for tumor proliferation [24]

activity. Additionally, they modulate the RANK/RANKL/OPG axis, which is central to bone resorption.

miRNAs critically influence these pathways. miR-214 promotes osteoclast differentiation, enhancing bone destruction. miR-141 and miR-219 inhibit osteoclastogenesis and serve as potential anti-resorptive agents. In osteoblasts, miR-29b supports differentiation and matrix formation, while miR-138 suppresses osteogenesis, contributing to impaired bone healing [21–23].

These cellular adaptations are influenced by finely tuned gene regulation an area where microRNAs have a profound impact, as we'll now explore.

MicroRNA and Its Role in the Regulation of Gene Expression

MicroRNA (miRs/miRNAs) is a family of small noncoding RNA molecules that regulate gene expression, influencing everything from cell survival to immune response in cancer. Its size is 21–25 bp. miRNAs can be expressed more broadly or in a manner unique to a particular cell type or tissue. They can also be released into extracellular fluids either actively or passively as free miRNAs, bound to lipoproteins and proteins, or encapsulated in vesicles [25]. miRNAs can act on various cellular target and are found in the blood, where they are rather plentiful (as well as in almost all bodily fluids). They serve as biomarkers for tracking biological activity, and as their expression is remarkably reliant on a variety of stimuli, they may probably be utilized for customized monitoring. Most cancer cells have upregulated miRNA, promoting proliferation and cell division.

MicroRNA is transcribed majorly from DNA sequences with the help of RNA polymerases II or III. In the nucleus, the gene codes for pri-miRNA that have a loop with double strands, which is further trimmed by enzymes Drosha and Pasha/DGCR8 which cleaves stem-loop/sequences. Activated pri-miRNA migrates out of the nucleus to the cytoplasm with the help of the exportin 5 complex. In the cytoplasm, miRNAs are assembled with dicer protein that cleaves out loops and miRNA convert into single-stranded miRNA. These strands individually combine with argonaute protein to form the RISC (RNA induced silencing complex), as illustrated in Fig. 15.2 [18, 26]. This complex binds to specific regions of target mRNA (mostly miRNA binds to 3′ UTR of target mRNA or can interact with other areas including 5′ UTR), and degrade target mRNA, inhibiting translation. Numerous conditions, such as tumors, inflammatory and autoimmune diseases, and others, have been shown to exhibit aberrant miRNA expression.

Since miRNAs inhibit gene expression, their expression is generally downregulated in tumors as opposed to non-tumor tissues. A particular neoplasia can be identified by its equivalents and a few miRNAs. Evey stage of the neoplastic transformation, including cancer cell growth, metastasis, progression, and treatment

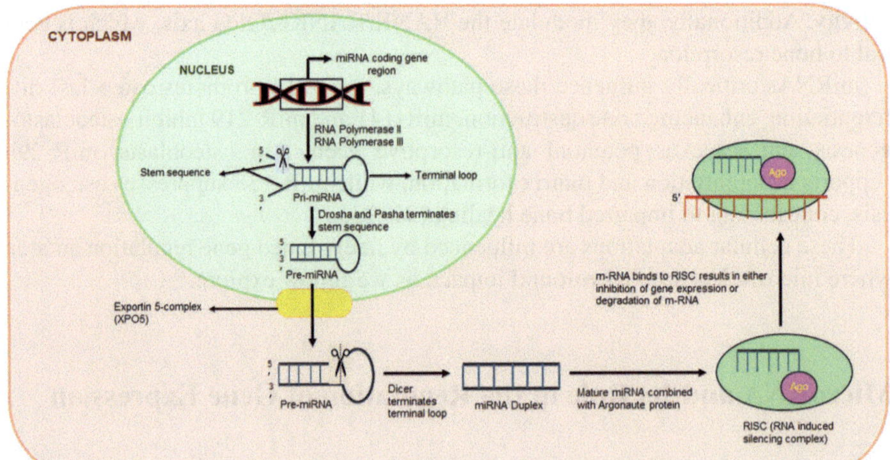

Fig. 15.2 Schematic overview of microRNA (miRNA) biogenesis and gene regulatory function. miRNA genes are transcribed by RNA polymerase II or III to form primary miRNAs (pri-miRNA), which are processed by Drosha and Pasha in the nucleus to produce precursor miRNAs (pre-miRNAs). These are exported to the cytoplasm by Exportin-5 (XPO5) and further cleaved by Dicer into mature miRNA duplexes. One strand is incorporated into the RNA-induced silencing complex (RISC) with Argonaute (Ago), which guides the complex to target mRNAs for translational inhibition or deregulation [4, 26]

resistance ontogeny, is linked to the dysregulation of the expression of particular miRs as well as the advancement regarding each distinct level. Additionally, cancer cells may eject miRs that focus on different organs that choose the ideal environment for a metastasis, for instance [27, 28]. We now turn our attention to how these miRNAs specifically regulate the activity of bone cells like osteoblasts and osteoclasts.

miRNAs in Osteoblast/Osteoclast Lineage and Bone Formation/Resorption

There are several targets of miRNAs involved in the inhibition and promotion of osteogenesis, making them key players in bone homeostasis; some of them are:

(a) Runt-related transcription factor 2 (Runx2) and miRNAs associated with osteogenesis: also known as acute myeloid leukemia gene 3 (*AML3*)-acts, polyomavirus enhancer binding protein (PEBP2a2), osteoblast-specific factor 2 (Osf2), and core-binding factor a1 (Cbfa1), is a significant transcription factor in osteogenesis regulation. Runx2 attaches itself to osteoblast-specific cis-acting element (OSE2), which is found in genes at promoter region, linked to osteogenesis, including alkaline phosphatase (*ALP*), secreted phosphoprotein 1 (*SPP1*), bone sialoprotein (BSP) collagen type 1 alpha 1 (*COL1A1*), and bone gamma-carboxyglutamic acid-containing protein (BGLAP) [29]. It is mostly needed to induce osteoblast differentiation, not adipogenesis or chondrogenesis, and during osteoblast development when Runx2 is overexpressed, ossification is inhibited. miR-146a reverse osteogenic differentiation by targeting JMJD3 (jumonji domain containing 3), which acts as a histone demethylase results in upregulation of Runx2. There are several other miRs that are responsible for osteogenesis and are associated with Runx2. For example, miR-9 promotes expression of Runx2 followed by BMSCs (bone marrow derived mesenchymal stem cells) osteogenic differentiation, whereas miR-10 play a role antagonist to miR-9. miR-204 downregulates the expression of Runx2 resulting in the prevention of osteoblastic differentiation and calcification in vascular mesenchymal stem cells (VMSC), whereas stimulation of adipogenesis and inhibition of osteogenesis in mesenchymal stem cells (MSC) [30, 31]. Histone acetyltransferase (HAT) balances transcriptional activity and chromatin structure through the action of a class of enzymes called HDACs. In humans, miR-449a was found to be target HDAC1, whereas miR-29 and miR-26a/b focused to HDAC4 and CDK6 in unrestricted somatic stem cells (USSCs). Osteoblast differentiation is inhibited by a co-repressor complex comprising Smad3 and HDC4/5; HDACs also act as Runx2 inhibitors like HDAC1 suppresses histone acetylation levels. Runx2 can create an amplification loop by upregulating two additional promoters of Runx2 degradation, Hoxa2 and HDAC5, which are blocked by miR-3960 and miR-2861. Furthermore, miR-22 targets HDAC6 deficiency, which results in increased trabecular bone density in mice. Additionally, transgenic miR-188 targeted mice for HDAC9 exhibited higher fat accumulation and loss of bone, and HDAC9 plays a flexible role in determining the direction of stem cell differentiation, which reduces in age-dependent adipogenesis [32–34].

(b) Smads and microRNAs associated with bone formation: An essential intracellular receptor, Smad factor takes involvement in the classical pathways of the BMP superfamily and transforming growth factor beta (TGF-b). It also independently promotes RSmad2/3 and RSmad1/5/8 signals. In order to control responsive genes, activated RSmads translocate into nuclei and establish a complicated interaction with Co-Smad4. Through a variety of methods, including its interaction with Runx2, Smad functions as a common activator of mechanisms like osteogenesis and bone regeneration. In addition to being engaged in numerous other pathways, Smad itself causes Msx2 to promote expression of Osx. However, BMP-7 induced osteogenesis involves a non-Smad mechanism that controls osteoblast survival and proliferation [35–37]. In order to direct osteoblast differentiation, miR-26a target Smad1 and is also engaged in BMP-pathways. MSCs from patients with multiple myeloma (MM) displayed increased miR-135b and poor bone formation, which targets mesenchymal cells like C2C12, human umbilical cord blood-derived unrestricted somatic stem cell (USSC) differentiation, and Smad5 to regulate bone disease [38, 39]. In addition, miR-17-5p, miR-140-5p, miR-100, and miR-106 can directly target BMP2 and BMPR2, and Smad1/5 can reduce osteogenic Runx2, Msx2, and TAZ while promoting adipogenesis. On the other hand, miR-17-5p is crucial to this response and has recently been shown to focus Smad7. Additionally, it can raise *COL1A1* and *Runx2* levels and encourage b-catenin's nuclear translocation, which can partially preserve bone remodeling and prevent nontraumatic osteonecrosis. Further, BMP-7 targeting miR-542-3p, prevents both BMP-7/PI3K surviving non-Smad induced proliferation and Smad-mediated osteoblast differentiation. On the other hand, miR-203 suppression might restore stem cell activity and PI3Kexpression that has been reduced [40–42].

(c) BMP/TGF-b/activin receptors: Osteogenesis also involves the TGF-b/Smads pathway. USSCs transfected with anti-miR-221 exhibited elevated Smad3 and TGF-bR expression as well as upregulation of osteogenesis. By blocking the osteoblastic-negative regulators TGF-bi (TGF-b induced) and TGF-bRI, miR-181 inhibited TGF-b signaling. miR-210 targets AcvR1b and Smad2/3 are targeted by miR-146a results in stimulation of osteoblastogenesis and suppression of TGF-b signaling. TGF-b signaling is suppressed and osteoblastogenesis is stimulated when miR-210 targets AcvR1b and miR-146a targets Smad2/3. Furthermore, it has been documented that Smad-specific E3 ubiquitin protein ligase 1 (Smurf) binds to miR497 ~ 195 cluster, which belongs to the miR-15 family. Remarkably, miR-194 and miR-302a target the chicken ovalbumin upstream promoter-transcription factor II (COUP-TFII), which is key requirement for preventing BMP-2 induced osteogenesis [43–45].

(d) RANKL and CSF1 in bone health: When combined with RANK ligand (RANKL), receptor activator for nuclear factor kB (RANK), a type I membrane protein produced on surface of osteoclast precursors, stimulates osteoclast development through close cell-cell interaction through ligands and receptors on the surfaces of T cells, stroma cells, osteoblasts and osteocytes. The JNK/AP-1 pathway and NF-kB can be activated in the amplification mechanism by transduction factors TNF receptor associated factors (TRAFs), particularly RANK associated TRAF6. Osteoclast differentiation is promoted by another significant transcription factor c-Fos (induced by RANKL), belonging to osteoclast functioning. However, to reduce osteoclastogenesis, the c-Fos-mediated osteoclast associated receptor (OSCAR) and nuclear factor of activated T cell calcineurin-dependent 1 (NFATc1) may be attenuated by protein inhibitor of activated STAT 3 (PIAS3). In this instance, M-CSF (also known as CSF-1) can upregulate RANK, while osteoprotegrin (OPG), a natural RANKL decoy receptor, can contend with RANK to inhibit osteoclast development [46–49]. CSF1 (colony stimulating factor 1) binds to CSF1R and their interaction results in growth, survival, and differentiation. CSF-1 are released by various cells and expressed by several cells including stromal cells, osteoblasts, and tumor cells. It acts as a key factor in bone health by regulating resorption of bone and osteoclastogenesis process. It was observed that the high level of CSF-1 was related to bone dysregulation like arthritis and osteoporosis [50–52].

Having understood their impact on bone biology, we now examine how these miRNAs contribute to the initiation and progression of bone metastasis.

MicroRNA and Bone Metastasis

Certain miRNAs act as pro- or anti-metastatic factors by influencing cancer cell invasion, interaction with the bone matrix, pre-metastatic niche formation, and establishing the basis for the creation of new therapeutic techniques. Identifying tumor cells' chances of surviving can be increased by immune cell's expression of miRNAs that regulate immune function. miRNAs that control various phases of metastasis formation can also promote this process, particularly by downregulating tumor suppressor genes and simultaneously upregulating oncogenes. For instance, they can prepare tumor cells to migrate or even prime the metastatic site to receive circulating tumor cells in order to start and complete metastasis [53, 54]. There are several miRNA targets of bone metastasis:

(a) miR-218: During metastasis, surface receptor CXCR4, expressed over tumor cell that has CXCL12 (SDF1, stromal derived factor 1) as its ligand, is upregulated due to miR-218 and it was observed that this CXCR4 expression significantly elevated in breast tumor cells. CXCR4 is responsible for bone metastasis in breast and prostate cancer. CXCR4 also has a role in the expression of MMP9 and MMP13. It was observed that MMP9 expression is reduced by miR-218. miR-218 significantly regulates osteomimcry in breast tumors. miR-218 functions as a "osteo-miR" and regulates bone formation, in osteoblasts, it upregulates Runx2 and downregulates Wnt signaling inhibitors (like Dickkopf-2, secreted frizzled-related protein2 and sclerostin). In metastatic breast cancer, miR-218 functions as pro-metastatic miRNA, especially for metastasis to bone. In breast cancer cells, it downregulates the Wnt-inhibitory protein, followed by increased Wnt-signaling. This results into upregulation of Wnt-responsive genes, including osteopontin and bone sialoprotein, which stimulated tumor's bone homing and growth capabilities [55, 56].

(b) miR-126: Breast and prostate cancer bone metastasis is reduced due to expression of MMP13, which is regulated by miR-126. Vascular endothelial molecule 1 (VCAM1) is a target of miR-126, and it is expressed over osteotropic breast cancer cells. Osteoclast progenitors have integrins ($\alpha4\beta7$ and $\alpha4\beta1$) that bind to VCAM1 and result in osteolytic lesions and increased bone resorption [57, 58].

(c) miR-25: Metastatic spread is reduced by overexpression of miR-25, which states that it is a critical regulator of bone metastasis formation and cancer stem cell properties. miR-25 significantly suppresses the high osteotropic cancer stem/progenitor cells of human prostate cancer. Integrin $\alpha\nu$ expression is regulated directly by miR-25, $\alpha\nu$ integrins regulate bone colonization and osteoclastic bone resorption [59].

(d) miR-26 and miR-29: These two miRs suppress LOXL2 and act as tumor suppressors and indirectly bone metastasis suppressors. miR-29 also inhibits cancer cell migration in prostate cancer. LOXL2 is linked with cancer progression and promotes premetastatic niche for bone metastasis cells [60, 61].

(e) miR-10b: It is one of the well-known pro-metastatic miRNAs. Breast cancer, pancreatic cancer, and glioblastoma are among the actively metastatic and aggressive types of human cancer that overexpress this miRNA. It has been demonstrated mechanistically to downregulate a number of tumor suppressor and genes that inhibit metastasis. The transcription factor TBX5 is specifically known to be directly targeted by miR-10b, which subsequently suppress *DYRK1A* and *PTEN* that are tumor suppressor genes, *KLF4* known to encourage migration and invasion, and neurofibromin encoding gene, which negatively control expression of Ras results in promoting proliferation. All of these factors ultimately contribute to the development of tumorigenesis and metastases. By inhibiting the transcription factor homeobox D10 (HOXD10) and subsequently upregulating the pro-metastatic gene RhoC, miR-10b overexpression in breast cancer encourages tumor cell invasion and metastasis, both in vitro and in vivo. In contrast, in a mouse model of breast tumors, blocking miR-10b using an antagomir prevents the development of lung metastases [62–64].

(f) miR-21, miR-30, miR-218, miR-183: These four miRNAs are known as ostemirs as they promote bone metastasis by promoting pre-metastatic niche formation and providing optimized condition bone microenvironment. miR-183 suppresses HO-1 resulting in enhanced osteoclastogenesis. The other three expressed highly in cancer stem cells and prostate cancer cells, promote bone metastasis, and upregulate the expression of *MMP9*, *MMP2*, and *MMP13* [65].

(g) miR-200 s and miR-205 Family: TGFb stimulates transcription factors like Slug, Snail, and Twist, suppressors of potent E-cadherin. MicroRNAs target

CSCs, leading to loss in their properties and metastatic progression. ZEB1 and ZEB2 mRNAs are targeted by miR-200 family and miR205 family; miR-200c targets ZEBs, TGFb, and BMII, resulting in the inhibition of cell stemness and metastasis. Let-7 miRNAs have downregulated expression in metastatic cancer cells and are used to silence *Ras*, *MYC*, and *HMGA2*, leading to the inhibition of cell stemness and BRCA metastases. There are various steps involved in dissemination of metastasis to bone: the addition in enhanced migratory capacity and invasiveness, followed by generation of ECM components, angiogenesis, and high tolerance to programmed cell death and the bone colonization, are regulated by miRNAs as pro or anti-metastatic. It has been demonstrated that the miR-200 family changes the tumor secreted proteins by directly aiming Sec23 homolog A (SEC23A), which encourages metastatic colonization at distant regions. Additionally, it should be noted that experimental data indicate that disruption of the microenvironmental sustaining hinders the in vivo de novo development of bone metastases. Bone metastases may be hindered by miRNAs generated in the bone microenvironment. Indeed, miRNAs have the ability to affect DTC dormancy and exit the mesenchymal stem cells through exosomes or gap junctions to reach breast cancer cells. The exchange of miRs between mesenchymal stem cells and breast cancer cells via gap junction may be a significant factor in the cancer cells' dormancy, which gives the bone marrow a quiescent phenotype [6, 66–68].

(h) miR-33a and miR-336: According to reports, miR-33a directly targets PTHrP in lung cancer. Between lung cancer cell lines and human normal bronchial cell lines, there is inverse relationship between miR-33a levels and PTHrP expression decrease the arousal of lung cancer cells on osteoblasts' generation of RANKL (receptor activator of nuclear factor kappa-B ligand) and M-CSF (macrophage colony stimulating factor) while also lowering PTHrP and IL-8 expression. As a result, reduced miR-33a expression potentially possibly indicate an unfavorable outlook for patients with lung cancer and contributes to cancer-mediated bone deterioration. Given their great steadiness and reachability in serum patients' miRNAs are in fact attractive prospects as prognostic indicators in cancer. Accordingly, Valencia et al. compared serum miRNA levels with those of common biochemical markers of bone turnover, such as CTX (carboxyterminal telopeptide), PINP (procollagen 1 amino-terminal propeptide) and BGP (osteocalcin) and assess the accuracy of serum miRNA measurement in a mouse model of human lung cancer of bone metastases. There was a robust correlation between PINP and osteolytic lesions tumor burden in this lung cancer model at both the early and late stages of colonization. BGP and CTX, on the other hand, only showed a high link in the later phases. It is interesting to note that in mice model, blood vessels of miR-326 were substantially correlated with PINP and tumor burden in animals given with a vehicle, but not in animals attended with zoledronate, an inhibitor of bone resorption. These findings imply that miR-326 may be used as a biochemical marker to track the development of bone metastases in advance lung cancer [69, 70].

miRNAs play a crucial role in interactions between primary and metastatic tumor cells and various bone/bone marrow stromal cells, like TAM, osteoblast, osteoclast, MSC, and others. For example, miR-511-3p suppresses TAM oncopromotion, whereas miR-31 and miR-214 promote the same. Tgif2 (stimulated by TGF-β) reduces miR-34a, enhancing osteoclastogenesis, whereas nanoparticles of miR-34a reduce the same. Loss of miR-16 and miR-15, along with miR-21 [71] upregulation in prostate cancer activates TGF-β signaling, resulting in osteolysis and bone colonization. miR-34a inhibits Notch signaling and osteoclastogenesis, reducing breast cancer bone metastasis. miR-155 affects TGF-β signaling and promotes invasion and metastasis. miR-16 and miR-378 also correlate with bone metastasis making them potential biomarkers.

Given their central role in metastatic behavior, these miRNAs are being investigated for their potential in diagnosis and therapy. An overview of key miRNAs, their molecular targets, and biological outcomes in bone metastasis is provided in Table 15.1.

miRNAs Involved in the Diagnosis of Bone Metastases and Their Therapeutic Aspects

Emerging research shows that specific microRNAs can act as potential diagnostic and prognostic biomarkers in cancer, that help in cancer medications, drug delivery, and treatment. The clinical trials of therapies associated with miRNA are under primary clinical trials due to the challenge of specific delivery of small interfering RNA to its target site. The aim of using miRNAs is to modulate the expression of key oncogenes, but there are several issues with specificity, drug delivery system, and potential immune responses. With the help of nanoparticle and lipid-based drug delivery systems, scientists are trying to overcome certain challenges of miRNA-associated medication. The therapeutic potential of miRNA and siRNA can be improved by using these approaches. Till now only a few miRNAs can prevent bone metastasis in animal models. For example, miR-141 and miR-219 are found to decrease bone metastasis in breast cancer by reducing osteoclast activity. miR-203 and miR-135 target key regulator RUNX2 resulting in decreased breast cancer bone metastasis. Currently, MIRX34 is the only miR-based therapy present in clinical trials, to restore this tumor suppressor miR in liver cancer and other solid tumors, a liposomal formulation of miR-34 mimic is utilized. miR-34 is downregulated in breast tumors and reduces bone metastasis. Many other miR-based therapies are under research and trying to find potential drugs to target bone metastatic cells [85, 88, 89].

The course of the disease and the spread of metastases in breast carcinomas are correlated with abnormal serum concentrations of miRNAs. In lymph node-positive versus node-negative breast cancer patients, the plasma levels of miR-373 and miR-10b differ. Elevated levels of miR-373 were also linked to the original tumor's

Table 15.1 Regulatory impact of miRNA-target interactions in bone metastasis

miRNAs	Targets	Outcome of interaction	References
miR-218	CXCR4, MMP9, Runx2, Wnt inhibitors (sclerostin, DKK2, sFRP2)	Promotes bone metastasis; enhances Wnt signaling; regulates osteomimcry and bone formation	[55]
miR-126	MMP13, VCAM1	Suppresses bone metastasis (BM); reduces osteolytic lesions and bone resorption	[72]
miR-25	Integrin αv	Suppresses BM and stem/progenitor properties in prostate cancer	[73]
miR-26, miR-29	LOXL2	Tumor suppressors; inhibit BM and cancer cell migration	[74]
miR-10b	TBX5, HOXD10, PTEN, DYRK1A, KLF4, RhoC	Promotes metastasis and tumor proliferation	[62]
miR-21, miR-30, miR-218, miR-183	MMP9, MMP2, MMP13, HO-1	Promote BM and osteoclastogenesis	[75]
miR-200s, miR-205	ZEB1, ZEB2, TGFβ, BMII	Inhibit CSC properties, metastasis, and EMT; downregulate stemness	[76]
let-7 family	Ras, MYC, HMGA2	Suppresses BRCA metastasis and stemness	[77]
miR-33a	PTHrP, IL-8	Reduces bone deregulation in lung cancer; correlated with prognosis	[69]
miR-326	Not specified	Potential biomarker for BM in lung cancer	[70]
miR-511-3p	Tumor associated macrophage (TAM) activity	Suppresses pro-tumoral TAM activity	[78]
miR-31, miR-214	Promote TAM activity	Enhance tumor progression and metastasis	[79]
miR-34a	Notch signaling	Inhibits osteoclastogenesis and breast cancer BM	[80]
miR-15, miR-16	TGF-β signaling	Loss promotes bone colonization and osteolysis	[81]
miR-155	TGF-β signaling, immune modulation	Promotes invasion and immunosuppressive environment	[82]
miR-378	Not specified	Correlated with BM	[83]
miR-141, miR-219	Osteoclast activity	Reduce BM in breast cancer	[84]
miR-203, miR-135	Runx2	Reduce BM in breast cancer	[85]
miR-195, miR-497	CD274 (PD-L1) in TNBC	Suppress immune evasion and tumor progression	[86]
miR-19a-3p	Fra-1	Reduces M2 TAMs; inhibits metastasis	[87]
miR-240-5p	TNF, cytokine genes	Reprograms tumor microenvironment; suppresses tumor and metastasis	[63]

HER2 negative status. Because blood levels of miR-34a, miR-199, miR-155, miR-335, miR-126, and miR-21 are disrupted in patients with hormone receptors, while miR-17 is disrupted in patients with hormone receptors, the receptor's status can also be determined by miRNA signature. Furthermore, compared to patients with breast cancer in its early stages, patients with advanced-stage disease had much higher blood levels of miR-34a, and variations in serum levels of miR-155, miR-34a, and miR-10b were associated with the existence of metastases [90, 91].

A well-known oncogenic miRNA, miR-155, is frequently overexpressed in several types of tumors, including breast cancer. According to reports, miR-155 controls the antitumoral response in various immune population subsets related to breast cancer [92]. In a mouse model of spontaneous breast cancer, by hindering TAM activation, boosting pro-tumor cytokine production, and decreasing activation maker expression, significant increase in tumor growth of myeloid cells was observed by a knockdown of miR-155. This resulted in the transition of TAM from an M1-phenotype to an M2-phenotype followed by the creation of an immune-suppressive environment. Similarly, by modifying macrophage polarization, the miR-23a/27a/24-2 cluster aids in the advancement of breast cancer. CD274 (often referred to as PD-L1), which is a cluster of differentiation in triple negative breast cancer cells, was recently reported to be downregulated by miR-195 and miR-497, indicating that these factors affect tumor progression suppress the immune system and encourage the immunological escape of tumors [86, 93, 94]. Conversely, in breast cancer, some miRNAs function as tumor-suppressive agent like miR-19a-3p and miR-240-5p. miR-19a-3p inhibits the M2-like TAM population and regulates the change in TAM from the M2 to M1 phenotype, both in vitro and in vivo by aiming the Fra-1 protooncogene and its signal transduction pathways. This helps to stop metastasis from occurring. The expression of cytokines like tumor necrosis factor (TNF) and other important genes involved in the immunological pathways are regulated by miR-240-5p, which aids in the reprogramming and redesigning of the tumor microenvironment. A significant decrease in MDSCs was associated with the misexpression of miR-240-5p in the tumor microenvironment, there are more CD4+ T cells, regulatory T cells, and CD8+ T cells in addition to macrophages and NK cells. Furthermore, the misexpression of miR-240-5p suppressed tumor development and metastasis in vivo and significantly changed the metabolic characteristics of cancer cells [87, 95].

The unification of miRNA mimics and PNA-anti-miRNA together is newly formed. Folic acid (FA) forms a PNA (recently found as a new type of miRNA inhibitors) complex that acknowledges the cationic R9 peptide, which enables the self-assembly with the miRNA mimic, and the folic receptor (FR) on the cancer cell membrane and a nanocomposite (PMN-34a/21) was created by assembling the PNA anti-miR-21 (FA-R9-PNA) with the miRNA mimic miR-34a. Effective delivery of the PMN-34a/21 nanocomposite in FR positive HeLa cells lead to the over-expression of miR-34a and the corresponding reduced response of miR-21. It also modulated the downstream target gene of the two miRNAs, ultimately driving cell death. A more potent anticancer treatment is made possible by this novel approach, which enables the simultaneous elevation of miRNA mimics that reduce tumors and

PNA-anti-miRNAs that inhibit carcinogenic miRNAs [96]. All of these investigations showed that PNA-induced miRNA suppression is very promising for the treatment of cancer. Given the flexibility of PNA design and alteration, PNA-based techniques to target several miRNAs at once can be developed. Additionally, as recently discovered, miRNA replacement and inhibitory therapies can be combined in a single formulation to improve the effects. Their use in clinical settings would also create fresh opportunities for nonviral gene treatments.

Future Prospective

The future of cancer care is increasingly moving beyond DNA-based diagnostics toward RNA-based strategies, with microRNAs (miRNAs) emerging as promising tools in oncology. miRNAs regulate gene expression post-transcriptionally and are involved in key processes such as tumor progression, immune modulation, and metastasis. Their stability in blood and specificity to disease states make them attractive noninvasive biomarkers and therapeutic targets, particularly for cancers with a high tendency to metastasize to bone, such as breast and prostate cancer.

Several miRNAs discussed in this chapter, including miR-34a, miR-155, and miR-16, are being investigated for their roles in bone metastasis and cancer stem cell regulation. MRX34, a liposomal mimic of miR-34a, was the first miRNA therapy to enter clinical trials in solid tumors, although it was discontinued due to immune-related toxicities. This highlights the key challenge in miRNA-based therapy: ensuring delivery specificity and avoiding off-target effects.

To overcome these barriers, researchers are now developing advanced delivery systems such as exosome-mimetic vesicles, aptamer-conjugated nanoparticles, and bone-targeted carriers. In parallel, combing miRNA-based approaches with existing therapies, like bisphosphonates or immune checkpoint inhibitors, is under exploration for synergistic effects in bone metastasis management. Additionally, miRNA profiling is being tested as a prognostic tool to predict the risk of bone relapse and treatment resistance. As clinical translation progresses, miRNA-based diagnostics and therapeutics hold strong potential for personalized, minimally invasive cancer management, especially for detecting and treating bone metastases at earlier stages.

References

1. Roy PS, Saikia BJ (2016) Cancer and cure: a critical analysis. Indian J Cancer 53(3):441–442
2. Hausman DM (2019) What is cancer? Perspect Biol Med 62(4):778–784
3. Hiraga T (2019) Bone metastasis: interaction between cancer cells and bone microenvironment. J Oral Biosci 61(2):95–98
4. Lu TX, Rothenberg ME (2018) MicroRNA. J Allergy Clin Immunol 141(4):1202–1207
5. Croset M et al (2014) MicroRNAs and bone metastasis: a new challenge. Molecules 19(7):10115–10128

6. Browne G et al (2014) MicroRNAs in the control of metastatic bone disease. Trends Endocrinol Metab 25(6):320–327
7. Clézardin P et al (2021) Bone metastasis: mechanisms, therapies, and biomarkers. Physiol Rev 101(3):797–855
8. Esposito M, Guise T, Kang Y (2018) The biology of bone metastasis. Cold Spring Harb Perspect Med 8(6)
9. Yoneda T, Sasaki A, Mundy GR (1994) Osteolytic bone metastasis in breast cancer. Breast Cancer Res Treat 32(1):73–84
10. Fidler IJ, Poste G (2008) The seed and soil; hypothesis revisited. Lancet Oncol 9(8):808
11. Pantel K, Alix-Panabières C, Riethdorf S (2009) Cancer micrometastases. Nat Rev Clin Oncol 6(6):339–351
12. Eger A et al (2005) DeltaEF1 is a transcriptional repressor of E-cadherin and regulates epithelial plasticity in breast cancer cells. Oncogene 24(14):2375–2385
13. Al-Hajj M et al (2003) Prospective identification of tumorigenic breast cancer cells. Proc Natl Acad Sci USA 100(7):3983–3988
14. Ricci-Vitiani L et al (2007) Identification and expansion of human colon-cancer-initiating cells. Nature 445(7123):111–115
15. Turley EA et al (2008) Mechanisms of disease: epithelial-mesenchymal transition—does cellular plasticity fuel neoplastic progression? Nat Clin Pract Oncol 5(5):280–290
16. Yin T, Li L (2006) The stem cell niches in bone. J Clin Invest 116(5):1195–1201
17. Poswal J, Mandal C (2025) Lipid metabolism dysregulation for bone metastasis and its prevention. Expert Rev Anticancer Ther 25:657
18. Mandal CC (2020) Osteolytic metastasis in breast cancer: effective prevention strategies. Expert Rev Anticancer Ther 20(9):797–811
19. Le Gall C et al (2007) A cathepsin K inhibitor reduces breast cancer induced osteolysis and skeletal tumor burden. Cancer Res 67(20):9894–9902
20. Bellahcène A et al (2007) Transcriptome analysis reveals an osteoblast-like phenotype for human osteotropic breast cancer cells. Breast Cancer Res Treat 101(2):135–148
21. Santini D et al (2012) The role of bone microenvironment, vitamin D and calcium. Recent Results Cancer Res 192:33–64
22. Santini D et al (2007) New developments of aminobisphosphonates: the double face of Janus. Ann Oncol 18(Suppl 6):vi164–vi167
23. Zoccoli A et al (2012) Premetastatic niche: ready for new therapeutic interventions? Expert Opin Ther Targets 16(Suppl 2):S119–SS29
24. Mandal CC (2015) High cholesterol deteriorates bone health: new insights into molecular mechanisms. Front Endocrinol (Lausanne) 6:165
25. Macfarlane LA, Murphy PR (2010) MicroRNA: biogenesis, function and role in cancer. Curr Genomics 11(7):537–561
26. Sharma T, Radosevich JA, Mandal CC (2021) Dual role of microRNAs in autophagy of colorectal cancer. Endocr Metab Immune Disord Drug Targets 21(1):56–66
27. Di Leva G, Garofalo M, Croce CM (2014) MicroRNAs in cancer. Annu Rev Pathol 9:287–314
28. Lu J et al (2005) MicroRNA expression profiles classify human cancers. Nature 435(7043):834–838
29. Dalle Carbonare L, Innamorati G, Valenti MT (2012) Transcription factor Runx2 and its application to bone tissue engineering. Stem Cell Rev Rep 8(3):891–897
30. Huszar JM, Payne CJ (2014) MIR146A inhibits JMJD3 expression and osteogenic differentiation in human mesenchymal stem cells. FEBS Lett 588(9):1850–1856
31. Zheng L et al (2017) Runx2/DICER/miRNA pathway in regulating osteogenesis. J Cell Physiol 232(1):182–191
32. Jensen ED, Nair AK, Westendorf JJ (2007) Histone deacetylase co-repressor complex control of Runx2 and bone formation. Crit Rev Eukaryot Gene Expr 17(3):187–196
33. Zhang Y et al (2008) Mice lacking histone deacetylase 6 have hyperacetylated tubulin but are viable and develop normally. Mol Cell Biol 28(5):1688–1701

34. Li CJ et al (2015) MicroRNA-188 regulates age-related switch between osteoblast and adipocyte differentiation. J Clin Invest 125(4):1509–1522
35. Chen G, Deng C, Li YP (2012) TGF-β and BMP signaling in osteoblast differentiation and bone formation. Int J Biol Sci 8(2):272–288
36. Afzal F et al (2005) Smad function and intranuclear targeting share a Runx2 motif required for osteogenic lineage induction and BMP2 responsive transcription. J Cell Physiol 204(1):63–72
37. Kureel J et al (2014) miR-542-3p suppresses osteoblast cell proliferation and differentiation, targets BMP-7 signaling and inhibits bone formation. Cell Death Dis 5(2):e1050
38. Xu S et al (2013) Upregulation of miR-135b is involved in the impaired osteogenic differentiation of mesenchymal stem cells derived from multiple myeloma patients. PLoS One 8(11):e79752
39. Li Z et al (2008) A microRNA signature for a BMP2-induced osteoblast lineage commitment program. Proc Natl Acad Sci USA 105(37):13906–13911
40. Hwang S et al (2014) miR-140-5p suppresses BMP2-mediated osteogenesis in undifferentiated human mesenchymal stem cells. FEBS Lett 588(17):2957–2963
41. Jia J et al (2014) MiR-17-5p modulates osteoblastic differentiation and cell proliferation by targeting SMAD7 in non-traumatic osteonecrosis. Exp Mol Med 46(7):e107
42. Liu T et al (2014) Suppression of MicroRNA-203 improves survival of rat bone marrow mesenchymal stem cells through enhancing PI3K-induced cellular activation. IUBMB Life 66(3):220–227
43. Bakhshandeh B et al (2012) Down-regulation of miRNA-221 triggers osteogenic differentiation in human stem cells. Biotechnol Lett 34(8):1579–1587
44. Jeong BC et al (2014) MicroRNA-194 reciprocally stimulates osteogenesis and inhibits adipogenesis via regulating COUP-TFII expression. Cell Death Dis 5(11):e1532
45. Kang IH et al (2015) MicroRNA-302a stimulates osteoblastic differentiation by repressing COUP-TFII expression. J Cell Physiol 230(4):911–921
46. Boyce BF (2013) Advances in the regulation of osteoclasts and osteoclast functions. J Dent Res 92(10):860–867
47. Boyce BF, Xing L (2007) Biology of RANK, RANKL, and osteoprotegerin. Arthritis Res Ther 9(Suppl 1):S1
48. Mizoguchi F et al (2013) miR-31 controls osteoclast formation and bone resorption by targeting RhoA. Arthritis Res Ther 15(5):R102
49. Boyce BF et al (2005) Roles for NF-kappaB and c-Fos in osteoclasts. J Bone Miner Metab 23(Suppl):11–15
50. Yagiz K, Rittling SR (2009) Both cell-surface and secreted CSF-1 expressed by tumor cells metastatic to bone can contribute to osteoclast activation. Exp Cell Res 315(14):2442–2452
51. Saleh R et al (2018) CSF-1 in inflammatory and arthritic pain development. J Immunol 201(7):2042–2053
52. Chockalingam S, Ghosh SS (2014) Macrophage colony-stimulating factor and cancer: a review. Tumour Biol 35(11):10635–10644
53. Bottani M, Banfi G, Lombardi G (2019) Circulating miRNAs as diagnostic and prognostic biomarkers in common solid tumors: focus on lung, breast, prostate cancers, and osteosarcoma. J Clin Med 8(10)
54. Kashyap D, Kaur H (2020) Cell-free miRNAs as non-invasive biomarkers in breast cancer: significance in early diagnosis and metastasis prediction. Life Sci 246:117417
55. Hassan MQ et al (2012) miR-218 directs a Wnt signaling circuit to promote differentiation of osteoblasts and osteomimicry of metastatic cancer cells. J Biol Chem 287(50):42084–42092
56. Zoni E, van der Pluijm G (2016) The role of microRNAs in bone metastasis. J Bone Oncol 5(3):104–108
57. Tai HC et al (2014) Osteoblast-derived WNT-induced secreted protein 1 increases VCAM-1 expression and enhances prostate cancer metastasis by down-regulating miR-126. Oncotarget 5(17):7589–7598

58. Wu Z et al (2014) MiR-126-5p regulates osteoclast differentiation and bone resorption in giant cell tumor through inhibition of MMP-13. Biochem Biophys Res Commun 443(3):944–949
59. Zoni E, Kruithof-de Julio M, van der Pluijm G (2015) miR-25, integrin and cancer invasiveness. Oncoscience 2(8):663–664
60. Fukumoto I et al (2015) Tumor-suppressive microRNAs (miR-26a/b, miR-29a/b/c and miR-218) concertedly suppressed metastasis-promoting LOXL2 in head and neck squamous cell carcinoma. J Hum Genet 61
61. Foenna Y et al (2025) miR-29 family as epigenetic regulators of DNMTs in prostate cancer and benign prostatic hyperplasia. Indones Biomed J 17:134–142
62. Sheedy P, Medarova Z (2018) The fundamental role of miR-10b in metastatic cancer. Am J Cancer Res 8(9):1674–1688
63. Gomarasca M et al (2020) microRNAs in the antitumor immune response and in bone metastasis of breast cancer: from biological mechanisms to therapeutics. Int J Mol Sci 21(8)
64. Zhao FL et al (2012) Serum overexpression of microRNA-10b in patients with bone metastatic primary breast cancer. J Int Med Res 40(3):859–866
65. Puppo M, Valluru MK, Clézardin P (2021) MicroRNAs and their roles in breast cancer bone metastasis. Curr Osteoporos Rep 19(3):256–263
66. Vimalraj S et al (2013) Regulation of breast cancer and bone metastasis by microRNAs. Dis Markers 35(5):369–387
67. Lim PK et al (2011) Gap junction-mediated import of microRNA from bone marrow stromal cells can elicit cell cycle quiescence in breast cancer cells. Cancer Res 71(5):1550–1560
68. Yu F et al (2007) let-7 regulates self renewal and tumorigenicity of breast cancer cells. Cell 131(6):1109–1123
69. Kuo PL et al (2013) MicroRNA-33a functions as a bone metastasis suppressor in lung cancer by targeting parathyroid hormone related protein. Biochim Biophys Acta 1830(6):3756–3766
70. Valencia K et al (2013) miR-326 associates with biochemical markers of bone turnover in lung cancer bone metastasis. Bone 52(1):532–539
71. Mandal CC et al (2012) miR-21 is targeted by omega-3 polyunsaturated fatty acid to regulate breast tumor CSF-1 expression. Carcinogenesis 33(10):1897–1908
72. Yokoyama Y et al (2013) Cancer-associated upregulation of histone H3 lysine 9 trimethylation promotes cell motility in vitro and drives tumor formation in vivo. Cancer Sci 104(7):889–895
73. Zoni E et al (2015) miR-25 modulates invasiveness and dissemination of human prostate cancer cells via regulation of αv- and α6-integrin expression. Cancer Res 75(11):2326–2336
74. Fukumoto I et al (2016) Tumor-suppressive microRNAs (miR-26a/b, miR-29a/b/c and miR-218) concertedly suppressed metastasis-promoting LOXL2 in head and neck squamous cell carcinoma. J Hum Genet 61(2):109–118
75. Zoni E, Pluijm G (2016) The role of microRNAs in bone metastasis. J Bone Oncol:5
76. Wellner U et al (2009) The EMT-activator ZEB1 promotes tumorigenicity by repressing stemness-inhibiting microRNAs. Nat Cell Biol 11(12):1487–1495
77. Wang L, Wang J (2012) MicroRNA-mediated breast cancer metastasis: from primary site to distant organs. Oncogene 31(20):2499–2511
78. Squadrito ML et al (2012) miR-511-3p modulates genetic programs of tumor-associated macrophages. Cell Rep 1(2):141–154
79. Pan Z et al (2020) Role of microRNAs in remodeling the tumor microenvironment (review). Int J Oncol 56(2):407–416
80. Krzeszinski JY et al (2014) miR-34a blocks osteoporosis and bone metastasis by inhibiting osteoclastogenesis and Tgif2. Nature 512(7515):431–435
81. Bonci D et al (2016) A microRNA code for prostate cancer metastasis. Oncogene 35(9):1180–1192
82. O'Connell RM et al (2010) MicroRNA-155 promotes autoimmune inflammation by enhancing inflammatory T cell development. Immunity 33(4):607–619
83. Wang J et al (2022) Tumor-derived miR-378a-3p-containing extracellular vesicles promote osteolysis by activating the Dyrk1a/Nfatc1/Angptl2 axis for bone metastasis. Cancer Lett 526:76–90

84. Ell B et al (2013) Tumor-induced osteoclast miRNA changes as regulators and biomarkers of osteolytic bone metastasis. Cancer Cell 24(4):542–556
85. Taipaleenmäki H et al (2015) Targeting of Runx2 by miR-135 and miR-203 impairs progression of breast cancer and metastatic bone disease. Cancer Res 75(7):1433–1444
86. Yang L et al (2018) miR-195/miR-497 regulate CD274 expression of immune regulatory ligands in triple-negative breast cancer. J Breast Cancer 21(4):371–381
87. Yang J et al (2014) MicroRNA-19a-3p inhibits breast cancer progression and metastasis by inducing macrophage polarization through downregulated expression of Fra-1 proto-oncogene. Oncogene 33(23):3014–3023
88. Waning DL, Mohammad KS, Guise TA (2013) Cancer-associated osteoclast differentiation takes a good look in the miR(NA)ror. Cancer Cell 24(4):407–409
89. Haider MT, Taipaleenmäki H (2018) Targeting the metastatic bone microenvironment by MicroRNAs. Front Endocrinol (Lausanne) 9:202
90. Roth C et al (2010) Circulating microRNAs as blood-based markers for patients with primary and metastatic breast cancer. Breast Cancer Res 12(6):R90
91. Schwarzenbach H et al (2014) Clinical relevance of circulating cell-free microRNAs in cancer. Nat Rev Clin Oncol 11(3):145–156
92. Mattiske S et al (2012) The oncogenic role of miR-155 in breast cancer. Cancer Epidemiol Biomarkers Prev 21(8):1236–1243
93. Zonari E et al (2013) A role for miR-155 in enabling tumor-infiltrating innate immune cells to mount effective antitumor responses in mice. Blood 122(2):243–252
94. Ma S et al (2016) A double feedback loop mediated by microRNA-23a/27a/24-2 regulates M1 versus M2 macrophage polarization and thus regulates cancer progression. Oncotarget 7(12):13502–13519
95. Hong BS et al (2019) Tumor suppressor miRNA-204-5p regulates growth, metastasis, and immune microenvironment remodeling in breast cancer. Cancer Res 79(7):1520–1534
96. Xiao X et al (2017) Multi-functional peptide-MicroRNA Nanocomplex for targeted MicroRNA delivery and function imaging. Chem Eur J 24

Chapter 16
Understanding the Role of miRNA Modulators in Gastrointestinal Cancers and Exploring Novel Therapeutic Capabilities

Megala Jayaraman

Abstract Gastrointestinal (GI) cancers continue to be of significant concern to the public, amounting to 26% of all cancer diagnoses worldwide and 35% of all fatalities associated with cancer. The present-day treatment options for patients suffering from GI cancer including radiation therapy, chemotherapy, immunotherapy, and hormonal treatments have been the backbone of a patient's recovery stage for several decades. Despite advancements in these treatment procedures, an effective cure for the permanent eradication of GI cancer is yet to exist. Researchers have now shifted their focus to the field of microRNAs (miRNAs), a class of small non-coding RNAs that have compelling evidence demonstrating their role in suppressing cancers, and they hope to develop miRNA modulators as a potentially useful weapon in the fight against GI cancers. In this chapter, the dysregulation of miRNAs in GI malignancies and a detailed study on the complex regulatory roles of miRNAs in the course of cancer are provided. It further delves into the therapeutic potential of miRNA modulators in arresting tumour growth, prospects of miRNA biomarkers, tumour-associated macrophages (TAM) polarisation, tumour microenvironments, and apoptosis of tumour cells. Moreover, this chapter also offers an insight into the progress of designing the miRNA modulators, the obstacles faced so far, and future perspectives in the use of miRNA as potent treatments for GI malignancies.

Keywords Apoptosis · Gastrointestinal cancers · miRNA biomarkers · Small non-coding RNA · Tumour microenvironment

---

M. Jayaraman (✉)
Department of Genetic Engineering, SRM Institute of Science and Technology, Chennai, India
e-mail: megalaj@srmist.edu.in

S. Pathak, A. Banerjee (eds.), *MicroRNAs: Role in Cancer Progression, Diagnostics and Therapy*, Advances in Biochemistry in Health and Disease 34,
https://doi.org/10.1007/978-3-032-08824-6_16

Introduction

MicroRNAs (miRNAs) are a class of short, single-stranded, non-coding RNAs ranging up to the length of 19 to 23 nucleotides that regulate gene expression by hindering translation and triggering degradation of a targeted messenger RNA, post-transcriptionally [106]. These miRNAs inhibit the translation of target mRNA through one of two major mechanisms: the first mechanism includes the two-step cleavage of primary miRNA, also known as pri-miRNA, resulting in the production of mature miRNA. This mature miRNA then becomes part of the RNA-induced silencing complex (RISC), which is responsible for its regulatory function. In the second mechanism, the miRNA binds with mRNA sequences by matching through partial complementarity between the second to the eighth nucleotides from the 5′ end of the miRNA with complementary sequences found in the 3′ untranslated region (UTR) of the mRNA. While we have a broad comprehension of the functions of miRNA and its activity, the specific mechanisms of miRNA production and gene silencing remain ambiguous and are currently being closely examined. The modified expression patterns of miRNAs in different tumours suggest that these small molecules may play a role in the development of cancer and other malignancies. miRNA expression was dysregulated [126, 127] and observed in numerous forms of human neoplasms, inclusive of digestive tract malignancies such as gastrointestinal, colorectal, and liver cancers. All together, they make for approximately 1/4th of the total death rates of non-cutaneous malignancies globally. A significant proportion of miRNAs have distinct expression patterns in these types of malignancies. There was a surging amount of evidence illustrating the significance of miRNAs in the treatment of tumours and being potential biomarkers of GI cancer. Various delivery mechanisms have been concocted, proposing innovative and promising therapeutic techniques based on miRNA to address the limitations of conventional cancer treatments [77]. Over the past few decades, numerous studies have revealed the significance of miRNAs in regulating both normal and abnormal immune function. There is a growing body of research focused on examining the interrelation between miRNA, immunity, and cancer. miRNAs play a vital role in regulating immune responses, influencing the function and activity of immune cells such as natural killer cells (NK cells), macrophages, and dendritic cells (DCs). miRNAs are also known to guide T cells in forming an immune response by modulating downstream signal transduction pathways from the T cell receptors. Understanding these interactions that are essential for developing new immunotherapeutic strategies to manage GI cancer effectively. miRNAs are often regulated by small molecule modulators such as enzymes and proteins; one such enzyme was Dicer, which initiates RNA interference and suppresses gene expression through siRNA (small interfering RNAs) and miRNA [26]. There were 88 small molecule drugs functioning as oncogene inhibitors that have been FDA-approved and shown potential in miRNA therapy. Other small molecules like neomycin inhibit miRNAs in gastric cancer (GC), and various small molecules regulate miRNA activity, showing therapeutic potential in cancer treatment. The prospects of miRNA as the treatment for GI cancer were found to be more optimistic and more promising with increasing research on its functions, with a vast amount left to be discovered [9]. miRNA also shows good results as biomarkers for GI cancers and contributes to the increasing effectiveness of chemotherapy for patients. In this chapter, we delve further into the diverse functions and roles of miRNA, gain a better comprehension of its working, and discuss the viable treatments for GI cancers.

Regulation of Biological Processes That Influence Tumours by Dysregulated miRNA

Irregular expression of miRNA was common in many types of cancer in humans, including GI cancers (GC), colorectal cancer (CRC), and liver cancer (LC) were the most common types of GI cancers; together, they cause 23.5% of all non-cutaneous cancers and 27% of deaths worldwide [40]. A wide range of miRNAs are expressed in these cancers and their dysregulation leads to the expression of oncogenic phenotypes. These include uncontrolled cell proliferation, inhibition of apoptosis, and upregulated metastasis [51, 107]. So far, 16 miRNAs in GC, 28 miRNAs in CRC, and 34 miRNAs in liver cancer are dysregulated. Some are upregulated while the others are downregulated (miR-12, 19, and 17 are upregulated and miR-4, 9, and 17 are downregulated in GI cancer, CRC, and hepatocellular cancer tissues, respectively). Some miRNA such as miR-18a, miR-19a miR-106b, and miR-21 are upregulated in all three cancers, whereas miRNA such as miR-145 and miR-195 are downregulated in CRC and hepatocellular cancer tissues and let-7a in CRC and GI cancer tissues. 34 of the 58 dysregulated miRNAs are associated with different pathways in cancer. These dysregulated miRNAs directly influence cancer pathways such as Platelet-derived growth factor (PDGF) signalling, MAPK cascade, Wnt/β-catenin signalling, and transforming growth factor (TGF)-β/bone morphogenetic protein (BMP) signalling pathway, in all three types of cancers discussed above. Other than signalling pathways, several biological processes are deregulated by the miRNAs which include epidermal growth factor receptor signalling, insulin signalling, angiogenesis, and apoptosis in GI, CRC, and hepatocellular cancers [126, 127].

Biological Processes Regulated by miRNAs

Several pathways play an important role in the differential expression of cancer. Table 16.1 provides data on biological processes that are governed by the miRNAs in GI cancers, specifically gastric, colorectal, and liver cancers [34, 63].

Table 16.1 miRNA in cancer pathways and their regulation in GI cancers

miRNA	Upregulated in	Downregulated in	Pathway	References
miR-19a	GC, colon cancer (CC), LC, and oral cancers		Wnt/β-catenin signalling, axon guidance, neurotrophin/NGF signalling, endocytosis, and PDGF signalling	[1, 121, 126]
miR-20a	GC and CC		Wnt/β-catenin signalling, TGF-β/BMP, axon guidance signalling, Ras-Raf-MAPK cascade, and endocytosis	[110, 126]
miR-20b	GC		Wnt/β-catenin signalling, axon guidance, Ras-Raf-MAPK cascade, and endocytosis	[45, 126]
miR-34a	LC		Wnt/β-catenin signalling and axon guidance	[28, 126]
miR-93	GC and LC		Wnt/β-catenin signalling, neurotrophin/NGF signalling, axon guidance, TGF-β/BMP signalling, and endocytosis	[50, 126]
miR-106a	GC and CC		Wnt/β-catenin signalling	[19, 126]
miR-125a		LC	Wnt/β-catenin signalling	[54, 126]
miR-106b	GC, CC, and LC		Wnt/β-catenin signalling, axon guidance, TGF-β/BMP signalling, and endocytosis	[50, 126]
miR-145		CC and LC	Wnt/β-catenin signalling, axon guidance, TGF-β/BMP signalling, and endocytosis	[81, 126]
miR-182	GC, CC, and LC		Wnt/β-catenin signalling, neurotrophin/NGF signalling, axon guidance, and PDGF signalling	[83, 126]
miR-195		CC and LC	Ras-Raf-MAPK cascade, Wnt/β--catenin signalling and neurotrophin/NGF signalling	[126]
miR-214		LC	Wnt/β-catenin signalling, neurotrophin/NGF signalling, axon guidance, and endocytosis	[114, 115, 126]
miR-221	GC and LC		Wnt/β-catenin signalling, neurotrophin/NGF signalling, axon guidance, and PDGF signalling	[17, 126]
miR-101		LC	Ras-Raf-MAPK cascade	[61, 62, 126]
miR-1		CC	PDGF signalling	[33, 126]
let-7a		GC and CC	Ras-Raf-MAPK cascade and PDGF signalling	[80, 126]

(continued)

Table 16.1 (continued)

miRNA	Upregulated in	Downregulated in	Pathway	References
let-7 g		LC	Ras-Raf-MAPK cascade, TGF-β/BMP signalling, axon guidance, and PDGF signalling	[80, 126]
let-7c		LC	Ras-Raf-MAPK cascade, TGF-β/BMP signalling, and PDGF signalling	[80, 126]
miR-15b	LC		Ras-Raf-MAPK cascade and neurotrophin/NGF signalling	[126]
miR-17	GC and CC		Ras-Raf-MAPK cascade, endocytosis, and PDGF signalling	[84, 126]
miR-18a	GC, CC, and LC		Ras-Raf-MAPK cascade, endocytosis, and PDGF signalling	[87, 126]
miR-21	GC and CC		Ras-Raf-MAPK cascade and TGF-β/BMP signalling	[50, 126]
miR-130b	LC		TGF-β/BMP signalling, neurotrophin/NGF signalling, and axon guidance	[18, 126]
miR-137		GC and CC	Ras-Raf-MAPK cascade	[12, 126]
miR-139		LC	Ras-Raf-MAPK cascade and neurotrophin/NGF signalling	[111, 112, 126]
miR-181b	CC		Ras-Raf-MAPK cascade and TGF-β/BMP signalling	[116, 117, 126]
miR-224	CC and LC		Ras-Raf-MAPK cascade and neurotrophin/NGF signalling	[58, 126]
miR-19b	CC	GC	Neurotrophin/NGF signalling, endocytosis, and axon guidance	[105, 126]
miR-29a	CC		PDGF signalling and axon guidance	[72, 126]
miR-96	CCand LC		PDGF signalling and neurotrophin/NGF signalling	[126]
miR-125b		LC	PDGF signalling, neurotrophin/NGF signalling, and axon guidance	[126, 130]
miR-143		CC	PDGF signalling and axon guidance	[2, 126]
miR-185	LC		PDGF signalling, neurotrophin/NGF signalling, and axon guidance	[5, 126]
miR-203	CC		PDGF signalling, neurotrophin/NGF signalling, and TGF-β/BMP signalling	[126]
miR-222	LC		Axon guidance	[17, 126]

(continued)

Table 16.1 (continued)

miRNA	Upregulated in	Downregulated in	Pathway	References
miR-218		GC	PDGF signalling, neurotrophin/NGF signalling, endocytosis, and axon guidance	[93, 94, 126]
miR-148a		LC	TGF-β/BMP signalling and axon guidance	[6, 126]
miR-199a		LC	Neurotrophin/NGF signalling and endocytosis	[126]
miR-31	CC	GC	Neurotrophin/NGF signalling	[126, 130]
miR-96	CC and LC		Axon guidance	[37, 126]
miR-126		LC	Neurotrophin/NGF signalling	[109, 126]
miR-135b	CC		TGF-β/BMP signalling and neurotrophin/NGF signalling	[97, 126]
miR-150		LC	Neurotrophin/NGF signalling	[126]
miR-151	LC		Neurotrophin/NGF signalling	[126]
miR-200b		LC	TGF-β/BMP signalling and neurotrophin/NGF signalling	[111, 112, 126]
miR-183	CC and LC		Neurotrophin/NGF signalling	[57, 126]
miR-223	GC	LC	Neurotrophin/NGF signalling	[126]
miR-122	CC	LC	Endocytosis	[126]

Wnt/β-Catenin Signalling

The Wnt/β-catenin signalling pathway is essential in adult tissue homeostasis, regeneration, and embryonic development. There are two kinds of Wnt signalling pathways: canonical and non-canonical. The canonical Wnt/β-catenin pathway is dependent on the nuclear translocation of *β-catenin* and the activation of target genes by TCF/LEF transcription factors, but the non-canonical route is independent of TCF/LEF/β-catenin-T cell factor/lymphoid enhancer-binding factor. While the non-canonical pathway governs cell polarity and migration, the canonical pathway controls cell proliferation [67]. Rapid invasion of the surrounding tissues is a characteristic of GC, where Wnt/β-catenin signalling is a potential factor in the growth of tumours by increasing the invasiveness and proliferation of GC cells. Suppression of the Wnt/β-catenin pathway demonstrates anti-metastatic properties in GI cancer cells [15]. Zhang et al. conducted a bioinformatic analysis that utilised computational algorithms, such as DIANA-microT, miRanda, miRDB, miRwalk, PICTAR

(4-way and 5-way), PITA, RNA22, RNAhybrid, and TargetScan, to identify targets. Subsequent gene ontology studies were conducted using the DAVID database, which revealed that a total of 58 miRNAs directly regulate cancer signalling pathways in GI cancers, of which 12 miRNAs deregulate the Wnt/β-catenin signalling pathway. miRNAs such as miR-214, miR-320, miR-101, miR-1826, miR-548b, and miR-33a directly deregulate the pathway [78, 126, 127].

Ras-Raf-MAPK Cascade

The Ras-Raf-MAPK pathway transduces signals to the cell nucleus from the extracellular milieu. This is where genes required for growth, division, and differentiation of cells are activated and also has a role in cell migration, cell cycle regulation, wound healing, and tissue repair. It also promotes changes in particular gene expressions which drive the formation of fresh blood vessels. This process in turn stimulates angiogenesis [73]. Dysregulation of this pathway was most commonly observed in cancerous tissues. *RAS* mutations occur in roughly 25% of all cancers and over 50% of CRC patients [74]. Regulation of this pathway is being researched as a viable novel therapy for cancer. In GI cancers, the Ras-Raf-MAPK cascade is perturbed by 20 miRNAs that were observed in the bioinformatic analysis done by Zhang et al. and other independent studies cited in Table 16.1 [12, 116, 117, 126, 127]. Hence, modification of the pathway by regulating miRNA present in the pathway is a potential form of cancer therapy [126, 127].

PDGF Signalling

Platelet-derived growth factor (PDGF) is responsible for stimulating the growth, survival, and motility of mesenchymal and other types of cells. They also have a role in embryonic development and in controlling tissue homeostasis in adult humans. Overexpression of PDGF signalling can be the cause of the development of malignant and non-malignant tumours. 16 out of the 58 dysregulated miRNAs in GI cancers regulate the pathway. PDGF signalling plays a key role in biological processes that impact cancer tissues, such as proliferation of cells, cell death, angiogenesis, and metastasis [27, 126, 127]. Metastasis of cancer cells depends upon an important process called epithelial-mesenchymal transition (EMT). PDGF and other growth factors are necessary to trigger EMT. In GI cancer, the overexpression of the PDGF signalling pathway leads to the upregulation of EMT which in turn aids metastasis [118]. miR-200b can suppress EMT when it is used via targeted treatment. This can further prove to be a novel treatment option to suppress and treat GI cancers [111, 112].

TGF-β/BMP Signalling

TGF-β is an abundantly expressed cytokine. Its function is cardinal in various cellular mechanisms. The activation and deactivation of *TGF-β* and its downstream mediators are directly related to elevated carcinogenesis. TGF-β/BMP signalling pathway suppresses tumours in the early stages of GI cancers but promotes tumour growth in the later stages [4, 70]. The pathway is deregulated by 13 dysregulated miRNAs in GI cancer. Upregulation of the pathway in later stages of cancer is usually related to poor prognosis in patients. miRNAs such as mir-17, miR-20a, and miR-32 are some miRNAs that regulate this pathway in oesophageal cancer [52, 126, 127]. Understanding and regulating the function of these miRNAs in order to control the pathway can be used to innovate novel cancer treatments.

HER2 Signalling Pathway

HER2 is an oncogene that plays a major role in tumorigenesis and tumour progression. When *HER2* is present in abundance, it triggers the activation of downstream signalling pathways such as the MAPK/ERK1/2 and PI3K/AKT/mTOR pathways. This leads to elevated levels of tumour progression, differentiation, proliferation, and metastasis. 11.3% of all GCs have the HER2/HER3 oncogene[144]; hence, when the genes are overexpressed, tumour progression is amplified [36, 52]; miRNA present in the *HER2* pathway can be used as biomarkers to identify *HER2*-positive GC and prove to aid in suppressing the tumour when target specific drugs are used. miR-21 when suppressed in HER2-positive GC increases the effectiveness of the anticancer drug trastuzumab [24]. miR-337 and miR-302f are reportedly overexpressed, while miR-139 and miR-129 are underexpressed in GC and serve as biomarkers for HER2-positive GCs [42].

Axon Guidance, Nerve Growth Factor (NGF), and Endocytosis Pathways

miRNAs can also influence rather unexpected biological pathways such as axon guidance, nerve growth factor (NGF) signalling, and endocytosis in cancer. Axon guidance molecules are not only required for axon migration but also for innervating cancer tissues, but they can get inactivated due to the action of miRNAs in these tissues [39]. 23 of the 58 dysregulated miRNAs deregulate the axon guidance molecules. NGF signalling pathway prevents neuronal apoptosis and reduces neuronal degeneration in cancerous tissues. It is deregulated by 26 miRNAs. Endocytosis is significantly disrupted by miRNAs in cancerous tissues, and this biological mechanism is recognised for its dual tumour-promoting and suppressive characteristics [21]. It is regulated by 13 dysregulated miRNAs such as miR-122, miR-199a, miR-218, miR-19b, miR-18a, miR-17, and more in GI cancers [11, 126, 127].

Common miRNAs That Are Involved in Gastrointestinal Cancer Pathway Regulation

The most common miRNAs expressed in GI cancers are miR-17 and miR-145. Their dysregulation affects epidermal growth factor receptor (EGFR) endocytosis. But endocytosis regulation caused by these miRNAs is not well understood [93, 94]. SW1116, a KRAS-wild-type colon cancer cell line, was used to confirm the elevated levels of miR-17 and miR-145 in the cells after they were compared with control miRNA, miR-17, or miR-145. This was done by RT-qPCR amplification. It was concluded that miR-145 inhibits EGFR endocytosis and miR-17 stimulates the process when it is present in abundance [126, 127]. miR-145 prolongs the p-Akt and pERK1/2 levels which are responsible for endocytosis. This causes overexpression of EGFR. EGFR overexpression is directly related to the increased expression of aggressive phenotypes. It is also known to have drug and therapy-resistant properties. The effects of miR-145 on the EGFR endocytosis were studied by understanding its effects on EGFR-targeting drugs. Anti-EGFR endocytosis antibodies and tyrosine kinase inhibitors (TKIs) are known for their ability to downregulate the receptor and decrease the activity of downstream regulators like p-Akt and pERK1/2, respectively. This resulted in the reduced effectiveness of the anti-EGFR antibody and increased effectiveness of the TKIs. Through this, we can understand the modulating effects of miR-145 on EGFR [25, 126, 127, 143].

Suppression of Hepatocellular Carcinoma by MicroRNAs

Hepatocellular carcinoma (HCC), a tumour of the hepatocytes, appears in roughly about 80–90% of patients suffering from severe chronic liver disease and cirrhosis. HCC is reported to be the fifth leading cause of cancer and the second most common cause of cancer-related deaths in men worldwide. The five-year survival rate of HCC was estimated to be around a meagre rate of 18%, primary cause being that HCC is highly prone to metastasis and recurrence [14]. HCC includes risk factors such as hepatitis B and hepatitis C, alcoholic liver disease, and non-alcoholic fatty liver disease. The pre-existing treatment options for severe cases of patients with HCC are surgical removal, liver transplantation, chemotherapy, targeted therapy, and immunotherapy. Patients diagnosed with advanced unresectable HCC are recommended to be treated with tyrosine kinase inhibitors, Sorafenib, Lenvatinib, or Bevacizumab as the first stage therapy and Regorafenib for the second stage therapy, yet drug resistance limits the clinical efficacy of Sorafenib [48]. Regardless of these therapeutic methods, HCC still does not have an effective cure for its complete and permanent eradication. In order to develop a novel therapeutic treatment, small non-coding microRNAs (miRNAs) have shown promising potential as a way to eradicate malignancy in hepatocytes. MicroRNAs belong to a family of small non-coding RNAs, composed of 21 to 25 nucleotides, that can inhibit the expression of the targeted DNA sequence either through repression of mRNA translation or degradation of the mRNA [135]. MiRNAs can be used as suppressors to regulate cellular processes such as proliferation, invasion, and apoptosis, through intrinsic and extrinsic pathways in HCC.

Apoptosis of HCC Cells

Anillin regulated by miRNA 15a and 16, was necessary for the formation of tumour in HBV-related HCC, influencing the HCC cells' ability to undergo apoptosis. By targeting apoptosis-related factor *p53*, signal transducers and activators of transcription 3 (*STAT3*), miR-221 and miR-222 aid in the development of HCC, as mentioned in Table 16.2. By modifying the tumour suppressor phosphatase, tissue metalloproteinase inhibitor 3, and tensin homomorphism on chromosome 10 (*PTEN*), miR-221 and miR-222 promote TNF-related apoptosis and ligand resistance during hepatocellular carcinogenesis [60]. Moreover, these miRNAs may facilitate HCC cell motility through activating the Akt pathway and matrix metalloproteinase activity.

MicroRNA Interference in HCC Angiogenesis

Several miRNAs are linked to angiogenesis, including miR-15b, miR-125b, miR-423-3p, miR-424, miR-494, miR-497, miR-612, miR-637, and miR-1255b, as shown in Table 16.2, where there is a substantial decrease in HCC. MiR-130b-3p directly targets the *HOXA5* gene, causing the membrane-bound proteins such as CD31 and CD34 to be expressed at higher levels. This, in turn, promoted the

Table 16.2 Role of miRNA in hepatocellular carcinomas

microRNA	Metabolic functions	References
miR-15a, miR-16-1	Regulates the apoptosis process observed in hepatocytes	[55]
miR-221, miR-222	Regulation of the metabolic activities such as proliferation, migration, and invasion of hepatocytes	[60]
miR-15b, miR-125b, miR-423-3p, miR-424, miR-494, miR-497, miR-612, miR-637, and miR-1255b, miR-20b	Regulates angiogenesis in HCC	[20, 113]
miR-3064-5p	Suppression of angiogenesis in HCC	[129]
miR-223, miR-148a-3p, miR-28-5p, miR-144, miR-451a	Regulates tumour-associated macrophages (TAMs)	[35, 38, 133]
miR-148b	Downregulation of HCC metastasis	[44]
miR-150-3p, miR-21	Regulation of TME	[122, 135]

angiogenesis of HCC, resulting in increased tumour size and decreased the overall survival rate. The expression of endostatin was increased by miR-3064-5p through the inhibition of vascular endothelial growth factor A. To help HCC cells adapt to various settings, miR-20b controlled VEGF and the hypoxia-inducible factor-1α [20].

HCC Microenvironment

Tumour microenvironment cells are surrounded by a multitude of mesenchymal cells, such as neutrophils, endothelial cells, white blood cells, and cancer-associated fibroblasts. Because of their great heterogeneity, macrophages may be polarised into a variety of morphologies by stimuli found in their surroundings. M1-type macrophages that have been classically activated have pro-inflammatory and pro-tumour properties, while, M2-type macrophages that have been alternatively activated also exhibit these traits. The M2 phenotype was often seen by tumour-associated macrophages known as TAMs. Exosomes produced from tumours can control TAM polarisation by transferring miRNAs. Three key components that comprise the control of TAMs by miRNAs were cellular crosstalk, functional polarisation, and differentiation. Microvesicles generated from macrophages that contain miR-223 facilitate the differentiation of monocytes. Through the Notch signalling pathway, miR-148a-3p encourages monocytes to develop into white blood cells such as macrophages in the presence of a stimulating factor [35]. By inducing macrophage infiltration, miR-148b downregulation facilitates the HCC metastasis via Chemokine Colony-Stimulating Factor-1 as portrayed in Fig. 16.1 [44]. MiR-28-5p was shown to be considerably downregulated in HCC and to have an impact on the metastasis of HCC through TAM infiltration. These findings imply that the feedback loop between miR-28-5p and macrophages is known to regulate the metastasis of HCC [134]. By focusing on hepatocyte growth factor, the group of miR-144 and miR-451a enhance M1 polarisation and predicted a better prognosis for patients with HCC [133].

MicroRNA Biomarkers for HCC Diagnosis

Since miRNAs have a variety of roles in the human genome and exhibit steady expression, they might be used as early cancer detection indicators. In HCC, serum levels of miR-130b and miR-15b were both observed to be elevated. When it came for the identification of HCC, miR-130b had a 87.7% sensitivity rate and 81.4% specificity rate, whereas miR-15b demonstrated a high sensitivity of 98.3% but a low specificity rate of 15.3%. These miRNAs performing the role of biomarkers used in combination may be advantageous, particularly for individuals with early-stage HCC who have low AFP levels. High sensitivity and specificity were demonstrated by miR-16-2 and miR-21-5p in differentiating between individuals with HCC and those with chronic hepatitis C. Patients with early-stage HCC had

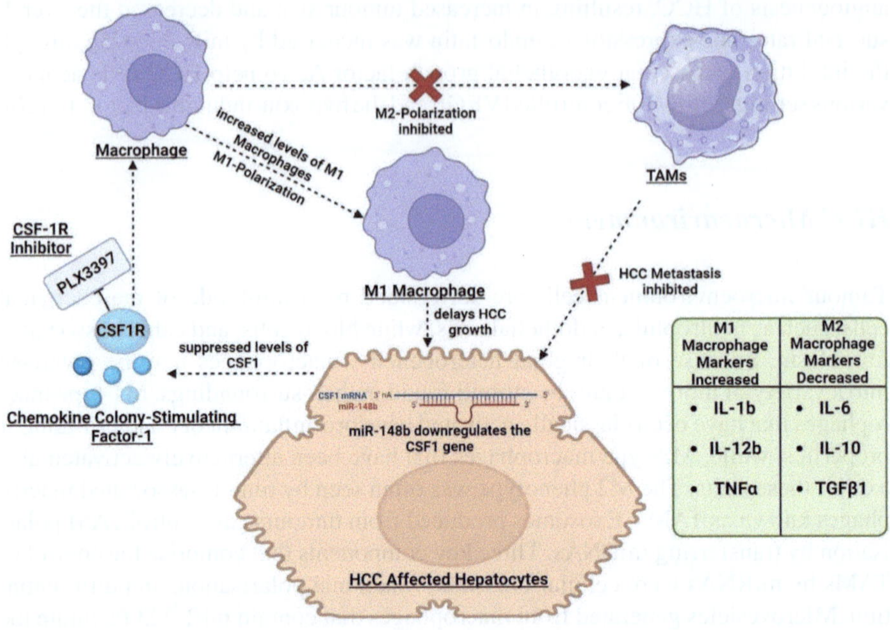

Fig. 16.1 Regulation of CSF1 gene by miRNA-148b and PLX3397 inhibitor in hepatocellular carcinoma

considerably higher expression levels of miRNA-16 and miRNA-122, which can be employed as biomarkers to diagnose early-stage HCC.

Table 16.2 mentions the role of multiple miRNAs in the suppression and removal of HCC.

Potential MicroRNA-Mediated Therapeutic Treatments of CRC

CRC, a type of human malignancy that was found primarily in the epithelial lining of the colon or rectum, is situated at the far end of the human digestive system. For decades, CRC diagnosis had been neglected and cast aside by cancer specialists, and since then, CRC has notoriously risen to the third rank in cancer incidence and second rank in cancer death worldwide, with an astonishing figure almost reaching a million cases annually. This surge in CRC cases was attributed to an alarming increase of the ageing population worldwide, obesity, lack of physical exercise, smoking, and other dietary habits [100].

The advent of CRC usually begins with a growth on the inner lining of the colon or the rectum, known as a polyp. They are mostly benign but some may later mature into cancer based on the type of polyp that has grown, as explained below:

- Adenomatous Polyps: They are the most commonly found type of polyp and change later into cancer, hence the name adenomatous. There are three subtypes: tubular, villous, and tubulovillous. The villous adenomas have the highest chances of turning into cancer.
- Hyperplastic Polyp: These polyps are common and smaller in size and are considered to have an extremely low risk of developing into cancer. They are removed surgically, and a biopsy test is done to confirm their benignancy.
- Sessile Serrated Polyp (SSP): Flat and slightly elevated in shape, these polyps have the highest risk of turning cancerous.

With only less than 10% survival rate for patients diagnosed with stage IV of CRC, early detection and abrogation of the cancer is of the highest priority. The foundation of CRC treatment comprises surgical excision, targeted therapy, radiotherapy, and adjuvant chemotherapy, although these treatment options come with certain specific adverse effects and complications, changing the patient's lifestyle. Albeit having innovated and made numerous advancements in these fields of treatment which have significantly raised the survival rate of patients, we are yet to find a sure-shot treatment or therapy to prevent CRC at its earliest stages. MicroRNA, the new subject of interest for researchers who are trying to demystify its functions and correlate it to pioneer a miRNA-regulated therapy for CRCs. miRNAs have been observed to have intervened in cancer growth and act as tumour suppressors. Henceforth, we highlight their role and the potential therapeutic methods of treating CRCs.

Downregulated Expression of CRC by miRNA Clusters

miRNAs that are in close proximity to one another in the human genome are found to form miRNA clusters or polycistronic miRNAs, amounting to half of the miRNA genes found in humans [79]. These clusters usually comprise two to eight members and are divided into two subtypes homologous and heterologous clusters. Homologous clusters contain miRNAs from the same family, whereas heterologous clusters contain miRNAs from different families. Another type of RNA, known as long non-coding RNA (Long ncRNA), 1000–10,000 nt long, has been shown to strongly regulate miRNA expression. Clusters of miR-99a/let7c, miR-100/let-7a-2/miR-125b-1, and miR-99b/let-7e/miR-125a are miRNAs that belong to the same family and are located on separate chromosomes. These clusters have been observed to show negative regulation through transcription factors in CRC models and correlated with improved survival and exhibit tumour-suppressing properties that are accomplished through targeting various genes. Long ncRNA let-7a-5p executes arrestation of the cancer cell cycle and decreases metastasis by targeting the ABC transporter ATP-binding cassette gene, *ABCC1*, which is responsible for the development of cell chemoresistance, hence increasing the effectiveness of chemotherapy [126, 127]. Up-regulation of let-7c-5p leads to a decrease in cell migration and inhibits cell growth as it targets *MMP11* (matrix metallopeptidase 11) and *PBX3* (PBX homeobox 3) genes [32]. Through the overexpression of let-7e-5p in CRC cell lines, results have shown that there is decreased cancer cell proliferation as it targets the *DCLK1* gene that codes for serine/threonine kinase and increases the sensitivity to 5-fluorouracil, which is a chemotherapy agent used in the treatment of CRC [46]. miR-1-3p, miR-133a-3p, miR-133a-3p, miR-133b-3p, and miR-206-3p are muscle-specific miRNAs that show an expression pattern specifically in CRC tissues. Their expression is observed to be restricted in CRC cell lines, and there is data indicating that increased expression or levels have been shown to enhance the survival rate of patients with CRC. Elevated levels of miR-1-3p show a negative correlation with tumour size, degree of differentiation, lymph node metastasis and their consequent stage. They also downregulate an apoptosis inhibitory protein known as NAIP, allowing apoptosis to occur freely [136]. Both miR-133b-3p and miR-206-3p mitigate the levels of cell invasion, proliferation, and migration by targeting the *MET* proto-oncogene which is involved in tumour growth [66, 86]. Furthermore, miR-133a-3p performs similar roles both in vitro and in vivo by inhibiting the *RFFL* gene expression, resulting in the degeneration of the mutated form of the p53 protein, whose presence is a characteristic feature of cancer cells [23].

Role of Apoptosis in CRC

Apoptosis, a homeostatic process that maintains the balance between the propagation of new cells and the removal of old, dying cells through programmed cell death. This process occurs through two distinct channels: extrinsic and intrinsic signalling pathways. The extrinsic pathway activates apoptosis by involving the *TNF* (tumour necrosis factor) gene whereas the intrinsic pathway triggers apoptosis by caspase activation, leading to DNA damage and disengagement of growth factor. The overall process of apoptosis is regulated by the *BCL-2* family proteins. Damaged cells are often seen to develop malignancy if they can evade apoptosis and survive. The p53 protein acts as a tumour suppressor by inducing the apoptotic death of incipient cancer cells. A multitude of miRNAs engage in the apoptotic pathway of CRC, enhancing the programmed cell death process [100]. Multiple studies show evidence of miR-92a present in the miR-17-92 cluster, highly expressed compared to the rest of the miRNAs in CRC cell lines [96]. It is known to enhance the levels in addition to targeting the anti-apoptotic molecule BCL-2-interacting mediator, making miR-92a an important factor in the initiation of CRC. Consequently, researchers have concluded that the downregulation of miR-92a results in an increase in apoptosis in CRC cells. The *BRAF* gene encodes the B-raf protein that was associated with controlled cell growth and differentiation. However, mutations found on the *BRAF* gene can cause uncontrolled cell growth, leading to cancer. More than 10% of CRCs are caused by the mutated *BRAF* gene [124, 125]. To counter this mutation, researchers have found that the increased expression of miR-378-5p targets and downregulates the *BRAF* gene through cancer cell cycle arrest and programmed cell death in CRC cell lines. miR-129 regulates the *BCL2* expression by activating the intrinsic apoptotic pathway and enhancing the sensitivity of human CRC cells to 5-fluorouracil chemotherapy [43]. miR-148a performs a similar function through an intrinsic mitochondrial pathway [123]. miR-769 was shown to be downregulated in CRC and acts on its target gene *CDK1* to control the proliferation, apoptosis, migration, and invasion of CRC cells. The restoration of miR-769 expression was observed to enhance programmed cell death in CRC cells [101].

Exosomal miRNA-Mediated Reversal of Chemotherapy Resistance in Colorectal Cancer

miRNAs interfere and regulate a magnitude of essential metabolic processes such as apoptosis, cell migration, differentiation, and invasion, some of which are carried out with the assistance of exosomes that form a complex known as exosomal miRNAs [59] Mature miRNAs are transported to exosomes within the nucleus through four different pathways:

- miRISC-related pathway,
- nSMase2-dependent pathway,
- miRNA sumoylated and motif hnRNPs-dependent pathway,
- 3'miRNA sequence-dependent pathway.

MRP5, a protein that is coded by the *ABCC5* gene, performs the role of an ATP-dependent transporter that transfers substrates across biological membranes in both directions. Elevated levels of MRP5 lead to increased resistance towards platinum drugs used in the treatment of CRC. Oxaliplatin is one such platinum-based chemotherapy drug prescribed for cases of advanced stages of CRC, that damages cellular DNA and prevents it from being copied. The exosomal miR-128-3p downregulates the expression of *MRP5* in CRC cells and reduces the export levels of oxaliplatin from the cancer cell. Hence the overexpression of miR-128-3p results in an increase of chemosensitivity in resistant cells, permitting oxaliplatin to damage the CRC cells [59].

miRNAs in GC

GC ranks fifth in terms of incidence and mortality globally [8]. It presents itself with lesions associated with atrophic gastritis and/or gastric intestinal metaplasia which are common precursors to GC and are difficult to treat.

The Multifaceted Role of Survivin in GC

Survivin, a component of the inhibitor of apoptosis protein (IAP) family, is integral to the regulation of mitosis and the inhibition of programmed cell death. It was largely absent from most mature, differentiated normal tissues but is highly expressed in various cancers, where it contributes to resistance against chemotherapeutic and radiotherapeutic interventions. Lowering survivin levels or inhibiting its function can lead to reduced tumour progression, enhanced cell death, and increased responsiveness to chemotherapy and radiation. Consequently, research is focused on developing small molecules that target survivin's gene expression and translation, as well as compounds that interact directly with the protein. Preliminary studies have identified several potential small-molecule survivin antagonists, although their clinical application may be limited by off-target effects. Additionally, survivin-targeted peptide vaccines are under development to stimulate a cytotoxic T-cell response against tumour cells expressing survivin [49]. miR-34 is present on human chromosome 1p36, a locus frequently subject to deletions in various malignancies. miR-34a, a known transcriptional target of p53, is frequently lost in numerous cancer types. The diminished expression of miR-34a may result from mutations that lead to p53 inactivation within tumour cells.

Weiguo Cao et al. [10] performed various experiments to investigate the effects of miR-34 on GC. Through cell line analysis, MTT assay, qRT-PCR, Western blotting, flow cytometry, and certain statistical analyses, they concluded that the expression levels of miR-34a were markedly reduced in the GC cell lines relative to the levels observed in the normal gastric epithelial cell line. miR-34a is also capable of downregulating survivin protein levels, thereby inhibiting both proliferation and invasion of GC cells. Consequently, therapeutic approaches aimed at increasing miR-34a levels or targeting survivin expression could offer potential benefits for patients with GC. As mentioned in Table 16.3, miRNAs play distinct roles in GC. miR-372 functions as an oncogene, impacting cell growth, cell cycle regulation, and apoptosis by downregulating the suppressor gene *LATS2*. miR-650 is intertwined with GC cell growth and development by directly targeting the *ING4* gene. Downregulation of miR-663 in GC contributes to the proliferation of aberrant cells. miR-126 is overexpressed in GC, promoting carcinogenesis by suppressing *SOX2* expression. miR-21 exhibits suppressive effects on excrescence in GC by targeting NF-kappaB1 and regulating cell growth. miR-148 b suppresses GC cell growth by targeting cholecystokinin B receptor (CCKBR). miR-141 and miR-451 are involved in GC progression. miR-29 slows down cell growth, migration, and irruption by targeting *MCL-1*. MiR-101 regulates cell growth, migration, and invasion in GC by modulating *FOS, CDC42, EZH2*, and *COX-2*. MiR-10b, miR-223, miRNA-30a-5p, and miRNA-338 are associated with relapse-free and overall survival in GC cases [7].

Table 16.3 Role of miRNAs as biomarkers in GI cancer

miRNA	Association with GC	Mechanism/target	References
miR-372	Controls cell proliferation, cycle, apoptosis	Downregulates *LATS2* tumour suppressor gene	[7]
miR-650	Associated with the increase and advancement of GC cells	Directly targets *ING4* gene	[7]
miR-663	Downregulation associated with GC development	Implicated in hyperplasia of aberrant cells	[7]
miR-126	Aberrant overexpression contributes to gastric carcinogenesis	Downregulates *SOX2*	[7]
miR-21	Regulates growth in GC cells	Targets NF-kappaB1, shows tumour suppressive activity in GC	[7]
miR-148b	Significantly inhibits the growth of GC cells	Targets CCKBR, suppresses GC cell growth	[7]
miR-141	Involved in GC progression	Inhibitory to cell proliferation	[7]
miR-451	Involved in GC progression	Inhibitory to cell proliferation	[7]
miR-29	Slows down cell growth, migration, invasion	Targets *MCL-1*	[7]
miR-101	Slows down cell growth, migration, invasion	Targets *FOS, CDC-42, EZH2, COX-2*	[7]
miR-10b	Directly related to improved survival without recurrence in GC patients		[7]
miR-223	Directly related to improved survival without recurrence in GC patients		[7]
miR-30a-5p	Directly related to improved survival without recurrence in GC patients		[7]
miR-338	Directly related to improved survival without recurrence in GC patients		[7]

The Potential of miRNAs: Drivers of Tumour Immune Response Modulation

Within the intricate and dynamic realm of cancer biology, the interplay between cancer cells and the immune system emerges as a pivotal determinant of disease progression and therapeutic outcomes. Research endeavours have increasingly directed their focus towards elucidating the complex mechanisms underpinning the tumour immune response, with particular attention to the regulatory roles of miRNAs. The TME constitutes a dynamic milieu wherein cancer cells are involved in intricate interactions with mesenchymal stem cells, immune effectors, and extracellular matrix components. Through precise modulation of immune-related gene expression, miRNAs exert profound influences on the infiltration, activation, and functional dynamics of various immune cell subsets, influencing tumour immune evasion and anti-tumour immunity. Accumulating evidence underscores the pivotal role of dysregulated miRNA expression in fostering an immunosuppressive TME, thereby facilitating tumour progression and metastasis. In this context, elucidating the intricate crosstalk between miRNAs and immune signalling pathways represents a fundamental imperative to developing innovative immunotherapeutic strategies for cancer management. Cancer frequently employs immune evasion mechanisms to promote its growth and hinder immunotherapeutic interventions. The effectiveness of the immune response against tumours depends on how various lymphocytes are activated and interact with each other. These cells play a dual role by enhancing and controlling antitumor immune responses. CD4+ helper T cells (Th cells) demonstrate flexibility in their characteristics, transforming into different subsets depending on environmental signals. The secretion of distinct cytokines and chemokines by these T cell subsets modulates the immune response in a context-dependent manner.

miRNA Regulation of Immune Responses Through Innate Immune Cells

The primary and first immune response was initiated by the body's innate immune system cells, which rapidly react to threats by recognising and eliminating tumour cells in a nonspecific manner. Changes in the expression of miRNAs within tumours can cause an impact on the functions and traits of innate immune cells, such as NK cells, macrophages, and dendritic cells, which are pivotal in the innate immune response [132]. As a result, compromised or modified activities of these immune effectors might allow tumour cells to evade detection by the innate immune system [64].

NK Cells

NK cells are primarily recognised for their ability to lyse target cells without the identification of tumour-specific antigens. Rather than altering their immunogenicity through reprogramming, tumour cells may evade elimination by adjusting their levels of miRNA expression to suppress the activity of cells. miRNAs could potentially aid tumour cells in evading immune surveillance by influencing lectin-like transmembrane activating receptor and its ligands which include major histocompatibility complex class I-chain-related molecules A, B, and UL16 binding proteins [132]. A hallmark characteristic of NK cells is their unique capacity to induce cytotoxicity against target cells without prior recognition of tumour-specific antigens. In GC, miRNAs play crucial roles in modulating the activity and function of NK cells, thereby influencing tumour progression and immune evasion mechanisms.

- Suppression of NK Cell Cytotoxicity: Certain miRNAs have been implicated in dampening NK cell cytotoxicity against GC cells. miR-20a targets the expression of NKG2D ligands on tumour cells, thereby reducing NK cell-mediated cytotoxicity and promoting tumour immune escape [138].

- Impairment of NK Cell Function: In GC, dysregulated miRNA expression in the TME leads to the impairment of the cell's function. miR-27a targets the expression of the receptor CD69 on NK cells, which diminishes their cytotoxic activity and cytokine production, aiding in tumour immune evasion [139].
- Modulation of NK Cell-Related Signalling Pathways: miRNAs can also regulate signalling pathways involved in NK cell activation and function. miRNA-155 has been reported to target the suppressor of cytokine signalling 1 (*SOCS1*), thus enhancing NK cell activation and cytotoxicity [13].

Macrophages

In GCs, miRNAs have been found to modulate macrophages, which are key components of the TME. miRNAs modulate the production of cytokines and chemokines by TAMs, which in turn influence the TME. miR-155 has been found to enhance the production of pro-inflammatory cytokines such as Interleukin-6 and tumour necrosis Factor-α by TAMs in CRC [132]. Moreover, TAMs can also express immune regulatory molecules such as *PD-L1*, which contribute to immune suppression. miRNAs are known to regulate the expression of these molecules in TAMs [102]. For instance, miR-34a has been identified as downregulating *PD-L1* expression in TAMs of GC, thereby enhancing anti-tumour immunity. In GCs, TAMs often exhibit a pro-tumorigenic phenotype, promoting tumour growth, angiogenesis, metastasis, and immune evasion. miRNAs can influence the phenotype and function of TAMs, thereby impacting tumour progression. TAMs can exhibit different polarisation states. Certain miRNAs have been identified to regulate TAM polarisation towards the M2 phenotype. For example, miRNA-21, miRNA-155, and miRNA-511 evidently regulate the polarisation of TAMs in GI cancers. Under specific conditions, macrophages endowed with phagocytic capabilities can differentiate into two distinct subsets, exerting contrasting effects: upon activation, M1 macrophages exhibit tumour-suppressive properties, and M2 macrophages foster tumour progression. Exploiting this plasticity of macrophages, tumour cells can induce a phenotypic switch of macrophages within the TME towards the M2 phenotype through the expression of specific microRNAs (miRNAs), thereby establishing an immunosuppressive milieu conducive to immune evasion [102]. miR-21 deficiency promotes polarisation of macrophages towards the M1 phenotype within the tumour milieu via activation of the signal transduction molecules and activator of transcription-1 signalling pathway and the Interferon-γ signalling mediator. Reduction of miR-21 levels enhances *STAT1* signalling, leading to increased expression of programmed death-ligand 1 (*PD-L1*) in macrophages, thus inhibiting their antitumor activities [132]. Additionally, miR-127 is implicated in promoting M2 polarisation of macrophages upon suppression of M1 marker genes while enhancing the functional transcripts of M2 markers [13]. Stimulation of miRNA127 expression due to lipopolysaccharide (LPS) decreases the expression of certain transcriptional repressors of Dual-specificity phosphatase (DUSP1). Decreased DUSP1 levels result in increased phosphorylation of Jun N-terminal kinase (JNK), thereby favouring the M1 phenotype [64].

Dendritic Cells (DC)

DCs serve as highly proficient antigen-presenting cells within the body, facilitating the essential linkage between the specific and non-specific immune responses. Nonetheless, within the (TME), dysregulated (miRNAs) frequently disrupt their maturation and functionality. Consequently, DCs often transform immune suppressive regulators, facilitating tumour evasion from immune surveillance. Aberrant miRNA expression patterns have been consistently observed using reverse transcription polymerase chain reaction and deep sequencing techniques, in both plasma and tissue samples obtained from patients with GI cancer. Notably, among these dysregulated miRNAs, miRNA-17-5p emerges as an important player in GC pathogenesis [132]. Specifically, miRNA-17-5p derived from GC cells can be internalised by immature DCs, leading to the inhibition of LPS-induced DC maturation and diminished endocytic activity. This dual effect ultimately promotes GC development by impairing the immune surveillance mechanisms mediated by DCs.

- miRNA-mediated DC maturation inhibition: Dysregulated miRNAs in the TMETumor microenvironment (TME) can impede the maturation process of DCs, rendering them less effective in antigen presentation and T cell activation. miRNAs such as miR-17-5p, often upregulated in GI cancers, have been found to inhibit DC maturation, thereby attenuating their ability to stimulate anti-tumour immune responses [137].
- Suppression of DC endocytosis and antigen presentation: miRNAs can down-regulate the expression of key molecules involved in DC endocytosis and antigen presentation pathways. miR-21 has been reported to target genes associated with antigen processing and presentation in DCs, leading to decreased antigen presentation to T cells and impaired anti-tumour immunity [64].
- Induction of immune tolerance by DCs: Certain miRNAs can promote the generation of tolerogenic DCs, which actively suppress anti-tumour immune responses. miRNAs such as miR-155 have been implicated in the induction of tolerogenic DCs in GI cancers, facilitating immune evasion by tumours [64].
- miRNA transfer from cancer cells to DCs: GI cancer cells can release miRNAs into the TME, which can be taken up by DCs, altering their function and phenotype. miRNA-29 and miRNA-155 [64, 76] have been detected in DCs within the TME, where they exert immunosuppressive effects by modulating DC function.

miRNA-Guided Specific Immune Response in T Cells

miRNAs play pivotal roles in the activation of T cells, influencing signal transduction pathways involving the T cell receptor (TCR) and costimulatory molecules [132]. Increasing evidence suggests that miRNA expression at distinct developmental phases serves as a regulatory mechanism in T cell development by modulating downstream signal transduction pathways from the T cell receptor [30]. miR-146a has been shown to promote Treg differentiation and function, contributing to immune modulation within the TME [102]. Modulation of T-cell-mediated cytotoxicity: miRNAs can modulate the cytotoxic function of T cells by targeting molecules involved in cytolytic granule release and target cell recognition. miR-181a has been implicated in regulating T cell cytotoxicity by targeting multiple genes involved in TCR signalling and cytolytic granule release [140]. Impact on immune checkpoint pathways: miRNAs can regulate immune checkpoint pathways that control T cell activation and function. miR-138, miR-34a, and miR-200 have been reported to target PD-L1 expression in tumour cells, thereby indirectly affecting T cell function within TME [141].

miRNA-Modulated Apoptotic Process in Tumour Cells

The apoptotic process in tumour cells is intricately regulated by microRNAs (miRNAs). Tumour cells often evade apoptosis, a programmed cell death mechanism controlled by genetic factors, thereby contributing to their survival and proliferation. miRNAs exert their influence by upregulating molecules such as *BCL2* oncogene which prevent apoptosis while downregulating molecules like death receptors that induce apoptosis. Examples of miRNAs with a promoting role include miR19a-3p, miR128a, miR200c, and miR204-3p, whereas miR23a, miR24a, miR15a, miR143a, miR186, miR125b, and miR590 are among those exhibiting an inhibitory role. As depicted in Fig. 16.2, within the extrinsic pathway of apoptosis, the activation occurs via ligands like FAS ligand (FASL) that induce death, which binds to the FAS receptor, leading to the recruitment of FAS-associated death domain protein (FADD) and inactive caspase-8. Subsequent cleavage and activation of pro-caspase-8 initiate a caspase cascade involving several caspases (−3, −6, and −7). Within the intrinsic pathway, various stimuli disrupt the balance between inductive and suppressive apoptotic proteins, triggering the exudation of cytochrome c from mitochondria. Cytochrome c subsequently associates with caspase-9 to form a complex, leading to its activation and subsequent cleavage of caspase-3, a key effector of apoptosis [132]. Ultimately, activation of caspase-3 results in the cleavage of numerous cytoplasmic proteins, orchestrating the apoptotic process through both extrinsic and intrinsic pathways.

Role of miRNA and Costimulatory Proteins in GC and CRC

Inducible costimulatory molecules (ICOSs), specific to activated T cells, bind selectively with the ligand B7-H2 on antigen-presenting cells (APCs), initiating ICOS/B7-H2 signalling to enhance the T-cell-mediated antitumour response. Wu et al. investigated single-nucleotide polymorphisms (SNPs) in the 3′-untranslated region of ICOS genes in a human case-control study of CRC, discovering that these miR-SNPs significantly attenuate the binding affinity between miRNAs and ICOS, consequently disrupting miRNA-mediated regulation of ICOS expression [142]. The identified miR-SNPs within ICOS disrupt interactions with miRNA-1279 and

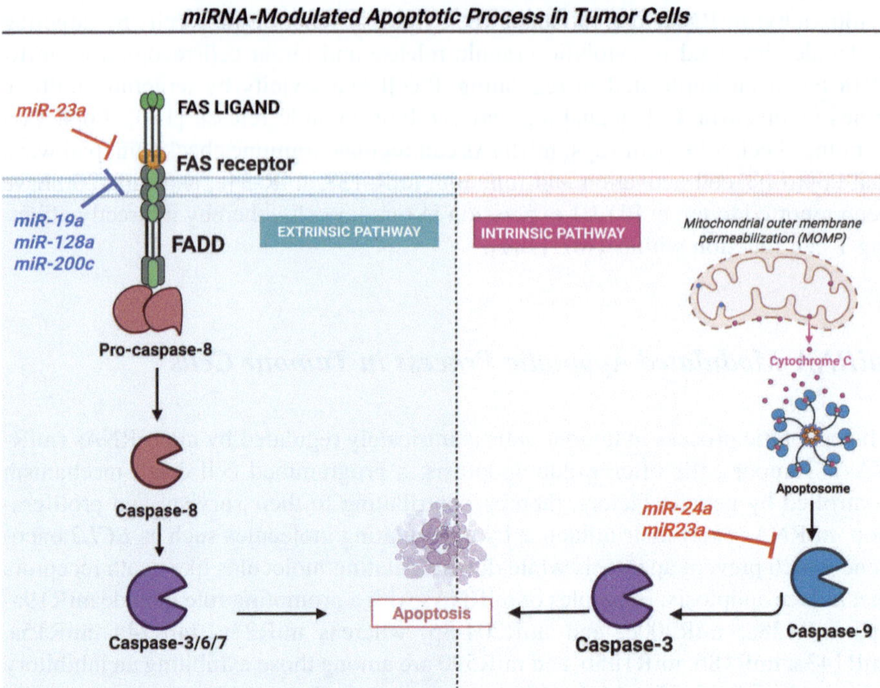

Fig. 16.2 miRNA-modulated apoptotic process in tumour cells

miRNA-2117, dampen the inhibitory effect of miR-186-5p on ICOS expression, and enhance T cell-mediated antitumour responses. B7-H2, an immunostimulatory molecule predominantly seen in macrophages, dendritic cells, and B cells, exhibits decreased expression in gastric and CRCs. The involvement of B7-H2 in immune evasion by cancer cells within tumours is linked, though the precise mechanism remains elusive [64]. Presently, miR-24 stands as the sole identified miRNA targeting B7-H2. The rs4819388 SNP in the B7-H2 3′-untranslated region disrupts miR-24 binding as demonstrated by dual-luciferase assay (in vitro assay), leading to an upregulated expression of B7-H2. Clinically, the A allele of this SNP may contribute to an increased incidence of GC, as substantiated by statistical analysis in a case-control genetic association study [114, 115].

Regulation of microRNAs by Small Molecule Modulators

The RNAi pathway initiated by the enzyme Dicer is a naturally occurring process in eukaryotes and prokaryotes which leads to sequence-specific suppression of gene expression, with siRNA and miRNA being the two central components of the pathway [92]. Macromolecules and small molecules are the two means of targeted drug treatment. Up until 2022, 88 small-molecule oncogene inhibitors have been approved by the US-FDA [61, 62]. Small molecules have been proven to exhibit therapeutic characteristics which could be exploited in miRNA therapy [26]. Enoxacin, a small molecule modulator belonging to the fluoroquinolone group, was first discovered by Shan et al. [85]. It was the first identified small molecule modifier found to be involved in miRNA activation and biogenesis. When tested for possible pleiotropic effects on the RNAi pathway, it was found that enoxacin tested negative, which led to the conclusion that it possessed interactions specific to the nucleic acids and proteins. miRNA TaqMan assays demonstrated that only 15/157 endogenous miRNAs in HEK293 cells displayed changes in expression and that it mediated the increased expression of miRNA in the presence of higher levels of precursor RNAs. Additionally, the levels of pri-miR125a and precursor miR-125a were observed to reduce the presence of enoxacin. These reports suggested that enoxacin possibly facilitated dicer-mediated processing. However, a series of experiments using siRNA duplexes and *TRBP* mRNA, using the levels of knockdown of *GAPDH* as the reference, proposed enoxacin to be TRBP dependent. Further studies by Melo et al. on the mechanism of action of enoxacin in promoting miRNA production aligned with the experimental conclusions by Shan et al. [85]. The former used Surface Plasmon Resonance and Isothermal Titration Calorimetry to study the effects on wild-type and mutant forms of TRBP protein. A frameshift mutation in the TRBP encoding gene, present in gastric, colorectal, and endometrial malignancies, known to have negative effects in miRNA processing, was analysed [71]. The TRBP-proficient cells displayed a higher miRNA biogenesisBiogenesis by enoxacin while it was found to be significantly lower in cells deficient in TRBP. Another crucial finding was that mutant cancer cells were inhibited by increased miRNA biogenesis in the presence of enoxacin [71]. Small molecules that directly bind to miRNAs are much easier to screen than the other modifiers mentioned above as the latter display a complex relationship with the genes they target [99]. These work again on the basic principle of inhibiting Dicer or Drosha-mediated cleavage. GC has a higher expression of miR-372 and miR-373, both of which have been identified to permit tumorigenic growth by suppressing the *p53* pathway in human germ cells [99]. Vo et al., using the FRET assay, showed that neomycin, one of the 7 aminoglycosides they screened, was able to inhibit Dicer. Further studies observed a reduction in the levels of miR-372 and miR-373 by the action of the small molecule modulator they developed, a neomycin–nucleobase conjugate. This newly discovered small molecule demonstrated the ability to bind to the pre-miRNAs and was proven to possess antiproliferative activity in GC [98]. The RISCs consist of proteins that belong to the Argonuate family [116, 117], with major roles in RNA biogenesis, establishment of anchor sites in the cytoplasm [3], dsDNA break repair processes [104] and target recognition. Other than the 3/2000 compounds discovered by Tan et al. [90] found to be involved in the loading of miRNA onto AGO2, Watashi et al. [103] discovered a small molecule modifier that significantly reduced tumour growth. Similarly, a small molecule, discovered by Naro et al. [75] was found to be specific to miR-21's inhibition. It functioned only by inhibiting the effects of miR-21 rather than affecting its synthesis and expression.

Apart from the modulators involved in biogenesis, small-molecule inhibitors and regulators specific to certain miRNAs were discovered and studied. Gumireddy et al. discovered an azobenzene-derived small molecule inhibitor of miR-21 using a luciferase-based cellular assay [31] while Young et al., using the same assay, screened a total of 1364 small molecules, out of which 3 small molecules were found to specifically regulate certain miRNAs [119]. Similarly, a small molecule inhibitor of cancer-related miRNAs was found to decrease the production of protein myoD while not altering the mRNA level. This mechanism was further studied by using the compound discovered as a marker and the action of miR-221/222 in muscle cells was detected to have a regulatory pathway in cells [91]. An endonuclease, Ago2, that exhibits formation of catalytically active compounds guided by RNA exhibits a key role in the biogenesis of siRNAs and miRNAs. It has been observed to show translation-inhibitory activity [16, 22, 47].

Future Prospects of miRNA in Gastrointestinal Cancers

miRNAs as Biomarkers in Cancer Genetics

miRNAs involved in cancer development, known as oncomirs, are concentrated within chromosomal fragile sites prone to cancer-associated genetic rearrangements. Their dysregulation can stem from various factors including transcriptional misregulation, epigenetic modifications such as DNA methylation and mutations, and abnormalities in miRNA processing pathways. Certain miRNA families control critical cellular processes like cell cycle regulation, senescence, differentiation, and proliferation. Mutations in these miRNAs can lead to cellular abnormalities and tumorigenesis; when present in germinal cells, they may predispose individuals to cancer. Several types of miRNAs can be found in various body fluids such as plasma, serum, tears, urine, amniotic fluid, and gastric juice. These circulating miRNAs exhibit diverse expression patterns in body fluids, possibly influenced by distinct cell types during specific physiological states. Based on these observations, miRNAs can potentially serve as valuable non-invasive biomarkers for diagnosing cancer [64].

miRNA's Influence on Resistance and Sensitivity to Chemotherapy

Chemotherapy is standard for treating advanced GC, improving patient survival but facing challenges like multi-drug resistance. MiRNAs play crucial roles in modulating drug sensitivity and resistance [128]. For instance, miR-29 family members enhance chemotherapy effectiveness by suppressing catenin-δ and RhoA signalling. MiR-148a-3p enhances cisplatin's cytotoxicity, while miR-143-3p restores cisplatin sensitivity by targeting bromodomain-containing protein 2. MiR-633 inhibition increases GC cell apoptosis induced by doxorubicin/cisplatin [108]. MiR-577 and miR-508-5p influence oxaliplatin resistance. Additionally, miR-1b-5p suppresses apoptosis and promotes cisplatin resistance. Various strategies like nanoparticle delivery of miRNAs and exosomal anti-miRs show promise in reversing drug resistance. MiR-204 enhances fluorouracil sensitivity, and miR-567 sensitises tumour cells to 5-fluorouracil and oxaliplatin [108]. These findings underscore the complex regulatory roles of miRNAs in GC chemotherapy resistance and sensitivity, suggesting potential therapeutic targets.

Conclusion

miRNAs, belonging to a group of noncoding RNAs, are being widely studied to understand mutation, cancer progression and metastasis. They have proven beneficial to understand the various cellular and biological processes occurring in the human body as well and have been a quintessential target of study in anticancer therapeutic treatments. In GI cancer, different pathways and processes determine the cardinal functions of cells such as proliferation and metastasis. These pathways are, in turn, regulated by miRNAs. Hence, identifying and modulating these miRNAs is a promising cancer therapy option. Wnt/β-catenin, TGF-β/BMP, Ras-Raf-MAPK cascade, HER2, and PDGF signalling pathways are some potential pathways that can be targeted using miRNA. Recent research highlights the potential of miRNAs as novel therapeutic targets in various cancers and these miRNAs can regulate gene expression involved in HCC progression, including apoptosis, angiogenesis, and TME interactions. For example, certain miRNAs influence apoptosis in HCC cells, while others affect angiogenesis and tumour growth. miRNAs also modulate the behaviour of TAMs, impacting HCC metastasis. In addition, CRC that affects the epithelial lining of the colon, has a mere survival rate below 10% and ranks third in cancer incidence and second in cancer deaths globally. miRNAs as potential therapeutic tools are being researched to help regulate gene expression involved in CRC growth. Further, dysregulation of a total of 34 miRNAs is discovered to be linked with the cancer pathways. Factors like polarisation of TAMs, inhibition of DC maturation and suppression of immune responses by the miRNAs have been linked to an increase in tumour progression. Early detection of cancer hugely determines the treatment success and the survival rate of the patient. miR-16 and miR-122 as biomarkers are found to be effective in diagnosing HCC in its early stages. GCs have displayed an elevated level of survivin, a member of the IAP family, which is responsible for the resistance conferred by the tumour cells against treatments. Various miRNAs also play a significant role in GC and its progression. Lower levels of miR34a, for example, are found to reduce the levels of survivin. Additionally, understanding the mechanisms of action of miRNAs offers valuable insights into tumour progression, thereby reinforcing their role as potential biomarkers in GC. For instance, miR-372 promotes tumour growth by downregulating LATS2, miR-148b suppresses cell proliferation by targeting CCKBR, and miR-21 has been shown to exert tumour-suppressive effects in GC by targeting NF-κB1. Moreover, in the tumour microenvironment, miRNAs (miR-20a and miR27a) play a crucial role in modulating immune responses by targeting the innate immune cells, thereby facilitating immune evasion. Certain miRNAs also regulate immune checkpoint pathways in the tumour microenvironment, which directly influence T cell functions. Various assays used to screen the small molecule modifiers have helped in the identification of potential effects of these molecules in the upregulation or downregulation of miRNAs. Neomycin, for example, shows inhibitory action against Dicer, while enoxacin is found to regulate the synthesis of miRNAs. The reduction in the levels of miR-372 and miR-373 by a specific, synthetic small molecule modulator, neomycin–nucleobase conjugate, was studied to have antiproliferative activity in GC. Research findings so far have provided valuable insights into miRNAs, a distinct class of non-coding RNAs, highlighting their significance in diagnosis, prognosis, and therapy. Further research in the field is expected to provide a better understanding of their roles, which could help in minimally invasive methods of diagnosis and better clinical-therapeutic applications.

References

1. Ardizzone A, Calabrese G, Campolo M, Filippone A, Giuffrida D, Esposito F, Colarossi C, Cuzzocrea S, Esposito E, Paterniti I (2021) Role of miRNA-19a in cancer diagnosis and poor prognosis. Int J Mol Sci 22(9):4697. https://doi.org/10.3390/ijms22094697. PMID: 33946718; PMCID: PMC8125123
2. Asghariazar V, Kadkhodayi M, Sarailoo M, Jolfayi AG, Baradaran B (2023) MicroRNA-143 as a potential tumor suppressor in cancer: An insight into molecular targets and signaling pathways. Pathol Res Pract 250:154792. https://doi.org/10.1016/j.prp.2023.154792. Epub 2023 Sept 6. PMID: 37689002
3. Azlan A, Dzaki N, Azzam G (2016) Argonaute: the executor of small RNA function. J Genet Genomics 43(8):481–494
4. Baba AB, Rah B, Bhat GR, Mushtaq I, Parveen S, Hassan R, Hameed Zargar M, Afroze D (2022) Transforming growth factor-Beta (TGF-β) signaling in cancer-A betrayal within. Front Pharmacol 13:791272. https://doi.org/10.3389/fphar.2022.791272. PMID: 35295334; PMCID: PMC8918694
5. Babaeenezhad E, Naghibalhossaini F, Rajabibazl M, Jangravi Z, Hadipour Moradi F, Fattahi MD, Hoheisel JD, Sarabi MM, Shahryarhesami S (2022) The roles of microRNA miR-185 in digestive tract cancers. Noncoding RNA 8(5):67. https://doi.org/10.3390/ncrna8050067. PMID: 36287119; PMCID: PMC9609348
6. Bao C, Guo L (2020) MicroRNA-148a-3p inhibits cancer progression and is a novel screening biomarker for gastric cancer. J Clin Lab Anal 34(10):e23454. https://doi.org/10.1002/jcla.23454. Epub 2020 Aug 12. PMID: 32785967; PMCID: PMC7595888
7. Bhat SA, Majid S, Rehman MU (2019) Scenario and future prospects of microRNAs in gastric cancer: a review. Iran J Basic Med Sci 22(4):345–352. https://doi.org/10.22038/ijbms.2019.32399.7765. PMID: 31168337; PMCID: PMC6535194
8. Bray F, Laversanne M, Sung H et al (2024) Global cancer statistics 2022: GLOBOCAN estimates of incidence and mortality worldwide for 36 cancers in 185 countries. CA Cancer J Clin 74(3):229–263
9. Budakoti M, Panwar AS, Molpa D, Singh RK, Büsselberg D, Mishra AP, Coutinho HDM, Nigam M (2021) Micro-RNA: the darkhorse of cancer. Cell Signal 83:109995. https://doi.org/10.1016/j.cellsig.2021.109995. Epub 2021 Mar 27. PMID: 33785398

10. Cao W, Fan R, Wang L, Cheng S, Li H, Jiang J et al (2012) Expression and regulatory function of miRNA-34a in targeting survivin in gastric cancer cells. Tumour Biol 34(2):963–971. https://doi.org/10.1007/s13277-012-0632-8
11. Chan ET, Kural C (2024) Targeting endocytosis to sensitize cancer cells to programmed cell death. Biochem Soc Trans 52(4):1703–1713. https://doi.org/10.1042/BST20231332
12. Chen Q, Chen X, Zhang M, Fan Q, Luo S, Cao X (2011) miR-137 is frequently down-regulated in gastric cancer and is a negative regulator of Cdc42. Dig Dis Sci 56(7):2009–2016. https://doi.org/10.1007/s10620-010-1536-3. Epub 2011 Jan 8. PMID: 21221794
13. Chen J, Zhang L, Jiang Q (2019) miR-155 in acute myeloid leukemia: from molecular pathogenesis to targeted therapy. Front Oncol 9:125
14. Chidambaranathan-Reghupaty S, Fisher PB, Sarkar D (2021) Hepatocellular carcinoma (HCC): epidemiology, etiology and molecular classification. Adv Cancer Res 149:1–61. https://doi.org/10.1016/bs.acr.2020.10.001. Epub 2020 Nov 28. PMID: 33579421; PMCID: PMC8796122
15. Chiurillo MA (2015) Role of the Wnt/β-catenin pathway in gastric cancer: An in-depth literature review. World J Exp Med 5(2):84–102. https://doi.org/10.5493/wjem.v5.i2.84. PMID: 25992323; PMCID: PMC4436943
16. Choe J, Cho H, Chi SG, Kim YK (2011) Ago2/miRISC-mediated inhibition of CBP80/20-dependent translation and thereby abrogation of nonsense-mediated mRNA decay require the cap-associating activity of Ago2. FEBS Lett 585(17):2682–2687
17. Chun-Zhi Z, Lei H, An-Ling Z, Yan-Chao F, Xiao Y, Guang-Xiu W, Zhi-Fan J, Pei-Yu P, Qing-Yu Z, Chun-Sheng K (2010) MicroRNA-221 and microRNA-222 regulate gastric carcinoma cell proliferation and radioresistance by targeting PTEN. BMC Cancer 10:367. https://doi.org/10.1186/1471-2407-10-367. PMID: 20618998; PMCID: PMC2914702
18. Colangelo T, Fucci A, Votino C, Sabatino L, Pancione M, Laudanna C, Binaschi M, Bigioni M, Maggi CA, Parente D, Forte N, Colantuoni V (2013) MicroRNA-130b promotes tumor development and is associated with poor prognosis in colorectal cancer. Neoplasia 15(9):1086–1099. https://doi.org/10.1593/neo.13998. PMID: 24027433; PMCID: PMC3769887
19. Cui ZJ, Xie XL, Qi W, Yang YC, Bai Y, Han Q, Ding J, Jiang HQ (2019) Cell-free miR-17-5p as a diagnostic biomarker for gastric cancer inhibits dendritic cell maturation. Onco Targets Ther 12:2661–2675. https://doi.org/10.2147/OTT.S197682
20. Daneshpour M, Ghadimi-Daresajini A (2023) Overview of miR-106a regulatory roles: from cancer to aging. Bioengineering (Basel) 10(8):892. PMID: 37627777; PMCID: PMC10451182. https://doi.org/10.3390/bioengineering10080892.
21. de Oliveira ARCP, Castanhole-Nunes MMU, Biselli-Chicote PM, Pavarino ÉC, da Silva RCMA, da Silva RF, Goloni-Bertollo EM (2020) Differential expression of angiogenesis-related miRNAs and VEGFA in cirrhosis and hepatocellular carcinoma. Arch Med Sci 16(5):1150–1157. https://doi.org/10.5114/aoms.2020.97967. PMID: 32864004; PMCID: PMC7444729
22. Di Fiore PP (2009) Endocytosis, signaling and cancer, much more than meets the eye. Preface. Mol Oncol 3(4):273–279. https://doi.org/10.1016/j.molonc.2009.06.002
23. Djuranovic S, Zinchenko MK, Hur JK, Nahvi A, Brunelle JL, Rogers EJ, Green R (2010) Allosteric regulation of Argonaute proteins by miRNAs. Nat Struct Mol Biol 17(2):144–150
24. Dong Y, Zhao J, Wu CW, Zhang L, Liu X, Kang W, Leung WW, Zhang N, Chan FK, Sung JJ et al (2013) Tumor suppressor functions of miR-133a in colorectal cancer. Mol Cancer Res 11:1051–1060
25. Dongjuan, Wu Rong, Tang Qinghua, Qi Xinru, Zhou Huan, Zhou Yong, Mao Rui, Li Cuiping, Liu Weipeng, Wang Dong, Hua Xueguang, Zhang (2015) Five functional polymorphisms of B7/CD28 co-signaling molecules alter susceptibility to colorectal cancer Cellular Immunology 293(1) 41-48 10.1016/j.cellimm.2014.11.006
26. Eto K, Iwatsuki M, Watanabe M, Ida S, Ishimoto T, Iwagami S, Baba Y, Sakamoto Y, Miyamoto Y, Yoshida N, Baba H (2014) The microRNA-21/PTEN pathway regulates the sensitivity of HER2-positive gastric cancer cells to trastuzumab. Ann Surg Oncol 21:343–350

27. Fan Z, Lu Y, Wu X, Mendelsohn J (1994) Antibody-induced epidermal growth factor receptor dimerization mediates inhibition of autocrine proliferation of A431 squamous carcinoma cells. J Biol Chem 269(44):27595–27602. PMID: 7961676
28. Fan R, Xiao C, Wan X, Cha W, Miao Y, Zhou Y, Qin C, Cui T, Su F, Shan X (2019) Small molecules with big roles in microRNA chemical biology and microRNA-targeted therapeutics. RNA Biol 16(6):707–718. Epub 2019 Apr 3. PMID: 30900502; PMCID: PMC6546413
29. Farooqi AA, Siddik ZH (2015) Platelet-derived growth factor (PDGF) signalling in cancer: rapidly emerging signalling landscape. Cell Biochem Funct 33(5):257–265. https://doi.org/10.1002/cbf.3120. Epub 2015 July 7. PMID: 26153649; PMCID: PMC5289298
30. Fatimat, Kipkeeva Tatyana, Muzaffarova Alexandra, Korotaeva Danzan, Mansorunov Pavel, Apanovich Maxim, Nikulin Olga, Malikhova Ivan, Stilidi Alexander, Karpukhin (2022) The Features of Immune Checkpoint Gene Regulation by microRNA in Cancer Int J Mol Sci 23(16) 9324. 10.3390/ijms23169324
31. Gao W, Zhou J, Morshedi M (2024) MicroRNA-34 and gastrointestinal cancers: a player with big functions. Cancer Cell Int 24(1):163. https://doi.org/10.1186/s12935-024-03338-w. PMID: 38725047; PMCID: PMC11084024
32. Gramantieri L, Giovannini C, Piscaglia F, Fornari F (2021) MicroRNAs as modulators of tumor metabolism, microenvironment, and immune response in hepatocellular carcinoma. J Hepatocell Carcinoma 8:369–385
33. Grigoryev YA, Kurian SM, Hart T, Nakorchevsky AA, Chen C, Campbell D, Head SR, Yates JR, Salomon DR (2011) MicroRNA regulation of molecular networks mapped by global MicroRNA, mRNA, and protein expression in activated T lymphocytes. J Immunol 187(5):2233–2243
34. Gumireddy K, Young DD, Xiong X, Hogenesch JB, Huang Q, Deiters A (2008) Small molecule inhibitors of microRNA miR-21 function. Angew Chem Int Ed Engl 47(39):7482
35. Han HB, Gu J, Zuo HJ, Chen ZG, Zhao W, Li M, Ji DB, Lu YY, Zhang ZQ (2012) Let-7c functions as a metastasis suppressor by targeting MMP11 and PBX3 in colorectal cancer. J Pathol 226:544–555
36. Han C, Yu Z, Duan Z, Kan Q (2014) Role of microRNA-1 in human cancer and its therapeutic potentials. Biomed Res Int 2014:428371. https://doi.org/10.1155/2014/428371. Epub 2014 May 18. PMID: 24949449; PMCID: PMC4052501
37. Hormati A, Hajrezaei Z, Jazi K, Aslani Kolur Z, Rezvan S, Ahmadpour S (2022) Gastrointestinal and pancratohepatobiliary cancers: a comprehensive review on epidemiology and risk factors worldwide. Middle East J Dig Dis 14(1):5–23. https://doi.org/10.34172/mejdd.2022.251
38. Huang F, Zhao JL, Wang L, Gao CC, Liang SQ, An DJ, Bai J, Chen Y, Han H, Qin HY (2017) miR-148a-3p mediates notch signaling to promote the differentiation and M1 activation of macrophages. Front Immunol 8:1327. https://doi.org/10.3389/fimmu.2017.01327. PMID: 29085372; PMCID: PMC5650608
39. Iqbal N, Iqbal N (2014) Human epidermal growth factor receptor 2 (HER2) in cancers: overexpression and therapeutic implications. Mol Biol Int 2014:852748. https://doi.org/10.1155/2014/852748. Epub 2014 Sept 7. PMID: 25276427; PMCID: PMC4170925
40. Iseki Y, Shibutani M, Maeda K, Nagahara H, Fukuoka T, Matsutani S, Hirakawa K, Ohira M (2018) MicroRNA-96 promotes tumor invasion in colorectal cancer via RECK. Anticancer Res 38(4):2031–2035. https://doi.org/10.21873/anticanres.12442. PMID: 29599320
41. Ismail N, Wang Y, Dakhlallah D, Moldovan L, Agarwal K, Batte K, Shah P, Wisler J, Eubank TD, Tridandapani S, Paulaitis ME, Piper MG, Marsh CB (2013) Macrophage microvesicles induce macrophage differentiation and miR-223 transfer. Blood 121(6):984–995. https://doi.org/10.1182/blood-2011-08-374793. Epub 2012 Nov 9. PMID: 23144169; PMCID: PMC3567345
42. Iyer AN, Bellon A, Baudet ML (2014) microRNAs in axon guidance. Front Cell Neurosci 8:78. https://doi.org/10.3389/fncel.2014.00078
43. Jemal A, Bray F, Center MM, Ferlay J, Ward E, Forman D (2011) Global cancer statistics. CA Cancer J Clin 61(2):69–90. https://doi.org/10.3322/caac.20107. Epub 2011 Feb 4. Erratum in: CA Cancer J Clin. 2011 Mar–Apr;61(2):134. PMID: 21296855

44. Ji J, Yin Y, Ju H, Xu X, Liu W, Fu Q (2017) Long non-coding RNA Lnc-Tim3 exacerbates CD8 T cell exhaustion via binding to Tim-3 and inducing nuclear translocation of Bat3 in HCC. Cell Mol Immunol 14(10):853–866
45. Kang HS, Jang SG, Kwon SY, Park YS, Green JE, Kim HK, Ro J (2014) MicroRNA signature for HER2-positive breast and gastric cancer. Anticancer Res 34(7):3807–3810. Erratum in: Anticancer Res. 2015;35(5):3109. Erratum in: Anticancer Res. 2019 Nov;39(11):6417. Kim J [removed]. PMID: 24982406
46. Karaayvaz M, Zhai H, Ju J (2013) miR-129 promotes apoptosis and enhances chemosensitivity to 5-fluorouracil in colorectal cancer. Cell Death Dis 4:e659
47. Ke M, Zhang Z, Cong L, Zhao S, Li Y, Wang X, Lv Y, Zhu Y, Dong J (2019) MicroRNA-148b-colony-stimulating factor-1 signaling-induced tumor-associated macrophage infiltration promotes hepatocellular carcinoma metastasis. Biomed Pharmacother 120:109523. https://doi.org/10.1016/j.biopha.2019.109523. Epub 2019 Oct 23. PMID: 31655310
48. Khalilian S, Abedinlou H, Hussen BM, Imani SZH, Ghafouri-Fard S (2022) The emerging role of miR-20b in human cancer and other disorders: pathophysiology and therapeutic implications. Front Oncol 12:985457. https://doi.org/10.3389/fonc.2022.985457. PMID: 36582800; PMCID: PMC9792503
49. Khodadadi Kohlan A, Saidijam M, Amini R, Samadi P, Najafi R (2019) Induction of let-7e gene expression attenuates oncogenic phenotype in HCT-116 colorectal cancer cells through targeting of DCLK1 regulation. Life Sci 228:221–227
50. Kim C, Ye Z, Weyand CM, Goronzy JJ. miR-181a-regulated pathways in T-cell differentiation and aging. Immun Ageing. 2021 Jun 15;18(1):28. 10.1186/s12979-021-00240-1. PMID: 34130717; PMCID: PMC8203492.
51. Kiriakidou M, Tan GS, Lamprinaki S, De Planell-Saguer M, Nelson PT, Mourelatos Z (2007) An mRNA m7G cap binding-like motif within human Ago2 represses translation. Cell 129(6):1141–1151
52. Komoll RM, Hu Q, Olarewaju O, von Döhlen L, Yuan Q, Xie Y, Tsay HC, Daon J, Qin R, Manns MP, Sharma AD, Goga A, Ott M, Balakrishnan A (2021) MicroRNA-342-3p is a potent tumour suppressor in hepatocellular carcinoma. J Hepatol 74(1):122–134. https://doi.org/10.1016/j.jhep.2020.07.039. Epub 2020 July 30. PMID: 32738449
53. Kondapuram SK, Ramachandran HK, Arya H, Coumar MS (2023) Targeting survivin for cancer therapy: strategies, small molecule inhibitors and vaccine based therapeutics in development. Life Sci 335:122260. https://doi.org/10.1016/j.lfs.2023.122260. Epub 2023 Nov 12. PMID: 37963509
54. LArki P, Ahadi A, Zare A, Tarighi S, Zaheri M, Souri M, Zali MR, Ghaedi H, Omrani MD (2018) Up-regulation of miR-21, miR-25, miR-93, and miR-106b in gastric cancer. Iran Biomed J 22(6):367–373. https://doi.org/10.29252/.22.6.367. Epub 2018 June 3. PMID: 29859516; PMCID: PMC6305817
55. Law PT, Wong N (2011) Emerging roles of microRNA in the intracellular signaling networks of hepatocellular carcinoma. J Gastroenterol Hepatol 26(3):437–449. https://doi.org/10.1111/j.1440-1746.2010.06512.x. PMID: 21332540
56. Lei ZN, Teng QX, Tian Q, Chen W, Xie Y, Wu K, Zeng Q, Zeng L, Pan Y, Chen ZS, He Y (2022) Signaling pathways and therapeutic interventions in gastric cancer. Signal Transduct Target Ther 7(1):358. https://doi.org/10.1038/s41392-022-01190-w. PMID: 36209270; PMCID: PMC9547882
57. Li Y, Jia R, Ge S, Wang J, Li S, Zhang Y, Li G (2016) Association of PD-1/PD-L axis expression with cytolytic activity, mutational load, and prognosis in melanoma and other solid tumors. Proc Natl Acad Sci 113(48):E7769–E7777
58. Li R, Hu Z, Wang Z, Zhu T, Wang G, Gao B, Wang J, Deng X (2021) miR-125a-5p promotes gastric cancer growth and invasion by regulating the Hippo pathway. J Clin Lab Anal 35(12):e24078. https://doi.org/10.1002/jcla.24078. Epub 2021 Oct 28. PMID: 34708891; PMCID: PMC8649339

59. Lian YF, Huang YL, Wang JL, Deng MH, Xia TL, Zeng MS, Chen MS, Wang HB, Huang YH (2018) Anillin is required for tumor growth and regulated by miR-15a/miR-16-1 in HBV-related hepatocellular carcinoma. Aging (Albany NY) 10(8):1884–1901. https://doi.org/10.18632/aging.101510. PMID: 30103211; PMCID: PMC6128427
60. Liao Y, Wang C, Yang Z, Liu W, Yuan Y, Li K, Zhang Y, Wang Y, Shi Y, Qiu Y, Zuo D, He W, Qiu J, Guan X, Yuan Y, Li B (2020) Dysregulated Sp1/miR-130b-3p/HOXA5 axis contributes to tumor angiogenesis and progression of hepatocellular carcinoma. Theranostics 10(12):5209–5224. https://doi.org/10.7150/thno.43640. PMID: 32373208; PMCID: PMC7196310
61. Lin J, Shen J, Yue H, Cao Z (2019) miRNA-183-5p.1 promotes the migration and invasion of gastric cancer AGS cells by targeting TPM1. Oncol Rep 42(6):2371–2381. https://doi.org/10.3892/or.2019.7354. Epub 2019 Oct 4. Erratum in: Oncol Rep. 2021 Feb;45(2):789–790. https://doi.org/10.3892/or.2020.7902. PMID: 31638242; PMCID: PMC6859460
62. Ling H, Pickard K, Ivan C, Isella C, Ikuo M, Mitter R, Spizzo R, Bullock M, Braicu C, Pileczki V, Vincent K, Pichler M, Stiegelbauer V, Hoefler G, Almeida MI, Hsiao A, Zhang X, Primrose J, Packham G, Liu K, Bojja K, Gafà R, Xiao L, Rossi S, Song JH, Vannini I, Fanini F, Kopetz S, Zweidler-McKay P, Wang X, Ionescu C, Irimie A, Fabbri M, Lanza G, Hamilton SR, Berindan-Neagoe I, Medico E, Mirnezami A, Calin GA, Nicoloso MS (2016) The clinical and biological significance of MIR-224 expression in colorectal cancer metastasis. Gut 65(6):977–989. https://doi.org/10.1136/gutjnl-2015-309372. Epub 2015 Mar 24. PMID: 25804630; PMCID: PMC4581915
63. Liu T, Zhang X, Du L et al. (2020) Correction to: Exosome-transmitted miR-128-3p increase chemosensitivity of oxaliplatin-resistant colorectal cancer. Mol Cancer 19:89. https://doi.org/10.1186/s12943-020-01211-8
64. Liu Z, Sun J, Liu B, Zhao M, Xing E, Dang C (2018) miRNA-222 promotes liver cancer cell proliferation, migration and invasion and inhibits apoptosis by targeting BBC3. Int J Mol Med 42(1):141–148. https://doi.org/10.3892/ijmm.2018.3637. Epub 2018 Apr 20. PMID: 29693134; PMCID: PMC5979783
65. Liu N, Yang C, Gao A, Sun M, Lv D (2022) MiR-101: an important regulator of gene expression and tumor ecosystem. Cancer 14:5861. https://doi.org/10.3390/cancers14235861
66. Liu GH, Chen T, Zhang X, Ma XL, Shi HS (2022) Small molecule inhibitors targeting the cancers. MedComm 3(4):e181
67. Loy LM, Low HM, Choi JY, Rhee H, Wong CF, Tan CH (2022) Variant hepatocellular carcinoma subtypes according to the 2019 WHO classification: an imaging-focused review. AJR Am J Roentgenol 219(2):212–223. https://doi.org/10.2214/AJR.21.26982
68. Lu Y, Deng M, Wang K, Peng Y, Ouyang M (2022) The regulatory effects of MicroRNAs on tumor immunity. Biomed Res Int 2022:2121993. https://doi.org/10.1155/2022/2121993. PMID: 35909469; PMCID: PMC9329000
69. Lu Y, Deng M, Wang K, Peng Y, Ouyang M (2022) The regulatory effects of MicroRNAs on tumor immunity. Biomed Res Int 2022:Article ID 2121993, 12 pages
70. Lv L, He L, Chen S, Yu Y, Che G, Tao X, Wang S, Jian Z, Zhang X (2019) Long non-coding RNA LINC00114 facilitates colorectal cancer development through EZH2/DNMT1-induced miR-133b suppression. Front Oncol 9:1383
71. MacDonald BT, Tamai K, He X (2009) Wnt/beta-catenin signaling: components, mechanisms, and diseases. Dev Cell 17(1):9–26. https://doi.org/10.1016/j.devcel.2009.06.016. PMID: 19619488; PMCID: PMC2861485
72. Maleki M, Golchin A, Javadi S, Khelghati N, Morovat P, Asemi Z et al (2021) Role of exosomal miRNA in chemotherapy resistance of colorectal cancer: a systematic review. Chem Biol Drug Des. https://doi.org/10.1111/cbdd.13947
73. Maleki M, Golchin A, Javadi S, Khelghati N, Morovat P, Asemi Z, Alemi F, Vaghari-Tabari M, Yousefi B, Majidinia M (2023) Role of exosomal miRNA in chemotherapy resistance of colorectal cancer: a systematic review. Chem Biol Drug Des 101(5):1096–1112. https://doi.org/10.1111/cbdd.13947. Epub 2021 Oct 6. PMID: 34480511

74. Matsuoka T, Yashiro M (2023) The role of the transforming growth factor-β Signaling pathway in gastrointestinal cancers. Biomolecules 13(10):1551. https://doi.org/10.3390/biom13101551. PMID: 37892233; PMCID: PMC10605301
75. Melo SA, Ropero S, Moutinho C, Aaltonen LA, Yamamoto H, Calin GA, Rossi S, Fernandez AF, Carneiro F, Oliveira C, Ferreira B (2009) A TARBP2 mutation in human cancer impairs microRNA processing and DICER1 function. Nat Genet 41(3):365–370
76. Mo WY, Cao SQ (2023) MiR-29a-3p: a potential biomarker and therapeutic target in colorectal cancer. Clin Transl Oncol 25(3):563–577. https://doi.org/10.1007/s12094-022-02978-6. Epub 2022 Nov 10. PMID: 36355327; PMCID: PMC9941256
77. Molina JR, Adjei AA (2006) The Ras/Raf/MAPK pathway. J Thorac Oncol 1(1):7–9. PMID: 17409820
78. Nandan MO, Yang VW (2011) An update on the biology of RAS/RAF mutations in colorectal cancer. Curr Colorectal Cancer Rep 7(2):113–120. https://doi.org/10.1007/s11888-011-0086-1. PMID: 21625338; PMCID: PMC3103074
79. Naro Y, Ankenbruck N, Thomas M, Tivon Y, Connelly CM, Gardner L, Deiters A (2018) Small molecule inhibition of MicroRNA miR-21 rescues chemosensitivity of renal-cell carcinoma to topotecan. J Med Chem 61(14):5900–5909. https://doi.org/10.1021/acs.jmedchem.7b01891. Epub 2018 July 11. PMID: 29993250
80. O'Connell RM, Kahn D, Gibson WSJ, Round JL (2017) MicroRNA-155 promotes autoimmune inflammation by enhancing inflammatory T cell development. Immunity 47(2):309–318.e5
81. Paladini L, Fabris L, Bottai G, Raschioni C, Calin GA, Santarpia L (2016) Targeting microRNAs as key modulators of tumor immune response. J Exp Clin Cancer Res 35:103. https://doi.org/10.1186/s13046-016-0375-2. PMID: 27349385; PMCID: PMC4924278
82. Peng Y, Zhang X, Feng X, Fan X, Jin Z (2017) The crosstalk between microRNAs and the Wnt/β-catenin signaling pathway in cancer. Oncotarget 8(8):14089–14106. https://doi.org/10.18632/oncotarget.12923
83. Pidíkova P, Reis R, Herichova I (2020) miRNA clusters with down-regulated expression in human colorectal cancer and their regulation. Int J Mol Sci 21(13):4633. https://doi.org/10.3390/ijms21134633. PMID: 32610706; PMCID: PMC7369991
84. Regis S, Caliendo F, Dondero A, Casu B, Romano F, Loiacono F, Moretta A, Bottino C, Castriconi R (2017) TGF-β1 downregulates the expression of CX3CR1 by inducing miR-27a-5p in primary human NK cells. Front Immunol 8:868. https://doi.org/10.3389/fimmu.2017.00868. PMID: 28791023; PMCID: PMC5524732.
85. Makoui RH, Fekri S, Ansari N, Makoui MH (2024) Investigating the co-expression rate of HER2 and HER3 biomarkers in cancer patients: A systematic review and meta-analysis. Asian Pac J Cancer Prev 25(9):2979–2990. https://doi.org/10.31557/APJCP.2024.25.9.2979
86. Roncarati R, Lupini L, Shankaraiah RC, Negrini M (2019) The importance of microRNAs in RAS oncogenic activation in human cancer. Front Oncol 9:988. https://doi.org/10.3389/fonc.2019.00988. PMID: 31612113; PMCID: PMC6777413
87. Sachdeva M, Mo YY (2010) MicroRNA-145 suppresses cell invasion and metastasis by directly targeting mucin 1. Cancer Res 70(1):378–387. https://doi.org/10.1158/0008-5472.CAN-09-2021. Epub 2009 Dec 8. PMID: 19996288; PMCID: PMC2805032
88. Salinas-Vera YM, Marchat LA, Gallardo-Rincón D, Ruiz-García E, Astudillo-De La Vega H, Echavarría-Zepeda R, López-Camarillo C (2019) AngiomiRs: MicroRNAs driving angiogenesis in cancer (review). Int J Mol Med 43(2):657–670. https://doi.org/10.3892/ijmm.2018.4003. Epub 2018 Nov 27. PMID: 30483765
89. Sameti P, Tohidast M, Amini M, Bahojb Mahdavi SZ, Najafi S, Mokhtarzadeh A (2023) The emerging role of MicroRNA-182 in tumorigenesis; a promising therapeutic target. Cancer Cell Int 23(1):134. https://doi.org/10.1186/s12935-023-02972-0. PMID: 37438760; PMCID: PMC10337107
90. Selven H, Andersen S, Pedersen MI et al (2022) High expression of miR-17-5p and miR-20a-5p predicts favorable disease-specific survival in stage I-III colon cancer. Sci Rep 12:7080. https://doi.org/10.1038/s41598-022-11090-2

91. Shan G, Li Y, Zhang J, Li W, Szulwach KE, Duan R, Faghihi MA, Khalil AM, Lu L, Paroo Z, Chan AW (2008) A small molecule enhances RNA interference and promotes microRNA processing. Nat Biotechnol 26(8):933–940
92. Shao HJ, Li Q, Shi T, Zhang GZ, Shao F (2019) LINC00707 promotes cell proliferation and invasion of colorectal cancer via miR-206/FMNL2 axis. Eur Rev Med Pharmacol Sci 23(9):3749–3759
93. Shen et al (2017) Silencing NKG2D ligand-targeting miRNAs enhances natural killer cell-mediated cytotoxicity in breast cancer. Cell Death Disease https://doi.org/10.1038/cddis.2017.158
94. Shen K, Cao Z, Zhu R, You L, Zhang T (2019) The dual functional role of MicroRNA-18a (miR-18a) in cancer development. Clin Transl Med 8(1):32. https://doi.org/10.1186/s40169-019-0250-9. PMID: 31873828; PMCID: PMC6928177
95. Shi T, Morishita A, Kobara H, Masaki T (2021) The role of long non-coding RNA and microRNA networks in hepatocellular carcinoma and its tumor microenvironment. Int J Mol Sci 22(19):10630. https://doi.org/10.3390/ijms221910630. PMID: 34638971; PMCID: PMC8508708
96. Taghehchian N, Samsami Y, Maharati A, Zangouei AS, Boroumand-Noughabi S, Moghbeli M (2023) Molecular biology of microRNA-342 during tumor progression and invasion. Pathol Res Pract 248:154672. https://doi.org/10.1016/j.prp.2023.154672. Epub 2023 July 3. PMID: 37413875
97. Tan GS, Chiu CH, Garchow BG, Metzler D, Diamond SL, Kiriakidou M (2012) Small molecule inhibition of RISC loading. ACS Chem Biol 7(2):403–410. https://doi.org/10.1021/cb200253h. Epub 2011 Nov 11. PMID: 22026461; PMCID: PMC3282558
98. Tan SB, Li J, Chen X, Zhang W, Zhang D, Zhang C, Li D, Zhang Y (2014) Small molecule inhibitor of myogenic microRNAs leads to a discovery of miR-221/222-myoD-myomiRs regulatory pathway. Chem Biol 21(10):1265–1270
99. Tang G (2005) siRNA and miRNA: an insight into RISCs. Trends Biochem Sci 30(2):106–114
100. Tang S, Wang D, Zhang Q, Li L (2016) miR-218 suppresses gastric cancer cell proliferation and invasion via regulation of angiopoietin-2. Exp Ther Med 12(6):3837–3842. https://doi.org/10.3892/etm.2016.3893. Epub 2016 Nov 11. PMID: 28105117; PMCID: PMC5228446
101. Tang S, Wu WK, Li X, Wong SH, Wong N, Chan MT, Sung JJ, Yu J (2016) Stratification of digestive cancers with different pathological features and survival outcomes by MicroRNA expression. Sci Rep 6:24466. https://doi.org/10.1038/srep24466. PMID: 27080237; PMCID: PMC4832245
102. Tavakoli Pirzaman A, Alishah A, Babajani B, Ebrahimi P, Sheikhi SA, Moosaei F, Salarfar A, Doostmohamadian S, Kazemi S (2024) The role of microRNAs in hepatocellular cancer: a narrative review focused on tumor microenvironment and drug resistance. Technol Cancer Res Treat 23:15330338241239188. https://doi.org/10.1177/15330338241239188. PMID: 38634139; PMCID: PMC11025440
103. Tsuchida A, Ohno S, Wu W, Borjigin N, Fujita K, Aoki T, Ueda S, Takanashi M, Kuroda M (2011) miR-92 is a key oncogenic component of the miR-17–92 cluster in colon cancer. Cancer Sci 102:2264–2271
104. Valeri N, Braconi C, Gasparini P, Murgia C, Lampis A, Paulus-Hock V, Hart JR, Ueno L, Grivennikov SI, Lovat F, Paone A, Cascione L, Sumani KM, Veronese A, Fabbri M, Carasi S, Alder H, Lanza G, Gafa' R, Moyer MP, Ridgway RA, Cordero J, Nuovo GJ, Frankel WL, Rugge M, Fassan M, Groden J, Vogt PK, Karin M, Sansom OJ, Croce CM (2014) MicroRNA-135b promotes cancer progression by acting as a downstream effector of oncogenic pathways in colon cancer. Cancer Cell 25(4):469–483. https://doi.org/10.1016/j.ccr.2014.03.006. PMID: 24735923; PMCID: PMC3995091
105. Vo DD, Staedel C, Zehnacker L, Benhida R, Darfeuille F, Duca M (2014) Targeting the production of oncogenic microRNAs with multimodal synthetic small molecules. ACS Chem Biol 9(3):711–721

106. Voorhoeve PM, Le Sage C, Schrier M, Gillis AJ, Stoop H, Nagel R, Liu YP, Van Duijse J, Drost J, Griekspoor A, Zlotorynski E (2006) A genetic screen implicates miRNA-372 and miRNA-373 as oncogenes in testicular germ cell tumors. Cell 124(6):1169–1181
107. Wang H (2020) MicroRNA, diabetes mellitus and colorectal cancer. Biomedicines 8(12):530. https://doi.org/10.3390/biomedicines8120530. PMID: 33255227; PMCID: PMC7760221.
108. Wang L, Xu M, Lu P, Zhou F (2018) microRNA-769 is downregulated in colorectal cancer and inhibits cancer progression by directly targeting cyclin-dependent kinase 1. Onco Targets Ther 11:9013
109. Wang J, Li D, Cang H, Guo B (2019) Crosstalk between cancer and immune cells: role of tumor-associated macrophages in the tumor microenvironment. Cancer Med 8(10):4709–4472
110. Watashi K, Yeung ML, Starost MF, Hosmane RS, Jeang KT (2010) Identification of small molecules that suppress microRNA function and reverse tumorigenesis. J Biol Chem 285(32):24707–24716. https://doi.org/10.1074/jbc.M109.062976. Epub 2010 June 7. PMID: 20529860; PMCID: PMC2915707
111. Wei W, Ba Z, Gao M, Wu Y, Ma Y, Amiard S, White CI, Danielsen JMR, Yang YG, Qi Y (2012) A role for small RNAs in DNA double-strand break repair. Cell 149(1):101–112
112. Wei Y, Guo S, Tang J, Wen J, Wang H, Hu X, Gu Q (2020) MicroRNA-19b-3p suppresses gastric cancer development by negatively regulating neuropilin-1. Cancer Cell Int 20:193. https://doi.org/10.1186/s12935-020-01257-0. PMID: 32508529; PMCID: PMC7249695
113. Wen D, Danquah M, Chaudhary AK, Mahato RI (2015) Small molecules targeting microRNA for cancer therapy: promises and obstacles. J Control Release 219:237–247. https://doi.org/10.1016/j.jconrel.2015.08.011. Epub 2015 Aug 6. PMID: 26256260; PMCID: PMC4749151
114. Wenxiu, Xu Yuting, Hua Fei, Deng Dandan, Wang Yang, Wu Wei, Zhang Jinhai, Tang (2020) MiR-145 in cancer therapy resistance and sensitivity: A comprehensive review Abstract Cancer Science 111(9) 3122-3131 10.1111/cas.14517
115. Wu WK, Lee CW, Cho CH, Fan D, Wu K, Yu J, Sung JJ (2010) MicroRNA dysregulation in gastric cancer: a new player enters the game. Oncogene 29(43):5761–5771. https://doi.org/10.1038/onc.2010.352. Epub 2010 Aug 30. PMID: 20802530
116. Wu SR, Wu Q, Shi YQ (2020) Recent advances of miRNAs in the development and clinical application of gastric cancer. Chin Med J 133(15):1856–1867. https://doi.org/10.1097/CM9.0000000000000921. PMID: 32649523; PMCID: PMC7469991
117. Wu S, Yuan W, Luo W, Nie K, Wu X, Meng X, Shen Z, Wang X (2022) miR-126 downregulates CXCL12 expression in intestinal epithelial cells to suppress the recruitment and function of macrophages and tumorigenesis in a murine model of colitis-associated colorectal cancer. Mol Oncol 16(19):3465–3489. https://doi.org/10.1002/1878-0261.13218. Epub 2022 Apr 11. PMID: 35363937; PMCID: PMC9533691
118. Xiao Z, Chen S, Feng S, Li Y, Zou J, Ling H, Zeng Y, Zeng X (2020) Function and mechanisms of microRNA-20a in colorectal cancer. Exp Ther Med 19(3):1605–1616. https://doi.org/10.3892/etm.2020.8432. Epub 2020 Jan 8. PMID: 32104211; PMCID: PMC7027132
119. Xu J, Zhang Y, Liu C, Yan P, Yang Z (2021) Roles of the miR-139-5p/CCT5 axis in hepatocellular carcinoma: a bioinformatic analysis. Int J Med Sci 18(15):3556–3564. https://doi.org/10.7150/ijms.57504. PMID: 34522182; PMCID: PMC8436101
120. Xu T, Xie F, Xu D, Xu W, Ge X, Lv S, Li S (2021) MiR-200b suppresses gastric cancer cell migration and invasion by inhibiting NRG1 through ERBB2/ERBB3 Signaling. J Oncol 2021:4470778. https://doi.org/10.1155/2021/4470778. PMID: 34531912; PMCID: PMC8440071
121. Xue TM, Tao LD, Zhang M, Zhang J, Liu X, Chen GF, Zhu YJ, Zhang PJ (2015) Clinicopathological significance of MicroRNA-20b expression in hepatocellular carcinoma and regulation of HIF-1α and VEGF effect on cell biological behaviour. Dis Markers 2015:325176. https://doi.org/10.1155/2015/325176. Epub 2015 Nov 3. PMID: 26612965; PMCID: PMC4646993

122. Yang TS, Yang XH, Wang XD, Wang YL, Zhou B, Song ZS (2013) MiR-214 regulate gastric cancer cell proliferation, migration and invasion by targeting PTEN. Cancer Cell Int 13(1):68. https://doi.org/10.1186/1475-2867-13-68. PMID: 23834902; PMCID: PMC3716801
123. Yang P, Tang R, Zhu J, Zou L, Wu R, Zhou H, Mao Y, Li R, Hua D, Wang W, Zhang H (2013) A functional variant at miR-24 binding site in B7-H2 alters susceptibility to gastric cancer in a Chinese Han population. Mol Immunol 56(1–2):98–103. https://doi.org/10.1016/j.molimm.2013.04.010. Epub 2013 May 18. PMID: 23688438
124. Yang X, Sun Y, Zhang Y, Han S (2020) Downregulation of miR-181b inhibits human colon cancer cell proliferation by targeting CYLD and inhibiting the NF-κB signaling pathway. Int J Mol Med 46(5):1755–1764. https://doi.org/10.3892/ijmm.2020.4720. Epub 2020 Sep 4. PMID: 32901872; PMCID: PMC7521473
125. Yang J, Cho WC, Zheng Y (2020) Argonaute proteins: structural features, functions and emerging roles. J Adv Res 24:317–324
126. Yin J, Guo Y, Li Z (2021) Platelet-derived growth factor-B signalling might promote epithelial-mesenchymal transition in gastric carcinoma cells through activation of the MAPK/ERK pathway. Contemp Oncol (Pozn) 25(1):1–6. https://doi.org/10.5114/wo.2021.103938. Epub 2021 Mar 22. PMID: 33911974; PMCID: PMC8063901
127. Young DD, Connelly CM, Grohmann C, Deiters A (2010) Small molecule modifiers of microRNA miR-122 function for the treatment of hepatitis C virus infection and hepatocellular carcinoma. J Am Chem Soc 132(23):7976–7981
128. Yu W, Liang X, Li X, Zhang Y, Sun Z, Liu Y, Wang J (2018) MicroRNA-195: a review of its role in cancers. Onco Targets Ther 11:7109–7123. https://doi.org/10.2147/OTT.S183600. PMID: 30410367; PMCID: PMC6200091
129. Yu FB, Sheng J, Yu JM, Liu JH, Qin XX, Mou B (2020) MiR-19a-3p regulates the Forkhead box F2-mediated Wnt/β-catenin signaling pathway and affects the biological functions of colorectal cancer cells. World J Gastroenterol 26(6):627–644. https://doi.org/10.3748/wjg.v26.i6.627. PMID: 32103872; PMCID: PMC7029353
130. Yugawa K, Yoshizumi T, Mano Y, Itoh S, Harada N, Ikegami T, Kohashi K, Oda Y, Mori M (2021) Cancer-associated fibroblasts promote hepatocellular carcinoma progression through downregulation of exosomal miR-150-3p. Eur J Surg Oncol 47(2):384–393. https://doi.org/10.1016/j.ejso.2020.08.002. Epub 2020 Aug 19. PMID: 32883551
131. Zhang H, Li Y, Huang Q, Ren X, Hu H, Sheng H, Lai M (2011) MiR-148a promotes apoptosis by targeting Bcl-2 in colorectal cancer. Cell Death Differ 18:1702–1710
132. Zhang Y, Wang X, Xu B, Wang B, Wang Z, Liang Y, Zhou J (2015) Epigenetic silencing of miR-203 upregulates SNAI2 and contributes to the invasiveness of malignant breast cancer cells. Genes Cancer 6(11–12):490–502
133. Zhang J, Xu X, Zhao S, Gong Z, Liu P, Guan W, He X, Wang T, Peng T, Teng J et al (2015) The expression and significance of the plasma Let-7 family in anti-N-methyl-d-aspartate receptor encephalitis. J Mol Neurosci 56:531–539
134. Zhang L, Zhang Y, Wong SH, Law PTY, Zhao S, Yu J, Chan MTV, Wu WKK (2018) Common deregulation of seven biological processes by MicroRNAs in gastrointestinal cancers. Sci Rep 8(1):3287. https://doi.org/10.1038/s41598-018-21573-w. PMID: 29459716; PMCID: PMC5818544
135. Zhang Z, Feng L, Liu P, Duan W (2018) ANRIL promotes chemoresistance via disturbing expression of ABCC1 by regulating the expression of Let-7a in colorectal cancer. Biosci Rep 38:BSR20180620
136. Zhang M, Yang Q, Zhang L, Zhou S, Ye W, Yao Q, Li Z (2019) MiR-20a targets BNIP2 and contributes chemotherapeutic resistance in colorectal adenocarcinoma SW480 and SW620 cell lines. Acta Cir Bras 34(3):e201900305
137. Zhang P, Ha M, Li L, Huang X, Liu C (2020) MicroRNA-3064-5p sponged by MALAT1 suppresses angiogenesis in human hepatocellular carcinoma by targeting the FOXA1/CD24/Src pathway. FASEB J 34(1):66–81. https://doi.org/10.1096/fj.201901834R. Epub 2019 Nov 15. PMID: 31914639

138. Zhang X, Li T, Han YN, Ge M, Wang P, Sun L, Liu H, Cao T, Nie Y, Fan D, Guo H, Wu K, Zhao X, Lu Y (2021) miR-125b promotes colorectal cancer migration and invasion by dual-targeting CFTR and CGN. Cancers (Basel) 13(22):5710. https://doi.org/10.3390/cancers13225710. PMID: 34830864; PMCID: PMC8616371
139. Zhang W, Zhu Y, Zhou Y, Wang J, Jiang P, Xue L (2022) miRNA-31 increases radiosensitivity through targeting STK40 in colorectal cancer cells. Asia Pac J Clin Oncol 18(3):267–278. https://doi.org/10.1111/ajco.13602. Epub 2021 June 25. PMID: 34170070; PMCID: PMC9291185
140. Zhang Z, Huang Q, Liuchunyang Y, Zhu D, Li Y, Xue Z, Hua Z, Luo X, Song Z, Cheng L, Zhao T, Liu Y (2022) The role of miRNA in tumor immune escape and miRNA-based therapeutic strategies. Front Immunol 12:807895
141. Zhao J, Li H, Zhao S, Wang E, Zhu J, Feng D, Zhu Y, Dou W, Fan Q, Hu J, Jia L, Liu L (2021) Epigenetic silencing of miR-144/451a cluster contributes to HCC progression via paracrine HGF/MIF-mediated TAM remodeling. Mol Cancer 20(1):46. https://doi.org/10.1186/s12943-021-01343-5. PMID: 33658044; PMCID: PMC7927270
142. Zhou SL, Hu ZQ, Zhou ZJ, Dai Z, Wang Z, Cao Y, Fan J, Huang XW, Zhou J (2016) miR-28-5p-IL-34-macrophage feedback loop modulates hepatocellular carcinoma metastasis. Hepatology 63(5):1560–1575. https://doi.org/10.1002/hep.28445. Epub 2016 Mar 7. PMID: 26754294
143. Zhou Y, Liu F, Ma C, Cheng Q (2022) Involvement of microRNAs and their potential diagnostic, therapeutic, and prognostic role in hepatocellular carcinoma. J Clin Lab Anal 36(10):e24673. https://doi.org/10.1002/jcla.24673. Epub 2022 Aug 29. PMID: 36036748; PMCID: PMC9551129
144. Zhu D, Sun Y, Zhang D, Dong M, Jiang G, Zhang X, Zhou J (2018) miR-1 inhibits the progression of colon cancer by regulating the expression of vascular endothelial growth factor. Oncol Rep 40:589–598

Chapter 17
Exploring MicroRNA-Mediated Pathways in Gynecological Cancer Progression and Therapeutic Resistance

Anshumi Banerjee, Srija Sadhukhan, Anushka Ghoshal, and Kousik Pramanick

Abstract MicroRNAs (miRNAs) are small noncoding RNAs that regulate gene expression by binding to target mRNAs, leading to their degradation or translational repression. Their dysregulation has been widely implicated in the development and progression of gynecological cancers, including ovarian, cervical, endometrial, and vulvar cancers. miRNAs influence crucial cellular processes such as cell cycle control, apoptosis, autophagy, angiogenesis, and metastasis by targeting key signaling pathways like PI3K/Akt, TGF-β, and MAPK. Recent findings also highlight the role of exosomal miRNAs in shaping the tumor microenvironment and contributing to resistance against chemotherapy and radiotherapy. This resistance is often mediated through altered expression of tumor suppressor genes and oncogenes, leading to therapeutic failure. This chapter explores the multifaceted roles of miRNAs in gynecological cancer biology, emphasizing their contribution to disease progression and therapeutic resistance, and discusses their potential as promising biomarkers and therapeutic targets.

Keywords MicroRNAs (miRNAs) · Gynecological cancers · Therapeutic resistance · Biomarkers and targets

---

A. Banerjee · K. Pramanick (✉)
Integrative Biology Research Unit (IBRU), Department of Life Sciences, Presidency University, Kolkata, India
e-mail: kousik.dbs@presiuniv.ac.in

S. Sadhukhan · A. Ghoshal
Department of Biotechnology, Amity University Kolkata, Kadampukur, West Bengal, India

S. Pathak, A. Banerjee (eds.), *MicroRNAs: Role in Cancer Progression, Diagnostics and Therapy*, Advances in Biochemistry in Health and Disease 34,
https://doi.org/10.1007/978-3-032-08824-6_17

Introduction

It was in the year 1993, when two scientists independently identified a noncoding RNA lin-4 and let-7 in *Caenorhabditis elegans* in two independent studies [48, 98]. Lee and colleagues had stated the presence of sequence complementarity of lin-4 with the multiple sites of 3′ untranslated region (3′UTR) of lin-14 mRNA. It was back then when they had seeded the idea of antisense RNA-RNA interaction as the occurrence of mRNA destabilization and transcriptional repression. This lin-4 RNA was 20–24 nucleotides long and would be placed in another large group of noncoding RNAs distinguished from other Dicer products, small temporal RNAs (stRNAs) and small interfering RNAs (siRNAs) based on their biogenesis and their distinct fold-back hairpin structures [6]. The new term coined for these tiny enodogenous RNAs as microRNAs, abbreviated miRNAs [44, 46, 47]. The RNA polymerase II-mediated transcription process of miRNA is evolutionarily conserved among eukaryotes. Complete functional miRNA processing is a result of endonucleolytic cleavage by two RNA polymerase III proteins: Drosha and Dicer. Drosha is acting in the nucleus upon the 60 to 100 nucleotide-long primary miRNA (pri-miRNA) transcript and chop them into a hairpin loop structure called precursor miRNA (pre-miRNA). The pre-miRNA is transported from the nucleus to the cytoplasm by nuclear exporting complex, exportin 5 and Ran-GTP, where Dicer cleaves it into ~22 nucleotides miRNA duplex. Here, the RNA-induced silencing complex (RISC) binds with the mature single-stranded miRNA and the other complementary fragment gets removed and degraded [19]. This miRNA-RISC has the potential to base pair nonspecifically with 3′UTR of target mRNA. This pairing will be mediated by incorporating RISC which is a combination of two protein components, AGO2 and GW182. In this way, a single miRNA-RISC complex can target hundreds to thousands of mRNA to its degradation or destabilization of posttranscriptional regulation [25]. Being an important regulator of gene expression, miRNA becomes crucial in diseases like cancer. When 13q14 was lost in patients with chronic lymphoid leukemia (CLL), only two noncoding RNAs, miR-15 and miR-16, were deleted instead of any protein-coding genes. This study was the first to confirm the connection between miRNA and cancer. That suggested that the cause of the CLL was the loss of miR15 and miR16 [8]. The onset and pathogenesis of cancer is also correlated with deregulation of miRNAs. Certain genetic and epigenetic modifications in malignancies influence the expression of components of the miRNA biogenesis machinery and, consequently, miRNA profiles. Overexpression and amplification of oncogenic miRNAs or the lower expression of tumor suppressor miRNAs are associated with cancer onset, metastasis, and prognosis [53]. Cancer cells show an expansive alteration in miRNA expressions that may be due to the aberration at the miRNA biogenesis via Dicer and Drosha. Studies indicate miRNA expression ideally repressed in most of the tumor cells with repression of Drosha [86]. As significantly lower expression of miRNA processing machinery components, Dicer and Drosha are observed with a suppressed expression of miRNA with a poor clinical outcome in patients with ovarian cancer [62, 67]. Similarly, a significant decrease of Dicer and Drosha mRNA has been observed in patients with endometrial cancer with higher invasiveness and worse prognosis [87]. Overexpression of Drosha can be seen in more than 50% of cervical squamous cell carcinoma (SCC). Increased Drosha causes altered cancer-related miRNA expression which in turn is responsible for affecting numerous protein-coding genes, causing invasion and metastasis in cervical SCC [65]. Solid metastases and effusions of serous ovarian carcinoma are associated with significantly greater expression of Dicer, Drosha, Argonaute1, and Argonaute2, which is typically a clear indicator of the role of miRNA in cancer progression [90]. The processing of miRNA is induced by several signal transduction pathways, and in turn, the miRNAs are found to regulate the effector molecules of these pathways [37]. Cancer onset and progression are a result of erroneous cell-cell communication that is governed by unusual cell signaling pathways [21]. Alterations in miRNA can lead to tumorigenesis, as well as compromised cell cycle progression, apoptosis, proliferation, metastasis, metabolic changes, immunological modulation, and resistance to treatment [89]. It has been found that microRNA can control cell cycle or cell signaling pathways at any level to alter its expression and induce or reduce cancer-related phenomenon.

Overview and Epidemiology of Gynecological Cancers

Ovarian cancer (OC) is the deadliest gynecological cancer which is characterized by poor prognosis with late diagnosis. It is the eighth most common cancer among women globally [77, 80]. In the United States, the lowest survival rate among the gynecological cancers was reported, with at least 21,000 new cases and around 14,000 deaths anticipated in 2022. The severe nature of the disease and its tendency to be diagnosed at advanced stages in 60% of cases leads to its poor prognosis, with a survival rate of roughly 48% [77]. OC is a complex disease comprising mainly of epithelial, stromal, and germ cells, with the majority being epithelial ovarian cancers. The epithelial type is further categorized into histotypes like high-grade serous, endometrioid, clear cell, mucinous, and low-grade serous carcinomas [9, 71, 80]. In addition, ovarian cancers are grouped into type I and II tumors depending on their molecular characteristics, in which type I tumors are typically low-grade, having mutations in *BRAF, KRAS,* and *PTEN*, whereas type II tumors are high-grade in nature, often mutated in *p53, BRCA1/2*, and overexpress *HER2/HER3* characterized by rapid progression and poor differentiation [5, 39, 72, 75]. Cervical cancer (CC) is the most common malignancy among worldwide women and a cause of cancer-related death in women in developing countries [78]. About 528,000 new cases and 266,000 deaths were reported approximately on 2010, among them 6000 cases had occurred in China. Mortality and high incidence are associated with lack of early detection procedures and prevention measures [15, 18]. Endometrial cancer (ECa) is the most widespread female genital tract cancer. Approximately, there were

52,630 new cases and 8590 deaths reported in the United States in 2014. Almost 10–20% endometrial cancers are categorized under type II tumors. Generally, endometrial cancers have two subtypes: type I and type II. Type I is often endometrioid adenocarcinomas, estrogen dependent, and have a better prognosis, while type II is non-endometrioid. Early-stage detection can promise a good prognosis but the survival rate might decrease due to effective therapies [41]. Vulvar cancer is a rare gynecological cancer that represents roughly 5% of all female genital tract cancers [36, 76]. The global incidence of vulvar cancer is closely connected to its pathogenesis that can be classified into two primary pathways. The first pathway is related to HPV-associated carcinogenesis which mostly affects younger women [16]. Another pathway is HPV-independent which is quite common in elderly women [17]. It is associated with inflammatory skin conditions like lichen sclerosis which in turn can lead to vulvar SCC [73].

MicroRNA in Gynecological Cancer Progression

A differential expression of miRNA has been observed in different tissues based on the type and stages of cancer, but in all the cases, they target diverse gene transcripts of cancer cell signaling pathways. Cancer metastasis is the result of the coordinated action of many gene products that can be targeted by multiple miRNAs and based upon their targets, miRNAs can be categorized as tumor suppressor or oncogenic (Fig. 17.1) [64].

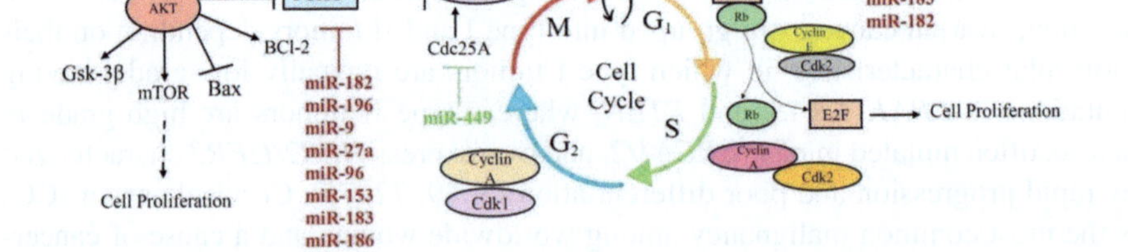

Fig. 17.1 Mechanism of action of a few microRNAs identified in different gynecological cancers. MicroRNAs having oncogenic activity are depicted in red, which are generally upregulated in cancer, and microRNAs having tumor suppressor activity are depicted in green, which are downregulated in cancer

MicroRNA-Mediated Unchecked Cell Cycle and Cell Growth

The altered regulation of the cell cycle shows tumor progression by uncontrolled proliferation rate. This uncontrolled cell proliferation enhances tumor mass expansion and results in the accumulation of genetic abnormalities. Cell cycle checkpoint disruption results in DNA damage, which amplifies mutation and chromosomal abnormalities [88]. Cell growth in cancer includes several processes like metabolism, cell proliferation mechanisms. There are several signaling pathways which regulate the cell growth and cell division, confirming that cell proliferation occurs when particular stimuli or response are observed. In cancer environment, these regulatory controls get damaged, which leads to tumor formation and uncontrolled cell growth [20]. FOXO1 is working on the upstream of CDK responses to regulate the cell cycle tightly. In cervical cancer, miR-182 and miR-196a have an upregulated response which causes the suppression of the FOXO1 gene. These oncogenic microRNAs promote cervical cancer development and progression by directly targeting FOXO1. The cell cycle regulator p27 is also found as a downstream target of miR-196a causes an elevated response of cyclin D1 [33]. Few miRNAs (miR-9, miR-27a, miR-96, miR-153, miR-183, miR-186) have been upregulated in endometrial tumorigenesis which might have promoted FOXO1 repression leading to uncontrolled cell cycle progression and altered cellular death machinery [66]. In ovarian cancer cell lines and among patients, miR-107 downregulation is noticed. Here, the miR-107 acts like a tumor suppressor as it can arrest the cell cycle at G1/S checkpoint by silencing cyclin E1 gene expression [85]. MiR-211 inhibits cell proliferation in epithelial ovarian carcinoma by inhibiting CDK6 and cyclin D1 and halting the cells at G1/S checkpoint [101]. Another protein, p16, controls the cell cycle G1-S checkpoint by preventing Rb phosphorylation. In cervical carcinoma, miR-29a causes p16 silencing by epigenetic methylation by targeting two DNA methyltranferases such as *DNMT3A, DNMT3B* [93, 94]. In endometrioid endometrial cancer (EEC), miR-503 is identified as such a tumor suppressor which has a negative regulation with a proto-oncogene cyclin D1. Suppressed expression of miR-503 regulates cell cycle proliferation and carcinogenesis in EEC [104]. Downregulation in the tumor suppressor gene p53 damages the responding ability of cells to DNA damage. This causes tumor evolution and genomic instability. Also, disruption in retinoblastoma protein (RB) shows activation of E2F transcription factors that promote uncontrolled progression from G1 phase to S phase [32]. Another transcription factor, E2F7, which promotes S phase cell cycle progression, is a direct target of miR-424. Endometrial cancer has a reduced miR-424 profile and increased E2F7 expression, which results in unchecked cell differentiation and tumorigenic qualities similar to loss of apoptosis. [51]. A member of the miR-17-92 family, miR-92a acts as an oncogene in cervical cancer progression. This miR-92a directly knocks down the expression of cyclin-dependent kinase (CDK) inhibitor p21, thus enhancing cancer cell progression through promoting uncontrolled cell cycle and proliferation [81]. Yin Yang-1 (YY1) is a transcription factor that regulates the expression profile of many oncogenes and tumor suppressor genes and plays a critical role in cancer cell growth and proliferation. In epithelial ovarian carcinoma (EOC), miR-381 expression is significantly decreased which is generally found to significantly inhibit tumorigenesis, cell proliferation, migration, and invasion by targeted suppression of YY1-mediated negative regulation of *p53* and wnt [102]. Tumor suppressor gene *TP53INP1* is identified as a direct target of miR-125b in type II endometrial cancer. An overexpressed miR-125b causes a loss in TP53INP1 protein which is a target protein of p53 and therefore the onset of an aberrant cell proliferation and migration [40]. In the endometrial cancer cell line, downregulated miR-152 induces cell proliferation by promoting the cells across G2/M checkpoint. Here, miR-152 acts as a tumor suppressor as it directly targets the cdc25B to knock down its expression [103]. In type II endometrial cancer, miR-449a expression loss is inversely correlated with CDC25A phosphatase activity. The onset of the oncogenic phenomena may be linked to the upregulation of *CDC25A* [110]. MicroRNA miR-302 inhibits cell proliferation and induces apoptosis by suppressing the expression of G2/M checkpoint controllers *CDK1* and *Cyclin B*, suppressed in endometrial cancer cells, thus promoting tumorigenesis [105]. In ovarian cancer, miR-1251-5p directly targets and downregulates the expression of the tumor suppressor gene tubulin-binding cofactor C (TBCC) involved in microtubule polymerization and cell cycle progression. Further, overexpression of miR-1251-5p enhances tumor growth [74].

In vulvar SCC, another miRNA, miR-4712-5p, directly targets the phosphatase and tensin homolog (*PTEN*) which eventually lowers the expression of *PTEN* and GSK3β and upregulates the expression of *Akt* and *cyclin D1* and therefore increases uncontrolled cell proliferation and invasion [107]. PI3/Akt/mTOR pathway supports increased cell growth rate and also metabolic processes that allow the cells to grow independently [29]. PI3k gets inhibited by *PTEN*. A decrease in *PTEN* function leads to the removal of inhibitory components on growth regulatory pathways that increase proliferation and metabolic activity. Increased expression of miR-205 suppresses the expression of *PTEN* while increasing the expression of phosphorylated Akt which in turn promotes cell signaling pathways and epithelial-mesenchymal transition (EMT) markers in cervical cancer cell [41, 43]. In ovarian cancer cell lines, elevated expression of miR-205 and miR-214 downregulates *PTEN*, thus removing the inhibitory effect on PI3K/Akt pathway that eventually increases cell viability and decreases apoptosis and promotes tumor progression [56, 57].

TGF-β signaling pathway or Transforming growth factor beta is defined as pleiotropic due to its different effects on cell types. It also shows diversification by dividing TGF-β into two different pathways: canonical and noncanonical. Canonical pathway is basically a Smad-dependent pathway and the noncanonical pathway is a Smad-independent pathway where TGF-β activates other signaling pathways (PI3K, Ras, MAPK pathways). In general, miR-520h binds to 3′UTR of Smad7 (negative regulator of TGF-β signaling) and thus further alleviates its inhibitory effect on TGF-β signaling pathway which eventually promotes epithelial ovarian cancer progression [115–117]. Another microRNA, miR-181a, allows epithelial-to-mesenchymal transition (EMT) that promotes ovarian cancer cell invasion, tumor progression and metastasis by suppressing Smad7 expression, a negative regulator of the TGF-β pathway [68]. Overexpression of miR-205 suppresses the expression of *PTEN* and *SMAD4*, the tumor suppressor genes, and promotes ovarian cancer cell growth and proliferation [14]. Similarly, miR-552 when overexpressed targets *PTEN*, the tumor suppressor gene and downregulates *PTEN* expression which facilitates ovarian cancer progression through increased cell proliferation and migration [120]. The transcriptional factors ZEB1 and ZEB2 are involved in TGF-β signaling as downstream effectors, which are targets of the miR-200 family in ovarian cancer. The reciprocal relation between miR-200 and ZEB shows a double negative feedback loop. An increase in miR-200 level decreases the ZEB1/2 level, which promotes epithelial to mesenchymal phenotype [7]. It has been found that human papillomavirus (HPV)-mediated cervical cancer correlates with upregulated miR-182 which is promoted by TGF-β signaling. The HPV oncogene E7 triggers TGF-κ signaling, which in turn causes increased Smad4 to attach to the miR-182 promoter and boost its production. This miR-182 in turn promotes further uncontrolled cell proliferation by targeting and suppressing *FOXO1* [11, 12]. In ovarian cancer cells, overexpressed miR-183 has a direct target to *SMAD4* which promotes the induction of TGF-β signaling leading to uncontrolled cell proliferation, migration, and invasion [120]. Downregulation of gremlin-1 (GREM1) shows downregulation of TGF-β1, Smad2, and Smad3 which result in suppression of EMT, cell proliferation, and invasion (Fig. 17.2). Overexpression of miR-137 shows decreased expression of GREM1 and inactivation of the Smad pathway that blocks EMT and decreases tumor cell invasiveness in cervical cancer [63]. Another data reveals that miR-362 acts as a tumor suppressor which targets B-cell receptor-associated protein 31 (BAP31), a protein involved in intracellular signal transduction and an oncogene in cervical cancer. Decreasing level of *BAP31* reduces interaction with protein Spectrin beta, non-erythrocytic 1 (SPTBN1) and leads to the binding of *SPTBN1* with SMAD3, thus activating the TGF-β/Smad signaling pathway (Table 17.1) [108]. Cytidine monophosphate kinase (CMPK) plays an important role in cell proliferation by controlling nucleic acid biosynthesis procedures. A class of onco-miR-130b-3p is found to be upregulated in the CMPK expression via the TGF-βsignaling pathway in ovarian cancer cells [122].

MicroRNA-Mediated Alteration in Autophagy and Apoptotic Cell Death

Autophagy plays a dual and complex role in cancer, depending on the stage of tumor development. It can limit the accumulation of defective organelles and genomic instability by inhibiting tumor initiation. For example, autophagy-related genes like Beclin1 and Bif-1 are tumor suppressors, and their deficiency contributes to increased tumorigenesis [69, 82]. MicroRNAs control these autophagic processes and possess the ability to influence this balance between autophagy's tumor-suppressing and tumor-promoting roles. In ovarian cancer, autophagy-related

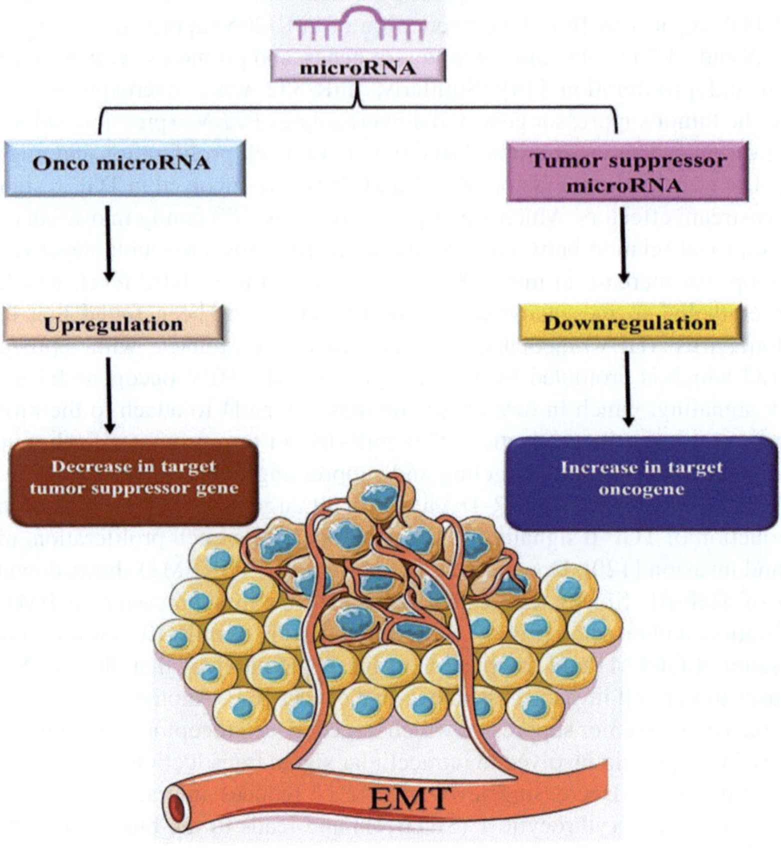

Fig. 17.2 MicroRNA found to cause cancers of two types—onco microRNA and tumor suppressor microRNA, depending upon their mode of action. Onco microRNAs are generally upregulated in cancers which cause the targeted repression of tumor suppressor mRNA. Tumor suppressor microRNAs are generally found to be downregulated in cancers, and as a result, the targeted suppression is withdrawn from onco-mRNAs

miRNAs play a crucial role by modulating autophagic processes which influence tumor development, progression, and their response to therapy. These small noncoding RNAs can either activate or inhibit autophagy, affecting cancer cell survival, apoptosis, chemoresistance, and metastasis. For example, under normal conditions, miR-342-3p acts as a tumor suppressor in ovarian cancer which targets F-box protein 11 (*FBX11*), an oncogene responsible for cancer progression which promotes ubiquitination and degradation of autophagy regulator Beclin-1. It binds to FBXL11 mRNA and downregulates it expression which facilitates autophagy. Thus, downregulation of miR-342-3p increases *FBXL11* expression and autophagy suppression leading to ovarian cancer progression [79]. TUG1 the long noncoding RNA confers paclitaxel resistance in ovarian cancer cells with the help of miR-29b-3p and autophagy. Autophagy-associated pathways are often suppressed by miR-29b-3p,

Table 17.1 Key microRNAs in various gynecological cancers and their targets

MicroRNA	Property	Targets	Type of cancer	Reference
miR-503	Promotes cell cycle progression	Cyclin D1	Endometrial	[104]
miR-152	Promotes cell proliferation	CDC25B	Endometrial	[103]
miR-302	Promotes tumor progression	Cyclin B and CDK1	Endometrial	[105]
miR-107	Promotes cell cycle progression	Cyclin E1	Ovarian	[85]
miR-520h, miR-183	Promote tumor growth and metastasis	SMAD 7 SMAD 4	Ovarian	[115–117] [123]
miR-200	Induces metastasis	ZEB1, ZEB 2	Ovarian	[7]
miR-182	Promotes uncontrolled cell proliferation	FOXO1	Cervical	[11, 12]
miR-130b-3p	Promotes uncontrolled cell proliferation	CMPK (cytidine monophosphate kinase)	Ovarian	[122]
miR-137	Inhibits cell migration and thus metastasis	GREM1	Cervical	[63]
miR-362	Induces apoptotic cell death	BAP31	Cervical	[108]
miRNA-342-3p	Induces autophagy suppression leading to cancer progression	FBX11	Ovarian	[79]
miR-150, 146a	Promotes metatstasis	Not identified	Ovarian	[92]
miR-182	Promotes metastasis and increases angiogenesis	MTSS1, HMGA2	Ovarian	[54, 55] [61]
miR-27a	Promotes tumor progression and angiogenesis	VEGF, Survivin, COX2	Ovarian	[45]
miR-205	Promotes cancer cell growth and proliferation	PTEN, SMAD4	Ovarian	[14]
miR-181a	Promotes tumor progression and metastasis	SMAD7	Ovarian	[68]
miR-99a-5p	Promotes ovarian cancer progression	FN1, VTN	Ovarian	[111]
miR-204-5p	Increases angiogenesis	THBS1	Ovarian	[11, 12]
miR-552	Increases cell proliferation and migration	PTEN	Ovarian	[120]
miR-20a	Promotes cancer progression and metastasis	ATG7, TIMP2	Cervical	[118]
miR-378	Promotes metastasis	ATG12	Cervical	[83]
miR-21	Promotes cell proliferation, migration, and metastasis	TIMP3	Cervical	[115–117]
miR-106a	Promotes cell migration and invasion	TIMP2, MMPs	Cervical	[52]
miR-150	Promotes cell proliferation and migration	PDCD4	Cervical	[115–117]

(continued)

Table 17.1 (continued)

MicroRNA	Property	Targets	Type of cancer	Reference
miR-152	Promotes cell proliferation, migration, and invasion	FoxF2	Endometrial	[109]
miR-30c	Promotes cell proliferation and migration	HMGA2	Vulvar	[1]

while TUG1 functions as a competitive endogenous RNA "sponging" miR-29b-3p. When TUG1 is overexpressed, it binds to miR-29b-3p and activates autophagy which allows cancer cell survival and increased resistance to paclitaxel [31]. Studies suggest that miR-20a promotes cervical cancer progression as well as metastasis by targeting autophagy-related protein 7 (*ATG7*) and tissue inhibitors of metalloproteinase 2 (*TIMP2*) [118]. In cervical cancer tissues, miR-378 expression is upregulated as compared to normal tissues. Autophagy-related protein 12 (*ATG12*) is a direct target of miR-378 which facilitates cervical cancer metastasis by downregulating *ATG12* expression [83].

In cancer cells, apoptotic or programmed cell death allows distorted cells to survive and proliferate uncontrollably. Studies showed miRNAs regulate apoptosis by targeting specific mRNAs which encode proteins associated with apoptotic pathways [42]. Generally, apoptosis occurs through two different pathways—intrinsic and extrinsic. Intrinsic (mitochondrial) and extrinsic (death receptor) pathways are involved with various proteins such as caspases, Bcl-2 family, tumor suppressor *p53*. MiRNA target oncogenes or anti-apoptotic proteins that promote apoptosis [59]. Overexpression of miR-886-5p shows a decrease in pro-apoptotic Bax level, reduced rate in apoptosis, and increased rate in cell proliferation [50]. Meanwhile, miR-143 acts as a tumor suppressor by binding to 3′UTR of *Bcl-2* and decreases *Bcl-2* expression level at both mRNA and protein levels. In consequence of this interaction, anti-apoptotic *Bcl-2* is reduced, which encourages apoptosis and prevents cervical cell line multiplication [54, 55]. In ovarian carcinoma, elevated miR-182 directly suppresses the expression of a tumor suppressor gene, programmed cell death 4 (*PDCD4*). This leads to the onset of multiple malignant properties including accumulation of faulty cells [92]. There are other miRNAs as well which are associated with cervical cancer like miR-150, which contributes to cancer cell proliferation and migration by downregulating the programmed cell death protein 4 (*PDCD4*) [115–117].

MicroRNA Promoted Metastasis and Angiogenesis

Angiogenesis plays a significant role in gynecological cancers because it promotes tumor growth, invasion, and metastasis. For cancer progression, tumors require a continuous supply of oxygen and nutrients from new blood vessels. Thus, as the disease progresses, gynecological malignancies such as ovarian, cervical, and endometrial tumors have increasing angiogenic activity. This entails the production of proangiogenic substances, such as vascular endothelial growth factor (VEGF), which encourages the development of new vessels and the proliferation of endothelial cells. Therefore, higher tumor stage, larger tumor size, and a worse prognosis are associated with increased levels of these variables. Moreover, angiogenesis allows tumor cells to invade surrounding tissues and metastasize to distant sites, making it a vital component of cancer progression [70].Varieties of miRNAs are differentially expressed in ovarian cancers and can target a repertoire of genes involved in cancer metastasis. In late-stage ovarian cancer, miR-22 expression is downregulated, showing an inhibitory effect in metastasis [49]. Loss of function mutation or suppression in the *BRCA1/BRCA2* gene is the hallmark of high-grade serous ovarian cancer. A tumorigenic role of miR-182 is determined in the pathogenesis of high-grade serous ovarian carcinoma by repressing the BRCA1 gene expression along with the potential repression of tumor suppressor gene FOXO3 and metastasis suppressor gene MTSS1. Overexpression of miR182 causes repression of BRCA1 expression, leading to the upregulation of oncogenic high mobility group A2 (HMGA2), a causative gene in high-grade serous ovarian carcinoma progression [54, 55, 61]. In cervical cancer patients, a decreased level of miR-497 is noticed which could control the expression of insulin-like growth factor 1 (IGf 1). Downregulation of this miRNA results in the occurrence of lymph node metastasis and uncontrolled cell migration in patients with cervical cancer [58]. RAS p21 protein activator 1 (RASA1) is acting as a controller of RAS/MAPK pathway. Mutation or loss of function of RASA1 promotes uncontrolled cell differentiation due to accumulation of phosphorylated Akt and Erk. Mutation or loss of function of RASA1 promotes uncontrolled cell differentiation due to accumulation of phosphorylated Akt and Erk. Akt and Erk are found in the downstream of the RAS signaling pathway. In cervical cancer patients, miR-21 upregulation results in cancer cell invasion through lymph node metastasis by reducing the RASA1 expression [113]. ELK1 is a member of the oncogene family and a target of the MAPK pathway contributes to cancer cell proliferation and abrupt cytoskeletal rearrangements has been identified as a target of miR-326 in cervical cancer. A lower miR-326 expression leads to cervical cancer cell migration and invasion, where ELK1 is upregulated [114]. The oncogene SOX4 is upregulated in endometrial cancer. The epigenetic methylation has silenced the expression of miR-129-2, working on the upstream of SOX4 [34]. FoxF2 has been shown to have tumor suppressor properties in endometrial cancer as it attenuates the epithelial-mesenchymal transition (EMT), thus inhibits cell invasion. Through miR-182 target binding, FoxF2 expression is downregulated in endometrial cancer, leading to unchecked cell proliferation and metastasis [109]. In ovarian epithelial cancer cells, miR-27a facilitates tumor development as well as angiogenesis. Overexpression of miR-27a upregulates proangiogenic and survival factors like VEGF, Survivin, and COX2. Apart from this, it also controls the expression of Sp1, ZBTB10 transcription factors that are involved in regulating genes associated with angiogenesis and tumor progression [45]. MiR-99a-5p affects the tumor microenvironment and accelerates the growth of ovarian cancer in serous ovarian cancer cells. It increases the expression of extracellular matrix proteins linked to cell invasion and metastasis, such as vitronectin and fibronectin [111]. MiR-204-5p regulates the expression of anti-angiogenic gene THBS1 which usually inhibits angiogenesis under normal conditions. Upregulation of miR-204-5p suppresses the expression of thrombospondin (THBS1), leading to increased angiogenesis [11, 12]. Another microRNA, miR-21, is upregulated in cervical cancer; it suppresses the activity of tissue inhibitor of metalloproteinase 3 (TIMP3) and promotes cell proliferation, migration, and metastasis [115–117]. Furthermore, by upregulating matrix metalloproteinases (MMPs) and downregulating tissue inhibitors of metalloproteinase 2 (TIMP2), miR-106a contributes to the cervical cancer cell migration and invasion [52]. High mobility group AT hook 1 (HMGA1) plays the role of an oncogene in cervical cancer as it upregulates miR-221/222. Here, the direct target of these miRNA is TIMP3, i.e., tissue inhibitor of metalloproteinase and thus expression of matrix metalloproteinase 2/9 (MMP2/MMP9) level is increased. Increased level of MMP2/MMP9 promotes cancer cell metastatic progression [30]. Higher expression of HMGA2 has been linked to vulva SCC, which promotes epithelial-mesenchymal transition (EMT) and cancer progression by favorably influencing the TGF-β pathway and transcription factors involved in EMT. Many microRNAs can target the 3′ UTR of HMGA2. It has been discovered that two microRNAs, miR-30c and let-7a, are downregulated in vulvar SCC and result in HMGA2 overexpression [1].

Therapeutic Resistance

In 1950, therapeutic resistance phenomenon was first documented in mouse leukemic cells that acquired resistance to the chemotherapeutic drug methotrexate.Therapeutic resistance has been claimed as one of the main causes of treatment failure. Several genomic and proteomic research have been carried out to bring some novel insights of therapeutic resistance [60]. The role of microRNAs (miRNAs) in contributing to therapeutic resistance in gynecological cancers is increasingly acknowledged, especially regarding their regulation of essential signaling pathways and cellular processes. Recent findings highlight the significance of miRNA-mediated regulation of autophagy in ovarian cancer, where altered autophagic flux enhances survival under chemotherapeutic stress, leading to treatment failure [22]. The interaction between N6-methyladenosine (m6A) RNA modifications and noncoding RNAs, such as miRNAs, contributes to the complexity of chemoresistance by affecting RNA stability, translation, and cellular stress responses [26]. Extracellular vesicles serve as vehicles for the transfer of miRNA between cells, facilitating the horizontal dissemination of resistance traits within the tumor microenvironment, particularly in cervical cancer [91]. There are number of literature showing the possible correlations of microRNA profile with the onset of therapeutic resistance, and the microRNAs are mostly exosomal derived from the surrounding tissues to the tumor microenvironment (TME). It has been found that after treatment with paclitaxel, miR-433 is overexpressed in the neighboring cells which further cause cellular senescence by inactivating the Rb and p53 targeted by cell-dependent kinases (CDKs). Senescent cells secrete pro-inflammatory cytokines into the tumor microenvironment in a paracrine manner which would eventually promote genetic instability and further cancer progression [96]. The hypoxic nature of TME contributes malignancy and drug resistance significantly. The hypoxic environment promotes macrophage infiltration and polarization into M2 macrophages and receives the exosomes from the surrounding donor tumor-associated macrophages (TAM) [97]. These M2 macrophages promote tumor progression and chemoresistance. An M2-derived miR-21 has a probable target to PI3k-Akt and itself could influence the population of M2 macrophages and further chemokine profiles, ultimately leading to chemoresistance [3]. Upregulation of miR-21 can also target the tumor suppressor programmed cell death 4 (PDCD4) in ovarian cancer and thus promote cisplatin resistance [23]. These exosome-derived microRNAs control the key regulators of cell proliferation pathways or cease the apoptotic pathway or autophagy and eventually drug resistance is noticed. Such a TAM-derived miR-223 is upregulated in paclitaxel-resistant epithelial ovarian carcinoma which targeted the tumor suppressor PTEN. This reduced activity of PTEN leads to activation of the PI3K/Akt signaling pathway that promotes cell survival, enabling the cancer cells to tolerate the drug sensitivity [124]. The miRNA-PTEN axis is significant as it modulates the PI3K/AKT pathway, which is known to contribute to resistance against platinum-based therapies and targeted agents in various malignancies [100]. Ovarian cancer cells resistant to cisplatin show an upregulation of miR-221/222 which also target 3′UTR of PTEN mRNA and confer survival capacity that can withstand the cytotoxic effects of cisplatin, thereby promoting chemoresistance [2]. Resistance in radiotherapy is also conferred by mi-214 in ovarian cancer by targeting PTEN. Cell survival gives the tolerance to the damaging effects of radiotherapy [114]. Multidrug resistance (MDR) is the ability of cancer cells to show resistance against different chemotherapeutic drugs. Generally, MDR is associated with overexpression of P-glycoprotein which is encoded by the MDR1 gene. In ovarian cancer cells, miR-130b expression is downregulated due to epigenetic silencing, thus leading to higher colony-stimulating factor 1 (CSF-1) expression and results in multidrug resistance like cisplatin and paclitaxel [106]. Copper transporter protein (CTR1) acts as a key transporter for cisplatin uptake into the cancer cells. MiR-130a shows cisplatin resistance in cervical cancer (CC) by targeting and inhibiting CTR1 as it directly binds to the 3′UTR of SLC31A1 (gene encoding CTR1), which leads to a decrease in the CTR1 protein level. A decrease in the CTR1 expression level also reduced cisplatin intracellular accumulation, which diminished the drug's cytotoxic effects and promoted chemoresistance [27]. Solute carrier family 40 member 1 (SLC40A1) acts as an iron transmembrane channel which is targeted by miR-194-5p in ovarian cancer. When miRNA-194-5p is upregulated, it causes a decrease in SLC40A1 levels, which promotes resistance to Cisplatin [99]. In malignant cells, an increased expression of a transmembrane glycoprotein CD44v6 appears which promotes the expression of a multidrug-resistant P-glycoprotein (P-gp). Elevation in miR-21 in ovarian cancer cells mediates the interaction between CD44v6 and P-gp, which leads to drug efflux and subsequently chemoresistance

[93, 94]. Under normal conditions, O-linked N-acetylglucosamine transferase (OGT) promotes KEAP1 (Kelch-like ECH-associated protein (1) glycosylation that facilitates ubiquitination and degradation of a transcription factor NRF2 (nuclear factor erythroid 2-related factor (2) associated with antioxidant response and chemoresistance. Overexpression of miRNA-181d targets and negatively regulates O-Linked N-acetylglucosamine transferase (OGT), thereby contributing to cisplatin resistance in ovarian cancer cells. This decrease in OGT levels reduces glycosylation of KEAP1, impairing its function. Therefore, NRF2 gets accumulated and its elevated levels facilitate the cell antioxidant defense, contributing to resistance to the oxidative stress induced by cisplatin and thereby, influencing tumor cell survival in spite of chemotherapy [35]. When miRNA-210 is overexpressed, it decreases the pro-apoptotic protein Bax and increases the anti-apoptotic protein Bcl-2. This causes a decrease in the apoptosis induced by radiation, reducing the sensitivity of the cancer cells to radiotherapy. In addition, miRNA-210 also decreases the activity of caspase 3/7, enzymes that are crucial for executing apoptosis thereby impairing the functionality of the process in response to radiation. These collectively enhance ovarian cancer cell survival and resistance against radiotherapy [121]. Two independent experiments show that an adaptor molecule of the apoptosis pathway, Apaf1, is inhibited by miR-27a and miR-21 and, in turn, promotes paclitaxel resistance in ovarian cancer cells. During hypoxic conditions, miRNA-27a expression is upregulated due to the exosomal transfer from the surrounding stromal cells. Overexpression of miRNA-27a suppresses APAF1 expression, reducing apoptosis conferred by paclitaxel, allowing cancer cells to survive despite chemotherapy [4, 28]. In cervical SCC, miR-181a induces cisplatin resistance by suppressing the expression of caspase 3 substrate protein kinase c δ (PRKCD). Inhibition of PRCKD reverts apoptosis, which in turn decreases apoptosis, allowing cancer cells to survive despite cisplatin action, leading to an increase in chemoresistance in cervical carcinoma cells [10]. Another one, miR-499a, targets SRY-Box Transcription Factor 6 (SOX6) and provides cisplatin resistance in cervical cancer. It binds to 3′UTR and decreases SOX6 expression, resulting in cancer cell migration, proliferation, invasion, and resistance to cisplatin. This shows chemoresistance by promoting cell survival and invasion [13]. Studies have shown that miR-551b promotes drug resistance in ovarian cancer by suppressing FOXO1 and TRIM31; both are tumor suppressor genes. This suppression leads to cisplatin resistance. It binds to 3′UTR of FOXO3 and TRIM31 transcripts that inhibit their expression and increase chemoresistance [95]. In further studies, miR-770-5p developed cisplatin resistance through excision repair cross-complementation group 2 (ERCC2) interaction. It directly binds to 3′UTR that leads to the degradation of mRNA and also reduces the ERCC2 level. Binding inhibits the translation and promotes its degradation, therefore influencing the level of ERCC2 in the cells [119]. MicroRNAs (miRNAs) in cervical cancer influence drug response by targeting genes associated with apoptosis, DNA repair, and immune evasion, positioning them as promising candidates for diagnostic and therapeutic strategies [24]. The insights collectively emphasize the critical function of miRNAs in mediating therapeutic resistance, indicating their potential as biomarkers sand modulators in addressing drug insensitivitys in gynecological cancers (Table 17.2).

Table 17.2 Some microRNAs and their respective target pathways associated to therapeutic resistance for quick reference

Name	Type of cancer	Effect	Target pathway	Resistance to treatment	References
miR-221/222	Ovarian	Downregulated	P13/AKT	Cisplatin	[2]
miR-214	Ovarian	Upregulated	PI3K/Akt	Radiotherapy	[114]
miR-130b	Ovarian	Downregulated	Colony stimulating factor 1	Cisplatin and paclitaxel	[106]
miR-130a	Cervical	Downregulated	CTR1	Cisplatin	[27]
miR-194-5p	Ovarian	Upregulated	SLC40A1	Cisplatin	[99]
miR 21	Ovarian	Upregulated	CD44v6/P-gp	Cisplatin	[93, 94]
miR-181d	Ovarian	Upregulated	O-linked N-acetylglucosamine transferase	Cisplatin	[35]
miR-210	Ovarian	Upregulated	Bcl-2	Radiotherapy	[121]
miR-27a	Ovarian	Upregulated	HIF-1α	Paclitaxel	[28]
miR-181a	Cervical	Downregulated	PRKCD	Cisplatin	[10]
miR-499a	Cervical	Upregulated	SOX6	Cisplatin	[13]
miR-551b	Ovarian	Upregulated	FOXO3 and TRIM31	Cisplatin	[95]
miR-770-5p	Ovarian	Upregulated	ERCC2	Cisplatin	[119]

Conclusion

In summary, microRNAs play an important role as regulators of gene expression. Its tiny yet powerful effects are observed when it binds to messenger RNAs and subsequently influences cellular processes. The binding site of miRNA is 3′UTR of target messenger RNAs. This binding initiates a cascade of events which leads to suppression of gene expression. MicroRNAs can be categorized as onco-miRNAs and tumor suppressor-miRNAs based upon their target mRNAs and expression level which results in the ensuing effects. Onco-miRNAs are microRNAs that are found to be increased under cancerous conditions and target one or many tumor suppressor genes. On the other hand, miRNAs, classified as tumor suppressor mi-RNAs, are observed to be downregulated in specific cancerous conditions where their probable target would be the oncogenes. Therefore, mi-RNAs influence gynecological cancers by influencing important pathways such as TGF-β, PI3/Akt-MAPK signaling, autophagy, apoptosis, angiogenesis, and metastasis by targeting essential genes. Additionally, mi-RNAs play a significant role in the phenomenon of therapeutic resistance, which has complex treatment outcomes. Recognizing the crucial functions of miRNA-mediated pathways not only increases our understanding of cancer pathways but also highlights promising targets for overcoming resistance and therapies. It is important to continue unravelling these mechanisms to pave the way for more productive interventions against gynecological cancers.

Acknowledgments We are thankful to the Department of Life Sciences, Presidency University, for the lab space. We acknowledge smart servier website for the art provided in the figure.

References

1. Agostini A, Brunetti M, Davidson B, Trope CG, Heim S, Panagopoulos I, Micci F (2016) Expressions of miR-30c and let-7a are inversely correlated with HMGA2 expression in squamous cell carcinoma of the vulva. Oncotarget 7(51):85058
2. Amini-Farsani Z, Sangtarash MH, Shamsara M, Teimori H (2018) MiR-221/222 promote chemoresistance to cisplatin in ovarian cancer cells by targeting PTEN/PI3K/AKT signaling pathway. Cytotechnology 70:203–213
3. An Y, Yang Q (2020) MiR-21 modulates the polarization of macrophages and increases the effects of M2 macrophages on promoting the chemoresistance of ovarian cancer. Life Sci 242:117162
4. Au Yeung CL, Co NN, Tsuruga T, Yeung TL, Kwan SY, Leung CS, Li Y, Lu ES, Kwan K, Wong KK, Schmandt R (2016) Exosomal transfer of stroma-derived miR21 confers paclitaxel resistance in ovarian cancer cells through targeting APAF1. Nat Commun 7(1):11150
5. Banerjee S, Kaye SB (2013) New strategies in the treatment of ovarian cancer: current clinical perspectives and future potential. Clin Cancer Res 19(5):961–968
6. Bartel DP (2004) MicroRNAs: genomics, biogenesis, mechanism, and function. Cell 116(2):281–297
7. Bendoraite A, Knouf EC, Garg KS, Parkin RK, Kroh EM, O'Briant KC, Ventura AP, Godwin AK, Karlan BY, Drescher CW, Urban N (2010) Regulation of miR-200 family microRNAs and ZEB transcription factors in ovarian cancer: evidence supporting a mesothelial-to-epithelial transition. Gynecol Oncol 116(1):117–125
8. Calin GA, Dumitru CD, Shimizu M, Bichi R, Zupo S, Noch E, Aldler H, Rattan S, Keating M, Rai K, Rassenti L (2002) Frequent deletions and down-regulation of micro-RNA genes miR15 and miR16 at 13q14 in chronic lymphocytic leukemia. Proc Natl Acad Sci 99(24):15524–15529
9. Chen VW, Ruiz B, Killeen JL, Coté TR, Wu XC, Correa CN, Howe HL (2003) Pathology and classification of ovarian tumors. Cancer 97(S10):2631–2642
10. Chen Y, Ke G, Han D, Liang S, Yang G, Wu X (2014) MicroRNA-181a enhances the chemoresistance of human cervical squamous cell carcinoma to cisplatin by targeting PRKCD. Exp Cell Res 320(1):12–20
11. Chen J, Deng Y, Ao L, Song Y, Xu Y, Wang CC, Choy KW, Chung KHT, Du Q, Sui Y, Yang T (2019a) The high-risk HPV oncogene E7 upregulates miR-182 expression through the TGF-β/Smad pathway in cervical cancer. Cancer Lett 460:75–85
12. Chen X, Mangala LS, Mooberry L, Bayraktar E, Dasari SK, Ma S, Ivan C, Court KA, Rodriguez-Aguayo C, Bayraktar R, Raut S (2019b) Identifying and targeting angiogenesis-related microRNAs in ovarian cancer. Oncogene 38(33):6095–6108
13. Chen Y, Song Y, Mi Y, Jin H, Cao J, Li H, Han L, Huang T, Zhang X, Ren S, Ma Q (2020) microRNA-499a promotes the progression and chemoresistance of cervical cancer cells by targeting SOX6. Apoptosis 25:205–216
14. Chu P, Liang A, Jiang A, Zong L (2018) miR-205 regulates the proliferation and invasion of ovarian cancer cells via suppressing PTEN/SMAD4 expression. Oncol Lett 15(5):7571–7578

15. Cohen PA, Jhingran A, Oaknin A, Denny L (2019) Cervical cancer. Lancet 393(10167):169–182
16. Creasman WT, Phillips JL, Menck HR (1997) The National Cancer Data Base report on early stage invasive vulvar carcinoma. Cancer 80(3):505–513
17. Crum CP (1992) Carcinoma of the vulva: epidemiology and pathogenesis. Obstet Gynecol 79(3):448–454
18. Daniyal M, Akhtar N, Ahmad S, Fatima U, Akram M, Asif HM (2015) Update knowledge on cervical cancer incidence and prevalence in Asia. Asian Pac J Cancer Prev 16(9):3617–3620
19. Davis-Dusenbery BN, Hata A (2010) Mechanisms of control of microRNA biogenesis. J Biochem 148(4):381–392
20. DeBerardinis RJ, Lum JJ, Hatzivassiliou G, Thompson CB (2008) The biology of cancer: metabolic reprogramming fuels cell growth and proliferation. Cell Metab 7(1):11–20
21. Dhillon AS, Hagan S, Rath O, Kolch W (2007) MAP kinase signalling pathways in cancer. Oncogene 26(22):3279–3290
22. Ding Y, Huang X, Ji T, Qi C, Gao X, Wei R (2024) The emerging roles of miRNA-mediated autophagy in ovarian cancer. Cell Death Dis 15(5):314
23. Echevarría-Vargas IM, Valiyeva F, Vivas-Mejía PE (2014) Upregulation of miR-21 in cisplatin resistant ovarian cancer via JNK-1/c-Jun pathway. PLoS One 9(5):e97094
24. Endale HT, Mariye YF, Negash HK, Hassen FS, Asrat WB, Mengstie TA, Tesfaye W (2024) MiRNA in cervical cancer: diagnosis to therapy: systematic review. Heliyon 10(3)
25. Fabian MR, Sonenberg N (2012) The mechanics of miRNA-mediated gene silencing: a look under the hood of miRISC. Nat Struct Mol Biol 19(6):586–593
26. Fan D, Shang Y, Cong Y, Jiao Y, Li N, Zhao H (2025) Reciprocal regulation between m6 A modifications and non-coding RNAs: emerging roles in cancer therapeutic resistance. Discov Oncol 16(1):1–26
27. Feng C, Ma F, Hu C, Ma JA, Wang J, Zhang Y, Wu F, Hou T, Jiang S, Wang Y, Feng Y (2018) SOX9/miR-130a/CTR1 axis modulates DDP-resistance of cervical cancer cell. Cell Cycle 17(4):448–458
28. Feng L, Shen F, Zhou J, Li Y, Jiang R, Chen Y (2020) Hypoxia-induced up-regulation of miR-27a promotes paclitaxel resistance in ovarian cancer. Biosci Rep 40(4):BSR20192457
29. Franke TF, Hornik CP, Segev L, Shostak GA, Sugimoto C (2003) PI3K/Akt and apoptosis: size matters. Oncogene 22(56):8983–8998
30. Fu F, Wang T, Wu Z, Feng Y, Wang W, Zhou S, Ma X, Wang S (2018) HMGA1 exacerbates tumor growth through regulating the cell cycle and accelerates migration/invasion via targeting miR-221/222 in cervical cancer. Cell Death Dis 9(6):594
31. Gu L, Li Q, Liu H, Lu X, Zhu M (2020) Long noncoding RNA TUG1 promotes autophagy-associated paclitaxel resistance by sponging miR-29b-3p in ovarian cancer cells. Onco Targets Ther:2007–2019
32. Harbour JW, Dean DC (2000) The Rb/E2F pathway: expanding roles and emerging paradigms. Genes Dev 14(19):2393–2409
33. Hou T, Ou J, Zhao X, Huang X, Huang Y, Zhang Y (2014) MicroRNA-196a promotes cervical cancer proliferation through the regulation of FOXO1 and p27Kip1. Br J Cancer 110(5):1260–1268
34. Huang YW, Liu JC, Deatherage DE, Luo J, Mutch DG, Goodfellow PJ, Miller DS, Huang TH (2009) Epigenetic repression of microRNA-129-2 leads to overexpression of SOX4 oncogene in endometrial cancer. Cancer Res 69(23):9038–9046
35. Huang W, Chen L, Zhu K, Wang D (2021) Oncogenic microRNA-181d binding to OGT contributes to resistance of ovarian cancer cells to cisplatin. Cell Death Discov 7(1):379
36. Hunter DJS (1975) Carcinoma of the vulva: a review of 361 patients. Gynecol Oncol 3(2):117–123
37. Ichimura A, Ruike Y, Terasawa K, Tsujimoto G (2011) miRNAs and regulation of cell signaling. FEBS J 278(10):1610–1618

38. Iorio MV, Croce CM (2012) microRNA involvement in human cancer. Carcinogenesis 33(6):1126–1133
39. Jayson GC, Kohn EC, Kitchener HC, Ledermann JA (2014) Ovarian cancer. Lancet 384(9951):1376–1388
40. Jiang F, Liu T, He Y, Yan Q, Chen X, Wang H, Wan X (2011) MiR-125b promotes proliferation and migration of type II endometrial carcinoma cells through targeting TP53INP1 tumor suppressor in vitro and in vivo. BMC Cancer 11:1–12
41. Jin C, Liang R (2015) miR-205 promotes epithelial–mesenchymal transition by targeting AKT signaling in endometrial cancer cells. J Obstet Gynaecol Res 41(10):1653–1660
42. Kanekura K, Nishi H, Isaka K, Kuroda M (2016) MicroRNA and gynecologic cancers. J Obstet Gynaecol Res 42(6):612–617
43. Karaayvaz M, Zhang C, Liang S, Shroyer KR, Ju J (2012) Prognostic significance of miR-205 in endometrial cancer. PLoS One 7(4):e35158
44. Lagos-Quintana M, Rauhut R, Lendeckel W, Tuschl T (2001) Identification of novel genes coding for small expressed RNAs. Science 294(5543):853–858
45. Lai Y, Zhang X, Zhang Z, Shu Y, Luo X, Yang Y, Wang X, Yang G, Li L, Feng Y (2013) The microRNA-27a: ZBTB10-specificity protein pathway is involved in follicle stimulating hormone-induced VEGF, Cox2 and survivin expression in ovarian epithelial cancer cells. Int J Oncol 42(2):776–784
46. Lau NC, Lim LP, Weinstein EG, Bartel DP (2001) An abundant class of tiny RNAs with probable regulatory roles in Caenorhabditiselegans. Science 294(5543):858–862
47. Lee RC, Ambros V (2001) An extensive class of small RNAs in Caenorhabditiselegans. Science 294(5543):862–864
48. Lee RC, Feinbaum RL, Ambros V (1993) The C. elegansheterochronic gene lin-4 encodes small RNAs with antisense complementarity to lin-14. Cell 75(5):843–854
49. Li J, Liang S, Yu H, Zhang J, Ma D, Lu X (2010) An inhibitory effect of miR-22 on cell migration and invasion in ovarian cancer. Gynecol Oncol 119(3):543–548
50. Li JH, Xiao X, Zhang YN, Wang YM, Feng LM, Wu YM, Zhang YX (2011) MicroRNA miR-886-5p inhibits apoptosis by down-regulating Bax expression in human cervical carcinoma cells. Gynecol Oncol 120(1):145–151
51. Li Q, Qiu XM, Li QH, Wang XY, Li L, Xu M, Dong M, Xiao YB (2015) MicroRNA-424 may function as a tumor suppressor in endometrial carcinoma cells by targeting E2F7. Oncol Rep 33(5):2354–2360
52. Li X, Zhou Q, Tao L, Yu C (2017) MicroRNA-106a promotes cell migration and invasion by targeting tissue inhibitor of matrix metalloproteinase 2 in cervical cancer. Oncol Rep 38(3):1774–1782
53. Lin S, Gregory RI (2015) MicroRNA biogenesis pathways in cancer. Nat Rev Cancer 15(6):321–333
54. Liu L, Yu X, Guo X, Tian Z, Su M, Long Y, Huang C, Zhou F, Liu M, Wu X, Wang X (2012a) miR-143 is downregulated in cervical cancer and promotes apoptosis and inhibits tumor formation by targeting Bcl-2. Mol Med Rep 5(3):753–760
55. Liu Z, Liu J, Segura MF, Shao C, Lee P, Gong Y, Hernando E, Wei JJ (2012) MiR-182 overexpression in tumourigenesis of high-grade serous ovarian carcinoma. J Pathol 228(2):204–215
56. Liu J, Chen W, Zhang H, Liu T, Zhao L (2017) miR-214 targets the PTEN-mediated PI3K/Akt signalling pathway and regulates cell proliferation and apoptosis in ovarian cancer. Oncol Lett 14(5):5711–5718
57. Liu HY, Zhang YY, Zhu BL, Feng FZ, Yan H, Zhang HY, Zhou B (2019) miR-21 regulates the proliferation and apoptosis of ovarian cancer cells through PTEN/PI3K/AKT. Eur Rev Med Pharmacol Sci 23(10)
58. Luo M, Shen D, Zhou X, Chen X, Wang W (2013) MicroRNA-497 is a potential prognostic marker in human cervical cancer and functions as a tumor suppressor by targeting the insulin-like growth factor 1 receptor. Surgery 153(6):836–847

59. Lynam-Lennon N, Maher SG, Reynolds JV (2009) The roles of microRNA in cancer and apoptosis. Biol Rev 84(1):55–71
60. Ma J, Dong C, Ji C (2010) MicroRNA and drug resistance. Cancer Gene Ther 17(8):523–531
61. McMillen BD, Aponte MM, Liu Z, Helenowski IB, Scholtens DM, Buttin BM, Wei JJ (2012) Expression analysis of MIR182 and its associated target genes in advanced ovarian carcinoma. Mod Pathol 25(12):1644–1653
62. Merritt WM, Lin YG, Han LY, Kamat AA, Spannuth WA, Schmandt R, Urbauer D, Pennacchio LA, Cheng JF, Nick AM, Deavers MT (2008) Dicer, Drosha, and outcomes in patients with ovarian cancer. N Engl J Med 359(25):2641–2650
63. Miao H, Wang N, Shi LX, Wang Z, Song WB (2019) Overexpression of mircoRNA-137 inhibits cervical cancer cell invasion, migration and epithelial–mesenchymal transition by suppressing the TGF-β/smad pathway via binding to GREM1. Cancer Cell Int 19:1–17
64. Mishra S, Yadav T, Rani V (2016) Exploring miRNA based approaches in cancer diagnostics and therapeutics. Crit Rev Oncol Hematol 98:12–23
65. Muralidhar B, Winder D, Murray M, Palmer R, Barbosa-Morais N, Saini H, Roberts I, Pett M, Coleman N (2011) Functional evidence that Droshaoverexpression in cervical squamous cell carcinoma affects cell phenotype and microRNA profiles. J Pathol 224(4):496–507
66. Myatt SS, Wang J, Monteiro LJ, Christian M, Ho KK, Fusi L, Dina RE, Brosens JJ, Ghaem-Maghami S, Lam EW (2010) Definition of microRNAs that repress expression of the tumor suppressor gene FOXO1 in endometrial cancer. Cancer Res 70(1):367–377
67. Pampalakis G, Diamandis EP, Katsaros D, Sotiropoulou G (2010) Down-regulation of dicer expression in ovarian cancer tissues. Clin Biochem 43(3):324–327
68. Parikh A, Lee C, Joseph P, Marchini S, Baccarini A, Kolev V, Romualdi C, Fruscio R, Shah H, Wang F, Mullokandov G (2014) microRNA-181a has a critical role in ovarian cancer progression through the regulation of the epithelial–mesenchymal transition. Nat Commun 5(1):2977
69. Qu X, Yu J, Bhagat G, Furuya N, Hibshoosh H, Troxel A, Rosen J, Eskelinen EL, Mizushima N, Ohsumi Y, Cattoretti G (2003) Promotion of tumorigenesis by heterozygous disruption of the beclin 1 autophagy gene. J Clin Invest 112(12):1809–1820
70. Rasila KK, Burger RA, Smith H, Lee FC, Verschraegen C (2005) Angiogenesis in gynecological oncology – mechanism of tumor progression and therapeutic targets. Int J Gynecol Cancer 15(5):710–726
71. Reid BM, Permuth JB, Sellers TA (2017) Epidemiology of ovarian cancer: a review. Cancer Biol Med 14(1):9–32
72. Rocco JW, Sidransky D (2001) p16 (MTS-1/CDKN2/INK4a) in cancer progression. Exp Cell Res 264(1):42–55
73. SCURRY (1999) Does lichen sclerosus play a central role in the pathogenesis of human papillomavirus negative vulvar squamous cell carcinoma? The itch-scratch-lichen sclerosus hypothesis. Int J Gynecol Cancer 9(2):89–97
74. Shao Y, Liu X, Meng J, Zhang X, Ma Z, Yang G (2019) MicroRNA-1251-5p promotes carcinogenesis and autophagy via targeting the tumor suppressor TBCC in ovarian cancer cells. Mol Ther 27(9):1653–1664
75. Shih IM, Kurman RJ (2004) Ovarian tumorigenesis: a proposed model based on morphological and molecular genetic analysis. Am J Pathol 164(5):1511–1518
76. Siegel R, Naishadham D, Jemal A (2013) Cancer statistics, 2013. CA Cancer J Clin 63(1)
77. Siegel RL, Miller KD, Fuchs HE, Jemal A (2021) Cancer statistics, 2021. CA Cancer J Clin 71(1):7–33
78. Small W Jr, Bacon MA, Bajaj A, Chuang LT, Fisher BJ, Harkenrider MM, Jhingran A, Kitchener HC, Mileshkin LR, Viswanathan AN, Gaffney DK (2017) Cervical cancer: a global health crisis. Cancer 123(13):2404–2412
79. Song W, Zeng Z, Zhang Y, Li H, Cheng H, Wang J, Wu F (2022) CircRNF144B/miR-342-3p/FBXL11 axis reduced autophagy and promoted the progression of ovarian cancer by increasing the ubiquitination of Beclin-1. Cell Death Dis 13(10):857

80. Stewart C, Ralyea C, Lockwood S (2019) Ovarian cancer: an integrated review. Semin Oncol Nurs 35(2):151–156
81. Su Z, Yang H, Zhao M, Wang Y, Deng G, Chen R (2017) MicroRNA-92a promotes cell proliferation in cervical cancer via inhibiting p21 expression and promoting cell cycle progression. Oncol Res 25(1):137
82. Takahashi Y, Coppola D, Matsushita N, Cualing HD, Sun M, Sato Y, Liang C, Jung JU, Cheng JQ, Mul JJ, Pledger WJ (2007) Bif-1 interacts with Beclin 1 through UVRAG and regulates autophagy and tumorigenesis. Nat Cell Biol 9(10):1142–1151
83. Tan D, Zhou C, Han S, Hou X, Kang S, Zhang Y (2018) MicroRNA-378 enhances migration and invasion in cervical cancer by directly targeting autophagy-related protein 12 Corrigendum in/10.3892/mmr. 2024.13366. Mol Med Rep 17(5):6319–6326
84. Tang T, Wong HK, Gu W, Yu MY, To KF, Wang CC, Wong YF, Cheung TH, Chung TKH, Choy KW (2013) MicroRNA-182 plays an onco-miRNA role in cervical cancer. Gynecol Oncol 129(1):199–208
85. Tang Z, Fang Y, Du R (2019) MicroRNA-107 induces cell cycle arrests by directly targeting cyclin E1 in ovarian cancer. Biochem Biophys Res Commun 512(2):331–337
86. Thomson JM, Newman M, Parker JS, Morin-Kensicki EM, Wright T, Hammond SM (2006) Extensive post-transcriptional regulation of microRNAs and its implications for cancer. Genes Dev 20(16):2202–2207
87. Torres A, Torres K, Paszkowski T, Jodłowska-Jędrych B, Radomański T, Książek A, Maciejewski R (2011) Major regulators of microRNAs biogenesis Dicer and Drosha are down-regulated in endometrial cancer. Tumour Biol 32:769–776
88. Tudzarova S, Trotter MW, Wollenschlaeger A, Mulvey C, Godovac-Zimmermann J, Williams GH, Stoeber K (2010) Molecular architecture of the DNA replication origin activation checkpoint. EMBO J 29(19):3381–3394
89. Uzuner E, Ulu GT, Gürler SB, Baran Y (2022) The role of MiRNA in cancer: pathogenesis, diagnosis, and treatment. In: miRNomics: MicroRNA biology and computational analysis. Humana Press, Totowa, pp 375–422
90. Vaksman O, Hetland TE, Trope CG, Reich R, Davidson B (2012) Argonaute, Dicer, and Drosha are up-regulated along tumor progression in serous ovarian carcinoma. Hum Pathol 43(11):2062–2069
91. Van der Merwe M, Myburgh K, Garnis C, Towle R, Engelbrecht AM (2025) Unravelling the role of extracellular vesicles in cervical cancer: mechanisms of progression, resistance, and emerging therapeutic strategies. Gene:149467
92. Wang YQ, Guo RD, Guo RM, Sheng W, Yin LR (2013) MicroRNA-182 promotes cell growth, invasion, and chemoresistance by targeting programmed cell death 4 (PDCD4) in human ovarian carcinomas. J Cell Biochem 114(7):1464–1473
93. Wang A, Xu Q, Sha R, Bao T, Xi X, Guo G (2021a) MicroRNA-29a inhibits cell proliferation and arrests cell cycle by modulating p16 methylation in cervical cancer. Oncol Lett 21(4):272
94. Wang Y, Chen G, Dai F, Zhang L, Yuan M, Yang D, Liu S, Cheng Y (2021b) miR-21 induces chemoresistance in ovarian cancer cells via mediating the expression and interaction of CD44v6 and P-gp. Onco Targets Ther:325–336
95. Wei Z, Liu Y, Wang Y, Zhang Y, Luo Q, Man X, Wei F, Yu X (2016) Downregulation of Foxo3 and TRIM31 by miR-551b in side population promotes cell proliferation, invasion, and drug resistance of ovarian cancer. Med Oncol 33:1–13
96. Weiner-Gorzel K, Dempsey E, Milewska M, McGoldrick A, Toh V, Walsh A, Lindsay S, Gubbins L, Cannon A, Sharpe D, O'Sullivan J (2015) Overexpression of the microRNA miR-433 promotes resistance to paclitaxel through the induction of cellular senescence in ovarian cancer cells. Cancer Med 4(5):745–758
97. Wen Z, Liu H, Li M, Li B, Gao W, Shao Q, Fan B, Zhao F, Wang Q, Xie Q, Yang Y (2015) Increased metabolites of 5-lipoxygenase from hypoxic ovarian cancer cells promote tumor-associated macrophage infiltration. Oncogene 34(10):1241–1252

98. Wightman B, Ha I, Ruvkun G (1993) Posttranscriptional regulation of the heterochronic gene lin-14 by lin-4 mediates temporal pattern formation in C. elegans. Cell 75(5):855–862
99. Wu J, Zhang L, Wu S, Yi X, Liu Z (2020) miR-194-5p inhibits SLC40A1 expression to induce cisplatin resistance in ovarian cancer. Pathol Res Pract 216(7):152979
100. Wu D, Huang C, Guan K (2024) Mechanistic and therapeutic perspectives of miRNA-PTEN signaling axis in cancer therapy resistance. Biochem Pharmacol 116406
101. Xia B, Yang S, Liu T, Lou G (2015) miR-211 suppresses epithelial ovarian cancer proliferation and cell-cycle progression by targeting Cyclin D1 and CDK6. Mol Cancer 14:1–13
102. Xia B, Li H, Yang S, Liu T, Lou G (2016) MiR-381 inhibits epithelial ovarian cancer malignancy via YY1 suppression. Tumour Biol 37:9157–9167
103. Xie D, Liang Y, Su Y, An Y, Qu P (2018) miR-152 inhibits proliferation of human endometrial cancer cells via inducing G2/M phase arrest by suppressing CDC25B expression. Biomed Pharmacother 99:299–305
104. Xu YY, Wu HJ, Ma HD, Xu LP, Huo Y, Yin LR (2013) Micro RNA-503 suppresses proliferation and cell-cycle progression of endometrioid endometrial cancer by negatively regulating cyclin D 1. FEBS J 280(16):3768–3779
105. Yan GJ, Yu F, Wang B, Zhou HJ, Ge QY, Su J, Hu YL, Sun HX, Ding LJ (2014) MicroRNA miR-302 inhibits the tumorigenicity of endometrial cancer cells by suppression of Cyclin D1 and CDK1. Cancer Lett 345(1):39–47
106. Yang C, Cai J, Wang Q, Tang H, Cao J, Wu L, Wang Z (2012) Epigenetic silencing of miR-130b in ovarian cancer promotes the development of multidrug resistance by targeting colony-stimulating factor 1. Gynecol Oncol 124(2):325–334
107. Yang S, Zhao Y, Wang L, Liu C, Lu Y, Fang Z, Shi H, Zhang W, Wu X (2019) MicroRNA-4712-5p promotes proliferation of the vulvar squamous cell carcinoma cell line A431 by targeting PTEN through the AKT/cyclin D1 signaling pathways. Oncol Rep 42(5):1689–1698
108. Yang S, Sun Y, Jiang D, Wang J, Dang E, Li Z, Zhou J, Lu Y, Shi J, Tao L, Wang J (2021) MiR-362 suppresses cervical cancer progression via directly targeting BAP31 and activating TGFβ/Smad pathway. Cancer Med 10(1):305–316
109. Yao H, Kong F, Zhou Y (2019) MiR-182 promotes cell proliferation, migration and invasion by targeting FoxF2 in endometrial carcinoma cells. Int J Clin Exp Pathol 12(4):1248
110. Ye W, Xue J, Zhang Q, Li F, Zhang W, Chen H, Huang Y, Zheng F (2014) MiR-449a functions as a tumor suppressor in endometrial cancer by targeting CDC25A. Oncol Rep 32(3):1193–1199
111. Yoshimura A, Sawada K, Nakamura K, Kinose Y, Nakatsuka E, Kobayashi M, Miyamoto M, Ishida K, Matsumoto Y, Kodama M, Hashimoto K (2018) Exosomal miR-99a-5p is elevated in sera of ovarian cancer patients and promotes cancer cell invasion by increasing fibronectin and vitronectin expression in neighboring peritoneal mesothelial cells. BMC Cancer 18:1–13
112. Zhang Q, Zhang S (2017) miR-214 promotes radioresistance in human ovarian cancer cells by targeting PETN. Biosci Rep 37(4):BSR20170327
113. Zhang L, Zhan X, Yan D, Wang Z (2016) Circulating microRNA-21 is involved in lymph node metastasis in cervical cancer by targeting RASA1. Int J Gynecol Cancer 26(5):810–816
114. Zhang P, Kong F, Deng X, Yu Y, Hou C, Liang T, Zhu L (2017) MicroRNA-326 suppresses the proliferation, migration and invasion of cervical cancer cells by targeting ELK1. Oncol Lett 13(5):2949–2956
115. Zhang J, Liu W, Shen F, Ma X, Liu X, Tian F, Zeng W, Xi X, Lin Y (2018a) The activation of microRNA-520h–associated TGF-β1/c-Myb/Smad7 axis promotes epithelial ovarian cancer progression. Cell Death Dis 9(9):884
116. Zhang Z, Wang J, Li J, Wang X, Song W (2018b) MicroRNA-150 promotes cell proliferation, migration, and invasion of cervical cancer through targeting PDCD4. Biomed Pharmacother 97:511–517
117. Zhang Z, Wang J, Wang X, Song W, Shi Y, Zhang L (2018c) MicroRNA-21 promotes proliferation, migration, and invasion of cervical cancer through targeting TIMP3. Arch Gynecol Obstet 297:433–442

118. Zhao S, Yao D, Chen J, Ding N, Ren F (2015) MiR-20a promotes cervical cancer proliferation and metastasis in vitro and in vivo. PLoS One 10(3):e0120905
119. Zhao H, Yu X, Ding Y, Zhao J, Wang G, Wu X, Jiang J, Peng C, Guo GZ, Cui S (2016) MiR-770-5p inhibits cisplatin chemoresistance in human ovarian cancer by targeting ERCC2. Oncotarget 7(33):53254
120. Zhao W, Han T, Li B, Ma Q, Yang P, Li H (2019) miR-552 promotes ovarian cancer progression by regulating PTEN pathway. J Ovarian Res 12:1–10
121. Zhao Y, Liu S, Wen Y, Zhong L (2021) Effect of microRNA-210 on the growth of ovarian cancer cells and the efficacy of radiotherapy. Gynecol Obstet Investig 86(1–2):71–80
122. Zhou D, Zhang L, Sun W, Guan W, Lin Q, Ren W, Zhang J, Xu G (2017) Cytidine monophosphate kinase is inhibited by the TGF-β signalling pathway through the upregulation of miR-130b-3p in human epithelial ovarian cancer. Cell Signal 35:197–207
123. Zhou J, Zhang C, Zhou B, Jiang D (2019) miR-183 modulated cell proliferation and apoptosis in ovarian cancer through the TGF-β/Smad4 signaling pathway. Int J Mol Med 43(4):1734–1746
124. Zhu X, Shen H, Yin X, Yang M, Wei H, Chen Q, Feng F, Liu Y, Xu W, Li Y (2019) Macrophages derived exosomes deliver miR-223 to epithelial ovarian cancer cells to elicit a chemoresistant phenotype. J Exp Clin Cancer Res 38:1–14

Chapter 18
MicroRNAs Involved in the Regulation of Critical Genes and Their Relevance in the Theragnostics of HPV-Driven Cervical Cancer

Sneha Krishnamoorthy, Vidya Balasubramanian, Jayapradha Gnanagurusamy, Rajalakshmi Sabanayagam, Ilangovan Ramachandran, and Sridhar Muthusami

Abstract Establishment of cervical cancer (CC) involves multiple steps such as binding of human papillomavirus (HPV) late (L) proteins L1 and L2 with key factors such as heparin sulphate proteoglycans (HSPG), epidermal growth factor receptor (EGFR), furin, and annexin A2 on the surface of epithelial cells in the transition zone of cervix. Upon internalization, the HPV early (E) proteins E5, E6, and E7 get integrated into host chromosomes and interfere with the actions of pRb and p53. Bromodomain-containing protein 4 (BRD4) is key in mediating the integration of viral proteins with host chromosomes. When these viral proteins are produced, they are processed into peptides and presented on the surface of infected cells, which invites cytotoxic T cells to ensue apoptosis. HPV downregulates the major histocompatibility complex (MHC)-I expression to minimize its presentation to immune cells and increases the programmed death (PD)-L1 expression, to exhaust the immune cells through PD-1. This chapter is an attempt to understand the microRNA (miRNA) signature of these critical molecules and to explore the possibilities of utilizing these miRNAs as markers in predicting the stages of diseases. The present chapter also explores the therapeutic potential of miRNA mimics and antagonists.

S. Krishnamoorthy · V. Balasubramanian · J. Gnanagurusamy · R. Sabanayagam · S. Muthusami (✉)
Department of Biochemistry, Karpagam Academy of Higher Education, Coimbatore, Tamil Nadu, India

Centre for Cancer Research, Karpagam Academy of Higher Education, Coimbatore, Tamil Nadu, India
e-mail: sridhar.m@kahedu.edu.in

I. Ramachandran
Department of Endocrinology, Dr ALM PG Institute of Basic Medical Sciences, University of Madras, Taramani Campus, Chennai, Tamil Nadu, India

S. Pathak, A. Banerjee (eds.), *MicroRNAs: Role in Cancer Progression, Diagnostics and Therapy*, Advances in Biochemistry in Health and Disease 34,
https://doi.org/10.1007/978-3-032-08824-6_18

Keywords Cervical cancer · miRNAs · PD-1 · PD-L1 · HPV

Introduction

The microRNAs (miRNAs) are small single-stranded, noncoding RNAs with 18 to 25 nucleotides, which bind to the complementary sequences in the 3′ untranslated regions (3′ UTRs) of target mRNAs. These miRNAs contribute in the regulation of various cellular processes such as expansion, differentiation, inflammation, apoptosis, and immune responses of human papillomavirus (HPV) infected cells. During the onset of several cancers, miRNAs target and alter the presence of proto oncogenes or tumor suppressor genes and are categorized into oncogenic miRNAs and tumor suppressor miRNAs. Interaction between miRNAs and progression of CC is mainly associated with the complex regulation of several signaling pathways such as PI3K/Akt, JAK/STAT, Notch, and Wnt/β catenin [6]. A strong correlation between miRNAs and CC tissue differentiation was identified, from which miR-126, miR-143, and miR-145 were downregulated and miR-15b, miR-16, miR-146a, and miR-155 were upregulated in CC. With the support of in vitro validation, miR-143 and miR-145 are known to exert tumor suppressive role and miR-146a is found to increase CC cell proliferation [51]. Several studies report that the dysregulated miRNAs in CC lead to disruption of these critical signaling cascades [35]. HPVs target the heparin sulphate proteoglycans (HSPG) and other secondary receptor like epidermal growth factor receptor (EGFR) for internalization in mucosal and cutaneous epithelium. Prolonged infection triggers the progression of CC from cervical intraepithelial neoplasia (CIN) I to CIN III. The integration of HPV into the host genome occurs mostly in the chromosomal fragile sites and 50% of the miRNAs associated with CC are found to be located in these sites [5]. Upon internalization, HPV interacts with host proteins, promoting endocytosis and trafficking to the golgi to suppress MHC-1 presentation and increase HPV E6 and E7 synthesis, which targets p53 and pRb tumor suppressor genes, respectively [36]. The transcription of miRNA involves cellular myelocytomatosis (c-Myc), early region 2 binding factor (E2F) and p53, where p53 loss enhances E2F activity, thereby promoting c-Myc driven miRNA transcription [56]. This common intersection of genes shows that there is a connection between miRNA synthesis and HPV infectivity. Various miRNAs get altered in respective CC stages, and targeting these miRNAs could help in the diagnostics and therapeutics of CC.

miRNAs Regulating HPV Internalization

The L1 protein of HPV initially binds with HSPG and this interaction alters the HPV conformation and exposes the L2 amino terminus. Furin, being a pro-protein convertase, gets recognized by L2 amino terminus containing Arg5 and Arg12 in their furin recognition site [43]. A successful HPV infection occurs only when the cleavage is done at Arg12 [60]. The L2 protein binds with secondary receptors such as EGFR, annexin A2 heterotetramer, α-6-integrin, and tetraspanin CD151 [23]. The high-risk (HR) HPV types such as HPV16, 18, 31, 33, 35, 39, 45, 51, 52, 56, 58, 59, 66, and 68 are known to interact with the host cell primary and secondary receptors when compared with low-risk (LR) HPV types. The miR-125b is known to inhibit the synthesis of HPV DNA and viral load, however, an increase in L2 protein reduced the expression of miR-125b during CIN-I [34]. As soon as the HPV L2 binds with their secondary receptors, it gets internalized via receptor-mediated endocytosis by involving the signaling of p21-activated kinase (PAK-1), PKC/PI3k/FAK, and forms uncoated vesicles inside the cytoplasm [10]. The overexpression of EGFR assists the persistence of HPV and transformation of CC cells. The increased expression of EGFR is associated with the reduced miR-133b in CC, as evidenced by EGFR-AS1 knockdown study [15]. Lal et al. [25] demonstrated that miR-34a inhibits basal ERK and AKR expression while mediating the upregulation of growth factor signaling in CC. EGFR is found to interact with 61 miRNAs through computational findings with respect to CC initiation, sustenance, and progression [32]. In a study conducted by Gunasekharan and Laimins [18], the miR-145 expression is modulated by HPV 31 E1 and E2 proteins. An upregulation of miR-21 is found in HPV-infected CC cells by activating EGFR/Akt/Wnt signaling, thereby aggravating the cervical carcinogenesis [55]. Once the HPV enters the cytoplasm via endocytosis, this later on forms the Rab5-dependent early endosomes under low pH conditions, leading to late endosome formation in the presence of lysosomotropic agents [46]. F-box and WD repeat domain containing 7 (FBXW7) is known to subdue the proteosomal degradation of c-Myc, cyclin E, Notch, and mTOR. miR-27a-3p is known to target FBXW7 and increase cervical carcinogenesis, which is suggested as a theragnostic marker in CC [2].

miRNAs Regulating HPV Integration in Host Genome

Once the HPV enters the cytoplasm, its genome will further be trafficked to nucleus from golgi or endoplasmic reticulum. Bromodomain-containing protein 4 (BRD4) plays an important role in transcriptional regulation and epigenetic modulation, through which it makes changes in the cellular function. BRD4 binds with acetylated histones and modulates the cell cycle, apoptosis, and metastasis [12]. HPV DNA replication begins in the host system by binding with BRD4 to E1 and E2 proteins. Once amplified, the BRD4 will be displaced from the HPV replication [45]. HPV E2 forms a complex with BRD4 and integrates into the host fragile site for its replication [22]. BRD4 in association with viral E2 protein interferes with various steps of HPV lifecycle including initiation, viral replication, and maintenance. Its carboxy-terminal motif and TopBP1 directly interact with HPV E2 and enhance plasmid segregation function in host genome [40]. BRD4 depletion could inhibit HPV 16-positive cell viability and it does not change the E6/E7 expression.

However, in HPV-negative cells, BRD4 does not inhibit cell viability, demonstrating the importance of BRD4 in HPV integration [4]. In HPV-positive CC, the miR-152-5p binds with BRD4 directly. Apigenin inhibited the viability, invasion, and EMT by regulating miR-152-5p/BRD4 axis [55]. EGFR inhibited by cetuximab could enhance BRD4 expression, leading to HPV E6 and E7 integration into the host genome. HPV E6/E7 expression was suppressed by EGFR and mediated through upregulation of miR-9-5p with silencing of BRD4 [33]. HPV integration in the loci of SOX2, MYC, and TP63 in several cancers can be targeted via BRD4 in these particular loci to benefit CC treatment. The miR-20a is upregulated by HPV 16 E6 through targeting and downregulating programmed cell death protein 6 (PDCD6) expression in Ca Ski CC cells [30]. Inhibition of BRD4 may cause HPV E6/E7 suppression with proliferation deployment and sensitization of HPV-positive cells. BRD4 inhibitors were used to suppress the HPV E6 expression in CC by administering limited dose of cisplatin to increase chemo-response [42]. BRD4 inhibitor, JQ1, significantly inhibited HPV-positive CC and affected the expression of super-enhancer-related IncRNAs and mRNAs in CC [57]. Once the HPV enters the nucleus, the viral protein L2 forms a complex with mitotic chromatin. This allows the formation of episomes and its replication through host DNA polymerase. E1 and E2 are involved in the replication and transcriptional activities, whereas E6 and E7 facilitate the oncogenesis and immune evasion [8].

miRNAs Regulating Immune Hijacking Mechanism

The HPV establishes an immune evasion mechanism involving reduction in L1 expression to bypass antibody recognition, enters through micro-abrasions to avoid immune cell exposure, and evades TLR9 detection in endosomal compartment. This results in neglecting the innate immune response of the host system. Further, the HPV oncoprotein E5 downregulates the MHC-I heavy chain expression on the cell membrane through SMAD-dependent actions [16]. The immune evasion strategy involves recruitment of T regulatory cells and cancer-associated fibroblasts (CAFs) through TGF-β and IL-10 overexpression [59]. Immune cells such as T lymphocytes and natural killer cells in CC decrease cytokine production and activate inhibitory receptors as a hallmark of T cell exhaustion in CC. Exhausted T cells have been categorized into two markers, PD-1 and transcription factor-1 (TCF-1) [3]. During HPV infection, the innate immune response facilitates and eliminates the infection through immune memory. PD-1 is a type I transmembrane protein belonging to Ig gene (CD279) superfamily, consisting of 288 amino acid residues with 55 kDa. The increased PD-1 in immune cells is capable of binding with PD-L1 of CC cells. Dong et al. [14] reported that suppression of miR-340, miR-383, miR-140, and miR-142 is associated with increased PD-L1 expression. Meanwhile, the Oct-4 upregulation increases the expression of miR-18a and targets SOX6, WNK2, and PTEN, leading to enhanced PD-L1. The miR-375 has tumor suppressor effects by inhibiting MELK and enhancing the apoptosis [1]. miR-20a-5p and miR-106b-5p tend to get upregulated during increased CC severity and miR-125b-5p is downregulated in CC samples. The differential expression of these miRNAs is found to target PD-L1 during CC using in silico analysis [26]. Upregulation of miR-9 in HPV-16-infected population compared with HPV-negative healthy subjects is reported. miR-21 supports the transformation of CC cells in response to prolonged HPV infection and activates monocytes to undergo M1 macrophage polarization via JAK2/STAT1 signaling cascade. The CC migration and invasion is enhanced by miR-214 by regulating UDP-N-acetyl-α-d-galactosamine: polypeptide N-acetylgalactosaminyl transferase 7 (GALNT7) [39], as GALNT7 is involved in immune cell infiltration by having an intricate association with immune-related genes [31].

miRNAs Regulating Cervical Carcinogenesis and Migration

The persistent HPV infection drives the cervical carcinogenesis via epithelial to mesenchymal transition (EMT), MMP-2 secretion, migration, and clonogenesis in HPV-positive CC cells when stimulated with or without EGF [24, 44]. The miRNAs such as miR-34a, miR-99a, miR-100, miR-143, miR-145, miR-199, and miR-218 were downregulated during CIN-1. The CIN 2–3 is associated with the downregulation of miR-26a and miR-29a. The invasive and advanced cervical carcinogenesis involves the upregulation of miR-20a, miR-21, miR-19a-3p, miR-92a, miR-93, miR-132, and miR-150, meanwhile, the miR-411, miR-195, miR-214, miR-125 a/b, miR-636, and miR-122 were downregulated [5, 9, 17, 20, 50, 54, 58]. Decreased miR-21 supports the increased apoptosis and antitumor effects by suppressing STAT1 signaling [29]. The E6 and E7 transcripts downregulate the expression of miR-29a, resulting in elevation of YY1 and CDK6, transcriptionally and translationally leading to the malignant transformation of CC cells [27]. During HPV-positive CC, miRNAs such as miR-143, miR-145, and miR-199a-3p were downregulated. miR-133a and miR-221 modify the therapeutic effectiveness of gefitinib via targeting PTEN and EGFR pathways [32]. miR-125 is a well-known tumor suppressor during CC and the inhibition of this miRNA targets VEGF, prevents invasion and migration of CC cells. Similarly, miR-125a-5p targets Abelson proto-oncogene (ABL-2) and its overexpression prevented the cervical carcinogenesis in Ca Ski and HeLa cells [41]. miR-218 is reported to be downregulated during CC by performing as a tumor suppressor gene and targeting roundabout guidance receptor 1 (ROBO1), a critical protein to drive CC progression, migration, and invasion [30]. Schematic representation of different stages of HPV-driven CC and their subsequent dysregulated miRNAs is depicted in Figs. 18.1, 18.2, and 18.3.

A conceptual overview depicts the different stages of HPV-driven CC and their subsequent dysregulated miRNAs. Five distinct stages of CC are included along with critical genes involved in the transformation (Fig. 18.1). The miRNAs which

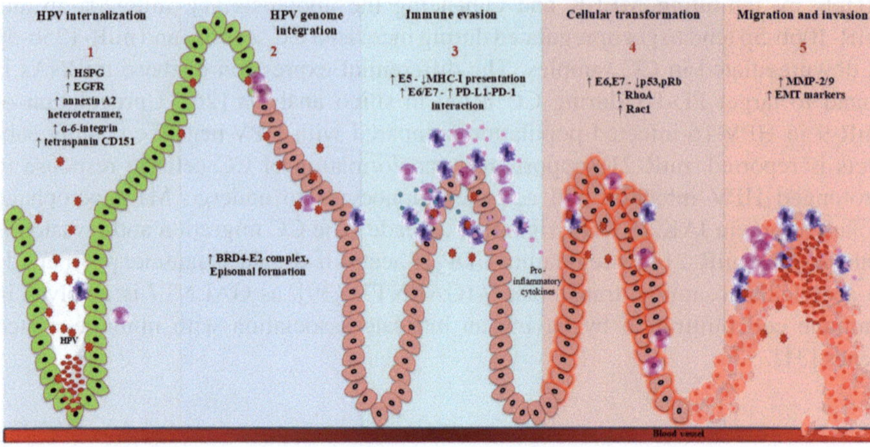

Fig. 18.1 The schematic illustrates the sequential steps of HPV-induced cancer progression: (*1*) HPV internalization, mediated by HSPG, EGFR, annexin A2 heterotetramer, α6-integrin, and tetraspanin CD151; (*2*) HPV genome integration, involving BRD4-E2 complex formation and episomal persistence; (*3*) immune evasion, driven by E5-mediated downregulation of MHC-I, E6/E7-induced PD-L1–PD-1 signaling, and suppression of pro-inflammatory cytokines; (*4*) cellular transformation, through E6/E7-mediated suppression of p53/pRb and activation of RhoA/Rac1 pathways; and (*5*) migration and invasion, facilitated by the upregulation of MMP-2/9 and EMT markers, ultimately promoting tumor cell dissemination

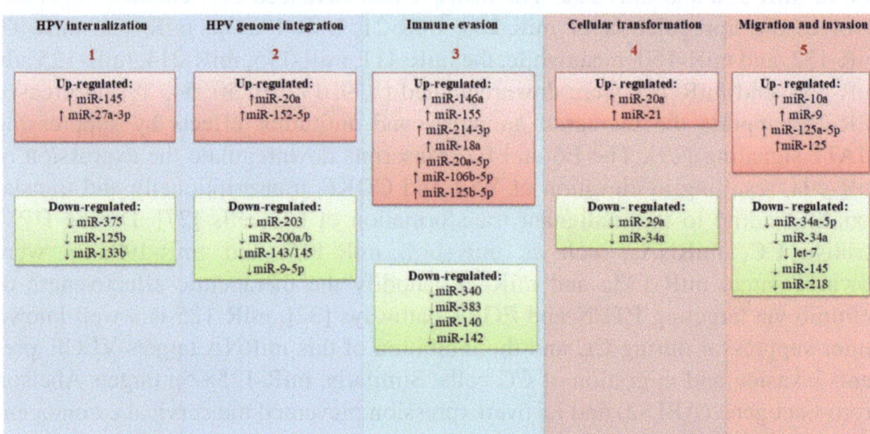

Fig. 18.2 The panel shows the stage-specific alterations in miRNA expression during HPV-mediated tumorigenesis: (*1*) HPV internalization—upregulation of miR-145 and miR-27a-3p; downregulation of miR-375, miR-125b, and miR-133b. (*2*) HPV genome integration—upregulation of miR-20a and miR-152-5p; downregulation of miR-203, miR-200a/b, miR-143/145, and miR-9-5p. (*3*) Immune evasion—upregulation of miR-146a, miR-155, miR-214-3p, miR-18a, miR-20a-5p, miR-106b-5p, and miR-125b-5p; downregulation of miR-340, miR-383, miR-140, and miR-142. (*4*) Cellular transformation—upregulation of miR-20a and miR-21; downregulation of miR-29a and miR-34a. (*5*) Migration and invasion—upregulation of miR-10a, miR-9, miR-125a-5p, and miR-125; downregulation of miR-34a-5p, miR-34a, let-7, miR-145, and miR-218

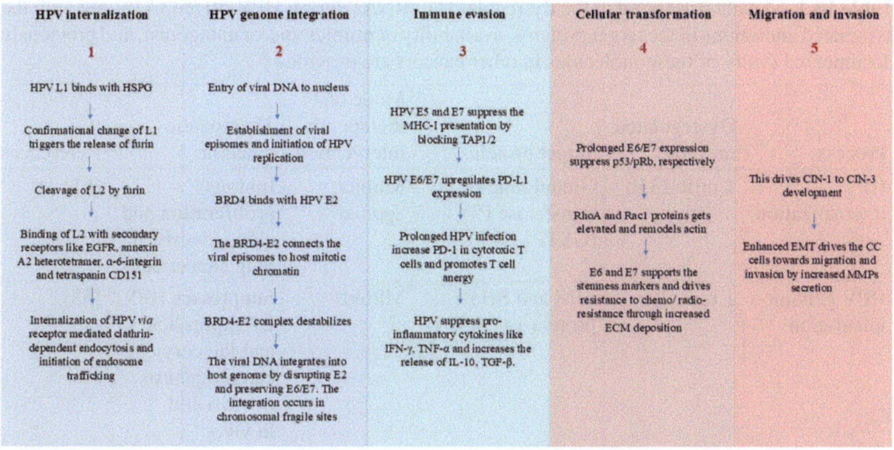

Fig. 18.3 This diagram outlines the mechanistic cascade of HPV infection and malignant progression: (*1*) HPV internalization—HPV L1 binds to HSPG, triggering conformational changes, furin-mediated cleavage of L2, binding to secondary receptors, and internalization via clathrin-dependent endocytosis. (*2*) HPV genome integration—entry of viral DNA into the nucleus, episome establishment via BRD4-E2 complex, destabilization of episomes, and viral DNA integration into fragile chromosomal sites. (*3*) Immune evasion—E5/E7-mediated suppression of MHC-I presentation, E6/E7-driven PD-L1 upregulation, induction of PD-1 in cytotoxic T cells, and suppression of IFN-γ and TNF-α with increased IL-10 and TGF-β release. (*4*) Cellular transformation—persistent E6/E7 expression suppressing p53/pRb, activation of RhoA and Rac1 to remodel actin, and enhanced stemness and resistance to therapy through ECM deposition. (*5*) Migration and invasion—progression from CIN1 to CIN3, epithelial-to-mesenchymal transition (EMT), and increased secretion of MMPs that drive cancer cell migration and invasion

are known to influence these genes are represented in Fig. 18.2. Key substages are detailed in Fig. 18.3 to emphasize the potential of specific miRNAs in intercepting distinct steps of carcinogenic progression.

Emerging Diagnostic and Therapeutic Agents Targeting miRNAs

For cell surface contact studies, early research employed noninfectious virus-like particles (VLPs); however, it was challenging to differentiate between infected and noninfectious particles [47]. Various miRNA mimics and antagonists have a potential impact in utilizing them as therapeutic interventions during CC. During CC, the malignant transformation and disease progression occurs by suppressing tumor-suppressive miRNAs such as miR-34a, miR-203, and miR-375, and upregulating onco-miRs like miR-21, miR-146a, and miR-155. Various miRNAs involved in different stages of HPV-driven carcinogenesis and their therapeutic interventions are elaborated in Table 18.1. The miR-34a is downregulated due to E6-driven p53

Table 18.1 Details pertaining to the dysregulated miRNAs during HPV-driven CC along with the associated alterations in the target proteins, availability of mimics and/or antagonist, and previously documented utility of these molecules in other cancers are provided

Process	Dysregulated miRNAs	Target protein	Mode of therapeutic intervention	Therapeutic outcome	References
HPV internalization	↓ miR-133b	Glutathione-S-transferase P1 (GSTP1)	Mimic/ agonist	Inhibits proliferation and clonogenesis of lung cancer cells	[61]
HPV genome integration	↓ miR-203	LIM and SH3 protein 1 (LASP1)	Mimic	Suppresses HPV E7 expression and the cervical carcinogenesis in vitro and in vivo	[38]
Immune evasion	↑ miR-146a	TNF-α	AntagomiR	Silences the pro-inflammatory genes induced by TLR-4	[62]
	↑ miR-125b	Transcriptional co-activator with PDZ-binding motif (TAZ) and transcriptional enhancer factor (TEAD)	Mimic	Suppresses the HPV E6 activity and enhanced apoptosis. Suppression of TAZ/TEAD improves innate immune response	[62]
	↑ miR-214-3p	β-catenin	Mimic	Reduces the expression of β-catenin and along with HOX transcript antisense intergenic RNA (HOTAIR) suppressed the CC proliferation and induced apoptosis	[63]
Cellular transformation	↓ miR-29a	DNA methyltransferase (DNMT)3A, DNMT3B	Mimic	Inhibits the gene methylation of p16, a tumor suppressor gene, and regains apoptotic potential of CC cells	[64]

(continued)

Table 18.1 (continued)

Process	Dysregulated miRNAs	Target protein	Mode of therapeutic intervention	Therapeutic outcome	References
Migration and invasion	↓ miR-34a	NOTCH1	Mimic	Reduce the tumor growth and improve the survival	[65]
	↓ miR-34a-5p	Murine double minute 4 (MDM4)	Mimic	Triggers the chemo-sensitivity of CC cells for oxaliplatin	[65]
	↓ miR-218	ROBO1	Mimic	Reduces CC cell viability, migration, and invasion along with reduced growth of xenograft	[66]

degradation and its overexpression via synthetic miR-34a mimics has shown to downregulate E6/E7 expression, reactivating p53 and pRb, inhibiting proliferation, and inducing apoptosis in CC cells [37]. miR-34a mimic is previously analyzed for multiple myeloma in vivo by delivering them via stable nucleic acid lipid particles (SNALPs). This approach significantly reduced the tumor growth in xenograft model and improved their survival [13]. miR-199b-5p encapsulated with SNALP delivery model impaired Hes-1 level and cancer stem cell markers in the different cell lines MDAMB-231T, MCF7 breast cancer cells, Colon cells SW480, HT-29, CaCo-2 and PC-3 prostate cells, U87 glioblastoma, and medulloblastoma cells ONS-76, UW-228 [11]. Encapsulation of SNALP with miR-34a has inhibited the tumor growth of multiple myeloma in mice models. SNALP with 2′-O-methylated miR-34a and SNALP with transferrin increased the life of the mice and their survival ability was higher compared with control [48]. Chlorotoxin-coupled SNALP with formulated anti-miR-21 reduced the tumor size and tumor cell proliferation and induced apoptosis in glioblastoma bearing mice [7].

Liposome-mediated delivery decreased the cell viability of CC and it mimics the efficient transfecting agent for cancer cells [21]. Liposome with folate encapsulated with cisplatin and miR-219a-5p (Lipo@DDP@miR-219a-5p@FA) was bound to folate receptors of A549 cells and enhanced the drug internalization efficacy. Lipo@DDP@miR-219a-5p@FA has inhibited the A549 proliferation and migration and induced apoptosis [53]. let-7 miRNA and siRNA of CDK4 encapsulated with Herceptin-mediated liposomes (Her-PEG-Lipo-mi/siRNA) have inhibited the growth and migration in SK-BR-3 breast cancer cells. Combination of miRNA and siRNA loaded with liposomes acts as a therapeutic model for carcinogenesis [49], including CC. MRX34, liposomal formulation of miR-34a mimic, is the first miRNA to be involved in the phase 1 clinical trial tested for advanced hepatocellular carcinoma [19]. Similarly, miR-203 is downregulated in HPV-infected keratinocytes and high-grade CC lesions and its mimic targets HPV E7-dependent LASP1 expression in order to suppress the cervical carcinogenesis in vitro and in vivo [38]. On the other hand, miRNA antagonists are chemically stabilized single-stranded oligonucleotides produced to bind and inhibit specific endogenous miRNAs. These antagonists suppress oncogenic miRNAs that are upregulated during HPV infection. For instance, miR-21, frequently upregulated in CC tissues, tends to target tumor suppressor genes such as PTEN and PDCD4. Inhibition of miR-21 using antisense oligonucleotides has demonstrated enhanced apoptosis and increased chemo-sensitivity to cisplatin in HPV-driven CC cell lines [52]. Similarly, blocking miR-155, which is known to elevate immune evasion, has a potential approach to resensitize tumors towards immune surveillance [6, 28].

Conclusion

The interplay between HPV and host miRNAs orchestrates critical events in cervical carcinogenesis such as viral genome integration, immune evasion, EMT, migration, and invasion. Differential miRNA expression is associated with different stages of HPV-driven progression, from initial HPV entry to invasive CC. This supports the conjuncture that these miRNAs could be the potential biomarkers for diagnosis, prognosis, and therapeutic interventions. The integration of miRNA-based strategies for the theragnostics of HPV-driven CC represents a significant leap towards personalized medicine. Future studies warrant the optimal delivery of these miRNAs, reducing the immunogenicity and combining miRNA-based treatment with the existing tyrosine kinase inhibitors or monoclonal antibodies.

Acknowledgments The constructive periodic discussions and support headed by the Managing Director, Dr. R. Vasanthakumar, Karpagam Educational Institutions, are responsible for this work. The authors thank the CEO, Vice Chancellor, and Registrar of KAHE for providing constructive support for the completion of this work.

Conflict of Interest The authors declare no conflicts of interest.

Funding Statement This study is supported by Department of Science and Technology (DST) in the form of NIDHI grant (KIIC/Incubatee/Ignition Grant/004/05.03.2024) to SM and INSPIRE fellowship (IF210086) to SM and SK.

References

1. Bañuelos-Villegas EG, Pérez-yPérez MF, Alvarez-Salas LM (2021) Cervical cancer, papillomavirus, and miRNA dysfunction. Front Mol Biosci 8:758337. https://doi.org/10.3389/fmolb.2021.758337. PMID: 34957212; PMCID: PMC8703027
2. Ben W, Zhang G, Huang Y, Sun Y (2020) MiR-27a-3p regulated the aggressive phenotypes of cervical cancer by targeting FBXW7. Cancer Manag Res 12:2925–2935. https://doi.org/10.2147/CMAR.S234897. PMID: 32431539; PMCID: PMC7198449
3. Brunell AE, Lahesmaa R, Autio A, Thotakura AK (2023) Exhausted T cells hijacking the cancer-immunity cycle: assets and liabilities. Front Immunol 14:1151632. https://doi.org/10.3389/fimmu.2023.1151632. PMID: 37122741; PMCID: PMC10140554
4. Canarte V, Munger K (2021) BRD4 downregulation inhibits the viability of cervical cancer cells without affecting viral oncoprotein expression. Curr Res Virol Sci 2:100010. https://doi.org/10.1016/j.crviro.2021.100010. ISSN 2666-478X
5. Causin RL, Freitas AJA, Hidalgo Filho CMT, Reis RD, Reis RM, Marques MMC (2021) A systematic review of MicroRNAs involved in cervical cancer progression. Cells 10(3):668. https://doi.org/10.3390/cells10030668. PMID: 33803022; PMCID: PMC8002658
6. Chauhan P, Pramodh S, Hussain A, Elsori D, Lakhanpal S, Kumar R, Alsaweed M, Iqbal D, Pandey P, Al Othaim A, Khan F (2024) Understanding the role of miRNAs in cervical cancer pathogenesis and therapeutic responses. Front Cell Dev Biol 12:1397945. https://doi.org/10.3389/fcell.2024.1397945. PMID: 39263322; PMCID: PMC11387185
7. Costa PM, Cardoso AL, Custódia C, Cunha P, Pereira de Almeida L, Pedroso de Lima MC (2015) MiRNA-21 silencing mediated by tumor-targeted nanoparticles combined with sunitinib: a new multimodal gene therapy approach for glioblastoma. J Control Release 207:31–39. https://doi.org/10.1016/j.jconrel.2015.04.002. Epub 2015 Apr 7
8. Coursey TL, McBride AA (2019) Hitchhiking of viral genomes on cellular chromosomes. Annu Rev Virol 6(1):275–296. https://doi.org/10.1146/annurev-virology-092818-015716. Epub 2019 July 5. PMID: 31283444; PMCID: PMC7415349
9. Cui F, Li X, Zhu X, Huang L, Huang Y, Mao C, Yan Q, Zhu J, Zhao W, Shi H (2012) MiR-125b inhibits tumor growth and promotes apoptosis of cervical cancer cells by targeting phosphoinositide 3-kinase catalytic subunit delta. Cell Physiol Biochem 30(5):1310–1318. https://doi.org/10.1159/000343320. Epub 2012 Oct 22
10. Day PM, Schelhaas M (2014) Concepts of papillomavirus entry into host cells. Curr Opin Virol 4:24–31. https://doi.org/10.1016/j.coviro.2013.11.002. Epub 2013 Dec 14. PMID: 24525291; PMCID: PMC3951680
11. de Antonellis P, Liguori L, Falanga A, Carotenuto M, Ferrucci V, Andolfo I, Marinaro F, Scognamiglio I, Virgilio A, De Rosa G, Galeone A, Galdiero S, Zollo M (2013) MicroRNA 199b-5p delivery through stable nucleic acid lipid particles (SNALPs) in tumorigenic cell lines. Naunyn Schmiedeberg's Arch Pharmacol 386(4):287–302. https://doi.org/10.1007/s00210-013-0837-4. Epub 2013 Jan 26
12. Dey A, Nishiyama A, Karpova T, McNally J, Ozato K (2009) Brd4 marks select genes on mitotic chromatin and directs postmitotic transcription. Mol Biol Cell 20(23):4899–4909. https://doi.org/10.1091/mbc.e09-05-0380. Epub 2009 Oct 7. PMID: 19812244; PMCID: PMC2785733
13. Di Martino MT, Campani V, Misso G, Gallo Cantafio ME, Gullà A, Foresta U, Guzzi PH, Castellano M, Grimaldi A, Gigantino V, Franco R, Lusa S, Cannataro M, Tagliaferri P, De Rosa G, Tassone P, Caraglia M (2014) In vivo activity of miR-34a mimics delivered by stable nucleic acid lipid particles (SNALPs) against multiple myeloma. PLoS One 9(2):e90005. https://doi.org/10.1371/journal.pone.0090005. PMID: 24587182; PMCID: PMC3937395
14. Dong P, Xiong Y, Yu J, Chen L, Tao T, Yi S, Hanley SJB, Yue J, Watari H, Sakuragi N (2018) Control of PD-L1 expression by miR-140/142/340/383 and oncogenic activation of the OCT4-miR-18a pathway in cervical cancer. Oncogene 37(39):5257–5268. https://doi.org/10.1038/

s41388-018-0347-4. Epub 2018 May 31. Erratum in: Oncogene. 2019;38(20):3972. https://doi.org/10.1038/s41388-019-0677-x. PMID: 29855617; PMCID: PMC6160397

15. Dong C, Dong R, Song J, Yu C, Zhuang Y, Guo Q (2024) Knockdown of lncRNA EGFR-AS1 promotes autophagy-mediated ferroptosis in cervical cancer via regulating EGFR expression through miR-133b. Mol Cell Toxicol 20(1):139–147. https://doi.org/10.1007/s13273-023-00332-7
16. Gnanagurusamy J, Krishnamoorthy S, Muthusami S (2024) Transforming growth factor-β micro-environment mediated immune cell functions in cervical cancer. Int Immunopharmacol 140:112837. https://doi.org/10.1016/j.intimp.2024.112837. Epub 2024 Aug 6
17. Granados López AJ, López JA (2014) Multistep model of cervical cancer: participation of miRNAs and coding genes. Int J Mol Sci 15(9):15700–15733. https://doi.org/10.3390/ijms150915700. PMID: 25192291; PMCID: PMC4200848
18. Gunasekharan V, Laimins LA (2013) Human papillomaviruses modulate microRNA 145 expression to directly control genome amplification. J Virol 87(10):6037–6043. https://doi.org/10.1128/JVI.00153-13. Epub 2013 Mar 6. PMID: 23468503; PMCID: PMC3648148
19. Hong DS, Kang YK, Borad M, Sachdev J, Ejadi S, Lim HY, Brenner AJ, Park K, Lee JL, Kim TY, Shin S, Becerra CR, Falchook G, Stoudemire J, Martin D, Kelnar K, Peltier H, Bonato V, Bader AG, Smith S, Kim S, O'Neill V, Beg MS (2020) Phase 1 study of MRX34, a liposomal miR-34a mimic, in patients with advanced solid tumours. Br J Cancer 122(11):1630–1637. https://doi.org/10.1038/s41416-020-0802-1. Epub 2020 Apr 2. PMID: 32238921; PMCID: PMC7251107
20. Hu T, Chang YF, Xiao Z, Mao R, Tong J, Chen B, Liu GC, Hong Y, Chen HL, Kong SY, Huang YM, Xiyang YB, Jin H (2016) miR-1 inhibits progression of high-risk papillomavirus-associated human cervical cancer by targeting G6PD. Oncotarget 7(52):86103–86116. https://doi.org/10.18632/oncotarget.13344. PMID: 27861141; PMCID: PMC5349900
21. Huang F, Zhao F, Liang LP, Zhou M, Qu ZL, Cao YZ, Lin C (2015) Optomizing transfection efficiency of cervical cancer cells transfected by cationic liposomes LipofectamineTM2000. Asian Pac J Cancer Prev 16(17):7749–7754. https://doi.org/10.7314/apjcp.2015.16.17.7749
22. Jang MK, Shen K, McBride AA (2014) Papillomavirus genomes associate with BRD4 to replicate at fragile sites in the host genome. PLoS Pathog 10(5):e1004117. https://doi.org/10.1371/journal.ppat.1004117. PMID: 24832099; PMCID: PMC4022725
23. Krishnamoorthy S, Muruganantham B, Yu JR, Park WY, Muthusami S (2023) Exploring the utility of FTS as a bonafide binding partner for EGFR: a potential drug target for cervical cancer. Comput Biol Med 167:107592. https://doi.org/10.1016/j.compbiomed.2023.107592. Epub 2023 Nov 4
24. Krishnamoorthy S, Vasudevan S, Muruganantham B, Muthusami S (2025) Epinephrine augments the phosphorylation of EGFR and promote the DNA synthesis and migration of cervical cancer cells. 3. Biotech 15(5):125. https://doi.org/10.1007/s13205-025-04285-7. Epub 2025 Apr 14. PMID: 40242059; PMCID: PMC11996747
25. Lal A, Thomas MP, Altschuler G, Navarro F, O'Day E, Li XL, Concepcion C, Han YC, Thiery J, Rajani DK, Deutsch A, Hofmann O, Ventura A, Hide W, Lieberman J (2011) Capture of microRNA-bound mRNAs identifies the tumor suppressor miR-34a as a regulator of growth factor signaling. PLoS Genet 7(11):e1002363. https://doi.org/10.1371/journal.pgen.1002363. Epub 2011 Nov 10. PMID: 22102825; PMCID: PMC3213160
26. Leventakou D, Gouloumi AR, Spathis A, Pouliakis A, Koufopoulos N, Pergialiotis V, Drakakis P, Panayiotides IG, Kottaridi C (2023) Expression profile of miRNAs computationally predicted to target PDL-1 in cervical tissues of different histology groups. Front Cell Dev Biol 11:1101041. https://doi.org/10.3389/fcell.2023.1101041. PMID: 36910137; PMCID: PMC9998664
27. Li Y, Wang F, Xu J, Ye F, Shen Y, Zhou J, Lu W, Wan X, Ma D, Xie X (2011) Progressive miRNA expression profiles in cervical carcinogenesis and identification of HPV-related target genes for miR-29. J Pathol 224(4):484–495. https://doi.org/10.1002/path.2873. Epub 2011 Apr 18

28. Li N, Cui T, Guo W, Wang D, Mao L (2019) MiR-155-5p accelerates the metastasis of cervical cancer cell via targeting TP53INP1. Onco Targets Ther 12:3181–3196. https://doi.org/10.2147/OTT.S193097. PMID: 31118671; PMCID: PMC6500876
29. Li C, Cang W, Gu Y, Chen L, Xiang Y (2023) The anti-PD-1 era of cervical cancer: achievement, opportunity, and challenge. Front Immunol 14:1195476. https://doi.org/10.3389/fimmu.2023.1195476. PMID: 37559727; PMCID: PMC10407549
30. Liu X (2018) Up-regulation of miR-20a by HPV16 E6 exerts growth-promoting effects by targeting PDCD6 in cervical carcinoma cells. Biomed Pharmacother 102:996–1002. https://doi.org/10.1016/j.biopha.2018.03.154. Epub 2018 Apr 5
31. Liu Y, Sun Y, Xiao M, Li S, Shi S (2024) Comprehensive pan-cancer analysis reveals the versatile role of GALNT7 in epigenetic alterations and immune modulation in cancer. Heliyon 10(11):e31515. https://doi.org/10.1016/j.heliyon.2024.e31515. PMID: 38845941; PMCID: PMC11153094
32. Muthusami S, Sabanayagam R, Periyasamy L, Muruganantham B, Park WY (2022) A review on the role of epidermal growth factor signaling in the development, progression and treatment of cervical cancer. Int J Biol Macromol 194:179–187. https://doi.org/10.1016/j.ijbiomac.2021.11.117. Epub 2021 Nov 27
33. Nantajit D, Presta L, Sauter T, Tavassoli M (2022) EGFR-induced suppression of HPV E6/E7 is mediated by microRNA-9-5p silencing of BRD4 protein in HPV-positive head and neck squamous cell carcinoma. Cell Death Dis 13(11):921. https://doi.org/10.1038/s41419-022-05269-8. PMID: 36333293; PMCID: PMC9636399
34. Nuovo GJ, Wu X, Volinia S, Yan F, di Leva G, Chin N, Nicol AF, Jiang J, Otterson G, Schmittgen TD, Croce C (2010) Strong inverse correlation between microRNA-125b and human papillomavirus DNA in productive infection. Diagn Mol Pathol 19(3):135–143. https://doi.org/10.1097/PDM.0b013e3181c4daaa. PMID: 20736742; PMCID: PMC4284817
35. Ortiz GGR, Mohammadi Y, Nazari A, Ataeinaeini M, Kazemi P, Yasamineh S, Al-Naqeeb BZT, Zaidan HK, Gholizadeh O (2023) A state-of-the-art review on the MicroRNAs roles in hematopoietic stem cell aging and longevity. Cell Commun Signal 21(1):85. https://doi.org/10.1186/s12964-023-01117-0. PMID: 37095512; PMCID: PMC10123996
36. Ozbun MA, Campos SK (2021) The long and winding road: human papillomavirus entry and subcellular trafficking. Curr Opin Virol 50:76–86. https://doi.org/10.1016/j.coviro.2021.07.010. Epub 2021 Aug 17. PMID: 34416595; PMCID: PMC8500947
37. Pal A, Kundu R (2020) Human papillomavirus E6 and E7: the cervical cancer hallmarks and targets for therapy. Front Microbiol 10:3116. https://doi.org/10.3389/fmicb.2019.03116. PMID: 32038557; PMCID: PMC6985034
38. Patterson MR, Meijers AS, Ryder EL, Wootton LM, Scarth JA, Evans D, Turner AL, Wasson CW, Darell JE, Theobald DA, Cogan JA, James CD, Wang M, Ladbury JE, Morgan IM, Samson A, Morgan EL, Macdonald A (2024) E7-mediated repression of miR-203 promotes LASP1-dependent proliferation in HPV-positive cervical cancer. Oncogene 43(28):2184–2198. https://doi.org/10.1038/s41388-024-03067-4. Epub 2024 May 24. PMID: 38789663; PMCID: PMC11226402
39. Peng RQ, Wan HY, Li HF, Liu M, Li X, Tang H (2012) MicroRNA-214 suppresses growth and invasiveness of cervical cancer cells by targeting UDP-N-acetyl-α-D-galactosamine:polypeptide N-acetylgalactosaminyltransferase 7. J Biol Chem 287(17):14301–14309. https://doi.org/10.1074/jbc.M111.337642. Epub 2012 Mar 7. PMID: 22399294; PMCID: PMC3340176
40. Prabhakar AT, James CD, Fontan CT, Otoa R, Wang X, Bristol ML, Hill RD, Dubey A, Morgan IM (2023) Human papillomavirus 16 E2 interaction with TopBP1 is required for E2 and viral genome stability during the viral life cycle. J Virol 97(3):e0006323. https://doi.org/10.1128/jvi.00063-23. Epub 2023 Feb 22. PMID: 36840558; PMCID: PMC10062148
41. Qin X, Wan Y, Wang S, Xue M (2015) MicroRNA-125a-5p modulates human cervical carcinoma proliferation and migration by targeting ABL2. Drug Des Devel Ther 10:71–79. https://doi.org/10.2147/DDDT.S93104. PMID: 26766902; PMCID: PMC4699546

42. Rataj O, Haedicke-Jarboui J, Stubenrauch F, Iftner T (2019) Brd4 inhibition suppresses HPV16 E6 expression and enhances chemoresponse: a potential new target in cervical cancer therapy. Int J Cancer 144(9):2330–2338. https://doi.org/10.1002/ijc.31986. Epub 2019 Jan 16
43. Richards KF, Mukherjee S, Bienkowska-Haba M, Pang J, Sapp M (2014) Human papillomavirus species-specific interaction with the basement membrane-resident non-heparan sulfate receptor. Viruses 6(12):4856–4879. https://doi.org/10.3390/v6124856. PMID: 25490765; PMCID: PMC4276933
44. Sabanayagam R, Krishnamoorthy S, Gnanagurusamy J, Muruganatham B, Muthusami S (2023) EGCG attenuate EGF triggered matrix abundance and migration in HPV positive and HPV negative cervical cancer cells. Med Oncol 40(9):261. https://doi.org/10.1007/s12032-023-02135-1
45. Sakakibara N, Chen D, Jang MK, Kang DW, Luecke HF, Wu SY, Chiang CM, McBride AA (2013) Brd4 is displaced from HPV replication factories as they expand and amplify viral DNA. PLoS Pathog 9(11):e1003777. https://doi.org/10.1371/journal.ppat.1003777. Epub 2013 Nov 21. PMID: 24278023; PMCID: PMC3836737
46. Schelhaas M, Shah B, Holzer M, Blattmann P, Kühling L, Day PM, Schiller JT, Helenius A (2012) Entry of human papillomavirus type 16 by actin-dependent, clathrin- and lipid raft-independent endocytosis. PLoS Pathog 8(4):e1002657. https://doi.org/10.1371/journal.ppat.1002657. Epub 2012 Apr 19. PMID: 22536154; PMCID: PMC3334892
47. Schiller JT, Day PM, Kines RC (2010) Current understanding of the mechanism of HPV infection. Gynecol Oncol 118(1 Suppl):S12–S17. https://doi.org/10.1016/j.ygyno.2010.04.004. PMID: 20494219; PMCID: PMC3493113
48. Scognamiglio I, Di Martino MT, Campani V, Virgilio A, Galeone A, Gullà A, Gallo Cantafio ME, Misso G, Tagliaferri P, Tassone P, Caraglia M, De Rosa G (2014) Transferrin-conjugated SNALPs encapsulating 2′-O-methylated miR-34a for the treatment of multiple myeloma. Biomed Res Int 2014:217365. https://doi.org/10.1155/2014/217365. Epub 2014 Feb 13. PMID: 24683542; PMCID: PMC3943297
49. Shin JH, Shin DH, Kim JS (2020) Let-7 miRNA and CDK4 siRNA co-encapsulated in Herceptin-conjugated liposome for breast cancer stem cells. Asian. J Pharm Sci 15(4):472–481. https://doi.org/10.1016/j.ajps.2019.03.001. Epub 2019 May 7. PMID: 32952670; PMCID: PMC7486552
50. Su Z, Yang H, Zhao M, Wang Y, Deng G, Chen R (2017) MicroRNA-92a promotes cell proliferation in cervical cancer via inhibiting p21 expression and promoting cell cycle progression. Oncol Res 25(1):137–145. https://doi.org/10.3727/096504016X14732772150262. PMID: 28081742; PMCID: PMC7840834
51. Wang X, Tang S, Le SY, Lu R, Rader JS, Meyers C, Zheng ZM (2008) Aberrant expression of oncogenic and tumor-suppressive microRNAs in cervical cancer is required for cancer cell growth. PLoS One 3(7):e2557. https://doi.org/10.1371/journal.pone.0002557. PMID: 18596939; PMCID: PMC2438475
52. Wang Y, Zhou S, Fan K, Jiang C (2019) MicroRNA-21 and its impact on signaling pathways in cervical cancer. Oncol Lett 17(3):3066–3070. https://doi.org/10.3892/ol.2019.10002. Epub 2019 Feb 1. PMID: 30867735; PMCID: PMC6396207
53. Wu Y, Zhang J, Zhao J, Wang B (2024) Folate-modified liposomes mediate the co-delivery of cisplatin with miR-219a-5p for the targeted treatment of cisplatin-resistant lung cancer. BMC Pulm Med 24(1):159. https://doi.org/10.1186/s12890-024-02938-6. PMID: 38561695; PMCID: PMC10986081
54. Yang Z, Chen S, Luan X, Li Y, Liu M, Li X, Liu T, Tang H (2009) MicroRNA-214 is aberrantly expressed in cervical cancers and inhibits the growth of HeLa cells. IUBMB Life 61(11):1075–1082. https://doi.org/10.1002/iub.252
55. Zhao S, Xie J, Zhao C, Cao W, Yu Y (2021) Ultrasound-targeted microbubble destruction enhances the inhibitive efficacy of miR-21 silencing in HeLa cells. Med Sci Monit 27:e923660. https://doi.org/10.12659/MSM.923660. PMID: 33606670; PMCID: PMC7901158

56. Zheng ZM, Wang X (2011) Regulation of cellular miRNA expression by human papillomaviruses. Biochim Biophys Acta 1809(11–12):668–677. https://doi.org/10.1016/j.bbagrm.2011.05.005. Epub 2011 May 17. PMID: 21616186; PMCID: PMC3175324
57. Zheng J, Huang B, Xiao L, Wu M (2024) Effects of BRD4 inhibitor JQ1 on the expression profile of super-enhancer related lncRNAs and mRNAs in cervical cancer HeLa cells. PeerJ 12:e17035. https://doi.org/10.7717/peerj.17035. PMID: 38410799; PMCID: PMC10896078
58. Zhou Q, Han LR, Zhou YX, Li Y (2016) MiR-195 suppresses cervical cancer migration and invasion through targeting Smad3. Int J Gynecol Cancer 26(5):817–824. https://doi.org/10.1097/IGC.0000000000000686
59. Zhou C, Tuong ZK, Frazer IH (2019) Papillomavirus immune evasion strategies target the infected cell and the local immune system. Front Oncol 9:682. https://doi.org/10.3389/fonc.2019.00682. PMID: 31428574; PMCID: PMC6688195
60. Bronnimann MP, Calton CM, Chiquette SF, Li S, Lu M, Chapman JA, Bratton KN, Schlegel AM, Campos SK (2016) Furin cleavage of L2 during papillomavirus infection: Minimal dependence on cyclophilins. J Virol 90(14):6224–6234. https://doi.org/10.1128/JVI.00038-16
61. Lin C, Xie L, Lu Y, Hu Z, Chang J (2018) miR-133b reverses cisplatin resistance by targeting GSTP1 in cisplatin-resistant lung cancer cells. Int J Mol Med https://doi.org/10.3892/ijmm.2018.3382
62. El Gazzar M, Church A, Liu T, McCall CE (2011) MicroRNA-146a regulates both transcription silencing and translation disruption of TNF-αduring TLR4-induced gene reprogramming. J Leukocyte Biol 90(3):509–519. https://doi.org/10.1189/jlb.0211074
63. Zhou Y, Wang Y, Lin M, Wu D, Zhao M (2021) LncRNA HOTAIR promotes proliferation and inhibits apoptosis by sponging miR-214-3p in HPV16 positive cervical cancer cells. Cancer Cell Int 21(1). https://doi.org/10.1186/s12935-021-02103-7
64. Wang A, Xu Q, Sha R, Bao T, Xi X, Guo G (2021) MicroRNA-29a inhibits cell proliferation and arrests cell cycle by modulating p16 methylation in cervical cancer. Oncol Lett 21(4). https://doi.org/10.3892/ol.2021.12533
65. He P, Liu X, Lou Y, Gong S, Cao L (2022) miR-34a-5p enhances the sensitivity of cervical cancer cells to oxaliplatin chemotherapy via targeting MDM4. Clin. Exp. Obstet Gynecol 49(2):54. https://doi.org/10.31083/j.ceog4902054
66. Liu Z, Mao L, Wang L, Zhang H, Hu X (2024) [Retracted] miR-218 functions as a tumor suppressor gene in cervical cancer. Mol Med Rep 30(5). https://doi.org/10.3892/mmr.2024.13326

Chapter 19
Role of MicroRNAs in Intestinal Epithelial Transport and Its Mechanism Leading to Colorectal Cancer

Rajesh Selvaraj, Sivakumari Kanakarajan, and Soudamani Singh

Abstract The function of microRNAs (miRNAs) as regulators of gene expression, their translational use as diagnostic biomarkers, and their potential as therapeutic targets for a number of diseases have all been the subject of much research in recent decades. By influencing the expression of genes involved in the production of tight junction proteins, miRNAs play a critical role in controlling intestinal epithelial transport. This directly affects the intestinal epithelium's barrier function and the movement of molecules across the intestinal lining; in other words, depending on which genes they target and regulate, miRNAs can either promote or inhibit the transport of substances. The onset and spread of colorectal cancer (CRC) are also significantly influenced by miRNAs. The miRNAs have two distinct roles in CRC, depending on the genes they target. They can either operate as oncogenes or tumor suppressors, depending on the cellular environment. Their influence also includes influencing important aspects of the pathophysiology of CRC, including as angiogenesis, apoptosis, chemoresistance, proliferation, and sensitivity to radiation. These miRNAs have the ability to influence pathways such as Wnt, TGF-β, PI3K-AKT, MAPK, and EMT, which makes them promising targets for therapy and clinical biomarkers. With a clear context, this chapter outlines current findings about miRNAs, including their function in intestinal transport of human epithelial cells, the process by which CRC progresses and spreads, and their potential as biomarkers and therapeutic targets in CRC. The roles of miRNAs in carcinogenesis and development were also integrated with CRC patient care, resulting in new insights into the diagnosis and management of CRC.

Keywords MicroRNA · Intestinal transport · Colorectal cancer · Exosomal miRNA

R. Selvaraj (✉) · S. Singh
Department of Clinical and Translational Sciences, Joan C. Edwards School of Medicine, Marshall University, Huntington, WV, USA

S. Kanakarajan (✉)
Department of Zoology, Presidency College, Chennai, Tamil Nadu, India

S. Pathak, A. Banerjee (eds.), *MicroRNAs: Role in Cancer Progression, Diagnostics and Therapy*, Advances in Biochemistry in Health and Disease 34,
https://doi.org/10.1007/978-3-032-08824-6_19

Abbreviations	
miRNAs	microRNAs
CRC	Colorectal Cancer
RNA	Ribonucleic acids
pre-miRNAs	precursor miRNAs
DNA	Deoxyribonucleic acid
pri-miRNAs	primary miRNAs
RISC	RNA-induced silencing complex
IECs	Intestinal epithelial cells
PTEN	Phosphatase and Tensin homolog
PDCD4	Programmed cell death 4
RECK	Ropivacaine, Epinephrine, Clonidine, and Ketorolac
SPRY2	Sprouty RTK signaling antagonist 2
TGFβR2	Transforming Growth Factor Beta Receptor 2
SOX9	SRY-Box Transcription Factor 9
DDK1	Dickkopf-related protein 1
PepT1	Peptide transporter 1
DRA	Downregulated in adenoma
CFTR	Cystic fibrosis transmembrane conductance regulator
PAT-1	Proton-coupled amino acid transporter 1
MCT1	Monocarboxylate transporter-1
UC	Ulcerative colitis
IBD	Inflammatory bowel disease
SCFA	Short-chain fatty acid transporters
SMCT1	Sodium-coupled MCT1
TMAO	Trimethylamine-N-oxide
OCT2	Organic cation transporter 2
OATP2B1	Organic anion transporting polypeptide 2B1 transporter
PAC	Pancreatic acinar cells
THTR-1	Thiamin transporter-1
CD	Crohn's disease
NOD2	Nucleotide-binding oligomerization domain-containing protein 2
MAPK	Mitogen-activated protein kinase
TGF-β	Transforming growth factor beta
SNP	Single nucleotide polymorphisms
NFκB	Nuclear factor kappa-light-chain-enhancer of activated B cells
EMT	Epithelial-mesenchymal transitions
MET	Mesenchymal epithelial transition
APC	Antigen-presenting cells
HT29	Human colorectal adenocarcinoma cell line
HCT-116	Human colorectal carcinoma cell line
HepG-2 and HuH 7	Human liver cancer cell line
HEK293	Human embryonic kidney 293 cells
TLR	Toll-like receptors

MyD88	Myeloid differentiation primary response 88
TRAF6	TNF receptor-associated factor 6
IRAK	Interleukin-1 receptor-associated kinase
TNF-α	Tumor necrosis factor alpha
IL	Interleukin
hnRNPA2B1	heterogeneous nuclear ribonucleoprotein A2/B1
METTL3	Methyltransferase-like 3
PD-L1	Programmed death-ligand 1
TAM	Tumor-associated macrophages
YTHDC2	YTH domain-containing protein 2
MYPT1	Myosin phosphatase-targeting protein 1
GIST	Gastrointestinal stromal tumors
PTPRO	Protein tyrosine phosphatase receptor type O
HMGB1	High mobility group box 1
ROS	Reactive oxygen species
GSH-Px	Glutathione peroxidase
HucMSC-Ex	Human umbilical cord mesenchymal stem cell-derived exosomes
ACSL4	Acyl-CoA synthetase long chain family member 4
STAT3	Signal transducer and activator of transcription 3
SPTBN2	Spectrin beta, nonerythrocytic 2
IKK-α	Inhibitor of κB kinase subunit alpha
DCLK1	Double cortin-like kinase 1
VMP1	Vacuole membrane protein 1
KRIT1	Krev interaction protein 1
TGFBI	Targeting transforming growth factor beta-induced protein
RhoB	Ras homolog gene family, member b
JNK	Jun N-terminal kinase
CDKN2B	Cyclin-dependent kinase inhibitor 2B
FIH-1	Factor inhibiting HIF-1α
RhoBTB1	Rho-related BTB domain containing 1
EGFR	Estimated glomerular filtration rate
CPM	Carboxypeptidase M
FAK	Focal adhesion kinase
c-Src	cellular Src; 7
CXCR	C-X-C motif chemokine receptor
CAF	Cancer-associated fibroblasts
PTK6	Protein tyrosine kinase 6
FoxO1	Forkhead box O1
TJ	Tight junction
DSS	Dextran sodium sulfate
HIF-1α	Hypoxia-inducible factor 1α
MLCK	Myosin light chain kinase
AID	Activation-induced cytidine deaminase
HP1	Heterochromatin protein 1

Overview and Function of miRNAs

MicroRNAs (miRNAs) are tiny, endogenous, evolutionarily conserved, noncoding single-stranded ribonucleic acids (RNA) with 19–24 nucleotides. Initially, they were thought to be developmental regulating genes that encoded short antisense RNA products [106]. Prior to becoming precursor miRNAs (pre-miRNAs) and mature miRNAs, the majority of miRNAs are translated from deoxyribonucleic acid (DNA) into primary miRNAs (pri-miRNAs) [14]. Mature miRNAs are a component of the RNA-induced silencing complex (RISC), which permits posttranscriptional control of gene expression. They bind to complementary sequences in the 3′ untranslated regions (3′UTRs) of their target messenger RNAs (mRNAs) and are processed by a particular machinery in the cytoplasm and nucleus [60]. Additionally, miRNAs can target proteins and DNA in addition to mRNAs [25]. According to Lewis et al. [107], miRNAs regulate almost one-third of human genes, and the synthesis of mature miRNAs is intricate. In the RISC, the mature and functional miRNA is loaded onto the Argonaute protein after pre-miRNAs are broken down by the RNase Dicer [14].

Since Dicer is required for miRNA processing, conditional gene ablation has been used to inactivate this gene, which has been used to investigate the function of miRNA in several organ systems [33, 68, 180]. The correct silencing mechanism-degradation of the mRNA target or translational silencing is determined by the complementarity between the miRNA and the mRNA target; these mechanisms ultimately lead to a reduction in the quantity of protein generated [14]. As a result, abnormal cellular activities, altered downstream gene regulation, and altered signaling cascades can result from poor miRNA expression. Although the majority of miRNAs work by interacting with the 3′-untranslated region (3′-UTR) of target mRNAs [64], they can also interact with coding sequences, gene promoters, and the 5′-UTR [20]. It has been demonstrated that miRNAs have a role in the control of gene expression by targeting mRNA, which results in mRNA degradation or posttranscriptional gene silencing, hence regulating the creation of proteins [146]. miRNAs have a significant impact on many biological processes, including signal transduction, cell death, cell proliferation, maturation and differentiation, modulation of chronic inflammation, carcinogenesis, and regulation of numerous cellular and metabolic pathways [5]. The development of a number of applications for mRNA/miRNA interaction prediction has improved the selection and interpretation of miRNA target prediction analytic tools and advanced our knowledge of how miRNAs exert their regulatory effects [160].

miRNAs are essential for a number of critical developmental and disease processes, from the development of Drosophila larvae to the control of human cancer growth. The involvement of miRNAs in CRC has been the primary focus of attention in the intestine thus far [2]. Certain miRNAs have also been linked to ulcerative colitis (UC) [212]. Technical constraints also resulted in a very restricted sequencing depth for the "colorectal microRNAome" [35]. Leukocytes express over 100 miRNAs [234], and many of these miRNAs have multiple targets within immune-related pathways, indicating that they play a significant role in the immune system [28]. Among the most extensively researched miRNAs in the immune system are *miR-155, miR-146a, miRs-17~92*, and *miR-181a* [13]. The development of therapeutic techniques based on miRNAs is being studied since intestinal disorders are significantly linked to the disruption of miRNA regulation and function. Recent research has demonstrated that miRNAs exhibit variable expression in autoimmune disorders; specifically, miRNAs have been linked to important inflammatory pathways that play a role in the pathophysiology of inflammatory bowel disease (IBD) [83, 94]. Biogenesis of miRNA is presented in Fig. 19.1.

Role of miRNAs Expression in the Intestinal Tract

According to the Tissue Atlas database, small intestine and colon express over 70% and 72% of all types of miRNAs, respectively [98]. Intestinal epithelial cells (IECs) are the source of the majority of fecal miRNAs. Some miRNAs, such as *miR-515-5p, miR-101, miR-325, miR-1253, miR-1226-5p, miR-876-5p, miR-1224-5p,* and *miR-623*, have been shown to target bacterial genes and control the makeup of bacterial communities in the gut [16, 119, 123]. In contrast, certain bacterial taxa and their by-products, including amyloids, lipopolysaccharide, and butyrate, control the expression of host genes [109, 154]. In CRC cells, butyrate therapy raised *miR-203* levels while inhibiting cell invasion, proliferation, and apoptosis [67]. Hu et al. [79] also demonstrated that butyrate produced from microorganisms suppressed *miR-92a* in CRC cells, while a certain mRNA can bind to many miRNAs [201]. Table 19.1 lists the number of predicted targets for each of the ten most prevalent intestinal miRNAs.

Role of miRNAs in Intestinal Transport

According to recent reports, certain miRNAs control the expression of intestinal transporters, including peptide transporter 1 (PepT1; an oligopeptide transporter) [39, 40], Downregulated in adenoma (*DRA*; Cl-/HCO3-exchanger) [7], Na-/K-/Cl cotransporter [58], and cystic fibrosis transmembrane conductance regulator (CFTR). *PAT-1* 3′UTR contains possible binding sites for many miRNAs, including *miR125a-5p, 339-5p, 423-5p, 485-5p,* and *501-3p,* according to miRNAs in the regulation of proton-coupled amino acid transporter 1 (*PAT-1*). One of these, *miR125a-5p,* was discovered to suppress PAT-1's mRNA and protein expression [8]. Furthermore, a recent study showed that *miR-29a, miR-29b,* and *miR-29c* mediated the regulation of monocarboxylate transporter-1 (MCT1) in pancreatic beta-cells [163] and that *miR-125a-5p* can use mRNA degradation to posttranscriptionally modify *PAT-1* expression [44].

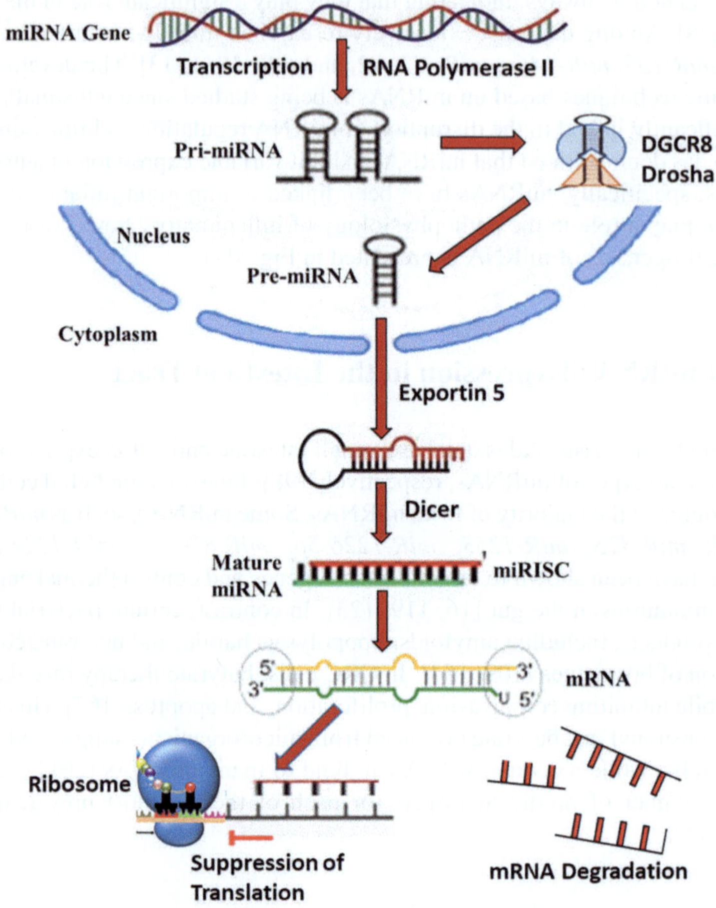

Fig. 19.1 Biogenesis of miRNA. In the nucleus, RNA polymerase II primarily transcribes miRNA genes to create pri-miRNA transcripts. The pre-miRNA precursor product is created by the microprocessor complex, which is made up of the dimeric RNA-binding protein DGCR8 and the RNase III enzyme Drosha, cleaving pri-miRNA. The cytoplasm receives the pre-miRNA through exportin 5. A complex consisting of the transactivation response element Dicer cleaves the pre-miRNA into a mature double-stranded miRNA. Following that, a mature miRNA is added to the miRNA-associated multiprotein RNA-induced silencing complex (mi-RISC). After binding to complementary areas of the target mRNA, the mature miRNA controls the production of the mRNA by acting as a guide through base pairing. The mature miRNA typically suppresses the translation of mRNAs into proteins by binding to the 30-untranslated regions (30-UTR) of particular mRNAs through partially complementary sequences. When the miRNA and mRNA have a strong complementarity, the target mRNA is cleaved

Dysregulation of PepT1 and intestinal inflammation have been linked to altered expression of *miR-92b* and *miR-193a-3p* [150]. PepT1 expression is suppressed by *miR-92b* by binding to the 3′UTR of PepT1 mRNA and encouraging its destruction. According to this research, intestinal epithelial cells' pro-inflammatory reactions

Table 19.1 Processes and functions of the most abundant miRNAs in the intestine

miRNA	Site of expression	Processes and functions	References
miR-21-5p	Small intestine, colon	Encourages the survival and growth of cancer cells by directly blocking its targets, such as *PTEN, PDCD4, RECK,* and *SPRY2*; in CRC; it encourages tumor invasion and metastasis through modifying the expression of several genes linked to cancer, such as *PTEN, PDCD4,* and *TGFβR2*	Nguyen et al. [143]
miR-24-3p	Small intestine, colon	Due to its impact on cell-cell junctions, it compromises the integrity and function of the intestinal barrier; in intestinal epithelial cell lines from patients with UC, its overexpression results in a reduction in the tight junction-associated protein cingulin and impaired barrier formation	Soroosh et al. [191]
miR-26a-5p	Small intestine, colon	Inhibits the intestinal inflammatory response in macrophages by lowering *NF-κB/STAT3* activation and interleukin 6 production; it plays a suppressive effect in CRC linked with colitis	Zhang et al. [231, 232]
miR-27a-3p	Small intestine, colon	Supports the growth, migration, and invasion of CRC cells; it influences the Wnt/β-catenin pathway through retinoid X receptor targeting; linked to distant metastases, histological differentiation, clinical stage, and survival of CRC patients	Liang et al. [111]
miR-29b-3p	Small intestine, colon	Suppresses cyclin-dependent kinase 2 translation, which inhibits intestinal mucosal expansion; it also suppresses menin mRNA translation, which modifies intestinal epithelial cell death and impacts intestinal epithelial homeostasis	Xiao et al. [217], Ouyang et al. [152]
miR-143-3p	Small intestine, colon	Has an impact on intestinal epithelial regeneration following dam age; it is expressed and only works in the intestinal mesenchymal compartment	Chivukula et al. [31]
miR-145-5p	Small intestine, colon	Impacts the expression of *SOX9*, a crucial transcription factor that is expressed and only functions in the intestinal mesenchymal compartment; it negatively regulates the expression of claudin 8, which is associated with decreased intestinal permeability and mucosal barrier homeostasis in Crohn's disease	Chivukula et al. [31], Zhuang et al. [244]
miR-192-5p	Small intestine, colon	Inhibits glycolysis by controlling the production of the protein X-linked 2, which contains sushi repeats, in colon cancer cells; it has an anticancer effect on CRC	Zhao et al. [236, 237]
miR-199a-3p	Small intestine, colon	By downregulating the interleukin-17A/interleukin-23 axis, it improves the intestinal barrier in UC, lowers ROS production, and increases the expression of junction protein in the intestinal tissue. Its loss exacerbates CRC by activating EMT-related signaling and targeting DDK1	Hu et al. [78]
let-7a-5p	Small intestine	It regulates cell cycle, cell proliferation, metabolism, apoptosis, and stemness	Mizuno et al. [135]

triggered by bacterial peptides are suppressed by *miR-92b* [38]. PepT1 and *miR-193a-3p* levels have also been shown to be inversely correlated in inflammatory colon tissues with active UC. Without altering the amounts of mRNA, this miRNA prevents the translation of PepT1 mRNA. According to Dai et al. [36], *miR-193a-3p* suppresses the following activation of inflammatory transcription factors by lowering PepT1 protein levels and activity. As regulators of colonic PepT1, *miR92b* and *miR-193a-3p* are both miRNAs that help preserve intestinal homeostasis [150].

In pancreatic beta-cells, a recent study demonstrated that the *miRNA-29* cluster controlled the regulation of *MCT1*, demonstrating that miRNA influenced the regulation of short-chain fatty acid transporters (SCFA). The control of *MCT1* expression by *miR-29a, miR-29b,* and *miR-29c* translated into a downregulation of sodium-coupled *MCT1* (SMCT1) expression [163]. It has been suggested that differently expressed miRNAs along the intestinal regions' length may control the expression of target genes in particular locations [52]. Additionally, the expression of *miR-29a* and *miR-29*c was considerably lower in the colon than in the small intestine. These miRNAs may have a role in controlling the expression of *SMCT-1* in specific regions. Since *miR-29a, miR-29b,* and *miR-29c* have been demonstrated to suppress *MCT1* expression, it is probable that the *miR-29* cluster also regulates the expression of *MCT1* in specific regions throughout the intestine [9].

Trimethylamine-N-oxide (TMAO) is primarily transported via the organic cation transporter 2 (OCT2); dietary TMAO intake raises the expression of *miR-21-5p*, which is known to play a role in inflammation, and *miR-30c-5p,* which is concerned with lipid metabolism [42]. Intestinal miRNAs may play a part in controlling lipid metabolism; according to a study, lipid intake changes the expression of miRNAs in the liver (*miR-30d-5p, -129-2-3p, -147-3p,* and *-202-5p*) and colon (*miR-30d-5p, -129-2-3p, -202-5p, -216b-5p-3p, 217-5p,* and *-218-2-3p*). In the duodenum and jejunum, *miR-218-2-3p* expression rose sharply in males during the second hour after dietary fat consumption and dropped back to baseline at the fourth hour, but in females, the expression level was steady. Two hours following dietary lipid intake, men jejunum showed a substantial increase in *miR-138-1-3p* levels, while female duodenum showed an increase in *miR-129-2-3p* levels [59].

Zinc finger proteins have roles in transcription, DNA repair, cell cycle progression, and miRNA degradation. The zinc finger SWIM region of zinc dinger SWIM-type containing 8 (ZSWIM8) contributes to target-directed degradation of *miR-7* [183]. Apart from its usual physiological functions, zinc also modifies the expression of miRNAs in pathological situations. Reduced expression of zinc transporters and lower zinc levels are seen in prostate cancer tumors. It has been discovered that a miRNA group controls zinc transporters. Zinc transporters are suppressed by *miR-183, miR-90,* and *miR-182,* which are overexpressed in prostate cancer tissue. Despite this, it is believed that eating more zinc reduces the incidence of cancer and its associated mortality [134]. The intestinal mouse epithelium's *miR-7027* directly controls the transporter Zip14, which binds to the 30 untranslated region (3′UTR) of the *STAT3* and interleukin-21 (IL-21) genes in response to pro-inflammatory stimuli [92]. Both crypts and villus exhibit variable expression of *miR-7027*, which has been linked to pathological conditions [148]. The intestinal riboflavin transporter-3 was shown to be posttranscriptionally regulated by *miR-423-5p*. It was discovered that intestinal epithelial cells had high levels of *miR-423-5p*, and that its mimic impacted the luciferase reporter activity of the pmirGLO-hRFVT3-3′UTR construct. Additionally, it inhibited the intestinal epithelial Caco-2 and HuTu-80 cells ability to absorb riboflavin [105].

miR-24 inhibits the transport function of organic anion transporting polypeptide 2B1 transporter (OATP2B1) and controls the production of the mRNA and protein of the OATP2B1. The distribution of OATP2B1 substrate medicines may be impacted by changes in *miR-24* expression. It may also play a crucial role in controlling OATP2B1 activity in diseases such as hyperlipidemia and cancer [120, 121]. Changes in OATP2B1's transport activity may affect the pharmacological effects of medications like statins and fexofenadine, since it is likely to aid in their absorption in the intestine [47, 139]. According to reports, CRC tissues exhibit decreased expression of *miR-24* when compared to their equivalent noncancerous tissues [51], whereas CRC tissues exhibit increased expression of OATP2B1 [103]. The correlation between OATP2B1 expression and changes in *miR-24* expression suggests that *miR-24* suppresses OATP2B1 expression in cancer cells, the liver, and the gut [46]. Additionally, increased expression of miR-24 explains decreased expression of serotonine transporter in intestinal epithelial cells seen in irritable bowel syndrome [112], supporting the functional expression of *miR-24* in the intestine to assess its role in regulating intestinal OATP2B1.

There are numerous water-soluble vitamin transporter genes (*SLC23A2, SLC19A2, SLC52A3, SLC26A3*, and *SLC15A1*) that are regulated posttranscriptionally by miRNAs [50, 86, 100, 105, 171]. A micronutrient that is necessary for healthy cellular growth, development, and function is vitamin C. One important factor in the intestinal absorption of amino acids is the apically localized intestinal vitamin C transporter-1 [192]. Both intestinal cell lines and the mouse jejunal mucosa exhibited significantly increased levels of mature *miR-103a* expression [193]. In mouse and human primary pancreatic acinar cells (PACs), *miR-200a-3p* significantly inhibits thiamin uptake and thiamin transporter-1 (THTR-1) expression (both at the protein and mRNA levels). This is because *miR-200a-3p* is involved in the membrane transporter of vitamin B1, also known as THTR-1 [166]. Recent studies have demonstrated that miRNA can have a significant impact on human physiology and transporters, potentially affecting transporters along the human intestine.

Role of miRNAs in IEC Pathophysiology

Variations in miRNA expression levels have been linked to the onset and progression of various diseases, including cancer [15, 34], neurological disorders [3, 11], cardiovascular diseases [140, 178], kidney diseases, and particularly cancers [200, 203]. miRNAs have been demonstrated to be essential for many cellular functions

[130], such as cell differentiation [219], proliferation [93], apoptosis [32], and metabolism [168]. Additionally, abnormal expression of miRNAs has been linked to a number of diseases, including Crohn's disease (CD) and UC [155, 214], as well as cancer [117].

A number of human illnesses, including allergies, different autoimmune disorders, and malignancies, are linked to disruption of the gut immune system and microbiota [95, 169, 179]. CD and UC are the two most prevalent forms of IBD, one of the most well-known and extensively researched classes of intestinal diseases [37]. IBD patients are far more likely to get CRC [141]. People with IBD have recently been demonstrated to have altered miRNA expression profiles. The miRNA profiles of UC and CD patients differ from those of healthy controls according to their disease stage [37, 84, 113, 155, 214]. During UC and CD, it was discovered that the following miRNAs had changed expression: *miR-16* [213], *miR-146a* and *miR-31* [113], *miR-340,* and *miR-199a-5p* [214].

Additionally, *miR-124* directly bonded to Stat3 and suppressed its expression. Only children with active UC exhibited high Stat3 levels and correspondingly low *miR-124* levels; not exhibited in children with CD, adults, or inactive UC or healthy individuals. In sick patients, hypermethylation of *miR-124* results in dysregulation of the Stat3 pathway [104]. Another study by Brain et al. [19] demonstrated that human CD patients with nucleotide-binding oligomerization domain-containing protein 2 (NOD2) gene polymorphisms could not increase *miR-29* expression. Consequently, *miR-29* was unable to suppress downstream cytokine signaling pathways linked to worse illness. Overall, these results confirm that miRNAs and their experimentally confirmed targets are significant contributors to IBD in humans.

Expression profiling has revealed that CRC tumors have downregulated levels of *miR-143, miR-145,* and *miR-451,* while increased levels of *miR-31, miR-21,* and *miR-191* [31, 116]. According to another study, the most highly expressed miRNAs in a CRC cohort were *miR-10a-5p, miR-21-5p, miR-22-3p, miR-143-3p,* and *miR-192-5* [176]. The results of a pathway analysis showed that these changed miRNAs may impact numerous cellular signaling pathways, such as those connected to oncogenesis, such as Wnt, MAPK, and TGF-β. Recent research has identified critical roles for *miR-143* and *miR-145* in pathways that can result in CRC, and animal studies employing genetic manipulation of miRNAs within the intestines are currently being conducted to understand cellular mechanisms of CRC-related miRNAs that have been found in human studies [31].

The *miR-146a* gene located on chromosome 5 has rs2910164 single nucleotide polymorphisms (SNPs) that have been demonstrated to reduce mature *miR-146a* levels and subsequently increase its target genes [88, 181]. Papillary thyroid carcinoma [88], sepsis [181], and other human disorders have all been linked to this miRNA SNP. Gazouli et al. [56] found that this *miR-146a* SNP was linked to CD in respect to the GI tract, but no significant correlation was found with UC patients [56, 149]. Intestinal metaplasia and dysplasia after *Helicobacter pylori* infection were similarly observed to be more likely in patients with this SNP mutation [189]. Furthermore, according to Chae et al. [26] and Ma et al. [125], this variation in *miR-146a* predicted cancer risk and disease-specific survival outcomes. The NFκB pathway is targeted by *miR-146a*, which reduces inflammation [18, 233, 234]. These studies indicate that a single nucleotide change within this miRNA may change the intestinal inflammatory state by dysregulating NFκB and other related pathways. In this chapter review, we will next focus on how miRNA can contribute to the pathophysiological regulation of tumor progression and metastasis.

Role of miRNAs in Tumor Progression and Metastasis in IECs

A key regulator of metastasis, miRNAs can either stimulate or inhibit metastatic processes through various mechanisms. These intricate biological processes encompass several key phenomena, including epithelial-mesenchymal transitions (EMT), which facilitate the transformation of stationary epithelial cells into more mobile mesenchymal cells. Additionally, they involve the unique characteristics of cancer stem cells, which are capable of self-renewal and driving tumor growth. Other critical aspects include the invasion of surrounding tissues, the colonization of distant sites in the body, and the dynamic migration of cancer cells. Furthermore, there is the manipulation of the tumor microenvironment, where cancer cells interact with and alter the surrounding stromal cells, blood vessels, and immune components to promote tumor survival and progression [101]. miRNAs play a crucial role in the intricate control of metastasis, coordinating various phases, including local invasion, anoikis resistance, extravasation, and colonization. miRNAs have a major impact on the metastatic cascade through their regulatory roles [27, 220]. Through mediating interactions between tumor and non-tumoral cells, miRNAs play a critical role in shaping the tumor microenvironment. Critical processes essential for metastasis, such as immunosuppression and angiogenesis, are influenced by this interplay [188].

Mesenchymal epithelial transition (MET), the reverse process of EMT, is precisely regulated by miRNAs to control the motility, invasiveness, and colonization of cancer cells at distant locations. According to Chan and Wang [27] and Solé and Lawrie [188], this dynamic control of EMT/MET processes demonstrates the importance of miRNAs in metastasis dissemination. Due to their crucial roles in controlling metastasis, several miRNAs have emerged as promising therapeutic targets for cancer treatment. Targeting these miRNAs may enable new therapeutic approaches to reduce metastasis progression [101, 132]. Additionally, dysregulated miRNAs have been linked to influencing invasion and metastasis, which are the two crucial elements of cancer progression, and may, therefore, serve as prognostic and diagnostic biomarkers for the disease. According to Peng and Croce [156], their aberrant expression profiles facilitate the design of individualized treatment plans and provide valuable insights into the prognosis of the condition. In CRC and other malignancies, data collectively demonstrate that miRNAs play a complex function in promoting tumor growth and metastasis. These studies shed light on potential directions for targeted treatments and diagnostic techniques to combat metastatic disease.

Role of miRNAs in CRC

Several types of cancers, such as breast, colon, pancreatic, and stomach cancers, demonstrate elevated levels of the pro-oncogenic gene *miR-21*. Understanding the role of *miR-21* in these cancers can provide valuable insights for developing targeted therapies and improving treatment outcomes. In recent years, it has been associated with the development of autoimmune diseases [198]. Although resting T cells and antigen-presenting cells (APCs) exhibit low levels of *miR-21* expression, cell stimulation results in noticeably increased levels of *miR-21* expression [172]. The intestinal tract serves as an important organ where the *miR-125a* is widely expressed, playing a crucial role in regulating various functional targets. This highlights the significance of *miR-125a* in intestinal health and its potential impact on related biological processes [161]. According to Potenza et al. [162], expression of *miR-125a-5p* typically increases with cell differentiation; however, it is downregulated in various tumors such as retinoblastoma, medulloblastoma, glioblastoma, neuroblastoma, hepatocellular carcinoma, and cancers of the breast, gastric, cervical, lung, ovarian, and colon. The progression of CRC from adenoma to carcinoma is strongly influenced by the regulation of miRNAs [34, 170, 194]. The role of miRNAs involved in CRC is explained in Fig. 19.2.

The EMT, apoptosis, metastasis, and cell survival are all significantly influenced by the dysregulated expression of miRNAs in CRC [30]. According to Michas et al. [133], these short noncoding RNAs play a critical role in CRC carcinogenesis by modulating key pathways such as Wnt/β-catenin and TGF-β signaling. They act as inflammatory mediators, tumor oncogenes, or tumor suppressors. Additionally, inflammation and EMT are closely related processes in the development of CRC, with miRNAs being vital for regulating both. miRNAs play a crucial role in targeting inflammatory signaling molecules, which in turn affects the transition of EMT, a critical stage in the development of CRC. By altering the expression of important molecules such as E-cadherin, miRNAs significantly influence EMT and the metastatic potential of CRC [30, 153]. The abnormal activation of the Wnt/β-catenin signaling pathway is a primary cause of CRC. Some miRNAs, like *miR-145*, specifically target genes within this pathway, including APC. This targeting leads to enhanced Wnt signaling, which is associated with a higher tumor grade and poorer prognosis in CRC patients [133]. For instance, restoring the expression of *miR-143* has been shown to suppress tumor cell proliferation and impact malignant transformation in colon cancer cells [30].

Tumor growth, metastasis, and the increased malignancy of tumor cells in CRC are significantly influenced by the dysregulation of miRNA expression [1]. Research has identified specific miRNAs that are dysregulated in CRC. Notably, the following miRNAs are downregulated: *miR-490-3p, miR-628-3p/-5p, miR-1297, miR-3151, miR-3163,* and *miR-3622a-5p*. In contrast, the following miRNAs are upregulated: *miR-105, miR-549, miR-1269, miR-1827, miR-3144-3p, miR-3177, miR-3180-3p,* and *miR-4326* [66]. In addition to serving as a strong training set for validation in larger cohort studies, these dysregulated miRNAs show promise as

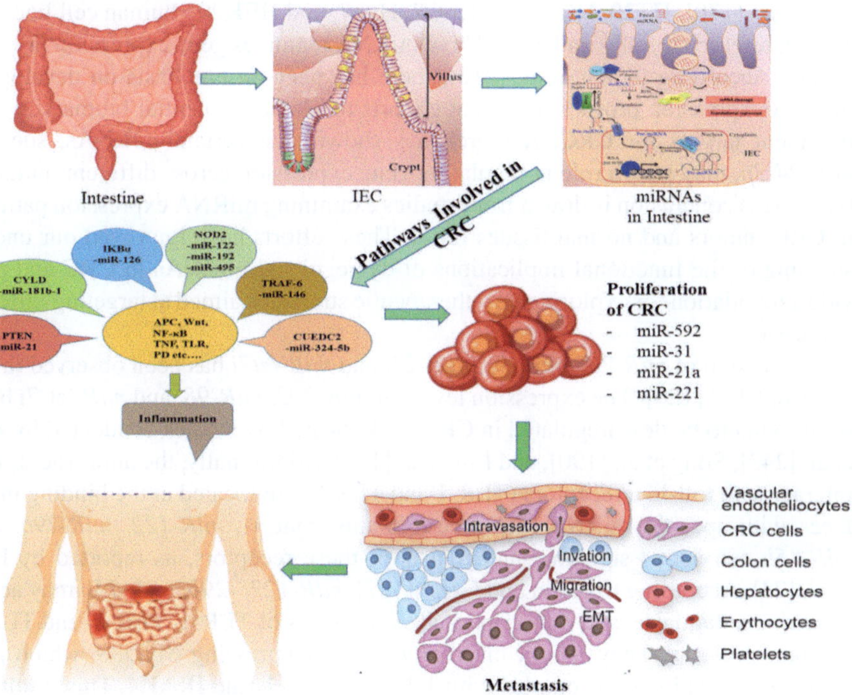

Fig. 19.2 Role of miRNAs involved in CRC**.** The figure illustrates the involvement of miRNAs in the multistep process of CRC development. Normal colonic epithelium undergoes transformation through crypt dysregulation and altered gene expression, where fecal and intracellular miRNAs modulate mRNA degradation and translational repression. Key signaling pathways including APC, Wnt, NF-κB, TNF, and TLR are influenced by dysregulated miRNAs (e.g., miR-21, miR-181b-1, miR-126, miR-122, miR-192, miR-495, miR-146, and miR-324-5b), contributing to inflammation and tumorigenesis. These changes promote uncontrolled proliferation, epithelial-mesenchymal transition (EMT), invasion, and migration, facilitating intravasation into the vasculature. The metastatic cascade is shown with circulating CRC cells interacting with vascular endothelial cells, hepatocytes, erythrocytes, and platelets, ultimately driving disease progression

biomarkers for both the diagnosis and prognosis of CRC [66]. Various dysregulated miRNAs have been clinically linked to the development, spread, and response to treatment of CRC. To accurately evaluate the efficacy of miRNAs in diagnosing and prognosing CRC, thorough testing across multiple cohorts is essential [43]. The oncogenes and tumor suppressors influenced by miRNAs in CRC affect critical cellular functions such as angiogenesis, metastasis, apoptosis, and proliferation. The dysregulation of miRNAs alters protein synthesis by disrupting the normal expression levels of target mRNAs [170].

The role of miRNAs in CRC is being explored through various experimental models to better understand how these molecules contribute to the development and progression of cancer. For instance, specific miRNAs, such as *miR-33a*, have been identified as potential tumor suppressors. Studies using animal models, particularly mice, have been essential for examining the effects of these miRNAs on tumor formation [170]. HT29, HCT116, HepG2, Huh7, and HEK293 human cell lines are used in research to fully understand how miRNA regulates gene expression in colon cancer. According to Wu et al. [216], there has been an emphasis on developing miRNA expression profiles in patient cohorts to identify potential biomarkers and treatment targets for CRC. Research has shown that certain miRNAs, such as *miR-24-3p*, may play a role in regulating gene expression across different mutation types. This conclusion is drawn from studies examining miRNA expression patterns in CRC tumors and normal tissues [204]. These efforts have improved our understanding of the functional implications of dysregulated miRNAs in CRC and provide a foundation for exploring new therapeutic strategies aimed at targeting miRNA expressions.

Expression of *miR-202, miR-98, miR-21,* and *miR-let7i* has been observed in the cases of CRC [129]. The expression levels of *miR-202, miR-98,* and *miR*-let-7i have been found to be downregulated in CRC, as indicated by studies conducted by Zhu et al. [243], Song et al. [190], and Lin et al. [114]. Additionally, the amino acids that interact with toll-like receptors (TLR-1 and TLR-8) are located at the binding interfaces with specific miRNA sequences. In this context, *miR-122, miR-29b*, and *miR-15b* can act as signaling antagonists of these receptors, as reported by Heil et al. [74]. According to Karin and Greten [97], *miR-122, -29b,* and *-15b* may act as paracrine antagonists and transcriptional suppressors of TLR-1, TLR-7, and TLR-8 receptor proteins. They inhibit macrophages by suppressing MyD88, which prevents the formation of a complex with TRAF6, IRAK1, and IRAK4. This results in the suppression of NF-kB and the secretion of pro-inflammatory cytokines like TNF-α and IL-6.

Research has shown that the *pri-miR-146b* (primary microRNA) interacts with heterogeneous nuclear ribonucleoprotein A2/B1 (hnRNPA2B1) and undergoes N6-methyladenosine (m6A) modification by methyltransferase-like 3 (*METTL3*). Inhibiting this interaction has been linked to an increased risk of developing CRC. Additionally, this knockdown enhances the effectiveness of anti-PD-1 immunotherapy and raises the expression of programmed death-ligand 1 (*PD-L1*) in tumor-associated macrophages (TAMs) through the p110β/PI3K/AKT pathway [73]. The binding of YTH domain-containing protein 2 (*YTHDC2*) inhibits the production of *pri-miR-17*, which is modified by the m6A methylation process mediated by *METTL14* and is found to be suppressed in CRC. When *YTHDC2* binding is reduced, there is an increase in the expression of *pre-miR-17* and *miR-17-5p*, along with a decrease in the m6A modification of *pri-miR-17*, particularly when *METTL14* expression is lowered [195]. In CRC, there is an overexpression of *METTL3*, and its growth is inhibited by knocking down *METTL3*. Additionally, *METTL3* promotes metastasis in cancer cells and enhances the expression of *miR-196b* by modifying *pri-miR-196b* with m6A [81]. The MAPK pathway is activated by the m6A modification of *pri-miR-1246* by *METTL3*, leading to increased expression of *miR-1246*, which promotes EMT and invasion in CRC [158]. *miR-675*, which is associated with a poor prognosis, is highly expressed in gastrointestinal stromal tumors (GISTs). By decreasing myosin phosphatase-targeting protein 1 (*MYPT1*), the m6A modification of *miR-675* by *METTL3* stabilizes *miR-675*, which contributes to cancer development by inhibiting the tumor-initiating RhoA/NF2/YAP1 signaling pathway [115]. Additionally, *METTL3* overexpression enhances the growth of GISTs by upregulating the expression of pre-*miR-25*, which subsequently increases the expression of *miR-25-3p* [164]. Such points demonstrate the dynamic nature of the miRNA and accentuate the need for further research.

Role of Exosomal miRNAs in CRC

Exosomal miRNAs are crucial in regulating inflammatory responses. By managing the expression of inflammation-related miRNAs, it can lead to reduce inflammation, minimize intestinal mucosal damage, and promote the growth and repair of IECs. This approach helps maintain the integrity of the intestinal barrier and lowers the risk of CRC, emphasizing the importance of targeting miRNA regulation for better gut health [70]. Exosomal miR-19a is significantly increased in the serum of CRC patients [127]. Colitis and colitis-associated CRC (CAC) are caused by *miR-19a*, which directly enhances NFκB signaling. TNF-α-induced protein 3 raises *miR-19a* levels, which can boost TNF-α production and activate NFκB. TNF-α activation, in turn, raises *miR-19a* expression. Furthermore, CAC can be lessened by blocking *miR-19a* [206]. Serum exosomes from CRC patients have been shown to have lower levels of miR-6869-5p. According to Yan et al. [221], Makarova et al. [126], Weil et al. [211], and Li et al. [110], miR-6869-5p directly targets Toll-like receptor 4 (*TLR4*) and inhibits cell proliferation and inflammatory cytokine production in CRC cells through the TLR4-NF-κB signaling pathway. This has protective effects on CRC and can be used as an inhibitor in CRC.

Protein tyrosine phosphatase receptor type O (PTPRO) negatively correlates with exosomal *miR-6803-5p*, which is elevated in CRC. By activating the TLR4/NF-κB signaling pathway, PTPRO exacerbates inflammation in UC [207]. Through the PTPRO/NF-κB axis, *miR-6803-5p* enhances oxidative stress and promotes the invasion and proliferation of cancer cells in CRC. When *miR-6803-5p* binds to the 3' UTR of the PTP receptor type 4 (PTPR4), it may promote the degradation of PTPR4 mRNA and inhibit its translation, resulting in reduced levels of PTPR4 protein expression [222].

Exosomal *miR-224-5p*, secreted by CRC cells, promotes cell proliferation, invasion, migration, and the malignant transformation of normal human colon epithelial cells through the downregulation of *CMTM4* expression. This process also contributes to tumorigenesis and malignant transformation [231, 232]. According to Qian et al. [165], *miR-216a-5p* regulates macrophage balance by modulating the high mobility group box 1 protein (*HMGB1*) and the TLR4/NF-κB signaling pathway. *TLR4* is activated by *HMGB1*, an inflammatory mediator, through the cluster of differentiation [96]. NF-κB can be activated through TLR signaling, with *TLR4* being a member of the TLR receptor family [12]. Almost all cases of CAC show increased TLR4 expression, which promotes reactive oxygen species (ROS) formation in IECs and activates microbiota [23, 48, 49, 110].

IECs play a crucial role in maintaining the integrity of the intestinal barrier and supporting innate immunity. Recent research indicates that patients with CRC exhibit elevated blood levels of exosomal *miR-222-3p*. One of the primary functions of *miR-222-3p* is the regulation of reactive oxygen species (ROS). In a mouse model of colitis, the effects of *miR-222-3p* on ROS in UC and CAC have been studied through both *in vitro* and *in vivo* experiments. Research has shown that inhibiting *miR-222-3p*, which targets the Brahma-related gene 1, activates the Nrf2/HO-1 signaling pathway in CAC IECs. This activation increases GSH-Px levels while reducing reactive ROS and MDA levels, leading to a decrease in the expression of TNF-α and IL-1β. Reducing ROS and inflammation in the damaged colon can help diminish colon inflammation and the risk of tumor development [54, 55, 110, 209]. Human umbilical cord mesenchymal stem cell-derived exosomes (HucMSC-Ex) release *miR-129-5p*, which targets and inhibits the expression of acyl-CoA synthetase long chain family member 4 (ACSL4) to reduce intestinal inflammation and repair damage, ACSL4 favorably modulates lipid peroxidation, which reduces IBD [110, 210]. More investigation is necessary on the mechanisms underlying this inhibition.

miRNAs as Diagnostic Biomarkers in CRC

The distinct expression patterns of miRNAs in cancerous and healthy cells make them promising candidates as potential diagnostic biomarkers for CRC. [177]. Numerous studies have identified specific miRNAs that may serve as prognostic and diagnostic markers for patients with CRC [102]. miRNAs in CRC have been extensively studied, mainly using stool and blood samples [172]. According to Zhao et al. [238], fecal miRNAs are considered reliable biomarkers for CRC screening due to their relatively high repeatability and stability in stool samples. Recent studies have investigated miRNAs in urine samples from CRC patients [87]. The most frequently examined miRNAs include *miR-21, miR-92a, miR-20a, miR-29a, miR-221, miR-23a, miR-223,* and *miR-150. miR-92a* has been associated with several other diseases, including liver cancer [184], breast cancer [186], and heart disease [90]. Therefore, its significance is not exclusive to CRC. The most well-studied and validated miRNAs, *miR-21* and *miR-92a*, have demonstrated significant potential as diagnostic markers.

The *miR-17-92* precursor cluster, which includes *miR-92a*, produces five miRNAs: *miR-17, miR-18a, miR-19a, miR-20a,* and *miR-19b* [199]. Research has suggested that *miR-20a* could serve as a diagnostic serum biomarker for several types of cancer, including prostate [21], gastric [157], and nasopharyngeal cancers [228]. It is notably found to be elevated in both solid tumors and hematologic cancers. While Yang *et al.* found that *miR-20a* was downregulated [223], research suggests that deregulation of *miR-29a* contributes to a number of different cancer types [144]. According to Mo and Cao [136], *miR-29a* can consistently function as an oncogene and a tumor suppressor in the development of CRC. Yang et al. [223]

found that *miR-20a* is downregulated in certain conditions. In contrast, research indicates that the deregulation of *miR-29a* plays a role in various cancer types [144]. Mo and Cao [136] suggest that *miR-29a* can act both as an oncogene and a tumor suppressor during the development of CRC. Huang et al. [80] and Shiosaki et al. [185] found that *miR-29a* acts as an oncogene and is expressed at elevated levels in CRC. Significant results for CRC detection were observed when combined with *miR-19a, miR-19b, miR-15b, miR-335,* and *miR-18a* [77]. A poor prognosis for patients is linked to higher levels of *miR-221*, which is recognized as an oncomiR [41]. In the study by Cai et al. [24], *miR-221* expression levels showed a notable change based on patient age, even though there were no significant variations related to gender, tumor type, or stage. Increased expression of *miR-23a* is linked to lymph node metastases, deeper invasion, and advanced tumor stages. This suggests that *miR-23a* could serve as a valuable biomarker for CRC, according to Tang et al. [196]. Ogata-Kawata et al. [147] reported that serum exosomal *miR-23a* has significant diagnostic potential for CRC and is particularly effective in detecting early-stage CRC.

The challenges of using miRNAs as universal biomarkers are highlighted by the varying expression patterns of miRNAs like *miR-150* and *miR-223* across different studies. Research analysis indicated that individuals with CRC may exhibit either upregulation [127, 147] or downregulation [173] of *miR-150.* In a study by Iwasaki et al. [87], it was found that patients with CRC had increased expression levels of *miR-566* and *miR-129-1-3p* compared to urine samples from healthy individuals. The tissue and serum samples from these patients demonstrated elevated expression levels of *miR-566* and *miR-129-1-3p*. This suggests that the overexpression of these miRNAs in CRC tissues leads to their secretion into the bloodstream and excretion into the urine [87].

Adenomas exhibited upregulation of *miR-20a, miR-29a,* and *miR-532-3p,* while CRC showed both up- and downregulation of the same genes. Upregulation of *miR-29b, miR-320a,* and *miR-423-5p* was demonstrated in adenomas, while CRC exhibited downregulation of these genes. Research indicates that *miR-145* and *miR-146a* are downregulated in adenomas, but their regulation may vary in CRC. *miR-151a-5p* is particularly noteworthy for clinical applications as it is downregulated in adenomas and upregulated in CRC. In addition, three other miRNAs, *miR-142-3p, miR-144-3p,* and *miR-193a-5p,* are downregulated in CRC but upregulated in adenomas [226]. According to Ždralević et al. [227], certain markers may be useful in identifying precancerous lesions, as their expression varies with the stage of malignancy. The oncogenic and tumor-suppressive effects of *miR-15b* have been observed in various cancers, including CRC [54, 55, 235]. *miR-15b* is involved in several mechanisms related to drug resistance, apoptosis, proliferation, and EMT. Specifically, it influences the NF-κB, STAT3, AKT/mTORC1, and β-catenin signaling pathways [57].

Emerging data indicates that *miR-15b* plays a significant role in the pathophysiology, development, and response to antitumor therapy in CRC. According to Zhao et al. [239], spectrin beta nonerythrocytic 2 (*SPTBN2*) has been identified as a direct target of *miR-15b*. Additionally, the expression levels of *miR-15b* in CRC cells were found to be significantly lower than in normal cells. The tumor-suppressive role of *miR-15b* is evident from its negative regulation of *SPTBN2*, as higher levels of SPTBN2 have been linked to poor prognosis in CRC patients. Additionally, studies using mouse models of CRC have demonstrated that *miR-15b* exerts its tumor-suppressive effects by inhibiting carcinogenesis, enhancing sensitivity to anti-PD-1 therapy, and reducing the expression of PD-L1 protein [122].

Drug resistance is a major cause of treatment failure, and *miR-15b* has been shown to significantly mediate 5-fluorouracil (5-FU) resistance in CRC. According to Zhao et al. [235], the overexpression of *miR-15b* specifically increases cell death by targeting NF-kB1 and its kinase complex, IKK-α, thereby making CRC cells more susceptible to 5-FU. Additionally, the overexpression of *miR-15b* targets double cortin-like kinase 1 (DCLK1), a potential marker for gastrointestinal stem cells. This activity suppresses the tumorigenic properties of tumor-initiating cells and enhances sensitivity to adjuvant chemotherapy and neoadjuvant radiotherapy in CRC patients [89].

The development of CRC is linked to *miR-21*, one of the most extensively studied miRNAs. The *miR-21* gene is located on chromosome 17, specifically within the vacuole membrane protein 1 (VMP1) locus. Research has shown that the prognosis for CRC is negatively associated with VMP1 expression; a lack of VMP1 expression is connected to a more aggressive form of the disease [61]. Inhibition of several well-known tumor suppressor genes, such as phosphatase and tensin homolog (*PTEN*) [215] and programmed cell death 4 (PDCD4) [10], leads to an oncogenic effect of *miR-21*. This enhances the invasion, intravasation, and metastasis of CRC cells by altering the expression of their downstream targets. Several studies have revealed the oncogenic mechanisms mediated by *miR-21* in CRC. For instance, it inhibited Krev interaction protein 1 (KRIT1) and activated the β-catenin signaling pathway in endothelial cells, leading to increased angiogenesis and vascular permeability [72]. Additionally, it induced pyroptosis in CRC cells by targeting transforming growth factor beta-induced protein (TGFBI) [91]. The knockdown of *miR-21* was associated with an increased expression of Sprouty2, a tumor suppressor gene, which led to a decrease in the proliferation rate of CRC cells. In contrast, the overexpression of *miR-21* promoted the proliferation and invasion of CRC cells and inhibited apoptosis by targeting the Ras homolog gene family, member b (RhoB) [118]. Additionally, miR-21 overexpression partially stimulated tumor growth by downregulating the expression of sec23a [108]. The importance of deregulated stromal miRNAs in the progression of CRC was highlighted by the finding that the expression of *miR-21* was associated with an increase in the stroma of CRC compared to normal tissue. Furthermore, the presence of ectopic stromal *miR-21* expression was linked to greater invasiveness [22]. Research indicates that *miR-21* regulates apoptosis, autophagy, glycolysis, the epithelial-mesenchymal transition, medication resistance, and radiation resistance across various cancer types [71].

As an oncogenic miRNA in CRC, *miR-31* plays a dual role in various human malignancies [225]. Through the p38/JNK pathways, IL-1β triggers the transcription of *miR-31*, which then binds to and directly targets E-selectin, influencing the metastatic process [240, 241]. Inhibition of *miR-31* has an antimetastatic effect by reducing the adhesion and trans-endothelial migration of colon cancer cells [240]. The significance of *miR-31's* role in CRC tumorigenesis is demonstrated by a large number of its targets, including cyclin-dependent kinase inhibitor 2B (CDKN2B), factor inhibiting HIF-1α (FIH-1) [29], paired box 6 (PAX6) [69], tensin 1 (TNS1) [131], Rho-related BTB domain containing 1 (RhoBTB1) [218], *NUMB* endocytic adaptor protein [158], STX12, eIF4EBP1, and eIF4EBP2 [197]. Compared to controls, *miR-31* was discovered to be increased in colorectal adenomas [138]. Furthermore, increased *miR-31* expression has been linked to deeper invasion of CRC tumors [205] as well as to more advanced tumor stages and poorer differentiation [17, 175]. Additionally, *miR-31* has emerged as a potential predictive biomarker for the effectiveness of anti-*EGFR* therapy in metastatic cases [6, 85].

Both the promotion and inhibition of CRC carcinogenesis have been demonstrated by *miR-146a* [124, 145, 174]. Specifically, *miR-146a* acts as a significant negative regulator of CRC development by modulating IL-17 responses, which in turn reduces tumorigenic inflammation. It is closely associated with both inflammatory signaling and the immune response [53]. *miR-146a* has been found to be poorly expressed in human CRC cells, and its ectopic expression seems to have a tumor-suppressive function by inhibiting the proliferation, migration, and invasion of CRC cells [174]. Research has shown that *miR-146a* has an oncogenic role and is upregulated in CRC tissues [124]. According to Lu et al. [124], *miR-146a* is believed to enhance cell migration and invasion by directly targeting carboxypeptidase M (CPM) and regulating the expression of focal adhesion kinase (FAK) and c-Src, a non-receptor tyrosine kinase. The activation of the Wnt signaling pathway was promoted by *miR-146a*, which consistently regulated the division of spheroid-derived CRC stem cells by targeting Numb, a tumor suppressor and segregation determinant [82]. The exosomal *miR-146a* and *miR-155* play a significant role in enhancing the levels of important inflammatory cytokines, such as IL-6, TNF-α, TNF-β, and CXCL12. This elevation of cytokines may contribute to the metastasis of CRC by activating the C-X-C motif chemokine receptor 7 (CXCR7) pathway. Understanding these mechanisms can provide valuable insights for developing targeted therapies to combat CRC metastasis [208]. Using tumor xenograft models, it was shown that the activation of cancer-associated fibroblasts (CAFs) by *miR-146a* and *miR-155-5p* enhances lung metastasis development and invasion *in vivo* [208].

Khorrami et al. [99] demonstrated that overexpression of *miR-146a* in HT-29 CRC cells leads to resistance to irinotecan, 5-FU, and chemotherapeutic agents. The prognosis and susceptibility to CRC have been associated with the *miR-146a* polymorphism rs2910164 [26]. Individuals diagnosed with CRC exhibit varying expression levels of *miR-146a* in their serum [159] as well as in tissue samples [151]. Additionally, Zeng et al. [229] discovered that individuals with higher levels of *miR-146a* tend to have better overall survival rates. Serum *miR-146a,* along with *miR-30e-3p* and *miR-148a-3p,* was found to have significant diagnostic potential for CRC when examined as part of a three-miRNA panel [159]. In patients with CRC, there was an increase in the expression of *miR-146a* in both tumor tissue and plasma.

Research indicated that *miR-146a* was both upregulated and downregulated in CRC, while it was consistently downregulated in adenomas [226]. These findings highlight the importance of miRNAs as diagnostic biomarkers in the induction and progression of CRC.

Therapeutic Use of miRNAs in Intestine

Research has shown that *miR-93* plays a protective role in maintaining the intestinal barrier by targeting protein tyrosine kinase 6 (PTK6). PTK6 is responsible for the nuclear accumulation of forkhead box O1 (FoxO1), which in turn decreases the expression of the tight junction (TJ) protein claudin-3. The authors reported that intestinal barrier function improved in PTK6-null mice. However, the use of *miR-93* mimics reduced TNF-α/IFN-γ mediated disruption of the intestinal epithelial barrier through PTK6 downregulation *in vitro* [65]. *miR-122a* is closely associated with TNF-α-dependent TJ permeability; its overexpression *in vivo* increases intestinal permeability, while its reduction *in vitro* limits TJ permeability [224]. Intestinal epithelial tissue with overexpressed *miR-122* exhibited lower levels of NOD2, which inhibited apoptosis and compromised the integrity of the intestinal barrier [28]. Additionally, the expression of occludin is negatively regulated by *miR-200c-3p* in both patients with UC and in experimental colitis induced by DSS. Antago*miR-200c*, which antagonizes *miR-200c-3p*, prevented the reduction of occludin both *in vitro* and in intestinal tissues of experimental colitis models, thereby maintaining the integrity of the TJ barrier [167].

miR-155 has been shown to reduce the expression of TJ proteins in DSS-induced colitis, which is associated with the failure of the intestinal barrier [120, 121]. In contrast, treatment with a *miR-155* antagomir led to an increase in HIF-1α levels. This indicates that *miR-155* downregulates hypoxia-inducible factor 1α (HIF-1α), a key player in the signaling pathways that regulate anti-inflammatory responses. The findings demonstrated that *miR-155* plays a significant role in DSS-induced colitis by suppressing the HIF-1α/TFF-3 axis, which leads to intestinal barrier dysfunction [120, 121]. Additionally, the regulation of TJs, including occludin, claudin-1, and myosin light chain kinase (MLCK), has been associated with *miR-155* [4]. High levels of *miR-155* have also been shown to downregulate IL-13R1, the main receptor subunit for IL-13, thereby regulating elevated IL-13 levels in UC [63]. Elevated IL-13 levels have been associated with an increased risk of apoptosis and alterations in claudin-2 expression, which contribute to epithelial barrier failure in Th2-mediated UC [75, 76]. According to Guz et al. [62], *miR-155* downregulates the expression of E-cadherin, an important adherens junction protein. This downregulation leads to decreased mucosal stability and increases the risk of cancer spread in patients with UC. Additionally, *miR-155* targets a key protein necessary for facilitating immunoglobulin (Ig) class switch recombination, known as activation-induced cytidine deaminase (AID). Blocking *miR-155* enhances AID levels and improves the gut's immunological barrier [45, 230].

Research indicates that *miR-200b* has a protective effect on gut barrier function by reducing gut permeability and inflammation through its targeting of myosin light chain kinase. This interaction decreases TNF-α-related tight junction dysfunction [182]. Both *hsa-miR-200c-3p* and immune-related IL-8 and CDH11, a crucial component of gut barrier function, were found to have a significant inverse relationship. *Hsa-miR-200c-3p* directly regulates the expression of IL-8 and CDH11 in inflamed mucosa of UC by either downregulating the NF-κB response to TLR4 activation or countering the inflammatory effects of IL-8 activity [202]. In IBD, several miRNAs have been associated with the regulation of nitric oxide synthase-2 (NOS2). Specifically, *miR-21, miR-126, miR-146a, miR-221,* and *miR-223* stimulate the nitric oxide (NO) pathway by increasing levels of heterochromatin protein 1 (HP1), which leads to senescence in adjacent epithelial cells [187]. Additionally, increasing *miR-320a* enhances barrier function in T84 cells, as HIF plays a crucial role in maintaining barrier integrity by responding to ambient oxygen levels in intestinal tissues [137]. These studies emphasize the complexity of miRNA interactions in CRC and highlight the necessity for further research in this field.

Conclusion

In conclusion, a number of significant findings were made by our detailed analysis of miRNA functions in intestinal transports. We have exposed several miRNAs that are up- and downregulated and linked to various physiological processes in intestinal disorders in humans. These miRNAs regulate important signaling pathways that are involved in immune evasion, invasion, and tumorigenesis. The potential of miRNAs as therapeutic targets and diagnostic indicators in the treatment of CRC is further highlighted by our investigation. Modified therapy plans for patients with CRC may also result from exploring the therapeutic potential of miRNA-based therapies in preclinical and clinical settings. Translating these findings into clinically meaningful interventions that eventually enhance patient outcomes in CRC will require close collaboration between researchers, doctors, and industry partners.

Conflicts of Interest The authors declare no conflict of interest.

Author Contributions All authors have made a substantial, direct, and intellectual contribution to the work and approved it for publication.

Funding This research received no external funding.

References

1. Ahadi A (2020) The significance of microRNA deregulation in colorectal cancer development and the clinical uses as a diagnostic and prognostic biomarker and therapeutic agent. Noncoding RNA Res 5:125–134. https://doi.org/10.1016/j.ncrna.2020.08.003
2. Akao Y, Nakagawa Y, Naoe T et al (2006) let-7 microRNA functions as a potential growth supressor in human colon cancer cells. Biol Pharm Bull 29:903–906
3. Alkhazaali-Ali Z, Sahab-Negah S, Boroumand AR et al (2024) MicroRNA (miRNA) as a biomarker for diagnosis, prognosis, and therapeutics molecules in neurodegenerative disease. Biomed Pharmacother 177:116899
4. Al-Sadi R, Engers J, Abdulqadir R (2020) Talk about micromanaging! Role of microRNAs in intestinal barrier function. Am J Physiol Gastrointest Liver Physiol 319:G170–G174
5. Ambros V (2004) The functions of animal microRNAs. Nature 431:350–355
6. Anandappa G, Lampis A, Cunningham D et al (2019) MiR-31-3p expression and benefit from anti-EGFR inhibitors in metastatic colorectal cancer patients enrolled in the prospective phase II PROSPECT-C trial. Clin Cancer Res 25:3830–3838
7. Anbazhagan AN, Priyamvada S, Kumar A et al (2014) Translational repression of SLC26A3 by miR-494 in intestinal epithelial cells. Am J Physiol Gastrointest Liver Physiol 306:G123–G131. https://doi.org/10.1152/ajpgi.00222.2013
8. Anbazhagan AN, Priyamvada S, Borthakur A et al (2019) miR-125a-5p: a novel regulator of SLC26A6 expression in intestinal epithelial cells. Am J Physiol Cell Physiol 317:C200–C208. https://doi.org/10.1152/ajpcell.00068.2019
9. Anbazhagan AN, Priyamvada S, Kumar A et al (2021) MiR-29a, b, and c regulate SLC5A8 expression in intestinal epithelial cells. Am J Physiol Gastrointest Liver Physiol 321:G223–G231. https://doi.org/10.1152/ajpgi.00148.2021
10. Asangani IA, Rasheed SAK, Nikolova DA et al (2008) MicroRNA-21 (MiR-21) post-transcriptionally downregulates tumor suppressor Pdcd4 and stimulates invasion, intravasation and metastasis in colorectal cancer. Oncogene 27:2128–2136
11. Azam HMH, Rößling RI, Geithe C et al (2024) MicroRNA biomarkers as next-generation diagnostic tools for neurodegenerative diseases: a comprehensive review. Front Mol Neurosci 17:1386735
12. Baker RG, Hayden MS, Ghosh S (2011) NF-κB, inflammation, and metabolic disease. Cell Metab 13:11–22
13. Baltimore D, Boldin M, O'Connell R et al (2008) MicroRNAs: new regulators of immune cell development and function. Nat Immunol 9:839–845. https://doi.org/10.1038/ni.f.209
14. Bartel D (2004) MicroRNAs: genomics, biogenesis, mechanism, and function. Cell 116:281–297
15. Baylie T, Kasaw M, Getinet M et al (2024) The role of miRNAs as biomarkers in breast cancer. Front Oncol 14:1374821
16. Bi K, Zhang X, Chen W et al (2020) MicroRNAs regulate intestinal immunity and gut microbiota for gastrointestinal health: a comprehensive review. Gene 11:1075
17. Bin H, Mei H, Hui W et al (2022) The correlation between miR -34a-3p, miR -31, PLEK2 and the occurrence, development and prognosis of colorectal cancer. Cell Mol Biol 68:192–200
18. Boldin MP, Taganov KD, Rao DS et al (2011) miR-146a is a significant brake on autoimmunity, myeloproliferation, and cancer in mice. J Exp Med 208:1189–1201. https://doi.org/10.1084/jem.20101823
19. Brain O, Owens BMJ, Pichulik T et al (2013) The intracellular sensor NOD2 induces microRNA-29 expression in human dendritic cells to limit IL-23 release. Immunity 39:521–536. https://doi.org/10.1016/j.immuni.2013.08.035
20. Broughton JP, Lovci MT, Huang JL et al (2016) Pairing beyond the seed supports MicroRNA targeting specificity. Mol Cell 64:320–333
21. Bryant RJ, Pawlowski T, Catto JWF et al (2012) Changes in circulating microRNA levels associated with prostate cancer. Br J Cancer 106:768–774

22. Bullock MD, Pickard KM, Nielsen BS et al (2013) Pleiotropic actions of miR-21 highlight the critical role of deregulated stromal microRNAs during colorectal cancer progression. Cell Death Dis 4:e684
23. Burgueño JF, Fritsch J, González EE et al (2021) Epithelial TLR4 signaling activates DUOX2 to induce microbiota-driven tumorigenesis. Gastroenterology 160:797–808
24. Cai K, Shen F, Cui JH et al (2015) Expression of miR-221 in colon cancer correlates with prognosis. Int J Clin Exp Med 8:2794–2798
25. Calin GA, Croce CM (2006) MicroRNA signatures in human cancers. Nat Rev Cancer 6(11):857–866. https://doi.org/10.1038/nrc1997
26. Chae YS, Kim JG, Lee SJ et al (2013) A miR-146a polymorphism (rs2910164) predicts risk of and survival from colorectal cancer. Anticancer Res 33:3233–3239
27. Chan SH, Wang LH (2015) Regulation of cancer metastasis by microRNAs. J Biomed Sci 22:9. https://doi.org/10.1186/s12929-015-0113-7
28. Chen Y, Wang C, Liu Y et al (2013) miR-122 targets NOD2 to decrease intestinal epithelial cell injury in Crohn's disease. Biochem Biophys Res Commun 438:133–139
29. Chen T, Yao LQ, Shi Q et al (2014) MicroRNA-31 contributes to colorectal cancer development by targeting factor inhibiting HIF-1α (FIH-1). Cancer Biol Ther 15:516–523
30. Chi Y, Zhou D (2016) MicroRNAs in colorectal carcinoma from pathogenesis to therapy. J Exp Clin Cancer Res 35:43. https://doi.org/10.1186/s13046-016-0320-4
31. Chivukula RR, Shi G, Acharya A et al (2014) An essential mesenchymal function for miR-143/145 in intestinal epithelial regeneration. Cell 157:1104–1116. https://doi.org/10.1016/j.cell.2014.03.055
32. Cimmino A, Calin GA, Fabbri M et al (2005) miR-15 and miR-16 induce apoptosis by targeting BCL2. Proc Natl Acad Sci 102:13944–13949. https://doi.org/10.1073/pnas.0506654102
33. Cobb BS, Nesterova TB, Thompson E (2005) T-cell lineage choice and differentiation in the absence of the RNase III enzyme dicer. J Exp Med 201:1367–1373
34. Coleman D, Kuwada S (2024) miRNA as a biomarker for the early detection of colorectal cancer. Genes (Basel) 15:338
35. Cummins JM, He Y, Leary RJ et al (2006) The colorectal microRNAome. Proc Natl Acad Sci 103:3687–3692
36. Dai X, Chen X, Chen Q et al (2015) MicroRNA-193a-3p reduces intestinal inflammation in response to microbiota via down-regulation of colonic PepT1. J Biol Chem 290:99–115
37. Dalal S, Kwon J (2010) The role of microRNA in inflammatory bowel disease. Gastroenterol Hepatol 6:714–722
38. Dalmasso G, Nguyen HT, Charrier-Hisamuddin L et al (2010) PepT1 mediates transport of the proinflammatory bacterial tripeptide L-Ala-γ-D-Glu-meso-DAP in intestinal epithelial cells. Am J Physiol Gastrointest Liver Physiol 299:G687–G696
39. Dalmasso G, Nguyen HTT, Yan Y et al (2011a) Microbiota modulate host gene expression via microRNAs. PLoS One 6:e19293. https://doi.org/10.1371/journal.pone.0019293
40. Dalmasso G, Nguyen HT, Yan Y et al (2011b) MicroRNA-92b regulates expression of the oligopeptide transporter PepT1 in intestinal epithelial cells. Am J Physiol Gastrointest Liver Physiol 300:G52–G59. https://doi.org/10.1152/ajpgi.00394.2010
41. Di Martino MT, Arbitrio M, Caracciolo D et al (2022) MiR-221/222 as biomarkers and targets for therapeutic intervention on cancer and other diseases: a systematic review. Mol Ther Nucleic Acid 27:1191–1224
42. Díez-Ricote L, Ruiz-Valderrey P, Micó V et al (2021) Trimethylamine n-Oxide (TMAO) modulates the expression of cardiovascular disease-related microRNAs and their targets. Int J Mol Sci 22:11145
43. Dong Y, Wu WK, Wu CW et al (2011) MicroRNA dysregulation in colorectal cancer: a clinical perspective. Br J Cancer 104:893–898. https://doi.org/10.1038/bjc.2011.57
44. Fabian MR, Sonenberg N, Filipowicz W (2010) Regulation of mRNA translation and stability by microRNAs. Annu Rev Biochem 79:351–379. https://doi.org/10.1146/annurev-biochem-060308-103103

45. Fairfax KA, Gantier MP, Mackay F et al (2015) IL-10 regulates Aicda expression through miR-155. J Leukoc Biol 97:71–78
46. Fisher CD, Lickteig AJ, Augustine LM et al (2009) Experimental non-alcoholic fatty liver disease results in decreased hepatic uptake transporter expression and function in rats. Eur J Pharmacol 613:119–127. https://doi.org/10.1016/j.ejphar.2009.04.002
47. Fujita D, Arai T, Komori H et al (2018) Apple derived nanoparticles modulate expression of organic-anion-transporting polypeptide (OATP) 2B1 in Caco-2 cells. Mol Pharm 15:5772–5780. https://doi.org/10.1021/acs.molpharmaceut.8b00921
48. Fukata M, Hernandez Y, Conduah D et al (2009) Innate immune signaling by Toll-like receptor-4 (TLR4) shapes the inflammatory microenvironment in colitis-associated tumors. Inflam Bow Dis 15:997–1006
49. Fukata M, Shang L, Santaolalla R et al (2011) Constitutive activation of epithelial TLR4 augments inflammatory responses to mucosal injury and drives colitis-associated tumorigenesis. Inflam Bow Dis 17:1464–1473
50. Gao Y, Schug J, McKenna LB et al (2011) Tissue-specific regulation of mouse microRNA genes in endoderm-derived tissues. Nucleic Acids Res 39:454–463
51. Gao Y, Liu Y, Du L et al (2015) Down-regulation of miR-24-3p in colorectal cancer is associated with malignant behavior. Med Oncol 32:362. https://doi.org/10.1007/s12032-014-0362-4
52. Gao X, Tyagi R, Magrini V et al (2017) Compartmentalization of functions and predicted miRNA regulation among contiguous regions of the nematode intestine. RNA Biol 14:1335–1352. https://doi.org/10.1080/15476286.2016.1166333
53. Garo LP, Ajay AK, Fujiwara M et al (2021) MicroRNA-146a limits tumorigenic inflammation in colorectal cancer. Nat Commun 12:2419
54. Gasparello J, Gambari L, Papi C et al (2020a) High levels of apoptosis are induced in the human colon cancer HT-29 cell line by co-administration of sulforaphane and a peptide nucleic acid targeting miR-15b-5p. Nucleic Acid Ther 30:164–174
55. Gasparello J, Papi C, Allegretti M et al (2020b) A distinctive microRNA (miRNA) signature in the blood of colorectal cancer (CRC) patients at surgery. Cancer 12:2410
56. Gazouli M, Papaconstantinou I, Stamatis K et al (2013) Association study of genetic variants in miRNAs in patients with inflammatory bowel disease: preliminary results. Dig Dis Sci 58:2324–2328. https://doi.org/10.1007/s10620-013-2640-y
57. Ghafouri-Fard S, Khoshbakht T, Hussen BM et al (2022) A comprehensive review on function of miR-15b-5p in malignant and non-malignant disorders. Front Oncol 12:870996
58. Gillen AE, Gosalia N, Leir SH et al (2011) MicroRNA regulation of expression of the cystic fibrosis transmembrane conductance regulator gene. Biochem J 438:25–32. https://doi.org/10.1042/BJ20110672
59. Gil-Zamorano J, Tomé-Carneiro J, Hazas MCLDL et al (2020) Intestinal miRNAs regulated in response to dietary lipids. Sci Rep 10:18921
60. Gregory RI, Chendrimada TP, Cooch N, Shiekhattar R (2005) Human RISC couples MicroRNA biogenesis and posttranscriptional gene silencing. Cell 123(4):631–640. https://doi.org/10.1016/j.cell.2005.10.022
61. Guo XZ, Ye XL, Xiao WZ et al (2015) Downregulation of VMP1 confers aggressive properties to colorectal cancer. Oncol Rep 34:2557–2566
62. Guz M, Dworzanski T, Jeleniewicz W et al (2020) Elevated miRNA inversely correlates with E-cadherin gene expression in tissue biopsies from Crohn Disease patients in contrast to ulcerative colitis patients. Biomed Res Int 2020:4250329
63. Gwiggner M, Martinez-Nunez RT (2018) MicroRNA-31 and MicroRNA-155 are overexpressed in ulcerative colitis and regulate IL-13 signaling by targeting interleukin 13 receptor α-1. Gene 9:85
64. Ha M, Kim VN (2014) Regulation of microRNA biogenesis. Nat Rev Mol Cell Biol 15:509–524

65. Haines RJ, Beard RS, Eitner RA et al (2016) TNFα/IFN mediated intestinal epithelial barrier dysfunction is attenuated by microRNA-93 downregulation of PTK6 in mouse colonic epithelial cells. PLoS One 11:e0154351
66. Hamfjord J, Stangeland AM, Hughes T et al (2012) Differential expression of miRNAs in colorectal cancer: comparison of paired tumor tissue and adjacent normal mucosa using high through put sequencing. PLoS One 7:e34150. https://doi.org/10.1371/journal.pone.0034150
67. Han R, Sun Q, Wu J et al (2016) Sodium butyrate upregulates miR-203 expression to exert anti-proliferation effect on colorectal cancer cells. Cell Physiol Biochem 39:1919–1929
68. Hand NJ, Master ZR, Le Lay J (2009) Hepatic function is preserved in the absence of mature microRNAs. Hepatology 49:618–626
69. Hao C, Gao C, Shang H et al (2020) MicroRNA-31 inhibits the growth and metastasis and enhances drug sensitivity of the human colon cancer cells by targeting PAX6. J BUON 25:1860–1865
70. Harrell CR, Jovicic N, Djonov V et al (2019) Mesenchymal stem cell-derived exosomes and other extracellular vesicles as new remedies in the therapy of inflammatory diseases. Cells 8:1605
71. Hashemi M, Mirdamadi MSA, Talebi Y et al (2023) Pre-clinical and clinical importance of miR-21 in human cancers: tumorigenesis, therapy response, delivery approaches and targeting agents. Pharmacol Res 187:106568
72. He Q, Ye A, Ye W et al (2021) Cancer-secreted exosomal miR-21-5p induces angiogenesis and vascular permeability by targeting KRIT1. Cell Death Dis 12:576
73. He S, Song W, Cui S et al (2023) Modulation of miR-146b by N6-methyladenosine modification remodels tumor-associated macrophages and enhances anti-PD-1 therapy in colorectal cancer. Cell Oncol (Dordr) 46:1731–1746
74. Heil F, Hemmi H, Hochrein H et al (2004) Species-specific recognition of single-stranded RNA via Toll-like receptor 7 and 8. Science 303:1526–1529
75. Heller F, Florian P, Bojarski C et al (2005) Interleukin-13 is the key effector Th2 cytokine in ulcerative colitis that affects epithelial tight junctions, apoptosis, and cell restitution. Gastroenterology 129:550–564
76. Heller F, Fromm A, Gitter AH et al (2008) Epithelial apoptosis is a prominent feature of the epithelial barrier disturbance in intestinal inflammation: effect of pro-inflammatory interleukin-13 on epithelial cell function. Mucosal Immunol 1(1):S58–S61
77. Herreros-Villanueva M, Duran-Sanchon S, Martín AC et al (2019) Plasma microRNA signature validation for early detection of colorectal cancer. Clin Transl Gastroenterol 10:e00003
78. Hu Y, Liu J, Jiang B et al (2014) MiR-199a-5p loss up-regulated DDR1 aggravated colorectal cancer by activating epithelial-to-mesenchymal transition related signaling. Dig Dis Sci 59:2163–2172
79. Hu S, Liu L, Chang EB et al (2015) Butyrate inhibits pro-proliferative miR-92a by diminishing c-Myc-induced miR-17-92a cluster transcription in human colon cancer cells. Mol Cancer 14:180
80. Huang Z, Huang D, Ni S et al (2010) Plasma microRNAs are promising novel biomarkers for early detection of colorectal cancer. Int J Cancer 127:118–126
81. Huang L, Liang D, Zhang Y et al (2023) METTL3 promotes colorectal cancer metastasis by promoting the maturation of pri-microRNA-196b. J Cancer Res Clin Oncol 149:5095–5108
82. Hwang WL, Jiang JK, Yang SH et al (2014) MicroRNA-146a directs the symmetric division of snail-dominant colorectal cancer stem cells. Nat Cell Biol 16:268–280
83. Iborra M, Bernuzzi F, Invernizzi P et al (2012) MicroRNAs in autoimmunity and inflammatory bowel disease: crucial regulators in immune response. Autoimmun Rev 11:305–314
84. Iborra M, Bernuzzi F, Correale C et al (2013) Identification of serum and tissue micro-RNA expression profiles in different stages of inflammatory bowel disease. Clin Exp Immunol 173:250–258. https://doi.org/10.1111/cei.12104

85. Igarashi H, Kurihara H, Mitsuhashi K et al (2015) Association of microRNA-31-5p with clinical efficacy of anti-EGFR therapy in patients with metastatic colorectal cancer. Ann Surg Oncol 22:2640–2648
86. Ikemura K, Yamamoto M, Miyazaki S et al (2013) MicroRNA-145 post-transcriptionally regulates the expression and function of P-glycoprotein in intestinal epithelial cells. Mol Pharmacol 83:399–405
87. Iwasaki H, Shimura T, Kitagawa M et al (2022) A novel urinary miRNA biomarker for early detection of colorectal cancer. Cancer 14:461
88. Jazdzewski K, Murray EL, Franssila K et al (2008) Common SNP in pre-miR-146a decreases mature miR expression and predisposes to papillary thyroid carcinoma. Proc Natl Acad Sci 105:7269–7274. https://doi.org/10.1073/pnas.0802682105
89. Ji D, Zhan T, Li M et al (2018) Enhancement of sensitivity to chemo/radiation therapy by using MiR-15b against DCLK1 in colorectal cancer. Stem Cell Rep 11:1506–1522
90. Jiang Y, Wang H, Li Y et al (2014) Peripheral blood miRNAs as a biomarker for chronic cardiovascular diseases. Sci Rep 4:5026
91. Jiang R, Chen X, Ge S et al (2021) MiR-21-5p induces pyroptosis in colorectal cancer via TGFBI. Front Oncol 10:610545
92. Jimenez-Rondan FR, Ruggiero CH, Cousins RJ (2022) Long noncoding RNA, microRNA, Zn transporter Zip14 (Slc39a14) and Inflammation in mice. Nutrients 14:5114. https://doi.org/10.3390/nu14235114
93. Johnson SM, Grosshans H, Shingara J et al (2005) RAS is regulated by the let-7 microRNA family. Cell 120:635–647. https://doi.org/10.1016/j.cell.2005.01.014
94. Kalla R, Ventham NT, Kennedy NA et al (2015) MicroRNAs: new players in IBD. Gut 64:504–517
95. Kamada N, Seo SU, Chen GY et al (2013) Role of the gut microbiota in immunity and inflammatory disease. Nat Rev Immunol 13:321–335. https://doi.org/10.1038/nri3430
96. Karampoga A, Tzaferi K, Koutsakis C et al (2022) Exosomes and the extracellular matrix: a dynamic interplay in cancer progression. Int J Dev Biol 66:97–102
97. Karin M, Greten FR (2005) NF-kappa B: linking inflammation and immunity to cancer development and progression. Nat Rev Immunol 5:749–759. https://doi.org/10.1038/nri1703
98. Keller A, Groger L, Tschernig T et al (2022) miRNA tissue Atlas 2: an update to the human miRNA tissue atlas. Nucleic Acids Res 50:D211–D221
99. Khorrami S, Zavaran Hosseini A, Mowla SJ et al (2017) MicroRNA-146a induces immune suppression and drug-resistant colorectal cancer cells. Tumour Biol 39:101042831769836
100. Kim S, Rhee JK, Yoo HJ et al (2015) Bioinformatic and metabolomics analysis reveals miR-155 regulates thiamine level in breast cancer. Cancer Lett 357:488–497
101. Kim J, Yao F, Xiao Z et al (2018) MicroRNAs and metastasis: small RNAs play big roles. Cancer Met Rev 37:5–15. https://doi.org/10.1007/s10555-017-9712-y
102. Kim JK, Qu X, Chen CT et al (2021) Identifying diagnostic microRNAs and investigating their biological implications in rectal cancer. JAMA Net Open 4:e2136913. https://doi.org/10.1001/jamanetworkopen.2021.36913
103. Kleberg K, Jensen GM, Christensen DP et al (2012) Transporter function and cyclic AMP turnover in normal colonic mucosa from patients with and without colorectal neoplasia. BMC Gastroenterol 12:78. https://doi.org/10.1186/1471-230X-12-78
104. Koukos G, Polytarchou C, Kaplan JL et al (2013) MicroRNA-124 regulates STAT3 expression and is downregulated in colon tissues of pediatric patients with ulcerative colitis. Gastroenterology 145:842–852. https://doi.org/10.1053/j.gastro.2013.07.001
105. Lakhan R, Subramanian VS, Said HM (2017) Role of MicroRNA-423-5p in posttranscriptional regulation of the intestinal riboflavin transporter-3. Am J Physiol Gastrointest Liver Physiol 313:G589–G598
106. Lee RC, Feinbaum RL, Ambros V (1993) The C. elegans heterochronic gene lin-4 encodes small RNAs with antisense complementarity to lin-14. Cell 75(5):843–854. https://doi.org/10.1016/0092-8674(93)90529-y. PMID: 8252621

107. Lewis BP, Burge CB, Bartel DP (2005) Conserved seed pairing, often flanked by adenosines, indicates that thousands of human genes are microRNA targets. Cell 120:15–20
108. Li C, Zhao L, Chen Y et al (2016) MicroRNA-21 promotes proliferation, migration, and invasion of colorectal cancer, and tumor growth associated with down-regulation of sec23a expression. BMC Cancer 16:605
109. Li M, Chen WD, Wang YD (2020) The roles of the gut microbiota-miRNA interaction in the host pathophysiology. Mol Med 26:101
110. Li Y, Li H, Cui M et al (2024) Interaction of exosomal MicroRNA and oxidative stress in the pathogenesis of colitis-associated cancer. Front Biosci (Landmark Ed) 29(8):276. https://doi.org/10.31083/j.fbl2908276
111. Liang J, Tang J, Shi H et al (2017) miR-27a-3p targeting RXR alpha promotes colorectal cancer progression by activating Wnt/beta-catenin pathway. Oncotarget 8:82991–83008
112. Liao XJ, Mao WM, Wang Q et al (2016) MicroRNA-24 inhibits serotonin uptake transporter expression and aggravates irritable bowel syndrome. Biochem Biophys Res Commun 469:288–2893. https://doi.org/10.1016/j.bbrc.2015.11.102
113. Lin J, Welker NC, Zhao Z et al (2014) Novel specific microRNA biomarkers in idiopathic inflammatory bowel disease unrelated to disease activity. Mod Pathol 27:602–608. https://doi.org/10.1038/modpathol.2013.152
114. Lin Y, Chen Z, Lin S et al (2019) MiR-202 inhibits the proliferation and invasion of colorectal cancer by targeting UHRF1. Acta Biochim Biophys Sin 51(6):598–606
115. Ling X, Wang R, Lin L et al (2023) N6-methyladenosine-modified microRNA-675 advances the development of gastrointestinal stromal tumors via inhibiting myosin phosphatase targeting protein 1. Genomics 115:110704
116. Liu M, Chen H (2010) The role of microRNAs in colorectal cancer. J Genet Genomics 37:347–358. https://doi.org/10.1016/S1673-8527(09)60053-9
117. Liu Y, Miyoshi H, Nakamura M (2007) Nanomedicine for drug delivery and imaging: a promising avenue for cancer therapy and diagnosis using targeted functional nanoparticles. Int J Cancer 120:2527–2537. https://doi.org/10.1002/ijc.22709
118. Liu M, Tang Q, Qiu M et al (2011) MiR-21 targets the tumor suppressor RhoB and regulates proliferation, invasion and apoptosis in colorectal cancer cells. FEBS Lett 585:2998–3005
119. Liu S, da Cunha AP, Rezende RM et al (2016) The host shapes the gut microbiota via fecal microRNA. Cell Host Microbe 19:32–43
120. Liu W, Nakano M, Nakanishi T et al (2020a) Post-transcriptional regulation of OATP2B1 transporter by a microRNA, miR-24. Drug Metab Pharmacokinet 35:515–521
121. Liu Y, Zhu F, Li H et al (2020b) MiR-155 contributes to intestinal barrier dysfunction in DSS-induced mice colitis via targeting HIF-1α/TFF-3 axis. Aging 12:14966–14977
122. Liu C, Liu R, Wang B et al (2021) Blocking IL-17A enhances tumor response to anti-PD-1 immunotherapy in microsatellite stable colorectal cancer. J Immunother Cancer 9:e001895
123. Liu T, Feng YL, Wang RY et al (2024) Long-term MNNG exposure promotes gastric carcinogenesis by activating METTL3/m6A/miR1184 axis-mediated epithelial-mesenchymal transition. Sci Total Environ 913:169752
124. Lu D, Yao Q, Zhan C et al (2017) MicroRNA-146a promote cell migration and invasion in human colorectal cancer via carboxypeptidase M/Src-FAK pathway. Oncotarget 8:22674–22684
125. Ma L, Zhu L, Gu D et al (2013) A genetic variant in miR-146a modifies colorectal cancer susceptibility in a Chinese population. Arch Toxicol 87:823–833. https://doi.org/10.1007/s00204-012-1004-2
126. Makarova J, Turchinovich A, Shkurnikov M et al (2021) Extracellular miRNAs and cell-cell communication: problems and prospects. Trends Biochem Sci 46(8):640–651. https://doi.org/10.1016/j.tibs.2021.01.007
127. Maminezhad H, Ghanadian S, Pakravan K et al (2020) A panel of six circulating miRNA signature in serum and its potential diagnostic value in colorectal cancer. Life Sci 258:118226

128. Masuda T, Hayashi N, Kuroda Y, Ito S, Eguchi H, Mimori K (2017) MicroRNAs as biomarkers in colorectal cancer. Cancers (Basel) 9:124. https://doi.org/10.3390/cancers9090124
129. Matboli M, Hossam N, Farag D et al (2024) miRNAs: possible regulators of toll like receptors and inflammatory tumor microenvironment in colorectal cancer. BMC Cancer 24:824. https://doi.org/10.1186/s12885-024-12417-0
130. McKenna LB, Schug J, Vourekas A et al (2010) MicroRNAs control intestinal epithelial differentiation, architecture, and barrier function. Gastroenterology 139:1654–1664. https://doi.org/10.1053/j.gastro.2010.07.040
131. Mi B, Li Q, Li T et al (2020) High miR-31-5p expression promotes colon adenocarcinoma progression by targeting TNS1. Aging 12:7480–7490
132. Michael IP, Saghafinia S, Hanahan D (2019) A set of microRNAs coordinately controls tumorigenesis, invasion, and metastasis. Proc Natl Acad Sci 116:24184–24195. https://doi.org/10.1073/pnas.1913307116
133. Michas A, Michas V, Anagnostou E et al (2023) The clinical significance of microRNAs in colorectal cancer signaling pathways: a review. Glob Med Genet 10:315–323. https://doi.org/10.1055/s-0043-1777094
134. Mihelich BL, Khramtsova EA, Arva N et al (2011) miR-183-96-182 cluster is overexpressed in prostate tissue and regulates zinc homeostasis in prostate cells. J Biol Chem 286:44503–44511
135. Mizuno R, Kawada K, Sakai Y (2018) The molecular basis and therapeutic potential of let-7 microRNAs against colorectal cancer. Can J Gastroenterol Hepatol 2018:5769591
136. Mo WY, Cao SQ (2022) MiR-29a-3p: a potential biomarker and therapeutic target in colorectal cancer. Clin Transl Oncol 25:563–577
137. Muenchau S, Deutsch R, de Castro IJ et al (2019) Hypoxic environment promotes barrier formation in human intestinal epithelial cells through regulation of microRNA 320a expression. Mol Cell Biol 39:e00553–e00518
138. Nagy ZB, Wichmann B, Kalmár A et al (2017) Colorectal adenoma and carcinoma specific miRNA profiles in biopsy and their expression in plasma specimens. Clin Epigenetics 9:22
139. Nakano M, Nakajima M (2018) Current knowledge of microRNA-mediated regulation of drug metabolism in humans. Expert Opin Drug Metabol Toxicol 14:493–504. https://doi.org/10.1080/17425255.2018.1472237
140. Nappi F, Singh SSA, Jitendra V et al (2023) The roles of microRNAs in the cardiovascular system. Int J Mol Sci 24:14277
141. Neurath MF (2014) Cytokines in inflammatory bowel disease. Nat Rev Immunol 14:329–342. https://doi.org/10.1038/nri3661
142. Nguyen HD, Kim M-S (2022) The role of mixed B vitamin intakes on cognitive performance: modeling, genes and miRNAs involved. J Psychiatr Res 152:38–56
143. Nguyen HT, Kacimi SEO, Nguyen TL et al (2021) MiR-21 in the cancers of the digestive system and its potential role as a diagnostic, predictive, and therapeutic biomarker. Biology 10:417
144. Nguyen TTP, Suman KH, Nguyen TB et al (2022) The role of miR-29s in human cancers-an update. Biomedicine 10:2121
145. Noorolyai S, Baghbani E, Shanehbandi D et al (2020) MiR-146a-5p and MiR-193a-5p synergistically inhibited the proliferation of human colorectal cancer cells (HT-29 Cell Line) through ERK signaling pathway. Adv Pharm Bull 11:755–764
146. O'Brien J, Hayder H, Zayed Y et al (2018) Overview of microRNA biogenesis, mechanisms of actions, and circulation. Front Endocrinol (Lausanne) 9:402
147. Ogata-Kawata H, Izumiya M, Kurioka D et al (2014) Circulating exosomal microRNAs as biomarkers of colon cancer. PLoS One 9:e92921
148. Ohsaka F, Sonoyama K (2018) Murine intestinal organoids resemble intestinal epithelium in their microRNA profiles. Biosci Biotechnol Biochem 82:1560–1567
149. Okubo M, Tahara T, Shibata T et al (2011) Association study of common genetic variants in pre-microRNAs in patients with ulcerative colitis. J Clin Immunol 31:69–73. https://doi.org/10.1007/s10875-010-9461-y

150. Olivo-Martínez Y, Martínez-Ruiz S, Cordero C et al (2024) Extracellular vesicles of the probiotic Escherichia coli Nissle 1917 reduce PepT1 levels in IL-1β-treated Caco-2 cells via upregulation of miR-193a-3p. Nutrients 16:2719. https://doi.org/10.3390/nu16162719
151. Omrane I, Kourda N, Stambouli N et al (2014) MicroRNAs 146a and 147b biomarkers for colorectal tumors localization. Biomed Res Int 2014:584852
152. Ouyang M, Su W, Xiao L et al (2015) Modulation by miR-29b of intestinal epithelium homoeostasis through the repression of menin translation. Biochem J 465:315–323
153. Pandey S, Jain A, Vagha S (2024) Insights into colorectal carcinoma: a comprehensive review of microRNA expression patterns. Cureus 16(3):e56739. https://doi.org/10.7759/cureus.56739
154. Peck BC, Mah AT, Pitman WA et al (2017) Functional transcriptomics in diverse intestinal epithelial cell types reveals robust MicroRNA sensitivity in intestinal stem cells to microbial status. J Biol Chem 292:2586–2600
155. Pekow JR, Kwon JH (2012) MicroRNAs in inflammatory bowel disease. Inflamm Bowel Dis 18:187–193. https://doi.org/10.1002/ibd.21691
156. Peng Y, Croce CM (2016) The role of microRNAs in human cancer. Signal Transduct Target Ther 1:15004. https://doi.org/10.1038/sigtrans.2015.4
157. Peng W, Liu YN, Zhu SQ et al (2018) The correlation of circulating pro-angiogenic MiRNAs expressions with disease risk, clinicopathological features, and survival profiles in gastric cancer. Cancer Med 7:3773–3791
158. Peng W, Li J, Chen R et al (2019) Upregulated METTL3 promotes metastasis of colorectal cancer via miR-1246/SPRED2/MAPK signaling pathway. J Exp Clin Cancer Res 38:393
159. Peng X, Wang J, Zhang C et al (2020) A three-miRNA panel in serum as a non-invasive biomarker for colorectal cancer detection. Int J Biol Markers 35:74–82
160. Peterson SM, Thompson JA, Ufkin ML et al (2014) Common features of microRNA target prediction tools. Front Genet 5:23
161. Potenza N, Russo A (2013) Biogenesis, evolution and functional targets of microRNA-125a. Mol Gen Genomics 288:381–389. https://doi.org/10.1007/s00438-013-0757-5
162. Potenza N, Panella M, Castiello F et al (2017) Molecular mechanisms governing microRNA-125a expression in human hepatocellular carcinoma cells. Sci Rep 7:10712. https://doi.org/10.1038/s41598-017-11418-3
163. Pullen TJ, da Silva Xavier G, Kelsey G et al (2011) miR-29a and miR-29b contribute to pancreatic beta-cell-specific silencing of monocarboxylate transporter 1 (Mct1). Mol Cell Biol 31:3182–3194. https://doi.org/10.1128/MCB.01433-10
164. Qian K, Xu W, Xia X et al (2022) Methyltransferase-like 3 (METTL3) mediated N6-methyladenosine (m6A) modifications facilitate mir-25-3p maturation to promote gastrointestinal stromal tumors (GISTs) progression. Genes Genomics 44:1519–1530
165. Qian W, Huang L, Xu Y et al (2023) Hypoxic ASCs-derived exosomes attenuate colitis by regulating macrophage polarization via miR-216a-5p/HMGB1 axis. Inflamm Bowel Dis 29:602–619
166. Ramamoorthy K, Anandam KY, Yasujima T et al (2020) Posttranscriptional regulation of thiamin transporter-1 expression by microRNA-200a-3p in pancreatic acinar cells. Am J Physiol Gastrointest Liver Physiol 319:G323–G332. https://doi.org/10.1152/ajpgi.00178.2020
167. Rawat M, Nighot M, Al-Sadi R et al (2020) IL1B increases intestinal tight junction permeability by up-regulation of MIR200C-3p, which degrades occludin mRNA. Gastroenterology 159:1375–1389
168. Rottiers V, Näär AM (2012) MicroRNAs in metabolism and metabolic disorders. Nat Rev Mol Cell Biol 13:239–250. https://doi.org/10.1038/nrm3313
169. Round JL, Mazmanian SK (2009) The gut microbiota shapes intestinal immune responses during health and disease. Nat Rev Immunol 9:313–323. https://doi.org/10.1038/nri2515
170. Sado AI, Batool W, Ahmed A et al (2024) Role of microRNA in colorectal carcinoma (CRC): a narrative review. Ann Med Surg 86:308–318. https://doi.org/10.1097/MS9.0000000000001494

171. Sangani R, Periyasamy-Thandavan S, Kolhe R et al (2015) MicroRNAs-141 and 200a regulate the SVCT2 transporter in bone marrow stromal cells. Mol Cell Endocrinol 410:19–26
172. Santos DAR, Gaiteiro C, Santos M et al (2023) MicroRNA biomarkers as promising tools for early colorectal cancer screening – a comprehensive review. Int J Mol Sci 24:11023
173. Sarlinova M, Halasa M, Mistuna D et al (2016) MiR-21, MiR-221 and MiR-150 are deregulated in peripheral blood of patients with colorectal cancer. Anticancer Res 36:5449–5454
174. Sathyanarayanan A, Chandrasekaran KS, Karunagaran D (2016) MicroRNA-146a inhibits proliferation, migration and invasion of human cervical and colorectal cancer cells. Biochem Biophys Res Commun 480:528–533
175. Schee K, Boye K, Abrahamsen TW et al (2012) Clinical relevance of microRNA miR-21, miR-31, miR-92a, miR-101, miR-106a and miR-145 in colorectal cancer. BMC Cancer 12:505
176. Schee K, Lorenz S, Worren MM et al (2013) Deep sequencing the microRNA transcriptome in colorectal cancer. PLoS One 8:e66165. https://doi.org/10.1371/journal.pone.0066165
177. Schou JV, Johansen JS, Nielsen D et al (2016) Circulating microRNAs as prognostic and predictive biomarkers in patients with colorectal cancer. Noncoding RNA 2:5. https://doi.org/10.3390/ncrna2020005
178. Searles CD (2024) MicroRNAs and cardiovascular disease risk. Curr Cardiol Rep 26:51–60
179. Sears CL, Garrett WS (2014) Microbes, microbiota, and colon cancer. Cell Host Microbe 15:317–328. https://doi.org/10.1016/j.chom.2014.02.007
180. Sekine S, Ogawa R, Ito R (2009) Disruption of Dicer1 induces dysregulated fetal gene expression and promotes hepatocarcinogenesis. Gastroenterology 136:2304–2315
181. Shao Y, Li J, Cai Y et al (2014) The functional polymorphisms of miR-146a are associated with susceptibility to severe sepsis in the Chinese population. Mediat Inflamm 916202. https://doi.org/10.1155/2014/916202
182. Shen Y, Zhou M, Yan J et al (2017) miR-200b inhibits TNF-α-induced IL-8 secretion and tight junction disruption of intestinal epithelial cells in vitro. Am J Physiol Gastrointest Liver Physiol 312:123–132
183. Shi CY, Kingston ER, Kleaveland B et al (2020) The ZSWIM8 ubiquitin ligase mediates target directed microRNA degradation. Science 370:eabc9359
184. Shigoka M, Tsuchida A, Matsudo T et al (2010) Deregulation of miR-92a expression is implicated in hepatocellular carcinoma development. Pathol Int 60:351–357
185. Shiosaki J, Tiirikainen M, Peplowska K et al (2020) Serum micro-RNA identifies early stage colorectal cancer in a multi-ethnic population. Asian Pac J Cancer Prev 21:3019–3026
186. Si H, Sun X, Chen Y et al (2013) Circulating microRNA-92a and microRNA-21 as novel minimally invasive biomarkers for primary breast cancer. J Cancer Res Clin Oncol 139:223–229
187. Sohn JJ, Schetter AJ, Yfantis HG et al (2012) Macrophages, nitric oxide and microRNAs are associated with DNA damage response pathway and senescence in inflammatory bowel disease. PLoS One 7:e44156
188. Solé C, Lawrie CH (2021) MicroRNAs in metastasis and the tumour microenvironment. Int J Mol Sci 22:4859. https://doi.org/10.3390/ijms22094859
189. Song M, Su H, Zhang L et al (2013) Genetic polymorphisms of miR-146a and miR-27a, H. pylori infection, and risk of gastric lesions in a Chinese population. PLoS One 8:e61250. https://doi.org/10.1371/journal.pone.0061250
190. Song J, Wang L, Ma Q et al (2018) Let-7i-5p inhibits the proliferation and metastasis of colon cancer cells by targeting kallikrein related peptidase 6. Oncol Rep 40(3):1459–1466
191. Soroosh A, Rankin CR, Polytarchou C et al (2019) MiR-24 is elevated in ulcerative colitis patients and regulates intestinal epithelial barrier function. Am J Pathol 189:1763–1774
192. Subramanian VS, Marchant JS, Boulware MJ et al (2004) A C-terminal region dictates the apical plasma membrane targeting of the human sodium-dependent vitamin C transporter-1 in polarized epithelia. J Biol Chem 279:27719–27728
193. Subramanian VS, Sabui S, Marchant JS et al (2019) MicroRNA-103a regulates sodium-dependent vitamin C transporter-1 expression in intestinal epithelial cells. J Nutr Biochem 65:46–53. https://doi.org/10.1016/j.jnutbio.2018.12.001

194. Sugai T, Osakabe M, Niinuma T et al (2022) Genome-wide analysis of microRNA and mRNA expression in colorectal intramucosal neoplasia and colorectal cancer with a microsatellite-stable phenotype based on adenoma-carcinoma sequences. Front Oncol 12:831100
195. Sun K, Chen L, Li Y et al (2023) METTL14-dependent maturation of pri-miR-17 regulates mitochondrial homeostasis and induces chemoresistance in colorectal cancer. Cell Death Dis 14:148
196. Tang H, Deng M, Liao Q et al (2012) Expression and clinical significance of miR-23a and metastasis suppressor 1 in colon carcinoma. Zhonghua Bing 41:28–32
197. Tang B, Lu X, Tong Y et al (2023) MicroRNA-31 induced by Fusobacterium nucleatum infection promotes colorectal cancer tumorigenesis. Sciences 26:106770
198. Tian SM, Wang J, Zhang FY et al (2022) Comparative analysis of microRNA binding site distribution and microRNA-mediated gene expression repression of oncogenes and tumor suppressor genes. Genes (Basel) 13:481
199. Tsuchida A, Ohno S, Wu W et al (2011) MiR-92 is a key oncogenic component of the MiR-17–92 cluster in colon cancer. Cancer Sci 102:2264–2271
200. Tsuji K, Nakanoh H, Fukushima K et al (2023) MicroRNAs as 9 biomarkers and therapeutic targets for acute kidney injury. Diagnostics 13:2893
201. Uhlmann S, Mannsperger H, Zhang JD et al (2012) Global microRNA level regulation of EGFR-driven cell-cycle protein network in breast cancer. Mol Syst Biol 8:570
202. Van der Goten J, Vanhove W, Lemaire K et al (2014) Integrated miRNA and mRNA expression profiling in inflamed colon of patients with ulcerative colitis. PLoS One 9:e116117
203. van Zonneveld AJ, Zhao Q, Rotmans JI et al (2023) Circulating non-coding RNAs 12 in chronic kidney disease and its complications. Nat Rev Nephrol 19:573–586
204. Villanueva JW, Kwong L, Han T et al (2022) Comprehensive microRNA analysis across genome-edited colorectal cancer organoid models reveals miR-24 as a candidate regulator of cell survival. BMC Genomics 23:792. https://doi.org/10.1186/s12864-022-09018-1
205. Wang CJ, Zhou ZG, Wang L et al (2009) Clinicopathological significance of microRNA-31, -143 and -145 expression in colorectal cancer. Dis Markers 26:27–34
206. Wang T, Xu X, Xu Q et al (2017) miR-19a promotes colitis-associated colorectal cancer by regulating tumor necrosis factor alpha-induced protein 3-NF-κB feedback loops. Oncogene 36:3240–3251
207. Wang J, Fry CME, Walker CL (2019) Carboxyl-terminal modulator protein regulates Akt signaling during skeletal muscle atrophy in vitro and a mouse model of amyotrophic lateral sclerosis. Sci Rep 9:3920
208. Wang D, Wang X, Song Y et al (2022) Exosomal miR-146a-5p and miR-155-5p promote CXCL12/CXCR7 - induced metastasis of colorectal cancer by crosstalk with cancer-associated fibroblasts. Cell Death Dis 13:380
209. Wang XJ, Zhang D, Yang YT et al (2023) Suppression of microRNA-222-3p ameliorates ulcerative colitis and colitis-associated colorectal cancer to protect against oxidative stress via targeting BRG1 to activate Nrf2/HO-1 signaling pathway. Front Immunol 14:1089809
210. Wei Z, Hang S, Wiredu Ocansey DK et al (2023) Human umbilical cord mesenchymal stem cells derived exosome shuttling mir-129-5p attenuates inflammatory bowel disease by inhibiting ferroptosis. J Nanobiotechnol 21:188
211. Weil PP, Reincke S, Hirsch CA et al (2023) Uncovering the gastrointestinal passage, intestinal epithelial cellular uptake, and AGO2 loading of milk miRNAs in neonates using xenomiRs as tracers. Am J Clin Nutr 117:1195–1210. https://doi.org/10.1016/j.ajcnut.2023.03.016
212. Wu F, Zikusoka M, Trindade A et al (2008) MicroRNAs are differentially expressed in ulcerative colitis and alter expression of macrophage inflammatory peptide-2a. Gastroenterology 135:1624–1635
213. Wu F, Zhang S, Dassopoulos T et al (2010) Identification of microRNAs associated with ileal and colonic Crohn's disease. Inflamm Bowel Dis 16:1729–1738. https://doi.org/10.1002/ibd.21267

214. Wu F, Guo N, Tian H (2011) Peripheral blood microRNAs distinguish active ulcerative colitis and Crohn'sdisease. Inflamm Bowel Dis 17:241–250. https://doi.org/10.1002/ibd.21450
215. Wu Y, Song Y, Xiong Y et al (2017) MicroRNA-21 (Mir-21) promotes cell growth and invasion by repressing tumor suppressor PTEN in colorectal cancer. Cell Physiol Biochem 43:945–958
216. Wu X, Yan F, Wang L et al (2020) MicroRNA: another pharmacological avenue for colorectal cancer? Front Cell Dev Biol 8:812. https://doi.org/10.3389/fcell.2020.00812
217. Xiao L, Rao JN, Zou T et al (2013) MiR-29b represses intestinal mucosal growth by inhibiting translation of cyclin-dependent kinase 2. Mol Biol Cell 24:3038–3046
218. Xu RS, Wu XD, Zhang SQ et al (2013) The tumor suppressor gene RhoBTB1 is a novel target of miR-31 in human colon cancer. Int J Oncol 42:676–682
219. Xu J, Xiao X, Yang D (2018) In vitro methods for analyzing miRNA roles in cancer cell proliferation, invasion, and metastasis. Methods Mol Biol 1733:159–171. https://doi.org/10.1007/978-1-4939-7601-0_13
220. Xu Q, Lu X, Li J et al (2022) Fusobacterium nucleatum induces excess methyltransferase-like 3-mediated microRNA-4717-3p maturation to promote colorectal cancer cell proliferation. Cancer Sci 113:3787–3800
221. Yan S, Liu G, Jin C et al (2018) MicroRNA-6869-5p acts as a tumor suppressor via targeting TLR4/NF-κB signaling pathway in colorectal cancer. J Cell Physiol 233:6660–6668
222. Yan S, Cheng M, Duan Q et al (2019) MiR-6803-5p promotes cancer cell proliferation and invasion via PTPRO/NF-κB axis in colorectal cancer. Mediat Inflamm 2019:8128501
223. Yang Q, Wang S, Huang J et al (2018) Serum MiR-20a and MiR-486 are potential biomarkers for discriminating colorectal neoplasia: a pilot study. J Cancer Res Ther 14:1572–1577
224. Ye D, Guo S, Al-Sadi R et al (2011) MicroRNA regulation of intestinal epithelial tight junction permeability. Gastroenterology 141:1323–1333
225. Yu T, Ma P, Wu D et al (2018) Functions and mechanisms of microRNA-31 in human cancers. Biomed Pharmacother 108:1162–1169
226. Ždralevic M, Raonic J, Popovic N et al (2023) The role of miRNA in colorectal cancer diagnosis: a pilot study. Oncol Lett 25:267
227. Ždralevic M, Radovic A, Raonic J et al (2024) Advances in microRNAs as emerging biomarkers for colorectal cancer early detection and diagnosis. Int J Mol Sci 25:11060. https://doi.org/10.3390/ijms252011060
228. Zeng X, Xiang J, Wu M et al (2012) Circulating MiR-17, MiR-20a, MiR-29c, and MiR-223 combined as non-invasive biomarkers in nasopharyngeal carcinoma. PLoS One 7:e46367
229. Zeng C, Huang L, Zheng Y et al (2014) Expression of miR-146a in colon cancer and its significance. Nan Fang Yi Ke Da Xue Xue Bao 34(3):396–400. PMID: 24670457
230. Zhang XY, Guan S, Zhang HF et al (2018) Activation of PD-1 protects intestinal immune defense through IL-10/miR-155 pathway after intestinal ischemia reperfusion. Dig Dis Sci 63:3307–3316
231. Zhang W, Fu X, Xie J et al (2021a) miR-26a attenuates colitis and colitis-associated cancer by targeting the multiple intestinal inflammatory pathways. Mol Ther Nucleic Acids 24:264–273
232. Zhang S, Li D, Zhao M et al (2021b) Exosomal miR-183-5p shuttled by M2 polarized tumor-associated macrophage promotes the development of colon cancer via targeting THEM4 mediated PI3K/AKT and NF-κB pathways. Front Oncol 11:672684
233. Zhao JL, Rao DS, Boldin MP et al (2011) NF-kappa B dysregulation in microRNA-146a deficient mice drives the development of myeloid malignancies. Proc Natl Acad Sci 108:9184–9189. https://doi.org/10.1073/pnas.1105398108
234. Zhao JL, Rao DS, O'Connell RM et al (2013) MicroRNA-146a acts as a guardian of the quality and longevity of hematopoietic stem cells in mice. elife 2:e00537. https://doi.org/10.7554/eLife.00537
235. Zhao C, Zhao Q, Zhang C et al (2017) MiR-15b-5p resensitizes colon cancer cells to 5-fluorouracil by promoting apoptosis via the NF-KB/XIAP axis. Sci Rep 7:4194

236. Zhao J, Xu J, Zhang R (2018) SRPX2 regulates colon cancer cell metabolism by miR-192/215 via PI3K-Akt. Am J Transl Res 10:483–490
237. Zhao K, Ye Z, Li Y et al (2020) LncRNA FTX contributes to the progression of colorectal cancer through regulating miR-192-5p/EIF5A2 Axis. Onco Targets Ther 13:2677–2688
238. Zhao Z, Zhu A, Bhardwaj M et al (2021) Fecal microRNAs, fecal microRNA panels, or combinations of fecal microRNAs with fecal hemoglobin for early detection of colorectal cancer and its precursors: a systematic review. Cancer 14:65
239. Zhao SY, Wang Z, Wu XB et al (2022) CERS6-AS1 contributes to the malignant phenotypes of colorectal cancer cells by interacting with miR-15b-5p to regulate SPTBN2. Kaohsiung J Med Sci 38:403–414
240. Zhong L, Simoneau B, Huot J et al (2017) P38 and JNK pathways control E-selectin-dependent extravasation of colon cancer cells by modulating miR-31 transcription. Oncotarget 8:1678–1687
241. Zhong L, Huot J, Simard MJ (2018) P38 activation induces production of miR-146a and miR-31 to repress E-selectin expression and inhibit trans endothelial migration of colon cancer cells. Sci Rep 8:2334
242. Zhou M, Kara H, Dai Y et al (2016) Circulating organ-specific microRNAs serve as biomarkers in organ-specific diseases: implications for organ allo and xeno-transplantation. Int J Mol Sci 17:1232
243. Zhu W, Huang Y, Pan Q et al (2017) MicroRNA-98 suppress Warburg effect by targeting HK2 in colon cancer cells. Dig Dis Sci 62(3):660–668
244. Zhuang X, Chen B, Huang S et al (2022) Hypermethylation of miR-145 promoter-mediated SOX9-CLDN8 pathway regulates intestinal mucosal barrier in Crohn's disease. EBioMedicine 76:103846

Chapter 20
The Role of MicroRNAs as Hepatocellular Carcinoma Biomarkers and Therapeutic Targets

Hadeer M. Morsy, Mohamed Y. Zaky, and Osama M. Ahmed

Abstract Hepatocellular carcinoma (HCC) constitutes a considerable worldwide health challenge, as evidenced by alarming statistics that indicate its increasing occurrence and elevated demise values. Notwithstanding advancements in care of medicine, the care of HCC continues to be problematic regardless of the late stage being diagnosed, few beneficial therapy choices, tumor diversity, and resistance to drugs. MicroRNAs (miRNAs) have garnered significant interest as critical regulators of HCC development. These small non-coding RNA molecules are crucial in regulating expression of genes and are associated with several cellular processes pertinent to cancer formation. Particular miRNAs have been found to exhibit specific expression in individuals with HCC, suggesting that they may serve as biomarkers for HCC diagnosis and prognosis. Comprehending the complex framework in HCC, miRNAs-mediated molecular pathways are crucial for elucidating the intricate techniques of HCC and formulating innovative cure strategies. Thus, this book chapter seeks to deliver an exhaustive analysis of the latest exploratory and health care results concerning the intricate function of miRNAs in affecting the principal characteristics of HCC, along with their prospective clinical applications as HCC diagnostic and prognostic biomarkers and therapeutic targets.

Keywords miRNAs · HCC · Prognosis · Diagnosis · Therapies

Introduction

Hepatocellular carcinoma (HCC) is one of the most frequent malignant tumors and it is responsible for 85–90% of primary liver cancer cases [22, 85].

H. M. Morsy · M. Y. Zaky · O. M. Ahmed (✉)
Physiology Division, Department of Zoology, Faculty of Science, Beni-Suef University, Beni-Suef, Egypt
e-mail: osama.ahmed@science.bsu.edu.eg

S. Pathak, A. Banerjee (eds.), *MicroRNAs: Role in Cancer Progression, Diagnostics and Therapy*, Advances in Biochemistry in Health and Disease 34,
https://doi.org/10.1007/978-3-032-08824-6_20

The miRNAs are short endogenous noncoding RNAs with an average of 22 nucleotides in length and they control gene expression at the post-transcriptional level. Most miRNAs are produced by transcription of DNA sequences into primary miRNAs, which are then converted into precursor miRNAs and, ultimately, mature miRNAs. The miRNAs have been reported to play important roles in regulating gene expression. It has been found that cancer cells exhibit substantial dysregulation of miRNAs [26, 47, 50].

Numerous studies have demonstrated that distinct miRNAs are expressed in the neoplastic tissues of patients with HCC and have been validated as important factors that can contribute to the control of essential signaling pathways in cancer cells. Because miRNAs are oncogenes or tumor suppressors, they can influence the prognosis, diagnosis, and therapy of HCC. Additionally, aberrant expression and regulation of miRNAs contribute to the development and progression of HCC. MiRNAs, therefore, have a lot of potential for the prognosis, diagnosis, and management of HCC [5, 22, 34, 38, 64].

Overview of Hepatocellular Carcinoma and Its Prevalence

Hepatocellular carcinoma is acknowledged to be one of the most common cancers globally, posing a significant health care challenge [51, 62]. The most common pathogenic form of liver cancer, HCC, makes up between 75% and 85% of all liver cancers [71]. It primarily arises in the setting of chronic liver illness and is marked by its aggressive characteristics and restricted therapeutic alternatives. HCC is a major worldwide health issue characterized by substantial regional disparities [81].

Prevalent risk factors encompass viral infections such as hepatitis B and C (HBV/HCV), chronic liver diseases including fatty liver and cirrhosis, alcohol misuse, and metabolic disorders such as diabetes [60]. Less prevalent risk factors include hemochromatosis, autoimmune hepatitis, and exposure to specific environmental contaminants. HCC frequently presents diagnostic challenges owing to the lack or vagueness of symptoms [11].

Initial symptoms generally include abdominal pain, weight loss, and exhaustion, but advanced stages may present with jaundice, ascites, and fever [6]. The subtle advancement of chronic liver illnesses exacerbates the surveillance of liver diseases [28]. Furthermore, the molecular and histological heterogeneity of HCC complicates diagnosis and treatment. In addition to the evident hepatitis virus, the existence of epigenetic modifications resulting from extended hepatic injury or hereditary factor mutations can result in HCC [58]. As patients generally do not receive genetic testing until there is a suspicion of liver disorders, these indicators are of diminished relevance in forecasting HCC [11].

Hepatocellular carcinoma etiology is intricate and encompasses several molecular problems that include cell-cycle deregulation, chromosomal instability, DNA methylation change, immunoregulation and the epithelial-to-mesenchymal transition (EMT), raised microRNA (miRNA), and HCC stem cell imbalance [33].

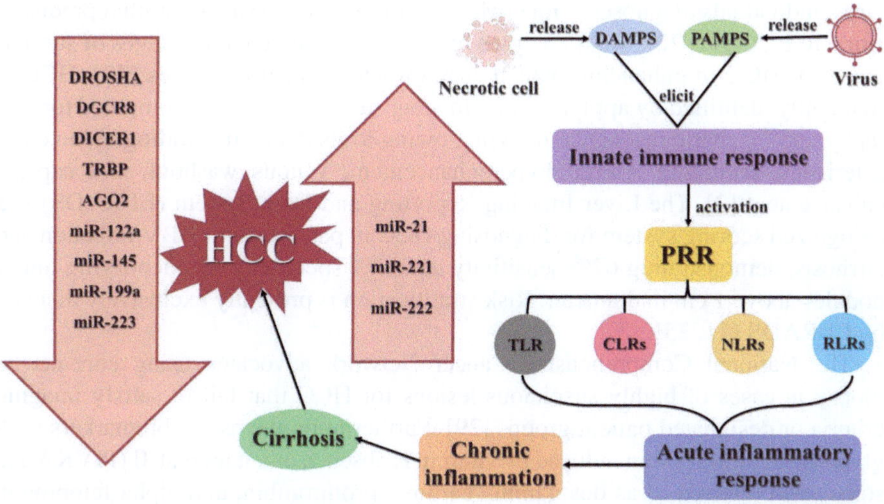

Fig. 20.1 Schematic representation of the role of miRNAs and innate immune signaling in hepatocellular carcinoma (HCC). Downregulation of key miRNAs such as miR-122a, miR-145, miR-199a, and miR-223, along with alterations in RNA-processing proteins (DROSHA, DGCR8, DICER1, TRBP, and AGO2), contributes to HCC initiation and progression. Conversely, upregulation of oncogenic miRNAs (miR-21, miR-221, and miR-222) promotes tumor development. Viral infections and necrotic cells release PAMPs and DAMPs, which activate the innate immune response through pattern recognition receptors (PRRs) including TLRs, CLRs, NLRs, and RLRs. This activation drives chronic and acute inflammatory responses, linking cirrhosis, chronic inflammation, and HCC progression

Although the disease mechanisms vary depending on underlying etiology, the usual sequence is liver injury, chronic inflammation, fibrosis, cirrhosis and HCC [51]. An innate immune response is triggered by the release of mediators of molecules through the activation of pattern recognition receptors (PRRs), such as Toll-like receptors (TLRs), nucleotide-binding oligomerization domain-like receptors (NLRs), C-type lectin receptors (CLRs), and retinoic acid-inducible gene I-like receptors (RLRs) present on different immune cells, such as damage-associated molecular patterns (DAMPs) from necrotic cells and pathogen-associated molecular patterns (PAMPs) by viruses (Fig. 20.1) [76]. An acute inflammatory reaction goes away once the outside trigger is gone. However, prolonged chronic inflammation can induce fibrosis and finally cirrhosis [68].

Importance of Early Detection

Surveillance programs and early detection tests are essential for the prompt identification and treatment of HCC. Screening initiatives predominantly focus on groups exhibiting numerous risk indicators, including confirmed carriers of the hepatitis virus, individuals diagnosed with cirrhosis, or those possessing a familial predisposition to HCC [11, 72]. Meta-analyses have corroborated the advantages of surveillance for HCC in enhancing patient survival and clinical outcomes [49]. HCC is frequently identified by applying safe imaging strategies, such as computed tomography and magnetic resonance imaging, owing to its distinctive radiographic characteristics, including arterial hyperenhancement, venous washout, and capsule enhancement [2]. The Liver Imaging Reporting and Data System (LI-RADS) is a recognized scoring system for diagnosing HCC in patients with HBV infection and cirrhosis, demonstrating 67% sensitivity and 93% specificity for identifying tumor nodules above 1 cm in diameter. Risk stratification is presently exclusively dictated by LI-RADS [11, 35].

The National Comprehensive Cancer Network advocates using core-needle biopsy in cases of highly suspicious lesions for HCC that fail to satisfy imaging criteria or designated patient groups [29]. Furthermore, the use of biomarkers such glypican-3, prothrombin induced by vitamin K absence or antagonist-II (PIVKA-II), sometimes referred to as des-gamma carboxy prothrombin, and alpha fetoprotein (AFP) has shown promise in improving the early diagnosis of HCC [36]. In HCC, accurate biomarkers are essential for predicting therapy response, tracking disease development, detecting possible therapeutic targets, and early identification [75].

Patients with early or intermediate-stage HCC may receive curative interventions including surgical resection, locoregional ablation, or liver transplantation [66]. Nonetheless, patients with advanced-stage HCC are restricted to chemotherapy or systemic therapies, including chemoembolization, sorafenib, and nivolumab [79]. Despite approximately 50% of HCC patients undergoing systemic therapy, the median overall survival remains constrained because of the scarcity of effective treatment alternatives [11].

MicroRNAs: Biological Function and Mechanisms

Definition and Role of microRNAs in Gene Regulation

The miRNAs are noncoding RNAs that serve as principal regulators of gene expression, precisely adjusting the activity of numerous genes in our cells by modifying gene expression at the post-transcriptional levels [14]. They are essential in regulating nearly all physiological functions, and their disruption can lead to the onset of various illnesses [32]. The majority of important biological processes, such as development, proliferation, differentiation, cell-fate determination, apoptosis, organ development, hematopoietic-lineage differentiation, host–viral interactions, nervous-system regulation, immunity, cardiac diseases, diabetes, angiogenesis, and tumorigenesis, are influenced by miRNAs, according to recent research (Fig. 20.2) [60, 61, 80].

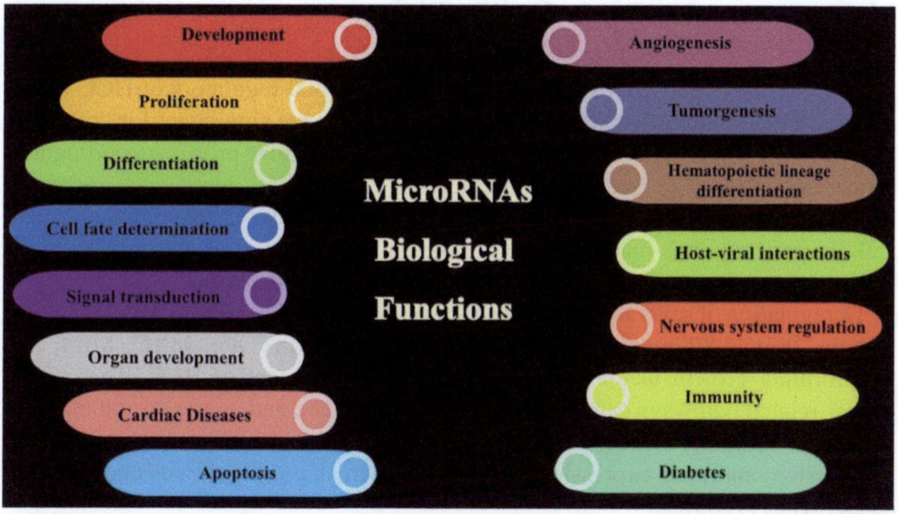

Fig. 20.2 Biological functions of miRNAs in diverse physiological and pathological processes. miRNAs regulate key cellular events including development, proliferation, differentiation, apoptosis, and organ development. They also influence angiogenesis, tumorigenesis, immunity, host–viral interactions, and nervous system regulation. Dysregulation of miRNAs contributes to diseases such as cardiac disorders and diabetes

MicroRNAs are formed by different genes found in intergenic tracts or areas that are intronic. RNA polymerase II or III transcribes these areas to create a primary miRNAs transcript, or pri-miRNAs, which exceeds a thousand nucleotides in extent. A complex of microprocessors in the nucleus processes the precursor miRNAs, comprising DiGeorge syndrome critical region 8 (DGCR8) and Drosha proteins, to produce a beginning miRNA (pre-miRNA) of around 85 nucleotides [37]. Next, the pre-miRNA is transported to the cytoplasm by means of Ran-GTP and exportin-5 complicated proteins. In the cytoplasm, the Dicer enzyme and the TRBP/PACT protein complex further cleave and process the pre-miRNA to produce a mature 20–22 nucleotide miRNA duplex consisting of passenger and guide strands [20]. The strand of guidance, exhibiting reduced stability of thermodynamics at the 5′ end, is preferentially selected for incorporation into the AGO 2 protein, leading to the establishment of a silencing complex produced by RNA [56]. RISC targets experience either translational suppression or mRNA destruction, chiefly via the base pairing of the 3′ untranslated region (UTR) of the target mRNA and the 5′ seed sequence of the miRNA [15].

The miRNA instability is ubiquitous in many malignancies, operating as tumor suppressors or oncomiRNAs, based on their expression levels and the individual disease setting. MiRNAs affect various essential biological processes, such as differentiation, angiogenesis, apoptosis, and cell proliferation, which are all pivotal in cancer initiation and advancement [37]. The dysregulation of miRNA expression may arise from pathways including gene amplifiers, elimination, alteration, and epigenetic modifications, resulting in aberrant miRNA expression that contributes to cancer [27].

Mechanisms of Action in Cancer Biology

MicroRNAs have garnered significant interest as pivotal regulators of HCC development. These diminutive noncoding RNA molecules are significant for governing expression of genes and are associated with several cellular processes pertinent to cancer formation [26, 37, 47]. Dysregulation of miRNAs, marked by increased oncomiRNA level and a decline in suppressive miRNAs, has a role in uncontrolled cell division, which is made possible by the multiplication of growth signals and a decreased reaction to pro-apoptotic and anti-growth signals [37].

Deregulation of several molecular pathways, such as p53, RAS/MAPK, WNT/β--catenin, PI3K/AKT/mammalian target of rapamycin (mTOR), MYC, MET, and transforming growth factor beta, leads to the development of HCC. The critical cancer-related pathways can be affected by both genetic and epigenetic changes, in addition to abnormal miRNA expression [67]. MiRNAs play crucial roles in cancer progression, and their expression profiles vary across normal tissues and different cancer types [7, 85].

Moreover, increased expression of miR-21 stimulates cancer-cell growth movement and penetration through the in vitro inhibition of Krüppel-like factor 5 [19]. Specifically, miR-550a-5p was demonstrated to enhance the migration and proliferation of HCC in a xenograft tumor model both in vivo and in vitro by using the Wnt/β-catenin signaling cascade to target glucosamine (UDPN-acetyl)-2-epimerase/N-acetylmannosamine kinase (GNE) [57]. Also, Morishita et al. [42] revealed that triggering AKT/PDK1 (pyruvate dehydrogenase kinase 1) signaling a pathway by releasing exosomal miR-21 in hematopoietic stem cells situated adjacent to HCC, it specifically targets *PTEN* as well as facilitating ongoing cancer by stimulating the release of angiogenesis factors, such as MMP9, MMP2, basic fibroblast growth factor (bFGF), vascular endothelial growth factor (VEGF), and transforming growth factor-β (TGF-β) by HCC cells. A separate study indicated that the expression levels of miR-221, miR-181b-1, miR-25, and miR-17-5p were markedly elevated in HCC. In Hep3B HCC and Huh-7 cell lines exhibiting consistent expression of the relevant miRNAs, miR-221 augmented hepatoma-cell proliferation, whereas miR-17-5p facilitated cell migration [67]. Furthermore, the overexpression of miR-155 enhances cellular spread via stimulation of the Wnt/β-catenin and AKT pathways, nuclear factor κB (NF-κB), and JAK2/STAT3 signaling pathways. Moreover, the overexpression of miR-221 promotes the proliferation of hepatic progenitor cells in tumorigenic mice by specifically targeting DNA damage-inducible transcript 4 (*DDIT4*), a regulatory factor of the mTOR pathway [57]. Moreover, miR-155 overexpression promotes cellular proliferation through the stimulation of the signaling pathways of JAK2/STAT3, AKT, and Wnt/β-catenin (Fig. 20.3) [46].

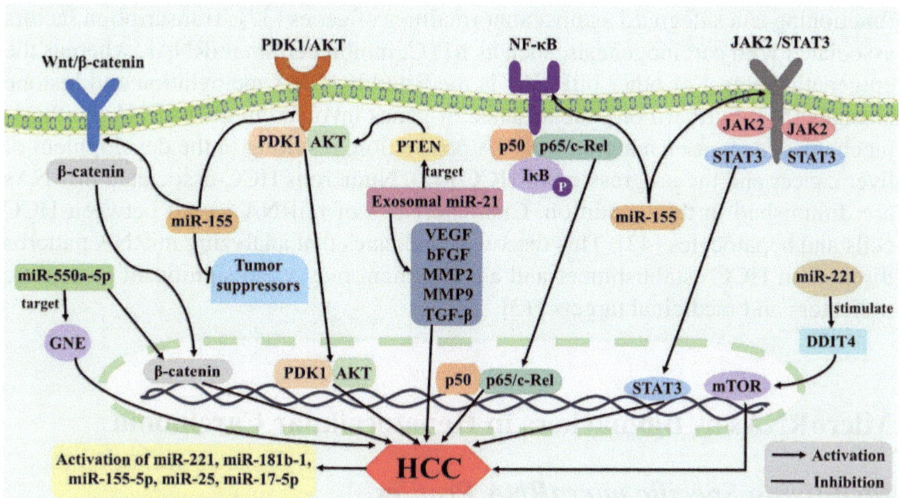

Fig. 20.3 Illustration of the molecular mechanisms by which dysregulated microRNAs (miRNAs) contribute to hepatocellular carcinoma (HCC) progression. The Wnt/β-catenin, PDK1/AKT, NF-κB, and JAK2/STAT3 signaling pathways are shown to be major oncogenic regulators in HCC. miR-155 promotes tumorigenesis by suppressing tumor suppressors and activating β-catenin, NF-κB, and STAT3 signaling. Exosomal miR-21 targets *PTEN* and enhances *VEGF, bFGF, MMP2, MMP9*, and *TGF-β* expression, promoting angiogenesis and invasion. miR-221 stimulates DDIT4, leading to *mTOR* activation, whereas miR-550a-5p targets GNE to regulate β-catenin signaling. Overall, activation of miR-221, miR-181b-1, miR-155-5p, miR-25, and miR-17-5p contributes to the oncogenic cascade, highlighting the central role of miRNA-mediated modulation of signaling pathways in HCC development and progression

Role of MicroRNAs in the Development and Pathogenesis of Hepatocellular Carcinoma

Hepatocellular carcinoma ranked as the seventh most common neoplasm and the fourth most important reason for death from cancer in the world. Significant progress in cure approaches notwithstanding, the alarm for the final stage of HCC remains bleak owing to the elevated prevalence rate. A comprehensive acknowledgement of the genesis of HCC is essential to enhance the effects [84]. MiRNAs stimulate the manifestation of the desired gene in the translational and post-transcriptional phases. Strange miRNA expression has often been identified in delicate areas or genetic regions linked to cancer across numerous human malignancies, including both HCC cells and tissues [42]. The specific patterns of abnormal miRNA expression vary according to disease etiology, encompassing several factors contributing to hepatocarcinogenesis, including alcoholic liver disease, non-alcoholic steatohepatitis, and viral hepatitis [16].

Although numerous tumor inhibitor miRNAs that inhibit cancer genes are reduced, other miRNAs such as oncomiRNAs that target genes that suppress tumors are increased in HCC. Genomic areas that encode miRNAs linked to dysfunctional functioning can safeguard against abnormalities of genes [37]. Transcription factors associated with carcinogenesis, such as MYC, inhibit certain miRNAs, whereas the epigenetic control of other miRNAs is mediated by DNA methylation and histone changes [54]. Furthermore, the absence of genes involved in the miRNA synthesis mechanism decreases matured miRNA production, resulting in the development of liver cancer and the progression of HCC [42]. Numerous HCC-associated miRNAs are diminished in this condition. Characteristics of miRNA varied between HCC cells and hepatocytes [42]. This discovery indicates that analyzing miRNA patterns throughout HCC establishment and advancement may yield significant diagnostic indicators and medicinal targets [13].

MicroRNAs as Biomarkers in Hepatocellular Carcinoma

Identifying Specific microRNA Profiles in Hepatocellular Carcinoma

The prognoses for patients with HCC are among the most unfavorable of all malignancies, with many patients presenting at an advanced stage upon diagnosis [69]. Tumor indicators, such as AFP and protein caused by PIVKA-II or lack of vitamin K, are extensively utilized to screen for HCC in high-risk individuals. Nonetheless, the assessment of HCC by traditional tumor indicators has produced inadequate outcomes. Consequently, supplementary complementary indicators are required, especially those linked to early HCC [69]. Also, the patient's sex, age, and body mass index may lead to considerable fluctuations in miRNA levels [53].

Circulating miRNAs are present in peripheral blood, making them more effective indicators for the assessment of HCC. miRNAs secreted by cancerous tissues and cells are enclosed in the exosomes, and the apoptotic bodies, or are associated with serum lipids or proteins [18]. Significantly, owing to the non-invasive and repetitive nature of blood-sample collection, along with the tissue specificity of miRNAs, miRNAs that circulate in serum or exosomes could serve as more effective indicators than those in tissues [19].

The prospective value of miR-122 and let-7 in the circulation in the detection of HCC in its early stages has been documented, demonstrating sensitivity similar to that of AFP according to blood levels in patients with HBV with early-stage HCC and precancerous nodules. miR-223-3p has shown promise as a trustworthy biomarker for HBV-related HCC [48]. Conversely, miR-122 may serve as a valuable biomarker in HCV-related HCC; the assessment of miR-221 and miR-22 indicated that miR-122 exhibited significant detection precision for identifying HCC in its early stages in individuals with cirrhosis of the liver [55]. Additionally, Megahed et al. [40] reported the effectiveness of using miR-125a-5p as a biomarker for the diagnosis of HCV-associated HCC. The degree to which particular miRNAs are expressed is contingent upon the progression of HCC; diminished miR-137 expression correlates with lymph node metastases or HCC vascular invasion [45].

Numerous miRNAs have shown capability as prognostic indicators for the prognosis of HCC. Current studies indicate that elevated levels of patients with HCC who have miR-92a, miR-221, miR-487a, or miR-1468 have a worse prognosis [48].

Diagnostic and Prognostic Potential

Numerous miRNAs may possess promise for both diagnostic and prognostic importance in HCC. For diagnostic purposes, it was demonstrated that miR-141, miR-200c, and miR-126, individually or collectively, can effectively differentiate primary carcinoma of the liver from other liver cancers that have spread with exceptional precision [25, 74]; Furthermore, the miR-205 to miR-194 ratio expression may serve to differentiate within gastrointestinal malignancies and metastases outside of the gastrointestinal tract, which is important because gastrointestinal tumors most often spread to the liver [9]. Owing to their varying concentrations in the serum or plasma of patients with diverse diseases compared with healthy individuals, miRNAs may serve as valuable biomarkers for monitoring HCC patients [83]. Numerous studies have indicated that the levels of miR-122 and miR-21 are significantly elevated in patients with HCC [70]. MiRNA panels or miRNAs with proteins such as AFP may offer a high diagnostic accuracy of HCC, and miRNAs may be an appropriate biomarker for early diagnosis of HCC. Additionally, aberrant circRNA expression has been found in a number of tumour types, including HCC [65].

The prognostic purposes of miRNAs lie in their ability to predict disease outcomes and progression. For instance, increased amounts of exosomal miR-32-5p, which focuses on PTEN, were unfavorable for patients with HCC. Expression of miR-638 in exosomes correlated with diminished general survival in patients with HCC [59]. Moreover, serum exosomes from patients with HCC had significantly higher expression levels of miR-122, miR-148a, and miR-1246 than patients with liver cirrhosis and healthy controls [63]. These data indicate that miRNAs may serve as possible biomarkers to enhance prognostic stratification and keep in touch with surveillance for individuals with HCC. They remain resilient in paraffin-embedded and formalin-fixed specimens, as well as in biological fluids such as serum or plasma, which is a crucial attribute for facilitating their identification and quantifying biological samples commonly utilized in clinical studies [78]. Remarkably, as fibrosis worsened in rats treated with diethylnitrosamine (DEN), miR-10b expression rose in both fibrotic and HCC livers. These findings point to the involvement of miRNA in the early stages of hepatic carcinogenesis. Moreover, it has been shown that miR-10b overexpression in HCC tissue maintains cancer-cell motility and invasion by modifying *RhoC, uPAR, MMP-2*, and *MMP-9* via the HOXD10 pathway. Through the regulation of CUB and Sushi multiple domains 1 (CSMD1), a protein implicated in EMT, miR-10b mimics were able to increase carcinogenesis in xenografted HCC mice [10, 86].

The downregulation of TLR3, which is linked to apoptosis evasion and a poor prognosis in these cancer cells, is facilitated by the often increased miR-155 in HCC. By promoting *TLR7/8* expression and triggering the anticancer activity of dendritic cells and natural killer cells, miR-1258 prevented tumor growth and metastasis in the HCC mouse model [77].

Therapeutic Applications of MicroRNAs in Hepatocellular Carcinoma

Two primary therapeutic techniques regarding miRNAs exist: restoring, replacement, or therapy based on overexpression, exemplified by miRNA-targeted therapeutics, and miRNA mimics, which encompass downregulation, inhibition, or decrease of miRNA, including antagomiRs [8]. MiRNA replacement treatments involve the introduction of chemically generated tumor suppressor miRNAs, which can be enhanced or recovered by miRNAs or miRNA mimics entering hepatocytes [22]. A report executed by Komoll et al. [30] demonstrated that injection into several mouse models of cancer, adeno-associated viral vector (AAV)-mediated miR-342-3p (AAV-miR-342-3p) can significantly slow tumor growth and increase overall survival in patients with HCC. Intravenously administered liposome-encapsulated MRX34, or miR-34, is the inaugural miRNA replacement treatment for individuals with sophisticated HCC. MiR-34a is crucial in impeding the advancement of HCC by downregulating hexokinase-1 expression, TNF receptor-associated factor 5, E2F Transcription Factor 1, and E2F Transcription Factor 3. Its Phase I study commenced in May 2013 [22]. Despite the initial data indicating an acceptable protection evaluation for individuals with severe HCC, few succumbed during MRX34 ersatz medication owing to severe immune-related side effects [21]. Nonetheless, it remains to be further investigated whether miR-34a possesses particular gene suppression action, whether the double-stranded RNA of MRX34 induces nonspecific inflammation effects, or whether successful transport among these RNA structures influences clinical outcomes [22].

MicroRNA-oriented methods are performed by altering miRNA expression that causes cancer using particular miRNA inhibitors. The first miRNA antagonist is miravirsen, which inhibits miR-122 [8]. The phase II clinical trial of miravirsen for the treatment of HCV was successfully concluded [31]. AC1MMYR2, a miR-21 small-molecule inhibitor, induces reversal EMT and suppresses invasion and growth of tumors and their metastases by obstructing miR-21 advancement without considerable tissue damage [23]. We were thrilled by a study conducted by Hassan et al. [24], who demonstrates significant potential for the inhibitor of miR-221 and miR-122 mimic combo as a novel medicinal approach for a mouse model of HCC DEN. The combination of miR-221 antagonist and miR-122 mimic is the most successful approach, as opposed to the use of either miR-221 inhibitor or miR-122 mimic alone. The combination of miR-221 inhibitor and miR-122 mimic drastically decreases the expression of *TGF-β, cyclin D1*, and *β-catenin* genes, which are crucial in regulating the cell cycle, EMT, and the growth of cells correspondingly [24].

MicroRNA Crosstalk with Immune Checkpoint Regulation in Cancer Immunotherapy

Immunotherapeutic drugs have shown promising initial results; the number of patients responding to them is low, indicating a need for greater efforts at drug development [39].

Immune-checkpoint proteins (ICPs), including programmed cell death-1 (*PD-1*) and its cognate programmed cell death ligand 1 (*PD-L1*), act as critical modulators of the immune system and are indispensable for preserving self-tolerance and regulating immune reactions. The PD-1/PD-L1 signaling pathway is essential for preventing autoimmune responses by inhibiting the activation and proliferation of T-cells, which are crucial components of the immune system [82]. Consequently, the inhibition of the PD-L1/PD-1 pathway has surfaced as an attractive therapeutic approach in cancer treatment, presenting new opportunities to improve patient outcomes. T-cell exhaustion can be reversed by disrupting PD-1/PD-L1 interactions. For instance, in clinical practice, monoclonal antibodies that target PD-1 (such as pembrolizumab and nivolumab) and PD-L1 (including atezolizumab and durvalumab) have shown considerable effectiveness across a range of malignancies [82].

MicroRNA-Based Therapies: Current Research and Trials

The RAS family is inhibited by tumor suppressor miRNAs such as the let-7 family [41]. Let-7 family members inhibit fibrosis by activating peroxisome proliferator-activated receptor gamma (PPARγ) at the level of transcription. These miRNAs modulate many target genes associated with fibrosis of the liver, thereby reducing carcinogenesis [46]. The expression of let-7 family members is significantly reduced in HCC linked to HCV. Among the let-7 family members [73], let-7 g levels are significantly reduced in metastatic HCC compared with nonmetastatic HCC, but elevated expression of let-7 g in HCC tissues is associated with improved prognosis in patients with HCC [1]. Members of the miR-15 family inhibit certain mRNAs and share a seed sequence. Members of this family are less effective tumor suppressors in HCC tissues, controlling inflammation by blocking TGF-β-activated kinase 1 binding protein 3 (TAB3) and IκB kinase (IKK) [4]. MiR-195 suppresses angiogenesis by targeting VEGF, cell-division cycle 42 (CDC42), and Vav guanine nucleotide exchange factor 2 (VAV2) in HCC cell lines [42]. A recent study found that miR-497, a member of the miR-15 family, influences the protein kinase B (PKB) pathway by targeting RPTOR independent companion of MTOR complex 2 (RICTOR) in HCC cell lines, and inhibits the growth of cells, and spread of cancer using the RICTOR/AKT signaling pathway [42]. In many types of cancer, miR-29 suppresses tumors and controls carcinogenic processes such as epigenetic changes, expansion, the cell death, fibrosis, and angiogenesis. The overexpression of miR-29 triggers apoptosis and suppresses HCC through the direct inhibition of anti-apoptotic proteins, encompassing myeloid leukemia 1 (MCL1) [52]. The absence of miR-101, which regulates MCL1, triggers apoptosis and suppresses carcinogenesis [43]. Additionally, miRNA-101 targets Rho-associated protein kinase (ROCK) in order to prevent EMT. ROCK, a down effector of RhoA GTPase, controls actomyosin bundles and focal adhesions [57]. miR-122, a prominent miRNA in the liver, is pivotal in lipid metabolism and HCV propagation HCC [12]. Consequently, miR-122 (Fig. 20.4) directly suppresses cyclin G1 (CCNG1), and insulin-like growth factor 1 (IGF-1R), A disintegrin and metalloproteinases 10 (ADAM10), and pyruvate kinase M2 (PKM2) function as tumor suppressors in HCC [3]. Notably, lack of miR-122 causes steatohepatitis and HCC in miR-122 knockout mice [17, 44]. Moreover, the reinstatement of miR-122 inhibits HCC in a murine model [10].

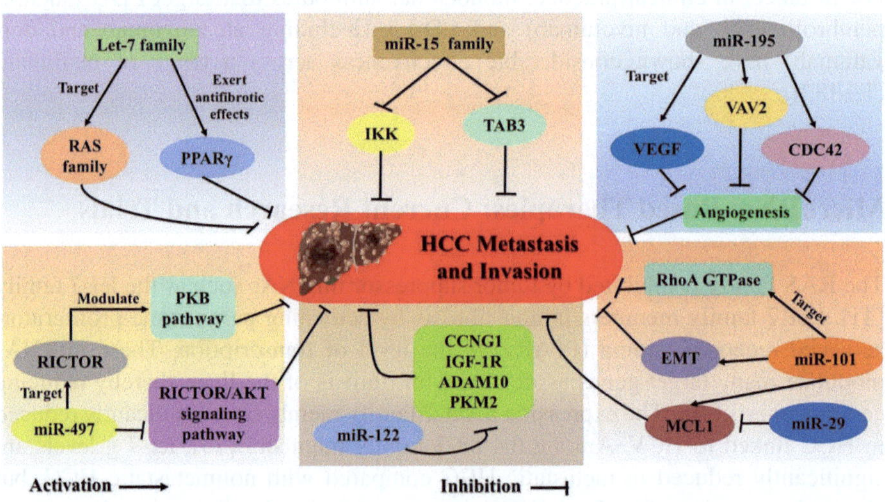

Fig. 20.4 Schematic representation of the regulatory roles of different microRNAs (miRNAs) in hepatocellular carcinoma (HCC) metastasis and invasion. The let-7 family targets the RAS family and PPARγ, exerting antifibrotic effects to inhibit HCC progression. The miR-15 family suppresses IKK and TAB3, reducing metastatic signaling. miR-195 targets VAV2, VEGF, and CDC42, thereby inhibiting angiogenesis. miR-497 targets RICTOR, modulating the PKB pathway and suppressing the RICTOR/AKT signaling cascade. miR-122 directly inhibits CCNG1, IGF-1R, ADAM10, and PKM2 to block invasion. miR-101 targets EMT via RhoA GTPase regulation, whereas miR-29 suppresses MCL1 expression

Conclusions and Future Perspectives

Hepatocellular carcinoma constitutes a substantial worldwide health challenge owing to its elevated frequency and prevalence, despite the existence of contemporary treatments. Breakthroughs in miRNA sequencing can significantly enhance our comprehension of the role of miRNAs in HCC and pharmacological reactions. These noncoding RNAs are essential in modulating target signaling pathways and genes, functioning as either tumor suppressors or oncogenes, thereby affecting numerous characteristics of cancer. The administration of miRNAs in HCC therapy is a fresh approach to molecular cancer treatment. Furthermore, serum miRNAs have become promising indicators for tracking the effectiveness of treatments and enhancing patient outcomes.

MicroRNAs have garnered attention recently for their potential as biomarkers in cancer detection and other disorders, as well as for predictive prognosis. Despite extensive study on miRNAs and their unique roles, aspects such as sample preparation procedures, analysis, and control selection remain under inquiry. Numerous variables must be considered while examining miRNAs and their function as biomarkers and mediators of disease. The patient's sex, age, and body mass index may lead to considerable fluctuations in miRNA levels.

Implications for Clinical Practice and Challenges

Our comprehension of the biological roles of miRNAs and their therapeutic and diagnostic significance in HCC is advancing swiftly. Numerous miRNAs have been recognized as playing a role in the regulation of biological processes during HCC development. Nonetheless, the translation of miRNA research into clinical practice faces several significant obstacles that must be overcome. A significant difficulty is achieving the correct control of miRNAs in vivo, given the complex dosage requirements and the likelihood of unforeseen side effects due to the multiple gene targets impacted by miRNAs. Moreover, guaranteeing the longevity of miRNA molecules in the circulation and their exact delivery to malignant cells while avoiding harmful effects on normal cells complicates therapeutic development further. Moreover, the discovery of key miRNAs involved in HCC could improve therapy aims, increase cure success, and lessen harmful adverse reactions. Therefore, it is critical to include these traits when analyzing the involvement of miRNAs in HCC and interpreting potentially contradicting results in the area. MiRNAs are effectively delivered when used for therapeutic purposes. miRNAs are hydrophilic and easily degraded by nucleases in the circulation, and they can potentially have off-target effects. To address these difficulties, numerous delivery techniques have been investigated, including liposomal carriers, which save miRNAs while improving their penetration into cells, and viral vectors, which provide efficient gene-transfer capabilities. These techniques are intended to increase miRNA stability, specificity, and therapeutic efficacy in clinical settings. Future research must concentrate on addressing these problems by creating effective separation procedures, standardizing detection technologies, and carrying out comprehensive clinical trials.

References

1. Abu-Shahba N, Hegazy E, Khan FM, Elhefnawi M (2023) In Silico analysis of MicroRNA expression data in liver cancer. Cancer Informat 22:11769351231171743
2. Afyouni S, Zandieh G, Nia IY, Pawlik TM, Kamel IR (2024) State-of-the-art imaging of hepatocellular carcinoma. J Gastrointest Surg 28(10):1717–1725
3. Al Ageeli E (2024) Dual roles of microRNA-122 in hepatocellular carcinoma and breast cancer progression and metastasis: a comprehensive review. Curr Issues Mol Biol 46(11):11975–11992
4. Antonakos N, Gilbert C, Theroude C, Schrijver IT, Roger T (2022) Modes of action and diagnostic value of miRNAs in sepsis. Front Immunol 13:951798
5. Aravalli RN (2013) Development of MicroRNA therapeutics for hepatocellular carcinoma. Diagnostics (Basel) 3(1):170–191
6. Asafo-Agyei KO, Samant H. (2025) Hepatocellular carcinoma. In: StatPearls [Internet]. StatPearls Publishing, Treasure Island. https://www.ncbi.nlm.nih.gov/books/NBK559177/
7. Assal RA, Abd El-bary RBED, Youness RA, Abdelrahman MM, Zahran H, Hosny KA, Esmat G, Breuhahn K, El-Ekiaby N, Fawzy IO, Abdelaziz AI (2024) OncomiR-181a promotes carcinogenesis by repressing the extracellular matrix proteoglycan decorin in hepatocellular carcinoma. BMC Gastroenterol 24(1):337
8. Brillante S, Volpe M, Indrieri A (2024) Advances in MicroRNA therapeutics: from preclinical to clinical studies. Hum Gene Ther 35(17–18):628–648
9. Callegari E, Elamin BK, Sabbioni S, Gramantieri L, Negrini M (2013) Role of microRNAs in hepatocellular carcinoma: a clinical perspective. Onco Targets Ther:1167–1178
10. Carpi S, Daniele S, de Almeida JFM, Gabbia D (2024) Recent advances in miRNA-based therapy for MASLD/MASH and MASH-associated HCC. Int J Mol Sci 25(22):12229
11. Chan YT, Zhang C, Wu J, Lu P, Xu L, Yuan H, Feng Y, Chen Z, Wang N (2024) Biomarkers for diagnosis and therapeutic options in hepatocellular carcinoma. Mol Cancer 23(1):189
12. Colaianni F, Zelli V, Compagnoni C, Miscione MS, Rossi M, Vecchiotti D, Padova MD, Alesse E, Zazzeroni F, Tessitore A (2024) Role of circulating microRNAs in liver disease and HCC: focus on miR-122. Genes 15(10):1313
13. D'Abundo L, Bassi C, Callegari E, Moshiri F, Guerriero P, Michilli A, Mora F, Gardini AC, Sangiovanni A, Piscaglia F, Sabbioni S, Gramantieri L, Negrini M (2024) Circulating microRNAs as biomarkers for stratifying different phases of liver cancer progression and response to therapy. Sci Rep 14(1):18551
14. Dayal S, Chaubey D, Joshi DC, Ranmale S, Pillai B (2024) Noncoding RNAs: emerging regulators of behavioral complexity. Wiley Interdiscip Rev RNA 15(3):e1847
15. Diener C, Keller A, Meese E (2024) The miRNA–target interactions: an underestimated intricacy. Nucleic Acids Res 52(4):1544–1557
16. El Hayek T, Alnaser-Almusa OA, Alsalameh SM, Alhalabi MT, Sabbah AN, Alshehri EA, Mir TA, Mani NK, Al-Kattan K, Chinnappan R, Yaqinuddin A (2024) Emerging role of exosomal microRNA in liver cancer in the era of precision medicine; potential and challenges. Front Mol Biosci 11:1381789
17. Fan W, Bradford TM, Török NJ (2024) Metabolic dysfunction– associated liver disease and diabetes: matrix remodeling, fibrosis, and therapeutic implications. Ann N Y Acad Sci 1538(1):21–33

18. Fekry B, Ugartemendia L, Esnaola NF, Goetzl L (2024) Extracellular vesicles, circadian rhythms, and cancer: a comprehensive review with emphasis on hepatocellular carcinoma. Cancer 16(14):2552
19. Gan L, Zheng L, Zou J, Luo P, Chen T, Zou J, Li W, Chen Q, Cheng L, Zhang F, Qian B (2024) MicroRNA-21 in urologic cancers: from molecular mechanisms to clinical implications. Front Cell and Dev Biol 12:1437951
20. Ghamlouche F, Yehya A, Zeid Y, Fakhereddine H, Fawaz J, Liu YN, Al-Sayegh M, Abou-Kheir W (2023) MicroRNAs as clinical tools for diagnosis, prognosis, and therapy in prostate cancer. Transl Oncol 28:101613
21. Grillone K, Caridà G, Luciano F, Cordua A, Di Martino MT, Tagliaferri P, Tassone P (2024) A systematic review of non-coding RNA therapeutics in early clinical trials: a new perspective against cancer. J Transl Med 22(1):731
22. Guo F, Li H, Wang J, Wang J, Zhang J, Kong F, Zhang Z, Zong J (2024) MicroRNAs in hepatocellular carcinoma: insights into regulatory mechanisms, clinical significance, and therapeutic potential. Cancer Manag Res:1491–1507
23. Han Q, Wang M, Dong X, Wei F, Luo Y, Sun X (2022) Non-coding RNAs in hepatocellular carcinoma: insights into regulatory mechanisms, clinical significance, and therapeutic potential. Front Immunol 13:985815
24. Hassan M, Elzallat M, Aboushousha T, Elhusseny Y, El-Ahwany E (2023) MicroRNA-122 mimic/microRNA-221 inhibitor combination as a novel therapeutic tool against hepatocellular carcinoma. Non-coding RNA Res 8(1):126–134
25. Hu H, Song H, Han B, Zhao H, He J (2024) Tumor-educated platelet RNA and circulating free RNA: emerging liquid biopsy markers for different tumor types. Front Biosci Landmark 29(2):80
26. Iacomino G (2023) miRNAs: the road from bench to bedside. Genes 2023(14):314
27. Improta-Caria AC, Ferrari F, Gomes JLP, Villalta PB, Soci ÚPR, Stein R, Oliveira EM (2024) Dysregulated microRNAs in type 2 diabetes and breast cancer: potential associated molecular mechanisms. World J Diabetes 15(6):1187
28. Jiang L, Meng Q, Liu L, Li W (2024) A comprehensive review on molecular mechanisms, treatments, and brief role of natural products in hepatocellular cancer. Nat Prod Commun 19(9):1–20
29. Kartik A, Armstrong VL, Stucky CC, Wasif N, Fong ZV (2024) Contemporary approaches to the surgical management of pancreatic neuroendocrine tumors. Cancer 16(8):1501
30. Komoll RM, Hu Q, Olarewaju O, von Döhlen L, Yuan Q, Xie Y, Tsay HC, Daon J, Qin R, Manns MP, Sharma AD, Goga A, Ott M, Balakrishnan A (2021) MicroRNA-342-3p is a potent tumour suppressor in hepatocellular carcinoma. J Hepatol 74(1):122–134
31. Kp A, Kaliaperumal K, Sekar D (2024) microRNAs and their therapeutic strategy in phase I and phase II clinical trials. Epigenomics 16(4):259–271
32. Levantini E, Rizzo M (2024) miRNAs: from master regulators of gene expression to biomarkers involved in intercellular communication. Biomedicine 12(4):721
33. Li A, Wang R, Zhao Y, Zhao P, Yang J (2024) Crosstalk between epigenetics and metabolic reprogramming in metabolic dysfunction-associated steatotic liver disease-induced hepatocellular carcinoma: a new sight. Meta 14(6):325
34. Lin HT, Castaneda AFA, Krishna SG, Mumtaz K (2024) MicroRNAs in hepatocellular carcinoma treatment: charting the path forward. World J Gastroenterol 30(11):1470–1474
35. Liu X, Ni X, Li Y, Yang C, Wang Y, Ma C, Zhou C, Lu X (2022) Diagnostic performance of LI-RADS version 2018 for primary liver cancer in patients with liver cirrhosis on enhanced MRI. Front Oncol 12:934045
36. Ma XL, Zhu J, Wu J, Tian L, Gao YY, Zhang CY, Zhou Y, Dai Q, Wang BL, Pan BS, Pan BS, Zhou J, Fan J, Yang XR, Guo W (2018) Significance of PIVKA-II levels for predicting microvascular invasion and tumor cell proliferation in Chinese patients with hepatitis B virus-associated hepatocellular carcinoma. Oncol Lett 15:8396–8404

37. Mahboobnia K, Beveridge DJ, Yeoh GC, Kabir TD, Leedman PJ (2024) MicroRNAs in hepatocellular carcinoma pathogenesis: insights into mechanisms and therapeutic opportunities. Int J Mol Sci 25(17):9393
38. Mallela VR, Rajtmajerová M, Trailin A, Liška V, Hemminki K, Ambrozkiewicz F (2024) miRNA and lncRNA as potential tissue biomarkers in hepatocellular carcinoma. Non-coding RNA Res 9(1):24–32
39. Marron TU, Luke JJ, Hoffner B, Jane Perlmutter J, Szczepanek C, Anagnostou V, Silk AW, Romero PJ, Garrett-Mayer E, Emens LA (2025) A SITC vision: adapting clinical trials to accelerate drug development in cancer immunotherapy. J Immunother Cancer 13:e010760
40. Megahed F, Tabll A, Atta S, Ragheb A, Smolic R, Petrovic A, Smolic M (2023) MicroRNAs: small molecules with significant functions, particularly in the context of viral hepatitis B and C infection. Medicina 59(1):173
41. Messina S (2024) The RAS oncogene in brain tumors and the involvement of let-7 microRNA. Mol Biol Rep 51(1):531
42. Morishita A, Oura K, Tadokoro T, Fujita K, Tani J, Masaki T (2021) MicroRNAs in the pathogenesis of hepatocellular carcinoma: a review. Cancer 13(3):514
43. Mu X, Wei Y, Fan X, Zhang R, Xi W, Zheng G, Yang AG (2024) Aberrant activation of a miR-101–UBE2D1 axis contributes to the advanced progression and chemotherapy sensitivity in human hepatocellular carcinoma. Cell Death Discov 10(1):422
44. Nakao K, Miyaaki H, Ichikawa T (2014) Antitumor function of microRNA-122 against hepatocellular carcinoma. J Gastroenterol 49:589–593
45. Nejati-Koshki K, Roberts CT, Babaei G, Rastegar M (2023) The epigenetic reader methyl-CpG-binding protein 2 (MeCP2) is an emerging oncogene in cancer biology. Cancer 15(10):2683
46. Nelaturi P, Kademani SP, Nallagangula KS, Ravikumar S (2024) Role of MicroRNAs in alcohol-related liver disease. Alcohol Treat Q 42(1):115–137
47. O'Brien J, Hayder H, Zayed Y, Peng C (2018) Overview of MicroRNA biogenesis, mechanisms of actions, and circulation. Front Endocrinol (Lausanne) 9:402
48. Oura K, Morishita A, Masaki T (2020) Molecular and functional roles of microRNAs in the progression of hepatocellular carcinoma—a review. Int J Mol Sci 21(21):8362
49. Pan CQ, Park AJ, Park JS (2024) New perspectives in hepatocellular carcinoma surveillance after hepatitis C virus eradication. Gastroenterology Report 12:goae085
50. Peng Y, Croce C (2016) The role of MicroRNAs in human cancer. Sig Transduct Target Ther 1:15004
51. Phoolchund AG, Khakoo SI (2024) MASLD and the development of HCC: pathogenesis and therapeutic challenges. Cancer 16(2):259
52. Pirzaman AT, Alishah A, Babajani B, Ebrahimi P, Sheikhi SA, Moosaei F, Salarfar A, Doostmohamadian S, Kazemi S (2024) The role of microRNAs in hepatocellular cancer: a narrative review focused on tumor microenvironment and drug resistance. Technol Cancer Res Treat 23:15330338241239188
53. Pointner A, Krammer UD, Tomeva E, Magnet U, Hippe B, Jacob U, Haslberger AG (2024) Lifestyle-driven variations in nutrimiromic MicroRNA expression patterns across and beyond genders. Life 14(3):390
54. Rac M (2024) Synthesis and regulation of miRNA, its role in oncogenesis, and its association with colorectal cancer progression, diagnosis, and prognosis. Diagnostics 14(13):1450
55. Rana M, Saini M, Das R, Gupta S, Joshi T, Mehta DK (2023) Circulating MicroRNAs: diagnostic value as biomarkers in the detection of non-alcoholic fatty liver diseases and hepatocellular carcinoma. MicroRNA 12(2):99–113
56. Ray A, Sarkar A, Banerjee S, Biswas K (2024) Non-CAnonical targets of MicroRNAs: role in transcriptional regulation, disease pathogenesis and potential for therapeutic targets. MicroRNA 13(2):83–95
57. Romeo M, Dallio M, Scognamiglio F, Ventriglia L, Cipullo M, Coppola A, Tammaro C, Scafuro G, Iodice P, Federico A (2023) Role of non-coding RNAs in hepatocellular carcinoma progression: from classic to novel clinicopathogenetic implications. Cancer 15(21):5178

58. Safri F, Nguyen R, Zerehpooshnesfchi S, George J, Qiao L (2024) Heterogeneity of hepatocellular carcinoma: from mechanisms to clinical implications. Cancer Gene Ther 31:1105–1112
59. Salehi M, Kamali MJ, Arab D, Safaeian N, Ashuori Z, Maddahi M, Latifi N, Jahromi AM (2024) Exosomal microRNAs in regulation of tumor cells resistance to apoptosis. Biochem Biophys Rep 37:101644
60. Sharma R, Simon TG, Hagström H, Lochhead P, Roelstraete B, Söderling J, Verna EC, Emond J, Ludvigsson JF (2024a) Statins are associated with a decreased risk of severe liver disease in individuals with noncirrhotic chronic liver disease. Clin Gastroenterol Hepatol 22(4):749–759
61. Sharma H, Kaushik M, Goswami P, Sreevani S, Chakraborty A, Ashique S, Pal R (2024b) Role of miRNAs in brain development. MicroRNA 13(2):96–109
62. Shen C, Jiang X, Li M, Luo Y (2023) Hepatitis virus and hepatocellular carcinoma: recent advances. Cancer 15:533
63. Shetti D, Mallela VR, Ye W, Sharif M, Ambrozkiewicz F, Trailin A, Liška V, Hemminki K (2024) Emerging role of circulating cell-free RNA as a non-invasive biomarker for hepatocellular carcinoma. Crit Rev Oncol Hematol 200:104391
64. Sohail SK (2025) Natural products as modulators of miRNA in hepatocellular carcinoma: a therapeutic perspective. J Gene Med 27(5):e70019
65. Song T, Li L, Wu S, Liu Y, Guo C, Wang W, Dai L, Zhang T, Wu H, Su B (2021) Peripheral blood genetic biomarkers for the early diagnosis of hepatocellular carcinoma. Front Oncol 11:583714
66. Susman S, Santoso B, Makary MS (2024) Locoregional therapies for hepatocellular carcinoma in patients with nonalcoholic fatty liver disease. Biomedicine 12(10):2226
67. Szilveszter RM, Muntean M, Florea A (2024) Molecular mechanisms in tumorigenesis of hepatocellular carcinoma and in target treatments – an overview. Biomolecules 14(6):656
68. Tanwar S, Rhodes F, Srivastava A, Trembling PM, Rosenberg WM (2020) Inflammation and fibrosis in chronic liver diseases including non-alcoholic fatty liver disease and hepatitis C. World J Gastroenterol 26(2):109–133
69. Teng W, Wang HW, Lin SM (2024) Management consensus guidelines for hepatocellular carcinoma: 2023 update on surveillance, diagnosis, systemic treatment, and posttreatment monitoring by the Taiwan liver cancer association and the gastroenterological society of Taiwan. Liver Cancer 13(5):468–486
70. Tobaruela-Resola AL, Milagro FI, Elorz M, Benito-Boillos A, Herrero JI, Mogna-Peláez P, Tur JA, Martínez JA, Abete I, Zulet MÁ (2024) Circulating miR-122-5p, miR-151a-3p, miR-126-5p and miR-21-5p as potential predictive biomarkers for metabolic dysfunction-associated steatotic liver disease assessment. J Physiol Biochem:1–14
71. Wang L, Wang T, Zhang W, Zheng S (2024) Survival and cardiovascular disease mortality among primary liver cancer patients: a population-based study. Heliyon 10(19):e37869
72. Yang JD, Hainaut P, Gores GJ, Amadou A, Plymoth A, Roberts LR (2019) A global view of hepatocellular carcinoma: trends, risk, prevention and management. Nat Rev Gastroenterol Hepatol 16(10):589–604
73. Yang C, Lv F, Yang J, Ding D, Cui L, Han Y (2025) Surveillance and management of hepatocellular carcinoma after treatment of hepatitis C with direct-acting antiviral drugs. Ann Hepatol 30:101582
74. Yin J, Bai Z, Song J, Yang Y, Wang J, Han W, Zhang J, Meng H, Ma X, Yang Y, Wang T, Li W, Zhang Z (2014) Differential expression of serum miR-126, miR-141 and miR-21 as novel biomarkers for early detection of liver metastasis in colorectal cancer. Chin J Cancer Res 26(1):95–103
75. Yu B, Ma W (2024) Biomarker discovery in hepatocellular carcinoma (HCC) for personalized treatment and enhanced prognosis. Cytokine Growth Factor Rev
76. Yu L, Gao F, Li Y, Su D, Han L, Li Y, Zhang X, Feng Z (2024) Role of pattern recognition receptors in the development of MASLD and potential therapeutic applications. Biomed Pharmacother 175:116724

77. Yuedong H, Yu H, Yuzhu Y (2022) miR-1258 enhances the anti-tumor effect of liver cancer natural killer (NK) cells by stimulating toll-liker receptor (TLR)7/8 to promote natural killer (NK)-dendritic cell (DC). Interact J Biomater Tissue Eng 12(6):1241–1246
78. Zakari S, Niels NK, Olagunju GV, Nnaji PC, Ogunniyi O, Tebamifor M, Israel EN, Atawodi SE, Ogunlana OO (2024) Emerging biomarkers for non-invasive diagnosis and treatment of cancer: a systematic review. Front Oncol 14:1405267
79. Zarlashat Y, Abbas S, Ghaffar A (2024) Hepatocellular carcinoma: beyond the border of advanced stage therapy. Cancer 16(11):2034
80. Zatterale F, Raciti GA, Prevenzano I, Leone A, Campitelli M, De Rosa V, Beguinot F, Parrillo L (2022) Epigenetic reprogramming of the inflammatory response in obesity and type 2 diabetes. Biomolecules 12(7):982
81. Zhan Z, Chen B, Huang R, Lin W, Lan S, Yao X, Huang S, Lin W, Xu S, Zhou S, Yu J, Wang X, Lin X, Guo Z (2025) Long-term trends and future projections of liver cancer burden in China from 1990 to 2030. Sci Rep 15(1):13120
82. Zhang H, Gong L, Yu L, Xian C, Ma Z, Wang X, Xia R (2025) Emerging roles of non-coding RNA derived from extracellular vesicles in regulating PD-1/PD-L1 pathway: insights into cancer immunotherapy and clinical applications. Cancer Cell Int 25(1):188
83. Zhao J, Hu Z, Zheng X, Lin Y, Liu X, Zhang J, Peng J, Gao H (2024) Blood biomarkers of hepatocellular carcinoma: a critical review. Front Cell Dev Biol 12:1489836
84. Zheng S, Chan SW, Liu F, Liu J, Chow PKH, Toh HC, Hong W (2024) Hepatocellular carcinoma: current drug therapeutic status. Adv Challenges Cancers 16(8):1582
85. Zhou Y, Liu F, Ma C, Cheng Q (2022) Involvement of microRNAs and their potential diagnostic, therapeutic, and prognostic role in hepatocellular carcinoma. J Clin Lab Anal 36(10):e24673
86. Zhu Q, Gong L, Wang J, Tu Q, Yao L, Zhang J, Han X, Zhu S, Wang S, Li Y, Zhang W (2016) miR-10b exerts oncogenic activity in human hepatocellular carcinoma cells by targeting expression of CUB and sushi multiple domains 1 (CSMD1). BMC Cancer 16:806

Chapter 21
The Role of MicroRNAs in Cancer Radiotherapy

Kshitija Aherkar, R. Mythreyi, and Kanthesh M. Basalingappa

Abstract Cancer remains one of the leading causes of death worldwide, thereby presenting an urgent need for reliable biomarkers and specific therapeutic targets. Despite the advancements in treatment strategies like radiotherapy, its effectiveness varies depending on the resistance or sensitivity of the cancer type. Thus, understanding the molecular interactions therein is vital for increasing the efficacy of radiotherapy. MicroRNAs (miRNAs) are central to the regulation of oncogenes and tumor suppressor genes, making them significant in the context of cancer. miRNAs are small, endogenous RNAs that serve to modulate gene expression. Recent research has highlighted the role of miRNAs in cancer as key regulators of radiation response. They serve a dual role as contributors to radiosensitizers and radioresistance, as per the cancer type and specific miRNA involved. Studies have also emphasized the function of miRNAs as a predictive biomarker in incident radiation or infrared exposure. This chapter explains how miRNAs function in the systems that control radiation sensitivity. Further investigation into miRNAs and their modulation could lead to more personalized and effective radiotherapy strategies. By analyzing these factors, this chapter seeks to provide a better understanding of how the role of miRNAs in cancer could transform the radiation treatment methods for the same.

Keywords Cancer · Radiotherapy · MicroRNAs · Biomarker · Treatment · Radiosensitization · Radioresistance

---

K. Aherkar · R. Mythreyi · K. M. Basalingappa (✉)
Division of Molecular Biology, School of Life Sciences, JSS Academy of Higher Education & Research, Mysuru, Karnataka, India
e-mail: kantheshmb@jssuni.edu.in

S. Pathak, A. Banerjee (eds.), *MicroRNAs: Role in Cancer Progression, Diagnostics and Therapy*, Advances in Biochemistry in Health and Disease 34,
https://doi.org/10.1007/978-3-032-08824-6_21

Introduction to MicroRNAs and Radiotherapy

One of the most important public health issues for several years has been cancer [1]. The evolution of cancer treatment has continued to be marked by significant milestones, from the early adoption of a standard anesthesia in surgical resections in the second half of the 1800s to X ray invention by Wilhelm Conrad Röntgen in the late nineteenth century, resulting in the era of combining radiation and surgery for cancer treatment [1, 2]. Further developments in chemotherapy during World War II and the most recent developments in gene therapy have contributed significantly to the multifaceted approach for cancer treatment [1, 2]. Given these advancements, the field of molecular biology has created new opportunities, especially in investigating miRNA and their potential to improve the effectiveness of radiation therapy or radiotherapy (RT) employed therein.

The miRNAs can be referred to as a group of single-stranded, nonprotein coding RNAs around 20 to 22 nucleotides long [3]. They have a role in the regulation of the gene expression posttranscriptionally [3]. The biogenesis of miRNAs begins with the specific sequence of DNA within miRNA genes or with a gene cluster that further undergoes transcription to form miRNA. There are two major pathways for their biogenesis: canonical pathway and noncanonical pathway [4–6].

miRNAs were formerly identified for their role in larvae of *Caenorhabditis elegans* in 1993 [7, 8]. Since then, they have continued to be a source for performing functional studies of RNA in various organisms, as most animal model systems have been found to include miRNAs, some of which are highly conserved throughout species [8, 9]. The human genome singularly encodes approximately more than 2500 miRNAs [9–11]. The functions of miRNAs in regulating genes are widely known, and new miRNAs are constantly being found in parallel. miRNAs are extensively connected to the etiology of a number of human illnesses, which include diseases like cancer [12]. The first evidence of a connection between miRNAs and cancer was presented two decades ago [13, 14]. It was suggested that miRNAs could enhance cancer treatment strategies by either restoring or inhibiting their function [14]. The abnormal levels of miRNAs in the case of a tumor have been known to cause serious consequences. For example, tumor suppressors like *PTEN* are downregulated by overexpression of miRNAs such as miR-21, which in turn leads to oncogenesis, as seen in the case of hepatocellular cancer [12, 13]. In contrast, certain tumor cells losing miRNAs lead to an increased expression of oncogenes. Thus, miRNAs can be considered to perform a dual role, acting as oncogenes for cancer progression and tumor suppressors depending on the specific pathway and gene expression pattern under consideration [9, 12, 13]. To understand the extensively observed variations in the expression of miRNA-associated genes involving tumor and normal cells, it takes an intricate combination of multiple factors, including oncogenes, tumor suppressor genes, epigenetic mechanisms, and exclusive genomic information about the location of miRNAs in regions associated with cancer [12–14].

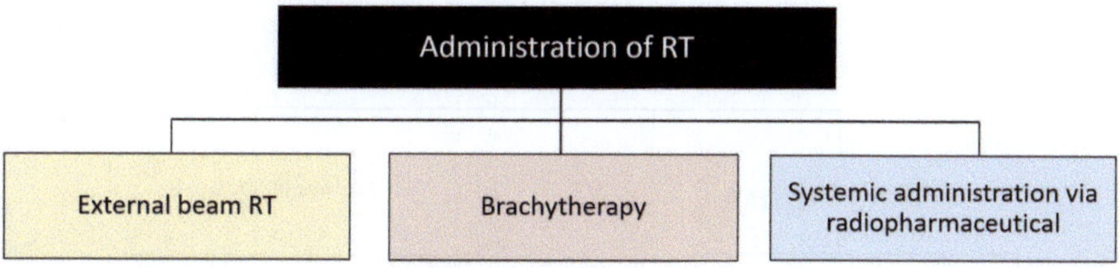

Fig. 21.1 Modes of radiotherapy (RT) administration

Approximately two-thirds of all cancer patients annually require RT [15]. Similar to the treatment modality of surgery, RT aims to target a specific location within the body as the principle behind the working of RT involves the use of ionizing radiation that serves to damage the DNA [16]. Specifically, the radiation causes breakage in the double-stranded structure of DNA present in cancer cells, which in turn prevents the restoration of the integrity of the genome, ultimately resulting in cell death [16]. So, RT involves exposure to a precise dosage of ionizing radiation that, in turn, can serve to cause damage to cancerous cells and facilitate their eradication [14, 15]. Thus, it is designed to bring maximum control over tumor cells and minimum damage to normal cells [15, 16]. As mentioned in Fig. 21.1, there are three ways in which RT can be administered: external beam RT, brachytherapy, and systemic administration of radiopharmaceutical substances [16, 17]. Conventional external RT uses gamma-ray photons from a source of radioactive decay, or more frequently, megavoltage X-rays from a linear accelerator [15–17].

In recent decades, there has been considerable development in RT treatments like intensity-modulated RT as well as imaging technologies to allow definite doses of radiation at the specific tumor dimension and reduce consequences on the normal cells [17]. Together with a refined comprehension of tumor biology at the molecular and physiological levels, these novel advances in technology have increased the effectiveness of RT as a treatment [18].

By itself, ionizing radiation upregulates certain miRNAs implicated in the DNA damage response, such as miR-34a and miR-143, while downregulating others, including miR-15b, and miR-181a [16, 17]. Interestingly, it has also been observed that a number of miRNAs involved in the DNA damage response (DDR) (e.g., let-7 family, miR-15a, miR-24, miR-155) are both down- and upregulated. This difference is postulated depending on the cell type, radiation dose, and time point of measurement after RT [16, 18]. As depicted in Fig. 21.2, certain pathways may be upregulated or downregulated in response to elevated miRNA levels, which may enhance radiosensitivity or radioresistance, respectively. On the other hand, a reduction in miRNA levels can also cause pathways to be upregulated or downregulated, which has the opposite impact of increasing radiosensitivity or causing radioresistance. In the end, these modifications affect how well radiation targets various cell types. Depending on the treatment objective, altering miRNA expression may make RT more effective by either protecting healthy cells from radiation or making cancer cells more sensitive to it.

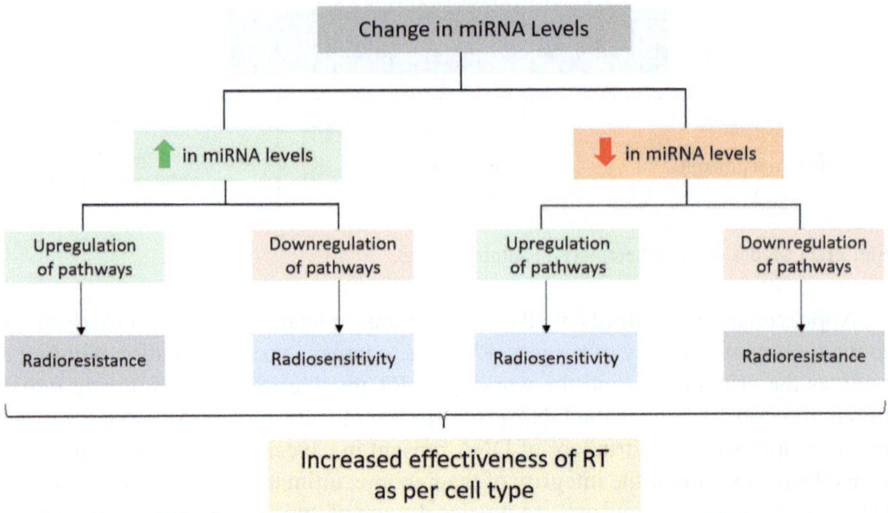

Fig. 21.2 Changes in miRNA level as per the cell type and radiotherapy

Challenges in Traditional Radiotherapy and Resistance Mechanisms

RT remains a cornerstone of cancer treatment, as it is employed for approximately 50% of cancer patients either primarily or in combination with other treatment regimens, including surgery and chemotherapy. While there are several problems that frequently impair RT's effectiveness, the development of resistance mechanisms in tumor cells is the most significant factor that could decrease its efficacy. This makes it essential to comprehend the underlying mechanisms in order to improve therapeutic results.

There is damage to healthy cells along with cancer cells being subjected to RT [18]. This leads to side effects and thus limits the tolerable radiation dosage. If continued, RT failure can even lead to the recurrence of cancer and thus is the prime barrier to optimizing therapeutic efficacy [18]. As summarized in Fig. 21.3, multiple mechanisms of RT resistance can develop in cancer cells as through the upregulation of certain proteins involved in DNA repair mechanisms, such as ATR, ATM, and those involved in double-stranded break repair (DSB) [19]. Also, the tumor cells could become resistant to RT [19]. This improvement enables them to effectively fix radiation-induced DNA double-strand breaks, preventing cell death. Another crucial factor in establishing resistance to RT is the tumor microenvironment. A prevalent characteristic of cancer is hypoxia, which stabilizes hypoxia-inducible factors (HIF) that increase angiogenesis and change cellular metabolism, improving tumor survival and treatment resistance [20].

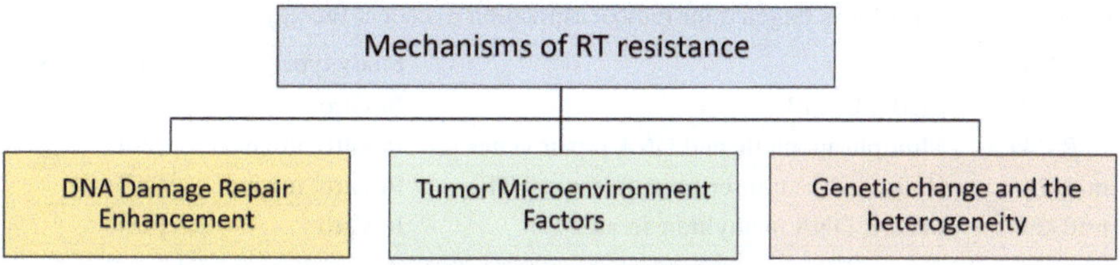

Fig. 21.3 Key mechanisms of radiotherapy resistance mediated by miRNAs

Additionally, numerous genetic changes and the heterogeneity of cancer cells, along with their surrounding microenvironment, are linked to RT resistance, which can arise in a variety of ways [21]. In this regard, traditional RT failure can arise because of resistance, thereby causing consequences to the normal functioning of cells via metastasis [21, 22]. With enhanced DNA repair ability, intrinsic and RT-induced radioresistant tumor cells are a major cause of radiation treatment failure [22]. To solve this issue, new radiosensitizers must be developed, and the processes of radiation resistance must be identified.

Targeting miRNAs for Radiosensitization

RT can be considered one of the key modes of cytotoxic therapy against cancer. With the increasing use of this treatment modality, the problem of radiation resistance or radioresistance is also increasing. Thus, the response brought about by tumor cells on exposure to radiation is crucial for RT as it is one of the major factors that ascertains if the cell will exhibit radioresistance or radiosensitivity [22, 23].

Numerous studies have shown that miRNAs can improve the effectiveness of RT by altering gene expression and cellular reactions to radiation. As a result, miRNAs have become attractive candidates for radiosensitization in cancer treatment [23]. A few groups of miRNAs have been recognized, in vitro, to bring radiosensitization after their alteration [24, 25]. In terms of research on radioresistance, as exhibited by tumor cells, it is inferred that inhibiting certain miRNAs could prove to be beneficial in countering the same. miRNAs can be significantly employed for their therapeutic potential since their effect results in tuning the expression of the target genes [3, 24, 25]. This is comparatively less detrimental to healthy cells than knocking out the genes, thereby offering a more targeted approach. The in vitro studies are further required to be validated for their applications in in vivo settings. Over the past years, the application of miRNAs for its function in increasing radiosensitivity is being recognized widely [3, 24]. It has been demonstrated that miR-205 targets *ZEB1* and *Ubc13* to prevent DNA damage repair, increasing tumor cells' radiosensitivity [22]. Also, miR-34a increases radiation response by regulating apoptosis and DNA repair processes [26]. Similar such studies have been done for key miRNAs that are targeted for radiosensitization, as summarized in Table 21.1.

Table 21.1 Key miRNAs targeted for radiosensitization in cancer therapy

miRNA	Targets	Study type	References
miR-205	ZEB1, Ubc13	In vitro	[22]
miR-34a	Multiple apoptotic and DNA repair genes	In vitro, in vivo	[26]
miR-145	Helicase-like transcription factor (HLTF)	In vitro, in vivo	[27]
miR-29b	MCL1, DNA methyltransferases	In vitro	[28]

Further, to increase the efficacy of RT, it is crucial to comprehend how miRNAs are regulated and function. It has been suggested that radiation-induced downregulation of certain miRNAs may increase radioresistance. While factors like the tumor microenvironment and epigenetic modifications are recognized to be simultaneously participating in the RT response, studies suggest that miRNA interactions are also a crucial aspect involved therein [22, 23, 25, 26].

miRNA-Based Biomarkers for Radiation Response

The current literature available on the application of miRNA is as per the studies conducted on the biopsy localization of miRNAs for their role as biomarkers for diagnosis [22, 24, 25]. The radiation-sensitive aspect of tumor cells is affected by miRNAs in various ways via the regulation of cellular components. These are majorly involved in cell cycle progression and cell signalling transduction pathways related to radiation. miRNAs are involved in various cellular processes involved for normal functioning [22]. This includes maintaining homeostasis and bringing response to cellular stress [3, 22, 24].

There is an urgent need for biomarkers that can quickly identify people who have been exposed to high levels of radiation and need immediate medical attention in mass-casualty disasters [29, 30]. Plasma miRNA expression profiles have been proposed as a predictor of varying radiation exposure levels and have been demonstrated to differentiate mice that underwent total body irradiation doses of 0.5 Gy, 2 Gy, and 10 Gy [29]. Ionizing radiation-induced miRNA profiles in mouse blood were found to be dose- and radiation-specific in another investigation [29]. Studies have also employed mouse blood-derived miRNA expression profiles that are used as biomarkers for radiation exposure [30, 31]. Likewise, to create biomarkers of radiation exposure, the levels of more than 600 miRNAs in the serum of mice exposed to radiation at doses ranging from 1 to 12 Gy were tracked at 24- and 48-h intervals [31]. It was found that the radiation exposure of both young and adult mice caused the induction of miR34a in numerous organs. After IR, miR-34a remains stable in serum and may be a sign of radiation damage [32–34]. A marker of lymphocyte depletion and bone marrow damage, miR-150, which is abundant in lymphocytes, showed a dose and time-dependent decrease in serum in patients undergoing total body radiation as part of a bone marrow transplant preparation regimen [34]. Evidently, it has been demonstrated that a number of miRNAs contribute to radioresistance in various tumor types. Additionally, following IR exposure, notable alterations in miRNA expression levels can be seen [16, 17, 34, 35]. The DDR following IR is one of the nearly all cellular pathways that miRNAs are known to regulate. Significantly, the development of customized radiotherapy depends on the discovery of such miRNA-based biomarkers since they aid in the prediction of unique tumor responses and radiation-induced toxicities, allowing for stratified patient treatment plans [35, 41, 81].

The need for newer molecular biomarkers per se exists for enhancing RT efficiency. In this context, miRNAs markedly could serve as new biomarkers for cancer diagnosis, recurrence monitoring, and survival prediction [16, 17, 35]. Besides, various circulating miRNAs in particular are a useful and readily available source. Given their persistence in bodily fluids, miRNAs are appealing options for noninvasive indicators [36]. Certain miRNA signatures have been found in serum or plasma that are correlated with radiation exposure levels [36]. A meta-analysis, for instance, showed that circulating miR-150 levels drop in a dose-dependent fashion after radiation exposure, indicating that it may be used as a biomarker for determining radiation exposure [36, 37]. Also, the potential of miR-21 as an indicator of prediction for treatment-related adverse outcomes is highlighted by the association between raised levels of the protein and increased genitourinary toxicity in patients with prostate cancer following radiation therapy [38]. In another systematic review and meta-analysis covering 19 tumor types, over 75 miRNAs were shown to be connected with patient response to RT. Other theragnostic biomarkers that could predict radioresistance or sensitivity were miR-21 and miR-125b [38, 88]. These realizations enable medical professionals to modify radiation dosage or combine with certain agents prior to the onset of resistance [38, 39, 88]. Circulating miRNAs detected during radiotherapy in prostate cancer have been demonstrated to predict early relapse as well as therapeutic success, allowing for prompt treatment regimen modification [14, 89]. Delivering the appropriate dosage to the appropriate patient at the appropriate time is supported by this dynamic monitoring [14, 89]. Correspondingly, such miRNA expression profiles specifically may indicate how a patient would react to radiation therapy and how likely they are to develop radiation-induced toxicities [38].

Therapeutic Potential of miRNAs in Radiotherapy

Despite the complexity of radioresistance, new studies have shown that epigenetic modifications, such as changes in miRNA, are crucial for RT resistance [16, 17, 28, 39]. Therefore, in order to explore miRNAs that can predict an individual cell's radiosensitivity, more thorough preclinical and clinical research examining miRNA expression profiles in relation to radioresistance is imperative. The molecular processes of the dysregulated circulating miRNAs that serve as biological markers for pancreatic cancer also need to be confirmed by additional research. Furthermore, miRNAs can aid in the development of novel cancer treatments, which is crucial for extending the survival of patients with cancer [16, 17, 39, 40].

Important signaling processes related to the radiation response, such as PI3-K/Akt, NF-κB, and MAPK, are regulated by miRNAs [41]. PTEN is regulated by miR-21, miR-26, miR-221/222, miR-216a/217, and miR-486, which also affect Akt activation [41, 42]. miR-221/222 targets PTEN to affect tumor development and radiosensitivity. miR-17-5p is a possible therapeutic target because it stimulates metastasis through the p38 MAPK-HSP27 axis [42, 43]. Numerous miRNAs influence NF-κB signaling, which in turn controls p50 activity and IKK subunits, hence promoting the growth of tumors [41, 44]. These interactions between miRNAs demonstrate their importance to the regulation and treatment of cancer [41]. As was previously indicated, miRNAs can be used in therapeutic techniques to either mimic or suppress translation-level gene expression. In the first strategy, miRNA can be added to restore underexpressed miRNA [14]. The second strategy involves adding synthetic anti-miRNAs to prevent miRNA from being overexpressed [14, 41]. According to preclinical research, using miRNAs is well-tolerated and does not result in serious side effects. Prior to treating patients, it is vital to enhance the targeted delivery and efficiency of the tumor [14]. As miRNAs can be customized to patient-specific tumor biology, reducing damage to healthy tissue and optimizing treatment efficacy, using them for radiosensitization is a significant step toward personalized radiotherapy.

Very few clinical trials exist that have studied the miRNA expression profile that RT causes in patients. In a study where patients received low-dose-rate prostate brachytherapy, miRNA-221 and miR-93 transcription were lower in patients' plasma after RT, making them radiosensitive [45, 46]. It is also reported that miR-93 drastically decreased in extracellular vesicles from patients' serum after RT [46]. Furthermore, miRNA-145 expression was examined in the tumor tissue of pancreatic cancer patients, and it was proposed that miRNA-145 may enhance responsiveness to radiation therapy by decreasing the effectiveness of DNA DSB repair [47]. For instance, because miR-99b regulates the mTOR pathway, its inhibition by radiation has been connected to greater radioresistance [48]. Additionally, a recent study used miRNA sequencing and indicated that miR-196b-5p and miR-194-5p were increased in exosomes taken from pancreatic cancer cell lines such as SW1990 following irradiation [48, 49]. Another study showed that several cancer cell lines showed increased expression of miR-194-5p following 10 Gy of radiation [49]. A recent systematic review and meta-analysis summarize the effects of RT on circulating miRNAs in humans, nonhuman primates, and mice. Radiation treatment in the included studies was heterogeneous, including total body irradiation and local tumor-specific irradiation with different fractionation schemes and irradiation doses [36, 41].

Personalized therapy has become more and more well-known in recent years, and it is even seen to be necessary for a number of clinical disorders. In particular, there is growing evidence that miRNAs can affect how cells react to ionizing radiation, increasing their radiosensitivity or radioresistance through a number of distinct mechanisms [50]. These include altering DNA repair pathways that disrupt apoptosis, the tumor microenvironment, and the activation of cell cycle checkpoints [50, 51]. miRNAs have generated a lot of interest in relation to the tumor microenvironment. MiRNAs are crucial in controlling the tumor radiation response, which includes DNA repair, and the epithelial-to-mesenchymal transition (EMT). But as radio resistance is a complicated phenomenon, further research is required to fully comprehend these mechanisms. The way radiation is administered, the dose administered, the stage or grade of the tumor, any complicating medical comorbidities, and the intrinsic tumor microenvironment affect the response rates to radiation [14, 17, 41, 52].

Limitations of miRNA Therapeutics in Radiotherapy

There has been a notable effect of miRNAs and their role as therapeutics for cancer treatment, specifically in RT enhancement. In contrast, irrespective of their potential, their clinical use in radiation is hindered by a number of restrictions. The studies involving applications on miRNA in clinical settings are limited, thereby making it challenging [4, 15, 24, 53]. Some of the challenges include the prevention of effects arising at off-target sites while developing highly specific delivery modes and standardized efficient doses. Possible ways to overcome these challenges are by carrying out research in preclinical and clinical areas with specific cell lines and animal models that can be further scaled up for human trials. Further, to verify the specificity of miRNAs for biomarkers, the patients treated with miRNAs could be followed and updated over a longer period. The miRNAs should be specific, stable, and sensitive [4, 53]. Multiple genes can be regulated by miRNAs, which may result in unexpected interactions and adverse effects. Maintaining specificity to prevent off-target gene silencing is still very difficult [54]. It is difficult to deliver miRNA-based treatments to tumor locations in an effective and focused manner. miRNAs' hydrophilic nature restricts their cellular absorption, and they are vulnerable to nuclease destruction [55]. It is crucial to create reliable delivery mechanisms that shield miRNAs and make it easier for them to enter target cells [55].

To overcome these obstacles, more research must be done on ways to improve miRNA delivery, increase specificity, reduce immunological responses, and carry out extensive clinical trials to determine safety and efficacy profiles.

Emerging miRNAs for the Enhancement of Radiotherapy

As evidenced by recent studies exploring the role of miRNAs, their modulation in tumor response to RT holds the potential for efficient treatment, thereby enhancing the same [23, 34, 56]. Specifically, miRNAs that target certain genes involved in the repair mechanism for DNA, such as miR-34a and miR-155, which serve as a tumor suppressor, have been studied in lung carcinoma cells [57]. They are also

Table 21.2 Influence of specific miRNAs on radiotherapy outcomes in different cancer types

Cancer type	miRNAs studied	Mechanism of action	References
Hepatoma	miR-210	Induced under hypoxia, targets AIFM3, increases cell cycle arrest at G phase	[60]
	miR-16	Downregulates Bcl2, increases radiosensitivity	[61]
Lung cancer	miR-9	Downregulates NFκB1, increases radiosensitivity	[62]
	miR-155	Overexpressed in lung cancer, protects against radiation	[63]
	miR-210	Stabilizes HIF-1, enhances DSB repair, radioresistant component	[64]
Glioblastoma	miR-17	Modulates autophagy, increases radiosensitivity	[65]
	miR-34	Regulates p53, Bcl2, increases radiosensitivity	[66]
Gastric cancer	miR-221	Regulates PTEN/AKT, explored for radiosensitization targets	[67]

recognized to enhance the radiosensitivity of *RAD50* [57]. Thus, the overexpression of miR-31a could decrease *RAD1* expression, thereby impairing the repair mechanism in cancer cells and thus increasing radiation-induced DNA damage to the same. miR-21 has been studied to target the cell cycle checkpoint Cdc25 gene. When overexpressed, miR-21 causes radioresistance in cancer cells [23]. A study on A549 cell culture indicated that by decreasing the expression of miR-21, the tumor migration was also decreased [23, 58]. The *PTEN* gene involved in cell growth, proliferation, and apoptosis is discovered to be targeted by miR-221 and miR-222 [23, 58]. On inhibiting the expression of these miRNAs, the underlying downstream signaling is blocked, as studied preliminary in the context of cell lines for prostate cancer [23, 58, 59]. Several such studies on miRNAs as per specific cancer type have been documented and are summarized in Table 21.2 below.

miRNAs have recently been shown to impact the immunological response after radiation therapy, in addition to DNA repair and apoptosis, opening up new avenues for collaboration with immune checkpoint blockade (ICB). Radiation-induced tumor cell death primes antitumor immunity by encouraging the release of damage-associated molecular patterns and tumor antigens. This postradiation immunological landscape has been demonstrated to be modulated by some miRNAs, including miR-155, miR-146a, and miR-16. The generation of IFN γ, memory formation, effector function, and cytotoxic T lymphocyte proliferation are enhanced by miR-155, a major activator of antitumor immunity that is essential for successful ICB outcomes. Following RT, miR-16 and miR-195 promote T-cell activation, which changes macrophages to the pro-inflammatory phenotype. The foundation for combined techniques improved by miRNA manipulation is laid by these miRNA-mediated effects, which improve tumor antigen presentation, encourage cytotoxic T lymphocyte infiltration, and reduce immunosuppressive signals [59, 82].

Finding potent miRNAs and reliably delivering them are equally essential for the clinical effectiveness of miRNA-based treatments. Natural extracellular vesicles called exosomes, which range in size from 30 to 150 nm, have the ability to transport miRNAs across biological boundaries [83, 84]. They are good delivery vehicles due to their biocompatibility, ability to target specific cells, and preservation of payload from enzymatic destruction [83–85]. For instance, in preclinical models, exosome-mediated transport of tumor-suppressive miRNAs, including miR-143, miR-146 b, and miR-122, has been demonstrated to improve chemosensitivity and inhibit tumor growth [84, 85]. Also, lipid nanoparticles (LNPs) work well for miRNA delivery and have been utilized successfully in mRNA vaccinations [85]. miRNAs are encapsulated by these carriers, which also shield them from harm while in circulation and promote cellular uptake. DOTMA-based LNPs are used to deliver miR-122 to liver cancer models, demonstrating tumor reduction without systemic damage [86]. By releasing miR-34a to cancer stem cells, small LNPs cause apoptosis and slow the formation of tumors [86, 87]. Adenovirus and lentivirus are examples of engineered viruses that provide great transfection efficiency and the possibility of long-term miRNA expression [87]. To lessen off-target effects in healthy tissues, miRNA response elements can be added to viral capsids [86, 87]. Viral vectors or Virus-like particles offer another viable platform for upcoming miRNA-based interventions by combining viral delivery effectiveness with reduced integration risk [87].

Challenges and Future Directions

miRNAs and their impact on RT are a promising tool for both patient stratification to enable individualized therapy and for efficient diagnosis due to their ability to be detected in blood samples [12, 23, 68, 69]. They represent critical regulators of tumor initiation, progression, and dissemination. Extensive evidence suggests that inhibition of overexpressed oncogenic miRNAs or substitution of tumor suppressive miRNAs could become a robust strategy for cancer therapy. [12, 70] The optimization of miRNA delivery systems, improvements in the stability of miRNAs, and a detailed understanding of the off-target effects of miRNA therapeutics are several challenges that need to be resolved for successful translation of miRNA therapeutics from bench to bedside [71].

It is important to take into account the difficulties associated with this approach. The standardization of protocols is vital to guarantee interlaboratory repeatability between collecting the samples and processing to miRNA detection [23, 68, 69, 72]. In this sense, RNA yield and purity can be impacted by sampling tubes, patient-related variables, sample processing, and sample storage [23, 73]. Choosing reliable biomarkers is made more difficult by the great heterogeneity of cancer, which could be considered an extra challenge therein.

With regard to the use of miRNAs in RT, it is required to be noted that little is currently known about how they function in various RT phenomenon, such as dosage fractionation. The usage of other radiation properties, like protons or carbon ions, as well as the tumor microenvironment, including hypoxia, is also not well documented [23, 73, 74]. Furthermore, it is currently difficult to translate preclinical results from xenograft or in vitro models into clinical practice employing patient materials. To build confidence in the chosen biomarkers, preclinical miRNA candidates are required to be examined in extensive prospective trials including multiple patient populations [23, 74]. In order to progress toward individualized RT, further studies could be planned to assess miRNA biomarkers across time by obtaining samples both before and after RT [23, 74, 75].

It is anticipated that in the near future, therapeutic medicines will take advantage of miRNA's regulatory function in tumor radiosensitivity [41, 74–76]. There may be several steps in the creative framework. In order to predict each patient's response to radiation, determine the customized dose of radiation for maximizing its therapeutic effects, and minimize both acute and latent damage to normal tissues, the gene expression range of radio-related miRNA in serum would first be assessed prior to tumor patients receiving RT [41, 75, 76]. Further, monitoring the expression spectrum of dominant and radio-specific miRNAs in serum during the radiation process and modifying the expression of particular miRNAs might also help to effectively accomplish the intended effect of RT and further enhance tumor radiosensitivity [75, 76].

To improve the overall effect of RT, the rate at which radiation kills tumor cells and thus the molecular instability of tumor cells could be increased. Likewise, RT can be used in conjunction with other chemotherapy medications, such as small molecule inhibitors, and other medications that target particular miRNAs during the therapeutic period [41, 42, 76]. Following RT, routinely measuring the level of expression of miRNAs in plasma may assist to track the beneficial effects of radiation and lower the risk of cancer recurrence and metastasis [76, 77]. The finding that miRNA regulates tumor radiosensitivity raises the prospect that miRNA could prove to be an appropriate target for clinical assessment and therapy. By employing radio-specific miRNAs as biomarkers in serum, it is possible to detect tumor radiosensitivity noninvasively [41, 76, 77]. Furthermore, there is a potential chance to enhance their radiotherapeutic effect by either activating or suppressing the expression of specific miRNAs and downstream target genes therein. Thus, proper therapeutics can be developed by understanding the interplay between miRNAs and the specific genes within the underlying pathway affected by radiation dose [41, 76].

Conclusion

Over the last few years, miRNAs have gained importance and thus have emerged as leading-edge tools with clinical relevance for prognosis, diagnosis, and personalized treatment [23, 77, 78]. Studies have shown that in some cancer entities, a number of miRNAs exhibit differential expression and are predictive of the outcome of treatment employed. In order to pave the path for future individualized treatment, this chapter has concentrated on the function of miRNAs in radiation-induced alterations such as radioresistance for cancer. The data suggest that, post-exposure to radiation, there is a differential expression of miRNAs that, in turn, alters the inherent radiosensitivity present in cells. This suggests that miRNAs are a component of the cellular response to the intrinsic response mechanism to radiation. The ubiquitin-mediated proteolysis, MAPK, thyroid cancer, oocyte meiosis, renal cell carcinoma, and other previously identified stress pathways have been validated by leading bioinformatics studies on the miRNA expression data [29, 79]. As precisely targeted and specialized treatments for cancer, miRNA-based therapies have a lot of potential [12, 14, 81]. In contrast, there is still a need to enhance their chemical designs, create better delivery systems, demonstrate sustained therapeutic efficacy, and assess the long-term reliability of these compounds in vivo in order to attain higher sensitivity and specificity and hasten their adoption in the clinic [12, 81]. Furthermore, before miRNAs are used in clinical settings, it is critical to comprehend the complex web of interactions that exist between them as well as the human genome, transcriptome, and proteome. Furthermore, a thorough evaluation of their toxicities must be carried out, and low-toxicity approaches such as low-dose combinations of miRNAs and siRNAs or the use of miRNA therapy as an adjuvant to conventional chemotherapy regimens should be considered [81]. It is possible to view miRNAs as prospective biomarkers that can predict radiation response and create a personalized treatment plan for individual patients [14, 81, 82]. This in turn would allow to create innovative therapeutic window for tailored intervention.

Our knowledge of miRNAs and their participation in specific malignancies is still developing, as is our ability to target or influence cells for more effective treatment. By investigating how miRNA affects biological pathways in particular cells, the mechanism to specifically change radiation sensitivity can be studied further [29, 79, 80]. In aggregate, a substantial amount of evidence points to miRNA as a possible therapeutic target, and it is anticipated that targeted suppression of certain miRNA expression in conjunction with RT could lead to a significant antitumor effect on human cancer cells [79, 80, 82]. Future research will help us better understand the radiobiology of cells and identify miRNA biomarkers for RT in humans.

References

1. Liu B, Zhou H, Tan L, Siu KT, Guan XY (2024) Exploring treatment options in cancer: tumor treatment strategies. Signal Transduct Target Ther 9(1):175
2. Frankel RI (1996) Centennial of Röntgen's discovery of x-rays. West J Med 164(6):497
3. Metheetrairut C, Slack FJ (2013) MicroRNAs in the ionizing radiation response and in radiotherapy. Curr Opin Genet Dev 23(1):12–19
4. Annese T, Tamma R, De Giorgis M, Ribatti D (2020) microRNAs biogenesis, functions and role in tumor angiogenesis. Front Oncol 10:581007
5. Han J, Lee Y, Yeom KH, Kim YK, Jin H, Kim VN (2004) The Drosha-DGCR8 complex in primary microRNA processing. Genes Dev 18(24):3016–3027
6. Ha M, Kim VN (2014) Regulation of microRNA biogenesis. Nat Rev Mol Cell Biol 15(8):509–524
7. Lee RC, Feinbaum RL, Ambros V (1993) The *C. elegans* heterochronic gene lin-4 encodes small RNAs with antisense complementarity to lin-14. Cell 75:843–854. https://doi.org/10.1016/0092-8674(93)90529-Y

8. Wightman B, Ha I, Ruvkun G (1993) Posttranscriptional regulation of the heterochronic gene lin-14 by lin-4 mediates temporal pattern formation in *C. elegans*. Cell 75:855–862. https://doi.org/10.1016/0092-8674(93)90530-4
9. O'Brien J, Hayder H, Zayed Y, Peng C (2018) Overview of microRNA biogenesis, mechanisms of actions, and circulation. Front Endocrinol 9:402
10. Pasquinelli AE, Reinhart BJ, Slack F, Martindale MQ, Kuroda MI, Maller B et al (2000) Conservation of the sequence and temporal expression of let-7 heterochronic regulatory RNA. Nature 408:86–89. https://doi.org/10.1038/35040556
11. Li SC, Chan WC, Hu LY, Lai CH, Hsu CN, Lin WC (2010) Identification of homologous microRNAs in 56 animal genomes. Genomics 96:1–9. https://doi.org/10.1016/j.ygeno.2010.03.009
12. Shah MY, Ferrajoli A, Sood AK, Lopez-Berestein G, Calin GA (2016) microRNA therapeutics in cancer—an emerging concept. EBioMedicine 12:34–42
13. Slaby O, Laga R, Sedlacek O (2017) Therapeutic targeting of non-coding RNAs in cancer. Biochem J 474(24):4219–4251. https://doi.org/10.1042/BCJ20170079
14. Soares S, Guerreiro SG, Cruz-Martins N, Faria I, Baylina P, Sales MG, Correa-Duarte MA, Fernandes R (2021) The influence of miRNAs on radiotherapy treatment in prostate cancer–a systematic review. Front Oncol 11:704664
15. Mehta SR, Suhag V, Semwal M, Sharma N (2010) Radiotherapy: basic concepts and recent advances. Med J Armed Forces India 66(2):158–162
16. Nguyen L, Schilling D, Dobiasch S, Raulefs S, Santiago Franco M, Buschmann D, Pfaffl MW, Schmid TE, Combs SE (2020) The emerging role of miRNAs for the radiation treatment of pancreatic cancer. Cancer 12(12):3703
17. Ahire V, Ahmadi Bidakhvidi N, Boterberg T, Chaudhary P, Chevalier F, Daems N, Delbart W, Baatout S, Deroose CM, Fernandez-Palomo C, Franken NA (2023) Radiobiology of combining radiotherapy with other cancer treatment modalities. In: Radiobiology textbook. Springer, Cham, pp 311–386
18. Chen HH, Kuo MT (2017) Improving radiotherapy in cancer treatment: promises and challenges. Oncotarget 8(37):62742
19. Wu Y, Song Y, Wang R, Wang T (2023) Molecular mechanisms of tumor resistance to radiotherapy. Mol Cancer 22(1):96
20. Barker HE, Paget JT, Khan AA, Harrington KJ (2015) The tumour microenvironment after radiotherapy: mechanisms of resistance and recurrence. Nat Rev Cancer 15(7):409–425
21. Larionova I, Rakina M, Ivanyuk E, Trushchuk Y, Chernyshova A, Denisov E (2022) Radiotherapy resistance: identifying universal biomarkers for various human cancers. J Cancer Res Clin Oncol 148(5):1015–1031
22. Zhang P, Wang L, Rodriguez-Aguayo C, Yuan Y, Debeb BG, Chen D, Sun Y, You MJ, Liu Y, Dean DC, Woodward WA (2014) miR-205 acts as a tumour radiosensitizer by targeting ZEB1 and Ubc13. Nat Commun 5(1):5671
23. Cellini F, Morganti AG, Genovesi D, Silvestris N, Valentini V (2014) Role of microRNA in response to ionizing radiations: evidences and potential impact on clinical practice for radiotherapy. Molecules 19(4):5379–5401
24. Moertl S, Mutschelknaus L, Heider T, Atkinson MJ (2016) MicroRNAs as novel elements in personalized radiotherapy. Transl Cancer Res 5(Suppl 6):S1262–S1269
25. West CM, Barnett GC (2011) Genetics and genomics of radiotherapy toxicity: towards prediction. Genome Med 3:1–5
26. Zhang L, Liao Y, Tang L (2019) MicroRNA-34 family: a potential tumor suppressor and therapeutic candidate in cancer. J Exp Clin Cancer Res 38:1–3
27. Ye C, Sun NX, Ma Y, Zhao Q, Zhang Q, Xu C, Wang SB, Sun SH, Wang F, Li W (2015) MicroRNA-145 contributes to enhancing radiosensitivity of cervical cancer cells. FEBS Lett 589(6):702–709
28. Mott JL, Kobayashi S, Bronk SF, Gores GJ (2007) mir-29 regulates Mcl-1 protein expression and apoptosis. Oncogene 26(42):6133–6140
29. Chaudhry MA (2014) Radiation-induced microRNA: discovery, functional analysis, and cancer radiotherapy. J Cell Biochem 115(3):436–449

30. Girardi C, De Pitta C, Casara S, Sales G, Lanfranchi G, Celotti L, Mognato M (2012) Analysis of miRNA and mRNA expression profiles highlights alterations in ionizing radiation response of human lymphocytes under modeled microgravity. PLoS One 7:e31293
31. Girardi C, De Pittà C, Casara S, Sales G, Lanfranchi G, Celotti L, Mognato M (2012) Analysis of miRNA and mRNA expression profiles highlights alterations in ionizing radiation response of human lymphocytes under modeled microgravity. PLoS One 7(2):e31293
32. Simone NL, Soule BP, Ly D, Saleh AD, Savage JE, DeGraff W, Cook J, Harris CC, Gius D, Mitchell JB (2009) Ionizing radiation-induced oxidative stress alters miRNA expression. PLoS One 4(7):e6377
33. Templin T, Amundson SA, Brenner DJ, Smilenov LB (2011) Whole mouse blood microRNA as biomarkers for exposure to γ-rays and 56Fe ions. Int J Radiat Biol 87(7):653–662
34. Li JJ, Xu L, Wang CL, Niu JW, Zou X, Feng XQ, Lu RJ (2022) Changes in patient peripheral blood cell microRNAs after total body irradiation during hematopoietic stem cell transplantation. Ann Transl Med 10(16):857
35. Dobiasch S, Goerig NL, Fietkau R, Combs SE (2018) Essential role of radiation therapy for the treatment of pancreatic cancer. Strahlenther Onkol 194(3):185–195
36. Małachowska B, Tomasik B, Stawiski K, Kulkarni S, Guha C, Chowdhury D, Fendler W (2020) Circulating microRNAs as biomarkers of radiation exposure: a systematic review and meta-analysis. Int J Radiat Oncol Biol Phys 106(2):390–402
37. Moher D, Shamseer L, Clarke M, Ghersi D, Liberati A, Petticrew M, Shekelle P, Stewart LA, PRISMA-P Group (2015) Preferred reporting items for systematic review and meta-analysis protocols (PRISMA-P) 2015 statement. Syst Rev 4:1–9
38. Singh J, Thachil T, Misir S, Altay DU, Yaman SO, Singh G, Eapen MS, McAlinden KD, Petrovic N, Sohal SS (2024) Expression of microRNAs following radiation therapy and association with severity of radiotherapy-induced toxicity among patients with prostate adenocarcinoma: a systematic review and meta-analysis. World Acad Sci J 6(3):27
39. Qu K, Zhang X, Lin T, Liu T, Wang Z, Liu S, Zhou L, Wei J, Chang H, Li K, Wang Z (2017) Circulating miRNA-21-5p as a diagnostic biomarker for pancreatic cancer: evidence from comprehensive miRNA expression profiling analysis and clinical validation. Sci Rep 7(1):1692
40. Madurantakam Royam M, Ramesh R, Shanker R, Sabarimurugan S, Kumarasamy C, Ramesh N, Gothandam KM, Baxi S, Gupta A, Krishnan S, Jayaraj R (2019) Mirna predictors of pancreatic cancer chemotherapeutic response: a systematic review and meta-analysis. Cancer 11(7):900
41. Zhao L, Bode AM, Cao Y, Dong Z (2012) Regulatory mechanisms and clinical perspectives of miRNA in tumor radiosensitivity. Carcinogenesis 33(11):2220–2227
42. Sayed D, Abdellatif M (2010) AKT-ing via microRNA. Cell Cycle 9(16):3233–3237
43. Yang F, Yin Y, Wang F, Wang Y, Zhang L, Tang Y, Sun S (2010) miR-17-5p promotes migration of human hepatocellular carcinoma cells through the p38 mitogen-activated protein kinase-heat shock protein 27 pathway. Hepatology 51(5):1614–1623
44. Ma X, Becker Buscaglia LE, Barker JR, Li Y (2011) MicroRNAs in NF-κB signaling. J Mol Cell Biol 3(3):159–166
45. Zedan AH, Hansen TF, Assenholt J, Madsen JS, Osther PJ (2019) Circulating miRNAs in localized/locally advanced prostate cancer patients after radical prostatectomy and radiotherapy. Prostate 79(4):425–432
46. Labbé M, Hoey C, Ray J, Potiron V, Supiot S, Liu SK, Fradin D (2020) microRNAs identified in prostate cancer: correlative studies on response to ionizing radiation. Mol Cancer 19:1–8
47. Gong P, Zhang T, He D, Hsieh JT (2015) MicroRNA-145 modulates tumor sensitivity to radiation in prostate cancer. Radiat Res 184(6):630–638
48. Sun D, Lee YS, Malhotra A, Kim HK, Matecic M, Evans C, Jensen RV, Moskaluk CA, Dutta A (2011) miR-99 family of MicroRNAs suppresses the expression of prostate-specific antigen and prostate cancer cell proliferation. Cancer Res 71(4):1313–1324
49. Jiang MJ, Chen YY, Dai JJ, Gu DN, Mei Z, Liu FR, Huang Q, Tian L (2020) Dying tumor cell-derived exosomal miR-194-5p potentiates survival and repopulation of tumor repopulating cells upon radiotherapy in pancreatic cancer. Mol Cancer 19:1–5

50. Pedroza-Torres A, Romero-Córdoba SL, Montaño S, Peralta-Zaragoza O, Vélez-Uriza DE, Arriaga-Canon C, Guajardo-Barreto X, Bautista-Sánchez D, Sosa-León R, Hernández-González O, Díaz-Chávez J (2024) Radio-miRs: a comprehensive view of radioresistance-related microRNAs. Genetics 227(4):iyae097
51. Biau J, Chautard E, Verrelle P, Dutreix M (2019) Altering DNA repair to improve radiation therapy: specific and multiple pathway targeting. Front Oncol 9:1009
52. El Bezawy R, Cominetti D, Fenderico N, Zuco V, Beretta GL, Dugo M, Arrighetti N, Stucchi C, Rancati T, Valdagni R, Zaffaroni N (2017) miR-875-5p counteracts epithelial-to-mesenchymal transition and enhances radiation response in prostate cancer through repression of the EGFR-ZEB1 axis. Cancer Lett 395:53–62
53. Hanahan D, Weinberg RA (2011) Hallmarks of cancer: the next generation. Cell 144(5):646–674
54. Seyhan AA (2024) Trials and tribulations of MicroRNA therapeutics. Int J Mol Sci 25(3):1469
55. Pagoni M, Cava C, Sideris DC, Avgeris M, Zoumpourlis V, Michalopoulos I, Drakoulis N (2023) miRNA-based technologies in cancer therapy. J Pers Med 13(11):1586
56. Oh JS, Kim JJ, Byun JY, Kim IA (2010) Lin28-let7 modulates radiosensitivity of human cancer cells with activation of K-Ras. Int J Radiat Oncol Biol Phys 76(1):5
57. Balca-Silva J, Sousa Neves S, Goncalves AC, Abrantes AM, Casalta-Lopes J, Botelho MF, Sarmento-Ribeiro AB, Silva HC (2012) Effect of miR-34b overexpression on the radiosensitivity of non-small cell lung cancer cell lines. Anticancer Res 32:1603–1609
58. Liu ZL, Wang H, Liu J, Wang ZX (2013) MicroRNA-21 (miR-21) expression promotes growth, metastasis, and chemo- or radioresistance in non-small cell lung cancer cells by targeting PTEN. Mol Cell Biochem 372:35–45. https://doi.org/10.1007/s11010-012-1443-3
59. Leung CM, Li SC, Chen TW, Ho MR, Hu LY, Liu WS, Wu TT, Hsu PC, Chang HT, Tsai KW (2014) Comprehensive microRNA profiling of prostate cancer cells after ionizing radiation treatment. Oncol Rep 31:1067–1078. https://doi.org/10.3892/or.2014.2988
60. Yang W, Sun T, Cao J, Liu F, Tian Y, Zhu W (2012) Downregulation of miR-210 expression inhibits proliferation, induces apoptosis and enhances radiosensitivity in hypoxic human hepatoma cells in vitro. Exp Cell Res 318(8):944–954
61. Cimmino A, Calin GA, Fabbri M, Iorio MV, Ferracin M, Shimizu M, Wojcik SE, Aqeilan RI, Zupo S, Dono M, Rassenti L (2005) miR-15 and miR-16 induce apoptosis by targeting BCL2. Proc Natl Acad Sci 102(39):13944–13949
62. Arora H, Qureshi R, Jin S, Park AK, Park WY (2011) miR-9 and let-7g enhance the sensitivity to ionizing radiation by suppression of NFκB1. Exp Mol Med 43(5):298–304
63. Babar IA, Czochor J, Steinmetz A, Weidhaas JB, Glazer PM, Slack FJ (2011) Inhibition of hypoxia-induced miR-155 radiosensitizes hypoxic lung cancer cells. Cancer Biol Ther 12(10):908–914
64. Grosso S, Doyen J, Parks SK, Bertero T, Paye A, Cardinaud B, Gounon P, Lacas-Gervais S, Noël A, Pouysségur J, Barbry P (2013) MiR-210 promotes a hypoxic phenotype and increases radioresistance in human lung cancer cell lines. Cell Death Dis 4(3):e544
65. Comincini S, Allavena G, Palumbo S, Morini M, Durando F, Angeletti F, Pirtoli L, Miracco C (2013) microRNA-17 regulates the expression of ATG7 and modulates the autophagy process, improving the sensitivity to temozolomide and low-dose ionizing radiation treatments in human glioblastoma cells. Cancer Biol Ther 14(7):574–586
66. Sasaki A, Udaka Y, Tsunoda Y, Yamamoto G, Tsuji M, Oyamada H, Oguchi K, Mizutani T (2012) Analysis of p53 and miRNA expression after irradiation of glioblastoma cell lines. Anticancer Res 32(11):4709–4713
67. Zhang C, Kang C, Wang P, Cao Y, Lv Z, Yu S, Wang G, Zhang A, Jia Z, Han L, Yang C (2011) MicroRNA-221 and-222 regulate radiation sensitivity by targeting the PTEN pathway. Int J Radiat Oncol Biol Phys 80(1):240–248
68. Kraemer A, Anastasov N, Angermeier M, Winkler K, Atkinson MJ, Moertl S (2011) MicroRNA-mediated processes are essential for the cellular radiation response. Radiat Res 176(5):575–586

69. Calin GA, Croce CM (2006) MicroRNA signatures in human cancers. Nat Rev Cancer 6(11):857–866
70. Chou J, Shahi P, Werb Z (2013) microRNA-mediated regulation of the tumor microenvironment. Cell Cycle 12(20):3262–3271
71. Cacheux J, Bancaud A, Leichlé T, Cordelier P (2019) Technological challenges and future issues for the detection of circulating microRNAs in patients with cancer. Front Chem 7:815
72. Korpela E, Vesprini D, Liu SK (2015) MicroRNA in radiotherapy: miRage or miRador? Br J Cancer 112(5):777–782
73. Pritchard CC, Kroh E, Wood B, Arroyo JD, Dougherty KJ, Miyaji MM, Tait JF, Tewari M (2012) Blood cell origin of circulating microRNAs: a cautionary note for cancer biomarker studies. Cancer Prev Res 5(3):492–497
74. Neesse A, Bauer CA, Öhlund D, Lauth M, Buchholz M, Michl P, Tuveson DA, Gress TM (2019) Stromal biology and therapy in pancreatic cancer: ready for clinical translation? Gut 68(1):159–171
75. Farsinejad S, Rahaie M, Alizadeh AM, Mir-Derikvand M, Gheisary Z, Nosrati H, Khalighfard S (2016) Expression of the circulating and the tissue microRNAs after surgery, chemotherapy, and radiotherapy in mice mammary tumor. Tumour Biol 37:14225–14234
76. Mladenov E, Iliakis G (2011) Induction and repair of DNA double strand breaks: the increasing spectrum of non-homologous end joining pathways. Mutat Res 711(1–2):61–72
77. Liu J, Zheng M, Tang YL, Liang XH, Yang Q (2011) MicroRNAs, an active and versatile group in cancers. Int J Oral Sci 3(4):165–175
78. Buscail E, Maulat C, Muscari F, Chiche L, Cordelier P, Dabernat S, Alix-Panabières C, Buscail L (2019) Liquid biopsy approach for pancreatic ductal adenocarcinoma. Cancer 11(6):852
79. Saleh AD, Savage JE, Cao L, Soule BP, Ly D, DeGraff W, Harris CC, Mitchell JB, Simone NL (2011) Cellular stress induced alterations in microRNA let-7a and let-7b expression are dependent on p53. PLoS One 6:e24429
80. Chaudhry MA (2008) Analysis of gene expression in normal and cancer cells exposed to γ-radiation. Biomed Res Int 2008(1):541678
81. Berindan-Neagoe I, Monroig PD, Pasculli B, Calin GA (2014) MicroRNAome genome: a treasure for cancer diagnosis and therapy. CA Cancer J Clin 64(5):311–336
82. Fabris L, Ceder Y, Chinnaiyan AM, Jenster GW, Sorensen KD, Tomlins S, Visakorpi T, Calin GA (2016) The potential of microRNAs as prostate cancer biomarkers. Eur Urol 70(2):312–322
83. Brinkley TE, Leng I, Register TC, Neth BJ, Zetterberg H, Blennow K, Craft S (2022) Changes in adiposity and cerebrospinal fluid biomarkers following a modified Mediterranean ketogenic diet in older adults at risk for Alzheimer's disease. Front Neurosci 16:906539
84. Tang WW, Bauer KM, Barba C, Ekiz HA, O'Connell RM (2022) miR-aculous new avenues for cancer immunotherapy. Front Immunol 13:929677
85. Forterre A, Komuro H, Aminova S, Harada M (2020) A comprehensive review of cancer MicroRNA therapeutic delivery strategies. Cancer 12(7):1852
86. Balaraman AK, Babu MA, Afzal M, Sanghvi G, Rekha MM, Gupta S, Rana M, Ali H, Goyal K, Subramaniyan V, Wong LS (2025) Exosome-based miRNA delivery: transforming cancer treatment with mesenchymal stem cells. Regen Ther 28:558–572
87. Chen Y, Gao DY, Huang L (2015) In vivo delivery of miRNAs for cancer therapy: challenges and strategies. Adv Drug Deliv Rev 81:128–141
88. Mohammadi C, Khoei SG, Fayazi N, Mohammadi Y, Najafi R (2021) miRNA as promising theragnostic biomarkers for predicting radioresistance in cancer: a systematic review and meta-analysis. Crit Rev Oncol Hematol 157:103183
89. Konoshenko MY, Bryzgunova OE, Laktionov PP (2021) miRNAs and radiotherapy response in prostate cancer. Andrology 9(2):529–545

Chapter 22
MicroRNAs: Innovative Clinical Tools for Diagnosis, Prognosis, and Therapy in Colorectal Cancer

Feiwu Long, Wei Wang, Ya Liu, Chuanhua Yang, Hailiang Ren, Cuo Leng, Peng Xiong, Xinyue Tang, Liyao Hu, Lin Liu, and Chuanwen Fan

Abstract The dysregulation of microRNAs (miRNAs), which can act as oncogenes or tumor suppressors, plays a pivotal role in the initiation and progression of colorectal cancer (CRC) by regulating key cancer hallmarks such as proliferation, invasion, and metastasis through pathways including Wnt/β-catenin and TGF-β signaling. Over the past two decades, accumulating evidence has shown that miRNAs exhibit higher stability and tissue specificity compared to conventional protein biomarkers and provide a novel strategy for noninvasive detection of liquid biopsy, prognostic stratification, and treatment targets. Specific miRNAs such as miR-21, miR-92a, and miR-29a show promise for early CRC diagnosis, while miR-21, miR-141, and miR-200c are related to prognostic stratification. Compared to individual miRNAs, blood-based miRNA panels (e.g., those incorporating miR-21 and

The authors Feiwu Long and Wei Wang have been contributed equally to this work.

F. Long · W. Wang · C. Yang · H. Ren · C. Leng · P. Xiong · X. Tang · L. Hu
Division of Gastrointestinal, Bariatric and Metabolic Surgery, Department of General Surgery, West China School of Public Health and West China Fourth Hospital, Sichuan University, Chengdu, China

Y. Liu
Department of Oncology and Department of Biomedical and Clinical Sciences, Linköping University, Linköping, Sweden

L. Liu (✉)
Department of Gastrointestinal Surgery, The Third Affiliated Hospital, Xinjiang Medical University, Urumqi, China

C. Fan (✉)
Division of Gastrointestinal, Bariatric and Metabolic Surgery, Department of General Surgery, West China School of Public Health and West China Fourth Hospital, Sichuan University, Chengdu, China

Department of Oncology and Department of Biomedical and Clinical Sciences, Linköping University, Linköping, Sweden
e-mail: chuanwen.fan@liu.se

S. Pathak, A. Banerjee (eds.), *MicroRNAs: Role in Cancer Progression, Diagnostics and Therapy*, Advances in Biochemistry in Health and Disease 34,
https://doi.org/10.1007/978-3-032-08824-6_22

miR-92a) have demonstrated higher diagnostic accuracy for CRC screening, risk prediction, and guiding personalized treatment. Moreover, multisource miRNA profiles, using samples from tumor tissue, serum/plasma, stool, urine, and exosomes, further enhance clinical utility by capturing both local and systemic tumor-derived signals. In addition to their role as biomarkers, miRNA-based therapeutic candidates such as miR-16 mimics (e.g., TargomiR) are being explored alone or in combination with standard chemoradiotherapy, providing novel options for CRC treatment. Despite their promise, challenges remain in standardizing detection methods, improving tumor specificity, and developing safe and efficient in vivo delivery systems. Thus, this review systematically summarized miRNA biogenesis, their functional roles in CRC, and recent advances in diagnostic, prognostic, and therapeutic applications. We highlight promising miRNA-based therapeutics under development or in early-phase clinical trials and discuss future directions, including the need for standardized detection, improved specificity, dynamic monitoring, and emerging delivery strategies to achieve efficient and safe in vivo administration.

Keywords MicroRNAs · Colorectal cancer · Biomarkers · Screening and diagnosis · Prognosis · Liquid biopsy · Therapy

Introduction

CRC remains a leading cause of cancer-related mortality worldwide, ranking among the top three most prevalent malignancies [1]. Noninvasive screening methods, such as fecal immunochemical testing (FIT), circulating tumor DNA (ctDNA) analysis, and carcinoembryonic antigen (CEA) detection, have improved early diagnosis and recurrence surveillance, but their utility is constrained by limited sensitivity for early-stage lesions, inconsistent specificity, and moderate prognostic performance [2]. The low 5-year survival rate in advanced CRC highlights the urgent need for robust biomarkers to facilitate early detection, recurrence risk stratification, and individualized treatment within a precision medicine framework.

MicroRNAs (miRNAs) are small noncoding RNAs (18–22 nucleotides), regulate gene expression by binding to the 3′ untranslated region (UTR) of target mRNAs, inducing posttranscriptional silencing. Through complex "one-to-many" (single miRNA regulating multiple genes) and "many-to-one" (multiple miRNAs targeting a single gene) networks, miRNAs influence critical cellular processes, such as proliferation, differentiation, and apoptosis. In CRC, dysregulation of these networks drives tumorigenesis, positioning miRNAs as critical nodes for clinical applications [2]. To understand their clinical potential, their biogenesis and roles in CRC must first be elucidated.

The clinical value of miRNAs lies in their high stability in biofluids, conferred by exosomal packaging, Argonaute protein binding, or lipoprotein association [3, 4]. Compared with FIT, ctDNA, and CEA, miRNAs demonstrate superior tissue specificity and dynamic reflection of disease progression and treatment response, making them ideal for noninvasive biomarker development in blood, stool, and other biofluids [2]. Over 2600 mature miRNAs have been identified in humans, providing a rich repertoire for translational research [5].

In CRC, aberrant miRNA expression contributes to tumor initiation, progression, and resistance [6]. For example, oncogenic miR-21 is overexpressed, while tumor-suppressive miR-145 is downregulated, modulating pathways like Wnt/β-catenin, EGFR/PI3K-AKT, and TGF-β, which drive proliferation, invasion, and therapeutic resistance [2, 7]. These alterations position miRNAs as versatile clinical tools with potential as (1) diagnostic biomarkers for early, noninvasive detection; (2) prognostic indicators for risk stratification and outcome prediction; (3) therapeutic targets for precision interventions via miRNA inhibition or replacement.

By bridging cancer biology with clinical applications, miRNAs hold significant promise for improving CRC management. Their validation in clinical settings will solidify their role in precision oncology.

Functional Mechanisms of miRNAs in CRC Pathogenesis

Canonical Pathway of miRNA Biogenesis

In the canonical pathway, the production of miRNAs is a tightly regulated, multistep process that begins in the nucleus and concludes in the cytoplasm of eukaryotic cells (Fig. 22.1). miRNA genes, derived from monocistronic loci, polycistronic clusters, or introns of protein-coding genes, are transcribed by RNA polymerase II into long primary transcripts (pri-miRNAs) with a hairpin structure (~35 base pair stem, terminal loop, and flanking regions) [8, 9].

In the nucleus, the microprocessor complex, comprising the RNase III enzyme Drosha and its cofactor DGCR8, processes pri-miRNAs. DGCR8 binds the hairpin's apical region, recognizing motifs like UGU, while Drosha cleaves the base of the stem, guided by UG and GHG motifs, yielding a ~ 70-nucleotide precursor miRNA (pre-miRNA) with a 3′ overhang and 5′ phosphate [10].

The pre-miRNA is exported to the cytoplasm via Exportin-5 in a Ran-GTP-dependent manner, with Exportin-5 binding the 3′ overhang for selective transport [11]. In the cytoplasm, Dicer binds the pre-miRNA's loop and overhang, cleaving it into a ~ 20 to 22-nucleotide duplex with 5′ phosphates and 3′ overhangs, regulated by cofactors TARBP2 and PACT [12].

The duplex is loaded onto an Argonaute (AGO) protein, assisted by chaperones HSC70 and HSP90, which ensure AGO's compatibility. The passenger strand (miRNA*) is degraded, while the guide strand forms the RNA-induced silencing complex (RISC). Strand selection favors the strand with a less stable 5′ end and A or U at the 5′ terminus, resulting in mature miRNA-5p or miRNA-3p based on hairpin orientation [13].

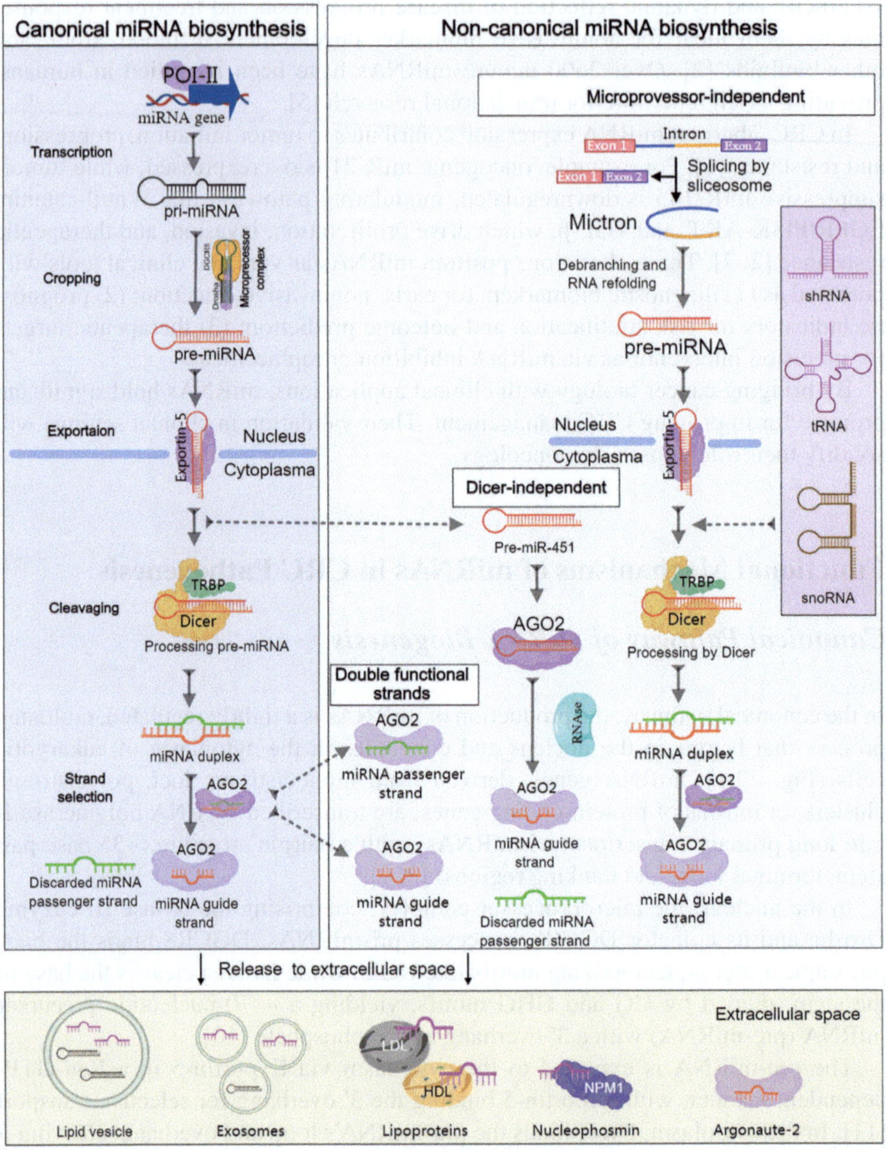

Fig. 22.1 Canonical and noncanonical pathways of miRNA biogenesis

Noncanonical Pathways of miRNA Biogenesis

In addition to the canonical route, several noncanonical biogenesis pathways generate functional miRNAs through alternative mechanisms that bypass one or more classical processing steps (Fig. 22.1). These pathways still result in mature miRNAs capable of entering AGO proteins and participating in gene silencing [9, 13].

One of the most studied noncanonical mechanisms is the mirtron pathway, in which spliced introns are debranched and fold into hairpin structures resembling pre-miRNAs [14]. These mirtrons bypass Drosha cleavage and are directly exported to the cytoplasm by Exportin-5, where Dicer processes them into mature duplexes for AGO loading [15].

Other Drosha-independent miRNAs can originate from tRNA-derived fragments (tRFs), snoRNA-derived RNAs (sdRNAs), and certain short hairpin RNAs (shRNAs). These precursors can enter the RNAi pathway via Dicer- or AGO-mediated routes [16].

Conversely, some miRNAs are generated through Dicer-independent mechanisms. A well-characterized example is miR-451, which, following Drosha-mediated cleavage, is directly loaded onto AGO2, the only member of the AGO protein family with catalytic slicing activity [17]. Unlike AGO1, AGO3, and AGO4, AGO2 uniquely enables the subsequent maturation of miR-451, which is further trimmed by the exoribonuclease PARN. This processing mechanism bypasses Dicer and relies on AGO2's catalytic function, highlighting an essential Dicer-independent miRNA biogenesis pathway [18].

These noncanonical pathways expand the versatility of miRNA maturation and increase the diversity of regulatory small RNAs, especially during specific tissue contexts, developmental stages, or stress conditions. Understanding these alternative mechanisms is essential for interpreting miRNA function in both physiological and pathological contexts.

Functional Roles of miRNAs in CRC Tumorigenesis

Under physiological conditions, miRNAs regulate cellular processes like proliferation, differentiation, apoptosis, and metabolism [19]. Through precise modulation of gene expression, they contribute to tissue homeostasis. In malignancies, however, this dysregulation is implicated in nearly all stages of tumor initiation and progression [6].

In CRC, aberrant miRNA expression has been extensively documented. Numerous miRNAs are significantly altered in tumor tissues compared to adjacent normal mucosa [19]. For instance, miR-21, a prototypical oncogenic miRNA (oncomiR), is frequently overexpressed in CRC and promotes tumor progression by targeting multiple tumor suppressor genes, such as *PTEN* (a negative regulator of the PI3K/AKT pathway), *PDCD4* (involved in apoptosis and inhibition of translation), and *TPM1* (a cytoskeletal regulator that suppresses cell migration and invasion). These targets collectively contribute to enhanced cell proliferation, survival, and metastasis in CRC [20]. Conversely, tumor-suppressive miRNAs like miR-34a and miR-143 are often downregulated, leading to the loss of negative regulation over critical signaling pathways, such as p53 and TGF-β [2, 7]. The disruption of the balance between oncogenic and tumor-suppressive miRNAs represents a key molecular mechanism in CRC development.

Beyond their cell-autonomous roles, miRNAs contribute to crosstalk between tumor cells and both the local microenvironment and distant premetastatic niches [4]. They are secreted into the extracellular space, often encapsulated in exosomes, and act as intercellular signaling molecules [21]. These exosomal miRNAs can enter the circulation and reach distant organs, where they condition the microenvironment to promote metastatic colonization an emerging mechanism particularly relevant to CRC metastasis [22]. By modulating immune responses, influencing inflammatory signaling, and reprogramming stromal cells toward tumor-supportive phenotypes, these extracellular miRNAs collectively facilitate tumor growth and dissemination [4, 23].

Given their central roles in CRC pathogenesis and their presence in accessible biological fluids, miRNAs are promising biomarkers for clinical application. Their expression profiles reflect underlying molecular changes and support noninvasive detection. These characteristics enable their use for early diagnosis, prognostic stratification, and targeted therapeutic strategies in colorectal cancer.

Sample Sources for miRNA Detection

Tissue

Since the publication of the first comprehensive mammalian miRNA atlas in 2007 [24], the diagnostic potential of tissue-derived miRNAs has become increasingly apparent. A growing body of evidence indicates that miRNAs are involved in both normal cellular function and the multistep progression of tumorigenesis, from precancerous lesions to invasive CRC [25]. This dual role emphasizes their value as indicators of disease progression and potential therapeutic targets in CRC [26]. Alterations in tissue-specific miRNA expression patterns support histopathological diagnosis, especially for poorly differentiated or morphologically ambiguous tumors.

Tissue miRNAs exhibit both spatial and cell-type specificity. In tumor specimens, miRNAs may originate from malignant epithelial cells and nonmalignant components of the tumor microenvironment (TME), including fibroblasts, endothelial cells, and infiltrating immune cells [27]. This complex cellular origin enables miRNA profiling to offer integrated insights into tumor biology and the surrounding stromal and immune context. However, both intertumoral and intratumoral heterogeneity can significantly influence the interpretation of miRNA expression data [26]. Apparent changes in miRNA levels may reflect shifts in the proportion of constituent cell types within the tissue, rather than true transcriptional alterations. Thus, integrating spatial localization with expression quantification may provide a more accurate understanding of their biological and clinical relevance, by revealing not only their expression levels but also their distribution within the tissue context—such as in tumor cells, immune infiltrates, or stromal compartments—which is essential for elucidating their functional roles and evaluating their potential as therapeutic targets or diagnostic biomarkers. Moreover, miRNAs also show greater stability than mRNAs in both fresh-frozen and formalin-fixed, paraffin-embedded (FFPE) tissues [28]. This biochemical robustness, combined with the availability of matched clinical follow-up and treatment data, makes tissue-based miRNA analysis a powerful approach for biomarker discovery and validation. For example, elevated expression of tissue miR-21 has been repeatedly shown to correlate with poor prognosis and resistance to chemotherapy in colorectal cancer, supporting its utility in risk stratification and treatment planning [29].

However, despite these advantages, tissue-based miRNA analysis faces several limitations. Bulk analysis often reflects miRNA expression from only a small portion of the tumor, making it difficult to account for intratumoral heterogeneity and potentially limiting its representativeness. In addition, current spatial transcriptomic approaches have limited scalability in large patient cohorts. Moreover, tissue sampling is typically restricted to a single time point, providing only a static snapshot and limiting longitudinal analysis.

Blood

Although tumor-derived DNA and RNA were detected in the plasma of cancer patients as early as the 1970s [30], it was assumed until the 2000s that RNA molecules were unsuitable as blood-based biomarkers due to their presumed rapid degradation by endogenous RNases [31]. This assumption was challenged by the findings that miRNAs remain intact and detectable even in FFPE tissues [32]. A subsequent landmark study by Chen et al. demonstrated that specific circulating miRNAs were stably detectable in human serum, laying the foundation for their use as noninvasive biomarkers in cancer and other diseases [33].

miRNAs are released into the extracellular environment, including the bloodstream, through mechanisms such as active secretion via exosomes, shedding of microvesicles, and protein-mediated export by binding to carriers like high-density lipoproteins (HDL) and Argonaute 2 (AGO2) protein complexes [34].

The high stability of circulating miRNAs is largely attributed to their association with these protective carriers. These complexes shield miRNAs from enzymatic degradation by extracellular nucleases and proteases [35], maintaining their integrity even under harsh sample handling conditions or repeated freeze-thaw cycles [36].

Stool

Blood samples provide a systemic overview of physiological and pathological states, whereas fecal samples offer detailed insight into the gastrointestinal environment [37]. This enables stool-based diagnostics to achieve higher sensitivity and specificity in the detection and study of colorectal diseases. Currently, widely adopted stool-based screening methods, such as the FIT and multitarget stool DNA test (mt-sDNA), identify patients requiring diagnostic colonoscopy [37].

In recent years, the role of miRNAs in the initiation and progression of CRC has gained attention [38]. In fecal samples, miRNAs may originate from exfoliated tumor cells, normal gastrointestinal epithelial cells, or precancerous lesions, or be secreted into the intestinal lumen via tumor-derived exosomes [39]. Protected by cellular membranes or vesicular structures, these miRNAs exhibit stability in the fecal matrix, remaining intact despite exposure to enzymatic degradation in the harsh digestive environment [39].

Fecal miRNA expression profiles correlate with those in CRC tumor tissues [40]. Beyond their tumor-associated functions, fecal miRNAs may regulate cross-kingdom gene expression, a process in which host-derived miRNAs influence the gene expression of gut microbiota, thereby modulating host-microbiota interactions [41]. This emerging mechanism has been implicated in shaping microbial composition, metabolism, and mucosal immune responses. Given the accessibility of stool samples and the availability of standardized detection platforms, fecal miRNAs are promising noninvasive biomarkers for early CRC detection.

Urine

Urine is a readily accessible biofluid that can be collected noninvasively. Traditionally, it has been used for the diagnosis of renal and urological diseases. However, as a filtrate of blood, urine can reflect systemic physiological and pathological changes beyond the urinary tract. Evidence shows that various circulating miRNAs are detectable in human urine, extending its utility for biomarker research for non-urological conditions [42].

Urinary miRNAs originate from exfoliated epithelial or tumor cells, freely circulating miRNAs filtered through the kidneys and extracellular vesicles (EVs) actively secreted by cells. Organoid models demonstrate that tumor-derived miRNAs, both those associated with vesicles and those free from vesicles, can be released into the culture medium. Notably, several of these miRNAs overlap with those identified in paired urine samples. This supports the notion that tumor cells contribute to the urinary miRNA profile [43]. Emerging studies suggest urinary miRNAs may reflect CRC-specific signatures [43]. Among these, miRNAs enclosed within EVs, particularly exosomes, are of significant interest. These vesicles protect their molecular cargo, including miRNAs, from enzymatic degradation and maintain integrity under variable storage and processing conditions [36]. EV-derived miRNAs partially reflect tissue-specific molecular signatures, such as those in pancreatic and prostate tumors, making them promising candidates for early cancer detection [43, 44].

However, despite this potential, several analytical challenges remain. Urinary miRNAs are present at lower concentrations than in plasma and are more susceptible to contamination by tRNA fragments [45]. Enrichment for EVs from urine offers greater feasibility and accuracy than direct measurement of free miRNAs

[43]. miRNAs, especially those encapsulated in extracellular vesicles or exosomes, represent a promising avenue for noninvasive cancer biomarker discovery.

Exosomes

Exosomes are membrane-bound vesicles (~40–100 nanometers), found in most biological fluids, including blood, urine, saliva, and cerebrospinal fluid [46]. They are actively released into the extracellular space by various cell types—such as epithelial cells, immune cells, and cancer-associated fibroblasts—through fusion of multivesicular bodies with the plasma membrane, entering the peripheral circulation [46].As intercellular cargo carriers, exosomes contain proteins and a variety of nucleic acids, including mRNAs, miRNAs, and other noncoding RNAs (ncRNAs) [23]. miRNAs are among the most abundant and functionally active RNA species within exosomes.

Circulating exosomes can be taken up by neighboring or distant cells, where encapsulated miRNAs modulate gene expression and cell behavior. This transfer plays a key role in shaping the pre-metastatic niche and supporting intercellular communication, enabling miRNAs to exert regulatory roles in both physiological and pathological conditions. The lipid bilayer of exosomes protects encapsulated miRNAs from degradation by circulating nucleases, enhancing their stability compared to freely circulating miRNAs. The molecular content of exosomes reflects the origin of the tumor cells, conferring tissue and disease specificity. These properties enhance the diagnostic performance of exosomal miRNAs over total plasma or serum miRNA profiling, as recent studies have demonstrated that isolating miRNAs specifically from exosomes improves both sensitivity and specificity in detecting early-stage colorectal cancer and distinguishing malignant from benign conditions [4].

Diagnostic Application of miRNAs in CRC

miRNAs play key regulatory roles in the initiation and progression of CRC and are recognized as potential noninvasive biomarkers. Circulating miRNAs are valued for early detection and screening due to their stability in body fluids (e.g., blood, urine, and stool), sensitivity, and ease of detection.

Single miRNA Biomarkers

Several miRNAs show diagnostic value in CRC, with miR-21 and miR-92a being the most validated, followed by miR-20a, miR-29a, miR-223, miR-221, miR-23a, and miR-150 (Table 22.1).

Table 22.1 Diagnostic performance of single miRNAs for colorectal cancer detection

miRNAs	Expression	AUC	Sensitivity	Specificity	Year	Country	Control	Cases	Biospecimen	PMID
miR-21	Upregulated	0.69	79.3	48.3	2019	Korea	29	29	Stool	PMID:31216561
miR-21	Upregulated	0.74	71.8	67.5	2016	Slovak Republic	80	71	Whole blood	PMID:27798914
miR-21	Upregulated	0.75	95.06	94.44	2021	China	150	117	Serum	PMID: 34457199
miR-21	Upregulated	0.75	/	/	2021	China	150	123	Exosomes from serum	PMID:34457199
miR-21	Upregulated	0.756	72.6	70.6	2020	USA	18	73	Serum	PMID:33112562
miR-21	Upregulated	0.783	86.05	72.97	2017	Iran	40	40	Serum	PMID:27432735
miR-21	Upregulated	0.798	61.4	90.9	2014	Japan	11	88	Exosomes from serum	PMID:24705249
miR-21	Upregulated	0.802	65	85	2013	China	80	200	Serum	PMID:23625654
miR-21	Upregulated	0.82	90	90	2012	USA	59	100	Plasma	PMID:22868372
miR-21	Upregulated	0.829	86.06	81.08	2017	Iran	40	40	Stool	PMID:27432735
miR-21	Upregulated	0.85	87.5	74.4	2012	China	39	32	Serum	PMID:22638884
miR-21	Upregulated	0.863	88.9	83.3	2020	China	20	40	Serum	PMID: 33054296
miR-21	Upregulated	0.87	77	78	2014	Iran	40	40	Serum	PMID:25178939
miR-21	Upregulated	0.877	90.3	75.2	2016	China	98	150	Stool	PMID:26621842
miR-21	Upregulated	0.893	82.9	97.1	2020	Egypt	35	35	Serum	PMID:33290159
miR-21	Upregulated	0.907	92	56	2025	Morocco	50	50	Whole blood	PMID: 40126350
miR-21	Upregulated	0.91	90	90	2012	USA	30	30	Plasma	PMID:22868372
miR-21	Upregulated	0.913	/	/	2020	Spain	45	27	Serum	PMID:32316350
miR-21	Upregulated	0.918	84	90	2020	Egypt	50	50	Serum	PMID:32862588
miR-21	Upregulated	0.927	82.8	90.6	2013	USA	53	186	Serum	PMID:23704278
miR-21	Upregulated	0.936	80.7	100	2020	Egypt	30	60	Serum	PMID:31512372
miR-21	Upregulated	0.94	95.8	91.7	2020	Egypt	48	48	Serum	PMID:31482406
miR-21	Upregulated	0.973	91.4	95	2018	Egypt	101	35	Serum	PMID:30357530
miR-21	Upregulated	0.977	90.4	100	2021	Egypt	20	52	Plasma	PMID:34132944
miR-21	Upregulated	/	65	85	2016	Russia	34	31	Plasma	PMID:27910062

miR-92a	Upregulated	0.506	/	/	2020	USA	18	73	Serum	PMID:33112562
miR-92a	Upregulated	0.672	66	68	2020	Egypt	50	50	Serum	PMID:32862588
miR-92a	Upregulated	0.739	/	/	2016	China	62	62	Stool	PMID:26848774
miR-92a	Upregulated	0.748	61	82	2016	China	247	76	Stool	PMID:26848774
miR-92a	Upregulated	0.751	/	/	2016	China	62	62	Plasma + stool	PMID:26848774
miR-92a	Upregulated	0.76	89.7	51.7	2019	Korea	29	29	Stool	PMID:31216561
miR-92a	Upregulated	0.766	71	67	2021	Egypt	40	44	Plasma	PMID:34132944
miR-92a	Upregulated	0.786	65.5	82.5	2013	China	80	200	Serum	PMID:23625654
miR-92a	Upregulated	0.809	/	/	2020	Spain	45	27	Serum	PMID:32316350
miR-92a	Upregulated	0.833	/	/	2016	China	47	187	Plasma	PMID:26848774
miR-92a	Upregulated	0.838	84	71.2	2010	China	59	100	Plasma	PMID: 19876917
miR-92a	Upregulated	0.844	84.4	100	2017	Egypt	27	64	Serum	PMID:28618945
miR-92a	Upregulated	0.991	94.2	100	2021	Egypt	20	52	Plasma	PMID:34132944
miR-92a	Upregulated	0.991	94.2	100	2025	Morocco	50	50	Whole blood	PMID: 40126350
miR-92a	Upregulated	0.994	98.1	93.9	2022	Egypt	15	54	Plasma	PMID:35225486
miR-92	Upregulated	0.885	89	70	2010	China	50	90	Plasma	PMID:19201770
miR-92a-3p	Upregulated	0.871	/	/	2014	China	94	160	Serum	PMID:25233400
miR-92b	Downregulated	0.793			2019		52	40	Exosome	PMID: 31588188
miR-20a	Upregulated	0.59	46	73.4	2015	China	79	100	Plasma	PMID:26261602
miR-20a	Downregulated	0.676	/	/	2018	China	33	26	Serum	PMID:30589041
miR-20a	Upregulated	0.73	55	82	2016	China	198	198	Stool	PMID:26621842
miR-20a	Upregulated	0.736	/	/	2018	China	50	50	Plasma	PMID:29731665
miR-20a	Upregulated	0.788	/	/	2016	Egypt	24	100	Serum	PMID:27135244
miR-20a	Upregulated	0.801	/	/	2018	China	40	40	Plasma	PMID:29731665
miR-20a	Upregulated	0.83	/	/	2020	Iran	4	45	Exosomes from serum	PMID:32771555

(continued)

Table 22.1 (continued)

miRNAs	Expression	AUC	Sensitivity	Specificity	Year	Country	Control	Cases	Biospecimen	PMID
miR-223	Downregulated	0.649	60	71	2016	China	51	80	Stool	PMID:26756616
miR-223	Upregulated	0.707	/	/	2016	China	121	153	Plasma + stool	PMID:26848774
miR-223	Upregulated	0.716	46.6	90.9	2014	Japan	11	88	Exosomes from serum	PMID:24705249
miR-223	Upregulated	0.734	/	/	2016	China	62	62	Plasma	PMID:26848774
miR-223	Upregulated	0.787	/	/	2016	China	51	80	Stool	PMID:26848774
miR-223	Upregulated	0.796	77	65	2016	China	62	62	Stool	PMID:26848774
miR-223	Upregulated	0.838	/	/	2016	Egypt	24	100	Serum	PMID:27135244
miR-221	Upregulated	0.606	86	41	2010	China	37	103	Plasma	PMID:20880178
miR-221	Upregulated	0.615	/	/	2020	USA	18	73	Serum	PMID:33112562
miR-221	Upregulated	0.73	62	74	2014	China	198	198	Stool	PMID:25233396
miR-221	Upregulated	0.754	71.8	68.8	2016	Slovak Republic	80	71	Whole blood	PMID:27798914
miR-221	Upregulated	0.882	/	/	2020	Spain	45	27	Serum	PMID:32316350
miR-23a	Upregulated	0.887	82.9	91.3	2020	Egypt	35	35	Serum	PMID:33290159
miR-23a	Upregulated	0.89	/	/	2019	Iran	5	13	Exosomes from serum	PMID:30913118
miR-23a	Upregulated	0.89	91	74.3	2021	China	35	100	Exosomes from serum	PMID:33550035
miR-23a	Upregulated	0.953	92	100	2014	Japan	11	88	Exosomes from serum	PMID:24705249
miR-27a	Upregulated	0.665	42.9	88.6	2020	Egypt	35	35	Serum	PMID:33290159
miR-27a	Upregulated	0.746	80	77.5	2018	China	40	40	Exosomes from plasma	PMID:29739748
miR-27a	Upregulated	0.773	75	77.5	2018	China	50	50	Exosomes from plasma	PMID:29739748
miR-27a	Upregulated	0.82	80	77.5	2018	China	50	50	Exosomes from plasma	PMID:29739748
miR-29a	Upregulated	0.696	/	/	2020	USA	18	73	Serum	PMID:33112562
miR-29a	Downregulated	0.777	85	61	2016	China	51	80	Stool	PMID:26756616
miR-29a	Upregulated	0.844	69	89.1	2010	China	59	100	Plasma	PMID: 19876917
miR-29a	Upregulated	0.898	76	94	2025	Morocco	50	50	Whole blood	PMID: 40126350
miR-29b	Downregulated	0.743	61.4	72.5	2016	China	400	200	Plasma	PMID:26466603

miR-29b	Downregulated	0.87	77	75	2015	Iran	55	55	Serum	PMID:26318304
miR-150	Downregulated	0.632	57.8	56.3	2016	Slovak Republic	80	71	Whole blood	PMID:27798914
miR-150	Upregulated	0.758	55.7	100	2014	Japan	11	88	Exosomes from serum	PMID:24705249
miR-150-5p	Downregulated	0.707	75.2	58.8	2019	China	153	165	Exosomes from serum	PMID:31750241
miR-150-5p	Upregulated	0.978	93.6	89.9	2019	Spain	22	19	EVs from PLF	PMID:31221189
miR-1246	Upregulated	0.681	64.2	68.2	2018	China	120	107	Serum	PMID:29780253
miR-1246	Upregulated	0.924	100	80	2020	Egypt	30	37	Serum	PMID:33402976
miR-1246	Upregulated	0.948	95.5	90.9	2014	Japan	11	88	Exosomes from serum	PMID:24705249
miR-1290	Upregulated	0.83	70.1	91.2	2016	Japan	57	211	Serum	PMID:27502702
miR-1290	Upregulated	0.92	85	88.6	2021	China	35	100	Exosomes from serum	PMID:33550035
miR-1290	Upregulated	0.96	78.79	93.33	2019	China	15	15	Serum	PMID: 30662524
miR-130a	Upregulated	0.697	70	80	2018	China	50	50	Exosomes from plasma	PMID:29739748
miR-130a	Upregulated	0.742	82.5	62.5	2018	China	40	40	Exosomes from plasma	PMID:29739748
miR-130a	Upregulated	0.787	70	80	2018	China	40	80	Exosomes from plasma	PMID:29739748
miR-592	Upregulated	0.844	82.8	78	2020	China	15	15	Serum	PMID:32724351
miR-592	Upregulated	0.88	86.6	73.4	2020	China	80	80	Serum	PMID:32724351
miR-592	Upregulated	0.884	/	/	2024	China	46	46	Serum	PMID:38571665
let-7a	Upregulated	0.67	50	90.9	2014	Japan	11	88	Exosomes from serum	PMID:24705249
let-7a	Upregulated	0.71	/	/	2020	Iran	4	45	Exosomes from serum	PMID:32771555
miR-106a	Upregulated	0.605	74	44.4	2015	China	79	100	Plasma	PMID:26261602
miR-106a	Upregulated	/	34.2	97.2	2013	China	10	117	Stool	PMID:23950216
miR-126	Downregulated	0.665	88.6	50.5	2018	Egypt	101	35	Serum	PMID:30357530
miR-126	Upregulated	0.94	84	88.6	2021	China	35	100	Exosomes from serum	PMID:33550035
miR-139-3p	Downregulated	0.726	/	/	2020	China	23	80	Exosomes from plasma	PMID:32626537
miR-139-3p	Downregulated	0.994	96.6	97.8	2017	China	90	117	Serum	PMID:28404964

(continued)

Table 22.1 (continued)

miRNAs	Expression	AUC	Sensitivity	Specificity	Year	Country	Control	Cases	Biospecimen	PMID
miR-146a	Downregulated	0.794	77.2	68.1	2016	China	98	150	Stool	PMID:26621842
miR-146a	Upregulated	0.805	78	74.1	2020	Egypt	30	60	Serum	PMID:31512372
miR-148a-3p	Downregulated	0.648	/	/	2020	China	88	80	Serum	PMID:32914665
miR-148a-3p	Downregulated	0.871	79.2	91.5	2019	China	47	48	Plasma	PMID:30925282
miR-15b	Upregulated	0.86	81.33	91.8	2021	China	150	117	Serum	PMID: 34457199
miR-15b	Upregulated	0.86	/	/	2021	China	4	45	Exosomes from serum	PMID:34457199
miR-17-3p	Upregulated	0.71	67.9	70.8	2019	Korea	29	29	Stool	PMID:31216561
miR-17-3p	Upregulated	0.717	64	70	2010	China	50	90	Plasma	PMID:19201770
miR-182	Upregulated	0.891	78	91	2018	China	40	40	Plasma	PMID:29731665
miR-182	Upregulated	0.929	/	/	2018	China	134	67	Plasma	PMID:29731665
miR-18a	Upregulated	0.67	61	69	2014	China	198	198	Stool	PMID:25233396
miR-18a	Upregulated	0.906	84	84	2020	Egypt	50	50	Serum	PMID:32862588
miR-19a	Upregulated	0.825	/	/	2016	Egypt	24	100	Serum	PMID:27135244
miR-19a	Upregulated	0.87	/	/	2020	Iran	4	45	Exosomes from serum	PMID:32771555
miR-206	Upregulated	0.705	/	/	2014	China	46	94	Plasma	PMID:25255814
miR-206	Downregulated	0.846	80	82.2	2017	China	45	73	Serum	PMID:31966596
miR-210	Upregulated	0.818	88.9	72.2	2020	China	20	40	Serum	PMID: 33054296
miR-210	Upregulated	0.934	88.6	90.1	2018	Egypt	101	35	Serum	PMID:30357530
miR-27a-3p	Downregulated	0.881	75	85	2019	China	47	48	Plasma	PMID:30925282
miR-27a-3p	Upregulated	0.922	88.7	81	2016	Czech Republic.	100	203	Serum	PMID:27485599
miR-375	Downregulated	0.749	76.92	64.63	2014	China	46	94	Plasma	PMID:25255814
miR-375	Downregulated	0.781	78.1	100	2017	Egypt	27	64	Serum	PMID:28618945
miR-760	Downregulated	0.788	80	72.4	2012	China	58	90	Plasma	PMID:22970209
miR-760	Downregulated	0.922	92.2	100	2017	Egypt	27	64	Serum	PMID:28618945
miR-92a-1	Upregulated	0.994	98.1	93.9	2020	China		84	Plasma	PMID: 35225486

miR-92a-1	Upregulated	0.914	81.8	95.6	2020	China	68	148	Serum	PMID:32562465
let-7c	Upregulated	0.855	77.6	96.2	2020	Egypt	30	60	Serum	PMID:31512372
let-7f-5p	Downregulated	0.709	/	/	2016	China	26	51	Stool	PMID:26793011
miR-103	Upregulated	0.662	55.9	75	2015	Japan	32	84	Serum	PMID:26134152
miR-103a-3p	Upregulated	0.759	/	/	2019	China	76	79	Exosomes from plasma	PMID:30458288
miR-10b-5p	Downregulated	0.957	87.5	86.6	2019	Spain	22	19	EVs from PLF	PMID:31221189
miR-1224-5p	Upregulated	0.61	31.8	100	2014	Japan	11	88	Exosomes from serum	PMID:24705249
miR-1229	Upregulated	0.614	22.7	100	2014	Japan	11	88	Exosomes from serum	PMID:24705249
miR-1229-3p	Upregulated	0.776	67.5	92.5	2018	China	120	107	Serum	PMID:29780253
miR-125a-3p	Upregulated	0.685	/	/	2017	China	50	50	Exosomes from plasma	PMID:28646161
miR-1260b	Downregulated	0.848	81.3	83.3	2019	China	47	48	Plasma	PMID:30925282
miR-127-3p	Upregulated	0.729	/	/	2019	China	26	30	Exosomes from plasma	PMID:30458288
miR-129-1-3p	Upregulated	0.856	/	/	2022	Japan	63	63	Urine	PMID:35053622
miR-135b	Upregulated	0.79	78	68	2014	China	109	104	Stool	PMID:24691020
miR-142-3p	Downregulated	0.71	/	/	2015	Iran	24	61	Plasma	PMID:26413249
miR-142-5p	Upregulated	0.891	/	/	2016	Czech Republic.	100	203	Serum	PMID:27485599
miR-143	Downregulated	0.76	/	/	2020	Iran	4	45	Exosomes from serum	PMID:32771555
miR-143-3p	Downregulated	0.777	72.9	78.7	2019	China	47	48	Plasma	PMID:30925282
miR-144	Upregulated	0.77	78.6	66.7	2019	Korea	29	29	Stool	PMID:31216561
miR-144-3p	Downregulated	0.887	93.8	78.7	2019	China	47	48	Plasma	PMID:30925282
miR-145	Downregulated	0.78	/	/	2020	Iran	4	45	Exosomes from serum	PMID:32771555
miR-145-5p	Downregulated	0.754	/	/	2020	China	80	80	Serum	PMID:32672054
miR-146a-5p	Upregulated	0.739	/	/	2020	China	88	80	Serum	PMID:32914665
miR150	Upregulated	0.75	/	/	2020	Iran	4	45	Exosomes from serum	PMID:32771555
miR-151a-5p	Upregulated	0.737	/	/	2019	China	26	30	Exosomes from plasma	PMID:30458288

(continued)

Table 22.1 (continued)

miRNAs	Expression	AUC	Sensitivity	Specificity	Year	Country	Control	Cases	Biospecimen	PMID
miR-1539	Upregulated	0.673	92.2	40.8	2021	China	49	51	Exosomes from serum	PMID:33680906
miR-155	Upregulated	0.776	58.2	95	2014	China	60	146	Serum	PMID:25528214
miR-16	Upregulated	0.58	/	/	2021	China	150	123	Exosomes from serum	PMID:34457199
miR-17	Upregulated	0.813	/	/	2016	Egypt	24	100	Serum	PMID:27135244
miR-17-5p	Upregulated	0.742	/	/	2019	China	26	30	Exosomes from plasma	PMID:30458288
miR-181a-5p	Upregulated	0.736	/	/	2019	China	26	30	Exosomes from plasma	PMID:30458288
miR-181c-5p	Downregulated	0.952	85.9	80.3	2019	Spain	22	19	EVs from PLF	PMID:31221189
miR-186-5p	Upregulated	0.655	/	/	2019	Spain	37	51	Saliva	PMID:31757017
miR-18a-5p	Upregulated	0.777	/	/	2019	China	26	30	Exosomes from plasma	PMID:30458288
miR-18b-5p	Upregulated	0.781	/	/	2019	China	26	30	Exosomes from plasma	PMID:30458288
miR-192-5p	Downregulated	0.652	/	/	2020	China	88	80	Serum	PMID:32914665
miR-193a-5p	Downregulated	0.962	85.2	89.7	2019	Spain	22	19	EVs from PLF	PMID:31221189
miR-194	Downregulated	0.85	72	80	2015	Iran	55	55	Serum	PMID:26318304
miR-196b	Upregulated	0.814	63	87.4	2017	China	100	103	Serum	PMID:28123705
miR-199a-3p	Downregulated	0.968	92	88.7	2019	Spain	22	19	EVs from PLF	PMID:31221189
miR-19a-3p	Upregulated	0.849	/	/	2014	China	60	146	Serum	PMID:25233400
miR-200c-3p	Downregulated	0.656	/	/	2020	China	80	80	Serum	PMID:32672054
miR-202-3p	Downregulated	0.815	69.2	88.3	2018	China	120	107	Serum	PMID:29780253
miR-203a-3p	Upregulated	0.712	/	/	2020	China	80	80	Serum	PMID:32672054
miR-211	Upregulated	0.794	71	90	2021	Egypt	40	44	Plasma	PMID:34132944
miR-21-3p	Downregulated	0.878	90.7	78.3	2018	China	120	107	Serum	PMID:29780253
miR-21-5p	Upregulated	0.928	90.6	86.2	2020	China	50	80	Serum	PMID:32293319
miR-218-5p	Downregulated	0.97	90.5	92.1	2019	Spain	22	19	EVs from PLF	PMID:31221189
miR-223-3p	Upregulated	0.89	/	/	2014	China	94	160	Serum	PMID:25233400
miR-224	Downregulated	0.744	75	63	2016	China	51	80	Stool	PMID:26756616

miR-23a-3p	Upregulated	0.922	88.7	81	2016	Czech Republic.	100	203	Serum	PMID:27485599
miR-24	Downregulated	0.839	78.4	83.9	2015	China	130	111	Plasma	PMID:26297223
miR-25	Upregulated	0.812	75	85	2021	Egypt	40	44	Plasma	PMID:34132944
miR-26a	Upregulated	0.918	77.6	96.2	2020	Egypt	30	60	Serum	PMID:31512372
miR-26a-5p	Downregulated	0.67	/	/	2015	Iran	24	61	Plasma	PMID:26413249
miR-29a-3p	Upregulated	0.631	/	/	2019	Spain	37	51	Saliva	PMID:31757017
miR-29c-3p	Upregulated	0.659	/	/	2019	Spain	37	51	Saliva	PMID:31757017
miR-29c-5p	Downregulated	0.973	94.3	94.4	2019	Spain	22	19	EVs from PLF	PMID:31221189
miR-301a	Upregulated	0.84	/	/	2019	Iran	5	13	Exosomes from serum	PMID:30913118
miR-30e-3p	Upregulated	0.731	/	/	2020	China	88	80	Serum	PMID:32914665
miR-31	Upregulated	0.75	/	/	2021	China	150	123	Exosomes from serum	PMID:34457199
miR-31-5p	Upregulated	0.669	/	/	2020	China	88	80	Serum	PMID:32914665
miR-320a	Downregulated	0.886	92.8	73.1	2015	China	130	111	Plasma	PMID:26297223
miR-320c	Upregulated	0.598	/	/	2017	China	50	50	Exosomes from plasma	PMID:28646161
miR-320d	Downregulated	0.89	93.94	73.33	2019	China	15	15	Serum	PMID: 30662524
miR-34b-3p	Upregulated	0.785	/	/	2020	China	88	80	Serum	PMID:32914665
miR-375-3p	Downregulated	0.715	/	/	2020	China	80	80	Serum	PMID:32672054
miR-376c-3p	Upregulated	0.697	/	/	2016	Czech Republic.	100	203	Serum	PMID:27485599
miR-378e	Downregulated	0.93	89	80	2020	China	90	110	Serum	PMID:32782563
miR-383-5p	Downregulated	0.968	94	93.8	2019	Spain	22	19	EVs from PLF	PMID:31221189
miR-422a	Downregulated	0.843	/	/	2014	China	94	160	Serum	PMID:25233400
miR-423-5p	Downregulated	0.833	91.9	70.8	2015	China	130	111	Plasma	PMID:26297223
miR-424-5p	Downregulated	0.919	79.2	93.6	2019	China	47	48	Plasma	PMID:30925282
miR-425-5p	Downregulated	0.91	83.3	91.5	2019	China	47	48	Plasma	PMID:30925282
miR-4316	Upregulated	0.744	83.9	60.9	2017	Poland	70	56	Plasma	PMID:28405738

(continued)

Table 22.1 (continued)

miRNAs	Expression	AUC	Sensitivity	Specificity	Year	Country	Control	Cases	Biospecimen	PMID
miR-451	Downregulated	0.757	73	80	2020	Egypt	30	37	Serum	PMID:33402976
miR-4516	Downregulated	0.958	94.4	89.8	2020	China	50	80	Serum	PMID:32293319
miR-486	Downregulated	0.629	/	/	2018	China	33	26	Serum	PMID:30589041
miR-491-5p	Upregulated	0.632	/	/	2019	Spain	37	51	Saliva	PMID:31757017
miR-506	Upregulated	0.747	60.7	76.8	2017	Poland	70	56	Plasma	PMID:28405738
miR-532-3p	Downregulated	0.743	60.8	96.3	2018	China	120	107	Serum	PMID:29780253
miR-549a	Upregulated	0.863	/	/	2024	China	46	46	Serum	PMID:38571665
miR-552	Upregulated	0.946	/	/	2024	China	46	46	Serum	PMID:38571665
miR-566	Upregulated	0.809	/	/	2022	Japan	63	63	Urine	PMID:35053622
miR-601	Downregulated	0.747	69.2	72.4	2012	China	58	90	Plasma	PMID:22970209
miR-622	Upregulated	/	87.8	67.5	2017	China	90	117	Serum	PMID:28404964
miR-720	Upregulated	0.63	58.3	56.3	2015	Japan	32	84	Serum	PMID:26134152
miR-766-3p	Upregulated	0.631	/	/	2019	Spain	37	51	Saliva	PMID:31757017
miR-940	Upregulated	0.88	90	77.1	2021	China	35	100	Exosomes from serum	PMID:33550035
miR-96	Upregulated	0.74	65.4	73.3	2016	USA	47	187	Plasma	PMID:26863633
miR-99a-3p	Downregulated	0.97	97.6	90	2019	Spain	22	19	EVs from PLF	PMID:31221189
miR-99b-5p	Downregulated	0.628	32.1	90.8	2019	China	153	165	Exosomes from serum	PMID:31750241
miRNA-199b-5p	Downregulated	1	96.8	96.4	2019	Spain	22	19	EVs from PLF	PMID:31221189

miR-21 is consistently upregulated in both CRC tissues and circulation (plasma or serum) compared to healthy controls. It not only distinguishes CRC from normal samples, but also shows moderate discriminative power between adenomas and carcinomas. Its expression has been associated with larger tumor size, distant metastasis, and poorer overall survival [29]. Notably, miR-21 levels significantly decrease in the postoperative serum of patients who undergo curative resection, suggesting potential utility for monitoring disease burden and surgical response [47]. A meta-analysis of 19 studies evaluating circulating miR-21 in CRC reported a sensitivity greater than 80% in 10 studies, specificity over 80% in 9, and an area under the ROC curve (AUC) exceeding 0.800 in 12, supporting its diagnostic value [2]. However, an exosome-based detection assay demonstrated poor performance (AUC = 0.5), likely reflecting methodological limitations [48]. Despite its promise, the clinical specificity of miR-21 remains limited due to its frequent upregulation in colorectal adenomas and multiple other cancer types, which may confound its use as a standalone CRC biomarker.

The miR-92a, from the miR-17–92 cluster (including miR-17, miR-18a, miR-19a, miR-20a, and miR-19b), is significantly overexpressed in the blood of CRC patients [49]. Its expression levels often decrease after surgical resection, suggesting correlation with tumor burden [50]. miR-92a can distinguish CRC from other gastrointestinal diseases and cancers, indicating possible disease specificity. For instance, Hassan et al. and Zaki et al. reported AUCs of 0.991 and 0.994, with sensitivities and specificities >90% [2]. In contrast, other studies have reported lower diagnostic performance of miR-92a with AUCs as low as 0.64, possibly due to sample variability [48]. Moreover, its elevated expression in other carcinomas diminishes its specificity for CRC diagnosis [2].

Other Promising miRNA Candidates

miR-20a While most studies report upregulation of miR-20a in colorectal cancer (CRC), some indicate downregulation, reflecting inconsistent expression across datasets [51, 52]. For example, a study on plasma samples from 100 CRC patients showed upregulation (74% sensitivity, 44.4% specificity, AUC 0.605), whereas another study on serum from 26 CRC patients reported downregulation (AUC 0.676) [51, 52]. This variability limits its reliability as a diagnostic biomarker.

miR-29 Expression of the miR-29 family in CRC varies by subtype. miR-29a is frequently upregulated and acts as an oncogene, while miR-29b is more often downregulated, functioning as a tumor suppressor [53, 54]. These context-dependent roles highlight the complexity of miR-29 as a biomarker, necessitating further validation to clarify its diagnostic utility.

miR-223 Similar to miR-20a, miR-223 exhibits inconsistent expression in CRC. It is predominantly upregulated in plasma, exosomes, and most stool samples, but downregulation has also been reported [55]. For instance, a study of 80 CRC stool samples found miR-223 downregulation (60% sensitivity, 71% specificity, AUC 0.649), underscoring the need for standardized protocols to establish its diagnostic reliability [55].

miR-221 As an oncomiR, miR-221 is consistently upregulated in CRC, with diagnostic accuracy (AUC ranging from 0.606 to 0.882) influenced by patient age but independent of sex or tumor stage [2]. Its potential as a biomarker is promising but requires larger cohort studies to confirm its clinical utility.

miR-23a Upregulated in CRC, miR-23a regulates metastasis-related genes. Notably, exosomal miR-23a demonstrates high diagnostic performance, with 92% sensitivity and 100% specificity in a study of 88 CRC patients, outperforming traditional markers like CEA and CA19–9 (AUC 0.953) [56]. This suggests miR-23a as a strong candidate for noninvasive CRC detection.

miR-150 miR-150 shows variable expression, with upregulation more commonly reported in exosome studies. A study of 88 CRC patients found exosomal miR-150 to have 55.7% sensitivity and 100% specificity (AUC 0.758), but its inconsistent expression across sample types reduces its diagnostic reliability [56].

miRNA Panels and Combined Detection Strategies

Unlike single miRNAs, miRNA panels leverage expression signatures across multiple miRNAs, reducing individual variability and improving diagnostic accuracy. Several panels, developed across diverse regions (e.g., China, Egypt, Spain, Morocco, Sweden, Japan), have demonstrated AUCs exceeding 0.90 (Table 22.2):

Two-miRNA panel (miR-21 + miR-92a): This panel achieved a sensitivity of 96.1%, specificity of 100%, and AUC of 0.981 in a study of 52 CRC patients [57]. However, the small sample size limits its generalizability.

Three-miRNA panel (miR-144-3p + miR-425-5p + miR-1260b): In a cohort of 48 CRC patients, this panel showed a sensitivity of 93.8%, specificity of 91.3%, and AUC of 0.954 [58].

Five-miRNA panel (miR-1246 + miR-1229-3p + miR-202-3p + miR-21-3p + miR-532-3p): Evaluated in 107 CRC patients, this panel achieved an AUC of 0.960, with 91.6% sensitivity and 91.7% specificity [59].

Seven-miRNA panel (miR-103a-3p + miR-127-3p + miR-151a-5p + miR-17-5p + miR-181a-5p + miR-18a-5p + miR-18b-5p): This panel was assessed in both plasma and plasma-derived exosomes from 30 CRC patients (validation cohort). Exosomal analysis showed higher diagnostic performance (AUC 0.895, sensitivity 76.9%, specificity 86.7%) compared to plasma (AUC 0.762, sensitivity 96.7%, specificity 53.3%) [60], suggesting exosomes may enhance specificity.

Table 22.2 Diagnostic performance of miRNA panels for colorectal cancer detection

miRNA	AUC	Sensitivity	Specificity	Year	Country	Control	Cases	Biospecimen	PMID
let-7, miR-16, miR-19, miR-21, miR-23, miR-29, miR-34, miR-92, miR-222, miR-451	0.9	65.2	95	2022	Germany	26	23	Serum	PMID: 35741223
let-7c + miR- 21 + miR-26a + miR-146a	0.95	82.1	100	2020	Egypt	30	60	Serum	PMID:31512372
let-7c, miR-21, miR-26a, miR-146a	0.95	91.8	91.7	2020	Egypt	45	84	Serum	PMID: 31512372
miR-103a-3p + miR-127-3p + miR-151a-5p + miR-17-5p + miR-181a-5p + miR-18a-5p + miR-18b-5p	0.895	76.9	86.7	2019	China	26	30	Exosomes from plasma	PMID:30458288
miR-103a-3p + miR-127-3p + miR-151a-5p + miR-17-5p + miR-181a-5p + miR-18a-5p + miR-18b-5p	0.824	85.3	35.1	2019	China	30	30	Exosomes from plasma	PMID:30458288
miR-103a-3p + miR-127-3p + miR-151a-5p + miR-17-5p + miR-181a-5p + miR-18a-5p + miR-18b-5p	0.762	96.7	53.3	2019	China	15	54	Plasma	PMID:30458288
miR-1246 + miR-1229-3p + miR-202-3p + miR-21-3p + miR-532-3p	0.96	91.6	91.7	2018	China	120	107	Serum	PMID:29780253
miR-126 + miR-1290 + miR-23a + miR-940	0.95	90	88.6	2021	China	35	100	Exosomes from serum	PMID:33550035
miR-126, miR-1290, miR-23a, miR-940	0.95	90	88.57	2021	China	35	100	Exosome	PMID: 33550035
miR-1290, miR-320d	0.98	90.91	93.33	2019	China	15	15	Plasma	PMID: 30662524
miR-129-1-3p + miR-566	0.868	88.9	76.2	2022	Japan	63	63	Urine	PMID:35053622
miR-139-3p + miR-431	0.829	91	57	2013	USA	26	45	Plasma	PMID:24022433
miR-144-3p + miR-425-5p + miR-1260b	0.954	93.8	91.3	2019	China	47	48	Plasma	PMID:30925282
miR-144-3p, miR-425-5p, miR-1260b	0.954	93.8	91.3	2019	China	47	48	Plasma	PMID: 30925282
miR-144-5p + miR-451a	0.89	66	95	2017	USA	115	29	Stool	PMID:28660489
miR-15b + miR-21 + miR-31	/	91.6	97.6	2021	China	150	123	Exosomes from serum	PMID:34457199

(continued)

Table 22.2 (continued)

miRNA	AUC	Sensitivity	Specificity	Year	Country	Control	Cases	Biospecimen	PMID
miR-15b + miR-21 + miR-31	/	95.1	94.4	2021	China	150	123	Exosomes from serum	PMID:34457199
miR-15b-5p, miR-18a-5p, miR-29a-3p, miR-335-5p, miR-19a-3p, miR-19b-3p	0.74	81	56	2019	Spain	94	75	Serum	PMID: 31614785
miR-15b-5p, miR-18a-5p, miR-29a-3p, miR-335-5p, miR-19a-3p, miR-19b-3p combined with fecal HB	0.88	81	78	2019	Spain	100	96	Plasma	PMID: 30702491
miR-16, miR-19, miR-21, miR-34, miR-222	0.87	56.5	95	2022	Germany	26	23	Serum	PMID: 35741223
miR-16, miR-19, miR-34	0.8	30.4	95	2022	Germany	26	23	Serum	PMID: 35741223
miR-16, miR-34	0.74	21.7	95	2022	Germany	26	23	Serum	PMID: 35741223
miR-182 + miR-20a	0.905	/	/	2018	China	40	40	Plasma	PMID:29731665
miR-182 + miR-20a	0.831	/	/	2018	China	50	50	Plasma	PMID:29731665
miR-186-5p + miR-29a-3p + miR-29c-3p + miR-766-3p + miR-491-5p	0.754	72	66.7	2019	Spain	37	51	Saliva	PMID:31757017
miR-18a + miR-21	0.966	88	92	2020	Egypt	50	50	Serum	PMID:32862588
miR-18a + miR-20a + miR-21 + miR-29a + miR-92a + miR-106b + miR-133a + miR-143 + miR-145 + miR-181b + miR-342-3p + miR-532-3p	0.745	/	/	2013	Germany	144	80	Plasma	PMID:23690963
miR-193a-5p, miR-210, miR-513a-5p, miR-628-3p	0.92	90	80	2022	Japan	110	149	Plasma	PMID: 35850198
miR19a + miR19b	0.82	78.6	77.4	2013	Spain	53	42	Plasma	PMID:23267864
miR19a + miR19b + miR15b	0.84	78.6	79.3	2013	Spain	53	42	Plasma	PMID:23267864
miR-19a + miR-19b + miR-15b + miR-29a + miR-335 + miR-18a	0.95	91	90	2019	Spain	100	96	Plasma	PMID:30702491
miR-19a, miR-19b, miR-15b, miR-29a, miR-335, miR-18a	0.95	91	90	2019	Spain	100	96	Plasma	PMID: 30702491

miR-19a, miR-19b, miR-15b, miR-29a, miR-335, miR-18a	0.95	91–95	90	2019	Spain	100	96	Plasma	PMID: 30702491
miR-19a-3p + miR-92a-3p + miR-223-3p + miR-422a	0.96	/	/	2014	China	94	160	Serum	PMID:25233400
miR-19a-3p + miR-92a-3p + miR-223-3p + miR-422a	0.951	84.3	91.6	2014	China	94	160	Serum	PMID:25233400
miR-19a-3p + miR-21-5p + miR-425-5p	0.886	/	/	2017	China	27	64	Serum	PMID:28177881
miR-19a-3p + miR-21-5p + miR-425-5p	0.83	/	/	2017	China	90	136	Serum	PMID:28177881
miR-19a-3p + miR-21-5p + miR-425-5p	0.768	/	/	2017	China	30	30	Serum	PMID:28177881
miR-203a-3p + miR-145-5p + miR-375-3p + miR-200c-3p	0.893	81.3	73.3	2020	China	80	80	Serum	PMID:32672054
miR-203a-3p, miR-145-5p, miR-375-3p, miR-200c-3p	0.893	81.25	73.33	2020	China	135	135	Plasma	PMID: 32672054
miR-20a + miR-135b	0.79	79	65	2016	China	198	198	Stool	PMID:26621842
miR-20a + miR-92a	0.77	57	84	2016	China	198	198	Stool	PMID:26621842
miR-21 + miR-146a	0.878	87	81.7	2016	China	98	150	Stool	PMID:26621842
miR-21 + miR-221 + miR-150	0.818	80	74	2016	Slovak Republic	80	71	Whole blood	PMID:27798914
miR-21 + miR-23a	0.908	82.9	97.1	2020	Egypt	35	35	Serum	PMID:33290159
miR-21 + miR-23a + miR-27a	0.908	82.9	97.1	2020	Egypt	35	35	Serum	PMID:33290159
miR-21 + miR-25 + miR-18a + miR-22	0.93	67	90	2018	Sweden	134	67	Plasma	PMID:29573205
miR-21 + miR-26a	0.953	91.8	91.7	2020	Egypt	30	60	Serum	PMID:31512372
miR-21 + miR-27a	0.899	88.6	85.7	2020	Egypt	35	35	Serum	PMID:33290159
miR-21 + miR-92	0.847	68	91.2	2013	China	80	200	Serum	PMID:23625654
miR-21 + miR-92a	0.981	96.1	100	2021	Egypt	20	52	Plasma	PMID:34132944

(continued)

Table 22.2 (continued)

miRNA	AUC	Sensitivity	Specificity	Year	Country	Control	Cases	Biospecimen	PMID
miR-21 + miR-92a + miR-221	0.891	/	/	2020	Spain	45	27	Serum	PMID:32316350
miR-21, miR-29a and miR-92a	0.968	85	80	2025	Morocco	50	50	Whole blood	PMID: 40126350
miR-211 + miR-25	0.898	82	90	2021	Egypt	40	44	Plasma	PMID: 34132944
miR-211, miR-25, miR-92a	0.945	91	93	2021	Egypt	40	44	Plasma	PMID: 34132944
miR-211, miR-25, TGF-β1	0.993	100	97	2021	Egypt	40	44	Plasma	PMID: 34132944
miR-221 + miR-18a	0.75	66	75	2014	China	198	198	Stool	PMID:25233396
miR-223 + miR-92a	/	71.7	79.9	2016	China	247	76	Stool	PMID:26848774
miR-23a-3p + miR-27a-3p + miR-142-5p + miR-376c-3p	0.922	87.5	81	2016	Czech Republic	24	100	Serum	PMID:27485599
miR-23a-3p + miR-27a-3p + miR-142-5p + miR-376c-3p	0.922	88.7	81	2016	Czech Republic	80	80	Serum	PMID:27485599
miR-24 + miR-320a + miR-423-5p	0.899	92.8	70.8	2015	China	130	111	Plasma	PMID:26297223
miR-27a + miR-130a	0.898	80	90	2018	China	40	80	Exosomes from plasma	PMID:29739748
miR-27a + miR-130a	0.846	82.5	75	2018	China	40	40	Exosomes from plasma	PMID:29739748
miR-27a + miR-130a	0.801	80	90	2018	China	50	50	Exosomes from plasma	PMID:29739748
miR-28-3p, let-7e-5p, miR-106a-5p, and miR-542-5p	0.86	88.9	86.7	2021	Brazil	27	41	Plasma	PMID: 33804927
miR-29a + miR-92a	0.883	83	84.7	2010	China	59	100	Plasma	PMID: 19876917
miR-30e-3p + miR-146a-5p + miR-148a-3p	0.932	84.6	86.1	2020	China	88	80	Serum	PMID:32914665
miR-30e-3p + miR-31-5p + miR-34b-3p + miR-146a-5p + miR-148a-3p + miR-192-5p	0.932	84.6	86.1	2020	China	88	80	Serum	PMID:32914665
miR-30e-3p, miR-31-5p, miR-34b-3p, miR-146a-5p, miR-148a-3p, miR-192-5p	0.932	85	86	2020	China	145	137	Serum	PMID: 32914665

miR-30e-3p, miR-31-5p, miR-34b-3p, miR-146a-5p, miR-148a-3p, miR-192-5p	0.932	85	86	2020	China	145	137	Serum	PMID: 32914665
miR-338-5p + miR-23a + miR-193a-3p	0.887	80	84.4	2013	Malaysia	32	70	Whole blood	PMID:23758639
miR-375 + miR-206	0.846	/	/	2014	China	46	94	Plasma	PMID:25255814
miR-421 + miR-27a-3p	0.74	96	33	2019	Spain	217	67	Stool	PMID:31622624
miR-4516 + miR-21-5p	0.943	92.1	87.6	2020	China	50	80	Serum	PMID:32293319
miR-506 + miR-4316	0.751	76.8	75	2017	Poland	70	56	Plasma	PMID:28405738
miR-627-5p, miR-199a-5p, CEA, CA19	0.98	98	100	2022	China	30	60	Serum	PMID: 35812667
miR-7 + miR-93 + miR-409-3p	0.897	82	89	2014	China	57	55	Plasma	PMID:23456911
miR-7 + miR-93 + miR-409-3p	0.866	91	88	2014	China	53	42	Plasma	PMID:23456911
miR-92a + miR-144	/	96.6	37.9	2019	Korea	29	29	Stool	PMID:31216561
miR-92a + miR-211 + miR-25	0.954	91	93	2021	Egypt	40	44	Plasma	PMID:34132944
miR-92a + miR-223	/	75.8	70.5	2016	China	121	153	Plasma + stool	PMID:26848774
miR-92a + miR-223	0.907	96.8	75	2016	China	121	153	Plasma + stool	PMID:26848774

Twelve-miRNA panel (miR-18a + miR-20a + miR-21 + miR-29a + miR-92a + miR-106b + miR-133a + miR-143 + miR-145 + miR-181b + miR-342-3p + miR-532-3p): In a study involving 144 CRC patients, this panel achieved an AUC of 0.745 [61]. Despite including a larger number of miRNAs, its diagnostic performance was not superior to panels with fewer miRNAs.

While increasing the number of miRNAs in a panel can enhance diagnostic performance, the marginal benefit diminishes beyond a certain threshold [59]. Combining miRNA panels with other diagnostic tools such as fecal immunochemical testing (FIT), tumor-associated antigens (e.g., CEA, CA19-9), or Tumor, Node, Metastasis (TNM) staging can further improve accuracy. For example, integrating miRNAs with FIT increased the AUC from 0.75 to 0.88 in a study of CRC patients [7].

Despite these advances, no miRNA-based panel has been approved for CRC screening. However, successful applications in other cancers provide a blueprint. For instance, GASTROClear®, a 12-miRNA panel approved in Singapore in 2019 for gastric cancer screening, achieved an AUC of 0.92, with 87.0% sensitivity and 68.4% specificity in clinical validation, offering a model for CRC screening development [62].

In summary, miRNA panels offer a promising noninvasive approach for early CRC detection, outperforming single miRNAs like miR-21 and miR-92a in diagnostic accuracy. Multimodal strategies integrating miRNA panels with conventional tools further enhance performance. However, the lack of standardized miRNA combinations across studies, evidenced by the diversity of panels, poses a challenge. Future research should prioritize standardizing sampling and detection protocols and validating biomarkers in large, prospective, multicenter cohorts to facilitate clinical translation.

Prognostic and Predictive Value of miRNAs in CRC

Integration of miRNAs with TNM Staging

The TNM staging system is the standard for assessing cancer burden and guiding treatment decisions in CRC. However, as a purely anatomical framework, TNM staging does not account for the molecular heterogeneity driving tumor behavior. Patients with the same TNM stage often have varied outcomes [63].

Integration of miRNA expression profiles with TNM staging adds molecular-level information to improve prognostic accuracy [64]. For example, high expression of miR-21 in advanced TNM stage shows superior prognostic stratification compared to TNM staging alone [65]. Similarly, low levels of miR-150, combined with anatomical staging, identify subgroups at high risk of recurrence and poor survival [66].

These integrated models use statistical methods like Cox proportional hazards regression or machine learning algorithms to derive risk scores and classify patients into low-, intermediate-, or high-risk categories. Advances in quantitative techniques, such as qRT-PCR and digital PCR, enable reliable measurement of miRNAs in tissue, serum/plasma, and extracellular vesicle samples, supporting clinical applicability through liquid biopsy approaches [67].

Circulating miRNAs, as minimally invasive biomarkers, are advantageous for dynamic disease monitoring and therapeutic response assessment. Studies show prognostic models incorporating miRNA expression outperform traditional models in metrics like AUC and identify molecularly aggressive tumors even at early stages, supporting risk stratification and individualized treatment planning.

Multivariate Prognostic Models Based on miRNAs

Although some miRNAs serve as independent prognostic indicators, CRC progression involves complex biological networks, and single markers are insufficient for comprehensive prognostic assessment. Multivariable models integrating multiple miRNAs, molecular biomarkers, genetic alterations, and clinical parameters are critical for personalized prognostic evaluation.

In addition, when applying machine learning or artificial intelligent algorithms, these models often employ high-dimensional data reduction methods, such as like LASSO regression, random forests, or support vector machines to identify the most informative variables. By assigning weighted contributions to each variable, these models generate patient-specific risk scores to guide clinical decision-making.

Robust bioinformatics workflows with feature selection, normalization, cross-validation, and model refinement enhance reproducibility and generalizability. As high-throughput technologies and new biomarkers emerge, these models can be refined to enhance predictive power. Multivariate models are useful for tailoring treatment intensity, selecting adjuvant therapies, and planning follow-up schedules. High-risk patients can benefit from intensified treatment or clinical trial enrollment, while low-risk patients can avoid overtreatment. Identifying regulatory miRNAs within these models offers opportunities for targeted therapy development.

Dynamic Monitoring and Early Detection of Recurrence

Prognostic Markers for Identifying High-Risk Populations

Studies show miRNA expression levels correlate with long-term survival in CRC. For instance, elevated miR-21 in tumor tissue correlates with shorter DFS and OS [68]. Circulating miRNAs like miR-141 and miR-181 are associated with liver metastasis and reduced survival, making them noninvasive prognostic markers [69,

70]. Their stability and ease of detection make them suitable for monitoring minimal residual disease and recurrence during follow-up.

Dynamic Recurrence Surveillance: "Molecular Sentinels" Preceding Imaging Detection

Fluctuations in circulating miRNA levels are early indicators of tumor activity. In postoperative CRC patients, elevated plasma miR-31 is linked to a threefold increased risk of recurrence compared to patients with stable levels [71]. Similarly, increased levels of miR-141 and miR-16 during follow-up correlate with disease relapse [2]. These liquid biopsy-based biomarkers can identify molecular recurrence earlier than conventional radiographic detection, providing a window for earlier therapeutic decision-making.

Predicting Therapeutic Response

Response to adjuvant chemotherapy and targeted therapies in metastatic CRC varies widely. miRNAs can predict outcomes, aiding treatment stratification. Liu et al. found that miR-21 levels predicted response to the FOLFOX regimen: using an optimal cut-off cycle threshold value of 10.32, high-expression groups showed lower remission rates and low expression linked to higher rates of pathological remission (sensitivity: 80.0%, specificity: 88.2%) [72]. However, some studies report inconsistent predictive value for miR-21 [4].

High expression of miR-20a and miR-130b correlates with resistance to 5-FU, while miR-200c downregulation promotes epithelial-mesenchymal transition (EMT) and oxaliplatin resistance [73]. In KRAS wild-type metastatic CRC, high miR-21 expression may enhance cetuximab efficacy by modulating EGFR signaling, though conflicting data exist [4].

Elevated miR-497, regulating ANG2, signals poor response to bevacizumab [74]. High plasma levels of miR-31 and miR-221 indicate resistance to anti-EGFR agents [75]. Pretreatment miRNA profiling supports regimen selection, while dynamic monitoring detects resistance, enabling therapeutic adjustments.

miRNAs are multifunctional biomarkers in CRC, aiding TNM integration, risk modeling, recurrence surveillance, and therapy response prediction. Standardized detection and validated models will enhance their role in personalized treatment and patient outcomes.

Therapeutic Potential and Delivery Strategies for miRNA-Based Therapy

miRNAs are key regulators in cancer development, and restoration of dysregulated miRNA levels is a promising strategy for CRC and other malignancies. miRNA-based therapies employ two primary approaches: (1) Inhibition of oncogenic miRNAs to block tumor-promoting activity using anti-miR molecules, and (2) Restoration of tumor-suppressive miRNAs to reactivate anticancer functions via miRNA mimics or gene delivery. These strategies target aberrant gene regulatory networks to suppress tumor growth.

Inhibiting Oncogenic miRNAs

Oncogenic miRNAs, overexpressed in tumors, drive cancer progression. Anti-miR strategies use antisense molecules to neutralize their function, including antisense oligonucleotides (AMOs), locked nucleic acids (LNAs), miRNA sponges, and antagomirs [76].

Antisense Oligonucleotides (AMOs)

AMOs arc short, single-stranded sequences (typically 17–22 nucleotides) that are chemically modified, for example, with 2′-O-methoxyethyl and phosphorothioate backbones—to enhance stability and affinity [77]. These oligonucleotides are designed to bind specifically to mature miRNAs, competitively hybridizing to them and thereby blocking their interaction with target mRNAs. This targeted inhibition leads to derepression of downstream target genes. For instance, AMOs fully modified with 2′-O-methoxyethyl and phosphorothioate backbones have been shown to effectively silence miR-122 in vivo, whereas unmodified AMOs typically exhibit minimal activity under physiological conditions [78].

Locked Nucleic Acids (LNAs)

LNAs represent a further optimization of antisense oligonucleotide design. By introducing an extra methylene bridge between the 2′-O and 4′-C atoms, the ribose ring is 'locked' into a C3′-endo conformation, which markedly improves thermal stability and binding affinity for the target miRNAs [79]. In addition, LNAs offer superior aqueous solubility and metabolic stability, enhancing their suitability for in vivo delivery. Preclinical studies have demonstrated that LNA-based anti-miR-21 can inhibit the proliferation and invasiveness of colorectal adenocarcinoma cells [80]. Notably, a recent first-in-human, phase 1, open-label, dose-escalation study of

LNA-i-miR-221 showed good tolerability with no grade 3–4 toxicities and evidence of clinical antitumor activity, providing initial proof-of-concept for miRNA-targeted strategies in cancer treatment and a framework for further development of this first-in-class LNA inhibitor [81].

Antagomirs

Antagomirs are chemically modified single-stranded RNA oligonucleotides, typically around 23 nucleotides long, that are fully complementary to their target miRNAs. They are designed with multiple stabilizing features, such as 2′-O-methyl and phosphorothioate backbone modifications, hydrophobic end groups, and cholesterol conjugation—to enhance cellular uptake, improve stability, protect against nuclease degradation, and improve in vivo stability. These modifications allow antagomirs to efficiently bind and neutralize mature miRNAs, thereby preventing their interaction with target mRNAs and restoring the expression of suppressed genes [82]. Preclinical studies have demonstrated that systemically delivered antagomirs can achieve potent and durable silencing of endogenous miRNAs across various tissues. Importantly, unlike some siRNA-based therapies, antagomirs have shown minimal induction of innate immune responses in animal models [82]. However, effective inhibition often requires relatively high doses to maintain sufficient bioavailability and distribution [83].

miRNA Sponges

These artificial synthetic transcripts contain multiple tandem binding sites complementary to specific miRNAs, thereby competitively sequestering homologous miRNAs and preventing them from interacting with their endogenous mRNA targets [84]. Compared to antisense oligonucleotides, these constructs generally exhibit lower sequence specificity because they rely on multiple partially complementary binding sites that primarily target the seed region, allowing them to sequester entire miRNA families rather than a single mature miRNA with perfect complementarity. In parallel, competitive endogenous RNAs (ceRNAs), such as pseudogenes, long noncoding RNAs (lncRNAs), and circular RNAs (circRNAs), act as natural sponges in physiological contexts. By competing for shared miRNAs, ceRNAs regulate miRNA availability and bioavailability, enabling coordinated regulation of multiple gene networks. Notably, ceRNAs often target modulate multiple or even entire miRNA clusters or families, offering broader regulatory effect than single-target oligonucleotide inhibitors [85]. Overall, this broader but less sequence-specific binding capacity can be advantageous for simultaneously suppressing whole oncogenic miRNA families or clusters in complex tumor contexts, but it may also increase the risk of off-target interactions compared to highly specific antisense oligonucleotides.

These strategies have demonstrated promising antitumor efficacy in preclinical models by effectively suppressing oncogenic miRNA activity, thereby inhibiting tumor growth. Notably, several LNA-based miRNA inhibitors have shown encouraging safety profiles and therapeutic potential in early clinical studies [86]. For example, Miravirsen, an LNA targeting miR-122, showed sustained viral suppression in a phase II trial for hepatitis C [86]. In CRC, a recent phase I open-label trial of LNA-i-miR-221 demonstrated good tolerability and preliminary signs of antitumor activity in patients with advanced solid tumors, including CRC, further supporting the translational potential of LNA-based anti-miR strategies in this setting [81].

Restoring Tumor-Suppressive miRNAs

Tumor-suppressive miRNAs, downregulated in cancers, inhibit oncogenes. Restoration using synthetic mimics or viral vectors reactivates anticancer pathways.

miRNA Mimics

These double-stranded RNA molecules, chemically modified (e.g., 2′-O-methoxy substitutions), are designed to mimic endogenous mature miRNAs. Delivery through liposomes or nanoparticles restores tumor-suppressive function of specific miRNA in cancer cells. MRX34, a miR-34a mimic in lipid nanoparticles, was the first miRNA-based therapeutic to enter clinical trials for advanced solid tumors. Phase I results confirmed biological activity evidenced by altered tumor gene expression, but the study was prematurely terminated following multiple severe immune-related adverse events, including cytokine release syndrome, hepatic failure, and other immune-mediated inflammatory toxicities [87]. However, in contrast, TargomiR, an improved synthetic miR-16 mimic formulated in EGFR-targeted bacterial minicells (EnGeneIC Dream Vectors) for the treatment of pleural mesothelioma, demonstrated acceptable safety and early signs of antitumor activity in a phase I study [88]. These contrasting outcomes suggest that while miRNA mimics hold promise as therapeutic agents in preclinic model, their clinical success depends critically on achieving robust antitumor efficacy while minimizing immune-mediated toxicities.

miRNA Expression in Viral Vectors

Adeno-associated viruses (AAVs) or lentiviruses can express genes encoding tumor-suppressive miRNAs for either temporary or sustained expression within target cells. This strategy has shown promise in preclinical cancer models by restoring the function of abnormally silenced tumor-suppressive miRNAs and reactivating associated downstream pathways. However, the use of viral vectors to express miRNAs also raises important safety concerns, including the risk of insertional mutagenesis—which can disrupt host genomic integrity—and potential host immune responses against viral components, both of which may limit repeated administration and long-term efficacy [89]. AAV vectors are generally favored for miRNA delivery because they combine low pathogenicity, minimal genomic integration, and customizable tissue tropism, making them an attractive option for safe and sustained in vivo expression. In contrast, lentiviral vectors carry a higher risk of genomic integration, which makes insertional mutagenesis a more significant concern in their clinical application.

Thus, these results highlight the importance of optimized careful patient selection, rational target miRNA design, and delivery systems to maximize safety and therapeutic benefit.

Delivery Systems for miRNA Therapeutics

miRNA therapies leverage multi-target regulation, providing advantages over traditional drugs. However, miRNAs are unstable, susceptible to nuclease degradation, and unable to cross membranes passively, necessitating advanced delivery systems. Strategies include nonviral and viral vectors, each with strengths and limitations.

Viral Delivery Systems

Viral vectors are widely used for delivering therapeutic miRNAs or anti-miRNA oligonucleotides to target tissues. Common systems include retroviruses, lentiviruses, adenoviruses, adeno-associated viruses (AAVs), and bacteriophage-based virus-like particles (VLPs) [90]. Depending on their integration ability, these vectors can provide transient or long-term expression with high efficiency. However, each vector type has unique benefits and inherent limitations that must be weighed for clinical application.

Retroviruses

Retroviral vectors can carry ~8 kb of genetic material and stably integrate therapeutic miRNAs or anti-miRNA cassettes into the host genome ensuring sustained expression, but only limited in dividing cells [91]. This stability, however, brings the risk of insertional mutagenesis and potential oncogenic effects [92].

Lentiviruses

A retrovirus subclass, lentiviral vectors can infect both dividing and nondividing cells with a somewhat lower risk of oncogenesis due to preferential integration in active transcription units compared with retroviruses [90]. Preclinical models have shown that lentiviruses can effectively deliver miRNA mimics or sponges, for example, restoring miR-15a/16 in chronic lymphocytic leukemia mice or suppressing tumor spread with a miR-494 sponge [93, 94].

Adenoviruses

Unlike retroviral and lentiviral vectors, adenoviral vectors are derived from non-enveloped DNA viruses with a double-stranded genome. Because they do not integrate into the host genome, the risk of insertional mutagenesis is minimized, which contributes to a relatively higher safety profile. Adenoviral vectors have been widely used in gene therapy, vaccine development, and functional genomic studies [95]. They offer high transduction efficiency and can infect a broad range of cell types, including nondividing cells, making them versatile for various in vivo and ex vivo applications. However, the transgene expression mediated by adenoviral vectors is typically transient, which can be advantageous for short-term miRNA-based interventions. A major limitation remains their high immunogenicity, especially in patients with preexisting immunity to common adenovirus serotypes, which may trigger strong host immune responses [3].

Adeno-Associated Viruses (AAVs)

Adeno-associated viral (AAV) vectors are considered one of the safest viral delivery systems for miRNA therapeutics because they are nonpathogenic and generally do not integrate into the host genome, minimizing risks to disturb genomic integrity. Meanwhile, same as Lentiviruses and Adenoviruses, AAVs can transduce both dividing and nondividing cells, which is critical for sustained therapeutic effects [89]. In addition, an important advantage of AAVs is that various serotypes exhibit distinct tissue tropism, enabling targeted delivery to specific organs and cell types [95]. Although their packaging capacity is relatively limited (approximately 4.8 kb), this size is generally sufficient for miRNA expression cassettes [89]. AAVs have therefore been successfully applied in the central nervous system and retinal gene delivery—for example, rAAV9 can cross the blood-brain barrier to deliver therapeutic miRNAs for amyotrophic lateral sclerosis (ALS), while rAAV2/8 vectors have been used to target retinal cells [96, 97]. Additionally, rAAV5-mediated delivery of engineered anti-miRNAs is currently under clinical investigation. Nonetheless, immune responses against AAVs can still occur and are influenced by factors such as vector dose, serotype, administration route, and preexisting neutralizing antibodies [98].

Bacteriophage-Based Virus-Like Particles (VLPs)

Bacteriophage-derived VLPs offer an alternative to eukaryotic viral systems, with lower cytotoxicity and reduced oncogenic risk. For example, MS2 and recombinant PP7 bacteriophage-based VLPs have been engineered to package miRNAs, often combined with CPP to enhance delivery [99, 100]. Applications have shown potential in autoimmune diseases, cancer, and myocarditis [101–103]. In addition, Phi29 pRNAs, with their unique oligomeric self-assembling structure, enable simultaneous cargo and targeting ligand delivery [104]. While promising, phage VLPs are still experimental, further studies are needed to address immunogenicity and large-scale production for clinical use.

Nonviral Delivery Systems

While viral vectors remain effective for miRNA delivery due to their high transduction efficiency and capacity for sustained expression, concerns about immunogenicity and potential cytotoxicity limit their clinical application; nonviral systems are therefore increasingly favored for their low immunogenicity, good biocompatibility, and flexible design. Main nonviral approaches include lipid-based nanoparticles, polymer-based carriers, inorganic nanoparticles, and naturally derived vesicles.

Lipid Nanoparticles (LNPs)

Liposomes, with a lipid bilayer that encapsulates nucleic acids, include cationic, anionic, and neutral. Cationic liposomes are widely used because their positive charge enables strong electrostatic interactions with high cell membrane affinity. In CRC models, cationic liposomes have demonstrated promising targeting and therapeutic efficacy [3]. However, challenges such as limited in vivo efficiency, nonspecific protein interactions, and scale-up difficulties persist. Modifications like PEGylation can improve circulation and stability; for example, PEG-liposomes have delivered miR-126 [105], while HDL-NPs carrying miR-34a have enhanced radiotherapy sensitization [106]. Anionic or neutral liposomes mitigate instability from serum protein interactions, for example, neutral DOPC liposomes have successfully delivered miR-506 or miR-520, resulting in tumor suppression in ovarian cancer models [107]. Solid lipid nanoparticles (SLNs) and nanostructured lipid carriers (NLCs) offer solvent-free production, better scalability, and improved stability and release profiles [108].

Polymeric Nanoparticles

Polymeric nanoparticles, including natural polymers (e.g., cell-penetrating peptides, CPPs) and synthetic polymers (e.g., polyethyleneimine, PEI; poly(lactic-co-glycolic acid), PLGA; polyamidoamine, PAMAM), have been widely studied for miRNA delivery due to their tunable properties and favorable biocompatibility. CPPs offer low toxicity and good membrane penetration but still face limitations such as low targeting specificity, inefficient endosomal escape, rapid clearance from the body, and high production costs, all of which restrict their large-scale clinical translation [109]. Although both low- and high-molecular-weight PEIs can deliver miRNAs, low-molecular-weight forms better balance cytotoxicity, cellular uptake, and endosomal release efficiency, effectively transporting miR-33a mimics and miR-145 into colon cancer xenografts and inhibiting tumor growth [110]. PEGylation can further improve its performance (e.g., PEG-PEI/miR-221/222 for prostate cancer) [111]. Chitosan is naturally mucoadhesive and biocompatible, protecting miRNAs from degradation with minimal toxicity. But its low transfection efficiency in some cells and batch variability can hinder consistency. PLGA, FDA-approved and biodegradable, is well suited for controlled drug release but may require combination with cationic polymers to improve nucleic acid loading. It can also suffer from burst release or incomplete cargo release [112]. PAMAM's cationic nature provides high transfection efficiency and can be ligand-modified for cell-specific delivery, though they may cause cytotoxicity, hemolysis, or complement activation at higher doses [113]. Hybrid lipopolyplexes (LPPs) combine lipid and polymer benefits, offering improved stability, uptake, and endosomal escape; however, their complex formulation can complicate reproducibility and scale-up [108].

Inorganic Nanoparticles

Gold nanoparticles (AuNPs) and Fe_3O_4-based and silica nanoparticles are valued for their structural stability and high nucleic acid loading capacity [110]. AuNPs offer excellent chemical stability, high miRNA payload density, and potential for combined therapeutic and imaging applications. Functionalization with thiol or amine groups enhances nucleic acid binding [114]. For example, thiolated AuNPs carrying anti-miR-155 improved cardiac function in diabetic mice [115], while PEGylated AuNPs loaded with miR-1 achieved high transfection efficiency and suppressed breast cancer xenografts [116]. However, slow degradation, potential liver and spleen toxicity, and high costs limit clinical expansion. Iron oxide (Fe_3O_4) nanoparticles combined with polymers like polyglutamic acid or PEI show systemic delivery potential; for example, Fe_3O_4 nanocomposites loaded with miR-100 coadministered with docetaxel significantly inhibited tumor growth in patient-derived xenografts [117]. However, they are prone to magnetic aggregation and metal ion release, which may induce oxidative stress or local inflammation if clearance is inadequate. Silica nanoparticles provide good thermal stability, biocompatibility, and a large surface area for payload loading [118]. Dopamine-aptamer-modified silica nanoparticles carrying anti-miR-155 have suppressed colorectal tumor growth [119]. However, slow degradation and the need for complex surface modifications pose risks of long-term tissue accumulation.

Exosomes and Outer Membrane Vesicles (OMVs)

Exosomes are natural nanovesicles with inherent biocompatibility, low immunogenicity, and innate targeting capability. Engineered tumor-derived exosomes have been shown to deliver therapeutic miRNAs and induce apoptosis in cancer cells. Two enrichment strategies include engineering donor cells to overexpress target miRNAs or loading isolated exosomes with synthetic miRNAs. For example, mesenchymal stem cell-derived exosomes carrying miR-34a-5p have suppressed CRC cell growth and induced apoptosis [120]. Nevertheless, large-scale GMP-compliant production, source standardization, and robust immunogenicity assessments remain major challenges for clinical translation. OMVs derived from bacterial outer membranes offer unique immunostimulatory properties that can enhance antitumor immune responses [121]. Preclinical studies have shown that OMVs can be engineered to encapsulate therapeutic nucleic acids such as miRNAs, but challenges remain in optimizing loading efficiency, minimizing endotoxin activity, and achieving scalable production.

These nonviral delivery systems offer low toxicity and design flexibility and greater versatility, but still require optimization to achieve higher transfection efficiency.

In summary, the therapeutic efficacy of miRNA-based strategies critically depends on achieving efficient and safe delivery. Both viral and nonviral systems have inherent limitations that must be carefully weighed. While ongoing technological advances have improved biocompatibility, reduced immunogenicity, and enhanced delivery efficiency, each platform still presents distinct advantages and challenges that require context-specific selection. However, challenges such as off-target effects, insertional mutagenesis risk, vector-related immunogenicity, and especially insufficient tissue or cell specific remain major barriers to the successful clinical translation of miRNA therapeutics to clinical success in precision medicine.

Challenges and Future Perspectives

Standardization and Validation

Significant inconsistencies across miRNA studies in CRC highlight the urgent need for standardized protocols and robust validation. Although several individual miRNAs, such as miR-21 and miR-92a, have shown diagnostic or prognostic potential in CRC, none have demonstrated consistent and universally validated performance across studies. Consequently, combined miRNA panels generally provide superior and more reproducible diagnostic and prognostic accuracy. For example, panels including miR-21, miR-92a, and miR-29a have shown improved sensitivity and specificity compared to single markers [20]. Large-scale, multicenter studies are essential to validate these miRNA signatures. In addition, establishing standardized workflows for sample collection, preprocessing, reference gene selection, and consistent result reporting will ensure cross-study comparability and facilitate integration into clinical practice.

Specificity and False Positives

Circulating miRNAs can originate from noncancerous processes such as inflammation or exercise, which may cause false-positive results and thereby lack tumor specificity. As discussed in diagnostic applications (see section "Diagnostic application of miRNAs in CRC"), analyzing miRNAs in tumor-derived exosomes (e.g., *EpCAM*-positive exosomes) can help enrich tumor-specific miRNA profiles [122]. Moreover, combining miRNAs with other biomarkers (e.g., ctDNA, or protein markers) into multi-marker panels can further enhance accuracy. For instance, several cohort studies have demonstrated that integrating miRNA panels with CEA levels or KRAS mutation status enhances early CRC detection compared to using either marker alone.

Technical Innovations for Detection

Accurate miRNA detection requires high sensitivity, specificity, and reproducibility. Quantitative PCR (qPCR) remains the gold standard for many studies due to its high sensitivity, but it relies on robust reference gene selection to ensure accurate quantification. Digital PCR offers absolute quantification with improved precision but at the cost of low throughput. High-throughput sequencing provides comprehensive miRNA profiling, but it remains costly and technically demanding for routine clinical implementation. Sample-specific protocols (e.g., for serum, plasma, tissue) and stable reference miRNAs (e.g., miR-191-5p) or synthetic standards can help normalize results and improve consistency. Innovations of technologies like microfluidic platforms, digital PCR, and cost-efficient sequencing technologies are driving improvements in high-throughput and reliability for routine clinical use.

Clinical Integration and Personalized Medicine

The clinical translation of miRNA-based therapies faces substantial challenges in delivery, stability, specificity, and safety. Due to their inherent mechanism of action—regulating multiple targets and pathways—some miRNAs often lack acceptable specificity. This makes the selection of target miRNAs particularly critical: they must exert sufficient functional effects while minimizing off-target risks, which remains a key barrier for clinical application.

Furthermore, naked miRNA molecules are highly susceptible to degradation by nucleases and rapid clearance in vivo, resulting in poor stability. Therefore, effective delivery systems are essential to protect therapeutic miRNAs and ensure their targeted accumulation within tumor tissue. However, while advanced carriers improve stability and delivery efficiency, they may introduce additional safety concerns, such as cytotoxicity or abnormal immune activation. For example, MRX34, the first-in-human miRNA therapy trial, was terminated due to severe immune-related adverse events, highlighting the need for low-immunogenicity carriers, such as exosomes or optimized nanoparticles [87].

From a pharmacokinetic perspective, achieving sufficient tumor-specific accumulation while minimizing system off-target effects remains a major challenge. Advances in nanotechnology, such as pH-responsive or tumor-microenvironment-sensitive carriers, as well as systems that rely on tumor-specific surface biomarkers, offer promising solutions for controlled and site-specific release.

Beyond therapeutics, miRNAs also enable dynamic monitoring across CRC management, including staging, minimal residual disease detection, treatment response evaluation, and recurrence surveillance. For example, persistently elevated miRNA levels post-surgery may indicate minimal residual disease or recurrence, allowing earlier intervention. Integration of miRNA profiles with patient-specific molecular data, such as tumor mutation burden or microsatellite instability (MSI) status, enables further refined personalized treatment strategies to overcome resistance. For example, combining miR-200 family profiling with EMT marker analysis could help tailor antimetastatic approaches for patients with advanced CRC, aligning with the goals of precision medicine.

Future Perspectives

Prospective studies and randomized trials will be essential to validate the clinical utility of miRNA biomarkers for CRC screening and prognosis, providing robust evidence for their integration into routine diagnostics. miRNA-based tests can complement colonoscopy for early detection, particularly for individuals unwilling to undergo invasive procedures or as follow-up testing after initial screening. Advances in liquid biopsy technologies and multi-omics integration are expected to enable the combined use of miRNAs with ctDNA, circulating tumor cells (CTCs), exosomes, and protein markers, thereby significantly enhancing the sensitivity and specificity of early CRC detection.

Therapeutically, advanced delivery systems such as next-generation LNPs, patient-derived personalized exosomes, OMVs, and pH- or TME-responsive carriers—hold promise for achieving tissue- and cell-specific accumulation, while minimizing off-target effects and systemic toxicity. The integration of AI algorithms and bioinformatics tools will further support individualized target selection and delivery strategy design, improving treatment precision and synergy with existing therapies.

In addition, miRNAs may play a pivotal role in the dynamic monitoring of treatment response, detection of minimal residual disease in high-risk patients, and recurrence risk assessment, becoming a key factor to CRC management. For example, combining postoperative circulating miRNA levels with patient-specific molecular features—such as tumor mutation burden (TMB), ctDNA, MSI status, exosomes, or EMT profiles—detects early minimal residual disease and guides timely personalized interventions.

In summary, the integration of optimized delivery platforms, scalable GMP-compliant manufacturing processes, systematic safety validation, and large multicenter clinical studies will be crucial for translating miRNA-based strategies from bench to bedside in precision oncology. Ultimately, miRNAs are poised to become an integral component of CRC management across the entire disease continuum, offering forward-looking solutions for noninvasive early detection, individualized therapy, and real-time recurrence monitoring.

Conclusion

miRNAs, as key regulators of gene expression, hold substantial promise for advancing CRC diagnosis and treatment. Their demonstrated accuracy in early diagnosis, combined with their capacity to predict recurrence, drug resistance, and potential, highlights their versatility as both biomarkers and therapeutic targets. Their remarkable stability in various biological samples, including blood, stool, urine, and exosomes, supports noninvasive screening, real-time risk stratification, and dynamic treatment monitoring.

However, several key challenges remain, especially in variability in detection platforms, lack of assay standardization, limited tumor specificity due to expression overlap with noncancerous processes, and the need for efficient, safe delivery systems, which continue to impede large-scale clinical translation. Addressing these limitations will require sustained progress in high-sensitivity detection methods, robust multi-marker panels, personalized delivery strategies, and the integration of AI-driven analytics and multi-omics profiling.

Looking ahead, the integration of optimized delivery systems, scalable GMP manufacturing, and well-designed multicenter clinical trials will be essential to move miRNA-based diagnostics and therapeutics from bench to bedside. If these hurdles are overcome, miRNAs are poised to become a critical component of precision oncology, offering a forward-looking solution for noninvasive early detection, individualized treatment strategies, and improved survival outcomes for patients with CRC.

References

1. Hossain MS, Karuniawati H, Jairoun AA, Urbi Z, Ooi J, John A, Lim YC, Kibria KMK, Mohiuddin AKM, Ming LC et al (2022) Colorectal cancer: a review of carcinogenesis, global epidemiology, current challenges, risk factors, preventive and treatment strategies. Cancer 14(7):1732
2. Ždralević M, Radović A, Raonić J, Popovic N, Klisic A, Vučković L (2024) Advances in microRNAs as emerging biomarkers for colorectal cancer early detection and diagnosis. Int J Mol Sci 25(20):11060
3. Yang N (2015) An overview of viral and nonviral delivery systems for microRNA. Int J Pharm Investigat 5(4):179–181
4. Shi X, Zhao X, Xue J, Jia E (2024) Extracellular vesicle biomarkers in circulation for colorectal cancer detection: a systematic review and meta-analysis. BMC Cancer 24(1):623
5. Marceca GP, Distefano R, Tomasello L, Lagana A, Russo F, Calore F, Romano G, Bagnoli M, Gasparini P, Ferro A et al (2021) MiREDiBase, a manually curated database of validated and putative editing events in microRNAs. Scientific data 8(1):199
6. Fan CW, Lu R, Fang C, Zhang XL, Lv ZY, Li Y, Zhang H, Zhou ZG, Mo XM, Sun XF (2021) Expression profile, molecular functions, and prognostic significance of miRNAs in primary colorectal cancer stem cells. Aging 13(8):12067–12085
7. AlZaabi A, Shalaby A (2024) A systematic review of diagnostic performance of circulating MicroRNAs in colorectal cancer detection with a focus on early-onset colorectal cancer. Int J Mol Sci 25(17):9565
8. Miyoshi K, Miyoshi T, Siomi H (2010) Many ways to generate microRNA-like small RNAs: non-canonical pathways for microRNA production. MGG 284(2):95–103
9. Santovito D, Weber C (2022) Publisher correction: non-canonical features of microRNAs: paradigms emerging from cardiovascular disease. Nat Rev Cardiol 19(9):639
10. Yeom KH, Lee Y, Han J, Suh MR, Kim VN (2006) Characterization of DGCR8/pasha, the essential cofactor for Drosha in primary miRNA processing. Nucleic Acids Res 34(16):4622–4629
11. Yi R, Qin Y, Macara IG, Cullen BR (2003) Exportin-5 mediates the nuclear export of pre-microRNAs and short hairpin RNAs. Genes Dev 17(24):3011–3016
12. Pinhal D, Gonçalves LB, Campos VF, Patton JG (2025) Decoding microRNA arm switching: a key to evolutionary innovation and gene regulation. Cell Mol Life Sci 82(1):197
13. Medley JC, Panzade G, Zinovyeva AY (2021) microRNA strand selection: unwinding the rules. WIREs RNA 12(3):e1627
14. Okamura K, Hagen JW, Duan H, Tyler DM, Lai EC (2007) The mirtron pathway generates microRNA-class regulatory RNAs in drosophila. Cell 130(1):89–100
15. Shapiro JS, Varble A, Pham AM, Tenoever BR (2010) Noncanonical cytoplasmic processing of viral microRNAs. RNA 16(11):2068–2074
16. Johanson TM, Lew AM, Chong MM (2013) MicroRNA-independent roles of the RNase III enzymes Drosha and dicer. Open Biol 3(10):130144
17. Meister G, Landthaler M, Patkaniowska A, Dorsett Y, Teng G, Tuschl T (2004) Human Argonaute2 mediates RNA cleavage targeted by miRNAs and siRNAs. Mol Cell 15(2):185–197
18. Yoda M, Cifuentes D, Izumi N, Sakaguchi Y, Suzuki T, Giraldez AJ, Tomari Y (2013) Poly(a)-specific ribonuclease mediates 3′-end trimming of Argonaute2-cleaved precursor microRNAs. Cell Rep 5(3):715–726
19. Rac M (2024) Synthesis and regulation of miRNA, its role in oncogenesis, and its association with colorectal cancer progression, diagnosis, and prognosis. Diagnostics (Basel) 14(13):1450
20. Fathi S, Aazzane O, Guendaoui S, Tawfiq N, Sahraoui S, Guessous F, Karkouri M (2025) A miRNA signature for non-invasive colorectal cancer diagnosis in Morocco: miR-21, miR-29a and miR-92a. Non-coding. RNA 11(2):26

21. Neviani P, Fabbri M (2015) Exosomic microRNAs in the tumor microenvironment. Front Med (Lausanne) 2:47
22. Bayat M, Sadri Nahand J (2024) Exosomal miRNAs: the tumor's trojan horse in selective metastasis. Mol Cancer 23(1):167
23. Sato-Kuwabara Y, Melo SA, Soares FA, Calin GA (2015) The fusion of two worlds: non-coding RNAs and extracellular vesicles – diagnostic and therapeutic implications (review). Int J Oncol 46(1):17–27
24. Landgraf P, Rusu M, Sheridan R, Sewer A, Iovino N, Aravin A, Pfeffer S, Rice A, Kamphorst AO, Landthaler M et al (2007) A mammalian microRNA expression atlas based on small RNA library sequencing. Cell 129(7):1401–1414
25. Shi X, Zhang Y, Cao B, Lu N, Feng L, Di X, Han N, Luo C, Wang G, Cheng S et al (2013) Genes involved in the transition from normal epithelium to intraepithelial neoplasia are associated with colorectal cancer patient survival. Biochem Biophys Res Commun 435(2):282–288
26. Gustafson D, Tyryshkin K, Renwick N (2016) microRNA-guided diagnostics in clinical samples. Best Pract Res Clin Endocrinol Metab 30(5):563–575
27. Kohlhapp FJ, Mitra AK, Lengyel E, Peter ME (2015) MicroRNAs as mediators and communicators between cancer cells and the tumor microenvironment. Oncogene 34(48):5857–5868
28. Dijkstra JR, Mekenkamp LJ, Teerenstra S, De Krijger I, Nagtegaal ID (2012) MicroRNA expression in formalin-fixed paraffin embedded tissue using real time quantitative PCR: the strengths and pitfalls. J Cell Mol Med 16(4):683–690
29. Toiyama Y, Takahashi M, Hur K, Nagasaka T, Tanaka K, Inoue Y, Kusunoki M, Boland CR, Goel A (2013) Serum miR-21 as a diagnostic and prognostic biomarker in colorectal cancer. J Natl Cancer Inst 105(12):849–859
30. Stroun M, Anker P, Maurice P, Gahan PB (1977) Circulating nucleic acids in higher organisms. Int Rev Cytol 51:1–48
31. Kamm RC, Smith AG (1972) Ribonuclease activity in human plasma. Clin Biochem 5(4):198–200
32. Xi Y, Nakajima G, Gavin E, Morris CG, Kudo K, Hayashi K, Ju J (2007) Systematic analysis of microRNA expression of RNA extracted from fresh frozen and formalin-fixed paraffin-embedded samples. RNA 13(10):1668–1674
33. Chen X, Ba Y, Ma L, Cai X, Yin Y, Wang K, Guo J, Zhang Y, Chen J, Guo X et al (2008) Characterization of microRNAs in serum: a novel class of biomarkers for diagnosis of cancer and other diseases. Cell Res 18(10):997–1006
34. Metcalf GAD (2024) MicroRNAs: circulating biomarkers for the early detection of imperceptible cancers via biosensor and machine-learning advances. Oncogene 43(28):2135–2142
35. Elkayam E, Kuhn CD, Tocilj A, Haase AD, Greene EM, Hannon GJ, Joshua-Tor L (2012) The structure of human argonaute-2 in complex with miR-20a. Cell 150(1):100–110
36. Chevillet JR, Kang Q, Ruf IK, Briggs HA, Vojtech LN, Hughes SM, Cheng HH, Arroyo JD, Meredith EK, Gallichotte EN et al (2014) Quantitative and stoichiometric analysis of the microRNA content of exosomes. Proc Natl Acad Sci USA 111(41):14888–14893
37. Chen CC, Chang PY (2025) Fecal miRNAs as potential biomarkers for early detection of colorectal cancer: An updated review. Biom J 48(1):100769
38. Sado AI, Batool W, Ahmed A, Zafar S, Patel SK, Mohan A, Zia U, Aminpoor H, Kumar V, Tejwaney U (2012) Role of microRNA in colorectal carcinoma (CRC): a narrative review. Ann Med Surg 86(1):308–318
39. Koga Y, Yasunaga M, Moriya Y, Akasu T, Fujita S, Yamamoto S, Matsumura Y (2011) Exosome can prevent RNase from degrading microRNA in feces. J Gastrointest Oncol 2(4):215–222
40. Chang PY, Chen CC, Chang YS, Tsai WS, You JF, Lin GP, Chen TW, Chen JS, Chan EC (2016) MicroRNA-223 and microRNA-92a in stool and plasma samples act as complementary biomarkers to increase colorectal cancer detection. Oncotarget 7(9):10663–10675

41. Liu S, da Cunha AP, Rezende RM, Cialic R, Wei Z, Bry L, Comstock LE, Gandhi R, Weiner HL (2016) The host shapes the gut microbiota via fecal MicroRNA. Cell Host Microbe 19(1):32–43
42. Lin JT, Tsai KW (2021) Circulating miRNAs act as diagnostic biomarkers for bladder cancer in urine. Int J Mol Sci 22(8):4278
43. Baba S, Kawasaki T, Hirano S, Nakamura T, Asano T, Okazaki R, Yoshida K, Kawase T, Kurahara H, Oi H et al (2024) A noninvasive urinary microRNA-based assay for the detection of pancreatic cancer from early to late stages: a case control study. EClinicalMedicine 78:102936
44. Jain G, Das P, Ranjan P et al (2023) Urinary extracellular vesicles miRNA-A new era of prostate cancer biomarkers. Front Genet 14:1065757
45. Srinivasan S, Yeri A, Cheah PS, Chung A, Danielson K, De Hoff P, Filant J, Laurent CD, Laurent LD, Magee R et al (2019) Small RNA sequencing across diverse biofluids identifies optimal methods for exRNA isolation. Cell 177(2):446–462
46. Zhang J, Li S, Li L, Li M, Guo C, Yao J, Mi S (2015) Exosome and exosomal microRNA: trafficking, sorting, and function. Genomics Proteomics Bioinformatics 13(1):17–24
47. Fukada M, Matsuhashi N, Takahashi T, Sugito N, Heishima K, Yoshida K, Akao Y (2021) Postoperative changes in plasma miR21-5p as a novel biomarker for colorectal cancer recurrence: a prospective study. Cancer Sci 112(10):4270–4280
48. Dohmen J, Semaan A, Kobilay M, Zaleski M, Branchi V, Schlierf A, Hettwer K, Uhlig S, Hartmann G, Kalff JC et al (2022) Diagnostic potential of Exosomal microRNAs in colorectal cancer. Diagnostics (Basel) 12(6):1413
49. Mogilyansky E, Rigoutsos I (2013) The miR-17/92 cluster: a comprehensive update on its genomics, genetics, functions and increasingly important and numerous roles in health and disease. Cell Death Differ 20(12):1603–1614
50. Ng EK, Chong WW, Jin H, Lam EK, Shin VY, Yu J, Poon TC, Ng SS, Sung JJ (2009) Differential expression of microRNAs in plasma of patients with colorectal cancer: a potential marker for colorectal cancer screening. Gut 58(10):1375–1381
51. Chen WY, Zhao XJ, Yu ZF, Hu FL, Liu YP, Cui BB, Dong XS, Zhao YS (2015) The potential of plasma miRNAs for diagnosis and risk estimation of colorectal cancer. Int J Clin Exp Pathol 8(6):7092–7101
52. Yang Q, Wang S, Huang J, Xia C, Jin H, Fan Y (2018) Serum miR-20a and miR-486 are potential biomarkers for discriminating colorectal neoplasia: a pilot study. J Cancer Res Ther 14(7):1572–1577
53. Huang Z, Huang D, Ni S, Peng Z, Sheng W, Du X (2010) Plasma microRNAs are promising novel biomarkers for early detection of colorectal cancer. Int J Cancer 127(1):118–126
54. Basati G, Razavi AE, Pakzad I, Malayeri FA (2016) Circulating levels of the miRNAs, miR-194, and miR-29b, as clinically useful biomarkers for colorectal cancer. Tumour Biol 37(2):1781–1788
55. Zhu Y, Xu A, Li J, Fu J, Wang G, Yang Y, Cui L, Sun J (2016) Fecal miR-29a and miR-224 as the noninvasive biomarkers for colorectal cancer. Cancer Biomark 16(2):259–264
56. Ogata-Kawata H, Izumiya M, Kurioka D, Honma Y, Yamada Y, Furuta K, Gunji T, Ohta H, Okamoto H, Sonoda H et al (2014) Circulating exosomal microRNAs as biomarkers of colon cancer. PLoS One 9(4):e92921
57. Radwan E, Shaltout AS, Mansor SG, Shafik EA, Abbas WA, Shehata MR, Ali M (2021) Evaluation of circulating microRNAs-211 and 25 as diagnostic biomarkers of colorectal cancer. Mol Biol Rep 48(5):4601–4610
58. Tan Y, Lin JJ, Yang X, Gou DM, Fu L, Li FR, Yu XF (2019) A panel of three plasma microRNAs for colorectal cancer diagnosis. Cancer Epidemiol 60:67–76
59. Guo S, Zhang J, Wang B, Zhang B, Wang X, Huang L, Liu H, Jia B (2018) A 5-serum miRNA panel for the early detection of colorectal cancer. Onco Targets Ther 11:2603–2614
60. Zhang H, Zhu M, Shan X, Zhou X, Wang T, Zhang J, Tao J, Cheng W, Chen G, Li J et al (2019) A panel of seven-miRNA signature in plasma as potential biomarker for colorectal cancer diagnosis. Gene 687:246–254

61. Luo X, Stock C, Burwinkel B, Brenner H (2013) Identification and evaluation of plasma microRNAs for early detection of colorectal cancer. PLoS One 8(5):e62880
62. Murano T, Chung KY, Tang YC, Kano Y, Takeuchi K, Sakamoto N, Kuwata T, Gao Y, Cheong JK, Cheng H et al (2025) Novel and effective blood-based miRNA diagnostic panel for gastric cancer: a pilot study in a Japanese population. Cancer Med 14(8):e70790
63. An Y, Gong J, Xiao A (2025) Development and validation of nomograms for predicting the prognosis of colorectal cancer patients. Transl Cancer Res 14(3):1651–1663
64. Saito M, Schetter AJ, Mollerup S, Kohno T, Skaug V, Bowman ED, Mathé EA, Takenoshita S, Yokota J, Haugen A et al (2011) The association of microRNA expression with prognosis and progression in early-stage, non-small cell lung adenocarcinoma: a retrospective analysis of three cohorts. Clin Cancer Res 17(7):1875–1882
65. Schetter AJ, Leung SY, Sohn JJ, Zanetti KA, Bowman ED, Yanaihara N, Yuen ST, Chan TL, Kwong DL, Au GK et al (2008) MicroRNA expression profiles associated with prognosis and therapeutic outcome in colon adenocarcinoma. JAMA 299(4):425–436
66. Ma Y, Zhang P, Wang F, Zhang H, Yang J, Peng J, Liu W, Qin H (2012) miR-150 as a potential biomarker associated with prognosis and therapeutic outcome in colorectal cancer. Gut 61(10):1447–1453
67. Cheong JK, Tang YC, Zhou L, Cheng H, Too HP (2022) Advances in quantifying circulatory microRNA for early disease detection. Curr Opin Biotechnol 74:256–262
68. Bautista-Sánchez D, Arriaga-Canon C, Pedroza-Torres A, De La Rosa-Velázquez IA, González-Barrios R, Contreras-Espinosa L, Montiel-Manríquez R, Castro-Hernández C, Fragoso-Ontiveros V, Álvarez-Gómez RM et al (2020) The promising role of miR-21 as a cancer biomarker and its importance in RNA-based therapeutics. Mol Ther Nucl acids 20:409–420
69. Peng Q, Yao W, Yu C, Zou L, Shen Y, Zhu Y, Cheng M, Feng Z, Xu B (2019) Identification of microRNA-181 as a promising biomarker for predicting the poor survival in colorectal cancer. Cancer Med 8(13):5995–6009
70. Yin J, Bai Z, Song J, Yang Y, Wang J, Han W, Zhang J, Meng H, Ma X, Yang Y et al (2014) Differential expression of serum miR-126, miR-141 and miR-21 as novel biomarkers for early detection of liver metastasis in colorectal cancer. Chin J Cancer Res 26(1):95–103
71. Yuan Z, Baker K, Redman MW, Wang L, Adams SV, Yu M, Dickinson B, Makar K, Ulrich N, Böhm J et al (2017) Dynamic plasma microRNAs are biomarkers for prognosis and early detection of recurrence in colorectal cancer. Br J Cancer 117(8):1202–1210
72. Liu K, Li G, Fan C, Zhou X, Wu B, Li J (2011) Increased expression of microRNA-21and its association with chemotherapeutic response in human colorectal cancer. J Int Med Res 39(6):2288–2295
73. Xie T, Huang M, Wang Y, Wang L, Chen C, Chu X (2016) MicroRNAs as regulators, biomarkers and therapeutic targets in the drug resistance of colorectal cancer. Cell Physiol Biochem 40(1–2):62–76
74. Soheilifar MH, Masoudi-Khoram N, Madadi S, Nobari S, Maadi H, Keshmiri Neghab H, Amini R, Pishnamazi M (2022) Angioregulatory microRNAs in breast cancer: molecular mechanistic basis and implications for therapeutic strategies. J Adv Res 37:235–253
75. Pugh S, Thiébaut R, Bridgewater J, Grisoni ML, Moutasim K, Rousseau F, Thomas GJ, Griffiths G, Liebaert F, Primrose J et al (2017) Association between miR-31-3p expression and cetuximab efficacy in patients with KRAS wild-type metastatic colorectal cancer: a post-hoc analysis of the new EPOC trial. Oncotarget 8(55):93856–93866
76. Shah MY, Ferrajoli A, Sood AK, Lopez-Berestein G, Calin GA (2016) microRNA therapeutics in cancer – an emerging concept. EBioMedicine 12:34–42
77. Garzon R, Marcucci G, Croce CM (2010) Targeting microRNAs in cancer: rationale, strategies and challenges. Nat Rev Drug Discov 9(10):775–789
78. Hutvágner G, Simard MJ, Mello CC, Zamore PD (2004) Sequence-specific inhibition of small RNA function. PLoS Biol 2(4):E98
79. Lundin KE, Højland T, Hansen BR, Persson R, Bramsen JB, Kjems J, Koch T, Wengel J, Smith CI (2013) Biological activity and biotechnological aspects of locked nucleic acids. Adv Genet 82:47–107

80. Nedaeinia R, Sharifi M, Avan A, Kazemi M, Rafiee L, Ghayour-Mobarhan M, Salehi R (2016) Locked nucleic acid anti-miR-21 inhibits cell growth and invasive behaviors of a colorectal adenocarcinoma cell line: LNA-anti-miR as a novel approach. Cancer Gene Ther 23(8):246–253
81. Tassone P, Di Martino MT, Arbitrio M, Fiorillo L, Staropoli N, Ciliberto D, Cordua A, Scionti F, Bertucci B, Salvino A et al (2023) Safety and activity of the first-in-class locked nucleic acid (LNA) miR-221 selective inhibitor in refractory advanced cancer patients: a first-in-human, phase 1, open-label, dose-escalation study. J Hematol Oncol 16(1):68
82. Krutzfeldt J, Rajewsky N, Braich R, Rajeev KG, Tuschl T, Manoharan M, Stoffel M (2005) Silencing of microRNAs in vivo with 'antagomirs'. Nature 438(7068):685–689
83. Longo F, Gattuso G, Spoto G, Ricci D, Vitale ACV, Lavoro A, Candido S, Libra M, Falzone L (2025) The multifaceted role of microRNAs in colorectal cancer: pathogenesis and therapeutic implications. Non-coding RNA Res 14:65–95
84. Pagoni M, Cava C, Sideris DC, Avgeris M, Zoumpourlis V, Michalopoulos I, Drakoulis N (2023) miRNA-based technologies in cancer therapy. J Personal Med 13(11)
85. Bak RO, Mikkelsen JG (2014) miRNA sponges: soaking up miRNAs for regulation of gene expression. Wiley Interdiscipl Rev RNA 5(3):317–333
86. Lindow M, Kauppinen S (2012) Discovering the first microRNA-targeted drug. J Cell Biol 199(3):407–412
87. Hong DS, Kang YK, Borad M, Sachdev J, Ejadi S, Lim HY, Brenner AJ, Park K, Lee JL, Kim TY et al (2020) Phase 1 study of MRX34, a liposomal miR-34a mimic, in patients with advanced solid tumours. Br J Cancer 122(11):1630–1637
88. van Zandwijk N, Pavlakis N, Kao SC, Linton A, Boyer MJ, Clarke S, Huynh Y, Chrzanowska A, Fulham MJ, Bailey DL et al (2017) Safety and activity of microRNA-loaded minicells in patients with recurrent malignant pleural mesothelioma: a first-in-man, phase 1, open-label, dose-escalation study. Lancet Oncol 18(10):1386–1396
89. Schultz BR, Chamberlain JS (2008) Recombinant adeno-associated virus transduction and integration. Mol Ther 16(7):1189–1199
90. Dasgupta I, Chatterjee A (2021) Recent advances in miRNA delivery systems. Methods Protocols 4(1):10
91. Herrera-Carrillo E, Liu YP, Berkhout B (2017) Improving miRNA delivery by optimizing miRNA expression cassettes in diverse virus vectors. Human Gene Ther Methods 28(4):177–190
92. Hacein-Bey-Abina S, Von Kalle C, Schmidt M, McCormack MP, Wulffraat N, Leboulch P, Lim A, Osborne CS, Pawliuk R, Morillon E et al (2003) LMO2-associated clonal T cell proliferation in two patients after gene therapy for SCID-X1. Science 302(5644):415–419
93. Kasar S, Salerno E, Yuan Y, Underbayev C, Vollenweider D, Laurindo MF, Fernandes H, Bonci D, Addario A, Mazzella F et al (2012) Systemic in vivo lentiviral delivery of miR-15a/16 reduces malignancy in the NZB de novo mouse model of chronic lymphocytic leukemia. Genes Immun 13(2):109–119
94. Liu Y, Lai L, Chen Q, Song Y, Xu S, Ma F, Wang X, Wang J, Yu H, Cao X et al (2012) MicroRNA-494 is required for the accumulation and functions of tumor-expanded myeloid-derived suppressor cells via targeting of PTEN. J Immunol 188(11):5500–5510
95. Daya S, Berns KI (2008) Gene therapy using adeno-associated virus vectors. Clin Microbiol Rev 21(4):583–593
96. Chen SK, Hawley ZCE, Zavodszky MI, Hana S, Ferretti D, Grubor B, Hawes M, Xu S, Hamann S, Marsh G et al (2023) Efficacy and safety of a SOD1-targeting artificial miRNA delivered by AAV9 in mice are impacted by miRNA scaffold selection. Mol Ther Nucl acids 34:102057
97. Carrella S, Di Guida M, Brillante S, Piccolo D, Ciampi L, Guadagnino I, Garcia Piqueras J, Pizzo M, Marrocco E, Molinari M et al (2022) miR-181a/b downregulation: a mutation-independent therapeutic approach for inherited retinal diseases. EMBO Mol Med 14(11):e15941
98. Vandenberghe LH, Wilson JM (2007) AAV as an immunogen. Curr Gene Ther 7(5):325–333

99. Pan Y, Zhang Y, Jia T, Zhang K, Li J, Wang L (2012) Development of a microRNA delivery system based on bacteriophage MS2 virus-like particles. FEBS J 279(7):1198–1208
100. Sun Y, Sun Y, Zhao R (2017) Establishment of MicroRNA delivery system by PP7 bacteriophage-like particles carrying cell-penetrating peptide. J Biosci Bioeng 124(2):242–249
101. Pan Y, Jia T, Zhang Y, Zhang K, Zhang R, Li J, Wang L (2012) MS2 VLP-based delivery of microRNA-146a inhibits autoantibody production in lupus-prone mice. Int J Nanomedicine 7:5957–5967
102. Wang G, Jia T, Xu X, Chang L, Zhang R, Fu Y, Li Y, Yang X, Zhang K, Lin G et al (2016) Novel miR-122 delivery system based on MS2 virus like particle surface displaying cell-penetrating peptide TAT for hepatocellular carcinoma. Oncotarget 7(37):59402–59416
103. Yao Y, Jia T, Pan Y, Gou H, Li Y, Sun Y, Zhang R, Zhang K, Lin G, Xie J et al (2015) Using a novel microRNA delivery system to inhibit osteoclastogenesis. Int J Mol Sci 16(4):8337–8350
104. Guo P (2005) Bacterial virus phi29 DNA-packaging motor and its potential applications in gene therapy and nanotechnology. Methods Mol Biol 300:285–324
105. Endo-Takahashi Y, Negishi Y, Nakamura A, Ukai S, Ooaku K, Oda Y, Sugimoto K, Moriyasu F, Takagi N, Suzuki R et al (2014) Systemic delivery of miR-126 by miRNA-loaded bubble liposomes for the treatment of hindlimb ischemia. Sci Rep 4:3883
106. Dehghankelishadi P, Maritz MF, Badiee P, Thierry B (2022) High density lipoprotein nanoparticle as delivery system for radio-sensitising miRNA: An investigation in 2D/3D head and neck cancer models. Int J Pharm 617:121585
107. Yang D, Sun Y, Hu L, Zheng H, Ji P, Pecot CV, Zhao Y, Reynolds S, Cheng H, Rupaimoole R et al (2013) Integrated analyses identify a master microRNA regulatory network for the mesenchymal subtype in serous ovarian cancer. Cancer Cell 23(2):186–199
108. Tenchov R, Bird R, Curtze AE, Zhou Q (2021) Lipid nanoparticles—from liposomes to mRNA vaccine delivery, a landscape of research diversity and advancement. ACS Nano 15(11):16982–17015
109. Yang D, Liu B, Sha H (2025) Advances and prospects of cell-penetrating peptides in tumor immunotherapy. Sci Rep 15(1):3392
110. Ibrahim AF, Weirauch U, Thomas M, Grünweller A, Hartmann RK, Aigner A (2011) MicroRNA replacement therapy for miR-145 and miR-33a is efficacious in a model of colon carcinoma. Cancer Res 71(15):5214–5224
111. Brillante S, Volpe M, Indrieri A (2024) Advances in MicroRNA therapeutics: from preclinical to clinical studies. Hum Gene Ther 35(17–18):628–648
112. Arora S, Swaminathan SK, Kirtane A, Srivastava SK, Bhardwaj A, Singh S, Panyam J, Singh AP (2014) Synthesis, characterization, and evaluation of poly (D,L-lactide-co-glycolide)-based nanoformulation of miRNA-150: potential implications for pancreatic cancer therapy. Int J Nanomedicine 9:2933–2942
113. Menon A, Abd-Aziz N, Khalid K, Poh CL, Naidu R (2022) miRNA: a promising therapeutic target in cancer. Int J Mol Sci 23(19):11502
114. Chen Y, Xianyu Y, Jiang X (2017) Surface modification of gold nanoparticles with small molecules for biochemical analysis. Acc Chem Res 50(2):310–319
115. Jia C, Chen H, Wei M, Chen X, Zhang Y, Cao L, Yuan P, Wang F, Yang G, Ma J (2017) Gold nanoparticle-based miR155 antagonist macrophage delivery restores the cardiac function in ovariectomized diabetic mouse model. Int J Nanomedicine 12:4963–4979
116. Ghosh R, Singh LC, Shohet JM, Gunaratne PH (2013) A gold nanoparticle platform for the delivery of functional microRNAs into cancer cells. Biomaterials 34(3):807–816
117. Sun S, Wang Y, Zhou R, Deng Z, Han Y, Han X, Tao W, Yang Z, Shi C, Hong D et al (2017) Targeting and regulating of an oncogene via Nanovector delivery of MicroRNA using patient-derived xenografts. Theranostics 7(3):677–693
118. Mamaeva V, Sahlgren C, Lindén M (2013) Mesoporous silica nanoparticles in medicine – recent advances. Adv Drug Deliv Rev 65(5):689–702
119. Li Y, Duo Y, Bi J, Zeng X, Mei L, Bao S, He L, Shan A, Zhang Y, Yu X (2018) Targeted delivery of anti-miR-155 by functionalized mesoporous silica nanoparticles for colorectal cancer therapy. Int J Nanomedicine 13:1241–1256

120. Zhao J, Lin H, Huang K (2022) Mesenchymal stem cell-derived extracellular vesicles transmitting MicroRNA-34a-5p suppress tumorigenesis of colorectal cancer through c-MYC/DNMT3a/PTEN Axis. Mol Neurobiol 59(1):47–60
121. Cui C, Guo T, Zhang S, Yang M, Cheng J, Wang J, Kang J, Ma W, Nian Y, Sun Z et al (2022) Bacteria-derived outer membrane vesicles engineered with over-expressed pre-miRNA as delivery nanocarriers for cancer therapy. Nanomed Nanotechnol Biol Med 45:102585
122. Wang K, Zhang S, Weber J, Baxter D, Galas DJ (2010) Export of microRNAs and microRNA-protective protein by mammalian cells. Nucleic Acids Res 38(20):7248–7259

Chapter 23
Artificial Intelligence-Enhanced Insights: Leveraging miRNA Sequencing Data for Cancer Prognosis and Treatment

Anand Thiyagaraj, Arul Jothi Nagarajan, and Satish Ramalingam

Abstract Cancer, characterized by its high heterogeneity and multifactorial nature, remains a leading cause of mortality worldwide. It results from a complex interaction of genetic mutations, epigenetic changes, and disrupted regulatory networks that collectively lead to uncontrolled cell growth, metastasis, and therapy resistance. MicroRNAs (miRNAs), a class of small noncoding RNAs that modulate gene expression at the posttranscriptional level, have recently been recognized as key players in both oncogenic and tumor-suppressive pathways driving cancer progression. Their capacity to affect multiple pathways at once makes them promising candidates for diagnostic, prognostic, and therapeutic biomarkers. With the introduction of next-generation sequencing (NGS), profiling miRNA expression has become more accessible and thorough. However, analyzing and interpreting these high-dimensional datasets necessitates computational frameworks that can capture complex biological interactions. Artificial intelligence (AI), especially machine learning (ML) and deep learning (DL), provides robust methods for managing, analyzing, and extracting meaningful insights from large-scale miRNA sequencing data. This chapter examines how AI-driven approaches are revolutionizing cancer prognosis and treatment by utilizing miRNA data, focusing on integrative models, data-driven biomarker discovery, and clinical decision support systems.

Keywords Cancer heterogeneity · MicroRNAs (miRNAs) · Next-generation sequencing (NGS) · Artificial intelligence (AI) · Machine learning (ML) · Deep learning (DL) · Biomarker discovery

A. Thiyagaraj (✉) · A. J. Nagarajan · S. Ramalingam
Department of Genetic Engineering, School of Bioengineering, SRM Institute of Science and Technology, Kattankulathur, Chennai, Tamil Nadu, India
e-mail: anandt@srmist.edu.in

S. Pathak, A. Banerjee (eds.), *MicroRNAs: Role in Cancer Progression, Diagnostics and Therapy*, Advances in Biochemistry in Health and Disease 34,
https://doi.org/10.1007/978-3-032-08824-6_23

Introduction

The integration of molecular biology and computational technologies in oncology has led to a deeper understanding of the complex mechanisms underlying cancer development and progression. miRNAs have emerged as crucial players in this landscape, demonstrating their ability to fine-tune gene expression through post-transcriptional regulation [1]. The versatility of miRNAs is evident in their capacity to target multiple mRNAs simultaneously, allowing them to modulate entire signaling pathways and cellular processes (Fig. 23.1). This property has established miRNAs as crucial regulators of multiple facets of cancer biology, such as cell proliferation, apoptosis, metastasis, and resistance to therapy [53, 4].

The potential of miRNAs as biomarkers extends beyond their stability and tissue specificity. Recent studies have shown that circulating miRNAs in blood and other bodily fluids can serve as "liquid biopsies," offering a noninvasive means to monitor cancer progression and treatment efficacy [15]. Moreover, the advent of high-throughput sequencing technologies has enabled the comprehensive profiling of miRNA expression patterns across different cancer types and stages, leading to the identification of miRNA signatures with diagnostic and prognostic value [79]. These advancements have paved the way for personalized medicine approaches in oncology, where miRNA profiles could guide treatment decisions and predict patient outcomes. As research in this field continues to evolve, the development of miRNA-based therapeutics, either through miRNA mimics or inhibitors, represents a promising avenue for novel cancer treatments that could potentially overcome the limitations of conventional therapies [64, 66].

Fig. 23.1 Workflow of miRNA biogenesis and functional targeting in cancer pathways

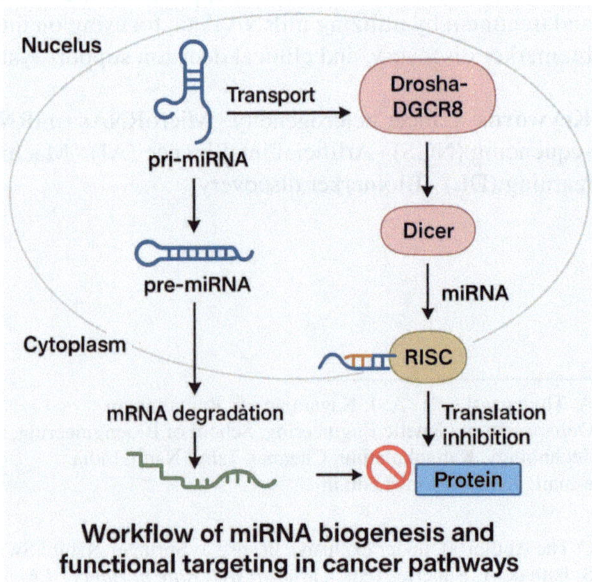

miRNAs have emerged as powerful molecular tools with significant potential in cancer research and clinical applications. However, their effective utilization has been hindered by challenges in data interpretation and the need for robust models that can be applied across diverse clinical scenarios [40]. The complexity of gene regulatory networks and the vast amount of sequencing data generated pose significant obstacles to deriving meaningful insights through conventional statistical approaches alone [89]. This is where artificial intelligence (AI)-driven techniques become crucial, offering advanced capabilities to analyze and interpret complex data [43]. Machine learning algorithms, a subset of AI, excel at identifying subtle, nonlinear patterns within large datasets. This capability enables researchers to classify patients, predict disease outcomes, and propose personalized therapeutic strategies based on miRNA profiles with greater accuracy and efficiency. The integration of AI and miRNA sequencing in cancer research has opened new avenues for improved diagnosis, prognosis, and treatment optimization [94]. By harnessing the analytical capabilities of AI, researchers can reveal previously undiscovered relationships between miRNA expression patterns and key elements of cancer biology, paving the way for the development of more precise and effective therapeutic strategies [67]. Furthermore, this convergence of technologies addresses the current challenges in miRNA analytics and paves the way for future advancements in clinical oncology, promising more personalized and effective cancer management strategies [23].

miRNA Profiling for Early Detection and Prognosis of Cancer

miRNAs have emerged as powerful biomarkers in cancer research and clinical diagnostics due to their unique characteristics and disease-specific expression patterns [33]. These small, noncoding RNA molecules demonstrate exceptional stability in bodily fluids such as blood, urine, and saliva, positioning them as promising candidates for noninvasive diagnostic and prognostic applications [19]. The presence of tumor-derived circulating miRNAs in serum, plasma, or exosomes has revolutionized early cancer detection, staging, and monitoring of therapeutic responses [44]. Extensive research has uncovered cancer-specific miRNA signatures, with certain miRNAs consistently exhibiting dysregulation across various cancer types. For instance, miR-21 is frequently upregulated in breast, colorectal, and pancreatic cancers, correlating with poor prognosis, while miR-143 and miR-145 act as tumor suppressors in various malignancies (Table 23.1) [31, 51].

The differential expression of miRNAs provides a molecular fingerprint that enables precise cancer subtyping and progression monitoring. This approach offers significant advantages over traditional diagnostic methods, as miRNA profiles can detect cancer at earlier stages and with greater accuracy [71]. However, the complexity of miRNA interactions and their multifaceted roles in cellular processes pose challenges for conventional statistical analyses [67]. To address these limitations, artificial intelligence (AI) methodologies have been increasingly employed to

Table 23.1 Selected miRNAs with validated roles in prognosis across various cancers

miRNA	Cancer type	Role	Prognostic value
miR-21	Breast, CRC, lung	Oncogenic	Poor survival [31]
miR-155	Lymphoma, breast	Oncogenic	Disease progression [51]
miR-143	Colon, bladder	Tumor suppressor	Better prognosis [51]
miR-210	Colon, bladder	Tumor suppressor	Better prognosis [51]

analyze miRNA datasets. AI-driven approaches, such as machine learning and deep learning algorithms, can uncover nonlinear interactions and multidimensional correlations within miRNA expression data [20]. These advanced computational techniques facilitate the discovery of novel miRNA biomarkers, enhance the accuracy of existing panels, and provide insights into the intricate biological networks underlying cancer development and progression [62]. The integration of AI with miRNA research holds immense potential for improving cancer diagnosis, prognosis, and personalized treatment strategies [55].

Artificial Intelligence and Machine Learning: A Primer for Systems Biologists

Artificial Intelligence (AI), along with Machine Learning (ML) and Deep Learning (DL), has transformed the landscape of systems biology by enabling more effective analysis and interpretation of complex biological data [82]. These computational approaches have become indispensable tools for extracting meaningful insights from large-scale omics datasets, such as miRNA sequencing. AI's ability to simulate human intelligence allows for the processing of vast amounts of biological information, while ML algorithms excel at identifying patterns and relationships within the data without the need for explicit programming. DL, with its multilayered neural networks, takes this a step further by modeling intricate nonlinear relationships that are often present in biological systems [90]. The application of these technologies in systems biology has led to significant advancements in our understanding of biological processes and disease mechanisms. For instance, AI-driven analysis of miRNA sequencing data has enabled researchers to identify novel biomarkers, predict disease outcomes, and uncover previously unknown regulatory networks [26, 72].

ML algorithms have proven particularly effective in integrating diverse omics datasets, allowing for a more holistic view of biological systems. DL models have shown remarkable success in tasks such as protein structure prediction and drug discovery, leveraging their ability to capture complex patterns in high-dimensional data [93]. With ongoing advancements, these computational technologies are expected to open novel pathways in precision medicine, enhance drug discovery efforts, and provide deeper insights into the sophisticated molecular frameworks of living organisms. Machine learning (ML) techniques in miRNA data analysis can

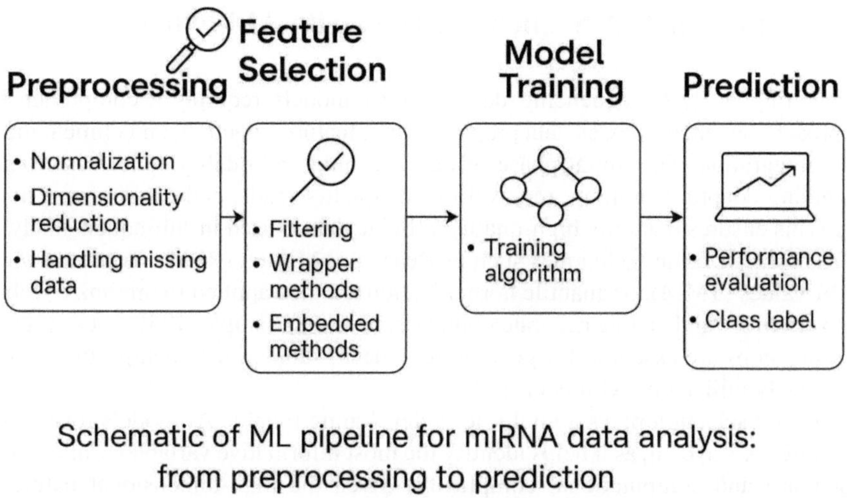

Fig. 23.2 Schematic of ML pipeline for miRNA data analysis: from preprocessing to prediction

be broadly categorized into supervised and unsupervised learning approaches (Fig. 23.2) [36]. Supervised learning methods utilize labeled datasets to learn how to predict defined outcomes, such as identifying cancer subtypes or estimating patient survival times. These models are trained on known examples to generalize their predictions to new, unseen cases. In contrast, unsupervised learning techniques operate on unlabeled data, aiming to uncover hidden patterns or groupings within the dataset. This approach is especially valuable for identifying previously unknown cancer subtypes or grouping patients based on their miRNA expression profiles without relying on predefined labels [49].

Several ML algorithms have proven effective in miRNA data analysis. Support Vector Machines (SVMs) excel in handling high-dimensional data and are particularly useful for classification tasks [75]. Random Forests offer a robust ensemble approach that combines multiple decision trees, providing both predictive power and interpretability through feature importance assessment [22]. Gradient boosting algorithms, such as XGBoost, have gained popularity due to their efficiency and superior performance in competitive environments. Deep neural networks, while requiring larger datasets, have the potential to uncover complex relationships within biological data [41]. To evaluate the performance of these models, various metrics are employed, including accuracy, precision, recall, F1-score, and Receiver Operating Characteristic (ROC) curves with corresponding Area Under the Curve (AUC) values [12]. These metrics are crucial for assessing the reliability and clinical relevance of the models in making accurate predictions and informing medical decisions [80].

Integrating miRNA Sequencing Data with AI Models

Integrating miRNA sequencing data with AI models requires a comprehensive approach that encompasses data preprocessing, feature selection, and clinical metadata integration. The initial phase of data cleaning and quality control is crucial, involving adapter trimming, removal of low-quality reads, and contaminant filtering. This ensures that only high-quality, reliable data is used in subsequent analyses [77]. Normalization techniques such as Reads per Million (RPM), Trimmed Mean of M-values (TMM), or quantile normalization are then applied to minimize technical variability and enable accurate comparisons across samples [95]. These preprocessing steps are essential for generating a clean, standardized dataset that can be effectively utilized by AI models [73].

Feature selection plays a vital role in developing reliable AI models for analyzing miRNA-seq data, as it helps identify the most informative variables while reducing noise and computational complexity. Given the high-dimensional nature of miRNA-seq data, where thousands of miRNAs may be present, it is crucial to identify the most informative features [27]. Algorithms such as Least Absolute Shrinkage and Selection Operator (LASSO), Minimum Redundancy Maximum Relevance (mRMR), and Recursive Feature Elimination (RFE) are commonly employed to select a subset of miRNAs that are most relevant to the biological question at hand [52]. This process not only reduces the risk of overfitting but also enhances model interpretability by focusing on the most significant miRNAs. Additionally, the integration of clinical metadata with miRNA expression data can provide valuable context and improve the predictive power of AI models, enabling more comprehensive insights into disease mechanisms and potential biomarkers [34, 83].

Open-access databases such as The Cancer Genome Atlas (TCGA), Gene Expression Omnibus (GEO), and ENCODE have significantly advanced bioinformatics and AI-powered genomic research by providing comprehensive, high-quality datasets for large-scale analysis and model development. These repositories offer a wealth of molecular and clinical data, including gene expression profiles, DNA methylation patterns, and patient outcomes, which are crucial for developing and validating sophisticated AI models [24, 91]. The integration of diverse data types allows for more comprehensive analyses, potentially uncovering complex relationships between molecular features and clinical phenotypes. However, researchers must navigate several challenges when utilizing these datasets [2]. Class imbalance, where certain patient subgroups (often high-risk cases) are underrepresented, can lead to biased model predictions and reduced generalizability. To overcome this challenge, strategies like oversampling, undersampling, and synthetic data generation are commonly applied to achieve a more balanced dataset, thereby improving model performance and reducing bias [16].

Another significant challenge in working with large-scale genomic data is the presence of batch effects, which arise from technical variations in sample processing, sequencing platforms, or laboratory conditions [60]. These effects can introduce systematic biases that confound biological signals and compromise the validity of downstream analyses. Tools like ComBat and Harmony have been developed to

Table 23.2 Public datasets with miRNA sequencing data and associated clinical annotations

Dataset	Cancer type	Sample size	Clinical data included
TCGA-BRCA	Breast cancer	>1000	Survival, therapy response
TCGA-LUAD	Lung adenocarcinoma	500+	Stage, mutation profiles
GEO-GSE48414	Multiple	300+	Treatment outcome, histology

mitigate these batch effects, allowing for more accurate integration of data from different sources [81]. In addition, dimensionality reduction techniques are essential for managing the high-dimensional complexity of genomic data, as they help simplify the dataset while preserving the most relevant biological information. Methods such as Principal Component Analysis (PCA) and t-distributed Stochastic Neighbor Embedding (t-SNE) not only facilitate visualization of complex datasets but also help in identifying underlying patterns and structures that may not be apparent in the raw data (Table 23.2) [30]. These techniques can reveal biologically meaningful clusters, guide feature selection for ML algorithms, and provide insights into the molecular basis of diseases, ultimately enhancing the interpretability and performance of AI models in genomic research [46].

AI-Based Prognostic Models Using miRNA Data

AI-based prognostic models leveraging miRNA data have demonstrated significant potential in oncology, particularly in breast cancer management [15]. These sophisticated algorithms analyze intricate relationships between miRNAs and clinical variables to generate accurate risk scores and survival probabilities [56]. By integrating miRNA expression profiles with traditional clinical parameters, these models offer a more comprehensive and personalized approach to patient stratification and treatment planning [87]. The success of random forest classifiers in identifying small yet powerful miRNA panels correlated with five-year survival outcomes underscores the clinical utility of these AI-driven approaches. The integration of miRNA-based biomarkers with established clinical features represents a significant advancement in prognostic modeling [21, 14].

By harnessing the synergistic capabilities of molecular biology and artificial intelligence, researchers can elucidate intricate regulatory mechanisms underlying cancer progression. The integration of miRNA and mRNA co-expression networks through deep learning methodologies not only enhances the resolution of disease characterization but also facilitates the identification of novel biomarkers and potential therapeutic targets [3]. This multifaceted analysis enables a more sophisticated understanding of tumor heterogeneity, which is essential for tailoring interventions to individual patient profiles. In the context of lung cancer, such approaches have demonstrated considerable potential in stratifying patients based on disease stage, thereby enabling earlier interventions and potentially improving prognostic outcomes [5]. Moreover, the ongoing refinement of AI algorithms, coupled with the increasing availability of high-throughput molecular data, is anticipated to further enhance the precision of prognostic tools. These advancements support a transition toward dynamic, data-driven clinical decision-making frameworks capable of adapting to evolving patient information over time [74]. As AI models become increasingly proficient at integrating diverse data types—including genomic, transcriptomic, and radiologic inputs—they are likely to become indispensable components of the clinical workflow. This evolution underscores the transformative potential of AI not only in breast and lung cancers but also across a broad spectrum of oncological diseases, paving the way for more efficacious and personalized patient care [[25, 45].

Deep learning models have significantly advanced cancer genomics research by providing unprecedented capabilities in analyzing complex biological data. These models, with their layered architecture, are adept at capturing complex regulatory patterns that traditional analytical approaches may overlook [71]. The efficacy of deep learning is attributed to its capacity to process and interpret high-dimensional data, capturing nonlinear relationships prevalent in cancer genomics. This capability enables researchers to obtain more comprehensive insights into the molecular mechanisms underlying cancer development and progression [92]. The incorporation of ensemble learning strategies, such as voting classifiers or stacked generalization, enhances the robustness and reliability of predictions by integrating multiple models or learning algorithms [57]. The transition from research to clinical application of these advanced machine learning techniques necessitates rigorous validation processes. Cross-validation, particularly k-fold validation, emerges as a crucial tool in this context [37]. This method involves partitioning the dataset into k subsets, utilizing k-1 subsets for training and the remaining subset for validation, iterating this process k times [9]. This strategy offers a thorough evaluation of the model's ability to generalize to unseen data, thereby mitigating the risk of overfitting. By evaluating the model's consistency across different subsets of the data, researchers can ensure its generalizability and reliability in real-world clinical settings [6]. This rigorous validation process is essential for developing robust predictive models that have the potential to transform cancer diagnosis, prognosis, and treatment strategies [10].

This enables researchers to derive a more reliable estimate of the model's generalizability and performance across varying data subsets. In parallel, interpretability tools such as SHAP (SHapley Additive exPlanations) and LIME (Local Interpretable Model-agnostic Explanations) have become indispensable in AI-driven diagnostics, particularly when transparency is critical for high-stakes clinical decision-making [37]. These tools offer granular insights into the contribution of individual miRNAs to the model's predictions, enhancing transparency and fostering greater trust among clinicians in AI-assisted diagnostic decisions. SHAP, for instance, uses game theory concepts to calculate the importance of each feature in the final prediction, while LIME creates a local linear approximation of the model's decision boundary to explain individual predictions [65]. By incorporating these interpretability techniques, researchers can address regulatory requirements and facilitate the integration of AI-driven diagnostics in lung cancer management, ultimately enhancing early detection, personalized treatment strategies, and patient outcomes [28].

Cross-validation techniques, especially k-fold validation, are essential for assessing a model's performance on independent data while helping to minimize the risk of overfitting and ensuring more reliable generalization.

This method entails partitioning the dataset into *k* subsets, where the model is trained on *k-1* of them and validated on the remaining one. The process is repeated *k* times, ensuring that each subset is used once as the validation set, providing a comprehensive evaluation of the model's performance. By doing so, researchers can obtain a more robust estimate of the model's generalizability and performance across different data partitions [13]. The k-fold cross-validation technique offers several advantages over traditional holdout methods. First, it ensures that every data point is used for both training and validation, maximizing the utilization of available data [85]. This is particularly beneficial when working with limited datasets, which is often the case in medical research. Second, it yields a more dependable estimate of the model's performance by averaging the results over multiple iterations, thereby minimizing the influence of random fluctuations in data partitioning. Moreover, cross-validation helps in identifying potential issues such as overfitting or underfitting [54, 35].

If a model shows strong performance on the training data but underperforms on the validation sets, it likely indicates overfitting, highlighting the need for strategies such as regularization or more refined feature selection. Conversely, if the model performs poorly on both training and validation sets, it might be underfitting, indicating the need for a more complex model or additional features [84]. Interpretability tools like SHAP (SHapley Additive exPlanations) and LIME (Local Interpretable Model-agnostic Explanations) have become essential in the context of AI-driven diagnostics, especially for high-stakes medical decisions [39]. These tools deliver in-depth explanations of how specific miRNAs influence the model's predictions, enhancing interpretability and fostering confidence among clinicians in adopting AI-based decision-making. SHAP, for instance, uses game theory concepts to calculate the importance of each feature in the final prediction, while LIME creates a local linear approximation of the model's decision boundary to explain individual predictions [85]. The importance of these interpretability techniques cannot be overstated in the medical field. They bridge the gap between complex machine learning models and human understanding, allowing healthcare professionals to gain insights into the reasoning behind AI-generated predictions [70].

In the context of genomics and miRNA studies, this transparency is crucial for building trust in AI systems and facilitating their adoption in clinical settings. SHAP values, derived from cooperative game theory, provide a unified measure of feature importance that is consistent and locally accurate for genomic data analysis. They offer both global interpretability, showing the overall impact of miRNA expression patterns across the entire dataset, and local interpretability, explaining the contribution of each miRNA to individual predictions [58]. This dual perspective is particularly valuable in personalized medicine, where understanding both population-level trends and individual patient miRNA profiles is essential [18]. LIME, on the other hand, focuses on local interpretability by creating simplified, interpretable models that approximate the behavior of the complex model in the vicinity of a specific instance of miRNA expression [47]. This approach is particularly useful for explaining individual predictions, helping clinicians understand why a particular diagnosis or prognosis was made for a specific patient based on their miRNA profile. By incorporating these interpretability techniques, researchers can address regulatory requirements and facilitate the integration of AI-driven diagnostics in genomic-based cancer management, particularly in the analysis of miRNA expression patterns and their role in disease progression [88, 61]. The integration of miRNA sequencing and interpretable AI models is revolutionizing lung cancer management. This approach offers improved diagnosis, treatment selection, and patient outcomes. miRNA sequencing provides detailed expression profiles, enabling AI-driven insights into lung cancer mechanisms [17]. The high-throughput technique allows comprehensive identification and quantification of miRNAs, detecting subtle changes indicative of early-stage cancer or treatment responses. Interpretable AI models analyzing miRNA expression profiles offer transparent insights, crucial for ethical and legal standards in medical applications [8]. They enable earlier detection, predict treatment responses, and guide personalized therapy. The collaboration between AI and human expertise enhances diagnostic accuracy and treatment planning [69]. Integrating miRNA-target interactions with drug databases through network-based approaches facilitates drug repurposing and combination therapy design [78]. This synergy advances personalized medicine for lung cancer. The transparency of interpretability tools improves patient education, engagement, and treatment adherence. Combining miRNA sequencing, robust cross-validation methods, and advanced AI interpretability techniques represents a significant advancement in lung cancer diagnostics and treatment, promising improved early detection [25, 42].

This method allows researchers to explore complex biological networks and predict potential therapeutic effects of drugs on specific targets or pathways [59]. By leveraging vast amounts of data from multiple sources, these AI-driven approaches can significantly accelerate the drug discovery process and potentially uncover unexpected connections between drugs and diseases (Fig. 23.3) [7]. Reinforcement learning, a subset of AI, has shown promise in adapting treatment regimens based on longitudinal patient data [50]. This approach enables the development of dynamic treatment strategies that can be tailored to individual patients' responses over time. As patients' conditions evolve, reinforcement learning algorithms can continuously optimize treatment plans, potentially improving outcomes and reducing adverse effects [76]. However, the successful implementation of these AI-driven approaches in clinical settings relies heavily on explainable AI techniques. These methods provide transparent and understandable model outputs, which are crucial for gaining clinician trust, ensuring patient acceptance, and meeting regulatory requirements set by organizations such as the FDA and EMA. By demystifying the decision-making process of AI models, explainable AI facilitates the responsible integration of these advanced technologies into healthcare systems, ultimately enhancing patient care and treatment efficacy.

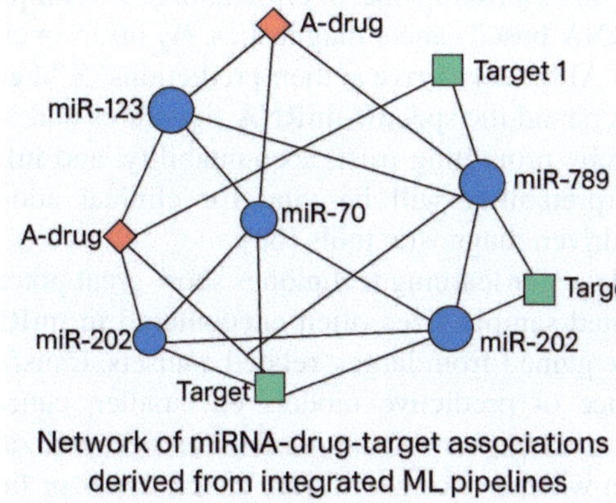

Fig. 23.3 Network of miRNA-drug-target associations derived from integrated ML pipelines

Future Directions and Challenges

Despite significant progress, numerous challenges remain in fully realizing the potential of AI-driven miRNA analysis for improving cancer diagnosis, prognosis, and treatment. Ethical and regulatory concerns, including bias in training data, lack of transparency in model design, and ambiguous accountability in AI-driven decisions, pose significant obstacles to clinical adoption [11]. Ensuring fairness, interpretability, and reproducibility is crucial for regulatory approval and public trust. The clinical translation of AI models requires rigorous validation through prospective trials, as current studies are predominantly retrospective and may not generalize across diverse patient populations [38]. Standardizing miRNA biomarkers and analytical workflows is essential to enhance reproducibility and interoperability across institutions. Data privacy represents another significant challenge in AI-enhanced miRNA analysis [29]. Federated learning presents a compelling approach by allowing AI models to be trained on decentralized datasets across multiple institutions without the need to exchange raw patient data, thereby safeguarding privacy while leveraging diverse and extensive data sources. Furthermore, embedding AI into real-time clinical decision support systems offers transformative potential to enhance precision and efficiency in cancer care. Embedding these models into electronic health records (EHRs) could facilitate dynamic, data-driven guidance for diagnosis, prognosis, and therapy [42]. This integration has the potential to usher in a new era of personalized medicine, where AI-driven insights can be seamlessly incorporated into clinical workflows, enabling more precise and timely interventions [32]. However, the effective deployment of AI-enhanced miRNA analysis in cancer care will depend on sustained collaboration among clinicians, data scientists, and regulatory authorities to navigate complex challenges and ensure its safe, ethical, and clinically meaningful integration into healthcare systems.

The future of miRNA research in cancer diagnostics and therapeutics holds promising avenues for advancement, particularly through the integration of cutting-edge technologies and methodologies. Explainable AI (XAI) is set to play a pivotal role in improving the interpretability of complex biological data, particularly in miRNA-based cancer diagnostics. By offering clear and transparent explanations of how AI models arrive at their predictions, XAI enables researchers and clinicians to understand the specific miRNA signatures and interactions driving those decisions, thereby promoting trust, accountability, and informed clinical use. This increased interpretability will be vital for clinical adoption and regulatory approval of AI-driven diagnostic tools [68].

Transfer learning techniques show great potential in addressing the challenge of limited sample sizes often encountered in miRNA studies. By leveraging knowledge gained from larger, related datasets, transfer learning can improve the performance of predictive models on smaller, cancer-specific miRNA datasets. This approach may enable more robust and generalizable miRNA-based diagnostic tools, even when working with rare cancer types or limited patient cohorts [86, 48]. The integration of multi-omics data represents another frontier in miRNA research. Integrating miRNA expression profiles with other molecular data types—such as genomics, transcriptomics, and proteomics—offers a more holistic view of cancer biology, enabling deeper insights into disease mechanisms and improving the precision of diagnostic and therapeutic strategies. AI-driven multi-omics integration has the potential to uncover complex relationships between miRNAs and other biomolecules, leading to more accurate cancer classification, improved treatment response predictions, and the identification of novel therapeutic targets.

References

1. Aigner A (2011) MicroRNAs (miRNAs) in cancer invasion and metastasis: therapeutic approaches based on metastasis-related miRNAs. J Mol Med 89(5):445–457. https://doi.org/10.1007/s00109-010-0716-0
2. Alameer A, Chicco D (2021) geoCancerPrognosticDatasetsRetriever: a bioinformatics tool to easily identify cancer prognostic datasets on gene expression omnibus (GEO). Bioinformatics 38(6):1761–1763. https://doi.org/10.1093/bioinformatics/btab852
3. Ali H (2023) Artificial intelligence in multi-omics data integration: advancing precision medicine, biomarker discovery and genomic-driven disease interventions. Int J Sci Res Arch 8(1):1012–1030. https://doi.org/10.30574/ijsra.2023.8.1.0189
4. Alkanli N, Ay A (2021) miRNAs AND miRNA gene variations in cancer development and progression. Gevher Nesibe J IESDR 6(13):38–45. https://doi.org/10.46648/gnj.226
5. Alosaimi HM, Bosbait JM, Alhazzaa AA, Alluqmani TA, Albeladi RS, Alraddadi RH, Al Qahtani SM, Al-Saif LK, Almohammadi SA, Alamri RA, Almutairi DR, Aljohani WK, Alshammari MK, Alkurdi AA, Alshilash AM (2025) AI models for the identification of prognostic and predictive biomarkers in lung cancer: a systematic review and meta-analysis. Front Oncol 15. https://doi.org/10.3389/fonc.2025.1424647
6. Aprihartha MA, Idham I (2024) Optimization of classification algorithms performance with k-fold cross validation. Eigen Mathematics J 7(2):61–66. https://doi.org/10.29303/emj.v7i2.212
7. Arjunan G (2021) Implementing explainable AI in healthcare: techniques for interpretable machine learning models in clinical decision-making. Int J Sci Res Manag (IJSRM) 9(05):597–603. https://doi.org/10.18535/ijsrm/v9i05.ec03

8. Azim SA, Tippu MSR, Khairi HF, Dawood MT, Abdullah MS, Al-yasiri NA, Shawky OM, Muthiah LA, Ali M (2024) Inverted maxillary third molar impaction: exploring capabilities of artificial intelligence (AI) versus human intelligence (HI) expertise in diagnosis and treatment planning. Open access. J Dental Sci 9(3). https://doi.org/10.23880/oajds-16000405
9. Azis H (2024) Assessing the performance of logistic regression in heart disease detection through 5-fold cross-validation. Int J Artificial Intell Med Issues 2(1):1–11. https://doi.org/10.56705/ijaimi.v2i1.137
10. Baba T, Hashimoto Y, Yasuda H, Araki M, Edahiro Y, Morishita S, Ochiai T, Shirane S, Ando J, Komatsu N (2022) Validation and reliability of current guidelines for the treatment of essential thrombocythemia under real-world clinical settings in Japan. Hematology 27(1):157–166. https://doi.org/10.1080/16078454.2021.2022848
11. Bahangulu J, Owusu-Berko L (2025) Algorithmic bias, data ethics, and governance: ensuring fairness, transparency and compliance in AI-powered business analytics applications. World J Adv Res Rev 25(2):1746–1763. https://doi.org/10.30574/wjarr.2025.25.2.0571
12. Barakat N, Bradley AP (2006) Rule extraction from support vector machines: measuring the explanation capability using the area under the ROC curve. In: 18th international conference on pattern recognition (ICPR'06), vol 2. IEEE, pp 812–815. https://doi.org/10.1109/icpr.2006.1021
13. Bates S, Hastie T, Tibshirani R (2023) Cross-validation: what does it estimate and how well does it Do it? J Am Stat Assoc 119(546):1434–1445. https://doi.org/10.1080/01621459.2023.2197686
14. Bello B, Varshney J, Bhave R, Bundey YN, Khotimchenko M, Chakravarty K, Baran SW (2023) Integrating AI/ML models for patient stratification leveraging omics dataset and clinical biomarkers from COVID-19 patients: a promising approach to personalized medicine. Int J Mol Sci 24(7):6250. https://doi.org/10.3390/ijms24076250
15. Biju TS, Francis AP, Veeraraghavan VP (2024) Theranostic potential of miRNAs in oral cancer: an emerging approach for precision oncology. Oral Sci Int 22(1). https://doi.org/10.1002/osi2.1277
16. Cao L, Yang J, Chen E (2021) Identification of methylation driven biomarkers for diagnosis and prognosis in colon cancer by integrative analysis of TCGA, GTEx, and GEO data. J Clin Trials. Research Square Platform LLC. https://doi.org/10.21203/rs.3.rs-179363/v1
17. Chapla D, Chorya HP, Ishfaq L, Khan A, Vr S, Garg S (2024) An artificial intelligence (AI)-integrated approach to enhance early detection and personalized treatment strategies in lung cancer among smokers: a literature review. Cureus 16(8). https://doi.org/10.7759/cureus.66688
18. Chen Y, Gelfond J, Shireman PK, Mcmanus LM (2010) MiRNA expression patterns during skeletal muscle regeneration. FASEB J 24(S1). https://doi.org/10.1096/fasebj.24.1_supplement.568.5
19. Clancy C, Joyce MR, Kerin MJ (2015) The use of circulating microRNAs as diagnostic biomarkers in colorectal cancer. Cancer Biomark 15(2):103–113. https://doi.org/10.3233/cbm-140456
20. Darvish M, Trask R, Tallon P, Khansari M, Ren L, Hershman M, Yousefi B (2024) AI-enabled lung cancer prognosis. https://doi.org/10.48550/arxiv.2402.09476
21. Davey MG, Davies M, Kerin MJ, Lowery AJ, Miller N (2021) The role of MicroRNA as clinical biomarkers for breast cancer surgery and treatment. Int J Mol Sci 22(15):8290. https://doi.org/10.3390/ijms22158290
22. Do T-N, Lenca P, Pham N-K, Lallich S (2010) Classifying very-high-dimensional data with random forests of oblique decision trees. Springer, Berlin/Heidelberg, pp 39–55. https://doi.org/10.1007/978-3-642-00580-0_3
23. Farooq M, Shahid MU (2023) Explainable AI and machine learning algorithms to predict treatment failures for patients with cancer. J Clin Oncol 41(Suppl 16):e13577. https://doi.org/10.1200/jco.2023.41.16_suppl.e13577
24. Gan T-Q, Chen W-J, Yang L-H, Li Z-Y, Pan L-J, Chen G, Huang S-N, Qin H, Fang Y-Y (2017) Clinical value and prospective pathway signaling of MicroRNA-375 in lung adenocarcinoma:

a study based on the cancer genome atlas (TCGA), gene expression omnibus (GEO) and bioinformatics analysis. Med Sci Monitor: Int Med J Exp Clin Res 23(1):2453–2464. https://doi.org/10.12659/msm.901460
25. Gandhi Z, Lokhandwala A, Koshiya H, Gurram P, Dewaswala N, Desai R, Mohammed A, Ganti S, Bhopalwala H, Amgai B, Lekkala SP, Surani S, Manne S (2023) Artificial intelligence and lung cancer: impact on improving patient outcomes. Cancer 15(21):5236. https://doi.org/10.3390/cancers15215236
26. Greener M (2017) Systems biology: transforming our understanding of disease. Prescriber 28(12):30–32. https://doi.org/10.1002/psb.1634
27. Hamada M, Hassan M, Tanimu JJ, Kakudi HA, Robert P (2021) Evaluation of recursive feature elimination and LASSO regularization-based optimized feature selection approaches for cervical cancer prediction, vol 106, pp 333–339. https://doi.org/10.1109/mcsoc51149.2021.00056
28. Hocking T, Thibault G, Bodine C, Arellano P, Shenkin A, Lindly O (2024) SOAK: same/other/all K-fold cross-validation for estimating similarity of patterns in data subsets. https://doi.org/10.48550/arxiv.2410.08643
29. Jeon K, Park WY, Kahn CE, Nagy P, You SC, Yoon SH (2024) Advancing medical imaging research through standardization: the path to rapid development, rigorous validation, and robust reproducibility. Investig Radiol 60(1). https://doi.org/10.1097/rli.0000000000001106
30. Jeong Y, Ronen J, Kopp W, Lutsik P, Akalin A (2024) scMaui: a widely applicable deep learning framework for single-cell multiomics integration in the presence of batch effects and missing data. BMC Bioinfo 25(1). https://doi.org/10.1186/s12859-024-05880-w
31. Joković SM, Korać A, Filipović L, Kotarac N, Brajušković G, Dobrijević Z, Popović M, Savić-Pavićević D, Vuković I (2022) MiR-375 and miR-21 as potential biomarkers of prostate cancer: comparison of matching samples of plasma and exosomes. Genes 13(12):2320. https://doi.org/10.3390/genes13122320
32. Journal I (2024) Generative AI for healthcare: applications, challenges, and ethical considerations. Interantional Journal of Scientific Research in Engineering and Management 08(12):1–6. https://doi.org/10.55041/ijsrem39600
33. Kartik Kumar S, Venkata Sasidhar M (2024) From fluids to forecasts: the promise of small extracellular vesicle miRNAs in Revolutionising cancer diagnostics. Intechopen. https://doi.org/10.5772/intechopen.1005059
34. Katiyar A, Kaur G, Rani L, Kaur P, Singh H, Gupta R (2021) Genome-wide identification of potential biomarkers in multiple myeloma using meta-analysis of mRNA and miRNA expression data. Res Square Platform LLC. https://doi.org/10.21203/rs.3.rs-132765/v1
35. Kh Arabo W, Malallah M, O. (2022) The effect of data splitting methods on classification performance in wrapper-based gene-selection model. Acad J Nawroz Univ 11(4):284–293. https://doi.org/10.25007/ajnu.v11n4a1424
36. Khandagale P (2024) Emerging trends in drug discovery: harnessing artificial intelligence and machine learning for drug development. Innov Pharm Planet 12(03):56–61. https://doi.org/10.31690/ipplanet.2024.v012i03.016
37. Kislay K, Singh S, Joshi S, Dutta R, Flint J, Zhu K (2024) Evaluating K-fold cross validation for transformer based symbolic regression models. https://doi.org/10.48550/arxiv.2410.21896
38. Krishna Pasupuleti M (2024) Quantum AI and precision genomics: redefining cancer care and personalized medicine. National Education Services, pp 40–55. https://doi.org/10.62311/nesx/46080
39. Kuzyurin NN (1977) Minimal coverings and maximal packings of (k−1)-subsets by k-subsets. Math Notes Acad Sci USSR 21(4):316–320. https://doi.org/10.1007/bf01787658
40. Lai X, Wolkenhauer O, Vera J (2016) Understanding microRNA-mediated gene regulatory networks through mathematical modelling. Nucleic Acids Res 44(13):6019–6035. https://doi.org/10.1093/nar/gkw550
41. Li J (2024) Evaluative comparison of machine learning algorithms for precision diagnosis in breast cancer. Highlight Sci Eng Technol 85:354–362. https://doi.org/10.54097/40fmfw48
42. Li F, Wang S, Gao Z, Qing M, Pan S, Liu Y, Hu C (2025a) Harnessing artificial intelligence in sepsis care: advances in early detection, personalized treatment, and real-time monitoring. Front Med 11. https://doi.org/10.3389/fmed.2024.1510792

43. Liang W, Cui X, Lu S, Xu Y, Li C (2024) Genome-wide identification and characterization of miRNAs and natural antisense transcripts show the complexity of gene regulatory networks for secondary metabolism in Aristolochia contorta. Int J Mol Sci 25(11):6043. https://doi.org/10.3390/ijms25116043
44. Liu X-G, Zhou S-Q, Huang Y-Y, Zhang Y-K, Zeng F, Zhou J-H, Wang Y-K, Ma L-N, Zhu W-Y (2011) High expression of serum miR-21 and tumor miR-200c associated with poor prognosis in patients with lung cancer. Med Oncol 29(2):618–626. https://doi.org/10.1007/s12032-011-9923-y
45. Lococo F, Evangelista J, Stefani A, Martino A, Boldrini L, Flamini S, Campanella A, Mohammed A, Ghaly G, Chiappetta M, Bria E, Margaritora S, Sassorossi C, Vita E (2024) Implementation of artificial intelligence in personalized prognostic assessment of lung cancer: a narrative review. Cancer 16(10):1832. https://doi.org/10.3390/cancers16101832
46. Lou RN, Therkildsen NO (2021) Batch effects in population genomic studies with low-coverage whole genome sequencing data: causes, detection, and mitigation. Authorea. https://doi.org/10.22541/au.162791857.78788821/v1
47. Lv Y, Huang Z, Lin Y, Fang Y, Chen Z, Pan L, Zhang Y, Xu Z (2019) MiRNA expression patterns are associated with tumor mutational burden in lung adenocarcinoma. Onco Targets Ther 8(10):e1629260. https://doi.org/10.1080/2162402x.2019.1629260
48. Ma T, Zhang A (2019) Integrate multi-omics data with biological interaction networks using multi-view factorization AutoEncoder (MAE). BMC Genomics 20(Suppl 11). https://doi.org/10.1186/s12864-019-6285-x
49. Mahfouz A, Mathe A, Ali A, Ahmed A, Gasmalla A (2025) Artificial intelligence in drug discovery and personalized medicine: transforming the future of pharmaceutical research. Int J Sci Res Arch 14(2):1394–1406. https://doi.org/10.30574/ijsra.2025.14.2.0495
50. Mahmud T, Sharmen N, Habiba SU, Hossain MS, Andersson K, Barua K (2024) An explainable AI paradigm for Alzheimer's diagnosis using deep transfer learning. Diagnostics 14(3):345. https://doi.org/10.3390/diagnostics14030345
51. Martínez-Espinosa I, Ortiz-Quintero B, Cabello-Gutiérrez C, Serrato JA, Carlos-Reyes Á (2024) Exosome-derived miRNAs in liquid biopsy for lung cancer. Life (Basel, Switzerland) 14(12):1608. https://doi.org/10.3390/life14121608
52. Mary PNV, Nagarajan R (2024) Feature selection in microarray using proposed hybrid minimum redundancy-maximum relevance (MRMR) and modified genetic algorithm (MGA). Int J Exp Res Rev 39(Spl Volume):82–91. https://doi.org/10.52756/ijerr.2024.v39spl.006
53. Mohammadi L, Alizadeh E, Imani M, Rahmati M, Mosayyebi B, Zarghami N (2020) Correlation between dexamethasone and miRNAs in the regulation of apoptosis, drug-resistance, and metastasis of cancer cell. Curr Mol Med 21(5):392–401. https://doi.org/10.2174/1566524020666200925155614
54. Morin K, Davis JL (2016) Cross-validation: what is it and how is it used in regression? Commun Stat - Theory Methods 46(11):5238–5251. https://doi.org/10.1080/03610926.2015.1099672
55. Muthamilselvan S, Palaniappan A, Baabu RS, P. (2023) Microfluidics for profiling miRNA biomarker panels in AI-assisted cancer diagnosis and prognosis. Technol Cancer Res Treat 22. https://doi.org/10.1177/15330338231185284
56. Namkung J, Kang MJ, Kim S, Kwon W, Park T, Han S, Choi Y, Jang J, Yi SG (2016) Molecular subtypes of pancreatic cancer based on miRNA expression profiles have independent prognostic value. J Gastroenterol Hepatol 31(6):1160–1167. https://doi.org/10.1111/jgh.13253
57. Nand Kumar EA (2023) Enhancing robustness and generalization in deep learning models for image processing. Power Sys Technol 47(4):278–293. https://doi.org/10.52783/pst.193
58. Niyazi M, Atkinson M, Schulze-Osthoff K, Niemöller OM, Tonn J-C, Mörtl S, Zehentmayr F, Kretzschmar H, Eigenbrod S, Belka C (2011) MiRNA expression patterns predict survival in glioblastoma. Radiat Oncol 6(1). https://doi.org/10.1186/1748-717x-6-153
59. Nwankwo E, Olaboye J, Maha C, Emeihe E, Ajegbile M (2024) AI in personalized medicine: enhancing drug efficacy and reducing adverse effects. Int Med Sci Res J 4(8):806–833. https://doi.org/10.51594/imsrj.v4i8.1453

60. Nygaard V, Rødland EA, Hovig E (2015) Methods that remove batch effects while retaining group differences may lead to exaggerated confidence in downstream analyses. Biostatistics 17(1):29–39. https://doi.org/10.1093/biostatistics/kxv027
61. Ocłoń E, Latacz A, Pierzchała-Koziec K, Zubel-Łojek J (2016) Hyperglycemia-induced changes in miRNA expression patterns in epicardial adipose tissue of piglets. J Endocrinol 229(3):259–266. https://doi.org/10.1530/joe-15-0495
62. Odah M (2024) Artificial intelligence (AI) and machine learning (ML) in diagnosing cancer: current trends. MDPI AG. https://doi.org/10.20944/preprints202403.0433.v1
63. Paliwal R (2024) Applications of AI (artificial intelligence) in diagnosis of breast cancer. Int J Multidiscipl Res 6(5). https://doi.org/10.36948/ijfmr.2024.v06i05.27787
64. Redova M, Slaby O, Sana J (2013) Circulating miRNAs as new blood-based biomarkers for solid cancers. Future Oncol 9(3):387–402. https://doi.org/10.2217/fon.12.192
65. Riazi Esfahani P, Patel R, Min M, Guirgus ME, Mazboudi P, Reddy AJ, Tak N, Farasat VP, Arakji GH (2023) Leveraging machine learning for accurate detection and diagnosis of melanoma and nevi: an interdisciplinary study in dermatology. Cureus 15(8). https://doi.org/10.7759/cureus.44120
66. Sabre L, Punga AR, Punga T (2020) Circulating miRNAs as potential biomarkers in myasthenia gravis: tools for personalized medicine. Front Immunol 11(e00653). https://doi.org/10.3389/fimmu.2020.00213
67. Sekhar G (2024) Advanced AI and machine learning techniques and algorithms in cancer research. Center for Open Science. https://doi.org/10.31219/osf.io/7skz3
68. Shah SAH, Shah STH, Khaled R, Buccoliero A, Shah SBH, Di Terlizzi A, Di Benedetto G, Deriu MA (2024) Explainable AI-based skin cancer detection using CNN, particle swarm optimization and machine learning. J Imaging 10(12):332. https://doi.org/10.3390/jimaging10120332
69. Sharma A (2023) AI powered identification of drug targets and pathways for diagnosis and treatment planning: a review. Nanomed Nanotechnol Open Access 8(2):1–6. https://doi.org/10.23880/nnoa-16000232
70. Shiroud Heidari B, Alihosseini A, Khamani S, Davachi SM, Hedayati Moghaddam A (2019) Optimization of process parameters in plastic injection molding for minimizing the volumetric shrinkage and warpage using radial basis function (RBF) coupled with the k-fold cross validation technique. J Polym Eng 39(5):481–492. https://doi.org/10.1515/polyeng-2018-0359
71. Shittu L, Olutimehin D, Ekundayo F (2024) Cancer diagnosis and prognosis using multi-omics data: a data science and machine learning approach. World J Adv Res Rev 24(3):2052–2069. https://doi.org/10.30574/wjarr.2024.24.3.3876
72. Singh R, Knight R (2009) Systems biology: understanding function from genes to networks. Bentham Sci 6(2):93–103. https://doi.org/10.2174/157016409788680974
73. Singh V, Song B-S, Lee S, Kirtipal N (2023) Normalization of RNA-Seq data using adaptive trimmed mean with multi-reference. Cold Spring Harbor Laboratory. https://doi.org/10.1101/2023.12.04.570016
74. Singh D, Singh L, Sachdeva D (2024) Advancing breast cancer drug delivery: the transformative potential of bioinformatics and artificial intelligence. Curr Cancer Therapy Rev 20. https://doi.org/10.2174/0115733947287709240229104857
75. Statnikov A, Wang L, Aliferis CF (2008) A comprehensive comparison of random forests and support vector machines for microarray-based cancer classification. BMC Bioinfo 9(1). https://doi.org/10.1186/1471-2105-9-319
76. Steging C, Renooij S, Verheij B (2021) Rationale discovery and explainable AI. Legal knowledge and information systems. In: Schweighofer E (ed) JURIX 2021: the thirty-fourth annual conference. IOS Press, Amsterdam, pp 225–234. https://doi.org/10.3233/faia210341
77. Sturm M, Schroeder C, Bauer P (2016) SeqPurge: highly-sensitive adapter trimming for paired-end NGS data. BMC Bioinfo 17(1). https://doi.org/10.1186/s12859-016-1069-7
78. Sun M, Cui C (2023) Transforming lung cancer care: synergizing artificial intelligence and clinical expertise for precision diagnosis and treatment. AIMS Bioengineering 10(3):331–361. https://doi.org/10.3934/bioeng.2023020

79. Tan IL, Castillejo G, Modderman R, Jonkers I, Weersma RK, Szajewska H, Shamir R, Withoff S, Barisani D, Wijmenga C, Stachurska A, Roca M, Troncone R, Zhernakova A, Visschedijk MC, Auricchio R, Kumar V, Koletzko S, Dekens J et al (2021) Circulating miRNAs as potential biomarkers for celiac disease development. Front Immunol 12(38). https://doi.org/10.3389/fimmu.2021.734763
80. Tanveer H, Ali MA, Khan MA, Shakoor A (2024) Analyzing the performance and efficiency of machine learning algorithms, such as deep learning, decision trees, or support vector machines, on various datasets and applications. Asian Bullet Big Data Manag 3(2):126–136. https://doi.org/10.62019/abbdm.v3i2.83
81. Tom JA, Graham RR, Bhangale TR, Hunkapiller J, Behrens TW, Reeder J, Forrest WF (2017) Identifying and mitigating batch effects in whole genome sequencing data. BMC Bioinfo 18(1). https://doi.org/10.1186/s12859-017-1756-z
82. Vogel R, Mück B (2024) Artificial intelligence-what to expect from machine learning and deep learning in hernia surgery. J Abdominal Wall Sur: JAWS 3. https://doi.org/10.3389/jaws.2024.13059
83. Wang J, Tao Y, Bian Q (2021) miRNA and mRNA expression profiling reveals potential biomarkers for metastatic cutaneous melanoma. Expert Rev Anticancer Ther 21(5):557–567. https://doi.org/10.1080/14737140.2021.1882860
84. Wu W, Maier HR, Dandy GC, May RJ (2013) A benchmarking approach for comparing data splitting methods for modeling water resources parameters using artificial neural networks. Water Resour Res 49(11):7598–7614. https://doi.org/10.1002/2012wr012713
85. Xu Y, Goodacre R (2018) On splitting training and validation set: a comparative study of cross-validation, bootstrap and systematic sampling for estimating the generalization performance of supervised learning. J Anal Test 2(3):249–262. https://doi.org/10.1007/s41664-018-0068-2
86. Yang S, Wang Z, Wang C, Li C, Wang B (2024) Comparative evaluation of machine learning models for subtyping triple-negative breast cancer: a deep learning-based multi-omics data integration approach. J Cancer 15(12):3943–3957. https://doi.org/10.7150/jca.93215
87. Yerukala Sathipati S, Ho S-Y (2017) Identifying the miRNA signature associated with survival time in patients with lung adenocarcinoma using miRNA expression profiles. Sci Rep 7(1). https://doi.org/10.1038/s41598-017-07739-y
88. Yin Y, Shan HQ, Huang W, Lu H, Jin Y, Wu YM (2014) Bioinformatic analysis of miRNA expression patterns in TFF2 knock-out mice. Genet Mol Res 13(4):8502–8510. https://doi.org/10.4238/2014.october.20.26
89. Yousef M, Allmer J, Trinh H (2014) Intersection of MicroRNA and gene regulatory networks and their implication in cancer. Curr Pharm Biotechnol 15(5):445–454. https://doi.org/10.2174/1389201015666140519120855
90. Yu C, Kobeissy F (2015) Systems biology application to decipher mechanisms and novel biomarkers in CNS trauma. CRC Press, pp 448–461. https://doi.org/10.1201/b18126-37
91. Yuan C, Gao Y, Cao K, Zhang N, Bai R, Xiang L, Xie C, Gong Y, Jiang X (2019) MiR-195 restrains lung adenocarcinoma by regulating CD4+ T cell activation via the CCDC88C/Wnt signaling pathway: a study based on the cancer genome atlas (TCGA), gene expression omnibus (GEO) and bioinformatic analysis. Annals Trans Med 7(12):263. https://doi.org/10.21037/atm.2019.05.54
92. Zamanitajeddin N, Jahanifar M, Xu K, Siraj F, Rajpoot N (2024) Benchmarking domain generalization algorithms in computational pathology. https://doi.org/10.48550/arxiv.2409.17063
93. Zhang T (2025) Integrating QSP and ML to facilitate drug development and personalized medicine. Handb Exp Pharmacol. https://doi.org/10.1007/164_2024_740
94. Zhang B, Shi H, Wang H (2023) Machine learning and AI in cancer prognosis, prediction, and treatment selection: a critical approach. J Multidiscip Healthc 16:1779–1791. https://doi.org/10.2147/jmdh.s410301
95. Zhao Y, Wong L, Goh WWB (2020) How to do quantile normalization correctly for gene expression data analyses. Sci Rep 10(1). https://doi.org/10.1038/s41598-020-72664-6

Chapter 24
Exosomal MicroRNAs in Cancer: Bridging Mechanistic Insights with Biomarker Potential

Sarwareddy Kartik Kumar, Supriya Thalakanti, Patnam Sreekanth, Neeharika Kannaparthi, and Manda Venkata Sasidhar

Abstract Exosome-derived microRNAs (Exo-miRNAs) have emerged as key signalling molecules in cancer progression, serving as a key link between mechanistic understanding and translational applications. These miRNAs, encapsulated within exosomes, are functionally active and play instrumental roles in various cancer hallmarks, including cell proliferation, immune evasion, angiogenesis, EMT, and metastasis. Compared to their free counterparts in blood, saliva, urine, and CSF, Exo-miRNAs exhibit enhanced stability, making them promising non-invasive biomarkers. Their cancer-specific expression patterns have been documented across multiple tumour types such as lung, breast, colorectal, brain, and prostate cancers. Moreover, alterations in exo-miRNA levels have been correlated with drug resistance, as validated by in vitro and in vivo functional studies. Therefore, clinically, exo-miRNAs have shown potential in differentiating between healthy and cancerous states, metastatic versus non-metastatic tumours, and drug-resistant versus responsive cases and are even associated with patient survival outcomes. Despite their promise, gaps remain in understanding their comprehensive functional roles, particularly in integrating diagnostic and prognostic value. This chapter provides a comprehensive overview of exo-miRNAs in cancer from their roles in tumour progression to their clinical utility. Furthermore, we highlight the synergistic value of

S. K. Kumar · S. Thalakanti · P. Sreekanth
Apollo Hospitals Educational and Research Foundation, Hyderabad, Telangana, India

N. Kannaparthi
Sephirah innovations Pvt Ltd, Apollo Hospitals Educational and Research Foundation, Hyderabad, Telangana, India

M. V. Sasidhar (✉)
Apollo Hospitals Educational and Research Foundation, Hyderabad, Telangana, India

Urvogelbio Private Ltd, Apollo Hospitals Educational and Research Foundation, Hyderabad, Telangana, India
e-mail: sasidhar@aherf.net

S. Pathak, A. Banerjee (eds.), *MicroRNAs: Role in Cancer Progression, Diagnostics and Therapy*, Advances in Biochemistry in Health and Disease 34,
https://doi.org/10.1007/978-3-032-08824-6_24

combining exo-miRNAs with traditional biomarkers like cfDNA, CTCs, and protein markers to enhance diagnostic sensitivity and specificity.

Keywords Exosomes · Cancer · Metastasis · Diagnosis · Prognosis

Introduction

Exosomes are small extracellular vesicles (30–150 nm in size) that have been recognized as vital mediators of intercellular communication in both physiological and pathological conditions. Extracellular vesicles are generally classified into three types: apoptotic bodies, microvesicles, and exosomes. Among these, exosomes are uniquely formed through the endosomal pathway and carry a variety of biomolecular cargo including proteins, lipids, mRNAs, and microRNAs (miRNAs) that reflect the physiological or pathological state of their parent cells [1, 2]. Notably, miRNAs, small non-coding RNAs that regulate gene expression, are selectively packaged into exosomes and can modulate tumour biology by influencing recipient cell behaviour [3, 4]. Studies have demonstrated that tumour-derived exosomal miRNAs contribute to cancer progression by promoting immune evasion, angiogenesis, epithelial-mesenchymal transition (EMT), and metastasis [5–8]. Their presence in various biological fluids such as blood, saliva, urine, and cerebrospinal fluid (CSF) along with cargo stability conferred by encapsulation within a protective lipid bilayer makes them highly attractive candidates for cancer diagnosis, prognosis, and targeted drug delivery [9–12].

Numerous studies have shown that exosomal miRNAs can function either as oncogenes or tumour suppressors, depending on the specific miRNA and the signalling pathways they affect [13–16]. For instance, miR-21, miR-155, and miR-1246 have been associated with tumour progression by targeting key tumour suppressors such as *PTEN, TP53*, and *DUSP3* [15, 17, 18]. Conversely, miR-143, miR-145, and miR-34a have demonstrated tumour-suppressive functions [19–21]. Clinical studies using high-throughput sequencing and subsequent validation in larger cohorts have identified disease-specific expression signatures, establishing a strong link between exosomal miRNA profiles and tumour stage, treatment response, recurrence, and patient survival [22–24]. Moreover, ongoing research efforts are aiming to enhance the diagnostic and prognostic utility of exosomal miRNAs by integrating them with other established biomarkers, including cell-free DNA (cfDNA), circulating tumour cells (CTCs), and serum proteins [25, 26]. Thus, understanding the mechanistic and translational relevance of exosomal miRNAs is critical for improving cancer diagnostics and patient outcomes. In this chapter, we highlight the functional roles of exosomal miRNAs in cancer progression and explore their clinical potential as biomarkers for diagnosis, prognosis, and therapeutic monitoring.

Exosome Biogenesis and miRNA Packaging

Exosome biogenesis begins in the endosomal pathway, where multivesicular bodies (MVBs) form intraluminal vesicles (ILVs) carrying proteins, lipids, and RNAs. MVBs either fuse with lysosomes for degradation or the plasma membrane to release exosomes. This process is regulated by factors such as lipid composition, signalling molecules, and Rab proteins. For example, Rab27a and Rab27b are required for MVB binding at the plasma membrane to support exosome secretion, whereas Rab7 directs MVBs to lysosomes for destruction [12].The central component of the biogenesis is the Endosomal Sorting Complex Required for Transport (ESCRT), composed of four conserved multi-protein complexes, ESCRT-0, ESCRT-I, ESCRT-II, and ESCRT-III, along with auxiliary proteins like ALIX (ALG-2-interacting protein X) and ATPase VPS4(2). ESCRT-0, made up of Hrs and STAM1/2, binds and aggregates ubiquitinated transmembrane proteins, recruiting them into ILVs. ESCRT-I and ESCRT-II assist membrane budding by stabilizing inward curvature, involving proteins such as TSG101, a known exosomal marker. ESCRT-III, the final complex, drives membrane scission by forming helical filaments that pinch off budding ILVs into the MVB lumen. VPS4 recycles ESCRT-III subunits, sustaining the sorting machinery. ALIX acts as a bridge between ESCRT-III, membrane cargo, and lipids, while also connecting to the syntenin-syndecan complex, integrating ESCRT-dependent and -independent pathways [27–29]. These mechanisms enable selective molecular packaging into exosomes, highlighting the complexity of their biogenesis and release.

The sorting of miRNAs into exosomes is another highly selective and regulated process involving RNA-binding proteins (RBPs), post-translational modifications, and lipid signalling pathways. One well-characterized RBP is heterogeneous nuclear ribonucleoprotein A2B1 (hnRNPA2B1), which recognizes specific sequences on target miRNAs, directing them into exosomes. Sumoylation enhances hnRNPA2B1's affinity for miRNAs such as miRNA-198, supporting their selective exosomal loading, a mechanism particularly relevant in cancer, where selective export of oncogenic miRNAs shapes the tumour microenvironment [30]. Similarly, Y-box protein 1 (YBX1) binds UCAGU-rich sequences on miRNAs, playing a critical role in their exosomal inclusion. Functional loss of YBX1 disrupts the miRNA composition of exosomes, confirming its key role in sorting [31]. Another significant player is Argonaute 2 (AGO2), a core RISC component, which has been identified in exosomes and contributes to miRNA sorting through the KRAS-MEK-ERK signalling pathway [32].

Beyond protein-mediated mechanisms, lipid signalling pathways also influence miRNA packaging. The ceramide pathway, regulated by Neural Sphingomyelinase 2 (nSMase2), promotes ILV formation and miRNA sorting into exosomes. Inhibition of nSMase2 reduces exosomal secretion of miR-16 and miR-146a, while its overexpression enhances their release, highlighting ceramide role in cargo selection [33]. Vacuolar Protein Sorting-Associated Protein 4 (Vps4A) is another key regulator. Studies by Jin et al. and Charles et al. reported that Vps4A overexpression boosts exosomal secretion of miR-27b-3p and miR-92a-3p, whereas its inhibition lowers miR-92a and miR-150 levels [34, 35]. Caveolin-1 (Cav-1), a structural protein in caveolae, also regulates membrane trafficking and miRNA sorting into exosomes. Under stress, Cav-1 enhances hnRNPA2B1 and its associated miRNA release into microvesicles, and its presence in miRNA-rich exosomes suggests a role in selective miRNA transport [36].

Despite advances, the precise mechanisms underlying miRNA sorting into exosomes, especially in cancer, remain incomplete. Evidence indicates cancer-specific pathways regulate this process. For example, Cha et al. (2015) demonstrated that *KRAS* mutation status influences miRNA sorting in colorectal cancer. Their study revealed that miR-10b is enriched in exosomes from wild-type cells, while miR-100 predominates in *KRAS*-mutant cells (12). Inhibiting nSMase2 in *KRAS*-mutant cells led to intracellular accumulation of miR-100, supporting a KRAS-dependent sorting mechanism [37]. Similarly, Sonia et al.'s study on breast cancer (BC) exosomes revealed the selective enrichment of Dicer, facilitated by *CD43*, for precursor miRNA processing inside exosomes [38]. Temoche et al. (2019) further highlighted both active and passive miRNA sorting, identifying Lupus La protein as a critical regulator of miRNA loading into cancer-derived exosomes [39].

These findings highlight the complex and context-dependent nature of miRNA packaging into exosomes, with significant implications for diagnostics and therapeutics. However, the underlying biological reasons for this selective packaging, especially in cancer, remain unclear. Whether miRNA sorting into exosomes serves primarily as waste disposal or as a mechanism to regulate key biological pathways is still under debate. A deeper understanding of this process is crucial, as its role in tumour growth and progression remains an open and important question.

Exosome Isolation and miRNA Detection

The method of exosome isolation plays a critical role in determining the quality and integrity of EV-associated miRNAs, which are predominantly used to study differential expression patterns such as upregulation or downregulation in disease contexts. Several EV isolation techniques have been reported in the literature, including precipitation, ultracentrifugation, density gradient centrifugation, and immunoaffinity-based methods [2]. Among these, commercial isolation kits based on precipitation, including total exosome isolation kits for serum, plasma, saliva, and urine, are widely adopted in biomarker studies due to their time efficiency, good yield, and reliable quality, which are particularly advantageous in diagnostic applications [40, 41]. Immunoaffinity-based isolation techniques utilize antibodies targeting exosome-specific surface markers like *CD63, CD81*, and *CD9*, offering high-purity vesicle populations [42]. For instance, Shimeng et al. demonstrated CD63-based immunocapture of exosomes, revealing that EpCAM-positive and HER2-positive exosome levels in plasma correlated well with tissue immunohistochemistry (IHC) findings in BC patients [43]. Interestingly, researchers have also developed tumour-specific immunocapture methods using tumour-associated antigens to distinguish tumour-derived exosomes from those of non-tumour origin, thereby enhancing diagnostic specificity [44]. A notable example is the study by Priyanka et al. (2018), who isolated melanoma-derived exosomes (MTEX) from melanoma patient plasma using tumour-associated antigens [45]. In parallel, microfluidic technologies have emerged as innovative platforms for EV isolation. Byeonggeol et al. (2023) introduced a magnetically labelled exosome isolation system (MEIS-chip) designed to selectively capture HER2-positive exosomes while minimizing the presence of non-target EVs, providing a promising tool for both diagnosis and treatment monitoring [46]. However, it remains unclear how each isolation method may specifically impact the miRNA content and profile of the isolated exosomes. Despite these advancements, it remains unclear how different isolation methods may impact the miRNA content and expression profiles of isolated exosomes. While such technologies pave the way for promising point-of-care applications in cancer diagnostics, further standardization and optimization are essential for their successful translation into clinical practice.

Once exosomes are isolated, miRNA detection methods are employed to analyse their content for cancer diagnostics. Quantitative PCR (qPCR) is a widely used technique that provides high sensitivity and specificity for detecting known miRNAs [47, 48]. However, it requires prior knowledge of target miRNAs. Microarrays allow the simultaneous screening of multiple miRNAs but have a limited dynamic range [49]. RNA sequencing (RNA-Seq) is a powerful technique that provides a comprehensive, unbiased analysis of exosomal miRNA profiles, though it is costly and computationally intensive [23]. Digital droplet PCR (ddPCR) offers absolute quantification of miRNAs with improved precision over qPCR and is increasingly utilized in cancer biomarker studies [50]. Emerging biosensor-based techniques, incorporating electrochemical, optical, or nanotechnology-based platforms, provide rapid and highly sensitive miRNA detection in biofluids, offering potential for real-time cancer diagnostics [51–53]. The integration of these miRNA detection methods with artificial intelligence (AI) and machine learning further enhances their diagnostic accuracy, paving the way for personalized cancer diagnostics and treatment strategies. Standardizing these methodologies will be critical for translating exosomal miRNA-based assays into routine clinical practice. The general workflow for utilizing exosomal miRNAs in cancer diagnosis and prognosis is illustrated in Fig. 24.1, outlining key steps from isolation to clinical interpretation.

Overview of Exosomal miRNAs in Various Cancers

Exosomal miRNAs have emerged as key players in the regulation of cancer biology, profoundly influencing tumour initiation, progression, metastasis, and therapeutic resistance across multiple malignancies [4]. In lung cancer, several exosomal miRNAs such as miR-21, miR-125b-5p, miR-126, and miR-5684 have been identified to drive oncogenic pathways. Among them, miR-21 is notably upregulated in

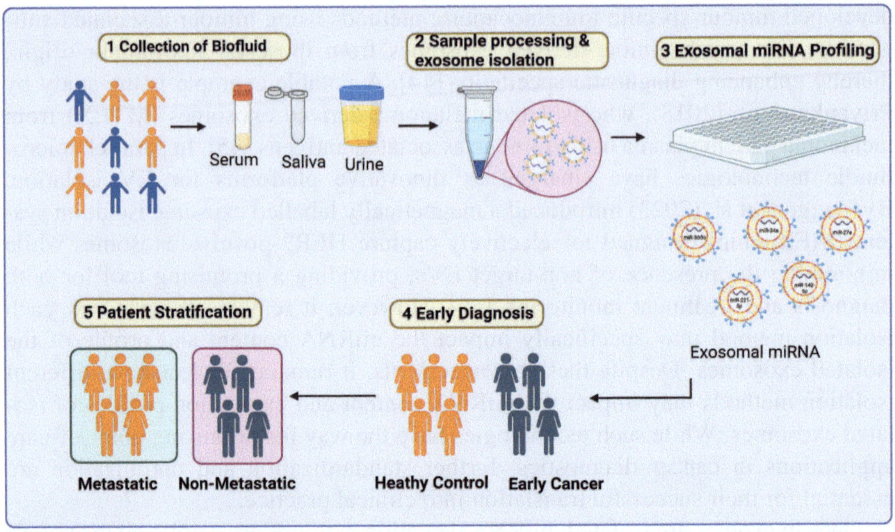

Fig. 24.1 Experimental workflow for utilizing exosomal mirnas in clinical cancer applications. This figure outlines the stepwise process starting from patient biofluid collection, followed by exosome isolation and miRNA profiling. Differential expression patterns of exosomal miRNAs enable distinction between healthy controls and cancer patients, as well as stratification of metastatic versus non-metastatic cases for improved diagnosis and prognosis

non-small cell lung cancer (NSCLC) and has been validated as a potential diagnostic and prognostic biomarker due to its strong correlation with tumour progression and chemoresistance [54–56]. In colorectal cancer (CRC), exosomal miRNAs also play a prominent role in modulating tumour behaviour. For example, miR-19a is significantly overexpressed in the serum exosomes of CRC patients and is associated with poor clinical outcomes [57]. Another key miRNA, miR-92a-3p, has been shown to activate the Wnt/β-catenin signalling pathway, thereby promoting tumour cell proliferation and inhibiting apoptosis [58]. Moreover, exosomal miRNAs such as miR-934, let-7a, miR-21, miR-23a, and miR-1229 have been implicated in shaping the tumour microenvironment (TME) by regulating macrophage polarization, a critical step in cancer metastasis [59, 60]. Notably, low levels of exosomal miR-4772-3p have been correlated with tumour recurrence in stage II and III CRC patients, making it a promising marker for disease monitoring and prognosis [61].

In brain tumours such as glioma, exosomal miRNAs have been increasingly recognized for their role in modulating the tumour microenvironment and enhancing tumour aggressiveness. Glioma-derived exosomes are thought to facilitate cellular communication within the TME by transferring oncogenic miRNAs to neighbouring cells [62, 63]. This intercellular exchange contributes to immune evasion, angiogenesis, and resistance to standard therapies. Several miRNAs, including miR-21, miR-301a, miR-210, and miR-222, have been found to be consistently upregulated in the plasma, serum, and CSF of glioma patients, suggesting their potential utility as non-invasive diagnostic biomarkers [63, 64]. In breast cancer, a wide spectrum of exosomal miRNAs has been reported, reflecting the heterogeneity of the disease. For instance, miR-16 and miR-150 are frequently dysregulated in BC patients [65]. Notably, miR-7641 expression is significantly higher in plasma exosomes from patients with distant metastasis compared to those without metastasis, highlighting its potential as a marker for disease progression [66]. A study by Aiko et al. (2017) identified a comprehensive panel of miRNAs associated with BC recurrence. In this panel, miR-338-3p, miR-340-5p, and miR-124-3p were found to be upregulated, while miR-29b-3p, miR-20b-5p, miR-17-5p, miR-130a-3p, miR-18a-5p, miR-195-5p, miR-486-5p, and miR-93-5p were downregulated in patients who experienced recurrence. Interestingly, these expression patterns were reversed in patients who remained disease-free, indicating their strong prognostic relevance and potential use in patient stratification [67].

In gastrointestinal and urogenital cancers, exosomal miRNAs have shown promise in both diagnosis and post-treatment monitoring. In gastric cancer, elevated levels of miR-21 and miR-1225-5p in peritoneal lavage fluid and malignant ascites have been linked to peritoneal dissemination and recurrence following curative surgery, indicating their predictive potential [68]. Similarly, in pancreatic cancer, upregulated exosomal miRNAs such as miR-1246, miR-4644, and miR-3976 have demonstrated a higher diagnostic accuracy than traditional serum markers like CA19-9, making them valuable tools for early detection [69]. In prostate cancer, exosomal miR-1246, miR-423-5p, and miR-1290 have been associated with therapy resistance and poor overall survival [70, 71]. Additionally, exosomal miR-21 has demonstrated strong potential in detecting aggressive forms of prostate cancer with high specificity and sensitivity [72]. Collectively, these findings underscore the multifaceted roles of exosomal miRNAs in cancer biology. Their involvement in oncogenic signalling, immune modulation, and therapy response not only enhances our understanding of cancer pathogenesis, but also supports their use in liquid biopsies for non-invasive cancer diagnostics. As a result, exosomal miRNAs hold substantial promise as integral components of precision oncology, offering new avenues for early detection, treatment stratification, and disease monitoring across various cancer types.

Functional Significance of Exosome miRNAs in Cancer Initiation and Progression

Exosome-transported miRNAs have emerged as key mediators of tumour progression by promoting cancer cell proliferation, a primary factor contributing to reduced patient survival. Numerous studies using both in vitro and in vivo models have shown that exosomal miRNAs facilitate uncontrolled growth in various cancer types [73–75]. For example, Xiu et al. (2017) demonstrated that exosomal miR-1246 from MDA-MB-231, BC cells enhanced cell viability [76], while Qian et al. (2018) reported a similar proliferative effect mediated by exosomal miR-148a through suppression of *CADM1* in T98G glioblastoma cells [77]. In lung cancer, overexpression of exosomal miR-499a-5p significantly stimulated tumour cell proliferation [78]. Yao et al. (2018) highlighted the role of paracrine exosomal miR-34a-5p in oral cancer xenograft models, where its levels were found to be lower in cancer-associated fibroblasts (CAFs); however, restoring its expression notably inhibited tumour growth, as evidenced by decreased cell viability and colony formation [79]. Moreover, Ming et al. (2020) provided novel insight by showing that even dying cancer cells release exosomal miR-194-5p under radiotherapy, which supports tumour repopulation by activating DNA damage response mechanisms [80]. Additional studies have reinforced the role of exosomal miRNAs in sustaining cancer cell proliferation, with examples including miR-1228-5p in small-cell lung cancer [81], miR-103a-3p in nasopharyngeal carcinoma [82], miR-20a-5p in prostate cancer [83], miR-200b in CRC [84], and miR-6876-5p in hepatocellular carcinoma (HCC) [85] through various signalling pathways. These findings collectively underscore the critical role of exosomal miRNAs in promoting cancer cell proliferation and stemness, positioning them as valuable targets for therapeutic strategies aimed at halting tumour growth and improving patient outcomes. Table 24.1 provides a comprehensive overview of functionally relevant exosomal miRNAs implicated in tumour initiation and progression across various cancer types.

Angiogenesis and immune escape are critical events that drive rapid tumour growth, leading to metastasis and poor clinical outcomes [102]. Emerging evidence highlights the role of exosomal miRNAs in orchestrating these processes, thereby extending their impact beyond tumour initiation to progression. For instance, exosomal miR-27a has been shown to promote angiogenesis in pancreatic cancer by targeting B-cell translocation gene 2 [96]. Similarly, studies in bladder cancer, melanoma, and HCC have identified miR-152-3p, miR-424-5p, and miR-210 as potent inducers of angiogenesis, validated through both in vitro and in vivo models [93–95]. In the context of immune escape, Xiaoli et al. (2020) demonstrated that exosomal miR-27a-3p facilitates immune evasion in BC via the MAGI2/PTEN/PI3K pathway by upregulating PD-L1 [90]. Further, Yu et al. (2021) and Shiquan et al. (2021) revealed that exosomal miR-155-5p and miR-222 contribute to immune suppression in colorectal cancer [103, 104]. Collectively, these findings emphasize the multifaceted roles of exosomal miRNAs in tumour progression and metastasis.

Exosomal miRNAs play a pivotal role in mediating cancer metastasis through the modulation of tumour microenvironment, immune evasion, and epithelial-mesenchymal transition (EMT) [105]. Several miRNAs, including miR-25-3p, miR-130b-3p, and miR-425-5p, are significantly upregulated in exosomes and are known to promote liver metastasis by inducing M2 polarization of macrophages, which in turn enhances cancer cell invasion via EMT and vascular endothelial growth factor (VEGF) secretion [92]. Liver is a common metastatic site for colorectal cancer (CRC), and exosomal miR-106b-3p has been implicated in this process by downregulating DLC-1, a tumour suppressor gene that restrains cancer cell migration and invasion [106]. Among the extensively studied oncogenic miRNAs, miR-21 stands out due to its frequent overexpression in exosomes from BC, lung cancer, and CRC. It facilitates proliferation, invasion, and resistance to

Table 24.1 Summary of functionally relevant exosomal miRNAs across various cancer types and their roles in tumour growth, progression, and therapy resistance

Cancer type	Exo-miRNA	Functional significance	References
Lung	miR-499a-5p, miR-1228-5p	Tumour growth	[78, 81]
	miR-194-5p	Tumour invasion	[80]
	miRNA-21-3p, miRNA-125b-5p, miRNA-181d-5p	Drug resistance	[86]
Breast	miR-21	Tumour growth	[87]
	miR-10b	Metastasis	[88]
	miR-9-5p	Drug resistance	[89]
	miR-27a-3p	Immune escape	[90]
Colorectal	miR-200b	Cell progression and metastasis	[84]
	miR-4326, miR-3615, miR-7974, miR-130b-3p, and miR-454-3p	Therapy resistance	[91]
Hepatocellular	miR-25-3p, miR-130b-3p, and miR-425-5p, miR-378b, miR-23a	Tumour metastasis and immune escape	[92]
	miR-152-3p, miR-424-5p, miR-210	Promotes angiogenesis	[93–95]
Prostate	miR-20a-5p	Cancer progression and metastasis	[83]
Brain	miR-1246	Promotes tumour progression	[17]
Pancreatic	miR-27a, miR-155	Promote tumour angiogenesis	[96, 97]
Ovarian	miR-205, miR-125b-5p, miR-222-3p, miR-21-3p, miR-181d-5p, miR-940	Promote cancer initiation, development, and angiogenesis	[98]
Bladder	miR-1247-3p	Promotes angiogenesis	[99]
Oral	miR-1246	Supports metastasis	[100]
Cervical	miR-221-3p	Promotes proliferation, migration, and angiogenesis	[101]

apoptosis by targeting tumour suppressors like *PTEN* and *PDCD4*. Moreover, it fosters a pro-tumourigenic microenvironment by activating fibroblasts and inducing inflammatory signals [87, 107, 108]. Intriguingly, miR-1246 has gained increasing attention for its strong association with invasion and metastasis in various cancers [109]. Studies by Xiaolong et al. (2025) and Wen et al. (2025) revealed its mechanistic involvement via TLR4/NF-κB/TGF-β and FSTL5/ERK-p38 MAPK pathways in colon and HCC [110, 111]. Similarly, Wei et al. (2021) demonstrated that miR-1246 promotes glioma progression by suppressing *DUSP3* and activating ERK signalling, while 2-Methoxyestradiol effectively inhibited tumour growth by downregulating miR-1246 and PD-L1 expression in MDSCs [17]. In oral cancer, miR-1246 was shown to enhance cell migration and invasion by targeting *DENND2D*, further supporting its role as a metastasis-promoting molecule and a potential biomarker for disease progression [100].

In addition to miR-1246, elevated levels of exosomal miR-155 have been linked to aggressive phenotypes in lymphomas and pancreatic cancers, primarily by targeting tumour suppressors such as *FOXO3a* and *SOCS1*, thus enabling immune evasion and tumour progression [97, 112]. In pancreatic cancer specifically, exosomal miR-155 also promotes M2 macrophage polarization, contributing to an immunosuppressive environment [103]. Likewise, miR-10b, often enriched in exosomes from metastatic BC cells, enhances metastasis by downregulating *HOXD10*, promoting EMT, and facilitating migration and invasion [113]. In line with this, Xin et al. (2019) demonstrated that exosomal transfer of miR-193a-3p, miR-210-3p, and miR-510 activates STAT3 signalling and induces EMT, leading to enhanced invasive capacity of cancer cells [88]. Notably, Vahid et al. (2017) showed that exosomal miRNAs from highly metastatic cells could induce malignant transformation in less aggressive BC cells, indicating their role in horizontal malignancy transfer [114]. Collectively, these findings underscore the profound influence of exosomal miRNAs in regulating metastatic potential and highlight their promise as biomarkers for early detection and targets for therapeutic intervention in metastatic cancer.

In addition to their roles in cancer initiation and progression, exosomal miRNAs are increasingly recognized for their involvement in drug resistance across various cancer types, although the underlying mechanisms remain incompletely understood [7, 115]. Shanliang et al. (2016) profiled exosomal miRNAs in BC drug-resistant sublines exposed to epirubicin, vinorelbine, and docetaxel, identifying the upregulation of several miRNAs out of 22 candidates after treatment [116]. Subsequent studies revealed functional insights; for instance, Jianhui et al. (2021) demonstrated that exosomal miR-9-5p modulated sensitivity to Tamoxifen (TAM) and promoted resistance in BC nude mouse models [89]. Moreover, exosomal uptake of miR-99b-3p was shown to confer resistance in BC cells through the PPP2CA/AKT/mTOR pathway [117]. In colorectal cancer, reduced levels of exosomal miR-224–5p were associated with 5-fluorouracil (5-FU) resistance via the S100A4 axis [118]. Additionally, in pancreatic cancer, gemcitabine-resistant stem cells were shown to transmit resistance to other pancreatic cells by transferring miR-210 through exosomes [119]. A recent study by Hoda et al. (2025) identified a panel of exosomal miRNAs, namely miR-4326, miR-3615, miR-7974, miR-130b-3p, and miR-454-3p, linked to oxaliplatin resistance in colorectal cancer using both in vitro and in silico models [91]. These findings underscore the critical role of exosomal miRNAs in modulating drug response, suggesting their potential as predictive biomarkers and therapeutic targets to overcome drug resistance in cancer treatment.

Overall, exosomal miRNAs play a pivotal role in cancer, extending from tumour initiation to drug resistance by modulating multiple oncogenic and tumour-suppressive pathways. Their ability to influence key mechanisms such as epithelial-mesenchymal transition, immune escape, angiogenesis, and therapy resistance highlights their functional significance in cancer progression (Fig. 24.2). Given their stability in biological fluids and disease-specific expression patterns, the analysis of exosomal miRNAs holds immense potential as non-invasive biomarkers for cancer diagnosis, prognosis, and prediction of therapeutic response.

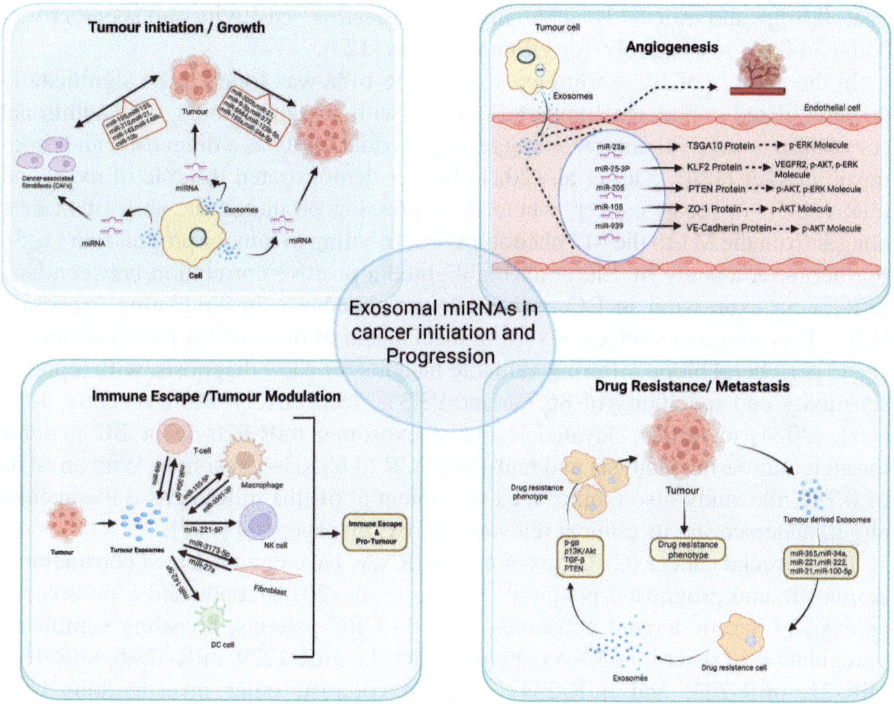

Fig. 24.2 Role of exosomal miRNAs in cancer initiation and progression. This figure illustrates how exosomal miRNAs influence key cancer hallmarks including tumour initiation, angiogenesis, immune escape, metastasis, and drug resistance by modulating intercellular communication and signalling pathways, ultimately contributing to tumour growth, microenvironment remodelling, therapy resistance, and cancer progression

Exosome miRNAs in Cancer Diagnosis

Exosomal miRNAs have emerged as promising non-invasive biomarkers for cancer diagnostics due to their functional relevance in tumour initiation and progression. Among them, exosomal miR-21 is one of the most extensively studied, identified as a potential universal biomarker across multiple cancer types. It demonstrates a sensitivity of 75% and specificity of 85%, with an AUC of 0.93, highlighting its robust diagnostic accuracy [120]. In epithelial ovarian cancer, plasma-derived exosomal miR-1260a, miR-7977, and miR-192-5p were significantly lower in patients compared to healthy controls, showing a strong diagnostic potential with an AUC of 0.8337 [121]. Debora et al. (2021) reported elevated levels of exosomal miR-21, miR-222, and miR-124-3p in glioblastoma patients [121]. Similarly, Lan et al. (2020) demonstrated the utility of serum exosomal miR-210 in distinguishing glioma patients from healthy individuals, reporting a sensitivity of 94.3%, specificity of 83.2%, and an AUC of 0.856 (95% CI: 0.795–0.917) [122]. Furthermore, Ebrahimkhani et al. (2018) applied random forest modelling on seven serum exosomal miRNAs (miR-182-5p, miR-328-3p, miR-339-5p, miR-340-5p, miR-485-3p, miR-486-5p, and miR-543), achieving a classification sensitivity and specificity of 0.86 and 0.87, respectively, for glioma patients [123].

In the context of BC, serum exosomal miR-148a was found to be significantly downregulated in patients compared to those with benign tumours, with additional correlation to survival outcomes suggesting its dual utility as a diagnostic and prognostic marker [124]. Xun et al. (2021) further demonstrated the role of exosomal miR-138-5p in breast cancer, where its expression promoted the shift of macrophages from the M1 to the M2 phenotype, contributing to tumour progression [125]. Furthermore, a study by Liu et al. highlighted a positive correlation between hsa-miR-21-5p expression in BC tissues and its abundance in circulating exosomes [126]. This suggests a selective sorting mechanism of the miRNA from tumour tissue to peripheral blood, offering valuable insights for early diagnosis, with reported sensitivity and specificity of 86.7% and 93.3%, respectively. More recently, Jung et al. (2024) identified elevated levels of exosomal miR-92b-5p in BC patients through microarray analysis and real-time PCR of CD63+ exosomes. With an AUC of 0.787, the study also emphasized the potential of this miRNA as a therapeutic target, underscoring its clinical relevance in BC management [127].

In colorectal cancer (CRC), exosomal miRNAs have demonstrated considerable diagnostic and prognostic potential. Hiroko et al. (2014) conducted a microarray analysis of serum-derived exosomes from 88 CRC patients, revealing significant upregulation of several miRNAs, including let-7a, miR-1229, miR-1246, miR-150, miR-21, miR-223, and miR-23a [128]. In contrast, other investigations have reported the downregulation of specific exosomal miRNAs during early-stage CRC. For instance, in a study involving 175 CRC patients, exosomal miRNA levels were found to be significantly reduced in early-stage disease when compared to healthy controls [129]. Another comprehensive analysis involving 125 CRC patients, 70 healthy individuals and 45 benign adenoma cases revealed significantly lower levels of serum exosomal miR-874 in cancer patients [130]. Notably, decreased miR-874 levels also correlated with poor overall survival, suggesting its dual utility as both diagnostic and prognostic markers. Collectively, these findings underscore the diagnostic versatility of exosomal miRNAs in CRC and their potential application in non-invasive liquid biopsy approaches for early cancer detection and monitoring.

Similarly, in lung cancer, exosomal miRNAs have emerged as valuable biomarkers for early diagnosis and disease stratification. A diagnostic panel comprising fucosylated exosomal miRNAs including miR-4732-5p, miR-451a, miR-486-5p, and miR-139-3p showed high diagnostic efficacy in early lung adenocarcinoma (LUAD), achieving an AUC of 0.85, with a sensitivity of 91% and specificity of 66% [131]. Chen et al. (2020) further identified elevated exosomal miR-7797 and reduced miR-98-3p levels in LUAD patients, with the combination of these two markers yielding an improved AUC of 0.816. Functional assays in A549 lung cancer cells confirmed that miR-7797 overexpression inhibited cell proliferation, suggesting a functional relevance beyond diagnostics [132]. In NSCLC, serum exosomal miR-155-5p and miR-658 were significantly decreased compared to healthy individuals, with respective AUC values of 0.71 and 0.73. When combined, these markers achieved an AUC of 0.78 and performed even better in early-stage NSCLC (stages 0 and I), with an AUC of 0.76 [133]. Additional studies reported by Zhijun et al. (2020) and Yi et al. (2019) identified miR-5684, miR-125b-5p, and miR-17-5p as effective components of diagnostic panels for NSCLC, with high sensitivity and specificity [54, 134]. These findings illustrate the dynamic shifts in exosomal miRNA profiles during lung cancer development and reinforce their utility as minimally invasive tools for early diagnosis and prognosis.

In addition to CRC and lung cancers, exosomal miRNAs have shown strong potential as diagnostic biomarkers in several other cancers, including prostate, HCC, gastric, oral, and bladder cancers. A study involving 31 prostate cancer (PCa) patients and 19 healthy individuals revealed a moderate increase in plasma exosomal miR-141-5p ($p = 0.085$) and a significant reduction in miR-125a-5p levels ($p = 0.032$) among PCa patients [135]. The respective AUC values of 0.65 and 0.69 suggest their potential in differentiating prostate cancer cases from healthy controls. Similarly, Zhuo et al. (2015) demonstrated a four-fold increase in exosomal miR-141 levels in PCa patients, with a sensitivity of 80% and specificity of 87%, further supporting its diagnostic value, especially when used alongside PSA testing [136]. In HCC, Jung Cho et al. (2020) analysed 178 serum samples including patients with HCC, chronic hepatitis B, liver cirrhosis, and healthy individuals and identified 49 significantly overexpressed miRNAs. Among these, exosomal miR-10b-5p, miR-18a-5p, miR-215-5p, and miR-940 were associated with HCC prognosis. Specifically, exosomal miR-10b-5p emerged as a promising early-stage biomarker, with an AUC of 0.93 (sensitivity: 90%; specificity: 75%) validated by qRT-PCR [137].

In gastric cancer (GC), Shi et al. (2020) reported elevated levels of serum exosomal miR-1246 in patients, particularly in early-stage GC. This marker demonstrated a sensitivity of 82% and specificity of 86%, highlighting its utility in early detection [138]. In bladder cancer, Andreu et al. (2017) performed microarray profiling of urinary EVs and identified 26 deregulated miRNAs in high-grade cases, 23 downregulated and 3 upregulated. Further validation using real-time PCR revealed miR-146a as a potential marker for low-grade bladder cancer and miR-375 as a specific indicator for high-grade disease [139]. Collectively, these findings illustrate the clinical significance of exosomal miRNAs in capturing cancer-specific signals from biofluids such as blood, urine, and cerebrospinal fluid. Their functional association with cancer initiation and progression underscores their reliability as non-invasive biomarkers, offering high sensitivity and specificity for early diagnosis and disease monitoring across multiple cancer types.

Prognostic Relevance of Exosomal miRNAs in Cancer

In continuation of their diagnostic utility, exosomal miRNAs have also demonstrated strong potential in the clinical prognosis of cancer, particularly in predicting treatment response and patient outcomes. For instance, Rie et al. (2018) reported that elevated levels of exosomal miR-451a in NSCLC patients were significantly associated with recurrence, metastasis, tumour invasion, and advanced staging and were directly correlated with poor prognosis across stages I to III [140]. Similarly, Dong et al. (2023) identified a prognostic panel comprising miR-200b-3p, miR-3124-5p, and miR-92b-5p in the serum-derived exosomes of small cell lung cancer (SCLC) patients [141]. In another study, Wulong et al. (2021) highlighted that exosomal miR-141 promotes angiogenesis and malignant behaviour in lung cancer via targeting the growth arrest-specific homeobox (GAX) gene [142]. Xinying et al. (2020) found that miR-484, miR-106b-3p, miR-192-5p, miR-10b-5p, and miR-151a-5p were significantly elevated in the plasma exosomes of LUAD patients and were associated with poor survival outcomes [143]. Notably, miR-10b-5p levels dropped post-tumour resection, reinforcing its dual role as a diagnostic and prognostic biomarker. Additionally, Xue et al. observed a similar trend with exosomal miR-484, which was markedly increased in LUAD patients and significantly decreased following surgery [137, 143]. Collectively, these findings underscore the prognostic value of exosomal miRNAs and their relevance in personalized cancer management.

In breast cancer, exosomal miRNAs have emerged as important prognostic and predictive biomarkers, with several also showing potential as therapeutic targets. Notably, miR-155 and miR-301 have demonstrated prognostic relevance, with miR-301 further identified as a candidate for targeted therapy [18, 144]. Exosomal miR-21, a well-characterized oncomiR, has been linked to treatment response in HER2-positive patients, showing reduced expression during trastuzumab therapy [145]. Additionally, Stevic et al. highlighted the role of exosomal miR-16 in promoting lymphatic metastasis and identified six exosomal miRNAs miR-382, miR-376a, miR-185, miR-628, miR-433, and miR-41 associated specifically with HER2-positive BC [146]. In triple-negative BC (TNBC), elevated levels of miR-374 were correlated with larger tumour size [146]. Notably, dynamic fluctuations in exosomal miR-376a levels, also observed in exosome-free plasma, further support their tumour-derived origin. After neoadjuvant therapy, the levels of miR-376a, miR-376c, miR-155, and miR-27a normalized, suggesting their release from the primary tumour and potential as indicators of therapeutic response [146]. Rodríguez et al. identified exosomal miR-1290, miR-146a, and miR-3662 as useful markers for monitoring chemotherapy and surgical outcomes [65]. Moreover, higher pretreatment levels of miR-21 and miR-105 were found in metastatic cases. Increased exosomal expression of miR-155, miR-222, and miR-21 also correlated significantly with circulating tumour cells (CTCs), underscoring their value in tracking disease progression [145].

Exosomal miRNAs play multifaceted roles in colorectal cancer (CRC) progression, metastasis, and chemotherapy resistance. Among them, miR-21 is one of the most studied oncomiRs, known for promoting poor prognosis and drug resistance by targeting tumour suppressor genes like *PTEN* and enhancing vascular permeability [147, 148]. Similarly, miR-135a/b contributes to tumour growth by inducing the increase of the stemness markers and self-renewal capacity [149]. In contrast, tumour-suppressor miRNAs such as miR-34 are well known to promote apoptosis and inhibit tumour progression, and decreased levels of this miRNA in cancer patients lead to reduced apoptosis and decreased chemotherapy efficacy [20]. Other exosomal miRNAs, including miR-203 and miR-21, have been implicated in shaping the pre-metastatic niche and modulating immune responses, further influencing metastasis and treatment outcomes [20, 150]. Clinically, studies by Yuki et al. (2017) and Mitsuo et al. (2017) demonstrated that these exosomal miRNAs are relevant biomarkers for tumour staging (II–IV) and metastasis detection using blood-derived exosomes, showing promising sensitivity and specificity [151, 152]. Additionally, Muwen et al. (2022) reported that downregulation of miR-431-5p is associated with poor prognosis, while functional assays confirmed that its overexpression inhibits tumour growth, highlighting its potential as a therapeutic target in CRC [153].

In pancreatic cancer, Lin et al. demonstrated that elevated levels of exosomal miR-1246 lead to downregulation of CCNG2, thereby promoting tumour growth, invasion, and chemoresistance [154]. These findings are consistent with earlier reports in breast cancer, where high miR-1246 levels were associated with aggressive tumour features and reduced chemotherapy response [76]. In colorectal cancer, Xiaoyi et al. (2016) reported that exosomal miR-1290 and miR-375 were significantly elevated in patients with poor overall survival, based on a screening cohort of 23 cancer patients [155]. In glioma, exosomal miR-21 detected in CSF has been linked to tumour metastasis and recurrence, supported by in vitro functional studies [156]. Additionally, Fengming et al. (2018) identified high levels of exosomal miR-301a in serum samples from patients with grade IV glioma. This miRNA was shown to enhance glioma cell invasion in functional assays. Notably, its expression peaked during tumour recurrence and decreased following surgical resection, highlighting its potential utility as a dynamic biomarker for monitoring disease progression and therapeutic response in glioma [157].

Exosomal miRNAs have also been explored in the prognosis of other cancers such as oral, HCC, and thyroid cancers, although the literature in these areas remains relatively limited. For example, a study by Aditi et al. (2022) revealed that exosomal miR-1307-5p is clinically correlated with chemoresistance and disease aggressiveness in oral cancer. Functional analyses further showed that this miRNA regulates the expression of key mRNAs including *THOP1, EHF, RNF4, GET4*, and *RNF*, providing mechanistic insight into its role in tumour progression [158]. Similarly, Tao et al. (2021) demonstrated that high expression of exosomal miR-130a is associated with poorer overall and recurrence-free survival outcomes [159]. In papillary thyroid carcinoma, high-throughput screening identified six exosomal miRNAs associated with prognosis, with their expression levels consistent with those found in tumour tissues [160]. In HCC, serum exosomal miR-665 was found to be significantly upregulated in high-grade (III and IV) and invasive tumours, and experimental models showed its involvement in activating the MAPK/ERK pathway [161].

Overall, beyond their role in early diagnosis and distinguishing cancer patients from healthy individuals, exosomal miRNAs hold significant potential in assessing tumour grade, invasiveness, drug response, recurrence, and survival outcomes across multiple cancer types, supported by functional evidence from both in vitro and in vivo studies. A comprehensive summary of diagnostically and prognostically relevant exosomal miRNAs across various cancer types is presented in Table 24.2.

Integration of Exosomal miRNAs with Other Biomarkers

The growing recognition of exosomal miRNAs as key regulators in cancer progression has fuelled efforts to integrate them with other established biomarkers, such as circulating tumour cells (CTCs), circulating tumour DNA (ctDNA), and protein biomarkers. This integration is crucial for enhancing the sensitivity, specificity, and overall accuracy of cancer diagnosis and prognosis [25, 26, 168]. Exosomal miRNAs, owing to their stability in biofluids and their ability to reflect the molecular landscape of tumours, offer a promising avenue for minimally or non-invasive cancer detection [25, 169]. However, challenges related to sensitivity and functional aspects must be addressed to optimize their clinical utility. The combination of exosomal miRNAs with CTC-based diagnostics provides real-time insights into tumour biology, as studies have demonstrated that specific exosomal miRNA signatures correlate with CTC counts and metastatic potential, improving the predictive value of CTC enumeration [26]. Similarly, ctDNA carries tumour-specific genetic alterations, such as mutations and methylation changes, but its fragmented nature and rapid clearance from circulation limit sensitivity. Integrating exosomal miRNAs with ctDNA analysis provides a stable and complementary molecular readout, improving early detection and prognosis [25, 170].

Traditional protein biomarkers such as carcinoembryonic antigen (CEA), alpha-fetoprotein (AFP), and prostate-specific antigen (PSA) are widely used in cancer screening and monitoring [171]. However, their specificity is often compromised by benign conditions that can cause elevated levels. The inclusion of exosomal miRNAs in biomarker panels can enhance diagnostic precision, as seen in HCC, where combining AFP with exosomal miR-122 and miR-192 significantly improves sensitivity compared to AFP alone [172]. Similarly, in prostate cancer, exosomal miRNA panels combined with PSA outperform PSA-based diagnosis in differentiating aggressive from indolent disease [173]. To further enhance diagnostic sensitivity, a multi-omics approach integrating genomic, transcriptomic, and proteomic data increases accuracy and enables better patient stratification. Machine learning and AI-based analysis allow the development of predictive algorithms with high sensitivity and specificity, while advanced detection techniques such as digital PCR (dPCR), next-generation sequencing (NGS), and droplet digital PCR (ddPCR) ensure reproducibility and precision in exosomal miRNA detection [174].

The integration of exosomal miRNAs with CTCs, ctDNA, and protein biomarkers represents a transformative approach to cancer diagnostics and prognostics. This multi-biomarker strategy enhances sensitivity, improves early detection, and provides comprehensive molecular insights into tumour biology. High-throughput multiplexing platforms that facilitate the simultaneous detection of exosomal miRNAs,

Table 24.2 Exosomal miRNAs as diagnostic and prognostic biomarkers in various cancers

Cancer type	ExomiRNA	Source	Clinical relevance	References
Diagnostic biomarkers				
Breast	miR-148a	Serum	Significantly downregulated in BC patients compared to benign tumours and healthy controls and correlates with BC stage	[162]
	miR-138-5p	Serum	Promotes macrophage, inhibits M1 polarization, and promotes M2, contributing to tumor progression	[125]
	hsa-miR-21-5p	Plasma	Significantly upregulated in BC patients compared to healthy individuals and effectively distinguishes BC cases with a good sensitivity	[126]
	miR-92b-5p	Serum	Significantly upregulated in BC patients, correlated with stage, and distinguishes BC cases with an AUC of 0.787	[127]
Colorectal	let-7a, miR-1229, miR-1246, miR-150, miR-21, miR-223, miR-23a	Serum	Significantly upregulated in patients compared to healthy controls and has high sensitivity to distinguish CRC patients from HC	[128]
	miR-874	Serum	Downregulated in patients compared to benign adenomas cases and healthy controls; correlated with poor overall survival	[130]
Lung	miR-4732-5p, miR-451a, miR-486-5p, miR-139-3p	Serum	Fucosylated exosomal miRNA panel able to predict LUAD diagnosis with good sensitivity and specificity	[131]
	miR-7797, miR-98-3p		Elevated exosomal miRNA expression demonstrated a diagnostic potential and miR-7797 overexpression inhibited cell proliferation	[132]
	miR-155-5p, miR-658	Serum	Significantly decreased in NSCLC patients versus healthy controls; distinguished patients with an AUC of 0.716 and 0.728 individually, and a combined AUC of 0.781	[163]
	miR-5684, miR-125b-5p,	serum	Significantly downregulated in NSCLC patients compared to healthy donors, demonstrating strong diagnostic potential, especially for early-stage detection	[54]
	miR-17-5p	Serum	Upregulated in NSCLC patients compared with the healthy controls	[134]

(continued)

Table 24.2 (continued)

Cancer type	ExomiRNA	Source	Clinical relevance	References
Prostate	miR-141-5p, miR-125a-5p	Plasma	miR-141-5p was moderately elevated, while miR-125a-5p was significantly reduced in patients compared to healthy controls; combined diagnostic AUC was 0.793	[135]
	miR-141	Serum	Elevated in PCa vs. BPH and controls; distinguishes metastatic from localized PCa	[136]
HCC	miR-10b-5p, miR-18a-5p, miR-215-5p, miR-940	Serum	Elevated miR-10b-5p shown potential for early diagnosis	[137]
Gastric	miR-1246	Serum	Elevated in early-stage GC; AUC 0.843 vs. controls	[138]
Bladder	miR-146a, miR-375	Urine	miR-146a was shown potential to diagnose low-grade and miR-375 was shown potential for high-grade bladder cancer detection	[139]
Prognostic biomarkers				
Breast				
	miR-155, miR-301	Plasma	Exosomal miR-155 and miR-301 serve as predictive biomarkers for therapy response in TNBC and HER2-positive BC types	[146]
	miR-382, miR-376a, miR-185, miR-628, miR-433, miR-410	Plasma	Associated with tumour size	[164]
	miR-374	Plasma	Elevated levels were correlated with tumour size in TNBC	[146]
	miR-27a, miR-155, miR-376a, miR-376c	Plasma	Show potential to analyse therapeutic response	[146]
	miR-21, miR-105	Serum	Higher pre-treatment levels found in metastatic BC cases than HC	[165]
	miR-155, miR-222, miR-21	Serum	Correlated significantly with CTCs; valuable for tracking disease progression	[165]
			Colorectal	
	miR-203,	Serum	Significantly upregulated in CRC patients; correlates with distant metastasis and serves as an independent predictor of poor prognosis	[151]

(continued)

Table 24.2 (continued)

Cancer type	ExomiRNA	Source	Clinical relevance	References
	miR-431-5p	hUCMSCs	Downregulation linked to poor prognosis; overexpression inhibits tumour growth	[153]
Others				
Pancreatic	miR-1246	Plasma	Significantly elevated in patients with intraductal papillary mucinous neoplasms (IPMN) and pancreatic cancer compared to healthy subjects, indicating its potential as a marker for localized pancreatic cancer	[166]
Glioma	miR-21	CSF	Significantly higher in glioma patients and associated with tumour metastasis, recurrence, and anatomical site preference	[156]
	Mir-301a	Serum	Significantly elevated in glioma patients compared and correlated with tumour grade	[157]
Oral	miR-1307-5p	Saliva	Significantly upregulated in oral cancer patients compared to non-cancerous controls, and associated with poor patient survival, disease progression, and chemoresistance	[158]
Thyroid	hsa-miR-129-2, hsa-miR-548j, hsa-miR-6734, hsa-miR-889, hsa-miR-129-2 and hsa-miR- 889	Serum	The exosomal miRNA expression patterns align with those observed in tumour tissues and are linked to prognosis	[167]
HCC	miR-665	Serum	Upregulated in HCC patients compared to healthy subjects, and correlated with tumour stage and size	[161]

ctDNA, and protein markers will reduce turnaround time and enhance diagnostic efficiency. Advances in bioinformatics, high-throughput detection methods, and functional studies will further refine this approach, paving the way for personalized and minimally invasive cancer management. Standardization of detection techniques, combined with the development of machine learning-based analytical frameworks, will be crucial in translating these integrated biomarker panels into clinical practice. As research progresses, exosomal miRNAs will continue to play a pivotal role in shaping the future of precision oncology, offering novel avenues for improved patient outcomes.

Conclusions

Exosomal miRNAs have significantly enhanced our understanding of cancer biology, revealing their functional roles in promoting tumour initiation and sustaining a pro-tumour microenvironment. These small regulatory RNAs, packaged within exosomes, influence various signalling pathways that contribute to drug resistance, immune evasion, and metastasis. Clinically, exosomal miRNAs isolated from biofluids such as blood, urine, saliva, and CSF have demonstrated strong correlations with tumour onset and progression across multiple cancer types. Their ability to reflect the molecular state of tumours makes them promising non-invasive biomarkers for early diagnosis and disease monitoring. Given the heterogeneity and complexity of tumours, integrating exosomal miRNAs with other biomarkers such as ctDNA, proteins, and CTCs could enhance diagnostic accuracy and prognostic potential. Recent studies suggest that such multimodal biomarker panels may provide a more precise depiction of tumour dynamics, which is crucial for advancing precision oncology. However, most current findings are based on limited sample sizes, highlighting the need for validation in larger cohorts. Additionally, the use of meta-analyses and AI-driven approaches may offer deeper insights into the diagnostic and predictive value of exosomal miRNAs, ultimately facilitating their translation from bench to bedside in clinical oncology.

Acknowledgements This work was supported by Apollo Hospitals Educational and Research Foundation (AHERF). We thank our AHERF President, Dr. N.K. Ganguly, Vice President; Ishita Shively, Clinical Director; Dr. Jayanthi Swaminathan for active organizational support.

Completing of Interests None.

References

1. Stefańska K et al (2023) The role of exosomes in human carcinogenesis and cancer therapy-recent findings from molecular and clinical research. Cells 12(3):356
2. Kumar SK, Sasidhar MV (2023) Recent trends in the use of small extracellular vesicles as optimal drug delivery vehicles in oncology. Mol Pharm 20(8):3829–3842
3. Manda S, Sarvareddy K (2024) From fluids to forecasts: the promise of small extracellular vesicle miRNAs in Revolutionising cancer diagnostics. In: Seifi M (ed) Beyond the blueprint – decoding the elegance of gene expression. Rijeka, IntechOpen
4. Dai J et al (2020) Exosomes: key players in cancer and potential therapeutic strategy. Signal Transduct Target Ther 5(1):145
5. Bhome R et al (2018) Exosomal microRNAs (exomiRs): small molecules with a big role in cancer. Cancer Lett 420:228–235
6. Nail HM et al (2023) Exosomal miRNA-mediated intercellular communications and immunomodulatory effects in tumor microenvironments. J Biomed Sci 30(1):69
7. Guo QR et al (2020) The role of exosomal microRNA in cancer drug resistance. Front Oncol 10:472
8. Mowla M, Hashemi A (2021) Functional roles of exosomal miRNAs in multi-drug resistance in cancer chemotherapeutics. Exp Mol Pathol 118:104592

9. Sarwareddy KK et al (2025) Harnessing tomato-derived small extracellular vesicles as drug delivery system for cancer therapy. Future Sci OA 11(1):2461956
10. Ghosh S et al (2024) Harnessing exosomes as cancer biomarkers in clinical oncology. Cancer Cell Int 24(1):278
11. Alotaibi F (2023) Exosomal microRNAs in cancer: potential biomarkers and immunotherapeutic targets for immune checkpoint molecules. Front Genet 14:1052731
12. Todorova VK et al (2022) Circulating Exosomal microRNAs as predictive biomarkers of Neoadjuvant chemotherapy response in breast cancer. Curr Oncol 29(2):613–630
13. Hu C et al (2020) Role of exosomal microRNAs in lung cancer biology and clinical applications. Cell Prolif 53(6):e12828
14. Yang S-S et al (2020) Breast cancer-derived exosomes regulate cell invasion and metastasis in breast cancer via miR-146a to activate cancer associated fibroblasts in tumor microenvironment. Exp Cell Res 391(2):111983
15. Donnarumma E et al (2017) Cancer-associated fibroblasts release exosomal microRNAs that dictate an aggressive phenotype in breast cancer. Oncotarget 8(12):19592–19608
16. Tan S et al (2020) Exosomal miRNAs in tumor microenvironment. J Exp Clin Cancer Res 39(1):67
17. Qiu W et al (2021) Exosomal miR-1246 from glioma patient body fluids drives the differentiation and activation of myeloid-derived suppressor cells. Mol Ther 29(12):3449–3464
18. Zhang Z et al (2020) Exosomal miR-1246 and miR-155 as predictive and prognostic biomarkers for trastuzumab-based therapy resistance in HER2-positive breast cancer. Cancer Chemother Pharmacol 86(6):761–772
19. Asghariazar V et al (2023) MicroRNA-143 as a potential tumor suppressor in cancer: an insight into molecular targets and signaling pathways. Pathol Res Pract 250:154792
20. Hosseini M et al (2021) Tumor-derived exosomes encapsulating miR-34a promote apoptosis and inhibit migration and tumor progression of colorectal cancer cells under in vitro condition. Daru 29(2):267–278
21. Cui SY, Wang R, Chen LB (2014) MicroRNA-145: a potent tumour suppressor that regulates multiple cellular pathways. J Cell Mol Med 18(10):1913–1926
22. Yang FK et al (2024) The value of urinary exosomal microRNA-21 in the early diagnosis and prognosis of bladder cancer. Kaohsiung J Med Sci 40(7):660–670
23. Bakhsh T et al (2024) Molecular detection of exosomal miRNAs of blood serum for prognosis of colorectal cancer. Sci Rep 14(1):8902
24. Uddin MH et al (2021) Exosomal microRNA in pancreatic cancer diagnosis. Prognosis Treatment Bench Bedside 13(11):2777
25. Yang Z et al (2020) A multianalyte panel consisting of extracellular vesicle miRNAs and mRNAs, cfDNA, and CA19-9 shows utility for diagnosis and staging of pancreatic ductal adenocarcinoma. Clin Cancer Res 26(13):3248–3258
26. Papakonstantinou D, Roumeliotou A, Pantazaka E, Shaukat AN, Christopoulou A, Koutras A, Dimitrakopoulos FI, Georgoulias V, Xagara A, Chantzara E, Koinis F (2025) Integrative analysis of circulating tumor cells (CTCs) and exosomes from small-cell lung cancer (SCLC) patients: a comprehensive approach. Mol Oncol 19(7):2038–2055
27. Juan T, Fürthauer M (2018) Biogenesis and function of ESCRT-dependent extracellular vesicles. Semin Cell Dev Biol 74:66–77
28. Wollert T, Hurley JH (2010) Molecular mechanism of multivesicular body biogenesis by ESCRT complexes. Nature 464(7290):864–869
29. Bissig C, Gruenberg J (2014) ALIX and the multivesicular endosome: ALIX in wonderland. Trends Cell Biol 24(1):19–25
30. Villarroya-Beltri C et al (2013) Sumoylated hnRNPA2B1 controls the sorting of miRNAs into exosomes through binding to specific motifs. Nat Commun 4:2980
31. Shurtleff MJ et al (2016) Y-box protein 1 is required to sort microRNAs into exosomes in cells and in a cell-free reaction. Elife:5:e19276
32. McKenzie AJ et al (2016) KRAS-MEK signaling controls Ago2 sorting into exosomes. Cell Rep 15(5):978–987

33. Kosaka N et al (2010) Secretory mechanisms and intercellular transfer of microRNAs in living cells. J Biol Chem 285(23):17442–17452
34. Wei JX et al (2015) Vps4A functions as a tumor suppressor by regulating the secretion and uptake of exosomal microRNAs in human hepatoma cells. Hepatology 61(4):1284–1294
35. Jackson CE et al (2017) Effects of inhibiting VPS4 support a general role for ESCRTs in extracellular vesicle biogenesis. Biophys J 113(6):1342–1352
36. Lee H et al (2019) Caveolin-1 selectively regulates microRNA sorting into microvesicles after noxious stimuli. J Exp Med 216(9):2202–2220
37. Cha DJ et al (2015) KRAS-dependent sorting of miRNA to exosomes. Elife 4:e07197
38. Melo SA et al (2014) Cancer exosomes perform cell-independent microRNA biogenesis and promote tumorigenesis. Cancer Cell 26(5):707–721
39. Temoche-Diaz MM et al (2019) Distinct mechanisms of microRNA sorting into cancer cell-derived extracellular vesicle subtypes. Elife 8:e47544
40. Jia Y et al (2022) Small extracellular vesicles isolation and separation: current techniques, pending questions and clinical applications. Theranostics 12(15):6548–6575
41. Omrani M et al (2024) Global trend in exosome isolation and application: an update concept in management of diseases. Mol Cell Biochem 479(3):679–691
42. Panwar D et al (2023) Efficient strategy to isolate exosomes using anti-CD63 antibodies conjugated to gold nanoparticles. AMB Express 13(1):90
43. Inubushi S et al (2024) Serum exosomes expressing CD9, CD63 and HER2 from breast-cancer patients decreased after surgery of the primary tumor: a potential biomarker of tumor burden. Cancer Genomics Proteomics 21(6):580–584
44. Castillo J et al (2018) Surfaceome profiling enables isolation of cancer-specific exosomal cargo in liquid biopsies from pancreatic cancer patients. Ann Oncol 29(1):223–229
45. Sharma P et al (2018) Immunoaffinity-based isolation of melanoma cell-derived exosomes from plasma of patients with melanoma. J Extracell Vesicles 7(1):1435138
46. Mun B et al (2023) An immuno-magnetophoresis-based microfluidic chip to isolate and detect HER2-positive cancer-derived exosomes via multiple separation. Biosens Bioelectron 239:115592
47. Gu Z et al (2022) Optimization of a method for the clinical detection of serum exosomal miR-940 as a potential biomarker of breast cancer. Front Oncol 12:956167
48. Giallombardo M et al (2016) Exosomal miRNA analysis in non-small cell lung cancer (NSCLC) patients' plasma through qPCR: a feasible liquid biopsy tool. J Vis Exp 111
49. Ueda H et al (2024) Preoperative treatment response prediction for pancreatic cancer by multiple microRNAs in plasma exosomes: optimization using machine learning and network analysis. Pancreatology 24(7):1097–1106
50. Wang C et al (2019) Droplet digital PCR improves urinary exosomal miRNA detection compared to real-time PCR. Clin Biochem 67:54–59
51. Luo L et al (2020) A ratiometric electrochemical DNA biosensor for detection of exosomal MicroRNA. Talanta 207:120298
52. Li X et al (2020) Electrochemical biosensor for ultrasensitive exosomal miRNA analysis by cascade primer exchange reaction and MOF@Pt@MOF nanozyme. Biosens Bioelectron 168:112554
53. Xia Y et al (2021) Colorimetric detection of exosomal microRNA through switching the visible-light-induced oxidase mimic activity of acridone derivate. Biosens Bioelectron 173:112834
54. Zhang Z et al (2020) Tumor-derived Exosomal miRNAs as diagnostic biomarkers in non-small cell lung cancer. Front Oncol 10:560025
55. Grimolizzi F et al (2017) Exosomal miR-126 as a circulating biomarker in non-small-cell lung cancer regulating cancer progression. Sci Rep 7(1):15277
56. Yang G et al (2020) Exosomal miR-21/let-7a ratio distinguishes non-small cell lung cancer from benign pulmonary diseases. Asia Pac J Clin Oncol 16(4):280–286
57. Matsumura T et al (2015) Exosomal microRNA in serum is a novel biomarker of recurrence in human colorectal cancer. Br J Cancer 113(2):275–281

58. Su Z et al (2024) Cancer-associated fibroblasts-secreted exosomal miR-92a-3p promotes tumor growth and stemness in hepatocellular carcinoma through activation of Wnt/β-catenin signaling pathway by suppressing AXIN1. J Cell Physiol 239(9):e31344
59. Hu HY et al (2019) Exosomal miR-1229 derived from colorectal cancer cells promotes angiogenesis by targeting HIPK2. Int J Biol Macromol 132:470–477
60. Jiang X et al (2017) Exosomal microRNA remodels the tumor microenvironment. PeerJ 5:e4196
61. Liu C et al (2016) Serum exosomal miR-4772-3p is a predictor of tumor recurrence in stage II and III colon cancer. Oncotarget 7(46):76250–76260
62. Figueroa J et al (2017) Exosomes from glioma-associated mesenchymal stem cells increase the Tumorigenicity of glioma stem-like cells via transfer of miR-1587. Cancer Res 77(21):5808–5819
63. Peng J et al (2021) Current understanding of exosomal MicroRNAs in glioma immune regulation and therapeutic responses. Front Immunol 12:813747
64. Olioso D et al (2021) Serum exosomal microRNA-21, 222 and 124-3p as noninvasive predictive biomarkers in newly diagnosed high-grade gliomas: a prospective study. Cancers (Basel) 13(12)
65. Linares-Rodríguez M, Blancas I, Rodríguez-Serrano F (2025) The predictive value of blood-derived exosomal miRNAs as biomarkers in breast cancer: a systematic review. Clin Breast Cancer 25(1):e48–e55
66. Shen S et al (2021) Cancer-derived exosomal miR-7641 promotes breast cancer progression and metastasis. Cell Commun Signal 19(1):20
67. Sueta A et al (2017) Differential expression of exosomal miRNAs between breast cancer patients with and without recurrence. Oncotarget 8(41):69934–69944
68. Tokuhisa M et al (2015) Exosomal miRNAs from peritoneum lavage fluid as potential prognostic biomarkers of peritoneal metastasis in gastric cancer. PLoS One 10(7):e0130472
69. Uddin MH et al (2021) Exosomal microRNA in pancreatic cancer diagnosis, prognosis, and treatment: from bench to bedside. Cancers (Basel) 13(11)
70. Wei J et al (2020) Serum miR-1290 and miR-1246 as potential diagnostic biomarkers of human pancreatic cancer. J Cancer 11(6):1325–1333
71. Shan G et al (2020) Cancer-associated fibroblast-secreted exosomal miR-423-5p promotes chemotherapy resistance in prostate cancer by targeting GREM2 through the TGF-β signaling pathway. Exp Mol Med 52(11):1809–1822
72. Joković SM et al (2022) MiR-375 and miR-21 as potential biomarkers of prostate cancer: comparison of matching samples of plasma and exosomes. Genes (Basel) 13(12)
73. Tian XP et al (2019) Acidic microenvironment up-regulates Exosomal miR-21 and miR-10b in early-stage hepatocellular carcinoma to promote cancer cell proliferation and metastasis. Theranostics 9(7):1965–1979
74. Reza AMMT et al (2016) Human adipose mesenchymal stem cell-derived exosomal-miRNAs are critical factors for inducing anti-proliferation signalling to A2780 and SKOV-3 ovarian cancer cells. Sci Rep 6(1):38498
75. Hao C et al (2023) Tumor-derived exosomal miR-148b-3p mediates M2 macrophage polarization via TSC2/mTORC1 to promote breast cancer migration and invasion. Thorac Cancer 14(16):1477–1491
76. Li XJ et al (2017) Exosomal MicroRNA MiR-1246 promotes cell proliferation, invasion and drug resistance by targeting CCNG2 in breast cancer. Cell Physiol Biochem 44(5):1741–1748
77. Cai Q, Zhu A, Gong L (2018) Exosomes of glioma cells deliver miR-148a to promote proliferation and metastasis of glioblastoma via targeting CADM1. Bull Cancer 105(7):643–651
78. He S et al (2019) Exosomal miR-499a-5p promotes cell proliferation, migration and EMT via mTOR signaling pathway in lung adenocarcinoma. Exp Cell Res 379(2):203–213
79. Li YY et al (2018) Cancer-associated fibroblasts contribute to oral cancer cells proliferation and metastasis via exosome-mediated paracrine miR-34a-5p. EBioMedicine 36:209–220
80. Jiang M-J et al (2020) Dying tumor cell-derived exosomal miR-194-5p potentiates survival and repopulation of tumor repopulating cells upon radiotherapy in pancreatic cancer. Mol Cancer 19(1):68

81. Mu X et al (2024) Exosomal miR-1228-5p down-regulates DUSP22 to promotes cell proliferation and migration in small cell lung cancer. Life Sci 351:122787
82. Shan Y et al (2024) Tumor-derived exosomal miR-103a-3p promotes vascular permeability and proliferation by targeting ZO-1 and ACOX-1 in nasopharyngeal carcinoma. Transl Cancer Res 13(9):4896–4912
83. Stoen MJ et al (2021) Overexpression of miR-20a-5p in tumor epithelium is an independent negative prognostic indicator in prostate cancer-a multi-institutional study. Cancers (Basel) 13(16)
84. Zhang Z et al (2018) Exosome-mediated miR-200b promotes colorectal cancer proliferation upon TGF-β1 exposure. Biomed Pharmacother 106:1135–1143
85. Zhang S et al (2025) CD63-high macrophage-derived exosomal miR-6876-5p promotes hepatocellular carcinoma stemness via PTEN/Akt-mediated EMT pathway. Hepatol Commun 9(1)
86. Li C et al (2022) The role of Exosomal miRNAs in cancer. J Transl Med 20:1–15
87. Ye B et al (2023) Hypoxic tumor-derived exosomal miR-21 induces cancer-associated fibroblast activation to promote head and neck squamous cell carcinoma metastasis. Cell Signal 108:110725
88. Zhang X et al (2019) Hypoxic BMSC-derived exosomal miRNAs promote metastasis of lung cancer cells via STAT3-induced EMT. Mol Cancer 18(1):40
89. Liu J et al (2021) Exosomes from tamoxifen-resistant breast cancer cells transmit drug resistance partly by delivering miR-9-5p. Cancer Cell Int 21(1):55
90. Yao X et al (2020) Endoplasmic reticulum stress-induced exosomal miR-27a-3p promotes immune escape in breast cancer via regulating PD-L1 expression in macrophages. J Cell Mol Med 24(17):9560–9573
91. Deli H et al (2025) The expression of exosomal and cellular miRNAs in predicting oxaliplatin resistance in colorectal cancer cells: an in silico and in vitro study. BMC Cancer 25(1):46
92. Kwon Y et al (2020) Exosomal MicroRNAs as mediators of cellular interactions between cancer cells and macrophages. Front Immunol 11:1167
93. Cao C et al (2025) Exosomes containing miR-152-3p targeting FGFR3 mediate SLC7A7-induced angiogenesis in bladder cancer. NPJ Precis Oncol 9(1):71
94. Du J et al (2025) Melanoma cell line-derived exosomal miR-424-5p: a key promoter of angiogenesis through LATS2 interaction. Oncol Res 33(2):357–367
95. Lin XJ et al (2018) Hepatocellular carcinoma cell-secreted Exosomal MicroRNA-210 promotes angiogenesis in vitro and in vivo. Mol Ther Nucleic Acids 11:243–252
96. Shang D et al (2020) Pancreatic cancer cell-derived exosomal microRNA-27a promotes angiogenesis of human microvascular endothelial cells in pancreatic cancer via BTG2. J Cell Mol Med 24(1):588–604
97. Zhou Z et al (2019) Exosomes carrying MicroRNA-155 target Forkhead box O3 of endothelial cells and promote angiogenesis in gastric cancer. Mol Ther Oncol 15:223–233
98. Xiong J et al (2023) Advances of exosomal miRNAs in the diagnosis and treatment of ovarian cancer. Discov Oncol 14(1):65
99. Liu Z et al (2024) Tumor-derived exosomal miR-1247-3p promotes angiogenesis in bladder cancer by targeting FOXO1. Cancer Biol Ther 25(1):2290033
100. Sakha S et al (2016) Exosomal microRNA miR-1246 induces cell motility and invasion through the regulation of DENND2D in oral squamous cell carcinoma. Sci Rep 6:38750
101. Zhang L et al (2019) Cervical cancer cells-secreted exosomal microRNA-221-3p promotes invasion, migration and angiogenesis of microvascular endothelial cells in cervical cancer by down-regulating MAPK10 expression. Cancer Manag Res 11:10307–10319
102. Lugano R, Ramachandran M, Dimberg A (2020) Tumor angiogenesis: causes, consequences, challenges and opportunities. Cell Mol Life Sci 77(9):1745–1770
103. Ma YS et al (2021) M2 macrophage-derived exosomal microRNA-155-5p promotes the immune escape of colon cancer by downregulating ZC3H12B. Mol Ther Oncol 20:484–498
104. Li S et al (2021) Extracellular vesicles-derived microRNA-222 promotes immune escape via interacting with ATF3 to regulate AKT1 transcription in colorectal cancer. BMC Cancer 21(1):349

105. Martinez-Espinosa I, Serrato JA, Ortiz-Quintero B (2023) The role of exosome-derived microRNA on lung cancer metastasis progression. Biomolecules 13(11):1574
106. Liu H et al (2020) Colorectal cancer-derived exosomal miR-106b-3p promotes metastasis by down-regulating DLC-1 expression. Clin Sci (Lond) 134(4):419–434
107. Bhattacharjee R et al (2022) Cellular landscaping of exosomal miRNAs in cancer metastasis: from chemoresistance to prognostic markers. Adv Cancer Biol Metast 5:100050
108. Hashemi M et al (2023) Pre-clinical and clinical importance of miR-21 in human cancers: tumorigenesis, therapy response, delivery approaches and targeting agents. Pharmacol Res 187:106568
109. Yang F et al (2019) MiR-1246 promotes metastasis and invasion of A549 cells by targeting GSK-3β–mediated Wnt/β-catenin pathway. Cancer Res Treat 51(4):1420–1429
110. Liu X et al (2025) Cancer-secreted exosomal miR-1246 promotes colorectal cancer liver metastasis by activating hepatic stellate cells. Mol Med 31(1):68
111. Chen WJ et al (2025) Exosomatic miR-1246 promotes hepatocellular carcinoma progression via FSTL5 and ERK/p38 MAPK pathway. J Biochem Mol Toxicol 39(3):e70148
112. Zhou X et al (2018) Melanoma cell-secreted exosomal miR-155-5p induce proangiogenic switch of cancer-associated fibroblasts via SOCS1/JAK2/STAT3 signaling pathway. J Exp Clin Cancer Res 37(1):242
113. Singh R et al (2014) Exosome-mediated transfer of miR-10b promotes cell invasion in breast cancer. Mol Cancer 13:256
114. Kia V et al (2019) Exosomal miRNAs from highly metastatic cells can induce metastasis in non-metastatic cells. Life Sci 220:162–168
115. Najminejad H et al (2019) Emerging roles of exosomal miRNAs in breast cancer drug resistance. IUBMB Life 71(11):1672–1684
116. Zhong S et al (2016) MicroRNA expression profiles of drug-resistance breast cancer cells and their exosomes. Oncotarget 7(15):19601–19609
117. Mao H et al (2024) Drug-resistant exosome miR-99b-3p induces macrophage polarization and confers chemoresistance on sensitive cells by targeting PPP2CA. Int Immunopharmacol 142(Pt B):113168
118. Yan Y-Y et al (2025) Low miR-224–5p in exosomes confers colorectal cancer 5-FU resistance by upregulating S100A4. Drug Resist Updat 79:101211
119. Yang Z et al (2020) Exosomes derived from cancer stem cells of gemcitabine-resistant pancreatic cancer cells enhance drug resistance by delivering miR-210. Cell Oncol (Dordr) 43(1):123–136
120. Shi J (2016) Considering exosomal miR-21 as a biomarker for cancer. J Clin Med 5(4)
121. Chen L et al (2022) Plasma exosomal miR-1260a, miR-7977 and miR-192-5p as diagnostic biomarkers in epithelial ovarian cancer. Future Oncol 18(26):2919–2931
122. Lan F, Yue X, Xia T (2020) Exosomal microRNA-210 is a potentially non-invasive biomarker for the diagnosis and prognosis of glioma. Oncol Lett 19(3):1967–1974
123. Ebrahimkhani S et al (2018) Deep sequencing of circulating exosomal microRNA allows non-invasive glioblastoma diagnosis. NPJ Precis Oncol 2:28
124. Li D et al (2020) Identification of serum exosomal miR-148a as a novel prognostic biomarker for breast cancer. Eur Rev Med Pharmacol Sci 24(13):7303–7309
125. Xun J et al (2021) Cancer-derived exosomal miR-138-5p modulates polarization of tumor-associated macrophages through inhibition of KDM6B. Theranostics 11(14):6847–6859
126. Liu M et al (2021) Exosomal hsa-miR-21-5p is a biomarker for breast cancer diagnosis. PeerJ 9:e12147
127. Kan JY et al (2024) Exosomal microRNA-92b is a diagnostic biomarker in breast cancer and targets survival-related MTSS1L to promote tumorigenesis. Int J Mol Sci 25(2)
128. Ogata-Kawata H et al (2014) Circulating exosomal microRNAs as biomarkers of colon cancer. PLoS One 9(4):e92921
129. Wang L et al (2022) Serum exosomal miR-377-3p and miR-381-3p as diagnostic biomarkers in colorectal cancer. Future Oncol 18(7):793–805
130. Zhang N et al (2020) Reduced serum exosomal miR-874 expression predicts poor prognosis in colorectal cancer. Eur Rev Med Pharmacol Sci 24(2):664–672

131. Chen X et al (2022) Fucosylated exosomal miRNAs as promising biomarkers for the diagnosis of early lung adenocarcinoma. Front Oncol 12:935184
132. Chen L et al (2020) Serum exosomal miR-7977 as a novel biomarker for lung adenocarcinoma. J Cell Biochem 121(5–6):3382–3391
133. Zhang Z et al (2020) Tumor-derived exosomal miR-155-5p and miR-658 in serum as diagnostic biomarkers for early-stage non-small cell lung cancer. Res Square
134. Zhang Y et al (2019) Detection of circulating exosomal miR-17-5p serves as a novel non-invasive diagnostic marker for non-small cell lung cancer patients. Pathol Res Pract 215(8):152466
135. Li W et al (2020) Plasma exosomal miR-125a-5p and miR-141-5p as non-invasive biomarkers for prostate cancer. Neoplasma 67(6):1314–1318
136. Li Z et al (2016) Exosomal microRNA-141 is upregulated in the serum of prostate cancer patients. Onco Targets Ther 9:139–148
137. Cho HJ et al (2020) Serum exosomal MicroRNA, miR-10b-5p, as a potential diagnostic biomarker for early-stage hepatocellular carcinoma. J Clin Med 9(1)
138. Shi Y et al (2020) Exosomal miR-1246 in serum as a potential biomarker for early diagnosis of gastric cancer. Int J Clin Oncol 25(1):89–99
139. Andreu Z et al (2017) Extracellular vesicles as a source for non-invasive biomarkers in bladder cancer progression. Eur J Pharm Sci 98:70–79
140. Kanaoka R et al (2018) Usefulness of plasma Exosomal MicroRNA-451a as a noninvasive biomarker for early prediction of recurrence and prognosis of non-small cell lung cancer. Oncology 94(5):311–323
141. Kim DH et al (2023) Identification of exosomal microRNA panel as diagnostic and prognostic biomarker for small cell lung cancer. Biomark Res 11(1):80
142. Wang W et al (2021) Tumor-derived exosomal miRNA-141 promote angiogenesis and malignant progression of lung cancer by targeting growth arrest-specific homeobox gene (GAX). Bioengineered 12(1):821–831
143. Xue X et al (2020) Exosomal miRNA profiling before and after surgery revealed potential diagnostic and prognostic markers for lung adenocarcinoma. Acta Biochim Biophys Sin Shanghai 52(3):281–293
144. Jia Y et al (2021) miR-301 regulates the SIRT1/SOX2 pathway via CPEB1 in the breast cancer progression. Mol Ther Oncolytics 22:13–26
145. Rodríguez-Martínez A et al (2019) Exosomal miRNA profile as complementary tool in the diagnostic and prediction of treatment response in localized breast cancer under neoadjuvant chemotherapy. Breast Cancer Res 21(1):21
146. Stevic I et al (2018) Specific microRNA signatures in exosomes of triple-negative and HER2-positive breast cancer patients undergoing neoadjuvant therapy within the GeparSixto trial. BMC Med 16(1):179
147. Sun LH et al (2020) Exosomal miR-21 promotes proliferation, invasion and therapy resistance of colon adenocarcinoma cells through its target PDCD4. Sci Rep 10(1):8271
148. He Q et al (2021) Cancer-secreted exosomal miR-21-5p induces angiogenesis and vascular permeability by targeting KRIT1. Cell Death Dis 12(6):576
149. Choi SA et al (2020) Extracellular vesicle-associated miR-135b and -135a regulate stemness in group 4 medulloblastoma cells by targeting angiomotin-like 2. Cancer Cell Int 20(1):558
150. Wang S et al (2020) Role of exosomal miR-21 in the tumor microenvironment and osteosarcoma tumorigenesis and progression. Int J Oncol 56(5):1055–1063
151. Takano Y et al (2017) Circulating exosomal microRNA-203 is associated with metastasis possibly via inducing tumor-associated macrophages in colorectal cancer. Oncotarget 8(45):78598–78613
152. Tsukamoto M et al (2017) Circulating exosomal MicroRNA-21 as a biomarker in each tumor stage of colorectal cancer. Oncology 92(6):360–370
153. Qu M et al (2022) The role of human umbilical cord mesenchymal stem cells-derived exosomal microRNA-431-5p in survival and prognosis of colorectal cancer patients. Mutagenesis 37(2):164–171

154. Lin SS et al (2018) miR-1246 targets CCNG2 to enhance cancer stemness and chemoresistance in oral carcinomas. Cancers (Basel) 10(8):272
155. Huang X et al (2015) Exosomal miR-1290 and miR-375 as prognostic markers in castration-resistant prostate cancer. Eur Urol 67(1):33–41
156. Shi R et al (2015) Exosomal levels of miRNA-21 from cerebrospinal fluids associated with poor prognosis and tumor recurrence of glioma patients. Oncotarget 6(29):26971–26981
157. Lan F et al (2018) Serum exosomal miR-301a as a potential diagnostic and prognostic biomarker for human glioma. Cell Oncol (Dordr) 41(1):25–33
158. Patel A et al (2022) Salivary Exosomal miRNA-1307-5p predicts disease aggressiveness and poor prognosis in oral squamous cell carcinoma patients. Int J Mol Sci 23(18)
159. He T et al (2021) Plasma-derived exosomal microRNA-130a serves as a noninvasive biomarker for diagnosis and prognosis of oral squamous cell carcinoma. J Oncol 2021:5547911
160. Maggisano V et al (2022) Identification of exosomal microRNAs and their targets in papillary thyroid cancer cells. Biomedicine 10(5):961
161. Qu Z et al (2017) Exosomal miR-665 as a novel minimally invasive biomarker for hepatocellular carcinoma diagnosis and prognosis. Oncotarget 8(46):80666–80678
162. Liao J et al (2014) Expression profiling of exosomal miRNAs derived from human esophageal cancer cells by Solexa high-throughput sequencing. Int J Mol Sci 15(9):15530–15551
163. Zhang Z et al (2020) Tumor-derived exosomal miR-155-5p and miR-658 in serum as diagnostic biomarkers for early-stage non-small cell lung cancer. Front Oncol 10:560025
164. Gupta I et al (2020) Circulating miRNAs in HER2-positive and triple negative breast cancers: potential biomarkers and therapeutic targets. Int J Mol Sci 21(18):6750
165. Rodríguez-Martínez A et al (2019) Exosomal miRNA profile as complementary tool in the diagnostic and prediction of treatment response in localized breast cancer under neoadjuvant chemotherapy. Breast Cancer Res 21:1–9
166. Xu YF et al (2017) Plasma exosome miR-196a and miR-1246 are potential indicators of localized pancreatic cancer. Oncotarget 8(44):77028–77040
167. Xin Y et al (2022) Exosomal hsa-miR-129-2 and hsa-miR-889 from a 6-microRNA signature might be a potential biomarker for predicting the prognosis of papillary thyroid carcinoma. Comb Chem High Throughput Screen 25(5):819–830
168. Madhavan B et al (2015) Combined evaluation of a panel of protein and miRNA serum-exosome biomarkers for pancreatic cancer diagnosis increases sensitivity and specificity. Int J Cancer 136(11):2616–2627
169. Wu J, Shen Z (2020) Exosomal miRNAs as biomarkers for diagnostic and prognostic in lung cancer. Cancer Med 9(19):6909–6922
170. Cheng F, Su L, Qian C (2016) Circulating tumor DNA: a promising biomarker in the liquid biopsy of cancer. Oncotarget 7(30):48832–48841
171. Hwang H et al (2019) Evaluation of MARK BTM for quantitative measurement of three tumor markers: prostate specific antigen, alpha fetoprotein, and carcinoembryonic antigen. Clin Lab 65(8)
172. Wang Y et al (2018) Serum exosomal micro RNA s combined with alpha-fetoprotein as diagnostic markers of hepatocellular carcinoma. Cancer Med 7(5):1670–1679
173. Barcelo M et al (2019) Semen miRNAs contained in exosomes as non-invasive biomarkers for prostate cancer diagnosis. Sci Rep 9:13772
174. Zhang B, Shi H, Wang H (2023) Machine learning and AI in cancer prognosis, prediction, and treatment selection: a critical approach. J Multidiscip Healthc 16:1779–1791

Chapter 25
Emerging Clinical Applications of MicroRNAs in Cancer

Mohamed Adil Althaf, Agalya Murugan, Afshan Sathar, Gautham Chengizkhan, Neha Guruswamy, Sridhar Muthusami, Satish Ramalingam, Lurdes Queimado, R. Ileng Kumaran, and Ilangovan Ramachandran

Abstract MicroRNAs (miRNAs/miRs) have become fundamental to modern cancer diagnostics and treatment, as they can regulate the expression of various genes, and their dysregulation can lead to several cancers. This book chapter integrates recent findings on miRNAs, highlighting their importance in diagnostic, prognostic, and therapeutic advancements in oncology and cancer biology. This chapter also delineates the dysregulation of miRNAs and its implications in the initiation and the progression of different cancers. In line with these insights, emerging technologies including long-read sequencing, single-cell sequencing, and next-generation sequencing have significantly strengthened the detection platforms that aid in the high-resolution profiling of miRNA expressions. Sensors and devices driven by nanotechnology offer increased sensitivity and specificity using materials such as graphene, carbon nanotubes, and gold nanoparticles. The CRISPR-based detection platforms, notably, the *CRISPR-Cas13* system have enabled researchers to accurately and rapidly detect the miRNAs through fluorescence and electrochemical techniques. Biomarker discoveries have benefited from machine learning and artificial intelligence (AI) algorithms, enabling rapid characterization of complex miRNA expression patterns. The adoption of non-intrusive liquid biopsy technologies and exosomal miRNA profiling has substantially improved the early cancer

M. A. Althaf · A. Murugan · I. Ramachandran (✉)
Department of Endocrinology, Dr. ALM PG Institute of Basic Medical Sciences, University of Madras, Taramani Campus, Chennai, Tamil Nadu, India
e-mail: ilangovan@unom.ac.in

A. Sathar
LIFE FIRST Multi Speciality Clinic, Chennai, Tamil Nadu, India

G. Chengizkhan
Department of Otolaryngology – Head and Neck Surgery, The University of Oklahoma Health Sciences Center, Oklahoma City, OK, USA

N. Guruswamy
University of Missouri – Kansas City, School of Medicine, Kansas City, MO, USA

S. Pathak, A. Banerjee (eds.), *MicroRNAs: Role in Cancer Progression, Diagnostics and Therapy*, Advances in Biochemistry in Health and Disease 34,
https://doi.org/10.1007/978-3-032-08824-6_25

detection using the circulating miRNA levels in blood, saliva, and urine. Portable point-of-care (PoC) diagnostic tools, such as microfluidic chips and smartphone-integrated biosensors, have expanded easy, cheap and rapid access to technologies for clinical applications. Collectively, these multidisciplinary strategies have significantly strengthened the clinical utility of miRNA-based technologies in cancer, improving the diagnosis, prognosis, treatment, and personalized cancer management.

Keywords MicroRNA (miRNA/miR) · Cancer diagnosis · Biomarker · Nanotechnology · *CRISPR-Cas13* · Next-generation sequencing (NGS) · Single-cell RNA sequencing (scRNA-seq) · Liquid biopsy

Introduction

Micro ribonucleic acids (miRNAs/miRs) are small noncoding RNA molecules that are 21 to 25 nucleotides long, which can regulate multiple genes, and their dysregulations are linked to numerous pathological conditions including cancer [1]. These important molecules, despite their small size, critically influence the gene expression because they regulate the RNA silencing processes [2]. MiRNAs are present in many different types of organisms, including animals, plants, and certain viruses [3]. Researchers have conducted numerous studies in medicine as well as molecular biology, identifying the critical molecular and cellular functions of miRNAs.

Experiments from Victor Ambros's and Gary Ruvkun's laboratories identified Lin-4 as the first miRNA in the nematode *Caenorhabditis elegans*. The Nobel Prize in Physiology or Medicine in 2024 was awarded to them for this discovery. MiRNAs interact with messenger RNAs (mRNAs) influencing the regulatory processes, but they do not encode the proteins [4], and thus miRNAs are classified as noncoding RNAs.

S. Muthusami
Department of Biochemistry, Karpagam Academy of Higher Education, Coimbatore, Tamil Nadu, India

Karpagam Cancer Research Centre, Karpagam Academy of Higher Education, Coimbatore, Tamil Nadu, India

S. Ramalingam
Department of Genetic Engineering, School of Bio-Engineering, SRM Institute of Science and Technology, Kattankulathur, Tamil Nadu, India

L. Queimado
Department of Otolaryngology – Head and Neck Surgery, The University of Oklahoma Health Sciences Center, Oklahoma City, OK, USA

Department of Cell Biology, The University of Oklahoma Health Sciences Center, Oklahoma City, OK, USA

R. I. Kumaran
Biology Department, Farmingdale State College, Farmingdale, NY, USA

MiRNAs regulate the gene expression at the posttranscriptional level [5]. MiRNAs can bind to mRNA complementary sequences, and thus silence the gene expression [6]. The initiation of gene silencing will be accomplished after the binding of miRNA to mRNA *via* several mechanisms. The primary mechanism involves cleaving the mRNA strand, and this cleavage results in two distinct fragments [7]. The mRNA cleavage ceases its translation into protein form completely, which is necessary for cellular functions [8]. MiRNAs also destabilize mRNA by shortening its poly(A) tail, which is crucial for mRNA stability and efficient protein translation. This reduction in poly(A) tail length facilitates mRNA degradation [9]. Ribosomal interference is another viable mechanism by which miRNAs can be used to stop protein synthesis, though it is not yet fully understood. This can be caused by various mechanisms that interfere with the ability of ribosomes to translate the mRNA into a functional protein [10, 217]. These effects of miRNAs highlight their importance in gene regulation. MiRNAs also provide cells with an influential tool so that cells can target protein synthesis when they respond to many cues.

An increasing focus on miRNAs by both researchers and clinicians has established their distinctive role, prompting cancer biologists to consider them as vital markers for diagnosis and therapy [11]. Due to their participation in numerous physiological and pathological processes, they are valuable biomarkers for identifying cancers, cardiovascular diseases, neurological disorders, etc [12]. Recently, the application of miRNAs in disease diagnostics and therapeutic research has expanded significantly. They may serve as potential biomarkers for cancer diagnosis, prognosis, and targeted therapies [14, 15]. Early detection and diagnosis of various cancers can be enhanced through several state-of-the-art techniques. Next-generation sequencing (NGS)-based profiling of miRNAs expression levels shows great potential for use in the drug development applications. This rapid advancement in the profiling of miRNAs expression levels, may enable the clinicians and researchers to design targeted therapeutics for a myriad of diseases including cancer [14]. Such advancements potentially bring revolutionary changes within the diagnostic methods, which could be personalized. MiRNAs can serve as cancer biomarkers and they can be considered for the development of targeted therapies in personalized medicine [16]. Numerous researchers are addressing the challenge of modifying miRNA activity through the use of miRNA mimics, inhibitors, and antagomiRs to regulate the dysregulated miRNA expression. These innovative approaches hold potential for future disease treatment and enhance the understanding of disease mechanisms [17]. This chapter delineates the recent progress made in the clinical research on miRNAs, with emphasis on their roles as potential diagnostic or prognostic biomarkers and therapeutics in the field of oncology. In this chapter, we analyzed the dysregulation of specific miRNAs in a spectrum of cancers and highlighted their clinical implications in cancer diagnosis, prognosis and therapy.

MiRNAs in Cancer

Alterations in inflammatory pathways, genetic or epigenetic regulation can contribute to the process of carcinogenesis [18]. Cancer is a multifaceted complex disease, which is triggered *via* changes to various genes. MiRNAs regulate several cellular processes and an aberrant miRNA signature can lead to development of various diseases, including cancer [19]. MiRNAs can function either as cancer-promoting molecules (oncomiRs) or as tumor suppressors depending on the specific genes they regulate. For example, miR-21 is frequently upregulated in many cancers, where it increases tumor growth since it inhibits the expression of tumor suppressor molecules like PTEN and PDCD4 [206–208]. In contrast, certain miRNAs (e.g., let-7 family) regulate oncogenes, including RAS and MYC. A decrease in the expression of these miRNAs is associated with the development of more aggressive cancer types [20, 209].

Multiple factors contribute to altered miRNA expression in cancer. These include changes in DNA structure and epigenetic modifications that affect miRNA gene regulation, mutations or dysregulation of key miRNA-processing enzymes such as *DROSHA* and *DICER*, and transcription factors [21, 210].

MiRNAs regulate cancer cells as well as cancer stem cells (CSCs). They also regulate the critical cell signaling pathways that are involved in self-renewal of CSCs and alter the expression of key stemness factors to increase the cancer stemness [22]. Targeting the key signaling pathway, namely WNT/β-catenin signaling in salivary gland carcinoma ex-pleomorphic adenoma (CaExPA) cancer cells modulated the expression of key miRNAs such as miR-200c as well as let-7a, which suppressed the cancer stemness [40].

The expression levels of different miRNAs/miRs that are frequently associated with the lung, cervical, breast, colorectal, prostate, pancreatic, brain, and liver cancers are depicted in Figure 25.1, which illustrates the upregulation and downregulation or expression patterns of different miRNAs/miRs in various types of cancer.

Lung Cancer

MiRNAs influence the development, progression, and therapy resistance of several cancers, including lung cancer. MiR-210 is an oncogenic miRNA that is often found to be highly expressed in lung cancer patients and is mostly overexpressed under hypoxic conditions. MiR-210 interacts with both *ISCU* 1/2 and *E2F3* genes, which promote the survival of lung cancer cells in hypoxic environments. This interaction facilitates cell proliferation and angiogenesis, ensuring an adequate supply of oxygen [23]. Among the miRNAs that are dysregulated in non-small cell lung cancer (NSCLC), miR-155 is frequently overexpressed. It aids in tumor cell proliferation and facilitates escaping immune surveillance *via* negatively regulating SOCS1 and

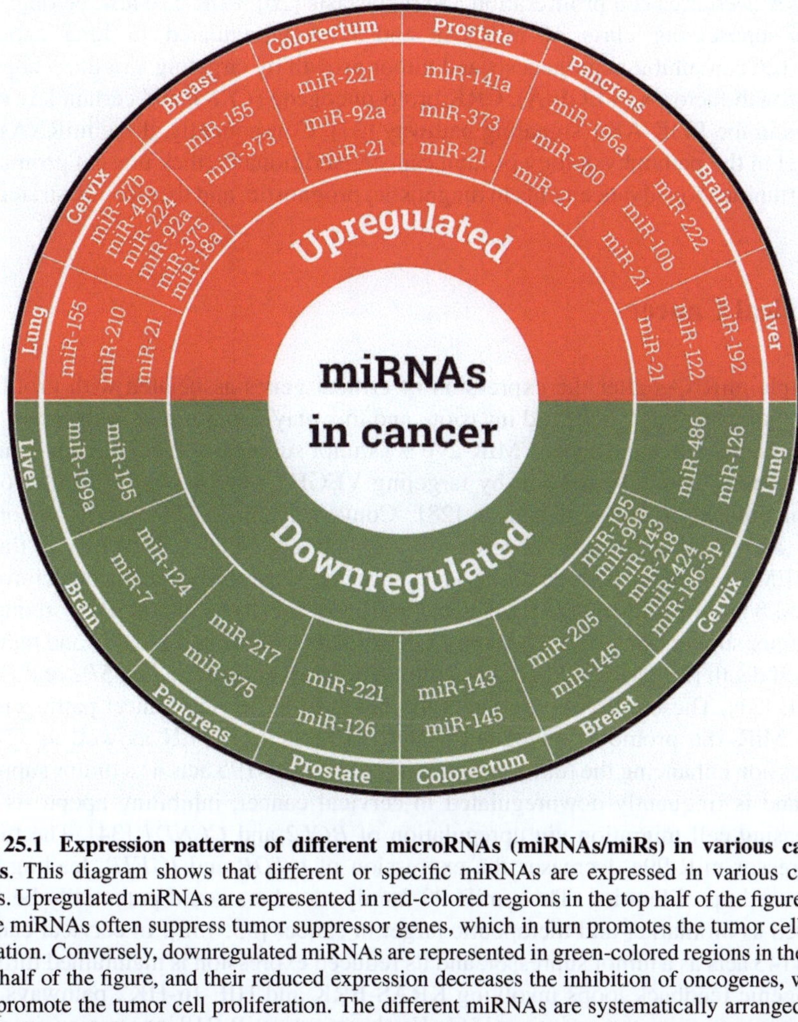

Fig. 25.1 Expression patterns of different microRNAs (miRNAs/miRs) in various cancer types. This diagram shows that different or specific miRNAs are expressed in various cancer types. Upregulated miRNAs are represented in red-colored regions in the top half of the figure, and these miRNAs often suppress tumor suppressor genes, which in turn promotes the tumor cell proliferation. Conversely, downregulated miRNAs are represented in green-colored regions in the bottom half of the figure, and their reduced expression decreases the inhibition of oncogenes, which can promote the tumor cell proliferation. The different miRNAs are systematically arranged corresponding to their various cancer types, which includes the lung, cervical, breast, colorectal, prostate, pancreatic, glioblastoma, and liver cancers. The arrangement or organization of different miRNAs in the figure, provides insight into the relationship between the different expression patterns of miRNAs and their corresponding various cancer types, and thereby highlighting the significance of miRNAs and their diagnostic as well as therapeutic potential in cancer treatment.

TP53INP1 tumor suppressor protein expression [24]. The overexpression of miR-21 facilitates the progression, invasion, and therapeutic resistance of lung cancer by modifying the crucial regulatory proteins, including *PTEN*, *PDCD4*, and *RECK* [25].

MiR-486 acts as a tumor suppressor and is often downregulated in lung cancer. By directly targeting insulin-like growth factor 1 (IGF1), Rho GTPase-activating protein 5 (ARHGAP5), and signaling other oncogenic pathways, this specific miRNA decreases cell proliferation and metastasis [26]. MiR-126 also belongs to a tumor-suppressing class of miRNAs and is downregulated in lung cancers. MiR-126 can inhibit angiogenesis and tumor growth by targeting vascular endothelial growth factor A (VEGF-A), CRK proto-oncogene (*CRK*), and certain key molecules in the PI3K/AKT signaling pathway [27]. Consequently, these miRNAs are pivotal in the pathophysiology of lung cancer. Additionally, they present promising opportunities for advancements in diagnostic, prognostic, and therapeutic strategies.

Cervical Cancer

Multiple miRNAs alter the expression of critical genes associated with proliferation, apoptosis, migration, and invasion, and this play a major role in the onset and progression of cervical cancer. MiR-27b is a tumor suppressor that typically inhibits angiogenesis and cell invasion by targeting VEGF-C, and its expression is often downregulated in cervical cancer [28]. Conversely, miR-224 acts as an oncogene and helps in cancer progression through epithelial–mesenchymal transition (EMT), is overexpressed in cervical cancer, and it promotes cell growth through RASSF8 and P21 targets [29]. Likewise, miR-92a acts as an oncomiR, by reducing the tumor suppressors like *PTEN* and *FOXF2*, increases the cell growth, and reduces the cell death [30]. MiR-375 acts as a tumor suppressor by acting on *SP1* and *PAX6* targets [31]. These miRNAs are integrally involved in cervical cancer pathogenesis [32]. MiR-18a promotes oncogenic traits *via* reducing PTEN as well as CTGF expression enhancing the tumor progression [33]. MiR-195 acts as a tumor suppressor, and is frequently downregulated in cervical cancer, inhibiting apoptosis and increasing cell migration *via* upregulation of *BCL2* and *CCND1* [34]. The tumor suppressor miR-99a decreases the expression of *mTOR* and *IGF1R*, leading to a decrease in proliferation. Thus, miR-99 family members could potentially be considered as biomarker and therapeutic target in cancer [35, 158]. In cervical cancer, miR-143 acts as a tumor supressor, and its reduced expression is maintained through oncogenic feedback loops involving KRAS-ERK and HIF-1α-HK2 pathways that further repress its transcription [36]. Furthermore, miR-218 expression is often downregulated in HPV-infected cervical tissues, which increases proliferation and metastasis through the regulation of *ROBO1* and *LAMB3* [37]. MiR-424 targets the *CDK6* and *CHAF1A* genes, which induce cell cycle arrest and act as tumor suppressors [38]. MiR-186-3p is another tumor suppressor miRNA that is downregulated in cervical cancer. It modulates *TWIST1* and *YAP1* expression to impede tumor progression and metastasis [39]. Comprehension of the biology of these miRNAs is critical. This understanding is needed, as knowledge of cervical cancer pathogenesis requires it. Recent studies suggest that miRNAs could greatly advance the biomarker diagnostics as well as targeted treatment.

Breast Cancer

MiR-373, miR-155, miR-205, and miR-145 are key miRNAs that play crucial roles in the initiation and development of breast cancer. Notably, in triple-negative breast cancer, the role of oncogenic miR-373 is critical. It promotes the invasion and metastasis of tumor cells by modulating the expression of *CD44* and *RAD52* genes [41]. MiR-155 is an oncomiR that is often overexpressed in breast cancer. It plays an important role in tumorigenesis because it stimulates proliferation of cells, evades the apoptosis and immunity. The promotion and evasion occur mainly through interactions with targets [42], which include the *TP53INP1* and *SOCS1* genes.

MiR-205 primarily functions as a tumor suppressor. However, it plays dual roles in breast cancer. It inhibits cell invasion and migration by modulating *ZEB1* and *ZEB2* expression, and also it suppresses EMT through the PI3K/AKT signaling pathway. Notably, reduced levels of miR-205 are closely associated with poor prognosis [44]. MiR-145 acts as a tumor suppresor and its expression is downregulated in breast carcinoma. It inhibits both cell proliferation and migration by targeting of FSCN1, SOX2, and OCT4 [43]. Thus, miRNAs could be used as biomarkers for prognosis, as well as potential therapeutic targets, as miRNAs are crucial regulatory molecules in breast cancer.

Colorectal Cancer

MiRNA mimics have the potential to activate the tumor suppressor genes or inhibit oncogenes, as well as pro-inflammatory cytokine genes [45]. While some miRNAs are classified as oncomiRs that can function as oncogenic regulators, others can act as anti-oncomiRs, similar to tumor suppressor miRNAs [46]. Several miRNAs, including miR-221, miR-92a, miR-21, miR-143, and miR-145, regulate multiple genes involved in colorectal cancer (CRC). These genes influence key cellular processes such as proliferation, apoptosis, migration, and neoangiogenesis, thereby driving CRC progression. Among them, miR-221 is frequently overexpressed in CRC and functions as an oncogenic miRNA. It supresses the tumor suppressors CDKN1B (p27) and CDKN1C (p57), and thereby, promotes the proliferation and metastasis [47]. Colorectal tumors overexpress miR-92a from the miR-17-92 cluster. This overexpression promotes tumor growth by suppressing both *PTEN* and *KLF4*, inhibiting the apoptosis, and increasing the invasive potential [48]. MiR-21 is a key miRNA in CRC and is associated with adverse outcomes. It promotes tumor growth and chemoresistance, as well as metastasis, through the downregulation of *PTEN*, *PDCD4*, and *TPM1* [49]. In CRC, there was a significant decrease in miR-143 and miR-145, the tumor suppressors; miR-143 targets *KRAS* and *ERK5* while inhibiting the cell proliferation and migration [50]. Stemness, invasion, and tumor growth are suppressed by miR-145 through alteration in *FSCN1*, *SOX2*, and

OCT4 gene expression [51]. Often, the downregulation of miR-143 and miR-145 is associated with tumor progression and poor CRC prognosis.

Prostate Cancer

MiRNAs play a crucical role in the progression of prostate cancer. MiR-141a is frequently upregulated in several cases of prostate cancer and it belongs to the miR-200 family. The increased expression of miR-141a is associated with metastatic behavior and can lead to more advanced disease stages, as it can promote EMT and increase the tumor growth by regulating the *PTEN* gene [52]. Moreover, by downregulating *LATS2* and *CD44* [53], miR-373 functions as an oncomiR in prostate cancer, promoting the cell migration and invasion. The miR-21 is a well studied oncomiR and is always found overexpressed in prostate cancer. Since it affects tumor suppressors like *RECK*, *PDCD4*, and *PTEN* [54], it helps the cells to proliferate, invade, and resist the chemotherapy. Likewise, miR-221 functions as an oncogenic miRNA and accelerates the cell cycle progression and increases the prostate cancer cell growth by inhibiting *p27Kip1* and *p57Kip2* [55]. In contrast, miR-128 functions as a tumor suppressor and is downregulated in prostate cancer. It targets the *BMI1* and *E2F3* genes, thereby inhibiting the tumor cell growth and survival [56]. Thus, miRNAs play a significant role in prostate cancer, making them as important biomarkers for diagnosis, prognosis and treatment.

Pancreatic Cancer

MiRNAs can modulate the critical genes that regulate cell growth, apoptosis, invasion, and chemotherapy resistance. Thus, changes in their expression can regulate the growth of pancreatic cancer cells. MiR-217 is a tumor suppressor miRNA in pancreatic ductal adenocarcinoma (PDAC). This miRNA inhibits the cancer cell growth and EMT by targeting *KRAS* and *EZH2* gene expressions [61]. Similarly, miR-375 can curtail *YAP1* and *JAK2* expression, restraining tumor advancement and spread. MiR-375 is frequently downregulated in pancreatic cancer cases [62]. MiR-196a, an oncogenic miRNA, is often overexpressed in pancreatic cancer, which can promote the tumor growth and can lead to poor prognosis in patients [63]. The tumor-suppressive genes *HOXB9* and *NFKBIA* are targeted by this miRNA. The miR-200 family includes miR-200a, -b, and -c and act as tumor suppresor miRNAs. They can inhibit the EMT and metastasis by downregulating the *ZEB1* and *ZEB2* gene expression [64]. The expression of miR-200 family varies according to the tumor stage. MiR-21 is an oncomiR that is frequently upregulated in pancreatic cancer. It targets *PDCD4*, *PTEN*, and *RECK,* and thus confers resistance to chemotherapy in pancreatic cancer [65].

Brain Cancer

MiRNAs can contribute to the initiation and progression of glioblastoma, as they can modulate key signaling pathways, and promote cancer cell survival and chemo resistance. MiR-124, is downregulated, in glioblastoma tumors which leads to increased cancer cell migration and proliferation since it targets several genes, including *SNAIL2*, *STAT3*, and *CDK6*, making it valuable for therapies [66]. The tumor suppressor miR-7 is also downregulated in glioblastoma and other cancers like breast and lung. It can suppress the EGFR/PI3K/AKT signaling pathway through EGFR and IRS-1, and thus reduce the tumor growth and invasion [67]. Importantly, miR-222 functions as an oncomiR in glioblastoma and other cancers, such as prostate, breast, and thyroid cancers. It can promote cell growth, migration, and resistance to chemotherapy by downregulating *p27Kip1* and *PTEN* [68]. Additionally, miR-10b is significantly overexpressed in glioblastoma as well as various high-grade cancers including breast and pancreatic cancers. It can target both *HOXD10* and *RhoC*, thus promoting the invasion and metastasis. Its expression is correlated with a poor prognosis [69]. Furthermore, miR-21 is one of the most well-studied oncomiRs, and it is overexpressed in glioblastoma and several other cancers, including colorectal, breast, and pancreatic cancers. It targets tumor suppressor genes like *PTEN*, *PDCD4*, and *RECK*, and thus enable tumors to grow, evade apoptosis, and confer therapeutic resistance [70]. These miRNAs act as key regulators of tumor behavior in diverse cancer types and therefore, they hold immense potential as diagnostic markers and therapeutic targets.

Liver Cancer

MiRNAs play a crucial role in the progression and development of liver cancer. MiR-195, which is often downregulated in hepatocellular carcinoma (HCC), acts as a tumor suppressor. This miRNA inhibits cell proliferation and induces apoptosis by targeting genes like *CCND1* and *BCL2* [34]. Similarly, miR-199a is also found to be downregulated in HCC and other liver diseases. It exhibits tumor-suppressive activity by targeting mTOR, CD44, and HIF-1α, and thus inhibits cell growth and angiogenesis [71]. Also, miR-192, a tumor suppressor miRNA is commonly downregulated in HCC. It can target the *ZEB2* and can alter the cell cycle, and inhibit the EMT. Importantly, the decreased expression of miR-192 is associated with poor prognosis [72]. On the other hand, miR-122 is recognized as liver-specific miRNA. Which can suppress tumor growth, by targeting important molecules like cyclin G1, ADAM10, and various oncogenic pathways. Its downregulation in HCC is linked to increased tumor cell proliferation, metastasis, and chemotherapeutic resistance [124]. The oncomiR-21 is found to be upregulated in liver cancer and it enhances tumor growth, invasion, and drug resistance by targeting the *PTEN*, *PDCD4*, and *RECK* genes [74].

Diagnostic Innovations

MiRNAs have garnered attention as valuable diagnostic/prognostic biomarkers due to their stability in biological fluids, distinct expression profiles in various tissues, and their involvement in a multitude of pathological states [57]. Their diagnostic potential across a wide range of diseases is clearly demonstrated by recent advancements in both research and clinical environments.

Key Innovations in miRNA-based Diagnostic Technologies

In recent years, the field of miRNA-based diagnostic technologies has undergone profound transformations, highlighting the convergence of advancements in molecular biology, nanotechnology, and computational methods and is represented in Table 25.1. Significant innovation have emerged, which can be categorized into several key areas:

Highly Sensitive Detection Methods

The sensitivity and specificity of miRNA detection methods have been markedly enhanced by the development of advanced techniques for the early detection of cancer.

Quantitative Reverse Transcription Polymerase Chain Reaction (qRT-PCR) has significantly enhanced the detection of low-abundance miRNAs in biological samples through primers and sensitive probes [58]. These refinements increase the sensitivity and specificity, underscoring qRT-PCR's essential role in molecular biology [59, 119]. Optimized primers boost amplification efficiency and minimize nonspecific binding for accurate miRNA quantification in complex samples [60]. Selecting primers based on sequence, melting temperature and binding affinity is crucial for ensuring specific hybridization and reducing cross-reactivity [75]. Such advancements enhance qRT-PCR accuracy, aiding in the analysis of miRNA expression and their implications in biological processes and diseases. qRT-PCR has been successfully used to quantify miRNAs like miR-21 and miR-155, which are critical regulators of genes namely *PTEN*, *PDCD4*, *RECK*, *TPM1*, *SOCS1*, *SHIP1*, *FOXO3a*, and *TP53INP1* in various cancers such as breast, lung, and colorectal cancer.

Droplet digital PCR (ddPCR) is an advanced technique for quantifiying nucleic acids, which provide improved sensitivity and specificity compared to conventional PCR systems [76]. This approach partitions samples into thousands of nanoliter-sized droplets, with each droplet serving as template for amplification [77]. By utilizing a droplet generator for emulsification, the system produces separate droplets

Table 25.1 Key innovations of miRNA-based cancer diagnostic technologies. In this table, key advancements made in the miRNA-based cancer diagnostics are summarized. It describes many diagnostic platforms or technologies as well as the respective critical mechanisms. This table also shows the different target miRNAs/miRs that are initially detected by the technologies and their corresponding various cancer types. Expression of different miRNAs in various cancer types have been shown to be of crucial relevance in clinical oncology and cancer research or biology. Therefore, these specific miRNAs expression or different miRNAs expression patterns can serve as important biomarkers for cancer diagnosis, prognosis, treatment, and identification of molecular residual disease.

S.No.	Technologies	Mechanisms of technologies	Target miRNAs/ miRs detected	Cancer types
1.	qRT-PCR	Reverse transcription followed by amplification and quantification of miRNAs	miR-21, miR-155	Breast, lung, colorectal
2.	Droplet digital PCR (ddPCR)	Partitioned amplification for precise quantification of low-abundance miRNAs	miR-210, miR-122	Liver, renal, pancreatic
3.	Isothermal amplification methods	Constant temperature amplification for rapid detection of miRNAs	miR-205, miR-141	Prostate, bladder
4.	Enhanced library preparation	Improved sequencing accuracy and coverage of miRNAs by reducing the ligation and size bias of miRNAs	miR-143, miR-200c	Colorectal, ovarian
5.	Single-cell RNA-seq (scRNA-seq)	Captures cell-to-cell heterogeneity in the miRNAs expression profiles	miR-1246, miR-451a	Breast, glioblastoma
6.	Long-read sequencing	Detects full-length miRNAs and isomiRs without fragmentation bias of miRNAs	miR-16, miR-375	Prostate, gastric
7.	Gold nanoparticle biosensors	Colorimetric detection *via* hybridization with target miRNAs	miR-21, miR-221	Breast, hepatocellular carcinoma
8.	Graphene/carbon nanotube sensors	Conductivity changes upon hybridization with miRNAs	miR-155, miR-10b	Breast, pancreatic
9.	SPR and electrochemical sensors	Detects binding to target miRNAs *via* real-time signal shifts	miR-126, miR-196a	Lung, gastric
10.	CRISPR-Cas13 systems	Cas13/Cas12 binds to target miRNAs and activates collateral cleavage to generate signals to detect specific target miRNAs	miR-21, miR-19a	Glioblastoma, colorectal

(continued)

Table 25.1 (continued)

S.No.	Technologies	Mechanisms of technologies	Target miRNAs/ miRs detected	Cancer types
11.	Cas-based fluorescence/ electrochemical sensors	MiRNAs detection signal amplification using Cas enzymes upon binding to miRNAs	miR-125b, miR-27a	Breast, leukemia
12.	Deep learning models	Predicts cancer-associated mRNAs patterns from large miRNA datasets	miR-34a, miR-200 family	Pancreatic, ovarian, lung
13.	AI-assisted pattern recognition algorithms	Identifies diagnostic miRNAs signatures across various cancer types	miR-203, miR-146a	Head and neck, cervical
14.	Circulating miRNAs detection	Quantifies free-floating miRNAs in body fluids	miR-21, miR-375	Lung, prostate, colorectal
15.	Exosomal miRNAs levels profiling	Isolates tumor-derived exosomes to analyze their miRNAs levels	miR-1246, miR-1290	Ovarian, pancreatic
16.	Portable microfluidic chips	Miniaturized platforms for on-site sample collection and miRNAs detection	miR-141, miR-195	Prostate, breast
17.	Smartphone-integrated biosensors	Mobile-linked sensors for remote real-time miRNAs detection	miR-21, miR-210	Lung, glioblastoma

that contain the target nucleic acids along with PCR reagents, enabling distinct amplification during the thermal cycling process [78]. In contrast to qPCR, which relies on threshold cycles to estimate concentrations, ddPCR offers absolute quantification by counting the number of positive and negative droplets after amplification, resulting in accurate measurements and enabling the detection of ultra-low copies of miRNAs [79]. This method has significantly impacted fields such as oncology, genetic research, and microbial detection, especially for detecting low-abundance targets in complex samples, including minimal residual disease in cancer patients and rare genetic variants [80]. Furthermore, ddPCR shows resilience against common biological inhibitors, making it a preferred choice for both research and clinical applications. Its increased sensitivity enhances quantification accuracy, contributing to the advancement of personalized medicine approaches [81]. Additionally, ddPCR represents a notable improvement in the absolute quantification of miRNAs, allowing for their evaluation without the necessity of reference genes, thereby streamlining analysis and comparisons [82]. Specifically, ddPCR has been successfully utilized to quantify miRNAs such as miR-210 and miR-122, which are important in liver, renal, and pancreatic cancers, thereby underscoring its significance in cancer diagnostics and treatment monitoring.

Isothermal amplification techniques are a significant innovation in molecular biology and diagnostics, enabling efficient nucleic acid amplification at a constant temperature without thermal cycling. The key techniques include Loop-Mediated Isothermal Amplification (LAMP) and Nucleic Acid Sequence-Based Amplification (NASBA) [83]. LAMP utilizes specially designed primers to create loop structures, facilitating rapid amplification through multiple primers, which is advantageous for point-of-care diagnostics requiring swift results [84]. Meanwhile, NASBA focuses on RNA amplification through several enzymatic steps, including reverse transcription and use of RNA polymerase functioning at a constant temperature [85, 86]. Both LAMP and NASBA have gained attention for their practicality in diverse settings and enhancing access to reliable nucleic acid testing, particulary resource-limited environments [87]. These isothermal methods have also been applied in the detection and quantification of specific miRNAs, such as miR-205 and miR-141, which can target genes such as *ZEB*, *VEGF*, *HER3*, *TGFβ2*, *DLK1*, and *PTEN*, and thus are important biomarkers associated with prostate and bladder cancers. Their integration into isothermal platforms supports the development of rapid, accessible, and sensitive diagnostic tools for cancer screening and monitoring.

Next-Generation Sequencing (NGS)

The realm of next-generation sequencing (NGS) has undergone notable improvements, leading to its significant implications in miRNA diagnostics. Improved library preparation techniques have significantly decreased sample bias and improved detection efficiency in sequencing. These techniques refine the extraction and amplification of nucleic acids, resulting in a more accurate representation of various miRNA species [88]. Consequently, researchers are able to capture a wider range of miRNA profiles during sequencing, which is crucial for both research and clinical applications [89]. The reduction of sample bias due to these advanced methods, enhances the reliability and reproducibility of results, essentials for both exploratory research and validation processes [90]. These enhancements have proven particularly beneficial in identifying and profiling specific miRNAs like miR-143 and miR-200c, that can regulate the genes *KRAS, ERK5, BCL-2, HK2, ZEB1/2, BMI1, TUBB3*, and *ETS1*, which play vital roles in colorectal and ovarian cancers. These innovation in miRNA sequencing improve the diagnostic accuracy and therapeutic outcomes for various health conditions, marking a noteworthy development in genomics and advancing the understanding of miRNA landscapes. NGS has demonstrated significant diagnostic utility in evaluating cancers driven by genetic mutations and molecular alterations.

Single-cell RNA sequencing (scRNA-seq) represents a major breakthrough in the field of genomics and molecular biology, facilitating the detailed examination of gene expression at the single-cell level and overcoming the constraints associated with bulk tissue analysis [91]. This method uncovers the transcriptomic variation present in diverse cell populations, providing essential insights into cellular functions, differentiation, and dynamics [92]. The scRNA-seq workflow includes isolating individual cell, extracting their RNA converting it to complementary DNA (cDNA) through reverse transcription, amplifying it, and sequencing it with next-generation technology [93]. The resulting data allow for the quantification of gene expression, help in the identification of unique cell types and elucidation of developmental pathways and molecular mechanisms related to biological functions and diseases [94]. scRNA-seq holds significant relevance across multiple disciplines, such as developmental biology, immunology, cancer research, and neuroscience, as it allows for the identification of distinct transcriptional profiles and previously unrecognized cellular subtypes and pathways [95]. Importantly, this technique has made it possible to detect and examine specific miRNAs, including miR-1246, which regulate genes like *DYRK1A, DR5, TP53INP1*, and *CCNG2* in cervical, lung, and colorectal cancers, and miR-451a, which regulates genes namely *MIF, CAB39*, and *YWHAZ* in gastric cancer and glioma. This method enhances our comprehension of intricate biological systems, particularly in scenarios where cell-type specific responses are vital, such as the immune reaction to pathogens and tumor microenvironments (TME).

Long-read sequencing technologies represent a significant breakthrough in molecular biology and genomics, allowing for the detection of long DNA sequences that may exceed what traditional short-read methods can achieve [96]. These technologies produce sequences that may encompass entire genes, regulatory elements, large structural variants leading to more comprehensive understanding of complex genomes [97]. By generating longer reads ranging from thousand to millions of base pairs, researcher are better equiped to overcome the challenges associated with short-read sequencing, such as difficulties in genome assembly and the analysis of repetitive region [98]. As a result long-read sequencing enables a more thorough investigation of genomic structure, variations, and functional components, providing deeper insight into genetic diversity, evolutionary patterns, and the molecular mechanism of diseases [99]. Recent advancements in this field include platforms such as Pacific Biosciences' Single Molecule Real-Time (SMRT) sequencing and Oxford Nanopore Technologies' (ONT) nanopore sequencing, each employing different methodologies to achieve the necessary read lengths for in-depth genomic studies [100]. Notably long read sequencing has also been utilized in examining miRNAs such as miR-16 and miR-375 that target genes such as *MTPN, SOX9, SOX17, HOXB13, SNAIL*, and *PAX6,* which can act as biomarkers for prostate and gastric cancers [204]. This application allows for a more precise characterization of their full-length precursors, isoforms and genomic contexts, enhancing diagnostic accuracy and propelling advancements in cancer genomics.

Nanotechnology-based miRNA Detection

Nanotechnology plays a pivotal role in advancing the miRNA detection methodologies through the development of highly specific and sensitive platforms. Gold nanoparticle (AuNP)-based biosensors are highly sensitive and innovative for the detection of biological analytes, owing to their unique optical, electronic, and catalytic characteristics [101]. Their favourable surface-to-volume ratio, biocompatibility and customizable surface chemistry enable the functionalization of AuNP with various biomolecules, ensuring specific binding to target analytes like proteins and pathogens [102]. These biosensors find widespread applications in medical diagnostics, environmental monitoring and food safety, necessitating precise biomolecule detection [103]. AuNPs' improve sensitivity and specificity by leveraging strong plasmonic properties that amplify signals the detection at picomolar concentrations through techniques such as colorimetry, fluorescence, and electrochemistry [104]. Advances in nanotechnology have further enhanced the capabilities of AuNP-based biosensors for multiplex detection, thereby increasing the diagnostic efficiency. The integration of advanced nanomaterials with microfluidics and lab-on-a-chip systems supports miniaturization and automation, making these biomarkers suitable for point-of-care testing and real-time monitoring [105]. AuNPs based biosensors have shown exceptional sensitivity in the detection of specific miRNAs, like miR-21 that can regulate genes such as *PTEN, PDCD4, RECK,* and *TPM1* are associated with breast cancer and miR-221 that target the genes *PTEN, STAT5a, IRF2*, and *ANXA*, which serve as crucial biomarkers for hepatocellular carcinoma [199]. Current research on the interactions between AuNPs and bimolecular targets aims to improve biosensor design, thereby advancing early disease detection and public health outcomes.

Graphene and carbon nanotube sensors mark a significant advancement in sensor technology due to their superior thermal, mechanical, and electrical properties. Graphene, a two-dimensional carbon material, is recognized for its excellent electrical conductivity, flexibility, and strength [106]. Simultaneously, carbon nanotubes derived from rolled graphene sheets exhibit high tensile strength, thermal conductivity, and advantageous surface area-to-volume ratio [107]. These traits make both materials ideal for application in chemical, biological and environmental sensing; their integration into sensors has produced devices that are highly sensitive and selective, capable of detecting low analyte concentrations [108]. Functionalized graphene develops sensors that target specific molecules for real-time data on environmental and health issues. Moreover, carbon nanotubes enhance a sensor sensitivity and response time through their high aspect ratio and conductivity, facilitating rapid electron transfer [109]. These sensor platforms have strong potential for detecting cancer-related miRNAs like miR-155, which target genes like *SHIP1, VHL, TP53, and SOX*, which are associated with breast cancer. MiR-10b, targets genes like *PTEN, RhoC*, and *KLF4*, and is associated with pancreatic cancer [200]. Their applications encompass environmental monitoring, medical diagnostics and industrial control, with ongoing research advancing innovations to enhance the accuracy and reliability [110].

Surface plasmon resonance (SPR) and electrochemical sensors are crucial biosensing techniques for the precise and sensitive detection of biomolecular interactions [111]. SPR is an optical method that leverages oscillations at the interface of metal and dielectric materials, where illumination triggers the formation of surface plasmons [112]. This allows for real-time tracking of molecular binding events, such as proteins interactions, without the need for labeling agents. Electrochemical sensors function through redox reactions using working, reference, and counter electrodes to identify specific analytes [113]. Variations in current from these electrochemical reactions facilitate accurate quantification, often improved by the incorporation of nanomaterials, enzymes, and antibodies [114]. These sensors are preferred for their quick response time, portability and minimal requirement for sample preparation in complex biological environments. Both SPR and electrochemical sensors play a vital role in medical diagnostics, environmental assessments and food safety, underscoring their importance in the field of analatical chemistry [115, 116]. These technologies have been utilized to identify cancer-associated miRNAs like miR-126 which target genes such as *VEGF, CRK, IRS-1* and *HOXA5*, can be used as biomarkers in lung cancer. The miR-196a targets genes like *FOXO1, HOXA9, ANXA1*, and *NOBOX*, and are reported in gastric cancers [201].

CRISPR-based miRNA Detection

The advent of CRISPR technology has opened new frontiers in the detection of miRNAs. CRISPR-Cas13 systems represent an advanced class of RNA-guided gene-editing technologies that have garnered significant interest in the field of genetic engineering and molecular biology [117]. Unlike the conventional *CRISPR-Cas9* systems, which targets DNA, Cas13 systems can specifically cleave and target RNA, providing unique capabilities for therapeutic use and gene regulation [118]. Originating from prokaryotic adaptive immune systems, these systems utilize RNA segments from previously encountered viruses as a molecular memory to defend against future infections [59, 119]. Class 2 CRISPR systems such as Cas13 proteins, function as ribonucleases that bind to RNA target sequences [120]. The versatality of CRISPR-Ca13 is exemplified by its applications in molecular diagnostics and therapeutic development, including its role in detecting viral pathogens like SARS-CoV-2 [121]. It induces RNA interference in eukaryotic cells, making it a valuable tool for modulating gene expression in diseases like cancer and genetic disorders [122]. *CRISPR-Cas13* has been used to target oncogenic miRNAs like miR-21, as it targets genes *PTEN, PDCD4* and *RECK*, which are critically involved in glioblastoma and miR-19a, which regulate genes including *FOXP1, TUSC2, SOX4*, and *TNFAIP3* are involved in CRC [202]. Cas13a and Cas13b provide different properties, allowing researchers to select systems tailored for specific applications [123]. Ongoing research into the mechanisms of *CRISPR-Cas13* continues to provide crucial insights into RNA targeting specificity and efficacy, cementing its importance in modern biotechnological advancements [124].

Caspase-mediated fluorescence and electrochemical biosensors play a vital role in the fields of biomedical research and diagnostics [125]. Caspases, are a group of cysteine proteases that are essential in the regulation of apoptosis, cellular growth and inflammation. Fluorescence biosensors operate with probes that emit light in response to specific biochemical interactions. When these probes bind to their targets, they undergo conformational changes, resulting in measureable fluctuations in fluorescene intensity. This mechanism provides valuable information about biological processes, including cell death and disease progression [126]. Electrochemical biosensors detect electrical signals produced by biochemical reactions and leverage the enzymatic activity of caspases for electron transfer. This capability allows for real-time monitoring of molecular interactions and enzyme activities, making these biosensors advantageous for live-cell assays and diagnostics [127]. These biosensors have demonstrated significant potential in detecting the cancer-related miRNAs, such as miR-125b, which targets several genes that are associated with breast cancer such as *ERBB2* (also known as HER-2), *BAK1, PUMA*, and *DRAM2* [128]. MiR-27a targets genes such as *EGFR, TGF-βRI, FSTL1*, and *KRAS*, which are associated with leukemia [129]. Owing to their rapid responsiveness and high specificity and sensitivity, biosensors are utilized in clinical diagnostics, environmental assessments, and research investigations. This versatality supports multiplexed assays, that enable the simultaneous detection of multiple targets [103].

Artificial Intelligence (AI) and Machine Learning (ML) for miRNA Biomarker Discovery

Artificial intelligence (AI) and machine learning (ML) have begun to play a crucial role in the discovery and characterization of miRNA biomarkers.

Deep learning models have become vital tools for the analysis of large datasets of miRNA expression [130]. These models surpass conventional techniques complex disease-related signatures within extensive datasets. MiRNA expression analysis is challenging due to complex biological data that is essential in regulating gene expression and cellular activities [131]. Dysregulation of miRNA is linked to a range of diseases, including cancer [132]. Therefore, the capability of deep learning to manage large datasets is vital for enhancing our comprehension of these diseases [133]. Such techniques allow researchers to explore disease-specific miRNA expression profiles, potentially leading to the discovery of diagnostic and prognostic biomarkers [134]. Deep learning algorithms can autonomously identify high-dimensional datasets, thus facilitating the effective recognition of clinically significant changes in miRNA expression indicative of disease onset or progression [135]. MiRNAs such as miR-34a, which targets *CDK4, MYB*, and *CCNE2*, and the miR-200 family, which regulates *ZEB1, CDK6, TGFβ2*, and *ZEB2*, have emerged as important biomarkers with implications in malignancies of pancreatic, ovarian, and lung cancers [203]. By providing innovative perspectives on the detection and interpretation of miRNA expression patterns, deep learning methods play a significant role in the advancement of targeted therapies and tumor biology.

Algorithms for pattern recognition driven by AI are crucial in enhancing the early detection of diseases, particularly in oncology and neurodegenerative disorders [136]. Through ML techniques and data analytics can uncover intricate patterns within large medical datasets, increasing the diagnostic precision of miRNA-based tests [137]. MiRNAs serve as critical biomarkers, given their role in regulating gene expression and their associations with various diseases. By refining the analysis of miRNA expression profiles, AI algorithms improve both the specificity and sensitivity of diagnostic tests, resulting in accurate and timely disease identification [138]. This advancement has the potential to transform diagnostic practices by enabling early disease detection; moreover, AI not only recognizes patterns but also supports personalized medicine by assisting in the development of tailored treatment plans, miRNAs such as miR-203 targets genes such *ZEB2, IRS-1, BMI-1* and *SOCS3*, which have been identified as biomarkers in head and neck cancer. MiR-146a regulates genes like *TRAF6, IRAK1* and *NUMB*, and it can be used as a biomarker of cervical cancer [139]. The AI-enhanced evaluation of these miRNAs supports more accurate cancer detection and enables personalized therapeutic approaches. The integration of AI in evaluating miRNA assay analysis significantly improves the ability to identify the early cancerous lesions [140].

Non-invasive Liquid Biopsy Technologies

The development of non-invasive liquid biopsy technologies marks a significant milestone in early disease detection. Circulating miRNAs detection in blood, saliva, and urine, represents a notable progress in diagnostic techniques [141]. This noninvasive approach allows for the early identification of diseases, thereby minimizing the discomfort or risk associated with conventional invasive techniques [142]. Investigating these small, noncoding RNA molecules provides critical information regarding both physiological and pathological conditions [143]. This strategy improves patient comfort by employing readily accessible samples and allows for rapid collection, alleviating the stress associated with standard blood collection practices [144]. Moreover, the availability of this diagnostic method makes it appropriate for those who are reluctant to undergo invasive procedures [145]. The incorporation of miRNA detection into clinical assessments overcomes the shortcomings of traditional diagnostic methods, which often detects diseases only at later stages, hindering timely treatment [146]. The high sensitivity and specificity of miRNA biomarkers facilitate earlier disease detection, resulting in improved patient outcomes through timely therapeutic interventions [147]. Circulating miRNAs, including miR-21 and miR-375, are crucial in the regulation of essential genes *PKD1, YAP1, YWHAZ,* and *CADM1*, which have strong diagnostic potential in lung, prostate, and colorectal cancers [204]. Their presence in biofluids supports creation of minimally invasive, highly accurate screening tools that could transform early cancer diagnosis and lead to more effective, personalized treatments.

Exosome-based miRNA profiling marks a notable progress in the detection of biomarkers, improving the stability and specificity of diagnostic indicators [148]. Exosomes are nano-sized extracellular vesicles released by various cell types and play an integral role in intercellular communication, carrying biomolecules such as miRNAs, proteins, and lipids [149]. The miRNAs found in exosomes provide insights into the physiological and pathological conditions of their source cells, which makes their profiling a promising tool for clinical applications in diagnosis and prognosis [150]. Exosomal miRNAs exhibit remarkable stability and are resistant to degradation in biological fluids such as urine and blood, which facilitates dependable biomarker identification [151]. This inherent stability is crucial for accurately identifying molecular markers associated to various diseases, including oncological and neurodegenerative conditions [152]. Furthermore, the examination of exosomal miRNAs reveals a level of specificity that distinguishes biomarkers linked to various diseases from baseline signals influenced by the cellular environment [153]. Noteworthy exosomal miRNAs, such as miR-1246 and miR-1290, have been recognized for their role in targeting genes such as *CCNG2, NAT1, NKD1, BCL2*, and *FOXA1*, which have shown considerate diagnostic potential in ovarian and pancreatic cancers. The identification of these miRNAs in exosomes supports the progress of targeted, noninvasive screening techniques, thereby facilitating early diagnosis and developing individualized therapeutic approaches. As a result, the profiling of exosomal miRNAs levels, is positioned as an essential element in the advancement of noninvasive diagnostic tools, which can ultimately improve the outcomes for patients through early or timely cancer detection and targeted cancer therapy strategies [154].

Point-of-Care (PoC) miRNA Diagnostic Devices

The evolution of point-of-care (PoC) diagnostic devices has transformed the landscape of miRNA diagnostics. Portable microfluidic chips represent a significant advancement in biomedical analysis, especially concerning the assessment of miRNAs [155]. These devices are designed for on-site application, allowing healthcare professionals to make prompt clinical decisions with minimal training required for complex testing procedures [156]. Their ability to provide real-time results is essential for effective patient management. Additionally, this technology improves diagnostic efficiency by minimizing the delays commonly encountered in traditional laboratory processing [157]. The combination of microfluidic technology with advanced detection techniques enhances the sensitivity and specificity of miRNA analysis, thus enabling the identification of low-abundance biomarkers in a variety of biological specimens [35, 158]. This enhanced capability plays a crucial role in supporting the early identification of diseases and the formulation of individualized treatment strategies, thereby improving patient outcomes [159]. MiRNAs such as miR-141 can regulate the genes such as *KLF9, YAP1* and *RAB32*, which are linked to prostate cancer and miR-141 regulates genes like *CCND3, E3F3, FGF2* and *SRC-3,* that are biomarkers in breast cancers. This shows their potential as diagnostic biomarkers when analyzed through microfluidic technologies [205]. These devices facilitate the rapid and sensitive detection of cancer-related miRNAs at the PoC, presenting significant opportunities for earlier intervention and tailored therapeutic stratergies. As the healthcare industry places increasing importance on PoC diagnostics, these portable chips improve clinical workflows and enable quicker, more accurate medical treatments, highlighting their essential role in modern healthcare practices [160].

Smartphone-integrated bisensors: Recent advancement in mobile technology and biosensing have led to the creation of biosensors that are embedded in smartphones, enabling users to have immediate access to accurate biological detection for real-time health monitoring [161]. These biosensors utilize the processing power, connectivity options, and user interfaces of smartphones to analyze biological samples, including blood, saliva, and sweat employing techniques like electrochemical detection, optical measurement and immunoassays [162]. Innovations in microfluidics and nanotechnology have improved sensitivity through miniaturization techniques, making these devices more portable when compared to traditional diagnostic methods [163]. Smartphone-based platforms are being designed to detect cancer-related miRNAs, such as miR-21 and miR-210, which are associated with lung cancer and glioblastoma related genes like *HIF-1α, EFNA3*, and *VEGF*. Such integrations facilitate rapid, non-invasive screening and continuous monitoring of high-risk individuals, thereby improving early detection and disease management. Additionally, these devices may revolutionize personal healthcare by encouraging self-monitoring and disease prevention while providing valuable data for public health studies [164].

MiRNAs as Diagnostic Markers

MiRNAs exhibit remarkable stability in noninvasive testing scenarios, outpacing conventional biomarkers in both resilience and precision of measurement [13]. The differential expression of particular miRNAs within tumor tissues correlates with various pathological conditions, including cancer, as their altered levels aid in disease identification and offer valuable perspectives on underlying biological mechanisms [165]. Additionally, miRNAs reflect disease-related molecular changes, assisting in monitoring treatment and risk assessment of relapse. By analyzing miRNA level variations, therapeutic strategies can be refined to promote the effective disease management [166]. These are strategies to improve the patient outcomes and reduce the side effects.

Circulating miRNAs as Diagnostic Indicators in Cancer Research

MiRNAs are significant, non-invasive biomarkers that are recognized for diagnosis, prognosis, and assessment of therapeutic responses in cancer [167]. Their resilience in biological fluids such as blood, saliva, and urine establishes them as promising candidates for applications in liquid biopsy. MiRNAs are present across various bodily fluids, indicating their extensive distribution [168]. The stability of miRNAs can be explained by several protective mechanisms, which include their encapsulation within exosomes, their association with protein complexes-particularly Argonaute, and their interactions with lipoproteins [169]. Furthermore, particular miRNAs express distinctive expression profiles, characterized by specific patterns of upregulation and downregulation across different cancer types [170]. Timely detection of these miRNAs is crucial to identify cancers at pre-symptomatic stages, often before tumors can be detected through imaging, thereby improving personalized medicine and targeted therapies [171]. Figure 25.2 depicts the biogenesis and secretion pathways of extracellular and endogenous miRNAs.

Advancement in miRNA-based Therapeutics in Cancer

The strategic development of miRNA-based therapies has become central to molecular medicine, impacting several diseases including cancer. These approaches consist of inhibition, modification and replacement [172]. MiRNA inhibition often uses antisense oligonucleotides (ASOs) including anti-miRs and antagomiRs to reduce the overexpression of disease-associated miRNAs [173]. Miravirsen, which targets miR-122, can target *IGF1R, CAT1*, and *PRKRA* genes to reduce hepatitis C virus replication [174]. RG-125 (AZD4076) enhances insulin sensitivity by therapeutically inhibiting the overexpressed miR-103/107, thereby improving insulin signalling pathways implicated in type 2 diabetes [175]. MiR-103/107 is also overexpressed in many cancers, and is associated with tumor progression, metastasis, and poor prognosis. In contrast, miRNA replacement therapy invloves the administration of synthetic miRNA mimics aimed at restoring the expression of tumor-suppressive miRNAs, which are frequently diminished in cancerous tissues [176]. For liver cancer, the therapeutic agent *MRX34* has been developed [177], while in idiopathic pulmonary fibrosis, the miR-29 mimic has demonstrated the capacity to regulate genes such as *CDK6* and *MCL1* [178]. Also, miR-29 has a complex role in cancer and its regulation could be critical. Methods such as miRNA sponges and decoys inhibit unwanted miRNAs through competitive binding. For instance, sequences targeting miR-155 (oncomiR) that regulate genes like *BCL6, SOCS1*, and *TNF-α*, have demonstrated efficacy in reducing the inflammation in autoimmune diseases [179, 180]. Genomic-editing tools like CRISPR enable the permanent modification of miRNA genes, allowing for the targeted degradation of oncogenic

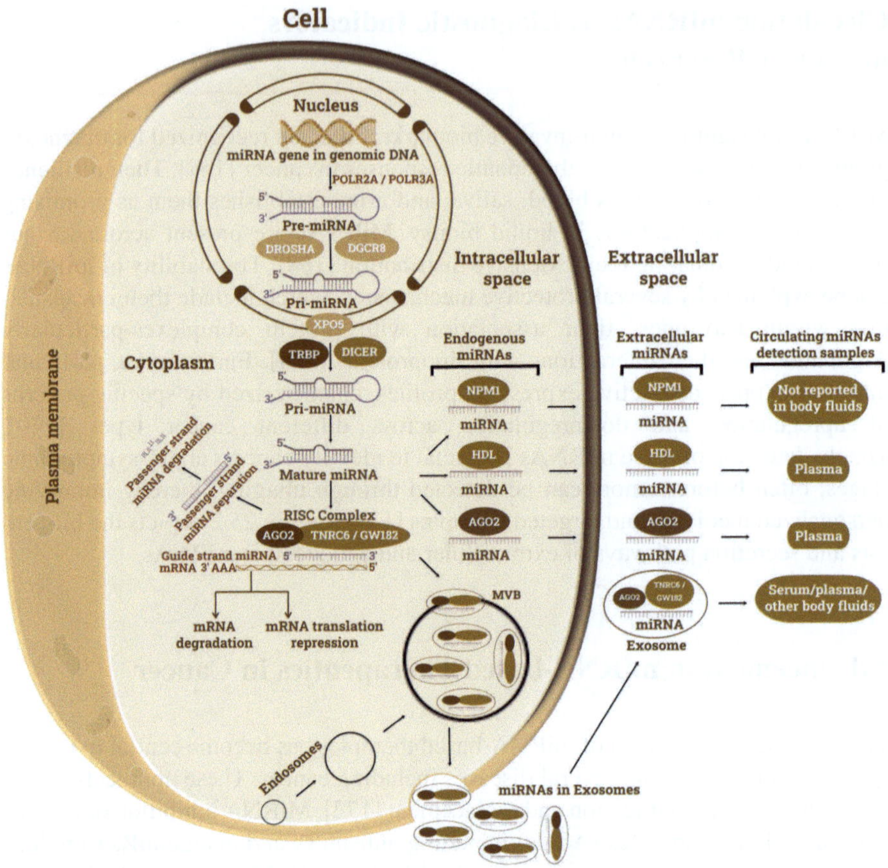

Fig. 25.2 Circulating miRNAs as diagnostic biomarkers. The figure depicts the biogenesis and secretion pathways of endogenous and extracellular miRNAs. In the nucleus, miRNA genes are transcribed by RNA polymerase II subunit A (POLR2A or Pol II) / RNA polymerase III subunit A (POLR3A or Pol III) into primary miRNAs (pri-miRNAs), which are processed by the DROSHA-DGCR8 complex into precursor miRNAs (pre-miRNAs). These pre-miRNAs are transported into the cytoplasm through the action of exportin-5 (XPO5), where they are further processed by DICER into mature miRNAs. The mature miRNA duplex separates into a guide strand, which is incorporated into the RNA-induced silencing complex (RISC), which comprises proteins such as Argonaute 2 (AGO2), TNRC6, and GW182. This complex is critical for the separation of the guide strand from the passenger strand, which is usually marked for degradation. This separation is crucial for the silencing of target genes, facilitated by mechanisms such as mRNA degradation or translational repression. Some mature miRNAs are packed into multivesicular bodies (MVB) and released into the extracellular space *via* exosomes. Extracellular miRNAs can also associate with transport proteins, such as high-density lipoprotein (HDL), AGO2, and nucleophosmin 1 (NPM1). While HDL-miRNAs and AGO2-miRNAs are found in plasma, NPM1-miRNAs have not been reported in body fluids. Exosomal miRNAs are present in plasma, serum, and other body fluids and extracellular miRNAs can be taken up by recipient cells through endocytosis

miRNAs, marking a significant milestone in the field of cancer treatment [181]. Strategies developed to inhibit the tumor growth by targeting the miR-21 and induce apoptosis in HCC *via* upregulation of miR-34 exemplify this innovative approach [182, 183].

Therapeutic Applications of miRNAs in Cancer

Advancements in understanding the role of miRNAs in cancer have paved the way for the development of novel therapeutic approaches. These therapies are designed either to suppress oncogenic miRNAs or to restore the activity of tumor-suppressive miRNAs using chemically engineered oligonucleotides.

MiRNA mimics are used to replace miRNAs that are absent or reduced in cancer. Another stratergy involves use of antisense molecules like locked nucleic acid (LNA) inhibitors and antagomiRs, to block the aberrantly expressed miRNAs that promote cancer progression [188]. These methods have yielded encouraging outcomes in laboratory studies and early human trials.

A notable example is *MRX34,* a synthetic molecule (mimic) of miR-34a, which can supress the tumors. *MRX34* was tested in clinical trials for liver cancer and targeted cancer-related genes like *MYC, MET*, and *BCL2*. Although immune-related side effects led to the early termination of the trial, it demonstrated that miRNAs can be delivered systemically and they can modulate the cancer-associated pathways [186].

TargomiRs, which can deliver miRNA mimic (for example, miR-16) using specially engineered bacterial minicells, have been tested in patients with mesothelioma and lung cancer. This approach was found to be safe and showed potential for slowing the disease progression [195]. Other miRNA-based treatments include MRG-106 (Cobomarsen) targeting miR-155 in blood cancers [212] and anti-miR-10b tested in animal studies to reduce the spread of breast cancer cells [211].

Although these treatments hold promise, several challenges remain, such as unwanted immune responses, preventing undesired side effects on healthy cells, and ensuring miRNAs reach the tumor effectively. The refining of these delivery systems and the overcoming of these obstacles are focuses of current research studies.

Table 25.2 outlines some of the preclinical and clinical approaches explored using miRNAs.

Conclusions and Future Perspectives

MiRNAs can function as oncogenes (oncomiRs) or tumor suppressors (anti-oncomiRs), and thereby impact the key cellular functions like growth, programmed cell death (PCD), metastasis, neoangiogenesis, and immune system evasion. Dysregulation of specific miRNAs can contribute to various types of cancer. Thus,

Table 25.2 This table summarizes key miRNAs that have been investigated in preclinical and clinical cancer research for therapeutic applications

Authors	miRNAs/ miRs used	Model systems/ Clinical Trials	Target organs	Study highlights
Yu et al. [184, 187]	miR-34a	Human bladder cancer cell lines	Bladder	miR-34a suppresses angiogenesis and metastasis by targeting CD44 mRNA/protein expression
Sun et al. [185]	miR-34a	Human nasopharyngeal carcinoma tissues and cell lines	Nasopharynx	miR-34a induces apoptosis by targeting lncRNA MCM3AP-AS1 mRNA expression
Beg et al. [186]	miR-34a (MRX34)	Phase I clinical trial in humans	Liver and other organs	Conducted the first clinical trial of a miRNA-based therapy (MRX34) in cancer patients
Yu et al. [184, 187]	miR-34a	Human renal carcinoma cell lines and xenograft mouse models	Kidney	miR-34a suppresses cell proliferation and metastasis by targeting CD44 mRNA/protein expression
Frères et al. [213]	miR-34a	Human breast cancer patients samples	Breast	Identified increased levels of miR-34a post-chemotherapy, suggesting its role as a biomarker
Wang and Wu. [214]	miR-155	Human breast cancer cell lines and xenograft mouse models	Breast	Inhibition of miR-155 reduces cancer stem cell formation and enhances chemosensitivity
Bouchie [188]	miR-34a (MRX34)	Preclinical mouse models	Liver	Reported that MRX34 induces tumor regression in HCC models
Di Martino et al. [47]	miR-34a	Multiple myeloma cell lines and xenograft mouse models	Bone marrow	Stable nucleic acid lipid particle (SNALP)-delivered miR-34a inhibits tumor growth with low toxicity
Li et al. [215]	miR-34a	Human glioblastoma cell lines and patient-derived cells cultures	Brain	miR-34a overexpression inhibits proliferation across glioblastoma subtypes
Chang et al. [209]	miR-375	HCC cell lines and xenograft mouse models	Liver	Overexpression of miR-375 suppressed liver cancer cell growth in vitro and in vivo by targeting AEG-1 mRNA/protein expression
Li et al. [10, 217]	miR-25	Human colon cancer tissues and xenograft mouse models	Colon	miR-25 was downregulated in colon cancer; its overexpression suppressed tumor growth by targeting Smad7 mRNA/protein expression

(continued)

Table 25.2 (continued)

Authors	miRNAs/ miRs used	Model systems/ Clinical Trials	Target organs	Study highlights
Ma et al. [190]	miR-10b	Mouse models of breast cancer metastasis	Breast	Systemic delivery of antagomiR-10b suppressed breast cancer metastasis without affecting primary tumor growth
Kota et al. [191]	miR-26a	Mouse models of hepatocellular carcinoma	Liver	Adeno-associated virus(AAV)-mediated delivery of miR-26a induced tumor-specific apoptosis and inhibited cancer cell proliferation
Wiggins et al. [192]	miR-34a	Human cancer cell lines and mouse models	Lung	Systemic delivery of miR-34a inhibited tumor growth by repressing c-Met and Bcl-2 mRNA/protein expression
Forrest et al. [193]	miR-155, miR-222, miR-424, miR-503	THP-1 human leukemia cell line	Blood	Overexpression of these miRNAs promoted monocytic differentiation and cell cycle arrest in leukemia cells
Xiao et al. [194]	miR-885-3p	HT-29 colon cancer cell xenografts in mice	Colon	miR-885-3p inhibited tumor angiogenesis and growth by targeting BMPR1A mRNA/ protein expression and thereby blocking BMP/Smad/Id1 signaling pathway
Tassone et al. [196]	miR-221	Clinical trial in humans	Advanced solid tumors	Showed promising antitumor activity and manageable toxicity
Querfeld et al. [197]	miR-155	Phase I/II clinical trial in humans	Cutaneous T cell lymphoma, DLBCL	Demonstrated tumor regression in some cancer patients, but clinical trial was halted before its completion
Ghosh et al. [198]	miR-10b	Phase 0 clinical trial in humans	Metastatic solid tumors	Early phase of human clinical trial results indicate the potential of miR-10b to reduce metastasis; but safety profiles of miR-10b are still being evaluated

altered expression levels of miRNAs can be used as biomarkers for early diagnosis/ detection, prognosis, and targeted therapies. Additionally, cancer-specific miRNAs expression profiles of patients can facilitate the advancement of their personalized medicine, as treatment plans for targeted cancer therapy can be tailored to individual miRNA profiles. Ongoing exploration of miRNA regulatory networks is expected to enhance our comprehension of cancer biology and drive the creation of innovative cancer diagnostic, prognostic, and therapuetic options.

Recent developments in miRNA-based diagnostic technologies have greatly improved the clinical accuracy and sensitivity, and access. Advanced detection techniques such as qRT-PCR, ddPCR, and isothermal amplification methods, allow the detection of low-abundance miRNAs, which are critical for the early diagnosis, treatment, and prognosis. The NGS methods, including the single-cell RNA sequencing and long-read sequencing, provide extensive high-resolution profiling of miRNAs gene expression levels, variants and regulatory regions, which can collaboratively support a more reliable cancer miRNAs biomarker discovery in the era of cancer genomics. Additionally, diagnostics based on nanotechnology, utilizing gold nanoparticles, graphene sensors, or SPR biosensors, have enhanced the specificity and real-time analysis of miRNA detection in cancer. Furthermore, innovative tools like the CRISPR-Cas13 systems have transformed miRNAs detection with improved sensitivity and precision. Importantly, the application or utilization of highly advanced and constantly evolving AI and ML models/algorithms/programs is reshaping the analysis of complex miRNA data, facilitating rapid discovery of miRNA biomarkers and early diagnosis, especially in cancer. Non-invasive liquid biopsy methods, which can be used for the identification of circulating miRNAs and exosome profiling, mark a significant advancements toward personalized, patient-friendly diagnostics and importantly, improving the prognostic assessment. This allows for early diagnosis and prognosis through periodical/continuous monitoring for cancer recurrence posttreatment, unlike the invasive tissue biopsy method, which is very cumbersome and inconvenient to monitor cancer patients. The rise of portable PoC devices, such as microfluidic chips and biosensors integrated with smartphones, is broadening the easy and affordable/cost-effective access to diagnostics, particularly in settings with limited resources. Together, these significant developments highlight the increasing clinical importance of miRNAs in cancer and have facilitated the creation of precise, swift, and noninvasive diagnostic tools. Therefore, future initiatives should prioritize standardization, regulatory clearance and cost-effectiveness to incorporate these innovations into standard healthcare practices.

From a therapeutic perspective, interventions based on miRNAs signify a groundbreaking development in molecular medicine. Preliminary clinical cancer research using miRNA mimics to restore tumor-suppressive functions, and antagomiRs to inhibit oncogenic miRNAs, has shown promising outcomes, including reduced tumor growth, decreased metastasis, and improved survival across multiple cancer types. The translation of miRNA therapies to clinical practice poses challenges, including efficient and specific delivery to tumor sites, reduction of their off-target effects and immune reactions, and addressing variations in miRNA expression because of cellular heterogeneity within and among tumors. To tackle these challenges, advancements in delivery systems such as lipid nanoparticles, exosome-based carriers, and targeted minicells are being developed to improve the stability, specificity, and biocompatibility. As our knowledge of miRNA biology advances, these therapeutic approaches are likely to become essential components of personalized cancer treatment and precision medicine, providing a new hope for effective cancer management.

Acknowledgments The authors express sincere thanks to Dr. Natarajan Bhaskaran, Center for Global Health Research (CGHR), Saveetha Medical College, Saveetha Institute of Medical and Technical Sciences (SIMATS), Thandalam, Chennai 602105, Tamil Nadu, India, for critically reading this manuscript and providing valuable comments.

Conflict of Interest The authors declare no conflict of interest.

References

1. Alizadeh M, Ghasemi H, Bazhan D, Mohammadi Bolbanabad N, Rahdan F, Arianfar N, Vahedi F, Khatami SH, Taheri-Anganeh M, Aiiashi S, Armand N (2025) MicroRNAs in disease states. Clin Chim Acta 569:120187. https://doi.org/10.1016/j.cca.2025.120187
2. Jonas S, Izaurralde E (2015) Towards a molecular understanding of microRNA-mediated gene silencing. Nat Rev Genet 16(7):421–433. https://doi.org/10.1038/nrg3965
3. Bhaskaran M, Mohan M (2013) MicroRNAs: history, biogenesis, and their evolving role in animal development and disease. Vet Pathol 51(4):759–774. https://doi.org/10.1177/0300985813502820
4. Bartel DP (2004) MicroRNAs: genomics, biogenesis, mechanism, and function. Cell 116(2):281–297. https://doi.org/10.1016/s0092-8674(04)00045-5
5. Catalanotto C, Cogoni C, Zardo G (2016) MicroRNA in control of gene expression: an overview of nuclear functions. Int J Mol Sci 17(10):1712. https://doi.org/10.3390/ijms17101712
6. Moreno-Moya JM, Vilella F, Simón C (2013) MicroRNA: key gene expression regulators. Fertil Steril 101(6):1516–1523. https://doi.org/10.1016/j.fertnstert.2013.10.042
7. Davis BN, Hata A (2009) Regulation of MicroRNA biogenesis: a miRiad of mechanisms. Cell Commun Signal 7:18. https://doi.org/10.1186/1478-811X-7-18
8. Alberts B, Johnson A, Lewis J, Raff M, Roberts K, Walter P (2002) Molecular biology of the cell, 4th edn. New York: Garland Science; From RNA to Protein. https://www.ncbi.nlm.nih.gov/books/NBK26829/
9. Eichhorn SW, Guo H, McGeary SE, Rodriguez-Mias RA, Shin C, Baek D, Hsu SH, Ghoshal K, Villén J, Bartel DP (2014) mRNA destabilization is the dominant effect of mammalian microRNAs by the time substantial repression ensues. Mol Cell 56(1):104–115. https://doi.org/10.1016/j.molcel.2014.08.028
10. Li S, Liu L, Zhuang X, Yu Y, Liu X, Cui X, Ji L, Pan Z, Cao X, Mo B, Zhang F, Raikhel N, Jiang L, Chen X (2013a) MicroRNAs inhibit the translation of target mRNAs on the endoplasmic reticulum in Arabidopsis. Cell 153(3):562–574. https://doi.org/10.1016/j.cell.2013.04.005
11. Ho PTB, Clark IM, Le LTT (2022) MicroRNA-based diagnosis and therapy. Int J Mol Sci 23(13):7167. https://doi.org/10.3390/ijms23137167
12. Paranjape T, Slack FJ, Weidhaas JB (2009) MicroRNAs: tools for cancer diagnostics. Gut 58(11):1546–1554. https://doi.org/10.1136/gut.2009.179531
13. Condrat CE, Thompson DC, Barbu MG, Bugnar OL, Boboc A, Cretoiu D, Suciu N, Cretoiu SM, Voinea SC (2020) miRNAs as biomarkers in disease: latest findings regarding their role in diagnosis and prognosis. Cells 9(2):276. https://doi.org/10.3390/cells9020276
14. Liu J, Jennings SF, Tong W, Hong H (2011) Next generation sequencing for profiling expression of miRNAs: technical progress and applications in drug development. J Biomed Sci Eng 4(10):666–676. https://doi.org/10.4236/jbise.2011.410083
15. Trang P, Weidhaas JB, Slack FJ (2008) MicroRNAs as potential cancer therapeutics. Oncogene 27(Suppl 2):S52–S57. https://doi.org/10.1038/onc.2009.353
16. Seyhan AA (2024) Trials and tribulations of MicroRNA therapeutics. Int J Mol Sci 25(3):1469. https://doi.org/10.3390/ijms25031469

17. Garreau M, Weidner J, Hamilton R, Kolosionek E, Toki N, Stavenhagen K, Paris C, Bonetti A, Czechtizky W, Gnerlich F, Rydzik A (2024) Chemical modification patterns for microRNA therapeutic mimics: a structure-activity relationship (SAR) case-study on miR-200c. Nucleic Acids Res 52(6):2792–2807. https://doi.org/10.1093/nar/gkae141
18. Chengizkhan G, Bisht B, Muthusami S, Anbalagan M, Ramachandran I, Kumaran RI (2020) Types of cancer stem cells. In: Pathak S, Banerjee A (eds) Cancer stem cells: new horizons in cancer therapies. Springer Nature Singapore, pp 15–50. (ISBN: 978-981-15-5119-2)
19. Visone R, Croce CM (2009) MiRNAs and cancer. Am J Pathol 174(4):1131–1138. https://doi.org/10.2353/ajpath.2009.080794
20. Johnson SM, Grosshans H, Shingara J, Byrom M, Jarvis R, Cheng A, Labourier E, Reinert KL, Brown D, Slack FJ (2005) RAS is regulated by the let-7 microRNA family. Cell 120(5):635–647. https://doi.org/10.1016/j.cell.2005.01.014
21. Croce CM (2009) Causes and consequences of microRNA dysregulation in cancer. Nat Rev Genet 10(10):704–714. https://doi.org/10.1038/nrg2634
22. Chengizkhan G, Thangavelu SK, Muthusami S, Banerjee A, Pathak S, Natarajan G, Ramalingam S, Queimado L, Kumaran RI, Ramachandran I (2024) Regulation of cancer stemness, cell signaling, reactive oxygen species, and microRNAs in cancer stem cells. In: Pathak S, Banerjee A (eds) Cancer stem cells and signaling pathways. Academic Press, An imprint of Elsevier, pp 243–263. https://doi.org/10.1016/B978-0-443-13212-4.00026-X. (ISBN: 978-0-443-13212-4)
23. Mondal I, Gupta N, Sharma V, Sarkar C, Mishra DP, Kulshreshtha R (2024) ALDH5A1/miR-210 axis plays a key role in reprogramming cellular metabolism and has a significant correlation with glioblastoma patient survival. Cancer Cell Int 24(1):259. https://doi.org/10.1186/s12935-024-03432-z
24. Hussen BM, Sulaiman SHA, Abdullah SR, Hidayat HJ, Khudhur ZO, Eslami S, Samsami M, Taheri M (2025) MiRNA-155: a double-edged sword in colorectal cancer progression and drug resistance mechanisms. Int J Biol Macromol 299:140134. https://doi.org/10.1016/j.ijbiomac.2025.140134
25. Moradi A, Bayat M, Pourmasoumi P, Chien S (2025) Unraveling the oncogenic potential of microRNAs in lung cancer: a narrative review article. Cancer Screen Prev 4(1):1120. https://doi.org/10.14218/csp.2025.00001
26. Moro M, Fortunato O, Bertolini G, Mensah M, Borzi C, Centonze G, Andriani F, Di Paolo D, Perri P, Ponzoni M, Pastorino U, Sozzi G, Boeri M (2022) MiR-486-5p targets CD133+ lung cancer stem cells through the p85/AKT pathway. Pharmaceuticals (Basel, Switzerland) 15(3):297. https://doi.org/10.3390/ph15030297
27. Chen Q, Chen S, Zhao J, Zhou Y, Xu L (2021) MicroRNA-126: a new and promising player in lung cancer. Oncol Lett 21(1):35. https://doi.org/10.3892/ol.2020.12296
28. Wang J, Chen L (2019) The role of miRNAs in the invasion and metastasis of cervical cancer. Biosci Rep 39(3):BSR20181377. https://doi.org/10.1042/bsr20181377
29. Chauhan P, Pramodh S, Hussain A, Elsori D, Lakhanpal S, Kumar R, Alsaweed M, Iqbal D, Pandey P, Othaim AA, Khan F (2024) Understanding the role of miRNAs in cervical cancer pathogenesis and therapeutic responses. Front Cell Develop Biol 12:1397945. https://doi.org/10.3389/fcell.2024.1397945
30. Yanshen Z, Lifen Y, Xilian W, Zhong D, Huihong M (2021 Dec) miR-92a promotes proliferation and inhibits apoptosis of prostate cancer cells through the PTEN/Akt signaling pathway. Libyan J Med 16(1):1971837. https://doi.org/10.1080/19932820.2021.1971837
31. Jayamohan S, Kannan M, Moorthy RK, Rajasekaran N, Jung HS, Shin YK, Arockiam AJV (2019) Dysregulation of MIR-375/AEG-1 axis by human papillomavirus 16/18-E6/E7 promotes cellular proliferation, migration, and invasion in cervical cancer. Front Oncol 9:847. https://doi.org/10.3389/fonc.2019.00847
32. Taghizadieh M, Kalantari M, Bakhshali R, Kobravi S, Khalilollah S, Baghi HB, Bayat M, Nahand JS, Akhavan-Sigari R (2025) To be or not to be: navigating the influence of MicroRNAs on cervical cancer cell death. Cancer Cell Int 25(1):153. https://doi.org/10.1186/s12935-025-03786-y

33. Shen K, Cao Z, Zhu R, You L, Zhang T (2019) The dual functional role of MicroRNA-18a (miR-18a) in cancer development. Clin Transl Med 8(1):32. https://doi.org/10.1186/s40169-019-0250-9
34. Xu S, He L, Chen Y, Lin T, Tang L, Wu Y, He Y, Sun X (2025) Clinical implications of miR-195 in cancer: mechanisms, potential applications, and therapeutic strategies. J Cancer Res Clin Oncol 151(4):148. https://doi.org/10.1007/s00432-025-06195-w
35. Wang Y, Huang D, Li M, Yang M (2025a) MicroRNA-99 family in cancer: molecular mechanisms for clinical applications. PeerJ 13:e19188. https://doi.org/10.7717/peerj.19188
36. Poli V, Seclì L, Avalle L (2020) The microRNA-143/145 cluster in tumors: a matter of where and when. Cancer 12(3):708. https://doi.org/10.3390/cancers12030708
37. Zhu L, Tu H, Liang Y, Tang D (2018) MiR-218 produces anti-tumor effects on cervical cancer cells in vitro. World J Surg Oncol 16(1):204. https://doi.org/10.1186/s12957-018-1506-3
38. Ghafouri-Fard S, Askari A, Hussen BM, Taheri M, Akbari DN (2024) Role of miR-424 in the carcinogenesis. Clin Transl Oncol 26(1):16–38. https://doi.org/10.1007/s12094-023-03209-2
39. Lu X, Song X, Hao X, Liu X, Zhang X, Yuan N, Ma H, Zhang Z (2021) miR-186-3p attenuates the tumorigenesis of cervical cancer via targeting insulin-like growth factor 1 to suppress PI3K-Akt signaling pathway. Bioengineered 12(1):7079–7092. https://doi.org/10.1080/21655979.2021.1977053
40. Ramachandran I, Ganapathy V, Gillies E, Fonseca I, Sureban SM, Houchen CW, Reis A, Queimado L (2014) Wnt inhibitory factor 1 suppresses cancer stemness and induces cellular senescence. Cell Death Dis 5(5):e1246. https://doi.org/10.1038/cddis.2014.219
41. Subramanian K, Sinha R (2024) Functions of differentially regulated MIRNAs in breast cancer progression: potential markers for early detection and candidates for therapy. Biomedicine 12(3):691. https://doi.org/10.3390/biomedicines12030691
42. Wang J, Wang Q, Guan Y, Sun Y, Wang X, Lively K, Wang Y, Luo M, Kim JA, Murphy EA, Yao Y, Cai G, Fan D (2022) Breast cancer cell-derived microRNA-155 suppresses tumor progression via enhancing immune cell recruitment and antitumor function. J Clin Invest 132(19):e157248. https://doi.org/10.1172/JCI157248
43. Ye D, Shen Z, Zhou S (2019) Function of microRNA-145 and mechanisms underlying its role in malignant tumor diagnosis and treatment. Cancer Manag Res 11:969–979. https://doi.org/10.2147/CMAR.S191696
44. Nair MG, Somashekaraiah VM, Ramamurthy V, Prabhu JS, Sridhar T (2021) miRNAs: critical mediators of breast cancer metastatic programming. Exp Cell Res 401(1):112518. https://doi.org/10.1016/j.yexcr.2021.112518
45. Muthusami S, Ramachandran IK, Babu KN, Krishnamoorthy S, Guruswamy A, Queimado L, Chaudhuri G, Ramachandran I (2021a) Role of inflammation in the development of colorectal cancer. Endocr Metab Immune Disord Drug Targets 21(1):77–90. https://doi.org/10.2174/1871530320666200909092908
46. Muthusami S, Ramachandran I, Krishnamoorthy S, Sambandam Y, Ramalingam S, Queimado L, Chaudhuri G, Ramachandran IK (2021b) Regulation of MicroRNAs in inflammation-associated colorectal cancer: a mechanistic approach. Endocr Metab Immune Disord Drug Targets 21(1):67–76. https://doi.org/10.2174/1871530320666200917112802
47. Di Martino MT, Arbitrio M, Caracciolo D, Cordua A, Cuomo O, Grillone K, Riillo C, Caridà G, Scionti F, Labanca C, Romeo C, Siciliano MA, D'Apolito M, Napoli C, Montesano M, Farenza V, Uppolo V, Tafuni M, Falcone F, Tassone P (2022) miR-221/222 as biomarkers and targets for therapeutic intervention on cancer and other diseases: a systematic review. Mol Ther Nucleic Acids 27:1191–1224. https://doi.org/10.1016/j.omtn.2022.02.005
48. Al-Nakhle HH (2024) Unraveling the multifaceted role of the MIR-17-92 cluster in colorectal cancer: from mechanisms to biomarker potential. Curr Issues Mol Biol 46(3):1832–1850. https://doi.org/10.3390/cimb46030120
49. Bautista-Sánchez D, Arriaga-Canon C, Pedroza-Torres A, De La Rosa-Velázquez IA, González-Barrios R, Contreras-Espinosa L, Montiel-Manríquez R, Castro-Hernández C, Fragoso-Ontiveros V, Álvarez-Gómez RM, Herrera LA (2020) The promising role of MIR-21

as a cancer biomarker and its importance in RNA-based therapeutics. Mol Ther Nucleic Acids 20:409–420. https://doi.org/10.1016/j.omtn.2020.03.003

50. Nabipoorashrafi SA, Shomali N, Sadat-Hatamnezhad L, Mahami-Oskouei M, Mahmoudi J, Shotorbani BS, Akbari M, Xu H, Shotorbani SS (2020) miR-143 acts as an inhibitor of migration and proliferation as well as an inducer of apoptosis in melanoma cancer cells in vitro. IUBMB Life 72(9):2034–2044. https://doi.org/10.1002/iub.2345
51. Zeinali T, Mansoori B, Mohammadi A, Baradaran B (2018) Regulatory mechanisms of miR-145 expression and the importance of its function in cancer metastasis. Biomed Pharmacother 109:195–207. https://doi.org/10.1016/j.biopha.2018.10.037
52. Klicka K, Grzywa TM, Mielniczuk A, Klinke A, Włodarski PK (2022) The role of miR-200 family in the regulation of hallmarks of cancer. Front Oncol 12:965231. https://doi.org/10.3389/fonc.2022.965231
53. Wei F, Cao C, Xu X, Wang J (2015) Diverse functions of miR-373 in cancer. J Transl Med 13:162. https://doi.org/10.1186/s12967-015-0523-z
54. Gan L, Zheng L, Zou J, Luo P, Chen T, Zou J, Li W, Chen Q, Cheng L, Zhang F, Qian B (2024) MicroRNA-21 in urologic cancers: from molecular mechanisms to clinical implications. Front Cell Develop Biol 12:1437951. https://doi.org/10.3389/fcell.2024.1437951
55. Abak A, Amini S, Sakhinia E, Abhari A (2018) MicroRNA-221: biogenesis, function and signatures in human cancers. Eur Rev Med Pharmacol Sci 22(10):3094–3117. https://doi.org/10.26355/eurrev_201805_15069
56. Budi HS, Younus LA, Lafta MH, Parveen S, Mohammad HJ, Al-Qaim ZH, Jawad MA, Parra RMR, Mustafa YF, Alhachami FR, Karampoor S, Mirzaei R (2023) The role of miR-128 in cancer development, prevention, drug resistance, and immunotherapy. Front Oncol 12:1067974. https://doi.org/10.3389/fonc.2022.1067974
57. Felekkis K, Papaneophytou C (2024) The circulating biomarkers league: combining miRNAs with cell-free DNAs and proteins. Int J Mol Sci 25(6):3403. https://doi.org/10.3390/ijms25063403
58. Fazzina M, Bergonzoni M, Massenzio F, Monti B, Frabetti F, Casadei R (2025) Author correction: selection of suitable reference genes for gene expression studies in HMC3 cell line by quantitative real-time RT-PCR. Sci Rep 15(1):3561. https://doi.org/10.1038/s41598-025-87066-9
59. Liu S, Chai X, Liu C, Bai J, Meng J, Tian H, Han X, Han G, Li Q, Xu X (2024a) Sensitivity analysis of RT-qPCR and RT-ddPCR for SARS-CoV-2 detection with mutations on N1 and E primer-probe region. Microbiol Spectrum 12(8):e0429223. https://doi.org/10.1128/spectrum.04292-23
60. He F, Ni N, Wang H, Zeng Z, Zhao P, Shi D, Xia Y, Chen C, Hu DA, Qin KH, Wagstaff W, Qin D, Hendren-Santiago B, Ho SH, Haydon RC, Luu HH, Reid RR, Shen L, Gan H, He T (2021) OUHP: an optimized universal hairpin primer system for cost-effective and high-throughput RT-qPCR-based quantification of microRNA (miRNA) expression. Nucleic Acids Res 50(4):e22. https://doi.org/10.1093/nar/gkab1153
61. Mortoglou M, Wallace D, Djordjevic AB, Djordjevic V, Arisan ED, Uysal-Onganer P (2021) MicroRNA-regulated signaling pathways: potential biomarkers for pancreatic ductal adenocarcinoma. Stress 1(1):30–47. https://doi.org/10.3390/stresses1010004
62. Wei J, Lu Y, Wang R, Xu X, Liu Q, He S, Pan H, Liu X, Yuan B, Ding Y, Zhang J (2021) MicroRNA-375: potential cancer suppressor and therapeutic drug. Biosci Rep 41(9):BSR20211494. https://doi.org/10.1042/BSR20211494
63. Prinz C, Fehring L, Frese R (2022) MicroRNAs as indicators of malignancy in pancreatic ductal adenocarcinoma (PDAC) and cystic pancreatic lesions. Cells 11(15):2374. https://doi.org/10.3390/cells11152374
64. Babaei G, Raei N, Milani T, Attabak & Gholizadeh, Shiva & Pourjabbar, Nima & Geravand, Faezeh. (2021) The emerging role of miR-200 family in metastasis: focus on EMT, CSCs, angiogenesis, and anoikis. Mol Biol Rep 48(10):6935–6947. https://doi.org/10.1007/s11033-021-06666-6

65. Chen C, Demirkhanyan L, Gondi CS (2024) The multifaceted role of miR-21 in pancreatic cancers. Cells 13(11):948. https://doi.org/10.3390/cells13110948
66. Behrooz AB, Latifi-Navid H, Nezhadi A, Świat M, Los M, Jamalpoor Z, Ghavami S (2023) Molecular mechanisms of microRNAs in glioblastoma pathogenesis. Biochimica Et Biophysica Acta (BBA)—Mol Cell Res 1870(6):119482. https://doi.org/10.1016/j.bbamcr.2023.119482
67. Korać P, Antica M, Matulić M (2021) MiR-7 in cancer development. Biomedicine 9(3):325. https://doi.org/10.3390/biomedicines9030325
68. Bassot A, Dragic H, Haddad SA, Moindrot L, Odouard S, Corlazzoli F, Marinari E, Bomane A, Brassens A, Marteyn A, Hibaoui Y, Petty TJ, Chalabi-Dchar M, Larrouquere L, Zdobnov EM, Legrand N, Tamburini J, Lincet H, Castets M, Yebra M et al (2023) Identification of a miRNA multi-targeting therapeutic strategy in glioblastoma. Cell Death Dis 14(9):630. https://doi.org/10.1038/s41419-023-06117-z
69. Charbit H, Lavon I (2024) Investigating expression dynamics of miR-21 and miR-10b in glioblastoma cells in vitro: insights into responses to hypoxia and secretion mechanisms. Int J Mol Sci 25(14):7984. https://doi.org/10.3390/ijms25147984
70. Si F, Wang Q, Chen F, Lu X (2025) Sophocarpine inhibits the proliferation and induces apoptosis of glioblastoma cells through regulating the miR-21/PTEN/PI3K/AKT axis. Discov Oncol 16(1):138. https://doi.org/10.1007/s12672-025-01839-2
71. Ma LN, Wu LN, Liu SW, Zhang X, Luo X, Nawaz S, Ma ZM, Ding XC (2024) miR-199a/b-3p inhibits HCC cell proliferation and invasion through a novel compensatory signaling pathway DJ-1\Ras\PI3K/AKT. Sci Rep 14(1):224. https://doi.org/10.1038/s41598-023-48760-8
72. Yang Y, Razak SRA, Ismail IS, Ma Y, Yunus MA (2025) Molecular mechanisms of miR-192 in cancer: a biomarker and therapeutic target. Cancer Cell Int 25(1):94. https://doi.org/10.1186/s12935-025-03666-5
73. Al Ageeli E (2024) Dual roles of microRNA-122 in hepatocellular carcinoma and breast cancer progression and metastasis: a comprehensive review. Curr Issues Mol Biol 46(11):11975–11992. https://doi.org/10.3390/cimb46110711
74. Yu H, Duan H, He R, Tian Y, Jiang J, Xiao F, Liu Q, Liu J, Li H, Yu X (2025) Integrated transcriptomics profile reveals the role of Gal-1 and miR-21 in intrahepatic cholangiocarcinoma progression. Biochim Biophys Acta Mol Basis Dis 1871(6):167882. Advance online publication. https://doi.org/10.1016/j.bbadis.2025.167882
75. Kohlberger M, Gadermaier G (2021) SELEX: critical factors and optimization strategies for successful aptamer selection. Biotechnol Appl Biochem 69(5):1771–1792. https://doi.org/10.1002/bab.2244
76. Spiteri S, Salamon I, Girolamini L, Pascale MR, Marino F, Derelitto C, Caligaris L, Paghera S, Ferracin M, Cristino S (2025) Surfaces environmental monitoring of SARS-CoV-2: loop mediated isothermal amplification (LAMP) and droplet digital PCR (ddPCR) in comparison with standard reverse-transcription quantitative polymerase chain reaction (RT-qPCR) techniques. PLoS One 20(2):e0317228. https://doi.org/10.1371/journal.pone.0317228
77. Jiang K, Zhang T, Liu C, Zhang Z, Li H (2025) An integrated droplet microfluidic chip for digital polymerase chain reaction with a fountain-like structure capable of removing bubble. Sensors Act B Chem 433:137540. https://doi.org/10.1016/j.snb.2025.137540
78. Hou Y, Chen S, Zheng Y, Zheng X, Lin J (2022) Droplet-based digital PCR (ddPCR) and its applications. TrAC Trends Anal Chem 158:116897. https://doi.org/10.1016/j.trac.2022.116897
79. McNair JN, Frobish D, Rediske RR, Hart JJ, Jamison MN, Szlag DC (2025) The theoretical basis of QPCR and DDPCR copy number estimates: a critical review and exposition. Water 17(3):381. https://doi.org/10.3390/w17030381
80. Semenkovich NP, Szymanski JJ, Earland N, Chauhan PS, Pellini B, Chaudhuri AA (2023) Genomic approaches to cancer and minimal residual disease detection using circulating tumor DNA. J Immunother Cancer 11(6):e006284. https://doi.org/10.1136/jitc-2022-006284

81. Nazir S (2023) Medical diagnostic value of digital PCR (dPCR): a systematic review. Biomed Eng Adv 6:100092. https://doi.org/10.1016/j.bea.2023.100092
82. Busato F, Ursuegui S, Deleuze J, Tost J (2024) Multiplex digital PCR for the simultaneous quantification of a miRNA panel. Anal Chim Acta 1335:343440. https://doi.org/10.1016/j.aca.2024.343440
83. Srivastava P, Prasad D (2023) Isothermal nucleic acid amplification and its uses in modern diagnostic technologies. *3*. Biotech 13(6):200. https://doi.org/10.1007/s13205-023-03628-6
84. Ruszova E, Vanek D, Stühmer W, Khaznadar Z, Subhashini N (2024) The utilization of the SALUX19-based loop-mediated isothermal amplification (LAMP) assay for the rapid and sensitive identification of minute amounts of a biological specimen. Life (Basel, Switzerland) 14(5):579. https://doi.org/10.3390/life14050579
85. Ngoc LTN, Lee Y (2024) Current trends in RNA virus detection via nucleic acid isothermal amplification-based platforms. Biosensors 14(2):97. https://doi.org/10.3390/bios14020097
86. Mao Z, Lei H, Chen R, Ren S, Liu B, Gao Z (2024) CRISPR/Cas13a analysis based on NASBA amplification for norovirus detection. Talanta 280:126725. https://doi.org/10.1016/j.talanta.2024.126725
87. Bellassai N, D'Agata R, Spoto G (2025) Advancements in the integration of isothermal nucleic acid amplification methods for point-of-care testing in resource-limited settings. Sensors Act Rep 100285. https://doi.org/10.1016/j.snr.2025.100285
88. Hess J, Kohl T, Kotrová M, Rönsch K, Paprotka T, Mohr V, Hutzenlaub T, Brüggemann M, Zengerle R, Niemann S, Paust N (2020) Library preparation for next generation sequencing: a review of automation strategies. Biotechnol Adv 41:107537. https://doi.org/10.1016/j.biotechadv.2020.107537
89. Nunes S, Bastos R, Marinho AI, Vieira R, Benício I, De Noronha MA, Lírio S, Brodskyn C, Tavares NM (2024) Recent advances in the development and clinical application of miRNAs in infectious diseases. Non-coding RNA Res 10:41–54. https://doi.org/10.1016/j.ncrna.2024.09.005
90. Nalbant E, Akkaya-Ulum YZ (2024) Exploring regulatory mechanisms on miRNAs and their implications in inflammation related diseases. Clin Exp Med 24(1):142. https://doi.org/10.1007/s10238-024-01334-y
91. Desta GM, Birhanu AG (2025) Advancements in single-cell RNA sequencing and spatial transcriptomics: transforming biomedical research. Acta Biochim Pol 72:13922. https://doi.org/10.3389/abp.2025.13922
92. Ma L, Luan Y, Lu L (2025) Analyze the diversity and function of immune cells in the tumor microenvironment from the perspective of single-cell RNA sequencing. Cancer Med 14(5):e70622. https://doi.org/10.1002/cam4.70622
93. Jovic D, Liang X, Zeng H, Lin L, Xu F, Luo Y (2022) Single-cell RNA sequencing technologies and applications: a brief overview. Clin Transl Med 12(3):e694. https://doi.org/10.1002/ctm2.694
94. Fu X, Mo S, Buendia A, Laurent AP, Shao A, Del Mar Alvarez-Torres M, Yu T, Tan J, Su J, Sagatelian R, Ferrando AA, Ciccia A, Lan Y, Owens DM, Palomero T, Xing EP, Rabadan R (2025) A foundation model of transcription across human cell types. Nature 637(8047):965–973. https://doi.org/10.1038/s41586-024-08391-z
95. Su J, Song Y, Zhu Z, Huang X, Fan J, Qiao J, Mao F (2024) Cell–cell communication: new insights and clinical implications. Signal Transduct Target Ther 9(1):196. https://doi.org/10.1038/s41392-024-01888-z
96. Oehler JB, Wright H, Stark Z, Mallett AJ, Schmitz U (2023) The application of long-read sequencing in clinical settings. Hum Genomics 17(1):73. https://doi.org/10.1186/s40246-023-00522-3
97. Satam H, Joshi K, Mangrolia U, Waghoo S, Zaidi G, Rawool S, Thakare RP, Banday S, Mishra AK, Das G, Malonia SK (2023) Next-generation sequencing technology: current trends and advancements. Biology 12(7):997. https://doi.org/10.3390/biology12070997

98. Espinosa E, Bautista R, Larrosa R, Plata O (2024) Advancements in long-read genome sequencing technologies and algorithms. Genomics 116(3):110842. https://doi.org/10.1016/j.ygeno.2024.110842
99. Blommaert J, Sandoval-Castillo J, Beheregaray LB, Wellenreuther M (2024) Peering into the gaps: Long-read sequencing illuminates structural variants and genomic evolution in the Australasian snapper. Genomics 116(5):110929. https://doi.org/10.1016/j.ygeno.2024.110929
100. Yu SCY, Deng J, Qiao R, Cheng SH, Peng W, Lau SL, Choy LYL, Leung TY, Wong J, Wong VW, Wong GLH, Jiang P, Chiu RWK, Chan KCA, Lo YMD (2022) Comparison of single molecule, real-time sequencing and nanopore sequencing for analysis of the size, end-motif, and tissue-of-origin of long cell-free DNA in plasma. Clin Chem 69(2):168–179. https://doi.org/10.1093/clinchem/hvac180
101. Kumalasari MR, Alfanaar R, Andreani AS (2024) Gold nanoparticles (AuNPs): a versatile material for biosensor application. Talanta Open 9:100327. https://doi.org/10.1016/j.talo.2024.100327
102. Georgeous J, AlSawaftah N, Abuwatfa WH, Husseini GA (2024) Review of gold nanoparticles: synthesis, properties, shapes, cellular uptake, targeting, release mechanisms and applications in drug delivery and therapy. Pharmaceutics 16(10):1332. https://doi.org/10.3390/pharmaceutics16101332
103. Bhatia D, Paul S, Acharjee T, Ramachairy SS (2023) Biosensors and their widespread impact on human health. Sensors Int 5:100257. https://doi.org/10.1016/j.sintl.2023.100257
104. Fu Y, Liu T, Wang H, Wang Z, Hou L, Jiang J, Xu T (2024) Applications of nanomaterial technology in biosensing. J Sci Adv Material Device 9(2):100694. https://doi.org/10.1016/j.jsamd.2024.100694
105. Kim C, Kang MS, Raja IS, Joung YK, Han D (2024) Advancements in nanobiosensor technologies for in-vitro diagnostics to point of care testing. Heliyon 10(22):e40306. https://doi.org/10.1016/j.heliyon.2024.e40306
106. Chowdhury MKH, Anik HR, Akter M, Hasan SMM, Tushar SI, Mahmud S, Nahar N, Tania IS (2024) Sensing the future with graphene-based wearable sensors: a review. Result Material 25:100646. https://doi.org/10.1016/j.rinma.2024.100646
107. Ali A, Koloor SSR, Alshehri AH, Arockiarajan A (2023) Carbon nanotube characteristics and enhancement effects on the mechanical features of polymer-based materials and structures—a review. J Mater Res Technol 24:6495–6521. https://doi.org/10.1016/j.jmrt.2023.04.072
108. Jalalvand AR, Karami MM (2025) Roles of nanotechnology in electrochemical sensors for medical diagnostic purposes: a review. Sensing Bio-Sensing Res 47:100733. https://doi.org/10.1016/j.sbsr.2024.100733
109. Song J, Luo Y, Hao Z, Qu M, Huang C, Wang Z, Yang J, Liang Q, Jia Y, Song Q, Zhang Q, Luo S (2025) Graphene-based wearable biosensors for point-of-care diagnostics: from surface functionalization to biomarker detection. Materials Today Bio 32:101667. https://doi.org/10.1016/j.mtbio.2025.101667
110. Hemdan M, Ali MA, Doghish AS, Mageed SSA, Elazab IM, Khalil MM, Mabrouk M, Das DB, Amin AS (2024) Innovations in biosensor technologies for healthcare diagnostics and therapeutic drug monitoring: applications, recent progress, and future research challenges. Sensors 24(16):5143. https://doi.org/10.3390/s24165143
111. Špringer T, Bocková M, Slabý J, Sohrabi F, Čapková M, Homola J (2025) Surface plasmon resonance biosensors and their medical applications. Biosens Bioelectron 278:117308. https://doi.org/10.1016/j.bios.2025.117308
112. Butt MA (2025) Surface Plasmon resonance-based biodetection systems: principles, Progress and applications—a comprehensive review. Biosensors 15(1):35. https://doi.org/10.3390/bios15010035
113. Zabitler D, Ülker E, Turan K, Erdoğan NÖ, Tığ GA (2025) Electrochemical sensor for biological samples monitoring. Top Catal 025(5):02080. https://doi.org/10.1007/s11244-025-02080-5

114. Curulli A (2023) Functional nanomaterials enhancing electrochemical biosensors as smart tools for detecting infectious viral diseases. Molecules (Basel, Switzerland) 28(9):3777. https://doi.org/10.3390/molecules28093777
115. Shanbhag MM, Manasa G, Mascarenhas RJ, Mondal K, Shetti NP (2023) Fundamentals of bio-electrochemical sensing. Chem Eng J Adv 16:100516. https://doi.org/10.1016/j.ceja.2023.100516
116. Hemmerová E, Homola J (2024) Combining plasmonic and electrochemical biosensing methods. Biosens Bioelectron 251:116098. https://doi.org/10.1016/j.bios.2024.116098
117. Aljabali AA, El-Tanani M, Tambuwala MM (2024) Principles of CRISPR-Cas9 technology: advancements in genome editing and emerging trends in drug delivery. J Drug Delivery Sci Technol 92:105338. https://doi.org/10.1016/j.jddst.2024.105338
118. Zhu G, Zhou X, Wen M, Qiao J, Li G, Yao Y (2024) CRISPR–CAS13: pioneering RNA editing for nucleic acid therapeutics. BioDesign Res 6:0041. https://doi.org/10.34133/bdr.0041
119. Liu S, Liu H, Wang X, Shi L (2024b) The immune system of prokaryotes: potential applications and implications for gene editing. Biotechnol J 19(2):e2300352. https://doi.org/10.1002/biot.202300352
120. Yang H, Patel DJ (2024) Structures, mechanisms and applications of RNA-centric CRISPR–Cas13. Nat Chem Biol 20(6):673–688. https://doi.org/10.1038/s41589-024-01593-6
121. Huang Z, Fang J, Zhou M, Gong Z, Xiang T (2022) CRISPR-Cas13: a new technology for the rapid detection of pathogenic microorganisms. Front Microbiol 13:1011399. https://doi.org/10.3389/fmicb.2022.1011399
122. Abosalha AK, Makhlouf S, Islam P, Mohapatra S, Prakash S (2024) Perspective chapter: mastering RNA interference (RNAI) delivery—strategies for effective targeting and gene silencing. IntechOpen 1005800. https://doi.org/10.5772/intechopen.1005800
123. Zilberzwige-Tal S, Altae-Tran H, Kannan S, Wilkinson ME, Vo SC, Strebinger D, Edmonds KK, Yao CJ, Mears KS, Shmakov SA, Makarova KS, Macrae RK, Koonin EV, Zhang F (2025) Reprogrammable RNA-targeting CRISPR systems evolved from RNA toxin-antitoxins. Cell 188(7):1925–1940. e20. https://doi.org/10.1016/j.cell.2025.01.034
124. Azeez SS, Hamad RS, Hamad BK, Shekha MS, Bergsten P (2024) Advances in CRISPR-Cas technology and its applications: revolutionising precision medicine. Front Genome Ed 6: 1509924. https://doi.org/10.3389/fgeed.2024.1509924
125. Kumaran A, Serpes NJ, Gupta T, James A, Sharma A, Kumar D, Nagraik R, Kumar V, Pandey S (2023) Advancements in CRISPR-based biosensing for next-gen point of care diagnostic application. Biosensors 13(2):202. https://doi.org/10.3390/bios13020202
126. Zhra M, Qasem RJ, Aldossari F, Saleem R, Aljada A (2024) A comprehensive exploration of caspase detection methods: from classical approaches to cutting-edge innovations. Int J Mol Sci 25(10):5460. https://doi.org/10.3390/ijms25105460
127. Saguin NS, Maulik G, Cao X, Luo X, Nag A, Gao J, Deng S, Wong JW (2024) CNTs-based biosensors for enzyme detection. Sensors Act Phy 377:115753. https://doi.org/10.1016/j.sna.2024.115753
128. Wang Y, Zeng G, Jiang Y (2020) The emerging roles of miR-125b in cancers. Cancer Manag Res 12:1079–1088. https://doi.org/10.2147/CMAR.S232388
129. Li X, Xu M, Ding L, Tang J (2019) MiR-27a: a novel biomarker and potential therapeutic target in tumors. J Cancer 10(12):2836–2848. https://doi.org/10.7150/jca.31361
130. Napoletano S, Dannhauser D, Netti PA, Causa F (2024) Integrative analysis of miRNA expression data reveals a minimal signature for tumour cells classification. Comput Struct Biotechnol J 27:233–242. https://doi.org/10.1016/j.csbj.2024.12.023
131. Byeon H, Gc P, Hannan SA, Alghayadh FY, Soomar AM, Soni M, Bhatt MW (2024) Deep neural network model for enhancing disease prediction using auto encoder based broad learning. SLAS Technol 29(3):100145. https://doi.org/10.1016/j.slast.2024.100145
132. Arunkumar G, Deva Magendhra Rao AK, Manikandan M, Prasanna Srinivasa Rao H, Subbiah S, Ilangovan R, Murugan AK, Munirajan AK (2018) Dysregulation of miR-200 family microRNAs and epithelial-mesenchymal transition markers in oral squamous cell carcinoma. Oncol Lett 15(1):649–657. https://doi.org/10.3892/ol.2017.7296

133. Aborode AT, Emmanuel OA, Onifade IA, Olotu E, Otorkpa OJ, Mehmood Q, Abdulai SI, Jamiu A, Osinuga A, Oko CI, Fakorede S, Mangdow M, Babatunde O, Olapade Z, Victoria AG, Salami A, Usman IA, Agboli VI, Adesola RO (2025) The role of machine learning in discovering biomarkers and predicting treatment strategies for neurodegenerative diseases: a narrative review. NeuroMarkers 2(1):100034. https://doi.org/10.1016/j.neumar.2024.100034
134. Metcalf GAD (2024) MicroRNAs: circulating biomarkers for the early detection of imperceptible cancers via biosensor machine-learning advances. Oncogene 43(28):2135–2142. https://doi.org/10.1038/s41388-024-03076-3
135. Danieli MG, Brunetto S, Gammeri L, Palmeri D, Claudi I, Shoenfeld Y, Gangemi S (2023) Machine learning application in autoimmune diseases: state of art and future prospectives. Autoimmun Rev 23(2):103496. https://doi.org/10.1016/j.autrev.2023.103496
136. Lei C, Dang K, Song S, Wang Z, Chew SP, Bian R, Yang X, Guan Z, De Abreu Lopes CIM, Wang MH, Choy RWC, Hu X, Lai KKH, Chong KKL, Pang CP, Song X, Su J, Ding X, Zhou H (2025) AI-assisted facial analysis in healthcare: from disease detection to comprehensive management. Patterns 6(2):101175. https://doi.org/10.1016/j.patter.2025.101175
137. Sayed GI, Solyman M, Gedawy GE, Moemen YS, Aboul-Ella H, Hassanien AE (2024) Circulating miRNA's biomarkers for early detection of hepatocellular carcinoma in Egyptian patients based on machine learning algorithms. Sci Rep 14(1):4989. https://doi.org/10.1038/s41598-024-54795-2
138. Wang Z, Wang H, Zhou S, Mao J, Zhan Z, Duan S (2024) miRNA interplay: mechanisms and therapeutic interventions in cancer. MedComm—Oncol 3(4). https://doi.org/10.1002/mog2.93
139. Nori LP, Lohitha M, Vadapalli RR, Bonthagarala B, Nagineni SR, Kalidindi VR (2025) Revolutionizing healthcare: the impact of AI on precision medicine. Int J Pharm Invest 15(2):334–343. https://doi.org/10.5530/ijpi.20250100
140. Gerede A, Stavros S, Danavasi M, Potiris A, Moustakli E, Machairiotis N, Zikopoulos A, Nikolettos K, Drakakis P, Nikolettos N, Eleftheriades M, Domali E (2025) MicroRNAs in preeclampsia: bridging diagnosis and treatment. J Clin Med 14(6):2003. https://doi.org/10.3390/jcm14062003
141. Zhang Z, Liu T, Dong M, Ahamed MA, Guan W (2024) Sample-to-answer salivary miRNA testing: new frontiers in point-of-care diagnostic technologies. Wiley Interdiscip Rev Nanomed Nanobiotechnol 16(3):e1969. https://doi.org/10.1002/wnan.1969
142. Shueng P, Shih K, Gambhir S, Kuo D, Chuang H (2022) Cancer detection using an artificial secretable MicroRNA found in blood and urine. Biomedicine 10(3):621. https://doi.org/10.3390/biomedicines10030621
143. Li X, Hallajzadeh J (2025) Circulating microRNAs and physical activity: impact in diabetes. Clin Chim Acta 569:120178. https://doi.org/10.1016/j.cca.2025.120178
144. Robotti M, Scebba F, Angeloni D (2023) Circulating biomarkers for cancer detection: could salivary microRNAs be an opportunity for ovarian cancer diagnostics? Biomedicine 11(3):652. https://doi.org/10.3390/biomedicines11030652
145. Prasad M, Sekar R, Priya MDL, Varma SR, Karobari MI (2024) A new perspective on diagnostic strategies concerning the potential of saliva-based miRNA signatures in oral cancer. Diagn Pathol 19(1):147. https://doi.org/10.1186/s13000-024-01575-1
146. Roi A, Boia S, Rusu L, Roi CI, Boia ER, Riviş M (2023) Circulating MIRNA as a biomarker in oral cancer liquid biopsy. Biomedicine 11(3):965. https://doi.org/10.3390/biomedicines11030965
147. Suvarnapathaki S, Serrano-Farias A, Dudley JC, Bettegowda C, Rincon-Torroella J (2024) Unlocking the potential of circulating miRNAs as biomarkers in glioblastoma. Life (Basel, Switzerland) 14(10):1312. https://doi.org/10.3390/life14101312
148. Jafari A, Karimabadi K, Rahimi A, Rostaminasab G, Khazaei M, Rezakhani L, Jouybari TA (2023) The emerging role of exosomal mIRNAs as biomarkers for early cancer detection: a comprehensive literature review. Technol Cancer Res Treat 22:15330338231205999. https://doi.org/10.1177/15330338231205999
149. Kumar MA, Baba SK, Sadida HQ, Marzooqi SA, Jerobin J, Altemani FH, Algehainy N, Alanazi MA, Abou-Samra A, Kumar R, Akil ASA, Macha MA, Mir R, Bhat AA (2024)

Extracellular vesicles as tools and targets in therapy for diseases. Signal Transduct Target Ther 9(1):27. https://doi.org/10.1038/s41392-024-01735-1
150. Umeche IE, Olaniyan MF (2023) Exosomes: emerging biomarkers unveiling cellular mysteries—a narrative review. J Bio-X Res 06(03):104–115. https://doi.org/10.1097/jbr.0000000000000146
151. Ljungström M, Oltra E (2025) Methods for extracellular vesicle isolation: relevance for encapsulated MIRNAs in disease diagnosis and treatment. Genes 16(3):330. https://doi.org/10.3390/genes16030330
152. Javdani-Mallak A, Salahshoori I (2024) Environmental pollutants and exosomes: a new paradigm in environmental health and disease. Sci Total Environ 925:171774. https://doi.org/10.1016/j.scitotenv.2024.171774
153. Alhenaky A, Alhazmi S, Alamri SH, Alkhatabi HA, Alharthi A, Alsaleem MA, Abdelnour SA, Hassan SM (2024) Exosomal MicroRNAs in Alzheimer's disease: unveiling their role and pioneering tools for diagnosis and treatment. J Clin Med 13(22):6960. https://doi.org/10.3390/jcm13226960
154. Bhadra M, Sachan M (2024) An overview of challenges associated with exosomal miRNA isolation toward liquid biopsy based ovarian cancer detection. Heliyon 10(9):e30328. https://doi.org/10.1016/j.heliyon.2024.e30328
155. Qiu J, Liang Y, Wang C, Yu Y, Zhang Y, Liu H, Han L (2024) Microfluidic barcode biochips for high-throughput real-time biomolecule and single-cell screening. Engineering 46(3):130–146. https://doi.org/10.1016/j.eng.2024.06.016
156. Rasheed S, Kanwal T, Ahmad N, Fatima B, Najam-Ul-Haq M, Hussain D (2024) Advances and challenges in portable optical biosensors for onsite detection and point-of-care diagnostics. TrAC Trends Anal Chem 173:117640. https://doi.org/10.1016/j.trac.2024.117640
157. Siavashy S, Soltani M, Rahimi S, Hosseinali M, Guilandokht Z, Raahemifar K (2024) Recent advancements in microfluidic based biosensors for detection of genes and proteins: applications and techniques. Biosensors Bioelectronics X 19:100489. https://doi.org/10.1016/j.biosx.2024.100489
158. Wang S, Guan X, Sun S (2025b) Microfluidic biosensors: enabling advanced disease detection. Sensors 25(6):1936. https://doi.org/10.3390/s25061936
159. Bakhshi A, Pandey A, Kharaba Z, Razlansari M, Sargazi S, Behzadmehr R, Rahdar A, Díez-Pascual AM, Fathi-Karkan S (2023) Microfluidic-based nanoplatforms for cancer theranostic applications: a mini-review on recent advancements. OpenNano 15:100197. https://doi.org/10.1016/j.onano.2023.100197
160. Lehnert T, Gijs M, a. M. (2024) Microfluidic systems for infectious disease diagnostics. Lab Chip 24(5):1441–1493. https://doi.org/10.1039/d4lc00117f
161. Bui H, Thangavel B, Sharipov M, Chen K, Shin JH (2023) Smartphone-based portable biochemical sensors: exploring recent advancements. Chem 11(9):468. https://doi.org/10.3390/chemosensors11090468
162. Zhang M, Cui X, Li N (2022) Smartphone-based mobile biosensors for the point-of-care testing of human metabolites. Materials Today Bio 14:100254. https://doi.org/10.1016/j.mtbio.2022.100254
163. Zhao F, Wu W, Zhao M, Ding S, Lin Y, Hu Q, Yu L (2022) Enzyme-like nanomaterials-integrated microfluidic technology for bioanalysis. TrAC Trends Anal Chem 158:116833. https://doi.org/10.1016/j.trac.2022.116833
164. Jafleh EA, Alnaqbi FA, Almaeeni HA, Faqeeh S, Alzaabi MA, Zaman KA (2024) The role of wearable devices in chronic disease monitoring and patient care: a comprehensive review. Cureus 16(9):e68921. https://doi.org/10.7759/cureus.68921
165. Kapplingattu SV, Bhattacharya S, Adlakha YK (2025) MiRNAs as major players in brain health and disease: current knowledge and future perspectives. Cell Death Discover 11(1):7. https://doi.org/10.1038/s41420-024-02283-x
166. Beylerli O, Shi H, Begliarzade S, Shumadalova A, Ilyasova T, Sufianov A (2024) MiRNAs as new potential biomarkers and therapeutic targets in brain metastasis. Non-coding RNA Res 9(3):678–686. https://doi.org/10.1016/j.ncrna.2024.02.014

167. Ghamlouche F, Yehya A, Zeid Y, Fakhereddine H, Fawaz J, Liu Y, Al-Sayegh M, Abou-Kheir W (2023) MicroRNAs as clinical tools for diagnosis, prognosis, and therapy in prostate cancer. Transl Oncol 28:101613. https://doi.org/10.1016/j.tranon.2022.101613
168. Gayosso-Gómez LV, Ortiz-Quintero B (2021) Circulating MicroRNAs in blood and other body fluids as biomarkers for diagnosis, prognosis, and therapy response in lung cancer. Diagnostics 11(3):421. https://doi.org/10.3390/diagnostics11030421
169. Li J, Zhang Y, Dong P, Yang G, Gurunathan S (2023) A comprehensive review on the composition, biogenesis, purification, and multifunctional role of exosome as delivery vehicles for cancer therapy. Biomed Pharmacother 165:115087. https://doi.org/10.1016/j.biopha.2023.115087
170. Verma VK, Beevi SS, Nair RA, Kumar A, Kiran R, Alexander LE, Kumar LD (2024) MicroRNA signatures differentiate types, grades, and stages of breast invasive ductal carcinoma (IDC): miRNA-target interacting signaling pathways. Cell Commun Signal 22(1):100. https://doi.org/10.1186/s12964-023-01452-2
171. Elmore LW, Greer SF, Daniels EC, Saxe CC, Melner MH, Krawiec GM, Cance WG, Phelps WC (2020) Blueprint for cancer research: critical gaps and opportunities. CA a Cancer J Clinicians 71(2):107–139. https://doi.org/10.3322/caac.21652
172. Diener C, Keller A, Meese E (2022) Emerging concepts of miRNA therapeutics: from cells to clinic. Trends Genet 38(6):613–626. https://doi.org/10.1016/j.tig.2022.02.006
173. Lima JF, Cerqueira L, Figueiredo C, Oliveira C, Azevedo NF (2018) Anti-miRNA oligonucleotides: A comprehensive guide for design. RNA Biol 15(3):338–352. https://doi.org/10.1080/15476286.2018.1445959
174. Panigrahi M, Palmer MA, Wilson JA (2022) MicroRNA-122 regulation of HCV infections: insights from studies of miR-122 independent replication. Pathogens 11(9):1005. https://doi.org/10.3390/pathogens11091005
175. Nigi L, Grieco GE, Ventriglia G, Brusco N, Mancarella F, Formichi C, Dotta F, Sebastiani G (2018) MicroRNAs as regulators of insulin signaling: research updates and potential therapeutic perspectives in type 2 diabetes. Int J Mol Sci 19(12):3705. https://doi.org/10.3390/ijms19123705
176. Bader AG, Brown D, Winkler M (2010) The promise of MicroRNA replacement therapy. Cancer Res 70(18):7027–7030. https://doi.org/10.1158/0008-5472.can-10-2010
177. Kim T, Croce CM (2023) MicroRNA: trends in clinical trials of cancer diagnosis and therapy strategies. Exp Mol Med 55(7):1314–1321. https://doi.org/10.1038/s12276-023-01050-9
178. Chioccioli M, Roy S, Newell R, Pestano L, Dickinson B, Rigby K, Herazo-Maya J, Jenkins G, Ian S, Saini G, Johnson SR, Braybrooke R, Yu G, Sauler M, Ahangari F, Ding S, DeIuliis J, Aurelien N, Montgomery RL, Kaminski N (2022) A lung targeted miR-29 mimic as a therapy for pulmonary fibrosis. EBioMedicine 85:104304. https://doi.org/10.1016/j.ebiom.2022.104304
179. Bernardo BC, Gregorevic P, Ritchie RH, McMullen JR (2018) Generation of MicroRNA-34 sponges and tough decoys for the heart: developments and challenges. Front Pharmacol 9:1090. https://doi.org/10.3389/fphar.2018.01090
180. Xu W, Feng S, Huang A (2022) Role of miR-155 in inflammatory autoimmune diseases: a comprehensive review. Inflamm Res 71(12):1501–1517. https://doi.org/10.1007/s00011-022-01643-6
181. Hussen BM, Rasul MF, Abdullah SR, Hidayat HJ, Faraj GSH, Ali FA, Salihi A, Baniahmad A, Ghafouri-Fard S, Rahman M, Glassy MC, Branicki W, Taheri M (2023) Targeting miRNA by CRISPR/Cas in cancer: advantages and challenges. Mil Med Res 10(1):32. https://doi.org/10.1186/s40779-023-00468-6
182. Rhim J, Baek W, Seo Y, Kim JH (2022) From molecular mechanisms to therapeutics: understanding MicroRNA-21 in cancer. Cells 11(18):2791. https://doi.org/10.3390/cells11182791
183. Fu J, Imani S, Wu M, Wu R (2023) MicroRNA-34 family in cancers: role, mechanism, and therapeutic potential. Cancer 15(19):4723. https://doi.org/10.3390/cancers15194723

184. Yu G, Yao W, Xiao W, Li H, Xu H, Lang B (2014a) MicroRNA-34a functions as an anti-metastatic microRNA and suppresses angiogenesis in bladder cancer by directly targeting CD44. J Exp Clin Cancer Res: CR 33(1):779. https://doi.org/10.1186/s13046-014-0115-4
185. Sun P, Feng Y, Guo H, Li R, Yu P, Zhou X, Pan Z, Liang Y, Yu B, Zheng Y, Shi Y, Wen L, Wei M, Chen Y (2020) MiR-34a inhibits cell proliferation and induces apoptosis in human nasopharyngeal carcinoma by targeting lncRNA MCM3AP-AS1. Cancer Manag Res 12:4799–4806. https://doi.org/10.2147/CMAR.S245520
186. Beg MS, Brenner AJ, Sachdev J, Borad M, Kang YK, Stoudemire J, Smith S, Bader AG, Kim S, Hong DS (2017) Phase I study of MRX34, a liposomal miR-34a mimic, administered twice weekly in patients with advanced solid tumors. Investig New Drugs 35(2):180–188. https://doi.org/10.1007/s10637-016-0407-y
187. Yu G, Li H, Wang J, Gumireddy K, Li A, Yao W, Tang K, Xiao W, Hu J, Xiao H, Lang B, Ye Z, Huang Q, Xu H (2014b) miRNA-34a suppresses cell proliferation and metastasis by targeting CD44 in human renal carcinoma cells. J Urol 192(4):1229–1237. https://doi.org/10.1016/j.juro.2014.05.094
188. Bouchie A (2013) First microRNA mimic enters clinic. Nat Biotechnol 31(7):577. https://doi.org/10.1038/nbt0713-577
189. Di Martino MT, Campani V, Misso G, Gallo Cantafio ME, Gullà A, Foresta U, Guzzi PH, Castellano M, Grimaldi A, Gigantino V, Franco R, Lusa S, Cannataro M, Tagliaferri P, De Rosa G, Tassone P, Caraglia M (2014) In vivo activity of miR-34a mimics delivered by stable nucleic acid lipid particles (SNALPs) against multiple myeloma. PLoS One 9(2):e90005. https://doi.org/10.1371/journal.pone.0090005
190. Ma L, Reinhardt F, Pan E, Soutschek J, Bhat B, Marcusson EG, Teruya-Feldstein J, Bell GW, Weinberg RA (2010) Therapeutic silencing of miR-10b inhibits metastasis in a mouse mammary tumor model. Nat Biotechnol 28(4):341–347. https://doi.org/10.1038/nbt.1618
191. Kota J, Chivukula RR, O'Donnell KA, Wentzel EA, Montgomery CL, Hwang HW, Chang TC, Vivekanandan P, Torbenson M, Clark KR, Mendell JR, Mendell JT (2009) Therapeutic microRNA delivery suppresses tumorigenesis in a murine liver cancer model. Cell 137(6):1005–1017. https://doi.org/10.1016/j.cell.2009.04.021
192. Wiggins JF, Ruffino L, Kelnar K, Omotola M, Patrawala L, Brown D, Bader AG (2010) Development of a lung cancer therapeutic based on the tumor suppressor microRNA-34. Cancer Res 70(14):5923–5930. https://doi.org/10.1158/0008-5472.CAN-10-0655
193. Forrest AR, Kanamori-Katayama M, Tomaru Y, Lassmann T, Ninomiya N, Takahashi Y, de Hoon MJ, Kubosaki A, Kaiho A, Suzuki M, Yasuda J, Kawai J, Hayashizaki Y, Hume DA, Suzuki H (2010) Induction of microRNAs, mir-155, mir-222, mir-424 and mir-503, promotes monocytic differentiation through combinatorial regulation. Leukemia 24(2):460–466. https://doi.org/10.1038/leu.2009.246
194. Xiao F, Qiu H, Cui H, Ni X, Li J, Liao W, Lu L, Ding K. MicroRNA-885-3p inhibits the growth of HT-29 colon cancer cell xenografts by disrupting angiogenesis via targeting BMPR1A and blocking BMP/Smad/Id1 signaling. Oncogene. 34(15):1968–1978. https://doi.org/10.1038/onc.2014.134
195. van Zandwijk N, Pavlakis N, Kao SC, Linton A, Boyer MJ, Clarke S, Huynh Y, Chrzanowska A, Fulham MJ, Bailey DL, Cooper WA, Kritharides L, Ridley L, Pattison ST, MacDiarmid J, Brahmbhatt H, Reid G (2017) Safety and activity of microRNA-loaded minicells in patients with recurrent malignant pleural mesothelioma: a first-in-man, phase 1, open-label, dose-escalation study. Lancet Oncol 18(10):1386–1396. https://doi.org/10.1016/S1470-2045(17)30621-6
196. Tassone P, Di Martino MT, Arbitrio M, Fiorillo L, Staropoli N, Ciliberto D, Cordua A, Scionti F, Bertucci B, Salvino A, Lopreiato M, Thunarf F, Cuomo O, Zito MC, De Fina MR, Brescia A, Gualtieri S, Riillo C, Manti F, Caracciolo D et al (2023) Safety and activity of the first-in-class locked nucleic acid (LNA) miR-221 selective inhibitor in refractory advanced cancer patients: a first-in-human, phase 1, open-label, dose-escalation study. J Hematol Oncol 16(1):68. https://doi.org/10.1186/s13045-023-01468-8

197. Querfeld C, Pacheco T, Foss FM, Halwani AS, Porcu P, Seto AG et al (2016) Preliminary results of a phase 1 trial evaluating MRG-106, a synthetic microRNA antagonist (LNA antimiR) of microRNA-155, in patients with CTCL. Blood 128(22):1829. https://doi.org/10.1182/BLOOD.V128.22.1829.1829
198. Ghosh S, Lazarus D, Robertson N, Crosier E, Liu P, Medarova Z (2023) The microRNA-10b targeted therapeutic, TTX-MC138, is effective in preclinical pancreatic adenocarcinoma. Cancer Res 83(7_Supplement):548–548
199. Lupini L, Bassi C, Ferracin M, Bartonicek N, D'Abundo L, Zagatti B, Callegari E, Musa G, Moshiri F, Gramantieri L, Corrales FJ, Enright AJ, Sabbioni S, Negrini M (2013) miR-221 affects multiple cancer pathways by modulating the level of hundreds messenger RNAs. Front Genet 4:64. https://doi.org/10.3389/fgene.2013.00064
200. Halim A, Al-Qadi N, Kenyon E, Conner KN, Mondal SK, Medarova Z, Moore A (2024) Inhibition of miR-10b treats metastatic breast cancer by targeting stem cell-like properties. Oncotarget 15:591–606. https://doi.org/10.18632/oncotarget.28641
201. Cui J, Yuan Y, Shanmugam MK, Anbalagan D, Tan TZ, Sethi G, Kumar AP, Lim LHK (2021) MicroRNA-196a promotes renal cancer cell migration and invasion by targeting BRAM1 to regulate SMAD and MAPK signaling pathways. Int J Biol Sci 17(15):4254–4270. https://doi.org/10.7150/ijbs.60805
202. Ouchida M, Kanzaki H, Ito S, Hanafusa H, Jitsumori Y, Tamaru S, Shimizu K (2012) Novel direct targets of miR-19a identified in breast cancer cells by a quantitative proteomic approach. PLoS One 7(8):e44095. https://doi.org/10.1371/journal.pone.0044095
203. Jiang JH, Liu C, Cheng H, Lu Y, Qin Y, Xu YF, Xu J, Long J, Liu L, Ni QX, Yu XJ (2015) Epithelial-mesenchymal transition in pancreatic cancer: is it a clinically significant factor? Biochim Biophys Acta 1855(1):43–49. https://doi.org/10.1016/j.bbcan.2014.11.004
204. Osmak G, Kiselev I, Baulina N, Favorova O (2020) From miRNA target gene network to miRNA function: miR-375 might regulate apoptosis and actin dynamics in the heart muscle via rho-GTPases-dependent pathways. Int J Mol Sci 21(24):9670. https://doi.org/10.3390/ijms21249670
205. Yu W, Liang X, Li X, Zhang Y, Sun Z, Liu Y, Wang J (2018) MicroRNA-195: a review of its role in cancers. Onco Targets Ther 11:7109–7123. https://doi.org/10.2147/OTT.S183600
206. Chan JA, Krichevsky AM, Kosik KS (2005) MicroRNA-21 is an antiapoptotic factor in human glioblastoma cells. Cancer Res 65(14):6029–6033. https://doi.org/10.1158/0008-5472.CAN-05-0137
207. Asangani IA, Rasheed SA, Nikolova DA, Leupold JH, Colburn NH, Post S, Allgayer H (2008) MicroRNA-21 (miR-21) post-transcriptionally downregulates tumor suppressor Pdcd4 and stimulates invasion, intravasation and metastasis in colorectal cancer. Oncogene 27(15):2128–2136. https://doi.org/10.1038/sj.onc.1210856
208. Wang LJ, He CC, Sui X, Cai MJ, Zhou CY, Ma JL, Wu L, Wang H, Han SX, Zhu Q (2015) MiR-21 promotes intrahepatic cholangiocarcinoma proliferation and growth in vitro and in vivo by targeting PTPN14 and PTEN. Oncotarget 6(8):5932–5946. https://doi.org/10.18632/oncotarget.3465
209. Chang TC, Yu D, Lee YS, Wentzel EA, Arking DE, West KM, Dang CV, Thomas-Tikhonenko A, Mendell JT (2008) Widespread microRNA repression by Myc contributes to tumorigenesis. Nat Genet 40(1):43–50. https://doi.org/10.1038/ng.2007.30
210. Lin S, Gregory RI (2015) MicroRNA biogenesis pathways in cancer. Nat Rev Cancer 15(6):321–333. https://doi.org/10.1038/nrc3932
211. Ma L, Teruya-Feldstein J, Weinberg RA. Tumour invasion and metastasis initiated by microRNA-10b in breast cancer. Nature. 449(7163):682–688. https://doi.org/10.1038/nature06174
212. Seto AG, Beatty X, Lynch JM, Hermreck M, Tetzlaff M, Duvic M, Jackson AL (2018) Cobomarsen, an oligonucleotide inhibitor of miR-155, co-ordinately regulates multiple survival pathways to reduce cellular proliferation and survival in cutaneous T-cell lymphoma. Br J Haematol 183(3):428–444. https://doi.org/10.1111/bjh.15547

213. Frères P, Josse C, Bovy N, Boukerroucha M, Struman I, Bours V, Jerusalem G (2015) Neoadjuvant Chemotherapy in Breast Cancer Patients Induces miR-34a and miR-122 Expression. J Cell Physiol 230(2):473–481. https://doi.org/10.1002/jcp.24730
214. Wang J, Wu J. Role of miR-155 in breast cancer. Front Biosci (Landmark Ed). 17(6):2350–2355. https://doi.org/10.2741/4056
215. Li Y, Guessous F, Zhang Y, Dipierro C, Kefas B, Johnson E, Marcinkiewicz L, Jiang J, Yang Y, Schmittgen TD, Lopes B, Schiff D, Purow B, Abounader R (2009) MicroRNA-34a inhibits glioblastoma growth by targeting multiple oncogenes. Cancer Res 69(19):7569–7576. https://doi.org/10.1158/0008-5472.CAN-09-0529
216. He XX, Chang Y, Meng FY, Wang MY, Xie QH, Tang F, Li PY, Song YH, Lin JS (2012) MicroRNA-375 targets AEG-1 in hepatocellular carcinoma and suppresses liver cancer cell growth in vitro and in vivo. Oncogene 31(28):3357–3369. https://doi.org/10.1038/onc.2011.500
217. Li Q, Zou C, Zou C, Han Z, Xiao H, Wei H, Wang W, Zhang L, Zhang X, Tang Q, Zhang C, Tao J, Wang X, Gao X (2013) MicroRNA-25 functions as a potential tumor suppressor in colon cancer by targeting Smad7. Cancer Lett 335(1):168–174. https://doi.org/10.1016/j.canlet.2013.02.029

Index

S. Pathak, A. Banerjee (eds.), *MicroRNAs: Role in Cancer Progression, Diagnostics and Therapy*, Advances in Biochemistry in Health and Disease 34, https://doi.org/10.1007/978-3-032-08824-6

N

O

P

MIX
Papier aus verantwortungsvollen Quellen
Paper from responsible sources
FSC® C105338

If you have any concerns about our products,
you can contact us on
ProductSafety@springernature.com

In case Publisher is established outside the EU,
the EU authorized representative is:
Springer Nature Customer Service Center GmbH
Europaplatz 3, 69115 Heidelberg, Germany

Printed by Libri Plureos GmbH
in Hamburg, Germany